Battery Management Systems

Volume III

Physics-Based Methods

For a listing of recent titles in the
Artech House Power Engineering and Power Electronics,
turn to the back of this book.

Battery Management Systems

Volume III

Physics-Based Methods

Gregory L. Plett
M. Scott Trimboli

ARTECH HOUSE
BOSTON | LONDON
artechhouse.com

Library of Congress Cataloging-in-Publication Data
A catalog record for this book is available from the U.S. Library of Congress.

British Library Cataloguing in Publication Data
A catalog record for this book is available from the British Library.

ISBN-13: 978-1-63081-904-0

Cover design by Joi Garron

685 Canton Street
Norwood, MA 02062

10 9 8 7 6 5 4 3 2 1

Contents

Preface

This book comprises the third and final volume in a series presenting battery-management systems with a particular emphasis on how to meet their algorithmic requirements. The first volume derived sets of mathematical equations (models) that describe how lithium-ion battery cells work, both internally (physics-based models) and as observed externally (empirical equivalent-circuit models). The second volume discussed state-of-the-art applications of equivalent-circuit models to solving problems in battery management and control. This third volume discusses how to use physics-based models of battery cells in a computationally efficient framework to optimize battery-pack management and control to maximize a battery pack's performance and extend its life.

The intention of the series is not to be encyclopedic; rather, it is to put forward only the current best practices with sufficient fundamental background to understand them thoroughly. This volume builds on the content in Vols. I and II. For the deepest understanding, the reader should have already studied these earlier volumes; however, we provide review material such that this third volume is generally self-contained.

Certainly, what is meant by "best practices" is at least somewhat subjective, and we may well have overlooked approaches and methodologies that are better in some applications than those described herein. Perhaps we should say, "best" from our own point of view, given what we happen to have tried, in application domains and problems we have attempted to address.

The majority of present battery-management systems still use equivalent-circuit models to aid with the computation of estimates of battery state of charge, state of health, available power, and available energy. While these methods can provide useful results, they are not able to predict the impact of battery usage on overall battery service life. Physics-based models and physics-based methods are needed to optimize the tradeoff between performance and life directly. This volume introduces and teaches the state of the art in physics-based methods for battery management. Since this field is still very young, it additionally points out where improvements beyond the present state-of-art can still be made.

This particular volume is organized in the following way:

- Chapter 1 introduces the need for physics-based battery management and reviews the equations comprising a baseline physics-based model of lithium-ion cells. It further shows how the parameter set of the model can be reduced to enable estimating the parameter values of the model to fit a particular physical cell.
- Chapter 2 augments the baseline model with additional terms to describe the electrical double layer and constant-phase-element behaviors observed in real cells. It then develops a frequency-domain impedance model of the cell, which is useful on its own but is further needed when estimating model parameter values.
- Chapter 3 describes how to estimate the parameter values of a physics-based model using carefully designed laboratory tests. In particular, it introduces the OCP, OCV, pulse-resistance, discharge, and EIS tests, each of which is useful for producing data that can be used to estimate a subset of model parameter values. The chapter teaches lab-test procedures, data-processing procedures, and parameter-optimization methods.
- Chapter 4 shows how to convert the physics-based models discussed so far—which comprise complicated coupled partial-differential equations—into computationally simple reduced-order physics-based models. Some aspects are similar to Chap. 6 in Vol. I, but this volume introduces new, better methods and results. It further describes how to simulate constant-voltage and constant-current events with these physics-based models and how to simulate battery packs comprising multiple cells.
- Chapter 5 develops state estimators for battery-management systems using physics-based reduced-order models. These methods can produce estimates of state of charge, but more importantly can also estimate cell internal electrochemical variables as well as corresponding confidence intervals on the estimates. These estimates are necessary to be able to optimize life and performance jointly.
- Chapter 6 introduces the ideas of diagnosis and prognosis as applied to aging battery packs. A concise survey of the field is presented and several different methods are described in more detail.
- Chapter 7 employs the reduced-order models developed in Chap. 4 within the framework of model-predictive control (MPC) in order to optimize battery fast charge. In particular, it demonstrates how to charge a cell as quickly as possible without violating electro-

chemical limits that would contribute to performance degradation (e.g., lithium plating).
- Chapter 8 explains the main ideas behind battery power utilization and shows how to use MPC to compute a dynamic state of power measure, and then concludes this volume.

The intended audience for this material is someone with an undergraduate degree in engineering—principally electrical or mechanical. Readers with different backgrounds may find some of the material too basic (because they have studied it before, whereas an engineering student would not), or not descriptive enough (because they are missing some background that would typically be encountered in an engineering degree program). Both problems have a remedy, although the solution to the second involves background research to become proficient in an unfamiliar discipline—not an easy undertaking. The book also assumes that Vols. I and II are at hand, to confer for some of the background review materials.

The content of this book has been taught to students of diverse backgrounds in *ECE5730: Physics-Based Battery Management* at the University of Colorado Colorado Springs. The feedback that we received from these students has been invaluable in helping us to refine our approach to presenting these topics.

We are also greatly indebted to our colleagues and both past and present students who have assisted us over the years to understand and develop the materials presented in this work. Dr. Shriram Santhanagopalan taught us the parameter-lumping method presented in Chap. 1. Drs. Albert Rodríguez, Zhengyu Chu, and Xiangdong Kong contributed to deriving and validating the new transfer functions of Chap. 2 and rewriting them in terms of lumped parameters. Dr. Ryan Jobman, Dr. Dongliang Lu, Mr. Brandon Guest, and Mr. Wesley Hileman pioneered the parameter-estimation methods presented in Chap. 3. Dr. Albert Rodríguez was the first to implement several new realization algorithms, including the HRA presented in Chap. 4; he also pioneered output blending versus the earlier model-blending methods. Mr. Kirk Stetzel implemented earlier versions of physics-based state-estimation methods; we followed his work when developing Chap. 5. Dr. Adam Smiley was the first to apply the interacting multiple model Kalman filter to the problem of model selection, presented in Chap. 6. Mr. Aloisio Kawakita de Souza was the first to implement the physics-based thermal model also presented in Chap. 6 and engineered the implementation of combined MPC-xKF using output blending, which forms a significant contribution to Chaps. 7 and 8. Drs. Marcelo Xavier, Kiana Karami, and Gustavo Florentino also contributed foundational material to the development of MPC methods for lithium-ion battery management presented in these two chapters.

Despite our best intentions, there are certain to be errors and confusing statements in this book. Please feel free to send us corrections and suggestions for improvements.

1 Redundant Parameter Elimination

At the time of writing, high-capacity battery systems appear to be at the threshold of a rapid acceleration in commercial adoption. Electric-vehicle market share in the United States remains low, around 2 %, but year-over-year growth is around 40 % globally at this point. Market share of plug-in vehicles in Norway is presently over 80 %, indicating that consumer acceptance of this technology can be very high. Many of the largest automakers have vowed to eliminate combustion vehicles from their product lineup this decade and many countries and regions have banned sales of new combustion vehicles by 2035 or earlier. The market for electric-drivetrain vehicles is about to be transformed.

Similarly, large-scale grid services are expanding. Utility-scale battery systems like the 150 MWh Hornsdale Power Reserve in South Australia provide grid stability and system security. Many big-box retail stores are investing in battery systems to implement peak shaving—storing electric energy during periods of the day when utility rates are low and then using the stored energy from the batteries during periods when the utility rates are high. This trend is likely to grow as the cost of batteries continue to drop and as more renewables—which have unpredictable generation capabilities—are added to the mix of power-generation means.

It is becoming increasingly more important to manage battery systems well, which is the focus of this series of volumes on battery-management systems (BMSs). The first volume[1] focused on developing models (i.e., sets of equations) that describe the dynamic behaviors of lithium-ion battery cells. These models divide into two main categories: empirical equivalent-circuit models (ECMs) and physics-based models (PBMs) developed from electrochemical and physical first principles. The second volume[2] showed how to use ECMs within battery-management algorithms to estimate cell state of charge (SOC), state of health (SOH), state of power (SOP, sometimes called state of function or SOF), and state of energy (SOE).[3] This vol-

[1] Gregory L. Plett, *Battery Management Systems, Volume 1: Battery Modeling*. Artech House, 2015.

[2] Gregory L. Plett, *Battery Management Systems, Volume 2: Equivalent-Circuit Methods*. Artech House, 2015.

[3] Collectively, these estimates are often termed "SOx."

ume shows how to use PBMs to achieve similar goals and also to enable fast charge.

Replacing ECMs with PBMs in BMS algorithms is vitally important to optimizing next-generation BMS. Why? Many battery systems that use ECM-based BMS operate very well already. Vol. II showed that BMS algorithms based on ECMs can give very good estimates of SOC, SOH, and SOE. However, ECMs predict only externally measurable cell quantities such as the terminal-voltage and surface-temperature response of a cell to an input-current stimulus. PBMs also predict these quantities but can additionally predict internal electrochemical variables such as electrical potentials, lithium concentrations, lithium fluxes, and temperatures at different spatial locations interior to a cell. This is important since the rate of cell aging depends directly on these internal variables and not on externally measurable quantities. We cannot predict the incremental degradation caused to a cell by a proposed load profile without knowing the cell's internal variables and therefore it is fundamentally impossible to optimize a tradeoff between cell performance and rate of aging directly and robustly using an ECM. If we wish to utilize a battery up to its physical limits of performance—but not beyond—then we must have good estimates of these internal variables. We must use PBM-based BMS.

This is why we believe that some form of PBM will form the basis for future BMS. As the market moves toward requirements for stricter guarantees of safety and service life and a demand for higher levels of performance, battery applications will require that the BMS compute battery power limits and charging profiles that are determined using PBMs and not empirical models.

The ultimate conclusion is that BMS based on ECMs can work well, as demonstrated by millions of existing products. But, they are not optimal. There remains room for improvement.

There have been a number of significant obstacles that have delayed adoption of PBMs in BMS. These include:

- It is more difficult for engineers having traditional educational backgrounds to learn the fundamentals of PBMs than ECMs. We devoted the majority of Vol. I in this series to developing PBM equations, but we needed only one chapter in that volume to develop ECMs. In the end, however, it is our hope that we were able to show that the processes occurring in a lithium-ion cell are fundamentally simple even if the equations appear daunting. The principal mechanism of lithium movement is diffusion, which is a fairly straightforward concept to understand. Lithium tends to move from high-concentration regions to low-concentration regions within the cell unless energy is applied to the cell from an external source that counteracts this natural movement.

- PBMs contain many more parameters than ECMs; values for these parameters must be determined for any physical cell we wish to model and manage. Historically, it has been necessary to disassemble cells and conduct special-purpose experiments to measure the values of the parameters. This requires highly trained scientists and expensive scientific equipment, neither of which are available to many companies wishing to build a BMS. Cell disassembly may also be forbidden by legal agreements made between a company and its cell supplier. Chap. 3 in this volume teaches methods that can be used to estimate PBM parameter values, most of which do not require cell disassembly.
- Solving PBM equations is computationally more demanding than when using ECMs. For example, the continuum-scale models developed in Vol. I comprise a set of four partial-differential equations (PDEs) coupled by a nonlinear algebraic closure term. A direct solution to this model requires multiphysics software that uses finite-element, finite-volume, or finite-difference methods. However, there are numerous ways to simplify the computational requirements of PBMs. We present our preferred method in Chap. 4 of this volume. The final physics-based reduced-order model (ROM) has computational complexity (number of floating-point operations per solution) similar to that of a high-fidelity ECM.
- BMS algorithms based on PBMs have not been proven in practice to the same extent as for ECMs. However, there is increasing evidence in the literature that research teams around the globe focusing on this topic are finding success in lab-scale implementations. We believe that it is only a matter of time before BMS methods based on PBMs attain a level of maturity that allows their use in commercial applications. Chaps. 5 to 8 in this volume share some methods that use PBMs to determine estimates of SOC, SOH, SOP, and optimized charging profiles.

When we published the first two volumes of this series in 2015, we thought that we were almost ready to write this third volume. We believed that we were about 90 % there. It has taken us a few years to realize that the last 10 % of the background research and development is much more difficult than the first 90 %. Now, we feel like we are 99 % of the way there. But it may well be that the remaining 1 % is more difficult than the first 99 %. Rather than delaying this volume any longer, we chose to move forward and share with you what we do know about the subject, hoping that it will inspire many of you to continue to move the field forward and accelerate the path toward 100 %. We don't pretend that we have the final answer to every question; in fact, we make mention in many chapters of the topics we know that we don't know enough about. We hope this will provide hints to researchers looking for important directions to pursue. That said, all topics presented in this book have been validated at least in simulation; most have undergone preliminary validation in a laboratory setting.

While there remains room to develop this field far beyond where it is now, we do consider the 99 % that we have to offer to be more than sufficient to build prototype PBM-based BMS for test and evaluation and perhaps even commercial BMS for some applications. The methods we present in detail in this volume have been developed exclusively by the research team at the University of Colorado Colorado Springs and, unlike most BMS algorithms based on ECMs, we have purposefully not sought patent protection for these methods. They are open source and we encourage adoption by the community.[4]

[4] We still recommend that the reader perform their own due diligence and search the patent literature; it may well be that other research teams around the globe have patented methods that are essentially the same.

In the remainder of this chapter, we first review some background topics from Vols. I and II upon which this volume builds. We then specifically review the continuum-scale PDEs that form the PDE PBM full-order model (FOM) that we use as a starting point for this volume. Finally, we transform this model into an equivalent mathematical form that will prove to be required later in the book.

1.1 Background topics and a roadmap to this book

The first volume in this series taught how to make sets of equations or *models* that describe how battery cells work, inside and out. The topics are illustrated iconically by the roadmap in Fig. 1.1.

We first studied empirical ECMs, which can describe input/output (current/voltage) behaviors of battery cells quite well. These models can be used by the BMS algorithms presented in Vol. II to estimate SOC, SOH, SOP, and SOE and to inform balancing decisions. We then looked at PBMs, which can additionally describe the internal electrochemical processes occurring in the cell. Different PBMs exist at different length scales. We noted that some research teams develop models that describe battery physics at molecular scales, but we did not discuss those models in detail. We began our in-depth derivation of cell models at the particle scale, where dynamics occurring inside electrode particles and inside the electrolyte are considered separately. These particle-scale models can be highly accurate but are extremely demanding computationally; it is not feasible to simulate more than a handful of particles using desktop computers. So,

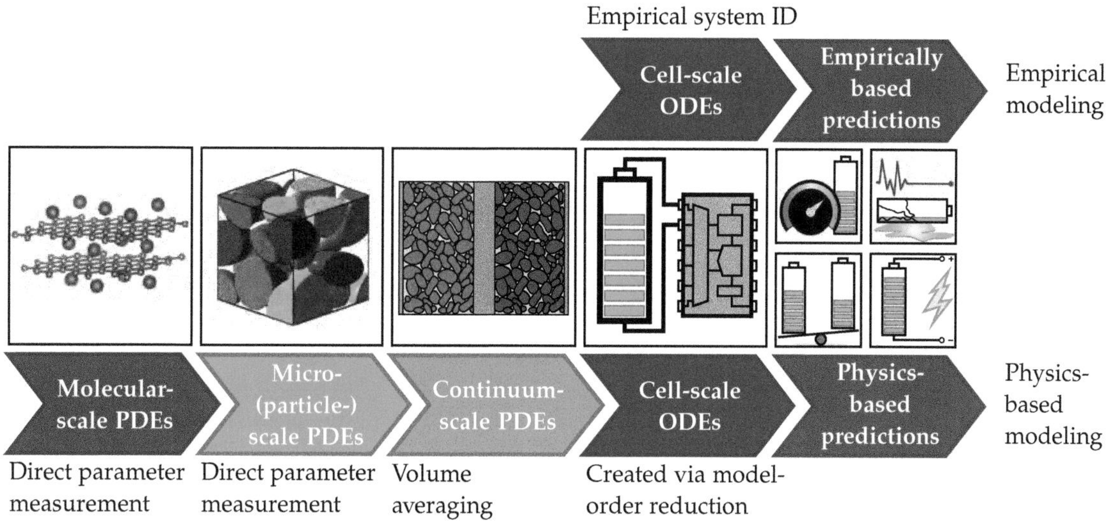

Figure 1.1: Different approaches to making models of lithium-ion cells presented in Vol. I.

we showed how volume-averaging theorems could be applied to the particle-scale equations to produce continuum-scale equations, the Doyle–Fuller–Newman (DFN) pseudo-two-dimensional (P2D) model. This model loses some fidelity by averaging out any existing inhomogeneities in microstructure details, but also greatly reduces the computational complexity of simulating the model—to the point where a desktop computer can often run cell-level simulations in near real time. This P2D model is still not well suited for BMS application, so we proposed the discrete-time realization algorithm (DRA) to approximate the P2D model using a low-order discrete-time state-space model form that has similar computational complexity to a high-fidelity ECM. This ROM is still able to compute predictions of all cell internal electrochemical variables at all spatial locations of interest interior to the cell, in addition to computing predictions of terminal voltage and temperature. This type of model is also suitable for estimating SOC, SOH, SOP, and SOE and to inform balancing decisions, as we will describe later in this volume.

Both kinds of model can be used inside of a BMS to regulate battery-pack usage. The second volume in this series taught how to use ECMs to enable battery-pack controls. The topics covered by that book are illustrated iconically by the roadmap in Fig. 1.2.

Vol. II first gave an overview of BMS requirements and then began to focus on the main algorithm control loop of the BMS. Every measurement interval, battery-pack current, module temperatures, and cell voltages are measured. These are used in combination with a cell model to produce estimates of SOC for all cells in the battery pack. Additional methods were developed to track SOH by updating

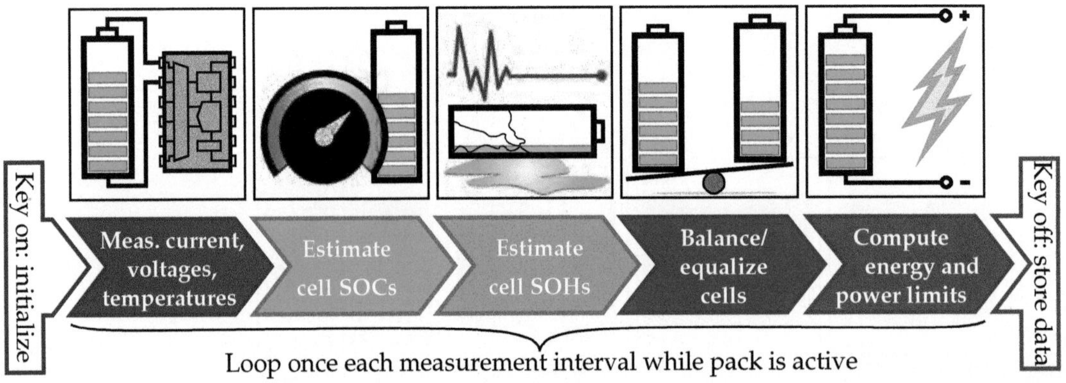

Figure 1.2: The battery-management-system main algorithm control loop presented in Vol. I.

estimates of cell total capacity and resistance as the pack ages. Cell balancing requirements and strategies and power-limits estimation were also covered in detail.

The focus of this third volume is to learn how to perform many of these same tasks, but using PBMs instead of ECMs. To do so, we will need to revisit the modeling and controls aspects of the first two volumes. The first four chapters of this volume are focused on modeling, and the remaining four chapters are focused on controls. The topics of the first four chapters are illustrated iconically by the roadmap figure presented in Fig. 1.3.

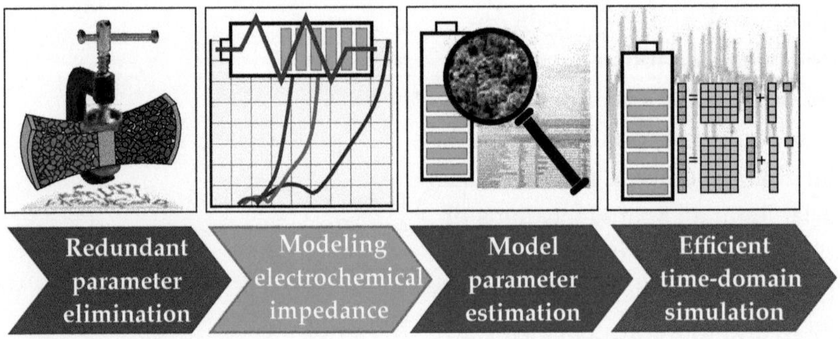

Figure 1.3: Topics in lithium-ion cell modeling that we cover in this volume.

This chapter reviews the continuum-scale DFN P2D model on which this volume is based. It then shows how to reformulate this model to eliminate redundant parameters, which will be necessary when we attempt to estimate the values of these parameters using experimental data. Chap. 2 shows how to enhance the model by adding detail—beyond that considered in the standard DFN model—to equations describing the processes occurring at the electrode/electrolyte interface. It also develops closed-form impedance models of the cell, which can describe frequency-dependent behaviors. Chap. 3 is one of the major contributions of this book, and shows how to estimate values for the parameters of a PBM using input/output (current/voltage) measurements of the cell in such a way that we frequently do not require cell teardown to characterize a PBM. Only standard types of equipment found in many battery laboratories are needed; specialized training and expensive scientific apparatus are not required. The parameter values determined by the methods of this chapter are used in a full-order PDE-based PBM that is still general; you may choose to adopt any of a number of methods in the literature to simplify this model further for BMS application. However, we present our preferred approach in Chap. 4. It is similar in many ways to the DRA methodology from Vol. I, except that we have shown in Chap. 2 how to improve the transfer functions used as a basis for the method, and we further show how to replace the DRA with a plug-in substitute named the hybrid realization algorithm (HRA), which executes much more quickly and is very robust. We also show how to use the resulting ROM in time-domain simulations across wide cell SOC and temperature operating ranges, and how to simulate constant-voltage and constant-power events.

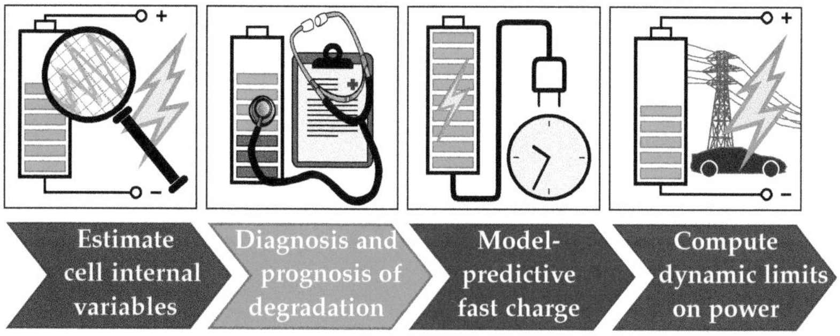

Figure 1.4: Topics in lithium-ion battery control that we cover in this volume.

The topics of the remaining four chapters are illustrated iconically by the roadmap figure presented in Fig. 1.4. Chap. 5 shows how to use the ROM to estimate the entire cell internal electrochemical state, including SOC but far more than that. Chap. 6 considers diagnostics and prognostics applied to aged cells. Chap. 7 teaches how to compute optimal fast-charge profiles using the ROM and a technology known as model-predictive control (MPC). Finally, Chap. 8 shows how to compute power-limit estimates that seek to optimize a tradeoff between battery long-term life and short-term performance.

1.2 *Lithium-ion cell models*

In Vol. I, we derived the equations for both ECMs and PBMs. There are advantages and disadvantages to using either inside BMS algorithms. ECMs are tuned by fitting measured current/voltage data to empirical relationships. In operation, the ECM is essentially performing an intelligent interpolation among previously seen data when using the model to predict the cell voltage at its present operating condition. We need to be cautious when using ECMs: interpolating among measured data can often be done reliably, but extrapolation beyond the measured data is hazardous. An ECM can make good predictions if the present cell operating condition resembles those under which the cell was exercised when collecting data to find the model parameter values. It can make very poor and even nonsensical predictions if the present operating condition is very different from those encountered in the training process. Therefore, large datasets must be collected to train the model and substantial model validation must be performed to ensure that BMS predictions using ECMs are reliable. As we have already mentioned, ECMs are also limited in that they can predict input/output (current/voltage) behaviors only. They cannot predict cell internal electrochemical states. But ECMs have the very positive feature of being numerically robust and computationally simple.

PBMs, on the other hand, tend to have opposite advantages and disadvantages. Unlike ECMs, their equations are based on first principles of electrochemistry and physics and so the models can predict cell behaviors over a wide range of operating conditions, even those not encountered in the laboratory when collecting data to estimate the model parameter values. PBMs also have the ability to predict a cell's internal electrochemical state, which is necessary for aging predictions and for developing fast-charge and power-limits algorithms that consider the impact of cell usage on its rates of degradation. However, traditional PBMs are computationally complex and the PDE solvers required for evaluating their equations are prone to numeric fragility when stimulated using rapidly changing input-current profiles.

Our preferred strategy develops ROMs from the PDEs of the DFN P2D model. These ROMs share all the advantages of both ECMs and PBMs and none of the disadvantages. That is, they can make high-fidelity predictions across a wide range of operating conditions, can predict values for all cell internal electrochemical variables, and have computational complexity similar to a high-accuracy ECM while also maintaining the numeric robustness of an ECM.

In this book, we do not redevelop ECMs or BMS algorithms based on ECMs. Instead, our focus is on using PBMs in BMS algorithms. So, we begin by reviewing the DFN P2D model upon which we base everything that follows.

1.3 *Review of DFN model*

Fig. 1.5 illustrates the geometry of a 1D slice through a lithium-ion cell, roughly to typical scale. The cell comprises three main sections: the negative-electrode, the separator, and the positive-electrode regions.[5] The electrodes themselves are not homogeneous blocks of material; instead, they comprise millions of microscopic particles, increasing the surface area of the electrode/electrolyte interface where the chemical reactions occur. In so doing, this has the effect of decreasing cell resistance and increasing its power capabilities. The figure shows the negative electrode in a green color and the positive electrode in a magenta color for illustrative purposes only; the actual materials are powders that tend to be gray or black in color.

[5] In the literature, the negative electrode is often called the anode and the positive electrode is often called the cathode. Strictly speaking, this terminology is correct only during discharge and must actually be reversed when describing a cell being charged. For this reason, we prefer to use the terms negative and positive electrode instead, to reduce confusion.

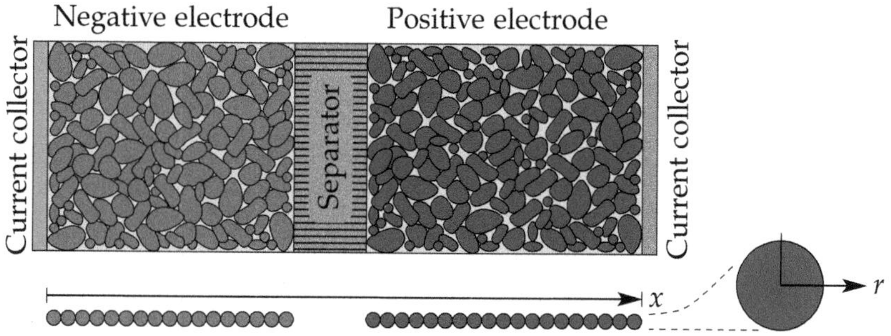

Figure 1.5: Illustration of the cross-sectional geometry of a lithium-ion cell.

The electrode particles are adhered to current collectors, which conduct electronic current to the cell's terminals. Negative-electrode current collectors are made from copper foils and positive-electrode current collectors are made from aluminum foils. The separator is a porous membrane that allows ions to pass through but blocks electron current, preventing the cell from having an internal short circuit. The voids between particles and in the porous separator are permeated with an electrolyte, which conducts ions from one cell region to another but is insulating to electron flow.

In the diagram, the thickness dimension is denoted by the variable x. The P2D model simplifies electrode-model equations by assuming that all electrode particles are spherical and have identical radius R_s. Then, the radial dimension inside any particle is denoted by the variable r. In the model, the length dimension is physical but the radial dimension is somewhat artificial because of this assumption: this is why the model is referred to as a "pseudo" two-dimensional model. The radial dimension is the pseudo dimension.

The P2D model describes the operation of a lithium-ion cell using four PDEs along with associated boundary conditions and initial values, as well as a single nonlinear algebraic equation that serves as a coupling closure term. This set of equations describes the following electrochemical variables of the cell at spatial location x and time t:

- The concentration of lithium in the solid electrode particles, denoted as $c_s(x,r,t)$, and particularly the concentration at the surface of the solid at the boundary between the solid and the electrolyte, denoted as $c_{s,e}(x,t)$;
- The concentration of lithium in the electrolyte, denoted as $c_e(x,t)$;
- The potential in the solid electrode particles, denoted as $\phi_s(x,t)$;
- The potential in the electrolyte, denoted as $\phi_e(x,t)$;
- The lithium flux (rate of lithium movement between solid electrode particles and the electrolyte), denoted as $j(x,t)$.

We now review the model equations. In the following, note that all variables are functions of x and t, but this dependence will be dropped from the notation unless it is required to clarify some specific point. Further, all parameters are functions of the cell region being described by the equation being reviewed, but their values are assumed to be uniform (constant) across that cell region.

EQUATION I: Charge conservation in the solid.

The first equation we review describes charge conservation within the solid particles that form an electrode. This is modeled as:[6]

$$\frac{\partial}{\partial x}\sigma_{\text{eff}}^{\text{r}}\frac{\partial}{\partial x}\phi_{\text{s}}^{\text{r}} = a_{\text{s}}^{\text{r}}Fj^{\text{r}}, \tag{1.1}$$

where $\sigma_{\text{eff}}^{\text{r}}$ is the effective conductivity of the electrode, a_{s}^{r} is the specific interfacial surface area of the electrode, and F is Faraday's constant.[7] This equation applies only to the negative-electrode and positive-electrode regions of the cell. It is a linear diffusion equation describing electron movement, with a forcing term that models flux of electrons, which is equal to the flux of lithium from the electrode to the electrolyte locally.

The boundary conditions for the PDE can be stated as:

$$\sigma_{\text{eff}}^{\text{n}}\frac{\partial}{\partial x}\phi_{\text{s}}^{\text{n}}\bigg|_{x=0} = \sigma_{\text{eff}}^{\text{p}}\frac{\partial}{\partial x}\phi_{\text{s}}^{\text{p}}\bigg|_{x=L^{\text{tot}}} = \frac{-i_{\text{app}}}{A}, \tag{1.2}$$

$$\frac{\partial}{\partial x}\phi_{\text{s}}^{\text{n}}\bigg|_{x=L^{\text{n}}} = \frac{\partial}{\partial x}\phi_{\text{s}}^{\text{p}}\bigg|_{x=L^{\text{n}}+L^{\text{s}}} = 0, \tag{1.3}$$

where L^{n} is the thickness of the negative electrode, L^{s} is the thickness of the separator, L^{p} is the thickness of the positive electrode, $L^{\text{tot}} = L^{\text{n}} + L^{\text{s}} + L^{\text{p}}$, A is the surface area of the current collector, and i_{app} is the electrical current measured at the terminals of the cell.[8]

The initial values for this PDE are:

$$\phi_{\text{s},0}^{\text{n}} = 0, \quad \text{and} \quad \phi_{\text{s},0}^{\text{p}} = U_{\text{ocp}}^{\text{p}}(\theta_{\text{s},0}^{\text{p}}) - U_{\text{ocp}}^{\text{n}}(\theta_{\text{s},0}^{\text{n}}), \tag{1.4}$$

[6] An appendix to this chapter, Sect. 1.A.2, presents a list defining the parameters of the DFN model for reference.

[7] The superscript "r" for "region" indicates that this equation is parameterized by different constants in different regions of the cell. When it is important to be more specific, we use superscripts "n" for the negative-electrode region, "s" for the separator region, and "p" for the positive-electrode region.

[8] In the sign convention that we use in this series of books, $i_{\text{app}} > 0$ for discharge currents and $i_{\text{app}} < 0$ for charge currents.

where $\theta_s^r = c_s^r / c_{s,max}^r$ is the stoichiometry of the electrode such that $0 \le \theta_s^r \le 1$, $\theta_{s,0}^r = c_{s,0}^r / c_{s,max}^r$ is the initial equilibrium stoichiometry of the electrode, and $U_{ocp}^r(\theta_s^r)$ is the open-circuit potential (OCP) of the electrode as a function of local stoichiometry.

EQUATION II: Mass conservation in the solid.

The second PDE describes mass conservation in the solid electrode materials. It is also valid only for the negative-electrode and positive-electrode regions and is modeled as:

$$\frac{\partial c_s^r}{\partial t} = \frac{1}{r^2}\frac{\partial}{\partial r}\left(D_s^r r^2 \frac{\partial c_s^r}{\partial r}\right), \tag{1.5}$$

where D_s^r is the solid diffusivity of the electrode.[9] This equation is a restatement of Fick's second law, formulated in spherical coordinates and assuming spherical symmetry. That is, lithium motion inside electrode particles is modeled as standard linear diffusion.

The boundary conditions imposed on this PDE are:

$$D_s^r \frac{\partial c_s^r}{\partial r}\bigg|_{r=R_s^r} = -j^r, \quad \text{and} \quad D_s^r \frac{\partial c_s^r}{\partial r}\bigg|_{r=0} = 0. \tag{1.6}$$

The initial values are:

$$c_{s,0}^r = c_{s,max}^r\left(\theta_0^r + z_0(\theta_{100}^r - \theta_0^r)\right), \tag{1.7}$$

where $0 \le z_0 \le 1$ is initial cell SOC, θ_0^r is the value of θ_s^r when the cell is resting at 0 % SOC, and θ_{100}^r is the value of θ_s^r when the cell is resting at 100 % SOC.

EQUATION III: Charge conservation in the electrolyte.

The third PDE describes charge conservation in the electrolyte and is valid for all cell regions. It is modeled as:

$$\frac{\partial}{\partial x}\kappa_{eff}^r\left(\frac{\partial}{\partial x}\phi_e^r + \frac{2RT\left(t_+^0 - 1\right)}{F}\left(1 + \frac{\partial \ln f_\pm}{\partial \ln c_e}\right)\frac{\partial \ln c_e^r}{\partial x}\right) + a_s^r F j^r = 0, \tag{1.8}$$

where κ_{eff}^r is the effective conductivity of the electrolyte, R is the universal gas constant, t_+^0 is the transference number of the positive ion in the electrolyte with respect to the solvent, and $f_\pm$ is the mean molar activity coefficient.[10] This equation is dominated by linear-diffusion terms, but the $\partial \ln c_e / \partial x$ term modifies the relationship somewhat to account for a nonlinear concentration dependence.

The boundary conditions for this equation enforce the physical constraint that all current at the current-collector boundaries must be electronic and all current at the separator boundaries must be ionic:

$$-\kappa_{eff}^r\left[\frac{\partial}{\partial x}\phi_e^r + \frac{2RT\left(t_+^0 - 1\right)}{F}\left(1 + \frac{\partial \ln f_\pm}{\partial \ln c_e}\right)\frac{\partial \ln c_e^r}{\partial x}\right|_{\substack{x=0\\x=L^{tot}}} = 0 \tag{1.9}$$

[9] For now, it is sufficient to think about D_s^r as being constant. However, this diffusivity is truly concentration-dependent. We incorporate a description of this dependence into the model in later chapters.

[10] We do not use the region superscript "r" on t_+^0 or on $\partial \ln f_\pm / \partial \ln c_e$ since these relationships are electrolyte properties that we assume to be constant over all regions.

$$-\kappa^{\mathrm{r}}_{\mathrm{eff}}\left[\frac{\partial}{\partial x}\phi^{\mathrm{r}}_{\mathrm{e}}+\frac{2RT\,(t^0_+-1)}{F}\left(1+\frac{\partial \ln f_\pm}{\partial \ln c_{\mathrm{e}}}\right)\frac{\partial \ln c^{\mathrm{r}}_{\mathrm{e}}}{\partial x}\right|_{\substack{x=L^{\mathrm{n}}\\ x=L^{\mathrm{n}}+L^{\mathrm{s}}}}=\frac{i_{\mathrm{app}}}{A}. \tag{1.10}$$

Initial values across the entire cell are:[11]

$$\phi^{\mathrm{r}}_{\mathrm{e},0}=-U^{\mathrm{n}}_{\mathrm{ocp}}(\theta^{\mathrm{n}}_{\mathrm{s},0}). \tag{1.11}$$

EQUATION IV: Mass conservation in the electrolyte.

The fourth PDE describes mass conservation in the electrolyte and is valid for all cell regions. It is modeled as:

$$\frac{\partial(\varepsilon^{\mathrm{r}}_{\mathrm{e}}c^{\mathrm{r}}_{\mathrm{e}})}{\partial t}=\frac{\partial}{\partial x}D^{\mathrm{r}}_{\mathrm{e,eff}}\frac{\partial}{\partial x}c^{\mathrm{r}}_{\mathrm{e}}+a^{\mathrm{r}}_{\mathrm{s}}(1-t^0_+)j^{\mathrm{r}}, \tag{1.12}$$

where $\varepsilon^{\mathrm{r}}_{\mathrm{e}}$ is the porosity (volume fraction of the electrolyte in the cell region), and $D^{\mathrm{r}}_{\mathrm{e,eff}}$ is the diffusivity of lithium in the electrolyte. This is again a linear diffusion equation with a forcing term that describes the addition of lithium due to flux of lithium from the electrode into the electrolyte locally. It applies to all regions of the cell.

The boundary conditions for this equation enforce continuity of both electrolyte concentration and flux of lithium across regions of the cell, and also enforce that there be no movement of lithium from inside the cell to the exterior of the cell:

$$\left.\frac{\partial c^{\mathrm{n}}_{\mathrm{e}}}{\partial x}\right|_{x=0}=\left.\frac{\partial c^{\mathrm{p}}_{\mathrm{e}}}{\partial x}\right|_{x=L^{\mathrm{tot}}}=0 \tag{1.13}$$

$$D^{\mathrm{n}}_{\mathrm{e,eff}}\left.\frac{\partial c^{\mathrm{n}}_{\mathrm{e}}}{\partial x}\right|_{x=(L^{\mathrm{n}})^-}=D^{\mathrm{s}}_{\mathrm{e,eff}}\left.\frac{\partial c^{\mathrm{s}}_{\mathrm{e}}}{\partial x}\right|_{x=(L^{\mathrm{n}})^+} \tag{1.14}$$

$$D^{\mathrm{s}}_{\mathrm{e,eff}}\left.\frac{\partial c^{\mathrm{s}}_{\mathrm{e}}}{\partial x}\right|_{x=(L^{\mathrm{n}}+L^{\mathrm{s}})^-}=D^{\mathrm{p}}_{\mathrm{e,eff}}\left.\frac{\partial c^{\mathrm{p}}_{\mathrm{e}}}{\partial x}\right|_{x=(L^{\mathrm{n}}+L^{\mathrm{s}})^+} \tag{1.15}$$

$$c^{\mathrm{n}}_{\mathrm{e}}\big|_{x=(L^{\mathrm{n}})^-}=c^{\mathrm{s}}_{\mathrm{e}}\big|_{x=(L^{\mathrm{n}})^+} \tag{1.16}$$

$$c^{\mathrm{s}}_{\mathrm{e}}\big|_{x=(L^{\mathrm{n}}+L^{\mathrm{s}})^-}=c^{\mathrm{p}}_{\mathrm{e}}\big|_{x=(L^{\mathrm{n}}+L^{\mathrm{s}})^+}. \tag{1.17}$$

Initial values across the cell are:

$$c_{\mathrm{e}}=c_{\mathrm{e},0}. \tag{1.18}$$

EQUATION V: Kinetics.

The final model equation is a nonlinear algebraic closure term known as the Butler–Volmer equation. It describes the kinetics of the cell—the rate of reaction and hence the rate of lithium flux from the electrode particles into the electrolyte:

$$j^{\mathrm{r}}=j^{\mathrm{r}}_0\left\{\exp\left(\frac{(1-\alpha^{\mathrm{r}})F}{RT}\eta^{\mathrm{r}}\right)-\exp\left(\frac{-\alpha^{\mathrm{r}}F}{RT}\eta^{\mathrm{r}}\right)\right\} \tag{1.19}$$

[11] To see this, notice from Eq. (1.21) that when the cell is in equilibrium, $\phi^{\mathrm{n}}_{\mathrm{s},0}-\phi^{\mathrm{n}}_{\mathrm{e},0}=U^{\mathrm{n}}_{\mathrm{ocp}}(\theta^{\mathrm{n}}_{\mathrm{s},0})$. Further, from Eq. (1.4) we know that $\phi^{\mathrm{n}}_{\mathrm{s},0}=0$ and since the cell is in equilibrium, Eq. (1.10) implies that $\nabla\phi^{\mathrm{r}}_{\mathrm{e}}=0$. Therefore, $\phi^{\mathrm{n}}_{\mathrm{e},0}=\phi^{\mathrm{p}}_{\mathrm{e},0}=-U^{\mathrm{n}}_{\mathrm{ocp}}(\theta^{\mathrm{n}}_{\mathrm{s},0})$.

$$j_0^{\mathrm{r}} = k_{\mathrm{norm},0}^{\mathrm{r}} \left(\frac{c_{\mathrm{e}}^{\mathrm{r}}}{c_{\mathrm{e},0}^{\mathrm{r}}} \right)^{1-\alpha^{\mathrm{r}}} \left(1 - \frac{c_{\mathrm{s,e}}^{\mathrm{r}}}{c_{\mathrm{s,max}}^{\mathrm{r}}} \right)^{1-\alpha^{\mathrm{r}}} \left(\frac{c_{\mathrm{s,e}}^{\mathrm{r}}}{c_{\mathrm{s,max}}^{\mathrm{r}}} \right)^{\alpha^{\mathrm{r}}} \tag{1.20}$$

$$\eta^{\mathrm{r}} = \phi_{\mathrm{s}}^{\mathrm{r}} - \phi_{\mathrm{e}}^{\mathrm{r}} - U_{\mathrm{ocp}}^{\mathrm{r}}(c_{\mathrm{s,e}}^{\mathrm{r}}/c_{\mathrm{s,max}}^{\mathrm{r}}) - FR_{\mathrm{f}}^{\mathrm{r}} j^{\mathrm{r}}, \tag{1.21}$$

where α^{r} is the charge-transfer coefficient, η^{r} is the local overpotential, j_0^{r} is the exchange-flux density, $k_{\mathrm{norm},0}^{\mathrm{r}}$ is a reaction-rate constant, and $R_{\mathrm{f}}^{\mathrm{r}}$ is the resistivity of the surface film that may exist on the electrode particles. This equation is valid for the negative- and positive-electrode regions and is a good description of reaction kinetics for constant-current events, but misses some critical cell behaviors at high frequencies. In Chap. 2, we will revisit this kinetics model and enhance it to describe more detail at the interface between the solid and electrolyte that will be important when modeling cell response to dynamic current inputs, especially at high frequencies.

1.4 *Reducing number of parameters: method*

The governing equations listed in Sect. 1.3 contain 36 parameter values (plus $U_{\mathrm{ocp}}^{\mathrm{r}}$ relationships) that must be determined to use the cell model.[12] These are listed in Table 1.1 and summarized in Sect. 1.A.2. If possible, we would like to find ways to estimate the values of these parameters for a specific cell without opening (or tearing down) the cell and performing invasive measurements. That is, we would prefer to use only input/output current/voltage measurements to find parameter values, which is the topic of Chap. 3.

[12] The content of this section has been adapted from: Gregory L. Plett and M. Scott Trimboli, "Process for lumping parameters to enable nondestructive parameter estimation for lithium-ion physics-based models," in *Proceedings of the 35th International Electric Vehicle Symposium and Exhibition (EVS35)*, Oslo, Norway, June 2022.

It turns out that some of the parameters in Table 1.1 cannot be estimated uniquely from input/output data. It is mathematically impossible to do so. Technically, we state that these parameters are *nonidentifiable*. However, the governing equations listed in Sect. 1.3 can be manipulated to combine together the nonidentifiable parameters into a smaller group of *lumped parameters* that are possible to estimate from input/output data, at least in principle. The resulting lumped-parameter model (LPM) retains the ability to predict voltage and (scaled) versions of the cell's internal variables, and can be used in BMS algorithms to optimize a tradeoff between performance and life.

Table 1.1: Parameters in the standard DFN model.

Negative electrode	Separator	Positive electrode
$\sigma_{\mathrm{eff}}^{\mathrm{n}}$		$\sigma_{\mathrm{eff}}^{\mathrm{p}}$
$a_{\mathrm{s}}^{\mathrm{n}}$		$a_{\mathrm{s}}^{\mathrm{p}}$
L^{n}	L^{s}	L^{p}
$\kappa_{\mathrm{eff}}^{\mathrm{n}}$	$\kappa_{\mathrm{eff}}^{\mathrm{s}}$	$\kappa_{\mathrm{eff}}^{\mathrm{p}}$
$D_{\mathrm{s}}^{\mathrm{n}}$		$D_{\mathrm{s}}^{\mathrm{p}}$
$R_{\mathrm{s}}^{\mathrm{n}}$		$R_{\mathrm{s}}^{\mathrm{p}}$
$\varepsilon_{\mathrm{e}}^{\mathrm{n}}$	$\varepsilon_{\mathrm{e}}^{\mathrm{s}}$	$\varepsilon_{\mathrm{e}}^{\mathrm{p}}$
$D_{\mathrm{e,eff}}^{\mathrm{n}}$	$D_{\mathrm{e,eff}}^{\mathrm{s}}$	$D_{\mathrm{e,eff}}^{\mathrm{p}}$
$k_{\mathrm{norm},0}^{\mathrm{n}}$		$k_{\mathrm{norm},0}^{\mathrm{p}}$
$c_{\mathrm{s,max}}^{\mathrm{n}}$		$c_{\mathrm{s,max}}^{\mathrm{p}}$
α^{n}		α^{p}
$R_{\mathrm{f}}^{\mathrm{n}}$		$R_{\mathrm{f}}^{\mathrm{p}}$
θ_0^{n}		θ_0^{p}
$\theta_{100}^{\mathrm{n}}$		$\theta_{100}^{\mathrm{p}}$
A, t_+^0, $\partial \ln f_\pm / \partial \ln c_{\mathrm{e}}$, $c_{\mathrm{e},0}$ span all		

The principal new contribution of this chapter is to demonstrate the process for taking the standard PBM equations and converting them to an LPM. This will be required by the parameter-estimation procedures developed in Chap. 3 and will factor into the model enhancements presented in Chap. 2. We will develop the LPM by modifying the standard PBM equation by equation. First, we normalize length scales and then we define new scaled variables to replace the original variables. By the end of this chapter, we will have reformulated all equations to eliminate redundant parameters, condensing the parameters we must determine from cell tests to characterize the dynamics of a cell to a minimum identifiable set.

1.4.1 *Intensive versus extensive quantities*

A quick example might help illustrate the issue we are trying to solve. Consider Newton's second law, which states: "force equals mass times acceleration." But this is equivalent to saying, "force equals density times volume times acceleration." So, both $F = ma$ and $F = \rho Va$ are true.

Suppose that we measure $F = 1\,\mathrm{N}$ and $a = 1\,\mathrm{m\,s^{-2}}$. We can easily calculate that $m = F/a = 1\,\mathrm{kg}$. However, based only on these measured F and a, we have *no idea* what are the distinct values of ρ and V.

- We could have $\rho = 1\,\mathrm{kg\,m^{-3}}$ and $V = 1\,\mathrm{m^3}$;
- We could have $\rho = 0.5\,\mathrm{kg\,m^{-3}}$ and $V = 2\,\mathrm{m^3}$;
- We could have $\rho = 2\,\mathrm{kg\,m^{-3}}$ and $V = 0.5\,\mathrm{m^3}$;

and infinitely many other combinations.

The issue is that ρ and V are joined together in Newton's model in a way that is *impossible* to separate mathematically using measurements of only $\{F, a\}$. It simply cannot be done. We would need to perform an independent *kind* of experiment (e.g., measure ρ or V directly) to determine both values with certainty.

Why do we have this problem with our PBMs? Recall from Vol. I that we say that a property is *intensive* if it is a normalized quantity. If everything (dimensions, moles, etc.) in a system is doubled, the value of an intensive property remains unchanged. Examples of intensive properties are pressure, temperature, concentration, and density. Alternately, we say that a property is *extensive* if it is a total quantity. If everything in a system is doubled, the value of an extensive property also doubles. Examples of extensive properties are internal energy, Gibbs free energy, volume, and mass.

In Vol. I we preferred to work with intensive properties. This allowed us to derive models that apply directly to any scale, whereby we scale intensive values to fit a particular application. But, for the parameter-estimation problem that we address in Chap. 3 of this volume, we *require* extensive properties! This is for the same reason that we could compute mass (extensive) in Newton's example but we could not find density (intensive, scaled by volume to compute mass).

So, we will now reformulate the physics-based-model PDEs to lump together groups of parameters that always appear together in equations and *cannot* be identified independently (although many researchers who do not recognize this problem have tried!). This is like lumping ρ and V by defining a new variable $m = \rho V$. The lumped parameters of the reformulated PBM are identifiable from input/output measurements, whereas the original parameters are not independently identifiable. In the majority of the cases, the effect of the reformulation is simply to convert intensive properties to extensive properties, but there are a few exceptions where more parameters combine to form a group.

1.4.2 *How are we going to do this?*

Before applying the method we will use to the P2D model, we will use it with a simple example to illustrate the process. Assume that we have pairs of measurements (x, y) where $0 \le x \le 1$. A possible dataset of such measurements is plotted in Fig. 1.6. For simplicity, we skip the step of normalizing x since it already has a normalized range.

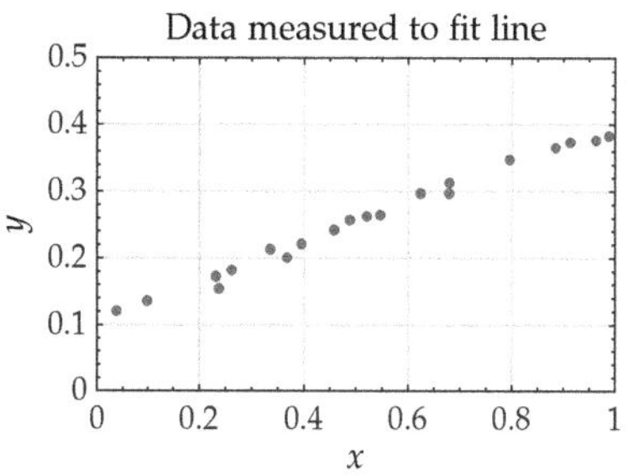

Figure 1.6: Noisy data to which we desire to fit a line.

Now suppose that we desire to fit these data to some physically derived relationship:

$$My = (A + B)x + C. \tag{1.22}$$

We define scaled versions of all of the variables in the equation: we let $\breve{y} = y/\bar{y}$ and $\breve{x} = x/\bar{x}$. In this notation, an overline $(\bar{\cdot})$ indicates a constant (not yet determined)—usually chosen to normalize the primary variable to make the $(\breve{\cdot})$ variable unitless—and $(\breve{\cdot})$ is a temporary notation that indicates a (possibly time-varying) variable that has been normalized and (often) made unitless.

When we make this substitution, we write

$$\begin{aligned} M\bar{y}\breve{y} &= (A + B)\bar{x}\breve{x} + C \\ \breve{y} &= \left[\frac{(A + B)\bar{x}}{M\bar{y}}\right]\breve{x} + \left[\frac{C}{M\bar{y}}\right]. \end{aligned}$$

Now let's look in more detail at the quantities in brackets. We choose to define $\bar{y} = C/M$ and $\bar{x} = C/(A + B)$. Then we have reformulated the original equation to a simpler form,

$$\breve{y} = \breve{x} + 1. \tag{1.23}$$

This equation itself has no more constants that can be estimated.

To implement the physical relationship, we must estimate two quantities from our measured (x, y) data: a modified slope (reciprocal) $\bar{x}$ and a modified y-intercept $\bar{y}$. These two parameters (or variations on these) are the only ones that can be estimated from input/output data. In terms of the original relationship of Eq. (1.22), it is impossible to find unique values for the parameters A, B, C, and M from input/output data to characterize this model. However, we can implement Eq. (1.23) and solve for $y = \check{y}\bar{y}$ and for $x = \check{x}\bar{x}$ to find the same variables. We have not lost any ability to predict the variables of the model—we have simply lumped constant parameter values together to make it possible to determine the coefficients of the equation via measured data.

1.5 *Reducing number of parameters: application*

We now apply this method to the PBM PDE model, equation by equation, starting with ϕ_s.

EQUATION 1: Charge conservation in solid.

The first step to applying the method is to normalize the cell's 1D thickness dimension by defining:

$$\tilde{x} = (x - x_0^{\mathrm{r}})/L^{\mathrm{r}} + x_1^{\mathrm{r}},$$

where x_0^{r} is the starting location of each region and x_1^{r} is the index of each region.[13] The result of this normalization is:

[13] Note that x_0^{r} is a dimension in meters and x_1^{r} is a dimensionless zero-based index: $x_1^{\mathrm{n}} = 0$, $x_1^{\mathrm{s}} = 1$, and $x_1^{\mathrm{p}} = 2$.

$0 \le \tilde{x} \le 1$ in the negative electrode;

$1 \le \tilde{x} \le 2$ in the separator region;

$2 \le \tilde{x} \le 3$ in the positive electrode.

Then we can also relate derivatives with respect to x to derivatives with respect to $\tilde{x}$:

$$\frac{\partial(\cdot)}{\partial x} = \frac{1}{L^{\mathrm{r}}}\frac{\partial(\cdot)}{\partial \tilde{x}}.$$

The second step to applying the method is to define scaled versions of all variables. For the charge-conservation equation, we let $\check{\phi}_s = \phi_s/\bar{\phi}_s$ and $\check{j} = j/\bar{j}$. Note that this also implies that $\partial\phi_s/\partial(\cdot) = \bar{\phi}_s\partial\check{\phi}_s/\partial(\cdot)$. So, starting with Eq. (1.1), we can write:

$$\begin{aligned}
\frac{\partial}{\partial x}\sigma_{\mathrm{eff}}^{\mathrm{r}}\frac{\partial}{\partial x}\phi_s^{\mathrm{r}} &= a_s^{\mathrm{r}}Fj^{\mathrm{r}}, \\
\frac{\bar{\phi}_s^{\mathrm{r}}\sigma_{\mathrm{eff}}^{\mathrm{r}}}{(L^{\mathrm{r}})^2}\frac{\partial^2}{\partial\tilde{x}^2}\check{\phi}_s^{\mathrm{r}} &= a_s^{\mathrm{r}}F\bar{j}^{\mathrm{r}}\check{j}^{\mathrm{r}} \\
\frac{\partial^2}{\partial\tilde{x}^2}\check{\phi}_s^{\mathrm{r}} &= F\left[\frac{a_s^{\mathrm{r}}(L^{\mathrm{r}})^2\bar{j}^{\mathrm{r}}}{\bar{\phi}_s^{\mathrm{r}}\sigma_{\mathrm{eff}}^{\mathrm{r}}}\right]\check{j}^{\mathrm{r}}.
\end{aligned}$$

We will ultimately define $\bar{j}^{\mathrm{r}}$ and $\bar{\phi}_s^{\mathrm{r}}$ to make the bracketed constant in this equation disappear, to help minimize the number of parameter values that must be determined.

Using the same approach, the nonzero boundary conditions of Eq. (1.2) are rewritten as:

$$\frac{\sigma^{\mathrm{n}}_{\mathrm{eff}}\bar{\phi}^{\mathrm{n}}_{\mathrm{s}}}{L^{\mathrm{n}}}\frac{\partial}{\partial\tilde{x}}\check{\phi}^{\mathrm{n}}_{\mathrm{s}}\bigg|_{\tilde{x}=0} = \frac{\sigma^{\mathrm{p}}_{\mathrm{eff}}\bar{\phi}^{\mathrm{p}}_{\mathrm{s}}}{L^{\mathrm{p}}}\frac{\partial}{\partial\tilde{x}}\check{\phi}^{\mathrm{p}}_{\mathrm{s}}\bigg|_{\tilde{x}=3} = \frac{-i_{\mathrm{app}}}{A}$$

$$\left[\frac{A\sigma^{\mathrm{n}}_{\mathrm{eff}}\bar{\phi}^{\mathrm{n}}_{\mathrm{s}}}{L^{\mathrm{n}}}\right]\frac{\partial}{\partial\tilde{x}}\check{\phi}^{\mathrm{n}}_{\mathrm{s}}\bigg|_{\tilde{x}=0} = \left[\frac{A\sigma^{\mathrm{p}}_{\mathrm{eff}}\bar{\phi}^{\mathrm{p}}_{\mathrm{s}}}{L^{\mathrm{p}}}\right]\frac{\partial}{\partial\tilde{x}}\check{\phi}^{\mathrm{p}}_{\mathrm{s}}\bigg|_{\tilde{x}=3} = -i_{\mathrm{app}}.$$

Finally, we consider the initial values from Eq. (1.4):

- $\phi^{\mathrm{n}}_{\mathrm{s},0} = \check{\phi}^{\mathrm{n}}_{\mathrm{s}} = 0$ in the negative electrode.
- $\phi^{\mathrm{p}}_{\mathrm{s},0} = U^{\mathrm{p}}_{\mathrm{ocp}}(\theta^{\mathrm{p}}_{\mathrm{s},0}) - U^{\mathrm{n}}_{\mathrm{ocp}}(\theta^{\mathrm{n}}_{\mathrm{s},0})$ in the positive electrode so,

$$\check{\phi}^{\mathrm{p}}_{\mathrm{s},0} = \left(U^{\mathrm{p}}_{\mathrm{ocp}}(\theta^{\mathrm{p}}_{\mathrm{s},0}) - U^{\mathrm{n}}_{\mathrm{ocp}}(\theta^{\mathrm{n}}_{\mathrm{s},0})\right)/\bar{\phi}^{\mathrm{p}}_{\mathrm{s}}.$$

The third step to applying the method is to make assignments to the scaling constants to simplify the equations. We choose to:

- Let $\bar{\phi}^{\mathrm{r}}_{\mathrm{s}} = \frac{L^{\mathrm{r}}}{A\sigma^{\mathrm{r}}_{\mathrm{eff}}}$ and define $\bar{\sigma}^{\mathrm{r}} = \frac{A\sigma^{\mathrm{r}}_{\mathrm{eff}}}{L^{\mathrm{r}}}$,[14] so $\bar{\phi}^{\mathrm{r}}_{\mathrm{s}} = 1/\bar{\sigma}^{\mathrm{r}}$.[15]
- Let $\bar{j}^{\mathrm{r}} = \frac{1}{a^{\mathrm{r}}_{\mathrm{s}}AL^{\mathrm{r}}}$.

This reduces the PDE, its boundary conditions, and initial values to:

$$\frac{\partial^2\check{\phi}^{\mathrm{r}}_{\mathrm{s}}}{\partial\tilde{x}^2} = F\check{j}^{\mathrm{r}}$$

$$\frac{\partial}{\partial\tilde{x}}\check{\phi}^{\mathrm{r}}_{\mathrm{s}}\bigg|_{\tilde{x}=0} = \frac{\partial}{\partial\tilde{x}}\check{\phi}^{\mathrm{r}}_{\mathrm{s}}\bigg|_{\tilde{x}=3} = -i_{\mathrm{app}}$$

$$\check{\phi}^{\mathrm{n}}_{\mathrm{s},0} = 0$$

$$\check{\phi}^{\mathrm{p}}_{\mathrm{s},0} = \left(U^{\mathrm{p}}_{\mathrm{ocp}}(\theta^{\mathrm{p}}_{\mathrm{s},0}) - U^{\mathrm{n}}_{\mathrm{ocp}}(\theta^{\mathrm{n}}_{\mathrm{s},0})\right)/\bar{\phi}^{\mathrm{p}}_{\mathrm{s}}.$$

This process has completely eliminated all cell-dependent parameters from the PDE and its boundary conditions, but we still must estimate $\bar{\phi}^{\mathrm{p}}_{\mathrm{s}}$ and specify $\theta^{\mathrm{n}}_{\mathrm{s},0}$ and $\theta^{\mathrm{p}}_{\mathrm{s},0}$ to be able to describe initial values. We must also determine the open-circuit-potential relationships $U^{\mathrm{r}}_{\mathrm{ocp}}(\theta^{\mathrm{r}}_{\mathrm{s}})$ for both electrodes.

EQUATION II: Mass conservation in solid.

We now apply the same process to the PDE describing mass conservation in the solid, Eq. (1.5). We normalize the radial pseudo dimension by defining $\tilde{r} = r/R^{\mathrm{r}}_{\mathrm{s}}$; then, $\tilde{r} = 0$ at the center of a particle and $\tilde{r} = 1$ at the surface of a particle. We also define $\check{c}^{\mathrm{r}}_{\mathrm{s}} = c^{\mathrm{r}}_{\mathrm{s}}/\bar{c}^{\mathrm{r}}_{\mathrm{s}}$.

Substituting these definitions into Eq. (1.5), we can write:

$$\frac{\partial c^{\mathrm{r}}_{\mathrm{s}}}{\partial t} = \frac{1}{r^2}\frac{\partial}{\partial r}\left(D^{\mathrm{r}}_{\mathrm{s}}r^2\frac{\partial c_{\mathrm{s}}}{\partial r}\right)$$

$$\bar{c}^{\mathrm{r}}_{\mathrm{s}}\frac{\partial\check{c}^{\mathrm{r}}_{\mathrm{s}}}{\partial t} = \frac{1}{(R^{\mathrm{r}}_{\mathrm{s}}\tilde{r})^2}\frac{1}{R^{\mathrm{r}}_{\mathrm{s}}}\frac{\partial}{\partial\tilde{r}}\left(D^{\mathrm{r}}_{\mathrm{s}}(R^{\mathrm{r}}_{\mathrm{s}}\tilde{r})^2\frac{\bar{c}^{\mathrm{r}}_{\mathrm{s}}}{R^{\mathrm{r}}_{\mathrm{s}}}\frac{\partial\check{c}^{\mathrm{r}}_{\mathrm{s}}}{\partial\tilde{r}}\right)$$

$$\frac{\partial\check{c}^{\mathrm{r}}_{\mathrm{s}}}{\partial t} = \frac{1}{\tilde{r}^2}\frac{\partial}{\partial\tilde{r}}\left(\left[\frac{D^{\mathrm{r}}_{\mathrm{s}}}{(R^{\mathrm{r}}_{\mathrm{s}})^2}\right]\tilde{r}^2\frac{\partial\check{c}^{\mathrm{r}}_{\mathrm{s}}}{\partial\tilde{r}}\right).$$

[14] The original parameter $\sigma^{\mathrm{r}}_{\mathrm{eff}}$ is the *intensive* effective solid conductivity of region r in [S m^{-1}] whereas the lumped parameter $\bar{\sigma}^{\mathrm{r}}$ is the *extensive* effective solid conductivity of region r in [S].

[15] Notice that the constant $\bar{\phi}^{\mathrm{r}}_{\mathrm{s}}$ has units of [Ω] and not [V]; in the procedure that we use, over-bar quantities do not need to have the same units as the variables that they normalize.

The nonzero boundary condition of Eq. (1.6) is:

$$\begin{aligned}
D_{\rm s}^{\rm r} \frac{\partial c_{\rm s}}{\partial r}\bigg|_{r=R_{\rm s}^{\rm r}} &= -j^{\rm r} \\
\frac{D_{\rm s}^{\rm r}\bar{c}_{\rm s}^{\rm r}}{R_{\rm s}^{\rm r}} \frac{\partial \breve{c}_{\rm s}^{\rm r}}{\partial \tilde{r}}\bigg|_{\tilde{r}=1} &= -\bar{j}^{\rm r}\breve{j}^{\rm r} \\
\frac{\partial \breve{c}_{\rm s}^{\rm r}}{\partial \tilde{r}}\bigg|_{\tilde{r}=1} &= -\left[\frac{\bar{j}^{\rm r} R_{\rm s}^{\rm r}}{D_{\rm s}^{\rm r}\bar{c}_{\rm s}^{\rm r}}\right]\breve{j}^{\rm r}.
\end{aligned}$$

The initial values of Eq. (1.7) can be rewritten as:

$$\begin{aligned}
c_{\rm s,0}^{\rm r} &= c_{\rm s,max}^{\rm r}\left(\theta_0^{\rm r} + z_0(\theta_{100}^{\rm r} - \theta_0^{\rm r})\right) \\
\breve{c}_{\rm s,0}^{\rm r} &= \bar{c}_{\rm s,max}^{\rm r}\left(\theta_0^{\rm r} + z_0(\theta_{100}^{\rm r} - \theta_0^{\rm r})\right).
\end{aligned}$$

We now make additional assignments, defining:

- $\bar{D}_{\rm s}^{\rm r} = \frac{D_{\rm s}^{\rm r}}{(R_{\rm s}^{\rm r})^2}$;
- $\bar{c}_{\rm s}^{\rm r} = \frac{\bar{j}^{\rm r} R_{\rm s}^{\rm r}}{D_{\rm s}^{\rm r}}$ and $\bar{c}_{\rm s,max}^{\rm r} = \frac{c_{\rm s,max}^{\rm r}}{\bar{c}_{\rm s}^{\rm r}} = \frac{a_{\rm s}^{\rm r} A L^{\rm r} c_{\rm s,max}^{\rm r} D_{\rm s}^{\rm r}}{R_{\rm s}^{\rm r}}$.

Note that cell total capacity (in Ah) can be written as:[16,17]

$$Q = \varepsilon_{\rm s}^{\rm r} A L^{\rm r} F c_{\rm s,max}^{\rm r} \left|\theta_{100}^{\rm r} - \theta_0^{\rm r}\right| / 3600.$$

If we assume that electrode particles are spherical, then $\varepsilon_{\rm s}^{\rm r} = a_{\rm s}^{\rm r} R_{\rm s}^{\rm r}/3$. So,

$$\begin{aligned}
Q &= \frac{a_{\rm s}^{\rm r} A L^{\rm r} F c_{\rm s,max}^{\rm r} R_{\rm s}^{\rm r}}{10\,800}\left|\theta_{100}^{\rm r} - \theta_0^{\rm r}\right| \\
&= \frac{\bar{c}_{\rm s,max}^{\rm r} F (R_{\rm s}^{\rm r})^2}{10\,800 D_{\rm s}^{\rm r}}\left|\theta_{100}^{\rm r} - \theta_0^{\rm r}\right|,
\end{aligned}$$

or

$$\bar{c}_{\rm s,max}^{\rm r} = \frac{10\,800 \bar{D}_{\rm s}^{\rm r} Q}{F\left|\theta_{100}^{\rm r} - \theta_0^{\rm r}\right|}. \tag{1.24}$$

The benefit of this analysis is that we have replaced two unknown constants, $\bar{c}_{\rm s,max}^{\rm n}$ and $\bar{c}_{\rm s,max}^{\rm p}$, with a single relationship in terms of a single (easily measurable) cell parameter Q and other constants that must be determined in any case when identifying a model of a lithium-ion cell. We can now express the PDE, its nonzero boundary condition, and its initial values as:

$$\begin{aligned}
\frac{\partial \breve{c}_{\rm s}^{\rm r}}{\partial t} &= \frac{1}{\tilde{r}^2}\frac{\partial}{\partial \tilde{r}}\left(\bar{D}_{\rm s}^{\rm r}\tilde{r}^2 \frac{\partial \breve{c}_{\rm s}^{\rm r}}{\partial \tilde{r}}\right) \\
\frac{\partial \breve{c}_{\rm s}^{\rm r}}{\partial \tilde{r}}\bigg|_{\tilde{r}=1} &= -\breve{j}^{\rm r} \\
\breve{c}_{\rm s,0}^{\rm r} &= \frac{10\,800 \bar{D}_{\rm s}^{\rm r} Q}{F\left|\theta_{100}^{\rm r} - \theta_0^{\rm r}\right|}\left(\theta_0^{\rm r} + z_0(\theta_{100}^{\rm r} - \theta_0^{\rm r})\right).
\end{aligned}$$

At this point, the list of unknown parameters that we must estimate comprises: Q, $\bar{\sigma}^{\rm p}$, $\theta_0^{\rm n}$, $\theta_0^{\rm p}$, $\theta_{100}^{\rm n}$, $\theta_{100}^{\rm p}$, $\bar{D}_{\rm s}^{\rm n}$, and $\bar{D}_{\rm s}^{\rm p}$.

[16] Note that electrochemists may be more comfortable expressing capacity in mol rather than Ah. However, we believe that most engineers will be more familiar with expressing capacity in Ah, and so this will be our convention.

[17] This equation relates cell-level measured total capacity to electrode-level utilized capacity. Each electrode actually has greater theoretical capacity than Q, but the utilized capacity is determined by the width of the stoichiometric operating window defined by $\theta_0^{\rm r}$ and $\theta_{100}^{\rm r}$.

EQUATION III: Charge conservation in the electrolyte.

We proceed by applying the same methodology to the PDE that describes charge conservation in the electrolyte, Eq. (1.8). We let $\check{\phi}_e^r = \phi_e^r/\bar{\phi}_e^r$ and $\check{c}_e^r = c_e^r/\bar{c}_e^r$. This gives:

$$\frac{\partial}{\partial x}\kappa_{\text{eff}}^r\left(\frac{\partial}{\partial x}\phi_e^r + \frac{2RT\,(t_+^0-1)}{F}\left(1+\frac{\partial \ln f_\pm}{\partial \ln c_e}\right)\frac{\partial \ln c_e^r}{\partial x}\right) + a_s^r F j^r = 0$$

$$\frac{1}{L^r}\frac{\partial}{\partial \tilde{x}}\left(\frac{\kappa_{\text{eff}}^r\bar{\phi}_e^r}{L^r}\frac{\partial}{\partial \tilde{x}}\check{\phi}_e^r + \frac{\kappa_{\text{eff}}^r}{L^r}\frac{2RT\,(t_+^0-1)}{F}\left(1+\frac{\partial \ln f_\pm}{\partial \ln c_e}\right)\frac{\partial \ln \check{c}_e^r}{\partial \tilde{x}}\right) + a_s^r F \bar{j}^r \check{j}^r = 0$$

$$\frac{\partial}{\partial \tilde{x}}\left(\left[\frac{\kappa_{\text{eff}}^r A\bar{\phi}_e^r}{L^r}\right]\frac{\partial}{\partial \tilde{x}}\check{\phi}_e^r + T\left[\frac{2R(t_+^0-1)}{F}\left(1+\frac{\partial \ln f_\pm}{\partial \ln c_e}\right)\frac{\kappa_{\text{eff}}^r A}{L^r}\right]\frac{\partial \ln \check{c}_e^r}{\partial \tilde{x}}\right) + \underbrace{a_s^r AFL^r\bar{j}^r}_{F}\check{j}^r = 0.$$

The nonzero boundary conditions from Eq. (1.10) are:

$$-\kappa_{\text{eff}}^r\left[\frac{\partial}{\partial x}\phi_e^r + \frac{2RT\,(t_+^0-1)}{F}\left(1+\frac{\partial \ln f_\pm}{\partial \ln c_e}\right)\frac{\partial \ln c_e^r}{\partial x}\right|_{\substack{x=L^n\\ x=L^n+L^s}} = \frac{i_{\text{app}}}{A}$$

$$-\left[\frac{\kappa_{\text{eff}}^r A\bar{\phi}_e^r}{L^r}\right]\frac{\partial}{\partial \tilde{x}}\check{\phi}_e^r - T\left[\frac{2R(t_+^0-1)}{F}\left(1+\frac{\partial \ln f_\pm}{\partial \ln c_e}\right)\frac{\kappa_{\text{eff}}^r A}{L^r}\right]\frac{\partial \ln \check{c}_e^r}{\partial \tilde{x}}\right|_{\tilde{x}=1,2} = i_{\text{app}}.$$

The initial values are reformulated from Eq. (1.11) as:

$$\phi_{e,0}^r = -U_{\text{ocp}}^n(\theta_{s,0}^n) \quad \text{and so}\ldots \quad \check{\phi}_{e,0}^r = -U_{\text{ocp}}^n(\theta_{s,0}^n)/\bar{\phi}_e^r,$$

where $\theta_{s,0}^r = c_{s,0}^r/c_{s,\max}^r$.

We define:

- $\bar{\phi}_e^r = \frac{L^r}{\kappa_{\text{eff}}^r A}$ and $\bar{\kappa}^r = \frac{A\kappa_{\text{eff}}^r}{L^r}$, so $\bar{\phi}_e^r = 1/\bar{\kappa}^r$;
- $\bar{\kappa}_D = \frac{2R(t_+^0-1)}{F}\left(1+\frac{\partial \ln f_\pm}{\partial \ln c_e}\right)$.

This reduces the PDE and its nonzero boundary conditions to:

$$\frac{\partial}{\partial \tilde{x}}\left(\frac{\partial}{\partial \tilde{x}}\check{\phi}_e^r + T\left[\bar{\kappa}_D\bar{\kappa}^r\right]\frac{\partial \ln(\check{c}_e^r)}{\partial \tilde{x}}\right) + F\check{j}^r = 0$$

$$-\frac{\partial}{\partial \tilde{x}}\check{\phi}_e^r - T\left[\bar{\kappa}_D\bar{\kappa}^r\right]\frac{\partial \ln(\check{c}_e^r)}{\partial \tilde{x}}\right|_{\tilde{x}=1,2} = i_{\text{app}}.$$

$$\check{\phi}_{e,0}^r = -U_{\text{ocp}}^n(\theta_{s,e}^n)/\bar{\phi}_e^r.$$

This PDE adds $\bar{\kappa}^n$, $\bar{\kappa}^s$, $\bar{\kappa}^p$, and $\bar{\kappa}_D$ to the list of values that must be identified.

EQUATION IV: Mass conservation in the electrolyte.

We now apply the method to the mass-conservation in electrolyte PDE, Eq. (1.12):

$$\frac{\partial(\varepsilon_e^r c_e^r)}{\partial t} = \frac{\partial}{\partial x}D_{e,\text{eff}}^r\frac{\partial}{\partial x}c_e^r + a_s^r(1-t_+^0)j^r$$

$$
\begin{aligned}
\varepsilon_e^r \bar{c}_e^r \frac{\partial \check{c}_e^r}{\partial t} &= \frac{1}{L^r}\frac{\partial}{\partial \tilde{x}} \frac{D_{e,\text{eff}}^r \bar{c}_e^r}{L^r} \frac{\partial}{\partial \tilde{x}} \check{c}_e^r + a_s^r (1-t_+^0) \bar{j}^r \check{j}^r \\
\left[\frac{\varepsilon_e^r \bar{c}_e^r A L^r}{1-t_+^0}\right] \frac{\partial \check{c}_e^r}{\partial t} &= \frac{\partial}{\partial \tilde{x}} \left[\frac{D_{e,\text{eff}}^r A \bar{c}_e^r}{L^r(1-t_+^0)}\right] \frac{\partial}{\partial \tilde{x}} \check{c}_e^r + \check{j}^r.
\end{aligned}
$$

The nonzero slope boundary condition at the negative-electrode/separator interface, Eq. (1.14), is:

$$
\begin{aligned}
D_{e,\text{eff}}^n \frac{\partial c_e^n}{\partial x}\bigg|_{x=(L^n)^-} &= D_{e,\text{eff}}^s \frac{\partial c_e^s}{\partial x}\bigg|_{x=(L^n)^+} \\
\left[\frac{D_{e,\text{eff}}^n A \bar{c}_e^n}{L^n(1-t_+^0)}\right] \frac{\partial \check{c}_e^n}{\partial \tilde{x}}\bigg|_{\tilde{x}=1^-} &= \left[\frac{D_{e,\text{eff}}^s A \bar{c}_e^s}{L^s(1-t_+^0)}\right] \frac{\partial \check{c}_e^s}{\partial \tilde{x}}\bigg|_{\tilde{x}=1^+}.
\end{aligned}
$$

The nonzero slope boundary condition at the separator/positive-electrode interface, Eq. (1.15), is:

$$
\left[\frac{D_{e,\text{eff}}^s A \bar{c}_e^s}{L^s(1-t_+^0)}\right] \frac{\partial \check{c}_e^s}{\partial \tilde{x}}\bigg|_{\tilde{x}=2^-} = \left[\frac{D_{e,\text{eff}}^p A \bar{c}_e^p}{L^p(1-t_+^0)}\right] \frac{\partial \check{c}_e^p}{\partial \tilde{x}}\bigg|_{\tilde{x}=2^+}.
$$

The continuity boundary conditions at the electrode/separator boundaries, Eqs. (1.16)–(1.17), are:

$$
\begin{aligned}
\check{c}_e^n\big|_{\tilde{x}=1^-} &= \check{c}_e^s\big|_{\tilde{x}=1^+} \\
\check{c}_e^s\big|_{\tilde{x}=2^-} &= \check{c}_e^p\Big|_{\tilde{x}=2^+}.
\end{aligned}
$$

The initial values from Eq. (1.18) are reformulated as:

$$
\begin{aligned}
c_e^r(x,0) &= c_{e,0} \\
\check{c}_e^r(\tilde{x},0) &= \frac{c_{e,0}}{\bar{c}_e^r}.
\end{aligned}
$$

We define $\bar{c}_e^r = c_{e,0}$, which makes $\check{c}_e^r(\tilde{x},0) = \check{c}_{e,0}^r = 1$. We further define $\bar{D}_e^r = D_{e,\text{eff}}^r A \bar{c}_e^r/(L^r(1-t_+^0))$. We temporarily define $\bar{n}_e^r = \varepsilon_e^r \bar{c}_e^r A L^r/(1-t_+^0)$. Then,

$$
\begin{aligned}
\bar{n}_e^r \frac{\partial \check{c}_e^r}{\partial t} &= \frac{\partial}{\partial \tilde{x}} \bar{D}_e^r \frac{\partial \check{c}_e^r}{\partial \tilde{x}} + \check{j}^r \\
\bar{D}_e^{\text{left}} \frac{\partial \check{c}_e}{\partial \tilde{x}}\bigg|_{\text{left}} &= \bar{D}_e^{\text{right}} \frac{\partial \check{c}_e}{\partial \tilde{x}}\bigg|_{\text{right}},
\end{aligned}
$$

where "left" indicates a point and its corresponding cell region immediately to the left of a separator/electrode interface and "right" indicates a point and its cell region immediately to the right of the same interface. That is, at the negative-electrode/separator interface, left $= 1^-$ for the position and left $=$ n for the region; right $= 1^+$ for the position and right $=$ s for the region. Similarly, at the separator/positive-electrode interface, left $= 2^-$ for the position and left = s for the region and right = 2^+ for the position and right = p for the region.

This set of definitions is perfectly correct, but we notice that the units of $\bar{n}_e^r$ are mol, and the parameter quantifies the total amount of lithium in the electrolyte in a certain cell region, scaled by the unitless constant $1 - t_+^0$. Ultimately, we will eliminate all units of mol from the model, retaining more common electrical units of amperes, volts, and ohms. As one step toward that goal, we define a new variable:[18]

$$\bar{q}_e^r = \frac{F}{3600}\bar{n}_e^r,$$

where $\bar{q}_e^r$ is now measured in ampere hours.[19] Then we can write

$$3600\bar{q}_e^r \frac{\partial \check{c}_e^r}{\partial t} = F\frac{\partial}{\partial \tilde{x}}\bar{D}_e^r \frac{\partial \check{c}_e^r}{\partial \tilde{x}} + F\check{j}^r.$$

It appears that we now need also to find $\bar{q}_e^n$, $\bar{q}_e^s$, $\bar{q}_e^p$, $\bar{D}_e^n$, $\bar{D}_e^s$, and $\bar{D}_e^p$ in order to characterize a lithium-ion cell. However, note that there is a relationship between $\bar{D}_e^r$ and $\bar{\kappa}^r$ that we can exploit. Recall from Vol. I the definitions of effective conductivity and diffusivity in terms of the intrinsic conductivity and diffusivity:

$$\begin{aligned}
\kappa_{\text{eff}}^n &= \kappa(\varepsilon_e^n)^{\text{brug}}, & D_{\text{e,eff}}^n &= D_e(\varepsilon_e^n)^{\text{brug}},\\
\kappa_{\text{eff}}^s &= \kappa(\varepsilon_e^s)^{\text{brug}}, & D_{\text{e,eff}}^s &= D_e(\varepsilon_e^s)^{\text{brug}},\\
\kappa_{\text{eff}}^p &= \kappa(\varepsilon_e^p)^{\text{brug}}, & D_{\text{e,eff}}^p &= D_e(\varepsilon_e^p)^{\text{brug}},
\end{aligned}$$

where "brug" is the Bruggeman exponent. Therefore,[20]

$$F\frac{D_{\text{e,eff}}^n}{\kappa_{\text{eff}}^n} = F\frac{D_{\text{e,eff}}^s}{\kappa_{\text{eff}}^s} = F\frac{D_{\text{e,eff}}^p}{\kappa_{\text{eff}}^p} = F\frac{D_e}{\kappa}.$$

Similarly,[21]

$$F\frac{\bar{D}_e^n}{\bar{\kappa}^n} = F\frac{\bar{D}_e^s}{\bar{\kappa}^s} = F\frac{\bar{D}_e^p}{\bar{\kappa}^p} = \bar{\psi}T, \qquad (1.25)$$

and if we have already identified the set of $\bar{\kappa}^r$, then we can compute the set of $\bar{D}_e^r$ by knowing only a single additional parameter, $\bar{\psi}$.

In summary, we rewrite the PDE, its boundary conditions, and initial condition as:

$$\begin{aligned}
3600\bar{q}_e^r \frac{\partial \check{c}_e^r}{\partial t} &= \bar{\psi}T\frac{\partial}{\partial \tilde{x}}\bar{\kappa}^r \frac{\partial \check{c}_e^r}{\partial \tilde{x}} + F\check{j}^r\\
\bar{\kappa}^{\text{left}} \frac{\partial \check{c}_e^r}{\partial \tilde{x}}\bigg|_{\text{left}} &= \bar{\kappa}^{\text{right}} \frac{\partial \check{c}_e^r}{\partial \tilde{x}}\bigg|_{\text{right}}\\
\check{c}_e^r\big|_{\text{left}} &= \check{c}_e^r\big|_{\text{right}}\\
\check{c}_{e,0}^r &= 1.
\end{aligned}$$

Therefore, we now need also to find only $\bar{q}_e^n$, $\bar{q}_e^s$, $\bar{q}_e^p$, and $\bar{\psi}$.

[18] We admit that the constant of 3600 is cumbersome in the equations, but we believe that the final set of equations is less error-prone than if we had expressed $\bar{q}_e^r$ in coulombs.

[19] A physical interpretation of $\bar{q}_e^r$ is that it is the quantity of lithium in the electrolyte in region "r," divided by $(1 - t_+^0)$. We use the symbol "$\bar{q}_e$" to represent capacity or *quantity* of lithium in the electrolyte in ampere hours, much like we use Q to represent the capacity of the cell in ampere hours.

[20] In this analysis, we are assuming that κ_{eff}^r is uniform across each cell region. This is not true in general since κ_{eff}^r is in fact concentration-dependent and the concentration of lithium in the electrolyte varies across the cell. However, since we will enforce this assumption with the lab tests that we will develop to estimate model parameter values in Chap. 3 and we will employ the same assumption when creating reduced-order models in Chap. 4 (as we did in Vol. I), it causes no more harm to make the assumption at this point. To relax this assumption, one would need to replace $\bar{\psi}$ in the model with $\bar{D}_e^r$, as appropriate.

[21] In Chap. 3 we will use the Nernst–Einstein relationship to show that $\bar{D}_e^r/\bar{\kappa}^r$ is expected to be proportional to temperature T. Therefore, we factor out this temperature dependence here such that $\bar{\psi}$ is expected to be independent of temperature.

EQUATION V: The reaction-kinetics.

Finally, we consider the closure term of the model, expressed via Eqs. (1.19)–(1.21). We will not dwell on the details in this chapter since we will refine this model in Chap. 2 to add detail at the electrode/electrolyte boundary. For the time being, we insert the definitions we have created to this point:

$$\breve{j}^{\mathrm{r}} = \breve{j}_0^{\mathrm{r}} \left\{ \exp\left(\frac{(1-\alpha^{\mathrm{r}})F}{RT} \eta^{\mathrm{r}} \right) - \exp\left(\frac{-\alpha^{\mathrm{r}} F}{RT} \eta^{\mathrm{r}} \right) \right\}$$

$$\breve{j}_0^{\mathrm{r}} = \frac{k_{\mathrm{norm},0}^{\mathrm{r}}}{\bar{j}^{\mathrm{r}}} \left(\breve{c}_{\mathrm{e}}^{\mathrm{r}} \right)^{1-\alpha^{\mathrm{r}}} \left(1 - \frac{\breve{c}_{\mathrm{s,e}}^{\mathrm{r}}}{\bar{c}_{\mathrm{s,max}}^{\mathrm{r}}} \right)^{1-\alpha^{\mathrm{r}}} \left(\frac{\breve{c}_{\mathrm{s,e}}^{\mathrm{r}}}{\bar{c}_{\mathrm{s,max}}^{\mathrm{r}}} \right)^{\alpha^{\mathrm{r}}}$$

$$\eta^{\mathrm{r}} = \bar{\phi}_{\mathrm{s}}^{\mathrm{r}} \breve{\phi}_{\mathrm{s}}^{\mathrm{r}} - \bar{\phi}_{\mathrm{e}}^{\mathrm{r}} \breve{\phi}_{\mathrm{e}}^{\mathrm{r}} - U_{\mathrm{ocp}}^{\mathrm{r}} (\breve{c}_{\mathrm{s,e}}^{\mathrm{r}} / \bar{c}_{\mathrm{s,max}}^{\mathrm{r}}) - F \left[R_{\mathrm{f}}^{\mathrm{r}} \bar{j}^{\mathrm{r}} \right] \breve{j}^{\mathrm{r}}.$$

We define $\bar{k}_0^{\mathrm{r}} = k_{\mathrm{norm},0}^{\mathrm{r}} / \bar{j}^{\mathrm{r}}$ and $\bar{R}_{\mathrm{f}}^{\mathrm{r}} = R_{\mathrm{f}}^{\mathrm{r}} \bar{j}^{\mathrm{r}}$. The additional parameter values we must then estimate are $\bar{k}_0^{\mathrm{n}}$, $\bar{k}_0^{\mathrm{p}}$, $\bar{R}_{\mathrm{f}}^{\mathrm{n}}$, $\bar{R}_{\mathrm{f}}^{\mathrm{p}}$, $\bar{\phi}_{\mathrm{s}}^{\mathrm{n}}$, α^{n}, and α^{p}.[22]

[22] Sometimes, we assume $\alpha^{\mathrm{r}} = 0.5$, which removes the requirement of finding α^{n} and α^{p}.

1.6 *Summary of reformulated model equations*

The process that we have followed has reduced the number of parameters that must be identified to characterize a physical lithium-ion cell from 36 down to 23 (21 if we assume $\alpha^{\mathrm{r}} = 0.5$). Thirteen "degrees of freedom" have been removed, making the parameter-estimation task mathematically possible in principle using only input/output current/voltage data. The lumped parameters of the reformulated model are summarized in Table 1.2 and in Sect. 1.A.4.

Table 1.2: Lumped parameters from the reformulated PDEs.

Negative electrode	Separator	Positive electrode
$\bar{\sigma}^{\mathrm{n}}$		$\bar{\sigma}^{\mathrm{p}}$
$\bar{\kappa}^{\mathrm{n}}$	$\bar{\kappa}^{\mathrm{s}}$	$\bar{\kappa}^{\mathrm{p}}$
$\bar{D}_{\mathrm{s}}^{\mathrm{n}}$		$\bar{D}_{\mathrm{s}}^{\mathrm{p}}$
$\bar{q}_{\mathrm{e}}^{\mathrm{n}}$	$\bar{q}_{\mathrm{e}}^{\mathrm{s}}$	$\bar{q}_{\mathrm{e}}^{\mathrm{p}}$
$\bar{k}_0^{\mathrm{n}}$		$\bar{k}_0^{\mathrm{p}}$
$\bar{R}_{\mathrm{f}}^{\mathrm{n}}$		$\bar{R}_{\mathrm{f}}^{\mathrm{p}}$
α^{n}		α^{p}
θ_0^{n}		θ_0^{p}
$\theta_{100}^{\mathrm{n}}$		$\theta_{100}^{\mathrm{p}}$
Q, $\bar{\kappa}_D$, and $\bar{\psi}$ span all regions		

Any similar formulation with the same number of parameters will also be minimal. Since some of the equations of the reformulated model are awkward, we slightly modify them to give a result that is easier to relate directly to the physical processes in the cell. The final notation we develop in this chapter is presented in Table 1.3 and is described below.

We consider first the equations that describe potential in the solid electrodes and the electrolyte. Analyzing the reformulated model shows that the constants $\bar{\phi}_{\mathrm{s}}^{\mathrm{r}}$ and $\bar{\phi}_{\mathrm{e}}^{\mathrm{r}}$ in all applicable cell regions must be estimated to solve for $\breve{\phi}_{\mathrm{s}}^{\mathrm{r}}$ and $\breve{\phi}_{\mathrm{e}}^{\mathrm{r}}$. Therefore, there is no benefit in converting the variables of the model from $\phi_{\mathrm{s}}^{\mathrm{r}}$ to $\breve{\phi}_{\mathrm{s}}^{\mathrm{r}}$ and from $\phi_{\mathrm{e}}^{\mathrm{r}}$ to $\breve{\phi}_{\mathrm{e}}^{\mathrm{r}}$. We choose to keep the original PDEs in terms of $\phi_{\mathrm{s}}^{\mathrm{r}}$ and $\phi_{\mathrm{e}}^{\mathrm{r}}$ but now written in terms of the new lumped-parameter values.

Considering the equations that describe lithium flux and concentrations in the solid electrodes and the electrolyte, we come to a different conclusion. After the reformulation, neither $\bar{c}_{\mathrm{s}}^{\mathrm{r}}$, $\bar{c}_{\mathrm{e}}^{\mathrm{r}}$, nor $\bar{j}^{\mathrm{r}}$ appear in the final PDEs, so we do not need to (nor can we) estimate these values from input/output data. So, for now, we will continue to write the corresponding PDEs as functions of $\breve{c}_{\mathrm{s}}$, $\breve{c}_{\mathrm{e}}$, and $\breve{j}$. However, we do note that the "breve" symbol in $\breve{j}$ becomes cumbersome,

Description	Governing equations	Boundary conditions	Initial conditions
Charge conservation in solid	$\bar{\sigma}^{\mathrm{r}} \frac{\partial^2 \phi_{\mathrm{s}}^{\mathrm{r}}}{\partial \tilde{x}^2} = i_{\mathrm{f}}^{\mathrm{r}}$	$\bar{\sigma}^{\mathrm{r}} \frac{\partial}{\partial \tilde{x}} \phi_{\mathrm{s}}^{\mathrm{r}} \Big\|_{\tilde{x}=0,3} = -i_{\mathrm{app}}$ $\bar{\sigma}^{\mathrm{r}} \frac{\partial}{\partial \tilde{x}} \phi_{\mathrm{s}}^{\mathrm{r}} \Big\|_{\tilde{x}=1,2} = 0$	$\phi_{\mathrm{s},0}^{\mathrm{p}} = U_{\mathrm{ocp}}^{\mathrm{p}}(\theta_{\mathrm{s},0}^{\mathrm{p}}) - U_{\mathrm{ocp}}^{\mathrm{n}}(\theta_{\mathrm{s},0}^{\mathrm{n}})$
Mass conservation in solid	$\frac{\partial \theta_{\mathrm{s}}^{\mathrm{r}}}{\partial t} = \frac{1}{\tilde{r}^2} \frac{\partial}{\partial \tilde{r}} (\bar{D}_{\mathrm{s}}^{\mathrm{r}} \tilde{r}^2 \frac{\partial \theta_{\mathrm{s}}^{\mathrm{r}}}{\partial \tilde{r}})$	$\bar{D}_{\mathrm{s}}^{\mathrm{r}} \frac{\partial \theta_{\mathrm{s}}^{\mathrm{r}}}{\partial \tilde{r}} \Big\|_{\tilde{r}=1} = -\frac{\|\theta_{100}^{\mathrm{r}} - \theta_0^{\mathrm{r}}\|}{10\,800 Q} i_{\mathrm{f}}^{\mathrm{r}}$ $\frac{\partial \theta_{\mathrm{s}}^{\mathrm{r}}}{\partial \tilde{r}} \Big\|_{\tilde{r}=0} = 0$	$\theta_{\mathrm{s},0}^{\mathrm{r}} = \theta_0^{\mathrm{r}} + z_0(\theta_{100}^{\mathrm{r}} - \theta_0^{\mathrm{r}})$
Charge conservation in electrolyte	$\frac{\partial}{\partial \tilde{x}} \left(\bar{\kappa}^{\mathrm{r}} \left(\frac{\partial}{\partial \tilde{x}} \phi_{\mathrm{e}}^{\mathrm{r}} + \bar{\kappa}_D T \frac{\partial \ln(\theta_{\mathrm{e}}^{\mathrm{r}})}{\partial \tilde{x}} \right) \right) + i_{\mathrm{f}}^{\mathrm{r}} = 0$	$\bar{\kappa}^{\mathrm{r}} \left(\frac{\partial}{\partial \tilde{x}} \phi_{\mathrm{e}}^{\mathrm{r}} + \bar{\kappa}_D T \frac{\partial \ln(\theta_{\mathrm{e}}^{\mathrm{r}})}{\partial \tilde{x}} \right) \Big\|_{\tilde{x}=1,2} = -i_{\mathrm{app}}$ $\bar{\kappa}^{\mathrm{r}} \left(\frac{\partial}{\partial \tilde{x}} \phi_{\mathrm{e}}^{\mathrm{r}} + \bar{\kappa}_D T \frac{\partial \ln(\theta_{\mathrm{e}}^{\mathrm{r}})}{\partial \tilde{x}} \right) \Big\|_{\tilde{x}=0,3} = 0$	$\phi_{\mathrm{e},0} = -U_{\mathrm{ocp}}(\theta_{\mathrm{s},0}^{\mathrm{n}})$
Mass conservation in electrolyte	$3600 \bar{q}_{\mathrm{e}}^{\mathrm{r}} \frac{\partial \theta_{\mathrm{e}}^{\mathrm{r}}}{\partial t} = \bar{\psi} T \frac{\partial}{\partial \tilde{x}} \bar{\kappa}^{\mathrm{r}} \frac{\partial}{\partial \tilde{x}} \theta_{\mathrm{e}}^{\mathrm{r}} + i_{\mathrm{f}}^{\mathrm{r}}$	$\bar{\kappa}^{\mathrm{n}} \frac{\partial \theta_{\mathrm{e}}^{\mathrm{n}}}{\partial \tilde{x}} \Big\|_{\tilde{x}=1^-} = \bar{\kappa}^{\mathrm{s}} \frac{\partial \theta_{\mathrm{e}}^{\mathrm{s}}}{\partial \tilde{x}} \Big\|_{\tilde{x}=1^+}$ $\bar{\kappa}^{\mathrm{s}} \frac{\partial \theta_{\mathrm{e}}^{\mathrm{s}}}{\partial \tilde{x}} \Big\|_{\tilde{x}=2^-} = \bar{\kappa}^{\mathrm{p}} \frac{\partial \theta_{\mathrm{e}}^{\mathrm{p}}}{\partial \tilde{x}} \Big\|_{\tilde{x}=2^+}$ $\bar{\kappa}^{\mathrm{r}} \frac{\partial \theta_{\mathrm{e}}^{\mathrm{r}}}{\partial \tilde{x}} \Big\|_{\tilde{x}=0,3} = 0$	$\theta_{\mathrm{e},0} = 1$
Kinetics	$i_{\mathrm{f}}^{\mathrm{r}} = i_0^{\mathrm{r}} \left(\exp\left(\frac{(1-\alpha^{\mathrm{r}})F}{RT} \eta^{\mathrm{r}} \right) - \exp\left(\frac{-\alpha^{\mathrm{r}} F}{RT} \eta^{\mathrm{r}} \right) \right)$ $i_0^{\mathrm{r}} = \bar{k}_0^{\mathrm{r}} (\theta_{\mathrm{e}}^{\mathrm{r}})^{1-\alpha^{\mathrm{r}}} (1 - \theta_{\mathrm{s,e}}^{\mathrm{r}})^{1-\alpha^{\mathrm{r}}} (\theta_{\mathrm{s,e}}^{\mathrm{r}})^{\alpha^{\mathrm{r}}}$ $\eta^{\mathrm{r}} = \phi_{\mathrm{s}}^{\mathrm{r}} - \phi_{\mathrm{e}}^{\mathrm{r}} - U_{\mathrm{ocp}}^{\mathrm{r}}(\theta_{\mathrm{s,e}}^{\mathrm{r}}) - \bar{R}_{\mathrm{f}}^{\mathrm{r}} i_{\mathrm{f}}^{\mathrm{r}}$	$\tilde{x}=\{0,1,2,3\}$ at neg/current-collector, neg/sep, sep/pos, and pos/current-collector boundaries; $\tilde{r}=\{0,1\}$ at particle center and surface. Concentration ratios in solid and electrolyte are $\theta_{\mathrm{s}}=c_{\mathrm{s}}/c_{\mathrm{s,max}}$ and $\theta_{\mathrm{e}}=c_{\mathrm{e}}/c_{\mathrm{e},0}$; potentials in solid and electrolyte are ϕ_{s} and ϕ_{e}; flux is $i_{\mathrm{f}}=a_{\mathrm{s}} ALFj$. "Neg," "sep," and "pos" denoted by "n," "s," and "p"; "r"∈{"n," "s," "p"}.	

Table 1.3: Summary of the lumped-parameter version of the DFN model. *This is a preliminary result. We will refine this model and present an updated table in Chap. 2.* See Table 2.1.

so we rename $i_{\mathrm{f}} = F\check{j}$, recognizing that $\check{j}$ has units of mol s^{-1} and so the "faradaic current" i_{f} has units of amperes.[23] Recapping, the charge-conservation in the solid equation is presented in the first row of Table 1.3, where $\phi_{\mathrm{s}}^{\mathrm{r}}$ is measured in volts and $\bar{\sigma}^{\mathrm{r}}$ in siemens.

[23] Subscript "f" stands for "faradaic"; more on this in Chap. 2. Note that i_{f} is different from i_{app}, i_{s}, and i_{e}.

For the mass conservation in solid equation, $\bar{D}_{\mathrm{s}}^{\mathrm{r}}$ has units of s^{-1}, which is manageable, but $\check{c}_{\mathrm{s}}^{\mathrm{r}}$ is in mol s^{-1}, which is an awkward unit for concentration. We choose to rewrite its PDE in terms of the already-defined unitless $\theta_{\mathrm{s}}^{\mathrm{r}} = \check{c}_{\mathrm{s}}^{\mathrm{r}}/\check{c}_{\mathrm{s,max}}^{\mathrm{r}}$, which does not require finding any additional parameter values. For cleaner notation in the charge conservation in electrolyte relationship, we define a unitless ratio of concentrations, $\theta_{\mathrm{e}}^{\mathrm{r}} = \check{c}_{\mathrm{e}}^{\mathrm{r}}$. In these equations, $\phi_{\mathrm{e}}^{\mathrm{r}}$ is measured in volts, $\bar{\kappa}^{\mathrm{r}}$ in siemens, and t_+^0 is unitless. For the relationship describing mass conservation in the electrolyte, $\bar{q}_{\mathrm{e}}$ is in ampere hours and $\bar{\psi}$ is in volts per kelvin. For the kinetics equation, $\bar{k}_0^{\mathrm{r}}$ is in amperes, $\bar{R}_{\mathrm{f}}^{\mathrm{r}}$ is in ohms, and η^{r} is in volts. These modified equations are listed in the remaining four rows of the table.

Note that we will refine this model in Chap. 2. This table of equations is a preliminary result.

1.7 *Recovering original electrochemical variables*

The reformulated set of PDEs allows us to simulate $\phi_{\mathrm{s}}^{\mathrm{r}}$, $\phi_{\mathrm{e}}^{\mathrm{r}}$, $\theta_{\mathrm{s}}^{\mathrm{r}}$, $\theta_{\mathrm{e}}^{\mathrm{r}}$, and $i_{\mathrm{f}}^{\mathrm{r}}$, which is sufficient to be able to compute cell voltage.[24] However,

[24] This is why it is mathematically possible, in principle, to estimate the required parameter values from current/voltage input/output data.

what if we also desire the ability to simulate c_s^r, c_e^r, and j^r?

This will turn out not to be as necessary as it may seem. For example, cell absolute performance limits $0 \le c_s^r \le c_{s,\max}^r$ automatically translate to $0 \le \theta_s^r \le 1$ without needing to simulate c_s^r directly or to estimate any new parameter values.

To recover all original parameter values and electrochemical variables from the modified set, we would need to be able to estimate ten additional coefficients: A, L^n, L^s, L^p, a_s^n, a_s^p, R_s^n, R_s^p, t_+^0, and $c_{e,0}$. It might seem that we should need to estimate thirteen values, since thirteen degrees of freedom have been removed from the model by the reduction process. However, two of those degrees of freedom were removed by noticing the relationship between $\bar{D}_e^r$ and $\bar{\kappa}^r$, which we summarized by replacing three $\bar{D}_e^r$ parameters with a single constant $\bar{\psi}$ in Eq. (1.25). The remaining degree of freedom was removed by noticing that operational capacity of both electrodes must be equal, which allowed us to replace two $\bar{c}_{s,\max}^r$ parameters with a relationship involving the single constant Q in Eq. (1.24). Note that most of the ten remaining unknown values are physical dimensions and relatively easy to measure via cell teardown if needed. No fancy electrochemical techniques are required.

To recover all original electrochemical variables, if we are not concerned with recovering all original parameter values themselves, we would need to determine only eight additional values: $c_{s,\max}^n$, $c_{s,\max}^p$, $c_{e,0}$, a_s^n, a_s^p, L^n, L^p, and A.[25] Then we could compute:

$$
\begin{aligned}
c_s^r &= c_{s,\max}^r \theta_s^r \\
c_e^r &= c_{e,0}^r \theta_e^r \\
j^r &= \frac{i_f^r}{a_s^r A L^r F}.
\end{aligned}
$$

[25] Note that we can combine the relationships from Sect. 1.5 to write:

$$
c_{s,\max}^r = \frac{10\,800 Q}{a_s^r A L^r R_s^r F |\theta_{100}^r - \theta_0^r|}.
$$

Therefore, we can replace the requirement of finding $c_{s,\max}^n$ and $c_{s,\max}^p$ with the equivalent requirement of finding R_s^n and R_s^p, should those quantities prove to be simpler to measure in practice.

1.8 *Where to from here?*

We have now reformulated the DFN P2D physics-based PDE model of a lithium-ion cell to eliminate redundant parameters. The final form has a minimal number of unknown parameter values that must be determined to characterize a physical lithium-ion cell.

For BMS purposes, we need to convert this PDE model into a reduced-order state-space model. We will follow the same general procedure we introduced in Vol. I, but with some enhancements: First, we find transfer functions for each electrochemical variable. Next, we look at how to find parameter values for the model from laboratory tests. We then use a subspace realization algorithm to convert the transfer functions into reduced-order models. When we have done so, we will be able to simulate cells and packs given known parameter values. Then we have fully parameterized ROMs that can be used for BMS tasks such as SOC estimation, SOH estimation, power-limits estimation, and so forth.

We look at these topics in that order.

1.A Summary of variables

1.A.1 Notation conventions for variables and parameters

In this volume, we have attempted to name variables using a consistent format:

$$\text{name}_{\text{phase}}^{\text{modifier,region}} \quad \text{or} \quad \text{name}_{\text{phase}}^{\text{region}}.$$

- "name" is the variable identifier, such as ϕ for potential, θ for normalized concentration, and so forth.
- "modifier" is a temporary descriptive tag sometimes added during a derivation, but is not permanent (i.e., the modifier is eventually removed from the name).
- "region" is the cell domain to which this variable applies: either n, s, or p, for the negative electrode, separator, and positive electrode, respectively.
  - We sometimes leave the region blank; this usually means that an equation applies to multiple regions.
  - Sometimes the identifier "r" is given, showing more explicitly that the equation applies to multiple regions.
- "phase" is the material to which this variable applies, such as "s" for solid, "e" for electrolyte, "f" for film, and so forth.

We have also attempted to name constant parameters using a similar consistent format:

$$\overline{\text{name}}_{\text{phase}}^{\text{modifier,region}} \quad \text{or} \quad \text{name}_{\text{phase}}^{\text{modifier,region}}.$$

The presence of an overbar denotes a lumped parameter; the absence of an overbar denotes a standard parameter.

- "name" is the parameter identifier, such as κ for electrolyte conductivity.
- "region" and "phase" have the same interpretation as for variables.
- "modifier" is again a temporary descriptive tag sometimes added during a derivation, but is not permanent.

1.A.2 *Variables in the standard DFN model reviewed in this chapter*

Table 1.A.1 lists the parameter values of the DFN model as developed in Vol. I. These are described briefly below. Note that all variables are at least potentially functions of space and time.

Table 1.A.1: Parameters of the DFN P2D model.

Negative electrode	Separator	Positive electrode
σ^n_{eff}		σ^p_{eff}
a^n_s		a^p_s
L^n	L^s	L^p
κ^n_{eff}	κ^s_{eff}	κ^p_{eff}
D^n_s		D^p_s
R^n_s		R^p_s
ε^n_e	ε^s_e	ε^p_e
$D^n_{\text{e,eff}}$	$D^s_{\text{e,eff}}$	$D^p_{\text{e,eff}}$
$k^n_{\text{norm},0}$		$k^p_{\text{norm},0}$
$c^n_{\text{s,max}}$		$c^p_{\text{s,max}}$
α^n		α^p
R^n_f		R^p_f
θ^n_0		θ^p_0
θ^n_{100}		θ^p_{100}
A, t^0_+, $\partial \ln f_\pm / \partial \ln c_e$, $c_{e,0}$ span all		

- α^r [unitless] is the asymmetric charge-transfer coefficient, $0 \le \alpha^r \le 1$.
- a^r_s [$\text{m}^2\,\text{m}^{-3}$] is the specific interfacial surface area: the area of the boundary between solid and electrolyte per unit volume.
- A [m^2] is the cell current-collector plate area.
- brug [unitless] is Bruggeman's exponent for computing effective material properties. Often, we approximate brug $= 1.5$.
- c [$\text{mol}\,\text{m}^{-3}$] is the concentration of lithium in the neighborhood of a given location. c^r_s is used to denote concentration in the solid, $c^r_{s,e}$ is the solid surface concentration, and $c^r_{s,\max}$ is the theoretic maximum lithium concentration in the solid. c^r_e is used to denote concentration in the electrolyte, and $c^r_{e,0}$ is the initial equilibrium electrolyte concentration.
- D [$\text{m}^2\,\text{s}^{-1}$] is a material-dependent diffusivity. D^r_s is used to denote diffusivity in the solid, and D^r_e is used to denote diffusivity in the electrolyte.
- $D^r_{\text{e,eff}}$ [$\text{m}^2\,\text{s}^{-1}$] is the effective diffusivity of the electrolyte, representing a volume-averaged diffusivity of the electrolyte phase in a porous media in the vicinity of a given point. We often model $D^r_{\text{e,eff}} \approx D^r_e(\varepsilon^r_e)^{\text{brug}} = D^r_e(\varepsilon^r_e)^{1.5}$.
- ε^r_e [unitless] is the volume fraction of the electrolyte in an electrode. It is also known as the porosity of the electrode.
- ε^r_s [unitless] is the volume fraction of the solid in an electrode.
- $F = 96\,485$ [$\text{C}\,\text{mol}^{-1}$] is Faraday's constant.
- $f_\pm$ [unitless] is the mean molar activity coefficient.
- η^r [V] is the reaction overpotential.
- j^r [$\text{mol}\,\text{m}^{-2}\,\text{s}^{-1}$] is the rate of positive charge flowing out of a particle across a boundary between the solid and the electrolyte.
- $\kappa^r_{\text{eff}}(x,t)$ [$\text{S}\,\text{m}^{-1}$] is the effective conductivity of the electrolyte, representing a volume-averaged conductivity of the electrolyte phase in a porous media in the vicinity of a given point. We often model $\kappa^r_{\text{eff}} \approx \kappa^r(\varepsilon^r_e)^{\text{brug}} = \kappa^r(\varepsilon^r_e)^{1.5}$.
- k^r_0 [$\text{mol}^{\alpha-1}\,\text{m}^{4-3\alpha}\,\text{s}^{-1}$] is the effective reaction-rate constant.
- $k^r_{\text{norm},0} = k_0(c_{e,0})^{1-\alpha}c_{s,\max}$ [$\text{mol}\,\text{m}^{-2}\,\text{s}^{-1}$] is the effective reaction rate constant with nicer units.
- L^r [m] is the thickness (length) of a cell region.
- ϕ [V] is the scalar field representing the electrostatic potential at a given point. ϕ^r_s is used to denote electric potential in the solid and ϕ^r_e is used to denote electric field in the electrolyte.

- Q [Ah] is the total capacity of a battery cell.
- $R = 8.314$ [J mol^{-1} K^{-1}] is the universal gas constant.
- R_s^r [m] is electrode average particle radius.
- R_f^r [Ω] is the resistance of a surface film that may exist on electrode particles.
- σ_{eff}^r [S m^{-1}] is an electrode-dependent parameter called the effective conductivity, representing a volume-averaged conductivity of the solid matrix in a porous media in the vicinity of a given point. We often model $\sigma_{\text{eff}}^r \approx \sigma^r(\varepsilon_s^r)^{\text{brug}} = \sigma^r(\varepsilon_s^r)^{1.5}$.
- T [K] is the temperature at a point.
- t_+^0 and t_-^0 [unitless] are the transference numbers of the cation and anion with respect to the solvent, respectively.
- θ [unitless] is a stoichiometry: $\theta_s^r = c_s^r/c_{s,\max}^r$. When the cell is resting at 0 % state of charge, then $\theta_s^r = \theta_0^r$; when the cell is resting at 100 % state of charge, then $\theta_s^r = \theta_{100}^r$.
- z [unitless] is cell state of charge.

1.A.3 *Variables and parameters used only in derivations in this chapter, but not thereafter*

The following variables and parameters were used in derivations in this chapter, but will not be used past this point.

- $\bar{c}_e = c_{e,0}$ [mol m^{-3}].
- $\check{c}_e = c_e/\bar{c}_e = \theta_e$ [unitless].
- $\bar{c}_s = \frac{\bar{j}R_s}{D_s} = \frac{R_s}{a_s ALD_s}$ [s m^{-3}].
- $\check{c}_{s,\max} = c_{s,\max}/\bar{c}_s = a_s ALD_s c_{s,\max}/R_s$ [mol s^{-1}].
- $\check{c}_s = c_s/\bar{c}_s = a_s ALD_s c_s/R_s$ [mol s^{-1}].
- $\bar{D}_e = \frac{D_{e,\text{eff}} A c_{e,0}}{L(1-t_+^0)}$ [mol s^{-1}].
- $\bar{\phi}_s = 1/\bar{\sigma}$ [Ω].
- $\bar{\phi}_e = 1/\bar{\kappa}$ [Ω].
- $\bar{j} = \frac{1}{a_s AL}$ [m^{-2}].
- $\check{j} = j/\bar{j} = i_f/F = a_s ALj$ [mol s^{-1}].

1.A.4 *Variables and parameters used continually after this point*

The following variables and parameters were either adopted from the DFN model without change or were defined in this chapter and will be used continually after this point.

- α^r [unitless] is the asymmetric charge-transfer coefficient (unchanged from the standard definition).
- $\bar{D}_s^r = D_s^r/(R_s^r)^2$ [s^{-1}] is the lumped solid diffusivity.
- ϕ_e^r [V], potential in the electrolyte (unchanged from the standard definition).

- ϕ_s^r [V], potential in the solid (unchanged from the standard definition).
- $\bar{\kappa}^r = \kappa_{\text{eff}}^r A/L^r$ [S] is the lumped electrolyte conductivity, where $\bar{\kappa}^r$ is treated as a constant.
- $\bar{\kappa}_D = \frac{2R(t_+^0-1)}{F}\left(1 + \frac{\partial \ln f_\pm}{\partial \ln c_e}\right)$ [V K^{-1}] is a short-hand constant from the electrolyte-potential PDE, where $\frac{\partial \ln f_\pm}{\partial \ln c_e}$ is treated as a constant.
- $\bar{k}_0^r = a_s^r A L^r F k_{\text{norm},0}^r$ [A] is the lumped reaction-rate constant.
- $i_f^r = a_s^r A L^r F j^r$ [A] is the lumped faradaic current from solid to electrolyte.
- $\bar{\psi} = \frac{F}{T}\frac{\bar{D}_e^n}{\bar{\kappa}^n} = \frac{F}{T}\frac{\bar{D}_e^s}{\bar{\kappa}^s} = \frac{F}{T}\frac{\bar{D}_e^p}{\bar{\kappa}^p}$ [V K^{-1}] is the scaled ratio between lumped electrolyte diffusivity and conductivity.
- $\bar{q}_e^r = \frac{\varepsilon_e^r c_{e,0} A L^r F}{3600(1-t_+^0)}$ [Ah] is the lumped value describing the total quantity of lithium in the electrolyte, scaled by $1 - t_+^0$.
- Q [Ah] is cell total capacity.
- $\bar{R}_f^r = \frac{R_f^r}{a_s^r A L^r}$ [Ω] is lumped film resistance.
- $\bar{\sigma}^r = \sigma_{\text{eff}}^r A/L^r$ [S] is lumped solid conductivity.
- $\theta_e^r = c_e^r/c_{e,0}$ [unitless] is the ratio of electrolyte concentration to its equilibrium value; note that $\theta_{e,0}^r = 1$ always.
- θ_s^r [unitless] is a stoichiometry: $\theta_s^r = c_s^r/c_{s,\max}^r$. Note that $\theta_{s,0}^r = \theta_0^r + z(\theta_{100}^r - \theta_0^r)$. When the cell is resting at 0 % state of charge, then $\theta_s^r = \theta_0^r$; when the cell is resting at 100 % state of charge, then $\theta_s^r = \theta_{100}^r$.
- z [unitless] is cell state of charge.

2

Modeling Electrochemical Impedance

In Chap. 1, we reviewed the DFN model and saw how to eliminate redundant parameters from its equations. With the resulting LPM it is mathematically possible, in principle, to estimate all model parameter values from current/voltage measurements.

We are now ready to proceed to the next step in the roadmap of Fig. 2.1. Our goal in this chapter is to develop impedance models of lithium-ion cells. These models will aid with parameter estimation in Chap. 3 and with converting PDEs to ROMs in Chap. 4. They also give insight into frequency-dependent aspects of cell dynamics.

Full-cell impedance models are based on Laplace-domain transfer functions (TFs) of individual electrochemical variables with respect to input current. These TFs are very similar to those presented in Chap. 6 of Vol. I; however, some important improvements have been made to the derivations since that volume was published and we incorporate those improvements here. Also, we add descriptions of several dynamic effects that were not considered in prior models.

We begin by updating the DFN model in five significant ways:

1. We replace the standard Butler–Volmer kinetics with the more general multi-site multi-reaction (MSMR) model.
2. We add a description of an electrical double layer to the electrode/electrolyte interface.

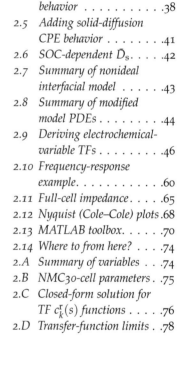

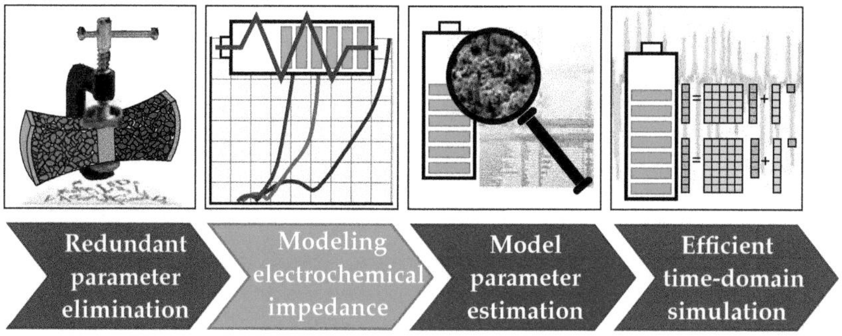

Figure 2.1: Topics in lithium-ion cell modeling that we cover in this volume.

3. We show how to incorporate constant-phase elements (CPEs) into the double-layer description.
4. We also incorporate CPEs into the solid-diffusion equations.
5. We add SOC-dependence to electrode solid diffusivity.

After we have added these new features to the model, we develop TFs for cell internal electrochemical variables. Finally, we combine TFs to achieve a closed-form full-cell impedance model.

2.1 *MSMR model*

The DFN model incorporates a Butler–Volmer kinetics equation to describe local lithium flux from the electrode into the electrolyte. The development of this equation assumes that the flux arises from a single electrochemical reaction whose rate depends on the electrode's OCP relationship. Further, this OCP may be specified as a completely general function of electrode surface stoichiometry.

In Chap. 3, we will find that data collected from physical cells do not always agree with these assumptions. Here, we introduce the MSMR model of interface kinetics, which builds the overall OCP function from a composite of multiple simpler functions.[1] Each simpler function corresponds to an independent reaction that fills and empties a particular *gallery* in the electrode host material.

The MSMR model is inspired by noticing the flat regions of some electrode OCP curves, as illustrated for graphite in Fig. 2.2. These correspond to a phase change between two single-phase solid-solution reactions. The sloped intervals bordering the flat regions correspond to the single-phase reactions.

[1] The most comprehensive treatment of the MSMR model of which we are aware is: Mark Verbrugge, Daniel Baker, Brian Koch, Xingcheng Xiao, and Wentian Gu, "Thermodynamic model for substitutional materials: Application to lithiated graphite, spinel manganese oxide, iron phosphate, and layered nickel-manganese-cobalt oxide," *Journal of The Electrochemical Society*, 164(11):E3243, 2017.

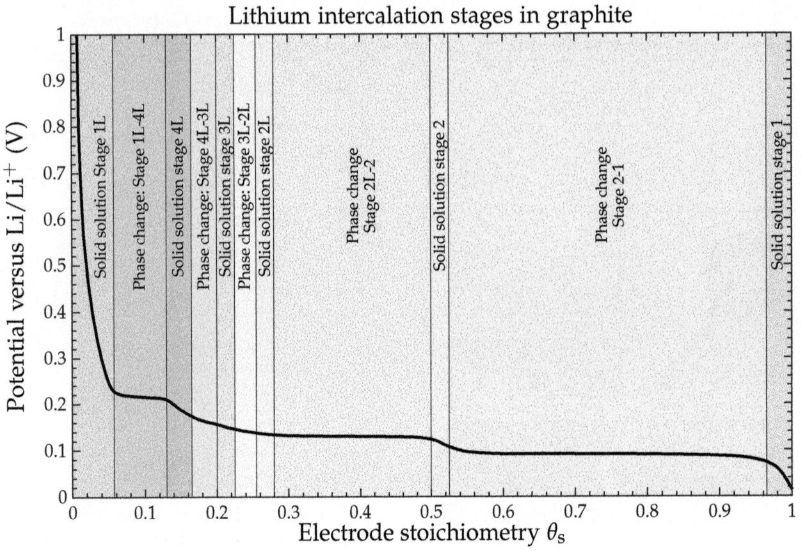

Figure 2.2: Negative-electrode OCP of NMC30 cell overlaid on staging diagram for graphite. (Adapted from Fig. 1 in: Michael Heß and Petr Novák, "Shrinking annuli mechanism and stage-dependent rate capability of thin-layer graphite electrodes for lithium-ion batteries," *Electrochimica Acta*, 106:149–158, 2013.)

From the figure, we observe six distinct solid-solution stages with intermediate phase-change regions. Fig. 2.3 motivates this result by showing how lithium self-organizes between graphene layers in graphite in different patterns for different levels of θ_s. The figure illustrates "stages" corresponding to "empty," "one quarter full," "one third full," "half full," and "full." Notice that patterns form; lithium does not tend to distribute randomly. Near any of these more stable patterns is a single-phase region; between these patterns is a "phase-change" region.

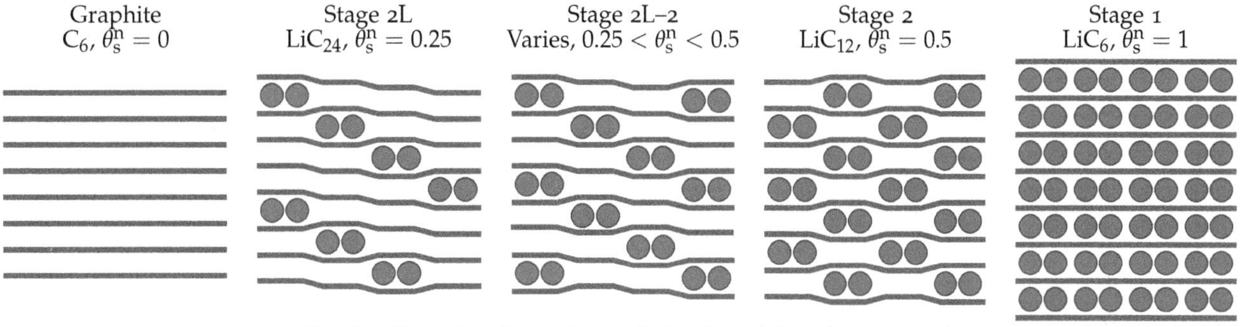

Figure 2.3: Patterns formed by lithium in graphite at different ranges of stoichiometry. (Adapted from Fig. 1 in: David Allart, Maxime Montaru, and Hamid Gualous, "Model of lithium intercalation into graphite by potentiometric analysis with equilibrium and entropy change curves of graphite electrode," *Journal of The Electrochemical Society*, 165(2):A380, 2018.)

The MSMR model describes this behavior by supposing that all phases may coexist locally in an electrode. The galleries of the model describe how full or empty each phase happens to be. Under dynamic conditions, all galleries become more or less full simultaneously, where the reaction for gallery j is expressed as:

$$\mathrm{Li}^+ + \mathrm{e}^- + \mathrm{Host}_j \rightleftharpoons (\mathrm{Li\text{-}Host})_j,$$

where "Host" is a vacant electrode site that could hold Li and (Li-Host) is a filled site. The equation states that every individual unfilled site in each gallery j is capable of holding one Li to become a filled site.

We define $0 < X_j \leq 1$ to be the total number of lithium host sites for gallery j divided by the total number of host sites for all J galleries. Thus, X_j is the fraction of (both filled and unfilled) available sites for gallery j and $\sum_{j=1}^{J} X_j = 1$. We further define $0 \leq x_j \leq X_j$ to be the fraction of filled sites (Li-Host) for gallery j.

The MSMR model then describes the equilibrium gallery potential using a generalized Nernst model (where $f = F/(RT)$):

$$U_j = U_j^0 + \frac{\omega_j}{f} \ln\left(\frac{X_j - x_j}{x_j}\right),$$

where U_j^0 is a concentration-independent potential and ω_j is a unitless shape factor. This model is illustrated by the solid lines in Fig. 2.4 for different ranges of ω; the dashed lines show the corresponding differential capacity of each gallery for the same set of ω.

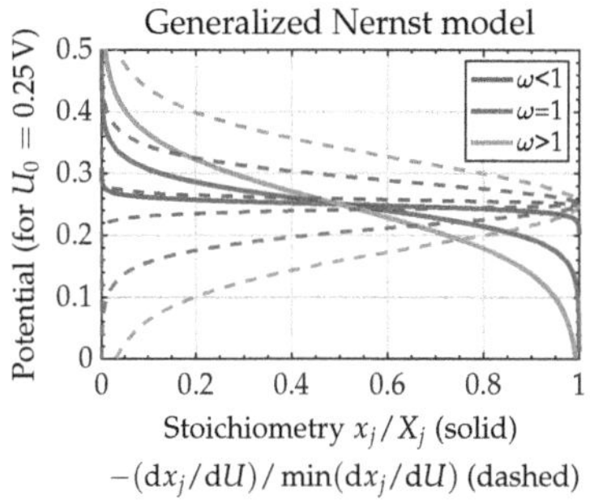

Figure 2.4: Nernst equations (and differential capacity) for different ranges of tuning variable ω.

The following are physical properties for these quantities:

- *In equilibrium*, the $\{x_j\}$ are fixed and the potentials of each gallery are equal (i.e., $U_{\text{ocp}} = U_j$). Also, $U_j = U_j^0$ when $x_j = X_j/2$.
- The logarithmic term predicts that the potential will approach positive infinity when a gallery is empty; the potential approaches negative infinity when a gallery is completely filled.[2]
- The parameter ω_j describes the degree to which the reaction pertains to two-phase versus single-phase behavior.
  - $\omega_j = 0$ represents ideal two-phase behavior (stair-like potential);
  - $\omega_j = 1$ represents ideal single-phase behavior (classic Nernst);
  - $\omega_j > 1$ represents disordered behaviors.

[2] We do not see such extreme limits in an experimental test due to the presence of additional electrochemical reactions to those of the OCP being modeled. For example, copper from the negative-electrode current collector oxidizes at around 3.4 V with respect to Li/Li$^+$ in common electrolytes, limiting maximum observed potential in a negative-electrode material to that level. Lithium plates at around 0 V, limiting minimum measured potential.

We can "read" approximate values for X_j, U_j^0, and ω_j from the OCP and differential-capacity curves.[3] Fig. 2.5 illustrates how we might do this from an OCP versus stoichiometry relationship: U_j^0 are OCP plateau voltages; X_j are roughly equal to plateau widths.

[3] In Chap. 3, we will see that this property helps when initializing optimizations to find MSMR parameter values to match data from an electrode having unknown composition.

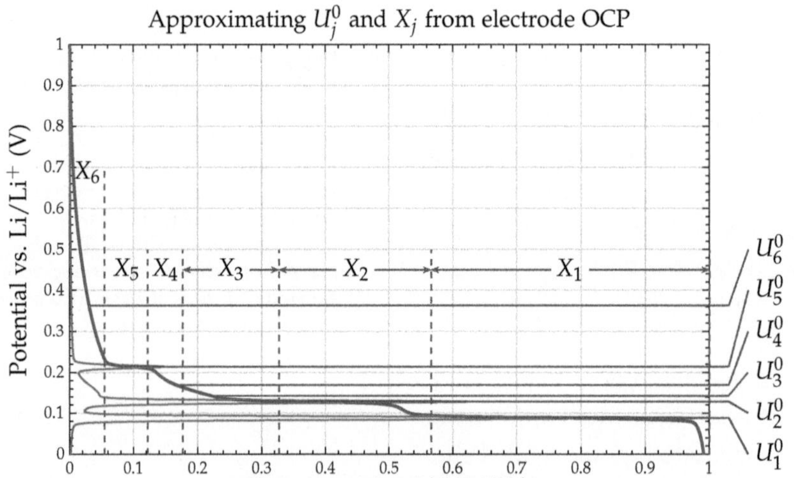

Figure 2.5: OCP of graphite (drawn as blue line), labeled with MSMR parameter values. Normalized differential capacity is drawn as a red line.

We can also read values from differential-capacity plots, as illustrated in Fig. 2.6. The number of galleries is equivalent to the number of peaks in the differential-capacity plot (it is sometimes difficult to see them all). Potentials U_j^0 are the peak locations in a $|\partial\theta_s/\partial U_{\text{ocp}}|$ against U_{ocp} plot and gallery sizes X_j are approximately the peak widths in a $|\partial\theta_s/\partial U_{\text{ocp}}|$ against θ_s plot. We can compute approximate ω_j by looking at peak heights in a differential-capacity plot:

$$\left.\frac{\partial x_j}{\partial U}\right|_{U=U_j^0, x_j=X_j/2} = -\frac{f}{4}\frac{X_j}{\omega_j}.$$

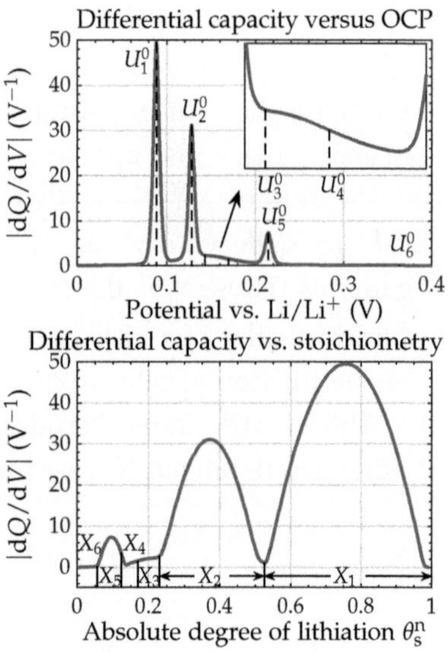

Figure 2.6: Differential capacity of graphite plotted versus potential and stoichiometry, with approximate U_j^0 and X_j marked.

The MSMR model can be manipulated to solve for some important properties in closed form. We can invert the potential equation to find the filled site fractions at the solid/electrolyte interface (denoted using subscript "s, e" as done in Vol. I) at any local potential U:

$$x_{\mathrm{s,e},j} = \frac{X_j}{1+\exp\left[f(U-U_j^0)/\omega_j\right]}.$$

The electrode surface stoichiometry $\theta_{\mathrm{s,e}}$ is then found by summing the fractional occupancies from all galleries:

$$\theta_{\mathrm{s,e}} = \sum_{j=1}^{J} x_{\mathrm{s,e},j} = \sum_{j=1}^{J} \frac{X_j}{1+\exp\left[f(U-U_j^0)/\omega_j\right]}. \tag{2.1}$$

Corresponding differential capacity can be computed analytically:

$$\frac{\partial \theta_{\mathrm{s,e}}}{\partial U} = -\sum_{j=1}^{J} \frac{fX_j\exp\left[f(U-U_j^0)/\omega_j\right]}{\omega_j\left\{1+\exp\left[f(U-U_j^0)/\omega_j\right]\right\}^2}.$$

In equilibrium, $U = U_{\mathrm{ocp}}$, $x_{\mathrm{s,e},j} = x_{0,j}$, and $\theta_{\mathrm{s,e}} = \theta_{\mathrm{s,0}}$. When the cell is not in equilibrium, each gallery fills and empties independently from the others, modeled by using a separate Butler–Volmer equation for each reaction. The fluxes out of each gallery are described as:[4]

$$i_{0,j} = \bar{k}_{0,j}(x_j)^{\omega_j\alpha_j}(X_j - x_j)^{\omega_j(1-\alpha_j)}\theta_{\mathrm{e}}^{(1-\alpha_j)} \tag{2.2}$$

$$i_{\mathrm{f}} = \sum_{j=1}^{J} i_{0,j}\left[\exp((1-\alpha_j)f\eta) - \exp(-\alpha_j f\eta)\right]. \tag{2.3}$$

Each gallery has its own solid-diffusion equation:

$$\frac{\partial x_j}{\partial t} = \frac{1}{\tilde{r}^2}\frac{\partial}{\partial \tilde{r}}\left(\bar{D}_{\mathrm{s}}\tilde{r}^2\frac{\partial x_j}{\partial \tilde{r}}\right) \tag{2.4}$$

$$\bar{D}_{\mathrm{s}}\left.\frac{\partial x_j}{\partial \tilde{r}}\right|_{\tilde{r}=1} = -\frac{|\theta_{100}-\theta_0|}{10\,800Q} i_{\mathrm{f},j}, \tag{2.5}$$

where we remember that the overall electrode stoichiometry at any point is $\theta_{\mathrm{s}} = \sum_{j=1}^{J} x_j$.

[4] Note that the overpotential $\eta = \phi_{\mathrm{s}} - \phi_{\mathrm{e}} - U_{\mathrm{ocp}}(\theta_{\mathrm{s,e}})$ in Eq. (2.3) has the same value for all MSMR reactions since neither ϕ_{s}, nor ϕ_{e}, nor $\theta_{\mathrm{s,e}}$ are a function of the gallery j.

2.2 *Electrical double layer*

We now consider a second model enhancement. Recall that the DFN model defines electrode particle surface overpotential η^{r} as:

$$\eta^{\mathrm{r}} = \phi_{\mathrm{s}}^{\mathrm{r}} - \phi_{\mathrm{e}}^{\mathrm{r}} - U_{\mathrm{ocp}}^{\mathrm{r}}(\theta_{\mathrm{ss}}^{\mathrm{r}}) - \bar{R}_{\mathrm{f}}^{\mathrm{r}} i_{\mathrm{f}}^{\mathrm{r}}. \tag{2.6}$$

Notice that we have renamed $\theta_{\mathrm{s,e}}(\tilde{x},t)$ to $\theta_{\mathrm{ss}}(\tilde{x},t) = \theta_{\mathrm{s}}(\tilde{x},1,t)$. The new subscript denotes solid surface stoichiometry. We can no longer call this variable $\theta_{\mathrm{s,e}}$ because we will be modeling more detail at the particle boundary, and the "s, e" subscript denoting the solid/electrolyte interface will no longer be entirely accurate.

We can rearrange Eq. (2.6) to get:

$$\phi_s^r = U_{ocp}^r(\theta_{ss}^r) + \eta^r + \bar{R}_f^r i_f^r + \phi_e^r. \tag{2.7}$$

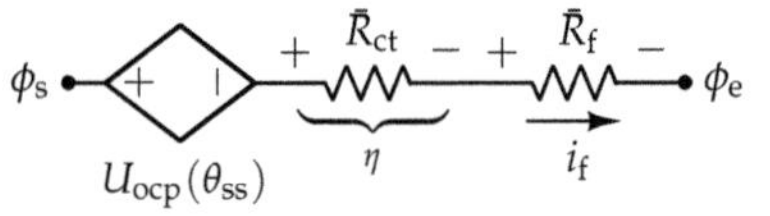

Figure 2.7: Visualizing the old-style interface equations using an equivalent circuit.

This interfacial model can be visualized with the equivalent circuit drawn in Fig. 2.7. Note that this is not the same as an equivalent-circuit model of a cell. We are not approximating dynamic behavior using an empirical relationship chosen in an attempt to describe observed dynamic behaviors; rather, the circuit in the figure is an exact schematic representation of the mathematical relationship of Eq. (2.7), using circuit elements to aid understanding of the equation. In the circuit, "charge-transfer resistance" $\bar{R}_{ct}$ is a virtual resistance that describes the voltage drop η^r when current i_f^r flows.

When seeking to match a PBM to the dynamics of commercial lithium-ion cells, this interfacial model is overly simplistic. This is because the original DFN model was derived to describe constant-current response.[5] To improve the ability of the model to predict transient response, we consider a more general model of the solid/electrolyte interface. This model includes the possibility of a particle surface film having resistance $\bar{R}_f$, as before. But, it adds a double layer of charges at different interfaces, describing the ordering of charged ions in the electrolyte near the electrode surface. The voltage dynamics of these double layers can be modeled as capacitive.

[5] Hence the adjective "galvanostatic" in: Marc Doyle, Thomas F. Fuller, and John Newman, "Modeling of galvanostatic charge and discharge of the lithium/polymer/insertion cell," *Journal of the Electrochemical Society*, 140(6):1526, 1993.

Many double-layer models have been proposed in the literature—there is no consensus regarding which is correct or which is best. We introduce one of the most general models,[6] and you will see that it is straightforward to substitute a different model if desired.

[6] Jeremy P. Meyers, Marc Doyle, Robert M. Darling, and John Newman, "The impedance response of a porous electrode composed of intercalation particles," *Journal of the Electrochemical Society*, 147(8):2930, 2000.

Fig. 2.8 illustrates the generalized interface. The particle surface has electric potential ϕ_s; the film at the particle/film interface has potential ϕ_{sf}; the film at film/electrolyte interface has potential ϕ_{fe}; and the electrolyte at the film interface has potential ϕ_e.

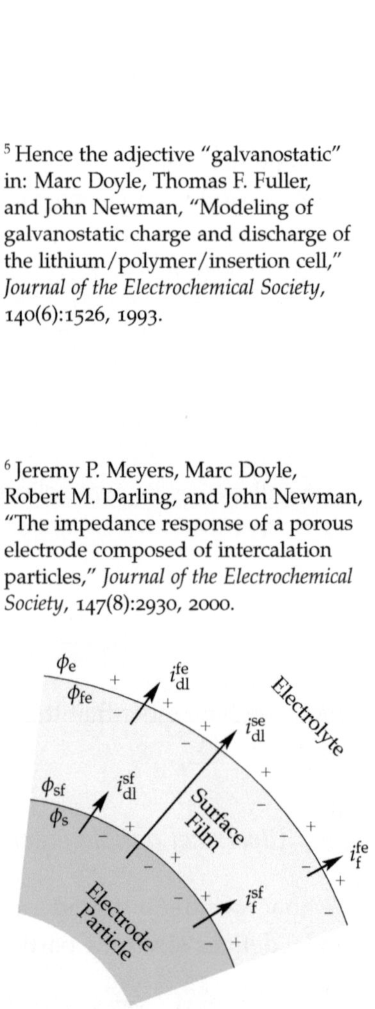

Figure 2.8: A more general model of the solid/electrolyte interface.

Both interfaces are considered to have *faradaic* (noncapacitive) lithium fluxes modeled by an MSMR relationship:[7]

$$i_f = \sum_{j=1}^{J} i_{0,j}\{\exp((1-\alpha_j)f(\phi_1-\phi_2-U_{ocp})) - \exp(-\alpha_j f(\phi_1-\phi_2-U_{ocp}))\},$$

where ϕ_1 is the potential of the phase closest to (or the same as) the particle, ϕ_2 is the potential of the phase closest to (or the same as) the electrolyte, and U_{ocp} is the OCP of the charge-transfer reaction.

[7] Ultimately, we will use a full MSMR model for the particle/film interface and a Butler–Volmer model for the film/electrolyte interface; the latter is simply an MSMR model with $J = 1$.

Interfaces between ϕ_s and ϕ_{sf}, ϕ_{fe} and ϕ_e, and across the film (between ϕ_s and ϕ_e) are also assumed to develop a *nonfaradaic* (capacitive) lithium flux that arises from the charging and discharging of an electrical double layer at that interface. If interface capacitance is considered ideal, we can model lithium flux generically as:

$$i_{\rm dl} = \bar{C}_{\rm dl}\frac{\partial(\phi_1 - \phi_2)}{\partial t}, \tag{2.8}$$

where double-layer capacitance $\bar{C}_{\rm dl}$ is measured in farads. However, we will later want to consider a more realistic double layer that also has a series resistance. In this case, the relationship becomes:[8]

$$\bar{C}_{\rm dl}\frac{\partial(\phi_1 - \phi_2)}{\partial t} = \bar{R}_{\rm dl}\bar{C}_{\rm dl}\frac{\partial i_{\rm dl}}{\partial t} + i_{\rm dl}. \tag{2.9}$$

This generalized model reduces to the ideal case when $\bar{R}_{\rm dl}$ (measured in ohms) is zero. Otherwise, $\bar{R}_{\rm dl}$ models electrolyte ionic friction that leads to noninstantaneous double-layer charge-carrier movement.

2.3 *Ideal interface impedance*

2.3.1 *Impedance at particle/film interface*

We seek to find the impedance of the solid/electrolyte interface when this double-layer model is assumed. We do so in steps, first investigating impedance across the particle/film interface. Recall, diffusion of lithium in a specific gallery has been modeled as having boundary condition Eq. (2.5), which is driven by $i_{{\rm f},j}$. We must now consider whether lithium flux through this boundary comprises only a faradaic component, or if it now has a nonfaradaic component as well. Following Meyers et al., we assume that it remains *only faradaic* since no lithium crosses the interface due to double-layer charging. Excess surface double-layer charge is due to electron flow only.

Because we now consider multiple interfaces, we (temporarily) rewrite $i_{{\rm f},j}$ as $i^{\rm sf}_{{\rm f},j}$. The superscript "sf" denotes "from 's' to 'f' phases." Next, we recall the Jacobsen–West TF from Chap. 5 of Vol. I, rewritten here in terms of the new lumped parameters and MSMR reactions (where $\tilde{x}_{\rm ss} = x_{\rm ss} - x_{{\rm s},0}$ debiases $x_{\rm ss}$):[9,10,11]

$$\widetilde{X}_{{\rm ss},j}(s) = \frac{|\theta_{100} - \theta_0|}{10\,800\bar{D}_{\rm s}Q}\left[\frac{1}{1 - \sqrt{s/\bar{D}_{\rm s}}\coth\left(\sqrt{s/\bar{D}_{\rm s}}\right)}\right] I^{\rm sf}_{{\rm f},j}(s).$$

Summing both sides of this equation over all galleries j gives (where $\tilde{\theta}_{\rm ss} = \theta_{\rm ss} - \theta_{{\rm s},0}$ debiases $\theta_{\rm ss}$):

$$\widetilde{\Theta}_{\rm ss}(s) = \frac{|\theta_{100} - \theta_0|}{10\,800\bar{D}_{\rm s}Q}\left[\frac{1}{1 - \sqrt{s/\bar{D}_{\rm s}}\coth\left(\sqrt{s/\bar{D}_{\rm s}}\right)}\right] I^{\rm sf}_{\rm f}(s). \tag{2.10}$$

We solve for $I^{\rm sf}_{\rm f}(s)$ by linearizing the electrode/electrolyte-interface kinetics equations and taking Laplace transforms of the result. We

[8] Consider the series RC circuit:

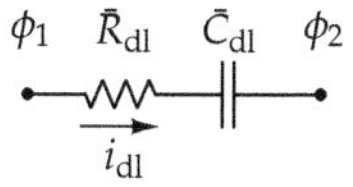

Analyzing in the Laplace domain gives:

$$\begin{aligned}\Phi_1(s) - \Phi_2(s) &= I_{\rm dl}(s)Z_{\rm dl}(s)\\ &= I_{\rm dl}(s)\left[\bar{R}_{\rm dl} + \frac{1}{s\bar{C}_{\rm dl}}\right],\end{aligned}$$

which can be rearranged as:

$$\begin{aligned}&s\bar{C}_{\rm dl}(s)(\Phi_1(s) - \Phi_2(s))\\ &\quad= I_{\rm dl}(s)\,[s\bar{R}_{\rm dl}\bar{C}_{\rm dl} + 1].\end{aligned}$$

Converting back to the time domain:

$$\begin{aligned}&\bar{C}_{\rm dl}\frac{\partial(\phi_1(t) - \phi_2(t))}{\partial t}\\ &\quad= \bar{R}_{\rm dl}\bar{C}_{\rm dl}\frac{\partial i_{\rm dl}(t)}{\partial t} + i_{\rm dl}(t).\end{aligned}$$

[9] The Jacobsen–West TF relates interfacial flux $J(s)$ to debiased solid surface concentration $\widetilde{C}_{\rm s,e}(s)$ as:

$$\frac{\widetilde{C}_{\rm s,e}(s)}{J(s)} = \frac{R_{\rm s}}{D_{\rm s}}\left[\frac{1}{1 - R_{\rm s}\sqrt{\frac{s}{D_{\rm s}}}\coth\left(R_{\rm s}\sqrt{\frac{s}{D_{\rm s}}}\right)}\right],$$

using standard (nonlumped) notation.

[10] As a reminder, we use lowercase symbols for time-domain variables and uppercase symbols for frequency-domain variables. We do not always include the time/frequency dependence in the notation. Therefore, for example, $i^{\rm sf}_{{\rm f},j}$ is a short form for $i^{\rm sf}_{{\rm f},j}(t)$ and $I^{\rm sf}_{{\rm f},j}$ is a short form for $I^{\rm sf}_{{\rm f},j}(s)$.

[11] As with Vol. I, we use the tilde symbol "~" to indicate a debiased variable; that is, a variable that has had a constant value subtracted from it to force its initial value (at $t = 0$) to be zero. This allows us to use the Laplace-transform derivative theorem when finding TFs.

Specifically, if we wish to find the Laplace transform of $\dot{x}_{{\rm ss},j}(t)$, we have:

$$\mathcal{L}\{\dot{x}_{{\rm ss},j}(t)\} = sX_{{\rm ss},j}(s) - x_{{\rm ss},j}(0),$$

which is equal to $sX_{{\rm ss},j}(s)$ only if $x_{{\rm ss},j}(0) = 0$. However, $x_{{\rm ss},j}(0) \neq 0$ and so proceeding to analyze this expression as-is does not lead to a TF solution for our battery model. But

$$\mathcal{L}\{\dot{\tilde{x}}_{{\rm ss},j}(t)\} = s\widetilde{X}_{{\rm ss},j}(s),$$

which does lead to a TF solution.

adopt the MSMR kinetics description, where $i_{0,j}$ for gallery j is expressed by Eq. (2.2) and where overpotential at the particle surface is $\eta = \phi_s - \phi_{sf} - U_{ocp}(\theta_s)$.[12] This gives us the overall faradaic flux:

$$i_f^{sf} = \sum_{j=1}^{J} i_{0,j} \left[\exp((1-\alpha_j)f\eta) - \exp(-\alpha_j f\eta) \right].$$

To find a TF, we linearize i_f^{sf} by forming a Taylor-series expansion around the equilibrium linearization setpoint (where $\theta_{s,0} = \sum_{j=1}^{J} x_{j,0}$):

$$p^* = \{x_j = x_{j,0}, \theta_e = \theta_{e,0} = 1, \phi_s - \phi_{sf} = U_{ocp}(\theta_{s,0}), i_f^{sf} = 0\},$$

and discarding second- and higher-order terms. To do so, we require:

$$\left.\frac{\partial i_f^{sf}}{\partial \phi_{sf}}\right|_{p^*} = f\bar{i}_0 \quad \text{and} \quad \left.\frac{\partial i_f^{sf}}{\partial x_j}\right|_{p^*} = -f[U_{ocp}]'\bar{i}_0,$$

where we have defined:[13]

$$\bar{i}_0 = \sum_{j=1}^{J} \bar{i}_{0,j} \tag{2.11}$$

$$\bar{i}_{0,j} = \bar{k}_{0,j}(x_{0,j})^{\omega_j\alpha_j}(X_j - x_{0,j})^{\omega_j(1-\alpha_j)} \tag{2.12}$$

$$[U_{ocp}]' = \left.\frac{\partial U_{ocp}(\theta_s)}{\partial \theta_s}\right|_{\theta_s=\theta_{s,0}}. \tag{2.13}$$

With these terms, we can write (where $\tilde{\phi}_s = \phi_s - U_{ocp}(\theta_{s,0})$):

$$\begin{aligned} i_f^{sf} &\approx f\bar{i}_0\left(\phi_s - \phi_{sf} - U_{ocp}(\theta_{s,0})\right) - f\bar{i}_0[U_{ocp}]'\sum_{j=1}^{J}\left(x_{ss,j} - x_{j,0}\right) \\ &= f\bar{i}_0\left(\tilde{\phi}_s - \phi_{sf}\right) - f\bar{i}_0[U_{ocp}]'\sum_{j=1}^{J} x_{ss,j} + f\bar{i}_0[U_{ocp}]'\sum_{j=1}^{J} x_{j,0} \\ &= f\bar{i}_0\left(\tilde{\phi}_s - \phi_{sf}\right) - f\bar{i}_0[U_{ocp}]'\left(\theta_{ss} - \theta_{s,0}\right). \end{aligned}$$

This can be rearranged in terms of debiased $\tilde{\phi}_s$ and ϕ_{sf} as:

$$\tilde{\phi}_s - \phi_{sf} = \bar{R}_{ct}^{sf} i_f^{sf} + [U_{ocp}]'\tilde{\theta}_{ss},$$

where we define charge-transfer resistance $\bar{R}_{ct}^{sf} = 1/(f\bar{i}_0)$.[14]

Taking Laplace transforms, we have:

$$\tilde{\Phi}_s - \Phi_{sf} = \bar{R}_{ct}^{sf} I_f^{sf} + [U_{ocp}]'\tilde{\Theta}_{ss}. \tag{2.14}$$

At this point, we rewrite Eq. (2.10) to define particle-diffusion impedance $\bar{Z}_s(s) = [U_{ocp}^r]'\tilde{\Theta}_{ss}(s)/I_f^{sf}(s)$:[15]

$$\bar{Z}_s(s) = \frac{|\theta_{100} - \theta_0|\,[U_{ocp}]'}{10\,800\bar{D}_s Q}\left[\frac{1}{1 - \sqrt{s/\bar{D}_s}\coth\left(\sqrt{s/\bar{D}_s}\right)}\right]. \tag{2.15}$$

[12] The observant reader will notice that this expression for overpotential does not include the voltage drop over the film resistance that we considered in Vol. I. This is because the overpotential being evaluated at this point is across the particle/film interface only, not the across the overall interface between the particle and the bulk electrolyte as was considered in Vol. I. We will include the contribution of the film resistance separately later in this development.

[13] This definition of $\bar{i}_0$ replaces the definition in Vol. I used for the Butler–Volmer model with a new definition for the MSMR model.

[14] Note that if we use a standard DFN kinetics model instead of an MSMR model, this definition of $\bar{R}_{ct}^{sf}$ still applies, and $\bar{i}_0$ is the exchange current density of the standard DFN kinetics from Vol. I instead of the redefinition in Eq. (2.11) for the MSMR model.

[15] Impedance is change in voltage divided by change in current, where the voltage is $U_{ocp}(\theta_{ss})$. The $[U_{ocp}]'$ term arises from linearizing this relationship.

Substituting this for $\widetilde{\Theta}_{ss}$ in Eq. (2.14) gives:

$$\widetilde{\Phi}_s - \Phi_{sf} = (\bar{R}_{ct}^{sf} + \bar{Z}_s) I_f^{sf}.$$

The interface also supports a lithium flux due to the double layer:[16]

$$\begin{aligned} \bar{C}_{dl}^{sf} \frac{\partial(\phi_s - \phi_{sf})}{\partial t} &= \bar{R}_{dl}^{sf} \bar{C}_{dl}^{sf} \frac{\partial i_{dl}^{sf}}{\partial t} + i_{dl}^{sf} \\ \bar{C}_{dl}^{sf} \frac{\partial(\tilde{\phi}_s + U_{ocp}(\theta_{s,0}) - \phi_{sf})}{\partial t} &= \bar{R}_{dl}^{sf} \bar{C}_{dl}^{sf} \frac{\partial i_{dl}^{sf}}{\partial t} + i_{dl}^{sf}. \end{aligned}$$

We take the Laplace transform of both sides of this equation and get:

$$s\bar{C}_{dl}^{sf} \left(\widetilde{\Phi}_s(s) - \Phi_{sf}(s) \right) = \left(s\bar{R}_{dl}^{sf} \bar{C}_{dl}^{sf} + 1 \right) I_{dl}^{sf}(s).$$

The overall solid/film interface supports a total lithium flux $I_{f+dl}^{sf}(s)$:

$$\begin{aligned} I_{f+dl}^{sf}(s) &= I_f^{sf}(s) + I_{dl}^{sf}(s) \\ &= \underbrace{\left(\frac{1}{\bar{R}_{ct}^{sf} + \bar{Z}_s(s)} + \frac{s\bar{C}_{dl}^{sf}}{s\bar{R}_{dl}^{sf}\bar{C}_{dl}^{sf} + 1} \right)}_{\bar{Y}_{sf}(s)} \left(\widetilde{\Phi}_s(s) - \Phi_{sf}(s) \right), \end{aligned}$$

where $\bar{Y}_{sf}(s)$ is overall admittance of the interface and the impedance of the interface can be expressed as $\bar{Z}_{sf}(s) = 1/\bar{Y}_{sf}(s)$.

2.3.2 *Impedance at film/electrolyte interface*

Now, we investigate the impedance at the film/electrolyte interface. We solve for $I_f^{fe}(s)$ by linearizing the Butler–Volmer equation for the interface and taking the Laplace transform of the result.[17] Following the same method as before, we find (where $\bar{R}_{ct}^{fe} = 1/(\bar{i}_0^{fe} f)$):

$$\Phi_{fe} - \Phi_e = \bar{R}_{ct}^{fe} I_f^{fe}.$$

The interface also supports a lithium flux due to the double layer:[18,19]

$$\begin{aligned} \bar{C}_{dl}^{fe} \frac{\partial(\phi_{fe} - \phi_e)}{\partial t} &= \bar{R}_{dl}^{fe} \bar{C}_{dl}^{fe} \frac{\partial i_{dl}^{fe}}{\partial t} + i_{dl}^{fe} \\ s\bar{C}_{dl}^{fe} (\Phi_{fe}(s) - \Phi_e(s)) &= \left(s\bar{R}_{dl}^{fe} \bar{C}_{dl}^{fe} + 1 \right) I_{dl}^{fe}(s). \end{aligned}$$

The overall film/electrolyte interface supports a total lithium flux:

$$I_{f+dl}^{fe}(s) = I_f^{fe}(s) + I_{dl}^{fe}(s) = \underbrace{\left(\frac{1}{\bar{R}_{ct}^{fe}} + \frac{s\bar{C}_{dl}^{fe}}{s\bar{R}_{dl}^{fe}\bar{C}_{dl}^{fe} + 1} \right)}_{\bar{Y}_{fe}(s)} (\Phi_{fe}(s) - \Phi_e(s)),$$

where $\bar{Y}_{fe}(s)$ is the overall admittance of the interface and the impedance of the interface is $\bar{Z}_{fe}(s) = 1/\bar{Y}_{fe}(s)$.

[16] This is another place where the discussion in sidenote 11 becomes important. Here, we require that $\tilde{\phi}_s(0) - \phi_{sf}(0) = 0$. This implies that the capacitor $\bar{C}_{dl}^{sf}$ is initially charged to the value $U_{ocp}(\theta_{s,0})$ volts so that $i_{dl}(0) = 0$ in Eqs. (2.8) or (2.9). It also means that $i_f^{sf}(0) = 0$, which causes $\eta(0) = 0$.

[17] Following Meyers et al. (see sidenote 6), we use a standard Butler–Volmer model of kinetics at this interface and ignore variation of OCP due to concentration of lithium in the film or in the solution of any reaction occurring at the interface. With these assumptions, $\bar{R}_{ct}^{fe}$ varies only with temperature per $f = F/(RT)$ and possible Arrhenius corrections applied to $\bar{i}_0^{fe}$ (see Chap. 3 for a review of the Arrhenius temperature-dependence model).

[18] We include this double-layer capacitance in the derivation at this point to be consistent with the development presented by Meyers et al. However, given that both the film and the electrolyte are ionically conductive, this capacitance tends to have a small effect. In Sect. 2.7 we will simplify the interface model to one that we feel presents the best tradeoff between being physically meaningful and identifiable. In the simplified model, the double-layer capacitance of the film/electrolyte interface is removed.

[19] This is another place where the discussion in sidenote 11 becomes important. Here, we require that $\phi_{fe}(0) - \phi_e(0) = 0$. This implies that the capacitor $\bar{C}_{dl}^{fe}$ is initially discharged to zero volts.

2.3.3 *Overall interfacial impedance model*

Finally, we add film resistance and the double layer between ϕ_s and ϕ_e. We start by recognizing that lithium flux $I^{sf}_{f+dl}(s) = I^{fe}_{f+dl}(s)$ since charge cannot accumulate in the film, and define that value to be $I^{f}_{f+dl}(s)$. We model film-layer resistance as being purely ohmic:

$$\Phi_{sf}(s) - \Phi_{fe}(s) = \bar{R}_f I^{f}_{f+dl}(s).$$

Now, we can rearrange equations for $I^{sf}_{f+dl}(s)$ and $I^{fe}_{f+dl}(s)$ and lithium flux through the film resistance as:

$$\begin{aligned}\widetilde{\Phi}_s(s) - \Phi_{sf}(s) &= \bar{Z}_{sf}(s) I^{f}_{f+dl}(s)\\ \Phi_{sf}(s) - \Phi_{fe}(s) &= \bar{R}_f I^{f}_{f+dl}(s)\\ \Phi_{fe}(s) - \Phi_e(s) &= \bar{Z}_{fe}(s) I^{f}_{f+dl}(s).\end{aligned}$$

Adding these three equations together, we find that:

$$\widetilde{\Phi}_s(s) - \Phi_e(s) = \left(\bar{Z}_{sf}(s) + \bar{R}_f + \bar{Z}_{fe}(s)\right) I^{f}_{f+dl}(s).$$

The interfacial impedance model is nearly complete: all that remains is to consider the double layer between ϕ_s and ϕ_e:[20]

$$\begin{aligned}\bar{C}^{se}_{dl}\frac{\partial(\phi_s - \phi_e)}{\partial t} &= \bar{R}^{se}_{dl}\bar{C}^{se}_{dl}\frac{\partial i^{se}_{dl}}{\partial t} + i^{se}_{dl}\\ \bar{C}^{se}_{dl}\frac{\partial(\tilde{\phi}_s + U_{ocp}(\theta_{s,0}) - \phi_e)}{\partial t} &= \bar{R}^{se}_{dl}\bar{C}^{se}_{dl}\frac{\partial i^{se}_{dl}}{\partial t} + i^{se}_{dl}\\ s\bar{C}^{se}_{dl}\left(\widetilde{\Phi}_s(s) - \Phi_e(s)\right) &= \left(s\bar{R}^{se}_{dl}\bar{C}^{se}_{dl} + 1\right) I^{se}_{dl}(s).\end{aligned}$$

[20] This is another place where the discussion in sidenote 11 becomes important. Here, we require that $\tilde{\phi}_s(0) - \phi_e(0) = 0$. This implies that the capacitor $\bar{C}^{se}_{dl}$ is initially charged to the value $U_{ocp}(\theta_{s,0})$ volts.

Now we have that the total lithium flux from solid to electrolyte is:

$$\begin{aligned}I^{se}_{f+dl}(s) &= I^{f}_{f+dl}(s) + I^{se}_{dl}(s)\\ &= \left(\frac{1}{\bar{Z}_{sf}(s) + \bar{R}_f + \bar{Z}_{fe}(s)} + \frac{s\bar{C}^{se}_{dl}}{s\bar{R}^{se}_{dl}\bar{C}^{se}_{dl} + 1}\right)\left(\widetilde{\Phi}_s(s) - \Phi_e(s)\right).\end{aligned}$$

Defining $\phi_{s\text{-}e} = \phi_s - \phi_e$ and $\tilde{\phi}_{s\text{-}e} = \tilde{\phi}_s - \phi_e$, interface impedance is:

$$\frac{\widetilde{\Phi}_{s\text{-}e}(s)}{I^{se}_{f+dl}(s)} = \bar{Z}_{se}(s) = \left(\frac{1}{\bar{Z}_{sf}(s) + \bar{R}_f + \bar{Z}_{fe}(s)} + \frac{s\bar{C}^{se}_{dl}}{s\bar{R}^{se}_{dl}\bar{C}^{se}_{dl} + 1}\right)^{-1}. \quad (2.16)$$

This interfacial model can be visualized using the equivalent circuit shown in Fig. 2.9. As before, the circuit illustrates the impedance across the solid/electrolyte interface that was developed directly from the linearized model differential equations.[21]

[21] That is, the equations imply the circuit; the circuit does not itself define the model equations.

2.4 *Adding double-layer constant-phase-element behavior*

The analysis performed so far in this chapter has focused on the solid/electrolyte interface for a single particle, not for an entire electrode. We have not considered the inhomogeneous nature of physical electrodes. For example, physical electrodes exhibit:

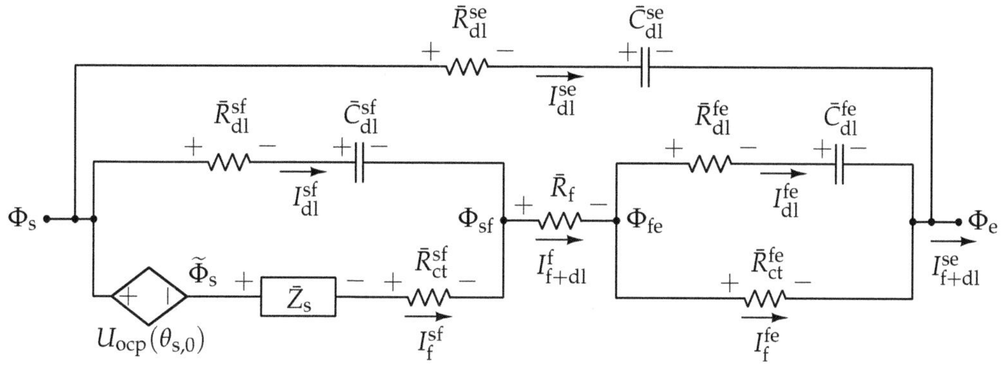

Figure 2.9: Visualizing the modified interface equations using an equivalent circuit. Initially, $\bar{C}_{dl}^{sf}$ and $\bar{C}_{dl}^{se}$ both have potential $U_{ocp}(\theta_{s,0})$ and $\bar{C}_{dl}^{fe}$ has potential of 0 V.

- Variations in surface reactivity;
- Variations in particle sizes, shapes, and surfaces of particles;
- Variations in the sizes and shapes of electrode pore openings.

When extending the solid/electrolyte interfacial model from particle-scale to a 1D continuum, these inhomogeneities are not always modeled well by the impedance $\bar{Z}_{se}(s)$ and its associated circuit. However, by replacing ideal capacitances with CPEs and by modifying particle impedance $\bar{Z}_s$, we can improve the ability of the interfacial model to describe observed behaviors.[22] We add these two features to the interfacial model in the next sections, beginning with a modification of the double-layer capacitance.[23]

The impedance of an ideal capacitance is $Z_c(s) = 1/(s\bar{C})$. A CPE replaces this with $Z_{\tilde{c}}(s) = 1/(s^{n_{dl}}\bar{C})$ for $0 < n_{dl} \leq 1$, where we use the "tilde" symbol ($\tilde{\cdot}$) in the subscript to denote CPE behavior. On a Bode plot, $Z_c(s)$ had a constant phase of $-90°$ and $Z_{\tilde{c}}(s)$ has a constant phase of $-90° \times n_{dl}$. So, in fact, the ideal capacitance is also a CPE, but the new terminology is used primarily when $n_{dl} \neq 1$.

Note that we can use Laplace analysis for circuits containing ideal capacitors to derive ODEs to simulate circuit behavior. This kind of analysis is covered in undergraduate courses in linear-circuit theory. However, we cannot use standard calculus or circuit theory to arrive at time-domain representations of CPEs. Instead, fractional-order calculus is required to do so, and this is not a topic covered in most undergraduate curricula. So, while the frequency-domain behaviors of CPE are very easy to see, the time-domain behaviors are less simple. However, the techniques presented in Chap. 4 to develop ROMs from TFs are able to work with models containing CPEs.

As an example illustrating how replacing an ideal capacitor with a CPE might change a circuit's behavior, consider Fig. 2.10 which shows the Bode magnitude and phase plots of a "ZARC" element.[24] This is a circuit comprising a resistor in parallel with a CPE. First, notice that the symbol for a CPE is similar to that of a capacitor,

[22] Jean-Baptiste Jorcin, Mark E. Orazem, Nadine Pébère, and Bernard Tribollet, "CPE analysis by local electrochemical impedance spectroscopy," *Electrochimica Acta*, 51(8–9):1473–1479, 2006.

[23] Some works in the literature also explore using transmission-line theory to investigate the impact of variations in electrode structures on impedance and may help to give insight regarding the magnitude of the effects to be discussed in these sections. These works include: Masayuki Itagaki, Yasunari Hatada, Isao Shitanda, and Kunihiro Watanabe, "Complex impedance spectra of porous electrode with fractal structure," *Electrochimica Acta*, 55(21):6255–6262, 2010, and Juhyun Song and Martin Z. Bazant, "Electrochemical impedance of a battery electrode with anisotropic active particles," *Electrochimica Acta*, 131:214–227, 2014.

[24] When this impedance is visualized in a Nyquist plot, it creates an arc shape, giving rise to the name of the circuit: "Z" for impedance, "arc" for its shape. See: J. Ross Macdonald and Robert L. Hurt, "Some simple equivalent circuits for ionic conductors," *Journal of Electroanalytical Chemistry and Interfacial Electrochemistry*, 200(1–2):69–82, 1986.

but uses parallel sets of *angled* lines "⟩⟩" instead of parallel *straight* lines "||". The ZARC element reduces to a standard Voigt (parallel resistor–capacitor) element when $n_{\mathrm{dl}} = 1$, which is shown by the blue lines in the Bode plots. ZARC and Voigt circuits have equivalent low-frequency gain and phase, but differ at high frequencies. By changing the value of n_{dl}, a ZARC circuit allows modeling a wider range of physical behaviors than does a fixed Voigt circuit.

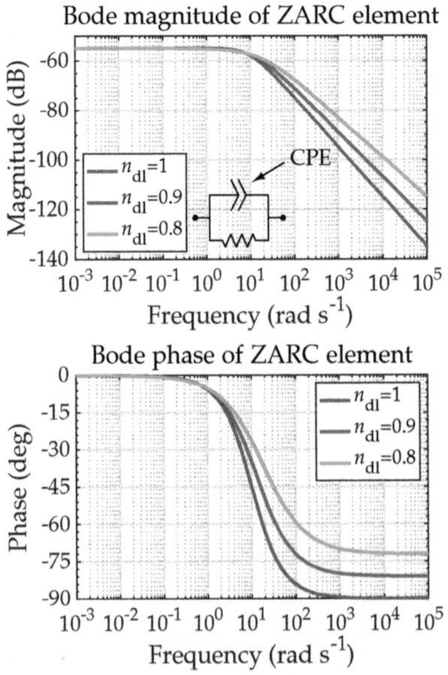

Figure 2.10: Bode magnitude and phase of ZARC (parallel R–CPE) elements.

We modify the interface model to describe a distribution of particle sizes by replacing all double-layer capacitances with CPEs, by replacing s with $s^{n_{\mathrm{dl}}}$ in all capacitor impedances. We write, where the tilde symbol ("~") in the subscript again denotes a CPE:

$$\bar{Z}_{\widetilde{\mathrm{sf}}}(s) = \left(\frac{1}{\bar{R}_{\mathrm{ct}}^{\mathrm{sf}} + \bar{Z}_{\mathrm{s}}(s)} + \frac{s^{n_{\mathrm{dl}}^{\mathrm{sf}}}\bar{C}_{\mathrm{dl}}^{\mathrm{sf}}}{s^{n_{\mathrm{dl}}^{\mathrm{sf}}}\bar{R}_{\mathrm{dl}}^{\mathrm{sf}}\bar{C}_{\mathrm{dl}}^{\mathrm{sf}} + 1}\right)^{-1}$$

$$\bar{Z}_{\widetilde{\mathrm{fe}}}(s) = \left(\frac{1}{\bar{R}_{\mathrm{ct}}^{\mathrm{fe}}} + \frac{s^{n_{\mathrm{dl}}^{\mathrm{fe}}}\bar{C}_{\mathrm{dl}}^{\mathrm{fe}}}{s^{n_{\mathrm{dl}}^{\mathrm{fe}}}\bar{R}_{\mathrm{dl}}^{\mathrm{fe}}\bar{C}_{\mathrm{dl}}^{\mathrm{fe}} + 1}\right)^{-1},$$

from which the overall solid/electrolyte impedance is:

$$\bar{Z}_{\widetilde{\mathrm{se}}}(s) = \left(\frac{1}{\bar{Z}_{\widetilde{\mathrm{sf}}}(s) + \bar{R}_{\mathrm{f}} + \bar{Z}_{\widetilde{\mathrm{fe}}}(s)} + \frac{s^{n_{\mathrm{dl}}^{\mathrm{se}}}\bar{C}_{\mathrm{dl}}^{\mathrm{se}}}{s^{n_{\mathrm{dl}}^{\mathrm{se}}}\bar{R}_{\mathrm{dl}}^{\mathrm{se}}\bar{C}_{\mathrm{dl}}^{\mathrm{se}} + 1}\right)^{-1}. \tag{2.17}$$

2.4.1 *Modified CPE for double layer*

You may recall from Vol. I that when converting TFs to ROMs we must take special steps when the TFs include integration dynamics. The integrator must first be removed, then the remainder of the TF converted to a discrete-time state-space model, and then the integration dynamics added back by augmenting the state-space model.[25]

As with integrators, a CPE violates this requirement for stability. So, we might consider subtracting out the effect of the CPE, converting the remaining portion of the TF to a reduced-order form, and then adding back the effect of the CPE by augmenting the model. However, this is impossible since there is no finite-order time-domain representation of a CPE that could be used to augment the model to add back in the CPE effects.

To address this problem, consider once again the Bode plot of a ZARC element in Fig. 2.10. Notice that the low-frequency behavior of a Voigt circuit containing an ideal capacitor is no different from that of a ZARC circuit containing a CPE. The only difference is at high frequencies. Our interface models will have similar characteristics to this. Therefore, if we could somehow replace the CPE with a relationship that blends between that of an ideal capacitor at low frequencies and a CPE at high frequencies, we would see no difference in the

[25] In Vol. I, we used the the discrete-time realization algorithm (DRA) to do this. In Chap. 4, we will present an updated method to perform the same task. Although this new method improves on the DRA, it still produces numerically better results when the TF being converted is strictly stable, so we impose the same constraint.

Bode plot. But we would solve our stability problem by being able to subtract out a pure integrator at low frequencies.

Here is how it can be done. Formerly, we defined the impedance of a CPE as $Z_{\tilde{c}} = 1/(s^{n_{\text{dl}}}\bar{C})$. Now, we modify this definition to be:[26]

$$Z_{\tilde{c}}(s) = \frac{(s\bar{C} + \omega_{\text{dl}})^{(1-n_{\text{dl}})}}{s\bar{C}^{2-n_{\text{dl}}}}. \tag{2.18}$$

This modified impedance behaves like an ideal capacitor for frequencies below ω_{dl} and like a CPE for frequencies above ω_{dl}.

Fig. 2.11 compares the Bode magnitude and phase plots of the unmodified and modified CPE impedances for $\omega_{\text{dl}} = 1\,\mu\text{Hz}$. At high frequencies (above ω_{dl}), the modified and unmodified impedances are indistinguishable, as desired. At low frequencies (below ω_{dl}), the modified impedance behaves like an integrator, having a phase of $-90°$, also as desired. Consequently, we can subtract a pure integrator from the modified impedance (shown as the green dotted lines) and later add back the integration dynamics after creating a ROM.

This is purely a mathematical exercise, but one that is required to make the numerics of Chap. 4 work. The behaviors of the TF are changed only at very low frequencies, so for all reasonable frequencies this modified model approximates an ideal CPE well.

Using these new definitions, the overall solid/electrolyte impedance of Eq. (2.17) may be rewritten as:

$$\bar{Z}_{\widetilde{\text{se}}}(s) = \left(\frac{1}{\bar{Z}_{\widetilde{\text{sf}}}(s) + \bar{R}_{\text{f}} + \bar{Z}_{\widetilde{\text{fe}}}(s)} + \frac{1}{\bar{R}^{\text{se}}_{\text{dl}} + Z^{\text{se}}_{\tilde{c}}(s)} \right)^{-1},$$

where we have substituted Eq. (2.18) for all CPE impedances.

[26] The integrator-removed version is:

$$Z^*_{\tilde{c}}(s) = Z_{\tilde{c}}(s) - \frac{1}{s}\frac{\omega_{\text{dl}}^{1-n_{\text{dl}}}}{\bar{C}^{2-n_{\text{dl}}}}.$$

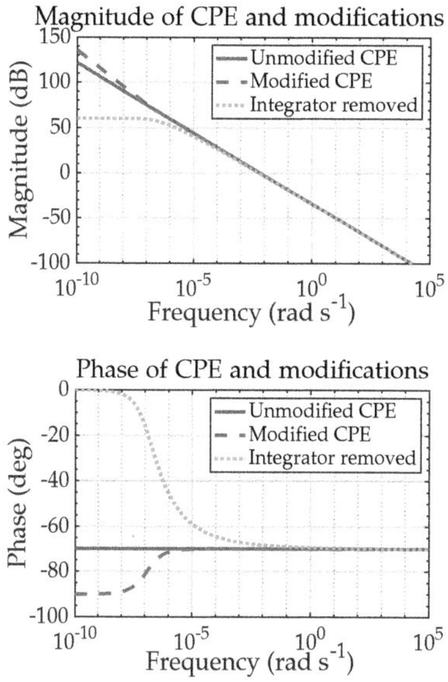

Figure 2.11: Bode magnitude and phase of modified CPE where $n_{\text{dl}} = 7/9$.

2.5 *Adding solid-diffusion CPE behavior*

We have replaced capacitances with CPEs, improving models for electrodes having nonuniform pore sizes and shapes. We now seek to modify the solid-diffusion impedance, improving models for electrodes having a distribution in particle sizes and shapes.[27]

Replacing a Voigt element with a ZARC element modified only high-frequency behaviors. Similarly, we desire to recast particle impedance $\bar{Z}_{\text{s}}$ to retain its low-frequency dynamics (particles must still integrate net lithium intercalated to account for SOC change) but modify high-frequency dynamics. To begin, recall $\bar{Z}_{\text{s}}(s)$ from Eq. (2.15). This impedance has a constant term (outside the brackets) and a dynamic term (inside the brackets). It also has a pole at $s = 0$ (which is not obvious by inspection): this integrator must be removed before applying the CPE. We examine the dynamic term

[27] This development uses CPE-like behavior to approximate the effects observed using transmission-line models in: Juhyun Song and Martin Z. Bazant, "Electrochemical impedance of a battery electrode with anisotropic active particles," *Electrochimica Acta*, 131:214–227, 2014.

and compute the residue of the integrator pole to be:

$$\text{res}_0 = \lim_{s\to 0} s\left(\frac{1}{1-\sqrt{s/\bar{D}_s}\coth\left(\sqrt{s/\bar{D}_s}\right)}\right) = -3\bar{D}_s.$$

With this knowledge, we can write the integrator-removed portion of $\bar{Z}_s(s)$ (i.e., $\bar{Z}_s^*(s)$) as:

$$\begin{aligned}\bar{Z}_s^*(s) &= \frac{|\theta_{100}-\theta_0|\,[U_{\text{ocp}}]'}{10\,800\bar{D}_sQ}\left[\frac{1}{1-\sqrt{s/\bar{D}_s}\coth\left(\sqrt{s/\bar{D}_s}\right)}-\frac{\text{res}_0}{s}\right]\\ &= \frac{|\theta_{100}-\theta_0|\,[U_{\text{ocp}}]'}{10\,800\bar{D}_sQ}\left[\frac{(s/\bar{D}_s)+3\left(1-\sqrt{s/\bar{D}_s}\coth\left(\sqrt{s/\bar{D}_s}\right)\right)}{(s/\bar{D}_s)\left(1-\sqrt{s/\bar{D}_s}\coth\left(\sqrt{s/\bar{D}_s}\right)\right)}\right].\end{aligned}$$

Inside $\bar{Z}_s^*(s)$, we replace $\sqrt{s/\bar{D}_s}$ with a constant-phase term:

$$\beta_{\tilde{s}}(s) = \sqrt{(s/\bar{D}_s)^{n_f}}, \tag{2.19}$$

resulting in integrator-removed and overall solid-diffusion impedances:

$$\begin{aligned}\bar{Z}_{\tilde{s}}^*(s) &= \frac{|\theta_{100}-\theta_0|\,[U_{\text{ocp}}]'}{10\,800\bar{D}_sQ}\left[\frac{\beta_{\tilde{s}}^2(s)+3\left(1-\beta_{\tilde{s}}(s)\coth\left(\beta_{\tilde{s}}(s)\right)\right)}{\beta_{\tilde{s}}^2(s)\left(1-\beta_{\tilde{s}}(s)\coth\left(\beta_{\tilde{s}}(s)\right)\right)}\right]\\ \bar{Z}_{\tilde{s}}(s) &= \frac{|\theta_{100}-\theta_0|\,[U_{\text{ocp}}]'}{10\,800\bar{D}_sQ}\left[\frac{\beta_{\tilde{s}}^2(s)+3\left(1-\beta_{\tilde{s}}(s)\coth\left(\beta_{\tilde{s}}(s)\right)\right)}{\beta_{\tilde{s}}^2(s)\left(1-\beta_{\tilde{s}}(s)\coth\left(\beta_{\tilde{s}}(s)\right)\right)}-\frac{3\bar{D}_s}{s}\right].\end{aligned} \tag{2.20}$$

Once again, note that the tilde symbol in the subscript denotes that this is a CPE term.

Fig. 2.12 shows the Bode magnitude and phase plots of integrator-included $\bar{Z}_{\tilde{s}}(s)$ for some sample parameter values. The low-frequency magnitude response has a slope of -20 dB per decade, as expected. The high-frequency slope is -10 dB per decade for the standard diffusion impedance, and we see that this slope becomes more shallow when we use a CPE. The high-frequency phase is $-45°$ for the standard impedance, and increases when we use a CPE.

Fig. 2.13 shows the Bode magnitude and phase plots for $\bar{Z}_{\tilde{s}}^*(s)$: this is the same impedance, but with the integrator removed. We see once again that the low-frequency behaviors of $\bar{Z}_s^*(s)$ and $\bar{Z}_{\tilde{s}}^*(s)$ are the same; the CPE changes only the high-frequency response. The nominal slope of the magnitude plot of -10 dB per decade becomes shallower; the phase also increases above the nominal value of $-45°$.

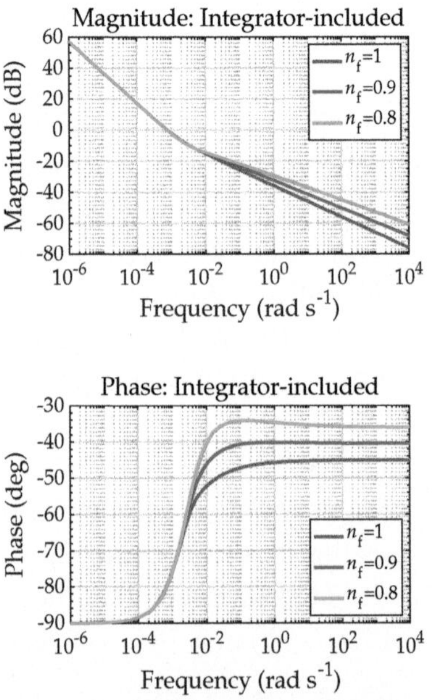

Figure 2.12: Bode magnitude and phase of solid diffusion with CPE (integrator included).

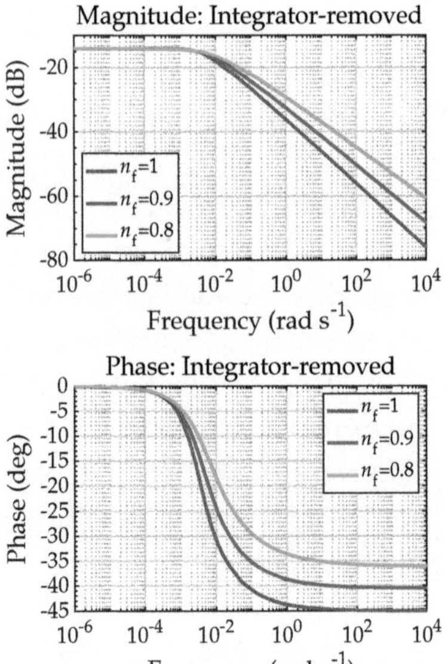

Figure 2.13: Bode magnitude and phase of solid diffusion with CPE (integrator removed).

2.6 *SOC-dependent solid diffusivity*

When seeking to estimate parameter values for a commercial cell, using a constant value of $\bar{D}_s^r$ in each electrode does not fit observed data well. Here, we briefly mention a simple model of SOC-dependent solid diffusivity that improves solid-diffusion modeling significantly by changing the definition of $\bar{D}_s^r$ to:[28]

$$\bar{D}_s^r(\theta_s^r) = \bar{D}_{s,\text{ref}}^r \underbrace{f\theta_s^r(\theta_s^r - 1)\,[U_{\text{ocp}}^r]'}_{\text{SOC dependence}}. \tag{2.21}$$

Fig. 2.14 illustrates this relationship for the NMC30 cell.[29]

[28] This model is developed in: Daniel R. Baker and Mark W. Verbrugge, "Intercalate diffusion in multiphase electrode materials and application to lithiated graphite," *Journal of The Electrochemical Society*, 159(8):A1341, 2012.

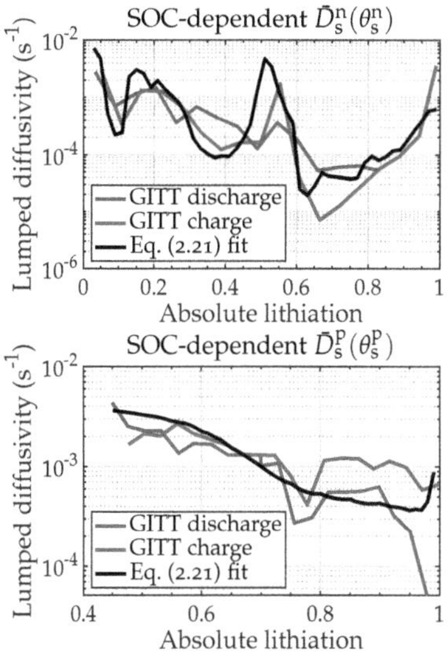

Figure 2.14: Fitting the SOC-dependent model of solid diffusivity from Eq. (2.21) to the data from Fig. 7 of: Johannes Schmalstieg, Christiane Rahe, Madeleine Ecker, and Dirk Uwe Sauer, "Full cell parameterization of a high-power lithium-ion battery for a physico-chemical model: Part I. Physical and electrochemical parameters," *Journal of The Electrochemical Society*, 165(16):A3799, 2018.

[29] The MATLAB toolbox to be introduced in Sect. 2.13 implements constant solid diffusivity when the user enters a value of `Ds` in the parameter spreadsheet; it implements SOC-dependent solid diffusivity using Eq. (2.21) when the user instead enters a value of `Dsref`.

2.7 *Summary of nonideal interfacial model*

We have now completed the development of a solid/electrolyte interface model that includes three double layers and includes CPE effects to help the TFs accommodate variation in particle and electrolyte pore-opening sizes and shapes and admits a description of SOC-dependent solid diffusivity. The equations describing the interfacial model can be visualized using the top circuit diagram in Fig. 2.15. This circuit is similar to the one drawn in Fig. 2.9; however, notice that capacitances have been replaced by CPEs and that the solid-diffusion impedance $\bar{Z}_s$ has been replaced by $\bar{Z}_{\tilde{s}}$.

This interface model is more general than we are likely to need in a BMS application. In particular, it is difficult and maybe impossible to estimate all of its parameter values for a commercial cell. So, instead, we will use a simplified version, removing the double layers across the film/electrolyte boundary and across the entire film, setting $\bar{C}_{\text{dl}}^{\text{fe}} = \bar{C}_{\text{dl}}^{\text{se}} = \bar{R}_{\text{ct}}^{\text{fe}} = 0$ and $\bar{R}_{\text{dl}}^{\text{se}} = \bar{R}_{\text{dl}}^{\text{fe}} = \infty$. This simplification also allows us to drop some superscripts from our notation since they are now unnecessary to distinguish between parameters. The simplified model is drawn as the middle circuit in Fig. 2.15.

For this simplified model, we write $\bar{Z}_{\widetilde{\text{se}}}(s) = \bar{Z}_{\widetilde{\text{sf}}}(s) + \bar{R}_f$, where:

$$\bar{Z}_{\widetilde{\text{sf}}}(s) = \left(\frac{1}{\bar{R}_{\text{ct}} + \bar{Z}_{\tilde{s}}(s)} + \frac{1}{\bar{R}_{\text{dl}} + Z_{\tilde{c}}(s)}\right)^{-1},$$

where charge-transfer resistance is defined as:

$$\bar{R}_{\text{ct}} = \frac{1}{f\bar{i}_0}, \tag{2.22}$$

and where $\bar{i}_0$ is defined in Eq. (2.11), $Z_{\tilde{c}}(s)$ is defined in Eq. (2.18) with $\bar{C} = \bar{C}_{\text{dl}}$, and $\bar{Z}_{\tilde{s}}(s)$ is defined in Eq. (2.20).[30]

[30] While we adopt the simplified model from this point forward, we elected to derive the full model in this chapter to demonstrate the method for doing so. Our hope is that by following this approach the reader will be able to adapt the interface description to any configuration of interest. However, in our experience we have not noticed any failure in the performance of BMS algorithms caused by the assumptions made by the simplified model.

We can compare this new interfacial model to the model we began with, redrawn in the lower circuit in Fig. 2.15 from Fig. 2.7 with the OCP now separated into two parts corresponding to the initial condition and the dynamic variation. The new interface model is largely the same, but now with inclusion of the double layer and CPE dynamics.

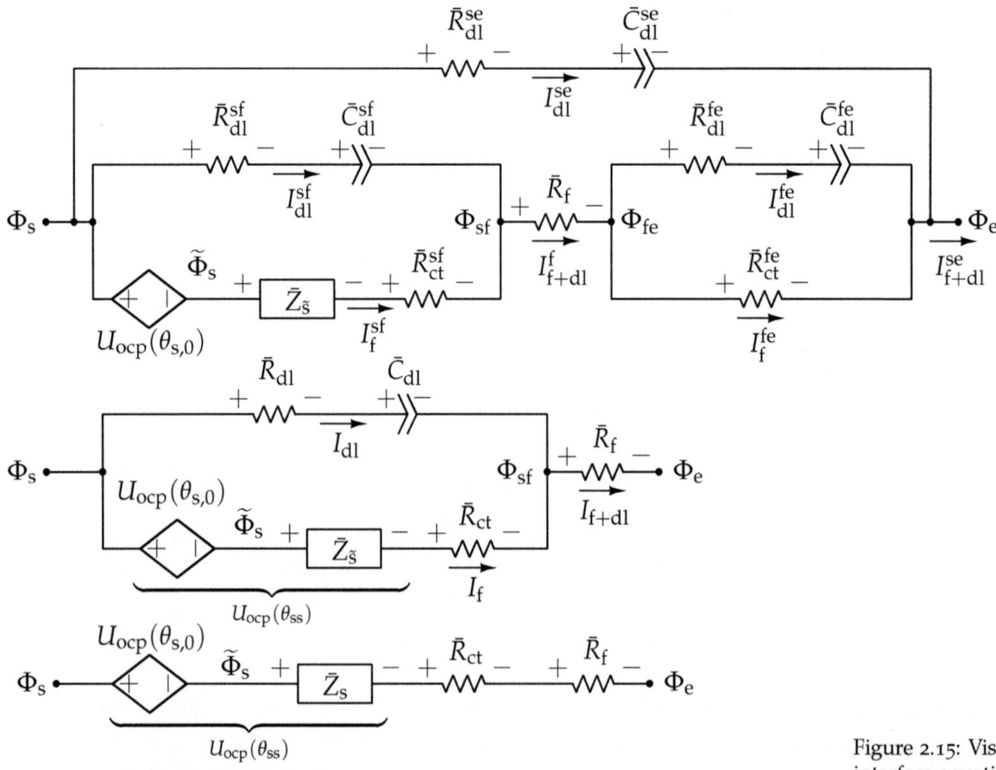

Figure 2.15: Visualizing the modified interface equations, including CPE, using an equivalent circuit. The top circuit is the full model, the middle circuit is the simplified model, and the lower circuit represents the original interface equations with OCP divided into constant and variable terms. Again, the diagram is developed from the equations as a means of visualizing the equations; the equations were not developed from the diagram.

2.8 *Summary of modified model PDEs*

We have now completed our modifications to the interfacial model. These modifications change somewhat the PDEs describing a cell. In particular, it is not possible any longer (without using fractional-order calculus, which is out of the scope of this book) to describe a model containing CPEs in the time domain using PDEs. The CPE behavior *can* be modeled perfectly in the Laplace domain for linearized TFs, which we do in Sect. 2.9. Here, we recap the PDEs for non-CPE behaviors to show how the generalized double layer (simple model of Fig. 2.15) changes them. Note that one subtle distinction between these equations and those in Chap. 1 is that previously we had only faradaic current i_{f}; now, we have faradaic current i_{f}, double-layer current i_{dl}, and total current $i_{\mathrm{f+dl}}$. It is important to be clear regarding which of these currents to use in which equations. All of these details are summarized for reference purposes in Table 2.1.

When using the standard (non-MSMR) model, the reformulated PDE for θ_{s} (for $0 \le \theta_{\mathrm{s}} \le 1$) is unchanged from Chap. 1, and is listed in Eq. (2.23). Its boundary conditions are given in Eqs. (2.24) and (2.25), where the forcing input is $i_{\mathrm{f}}^{\mathrm{r}}$. The initial condition is presented in Eq. (2.26). When we are using the MSMR model, the reformulated PDE for x_j (for $0 \le x_j \le X_j$) is listed in Eq. (2.27). Its

Description	Governing equations	Boundary conditions	Initial conditions
Mass conservation in solid (non-MSMR)	$\frac{\partial \theta_s^r}{\partial t} = \frac{1}{\tilde{r}^2}\frac{\partial}{\partial \tilde{r}}\left(\bar{D}_s^r \tilde{r}^2 \frac{\partial \theta_s^r}{\partial \tilde{r}}\right)$ (2.23)	$\bar{D}_s^r \frac{\partial \theta_s^r}{\partial \tilde{r}}\Big\vert_{\tilde{r}=1} = -\frac{\lvert\theta_{100}^r - \theta_0^r\rvert}{10\,800Q} i_f^r$ (2.24) $\frac{\partial \theta_s^r}{\partial \tilde{r}}\Big\vert_{\tilde{r}=0} = 0$ (2.25)	$\theta_{s,0}^r = \theta_0^r + z_0(\theta_{100}^r - \theta_0^r)$ (2.26)
Mass conservation in solid (MSMR)	$\frac{\partial x_j^r}{\partial t} = \frac{1}{\tilde{r}^2}\frac{\partial}{\partial \tilde{r}}\left(\bar{D}_s^r \tilde{r}^2 \frac{\partial x_j^r}{\partial \tilde{r}}\right)$ (2.27)	$\bar{D}_s^r \frac{\partial x_j^r}{\partial \tilde{r}}\Big\vert_{\tilde{r}=1} = -\frac{\lvert\theta_{100}^r - \theta_0^r\rvert}{10\,800Q} i_{f,j}^r$. (2.28) $\frac{\partial x_j^r}{\partial \tilde{r}}\Big\vert_{\tilde{r}=0} = 0$ (2.29)	$x_{j,0}^r = \frac{X_j^r}{1+\exp\left[f(U_{ocp}^r(\theta_{s,0}^r) - U_j^{0,r})/\omega_j^r\right]}$ (2.30)
Charge conservation in solid	$\bar{\sigma}^r \frac{\partial^2 \phi_s^r}{\partial \tilde{x}^2} = i_{f+dl}^r$ (2.31)	$\bar{\sigma}^n \frac{\partial}{\partial \tilde{x}}\phi_s\Big\vert_{\tilde{x}=0} = \bar{\sigma}^p \frac{\partial}{\partial \tilde{x}}\phi_s\Big\vert_{\tilde{x}=3} = -i_{app}$ (2.32) $\bar{\sigma}^n \frac{\partial}{\partial \tilde{x}}\phi_s\Big\vert_{\tilde{x}=1} = \bar{\sigma}^p \frac{\partial}{\partial \tilde{x}}\phi_s\Big\vert_{\tilde{x}=2} = 0$ (2.33)	$\phi_{s,0}^n = 0$ (2.34) $\phi_{s,0}^p = U_{ocp}^p(\theta_{s,0}^p) - U_{ocp}^n(\theta_{s,0}^n)$ (2.35)
Charge conservation in electrolyte	$\frac{\partial}{\partial \tilde{x}}\left(\bar{\kappa}^r\left(\frac{\partial}{\partial \tilde{x}}\phi_e^r + \bar{\kappa}_D T \frac{\partial \ln(\theta_e^r)}{\partial \tilde{x}}\right)\right) + i_{f+dl}^r = 0$ (2.36)	$-\bar{\kappa}^r\left[\frac{\partial}{\partial \tilde{x}}\phi_e^r + \bar{\kappa}_D T \frac{\partial \ln(\theta_e^r)}{\partial \tilde{x}}\right]_{\tilde{x}=0,3} = 0$ (2.37) $-\bar{\kappa}^r\left[\frac{\partial}{\partial \tilde{x}}\phi_e^r + \bar{\kappa}_D T \frac{\partial \ln(\theta_e^r)}{\partial \tilde{x}}\right]_{\tilde{x}=1,2} = i_{app}$ (2.38)	$\phi_{e,0}^r = -U_{ocp}^n(\theta_{s,0}^n)$ (2.39)
Mass conservation in electrolyte	$3600\bar{q}_e^r \frac{\partial \theta_e^r}{\partial t} = \bar{\psi} T \frac{\partial}{\partial \tilde{x}}\bar{\kappa}^r \frac{\partial}{\partial \tilde{x}}\theta_e^r + i_{f+dl}^r$ (2.40)	$\bar{\kappa}^r \frac{\partial \theta_e^r}{\partial \tilde{x}}\Big\vert_{\tilde{x}=0,3} = 0$ (2.41) $\bar{\kappa}^{left} \frac{\partial \theta_e}{\partial \tilde{x}}\Big\vert_{left} = \bar{\kappa}^{right} \frac{\partial \theta_e}{\partial \tilde{x}}\Big\vert_{right}$ (2.42) $\theta_e\vert_{left} = \theta_e\vert_{right}$ (2.43)	$\theta_{e,0}^r = 1$ (2.44)
Double layer	$\bar{R}_{dl}\bar{C}_{dl}\frac{\partial i_{dl}}{\partial t} + i_{dl} = \bar{C}_{dl}\frac{\partial(\phi_s - \phi_e - \bar{R}_f i_{f+dl})}{\partial t}$ (2.45) $i_{f+dl} = i_f + i_{dl}$ (2.46)		
Kinetics	$i_f^r = \sum_{j=1}^J i_{f,j}^r$ (2.47) $i_{f,j}^r = i_{0,j}^r\left(\exp\left((1-\alpha_j^r)f\eta^r\right) - \exp\left(-\alpha_j^r f\eta^r\right)\right)$ (2.48) $i_{0,j}^r = \bar{k}_{0,j}^r(\theta_e^r)^{1-\alpha_j^r}(X_j - x_{ss,j}^r)^{\omega_j^r(1-\alpha_j^r)}(x_{ss,j}^r)^{\omega_j\alpha_j^r}$ (2.49) $\eta^r = \phi_s^r - \phi_e^r - U_{ocp}^r(\theta_{ss}^r) - \bar{R}_f^r i_{f+dl}^r$ (2.50)		

$\tilde{x}=\{0,1,2,3\}$ at neg/current-collector, neg/sep, sep/pos, and pos/current-collector boundaries; $\tilde{r}=\{0,1\}$ at particle center and surface. Concentration ratios in solid and electrolyte are θ_s and θ_e; potentials in solid and electrolyte are ϕ_s and ϕ_e; faradaic, double-layer, and total fluxes are i_f, i_{dl}, and i_{f+dl}. $f = F/(RT)$. "Neg," "sep," and "pos" denoted by "n," "s," and "p"; "r"∈{"n," "s," "p"}.
In the MSMR model, an electrode is considered to support J reactions or "galleries." Gallery j has concentration ratio x_j. In many cases it can be convenient to note that $\theta_s = \sum_{j=1}^J x_j$ (so, e.g., $\theta_{ss} = \sum_{j=1}^J x_{ss,j}$).

Table 2.1: Summary of the final physics-based lumped-parameter model with enhanced interface.

boundary conditions are given in Eqs. (2.28) and (2.29), where the forcing input is $i^{\mathrm{r}}_{\mathrm{f},j}$. The initial condition is shown in Eq. (2.30).

The charge-conservation equations do not change except that their forcing inputs are $i^{\mathrm{r}}_{\mathrm{f+dl}}$. In the solid, the PDE is listed in Eq. (2.31). Its boundary conditions are presented in Eqs. (2.32) and (2.33). Its initial conditions for the negative-electrode and positive-electrode regions are given in Eqs. (2.34) and (2.35), respectively. In the electrolyte, the PDE is listed in Eq. (2.36). Its boundary conditions are presented in Eqs. (2.37) and (2.38). Its initial condition is shown in Eq. (2.39).

The electrolyte mass-conservation equations also do not change from Chap. 1 except that the forcing input is now $i^{\mathrm{r}}_{\mathrm{f+dl}}$. The PDE is listed in Eq. (2.40). Its gradient boundary conditions at the current collectors are presented in Eq. (2.41); its boundary conditions at the separator interfaces are given in Eqs. (2.42) and (2.43). Its initial condition is shown in Eq. (2.44).

We have added a description of the double layer, which includes an equation to compute $i_{\mathrm{f+dl}}$ from i_{f} and i_{dl}. We need to be careful to note that the double layer spans ϕ_{s} to ϕ_{sf}, so an intermediate result is:

$$\bar{R}_{\mathrm{dl}}\bar{C}_{\mathrm{dl}}\frac{\partial i_{\mathrm{dl}}}{\partial t} + i_{\mathrm{dl}} = \bar{C}_{\mathrm{dl}}\frac{\partial\left(\phi_{\mathrm{s}} - \phi_{\mathrm{sf}}\right)}{\partial t}.$$

We then recognize that $\phi_{\mathrm{sf}} = \phi_{\mathrm{e}} + \bar{R}_{\mathrm{f}} i_{\mathrm{f+dl}}$. Combining, we have the ODE describing the double layer in Eq. (2.45) and the equation defining $i_{\mathrm{f+dl}}$ in Eq. (2.46).

Finally, the kinetics equations now add the MSMR model details, where we must again be careful to distinguish between i_{f} and $i_{\mathrm{f+dl}}$. These equations are listed in Eqs. (2.47) through (2.50). Notice that in the notation we use, $\bar{i}_{0,j}$ (having the overbar symbol) from Eq. (2.12) is a *constant* evaluated at some linearization setpoint while $i_{0,j}$ (without the overbar symbol) from Eq. (2.49) is a *time-varying* quantity that depends on the present local value of θ_{e} and $x_{\mathrm{ss},j}$.

2.9 *Deriving TFs for all electrochemical variables*

We are now ready to begin developing a full-cell impedance model, which is the principal objective of this chapter. This model will be based on a linearized Laplace-domain interpretation of the PDEs, so we will need to find TFs of all variables of interest. Then we will combine the TFs to compute full-cell impedance.[31]

The approach is similar to the one presented in Vol. I to develop TFs for all electrochemical variables. However, two significant advances have been made since that volume was published. First, the new TFs incorporate the effects of the enhanced solid/electrolyte interfacial model we have just discussed. Second, the new TFs are

[31] We will follow the derivation that was first presented for standard parameters and variables in: Albert Rodríguez, Gregory L. Plett, and M. Scott Trimboli, "Improved transfer functions modeling linearized lithium-ion battery-cell internal electrochemical variables," *Journal of Energy Storage*, 20:560–575, 2018.

However, we will modify the derivation to accommodate the generalized interface model as well as to use lumped parameters and variables, similar to what is presented in: Zhengyu Chu, Gregory L. Plett, M. Scott Trimboli, and Minggao Ouyang, "A control-oriented electrochemical model for lithium-ion battery, part i: Lumped-parameter reduced-order model with constant phase element," *Journal of Energy Storage*, 25:100828, 2019.

The results have been visualized and validated in: Xiangdong Kong, Gregory L. Plett, M. Scott Trimboli, Zhendong Zhang, and Yuejiu Zheng, "An exact closed-form impedance model for porous-electrode lithium-ion cells," *Journal of The Electrochemical Society*, 167(1):013539, 2020.

derived in a different way that eliminates the need to make a simplifying assumption, and so the new TFs are now exact small-signal frequency-domain transformations of the model PDEs.

Recall from Vol. I that we made two basic assumptions when deriving TFs of model variables:[32]

A1. Linearity: We linearized nonlinear equations using Taylor series;

A2. We assumed that current $i_{\text{f+dl}}(\tilde{x},t)$ was decoupled from (not a function of) electrolyte concentration $\theta_{\text{e}}(\tilde{x},t)$.[33] This assumption allowed us to modify the electrolyte-potential equation from:

$$\frac{\partial}{\partial \tilde{x}}\left(\bar{\kappa}\left(\frac{\partial}{\partial \tilde{x}}\phi_{\text{e}} + \bar{\kappa}_D T \frac{\partial \ln(\theta_{\text{e}})}{\partial \tilde{x}}\right)\right) + i_{\text{f+dl}} = 0.$$

to

$$\frac{\partial}{\partial \tilde{x}}\left(\bar{\kappa}\frac{\partial}{\partial \tilde{x}}\phi_{\text{e}}\right) + i_{\text{f+dl}} \approx 0.$$

Note that Assumption A1 is necessary when making an impedance model. By definition, TFs *always* model linear systems. However, Assumption A2 was made simply because we did not know of any other way to find TFs from the PDE model. It was made out of pragmatic expedience rather than being motivated by physical insight.

Subsequent to the publication of Vol. I, we discovered a different way to derive TFs that does not require making Assumption A2, and so the resulting TFs are exact. We present the full development in this section; however, the derivation itself is not the main point of the chapter and so on a first reading we recommend skipping ahead to Sect. 2.10 and returning here later.

The derivation is lengthy and it is easy to get lost in its details. The flowchart in Fig. 2.16 shows the overall process, hoping to add perspective and clarity: boxes shaded blue represent time-domain equations, boxes shaded yellow represent frequency-domain steps leading to $\widetilde{\Theta}_{\text{e}}^{\text{r}}(\tilde{x},s)/I_{\text{app}}(s)$, and boxes shaded green represent TFs developed from $\widetilde{\Theta}_{\text{e}}^{\text{r}}(\tilde{x},s)/I_{\text{app}}(s)$.

2.9.1 *Finding the form of the $\widetilde{\Theta}_{\text{e}}^{\text{r}}(\tilde{x},s)/I_{\text{app}}(s)$ TF*

We begin by finding $\widetilde{\Theta}_{\text{e}}^{\text{r}}(\tilde{x},s)/I_{\text{app}}(s)$, from which all other TFs are later derived. For brevity, we omit superscripts $\text{r} \in \{\text{n},\text{s},\text{p}\}$, assuming a generic applicable region; when distinctions between regions become important, we will reintroduce the superscript notation.

The first step is to subtract Eq. (2.36) from Eq. (2.31),

$$\left[\frac{\partial^2 \phi_{\text{s}}(\tilde{x},t)}{\partial \tilde{x}^2} \qquad\qquad - \frac{1}{\bar{\sigma}} i_{\text{f+dl}}(\tilde{x},t) = 0\right]$$

[32] Simulations show that Assumption A2 is quite good for HEV-type simulations (random, charge-neutral). Assumption A1 is less good, but nonlinear corrections help improve linear predictions.

[33] In terms of standard variables, we assumed that $j(x,t)$ was not a function of $c_{\text{e}}(x,t)$.

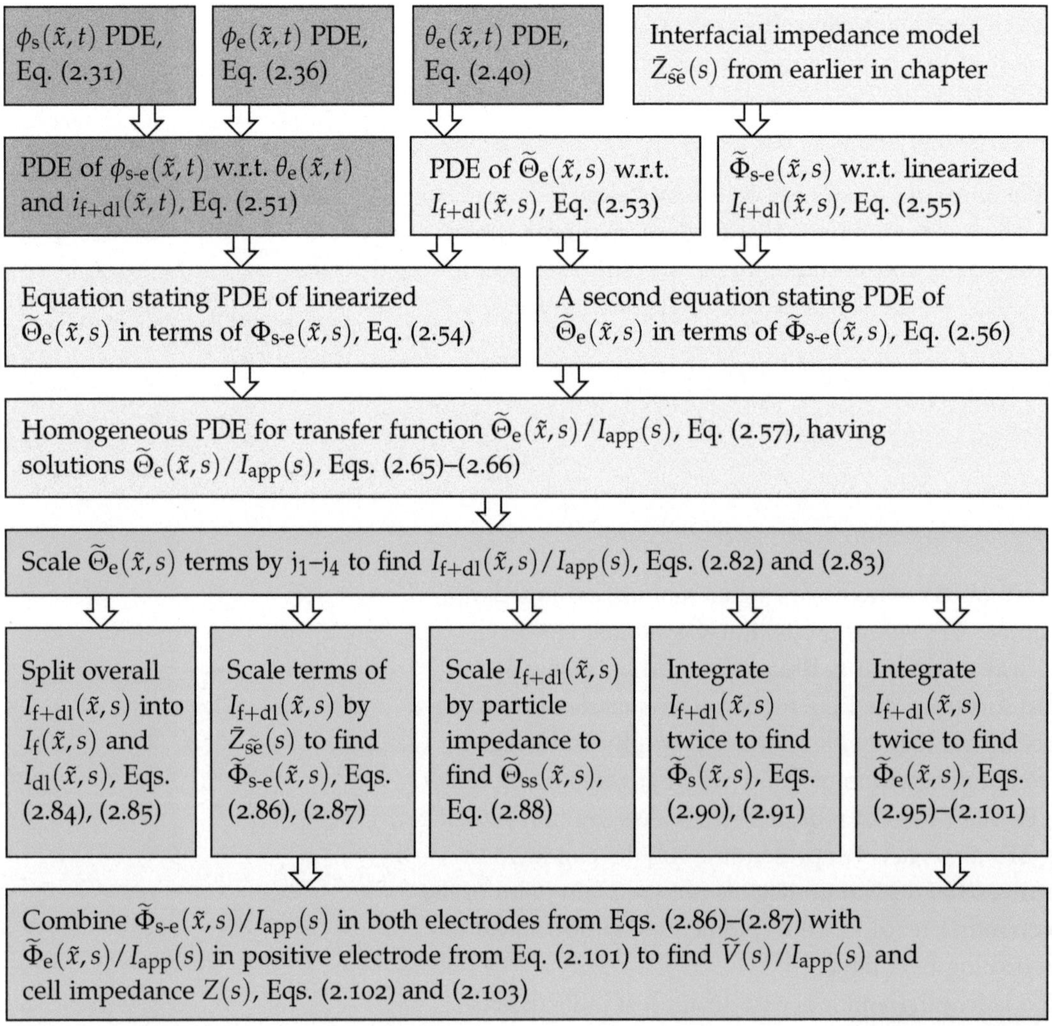

Figure 2.16: Flowchart of the method used to derive the model TFs (a number of the references are to equations in Table 2.1).

$$-\left[\frac{\partial^2 \phi_e(\tilde{x},t)}{\partial \tilde{x}^2} \quad + \bar{\kappa}_D T \frac{\partial^2 \ln \theta_e(\tilde{x},t)}{\partial \tilde{x}^2} \quad + \frac{1}{\bar{\kappa}} i_{f+dl}(\tilde{x},t) = 0\right],$$

to find a PDE for electrode/electrolyte phase potential difference $\phi_{s\text{-}e}(\tilde{x},t) = \phi_s(\tilde{x},t) - \phi_e(\tilde{x},t)$:

$$\frac{\partial^2 \phi_{s\text{-}e}(\tilde{x},t)}{\partial \tilde{x}^2} = \left(\frac{1}{\bar{\sigma}} + \frac{1}{\bar{\kappa}}\right) i_{f+dl}(\tilde{x},t) + \bar{\kappa}_D T \frac{\partial^2 \ln \theta_e(\tilde{x},t)}{\partial \tilde{x}^2}. \qquad (2.51)$$

We linearize this relationship around $\theta_e(\tilde{x},t) \approx \theta_{e,0} = 1$ and define $\tilde{\theta}_e(\tilde{x},t) = \theta_e(\tilde{x},t) - \theta_{e,0}$. To do so, we note that the natural logarithm can be approximated using its Taylor-series (recall that $\theta_{e,0} = 1$):

$$\begin{aligned}\ln(\theta_e(\tilde{x},t)) &= \ln(\theta_{e,0} + \tilde{\theta}_e(\tilde{x},t)) \\ &\approx \ln \theta_{e,0} + \tilde{\theta}_e(\tilde{x},t)/\theta_{e,0} = \tilde{\theta}_e(\tilde{x},t).\end{aligned}$$

Then,

$$\frac{\partial^2 \phi_{\text{s-e}}(\tilde{x},t)}{\partial \tilde{x}^2} \approx \left(\frac{1}{\bar{\sigma}} + \frac{1}{\bar{\kappa}}\right) i_{\text{f+dl}}(\tilde{x},t) + \bar{\kappa}_D T \frac{\partial^2 \tilde{\theta}_{\text{e}}(\tilde{x},t)}{\partial \tilde{x}^2}.$$

We take the Laplace transform of this relationship, giving:

$$\frac{\partial^2 \Phi_{\text{s-e}}(\tilde{x},s)}{\partial \tilde{x}^2} = \left(\frac{1}{\bar{\sigma}} + \frac{1}{\bar{\kappa}}\right) I_{\text{f+dl}}(\tilde{x},s) + \bar{\kappa}_D T \frac{\partial^2 \widetilde{\Theta}_{\text{e}}(\tilde{x},s)}{\partial \tilde{x}^2}. \tag{2.52}$$

We also take the Laplace transform of Eq. (2.40), giving:

$$I_{\text{f+dl}}(\tilde{x},s) = -\bar{\psi}\bar{\kappa}T \frac{\partial^2 \widetilde{\Theta}_{\text{e}}(\tilde{x},s)}{\partial \tilde{x}^2} + 3600\bar{q}_{\text{e}} s \widetilde{\Theta}_{\text{e}}(\tilde{x},s). \tag{2.53}$$

Substituting Eq. (2.53) into (2.52) produces:

$$\begin{aligned} \frac{\partial^2 \Phi_{\text{s-e}}(\tilde{x},s)}{\partial \tilde{x}^2} &= \left(-\bar{\psi}\bar{\kappa}T\left(\frac{1}{\bar{\sigma}} + \frac{1}{\bar{\kappa}}\right) + \bar{\kappa}_D T\right) \frac{\partial^2 \widetilde{\Theta}_{\text{e}}(\tilde{x},s)}{\partial \tilde{x}^2} \\ &\quad + 3600\bar{q}_{\text{e}}\left(\frac{1}{\bar{\sigma}} + \frac{1}{\bar{\kappa}}\right) s\widetilde{\Theta}_{\text{e}}(\tilde{x},s). \end{aligned} \tag{2.54}$$

At this point, we have expressed the PDE of phase potential difference $\Phi_{\text{s-e}}$ as a function of only the electrolyte concentration. We next seek another independent relationship between phase potential difference and electrolyte concentration: we will combine equations to find a homogeneous PDE in $\widetilde{\Theta}_{\text{e}}$ that can be solved. To do so, we start with the interfacial model found earlier, Eq. (2.16), which states $\widetilde{\Phi}_{\text{s-e}}(s) = \bar{Z}_{\widetilde{\text{se}}}(s) I_{\text{f+dl}}(s)$. Note, this TF is the same at every spatial location $\tilde{x}$, so we can also write:

$$\widetilde{\Phi}_{\text{s-e}}(\tilde{x},s) = \bar{Z}_{\widetilde{\text{se}}}(s) I_{\text{f+dl}}(\tilde{x},s). \tag{2.55}$$

From Eqs. (2.55) and (2.53), we can also find:

$$\widetilde{\Phi}_{\text{s-e}}(\tilde{x},s) = \bar{Z}_{\widetilde{\text{se}}}(s)\left(-\bar{\psi}\bar{\kappa}T \frac{\partial^2 \widetilde{\Theta}_{\text{e}}(\tilde{x},s)}{\partial \tilde{x}^2} + 3600\bar{q}_{\text{e}} s \widetilde{\Theta}_{\text{e}}(\tilde{x},s)\right). \tag{2.56}$$

Since $\partial^2 \Phi_{\text{s-e}}(\tilde{x},s)/\partial \tilde{x}^2 = \partial^2 \widetilde{\Phi}_{\text{s-e}}(\tilde{x},s)/\partial \tilde{x}^2$, we can rewrite Eq. (2.54):

$$\begin{aligned} -\bar{\psi}\bar{\kappa}T \frac{\partial^4 \widetilde{\Theta}_{\text{e}}(\tilde{x},s)}{\partial \tilde{x}^4} + 3600\bar{q}_{\text{e}} s \frac{\partial^2 \widetilde{\Theta}_{\text{e}}(\tilde{x},s)}{\partial \tilde{x}^2} &= \frac{3600\bar{q}_{\text{e}}}{\bar{Z}_{\widetilde{\text{se}}}(s)}\left(\frac{1}{\bar{\sigma}} + \frac{1}{\bar{\kappa}}\right) s\widetilde{\Theta}_{\text{e}}(\tilde{x},s) \\ &\quad + \left(\frac{-\bar{\psi}\bar{\kappa}T}{\bar{Z}_{\widetilde{\text{se}}}(s)}\left(\frac{1}{\bar{\sigma}} + \frac{1}{\bar{\kappa}}\right) + \frac{\bar{\kappa}_D T}{\bar{Z}_{\widetilde{\text{se}}}(s)}\right) \frac{\partial^2 \widetilde{\Theta}_{\text{e}}(\tilde{x},s)}{\partial \tilde{x}^2}. \end{aligned}$$

We divide both sides of this equation by $I_{\text{app}}(s)$, resulting in a fourth-order homogeneous PDE with unitless coefficients $\tau_1^{\text{r}}(s)$ and $\tau_2^{\text{r}}(s)$:

$$\frac{\partial^4}{\partial \tilde{x}^4} \frac{\widetilde{\Theta}_{\text{e}}^{\text{r}}(\tilde{x},s)}{I_{\text{app}}(s)} - \tau_1^{\text{r}}(s) \frac{\partial^2}{\partial \tilde{x}^2} \frac{\widetilde{\Theta}_{\text{e}}^{\text{r}}(\tilde{x},s)}{I_{\text{app}}(s)} + \tau_2^{\text{r}}(s) \frac{\widetilde{\Theta}_{\text{e}}^{\text{r}}(\tilde{x},s)}{I_{\text{app}}(s)} = 0, \tag{2.57}$$

where:

$$\tau_1^{\mathrm{r}}(s) = \frac{3600\bar{q}_{\mathrm{e}}^{\mathrm{r}} s}{\bar{\psi}\bar{\kappa}^{\mathrm{r}} T} + \frac{1}{\bar{Z}_{\widetilde{\mathrm{se}}}^{\mathrm{r}}(s)} \left(\frac{1}{\bar{\sigma}^{\mathrm{r}}} + \frac{1}{\bar{\kappa}^{\mathrm{r}}} \right) - \frac{\bar{\kappa}_D T}{\bar{\psi}\bar{\kappa}^{\mathrm{r}} T \bar{Z}_{\widetilde{\mathrm{se}}}^{\mathrm{r}}(s)} \tag{2.58}$$

$$\tau_2^{\mathrm{r}}(s) = \frac{3600\bar{q}_{\mathrm{e}}^{\mathrm{r}} s}{\bar{\psi}\bar{\kappa}^{\mathrm{r}} T \bar{Z}_{\widetilde{\mathrm{se}}}^{\mathrm{r}}(s)} \left(\frac{1}{\bar{\sigma}^{\mathrm{r}}} + \frac{1}{\bar{\kappa}^{\mathrm{r}}} \right). \tag{2.59}$$

The generic solution to the electrolyte concentration TF is then:[34]

$$\frac{\widetilde{\Theta}_{\mathrm{e}}^{\mathrm{r}}(\tilde{x}, s)}{I_{\mathrm{app}}(s)} = \gamma_1^{\mathrm{r}}(s) e^{\Lambda_1^{\mathrm{r}}(s)\tilde{x}} + \gamma_2^{\mathrm{r}}(s) e^{-\Lambda_1^{\mathrm{r}}(s)\tilde{x}} + \gamma_3^{\mathrm{r}}(s) e^{\Lambda_2^{\mathrm{r}}(s)\tilde{x}} + \gamma_4^{\mathrm{r}}(s) e^{-\Lambda_2^{\mathrm{r}}(s)\tilde{x}}, \tag{2.60}$$

where:

$$\Lambda_1^{\mathrm{r}}(s) = \sqrt{\frac{1}{2}\left(\tau_1^{\mathrm{r}}(s) - \sqrt{(\tau_1^{\mathrm{r}}(s))^2 - 4\tau_2^{\mathrm{r}}(s)} \right)} \tag{2.61}$$

$$\Lambda_2^{\mathrm{r}}(s) = \sqrt{\frac{1}{2}\left(\tau_1^{\mathrm{r}}(s) + \sqrt{(\tau_1^{\mathrm{r}}(s))^2 - 4\tau_2^{\mathrm{r}}(s)} \right)}. \tag{2.62}$$

$\Lambda_1^{\mathrm{r}}(s)$ and $\Lambda_2^{\mathrm{r}}(s)$ have nonnegative real parts that tend toward infinity as $s \to \infty$, so $\exp(\Lambda_1^{\mathrm{r}}(s))$ and $\exp(\Lambda_2^{\mathrm{r}}(s))$ have magnitudes that also approach infinity. For numeric robustness, we modify how we compute Eq. (2.60) by first defining spatial variable $0 \le z \le 1$:[35]

$$z = \begin{cases} \tilde{x}, & \text{in the negative electrode} \\ \tilde{x} - 1, & \text{in the separator region} \\ 3 - \tilde{x}, & \text{in the positive electrode.} \end{cases} \tag{2.63}$$

Second, we rewrite Eq. (2.60) as:[36,37]

$$\frac{\widetilde{\Theta}_{\mathrm{e}}^{\mathrm{r}}(z, s)}{I_{\mathrm{app}}(s)} = c_1^{\mathrm{r}}(s) e^{\Lambda_1^{\mathrm{r}}(z-1)} + c_2^{\mathrm{r}}(s) e^{-\Lambda_1^{\mathrm{r}} z} + c_3^{\mathrm{r}}(s) e^{\Lambda_2^{\mathrm{r}}(z-1)} + c_4^{\mathrm{r}}(s) e^{-\Lambda_2^{\mathrm{r}} z}. \tag{2.64}$$

The input to $\exp(\cdot)$ now always has a nonpositive real part; so, the $\exp(\cdot)$ function will always produce a finite result. In this new formulation, the $c_k^{\mathrm{r}}(s)$ coefficients are related to but different from $\gamma_k^{\mathrm{r}}(s)$. We will use this new formulation exclusively from this point forward.

This solution corresponds to a generic electrode region. The specific electrolyte-concentration solutions for the two regions will be

- Negative-electrode region (where $z = \tilde{x}$):

$$\frac{\widetilde{\Theta}_{\mathrm{e}}^{\mathrm{n}}(z, s)}{I_{\mathrm{app}}(s)} = c_1^{\mathrm{n}}(s) e^{\Lambda_1^{\mathrm{n}}(z-1)} + c_2^{\mathrm{n}}(s) e^{-\Lambda_1^{\mathrm{n}} z} + c_3^{\mathrm{n}}(s) e^{\Lambda_2^{\mathrm{n}}(z-1)} + c_4^{\mathrm{n}}(s) e^{-\Lambda_2^{\mathrm{n}} z}. \tag{2.65}$$

- Positive-electrode region (where $z = 3 - \tilde{x}$):

$$\frac{\widetilde{\Theta}_{\mathrm{e}}^{\mathrm{p}}(z, s)}{I_{\mathrm{app}}(s)} = c_1^{\mathrm{p}}(s) e^{\Lambda_1^{\mathrm{p}}(z-1)} + c_2^{\mathrm{p}}(s) e^{-\Lambda_1^{\mathrm{p}} z} + c_3^{\mathrm{p}}(s) e^{\Lambda_2^{\mathrm{p}}(z-1)} + c_4^{\mathrm{p}}(s) e^{-\Lambda_2^{\mathrm{p}} z}. \tag{2.66}$$

[34] An alternate generic form of the solution is (where $\gamma_k^{\mathrm{r}}(s) \neq \bar{\gamma}_k^{\mathrm{r}}(s)$):

$$\frac{\widetilde{\Theta}_{\mathrm{e}}^{\mathrm{r}}(\tilde{x}, s)}{I_{\mathrm{app}}(s)} = \bar{\gamma}_1^{\mathrm{r}}(s)\cosh(\Lambda_1^{\mathrm{r}}(s)\tilde{x}) + \bar{\gamma}_2^{\mathrm{r}}(s)\sinh(\Lambda_1^{\mathrm{r}}(s)\tilde{x}) + \bar{\gamma}_3^{\mathrm{r}}(s)\cosh(\Lambda_2^{\mathrm{r}}(s)\tilde{x}) + \bar{\gamma}_4^{\mathrm{r}}(s)\sinh(\Lambda_2^{\mathrm{r}}(s)\tilde{x}).$$

We have implemented both: while they are theoretically equivalent, we choose to present the exponential rather than the hyperbolic form since we find that its numeric performance at high frequencies is more robust in practice.

[35] This is the same definition for 1D spatial-location variable z that we used in Vol. I.

[36] We recognize that Λ_1^{r} and Λ_2^{r} are both functions of the Laplace variable s. However, for a more compact notation, we will usually omit the dependence on s from the equations from this point forward.

[37] The coefficients $c_k^{\mathrm{r}}(s)$ absorb scale factors during the conversion from Eq. (2.60) to Eq. (2.64); therefore $c_k^{\mathrm{r}}(s)$ are related to but different from $\gamma_k^{\mathrm{r}}(s)$, as mentioned in the main text.

In the separator, $i^{\rm s}_{\rm f+dl}(\tilde{x},t) = 0$ and $\phi^{\rm s}_{\rm s\text{-}e}(\tilde{x},t)$ does not exist. For this region then, directly from Eq. (2.40),

$$\frac{\partial^2}{\partial \tilde{x}^2}\frac{\widetilde{\Theta}_{\rm e}(\tilde{x},s)}{I_{\rm app}(s)} - \frac{3600\bar{q}^{\rm s}_{\rm e}s}{\bar{\psi}\bar{\kappa}^{\rm s}T}\frac{\widetilde{\Theta}_{\rm e}(\tilde{x},s)}{I_{\rm app}(s)} = 0.$$

The solution to this lower-order PDE is (where $z = \tilde{x} - 1$):

$$\frac{\widetilde{\Theta}^{\rm s}_{\rm e}(z,s)}{I_{\rm app}(s)} = c^{\rm s}_1(s)e^{\Lambda^{\rm s}_1(z-1)} + c^{\rm s}_2(s)e^{-\Lambda^{\rm s}_1 z}, \tag{2.67}$$

where there is no $\Lambda^{\rm s}_2(s)$ and

$$\Lambda^{\rm s}_1(s) = \sqrt{3600\bar{q}^{\rm s}_{\rm e}s/(\bar{\psi}\bar{\kappa}^{\rm s}T)}. \tag{2.68}$$

2.9.2 *Solving for coefficients in* $\widetilde{\Theta}^{\rm r}_{\rm e}(\tilde{x},s)/I_{\rm app}(s)$

To summarize to this point: we have found a generic Laplace-domain relationship for $\widetilde{\Theta}^{\rm r}_{\rm e}(\tilde{x},s)/I_{\rm app}(s)$, written in terms of ten undetermined functions: $c^{\rm n}_1(s)$, $c^{\rm n}_2(s)$, $c^{\rm n}_3(s)$, $c^{\rm n}_4(s)$, $c^{\rm s}_1(s)$, $c^{\rm s}_2(s)$, $c^{\rm p}_1(s)$, $c^{\rm p}_2(s)$, $c^{\rm p}_3(s)$, $c^{\rm p}_4(s)$.[38] Hence, to find $\widetilde{\Theta}^{\rm r}_{\rm e}(\tilde{x},s)/I_{\rm app}(s)$ at some point $\tilde{x}$ in the cell, we must first solve for these ten functions by using the ten boundary conditions (BCs) of the $\phi_{\rm s}$, $\phi_{\rm e}$, and $\theta_{\rm e}$ PDEs.

[38] We also recognize that $c^{\rm r}_k$ are all functions of the Laplace variable s. However, for a more compact notation, we will usually omit the dependence on s from the equations from this point forward.

If the solid- and electrolyte-potential BCs stated in Eqs. (2.32), (2.33), (2.37), and (2.38) are combined, we have:

$$\begin{aligned}\left.\frac{\partial \phi_{\rm s\text{-}e}(\tilde{x},t)}{\partial \tilde{x}}\right|_{\tilde{x}=0,3} &= -\frac{i_{\rm app}(t)}{\bar{\sigma}} + \underbrace{\bar{\kappa}_D T\left.\frac{\partial \tilde{\theta}_{\rm e}(\tilde{x},t)}{\partial \tilde{x}}\right|_{\tilde{x}=0,3}}_{0\text{ by Eq. (2.41)}}\\ \left.\frac{\partial \phi_{\rm s\text{-}e}(\tilde{x},t)}{\partial \tilde{x}}\right|_{\tilde{x}=1,2} &= \frac{i_{\rm app}(t)}{\bar{\kappa}} + \bar{\kappa}_D T\left.\frac{\partial \tilde{\theta}_{\rm e}(\tilde{x},t)}{\partial \tilde{x}}\right|_{\tilde{x}=1,2}.\end{aligned}$$

Recognizing that $\partial\phi_{\rm s\text{-}e}/\partial\tilde{x} = \partial\tilde{\phi}_{\rm s\text{-}e}/\partial\tilde{x}$, Laplace transforms give:

$$\begin{aligned}\left.\frac{\partial \widetilde{\Phi}_{\rm s\text{-}e}(\tilde{x},s)}{\partial \tilde{x}}\right|_{\tilde{x}=0,3} &= -\frac{I_{\rm app}(s)}{\bar{\sigma}}\\ \left.\frac{\partial \widetilde{\Phi}_{\rm s\text{-}e}(\tilde{x},s)}{\partial \tilde{x}}\right|_{\tilde{x}=1,2} &= \frac{I_{\rm app}(s)}{\bar{\kappa}} + \bar{\kappa}_D T\left.\frac{\partial}{\partial \tilde{x}}\widetilde{\Theta}_{\rm e}(\tilde{x},s)\right|_{\tilde{x}=1,2}.\end{aligned}$$

To find the unknown coefficients, we must express all BCs in terms of the electrolyte concentration. We reuse Eq. (2.56) and differentiate:

$$\begin{aligned}\widetilde{\Phi}_{\rm s\text{-}e}(\tilde{x},s) &= \bar{Z}_{\rm \tilde{s}\tilde{e}}(s)\left(-\bar{\psi}\bar{\kappa}T\frac{\partial^2\widetilde{\Theta}_{\rm e}(\tilde{x},s)}{\partial\tilde{x}^2} + 3600\bar{q}_{\rm e}s\widetilde{\Theta}_{\rm e}(\tilde{x},s)\right)\\ \frac{\partial\widetilde{\Phi}_{\rm s\text{-}e}(\tilde{x},s)}{\partial\tilde{x}} &= \bar{Z}_{\rm \tilde{s}\tilde{e}}(s)\left(-\bar{\psi}\bar{\kappa}T\frac{\partial^3\widetilde{\Theta}_{\rm e}(\tilde{x},s)}{\partial\tilde{x}^3} + 3600\bar{q}_{\rm e}s\frac{\partial\widetilde{\Theta}_{\rm e}(\tilde{x},s)}{\partial\tilde{x}}\right).\end{aligned}$$

At boundaries $\tilde{x} = 0$ and 3, the first-derivative term disappears, so:

$$\left.\frac{\partial\widetilde{\Phi}_{\text{s-e}}(\tilde{x},s)}{\partial\tilde{x}}\right|_{\tilde{x}=0,3} = -\bar{\psi}\bar{\kappa}T\bar{Z}_{\tilde{\text{se}}}(s)\frac{\partial^3\widetilde{\Theta}_{\text{e}}(\tilde{x},s)}{\partial\tilde{x}^3} = -\frac{I_{\text{app}}(s)}{\bar{\sigma}}$$

$$\left.\frac{\partial^3}{\partial\tilde{x}^3}\frac{\widetilde{\Theta}_{\text{e}}(\tilde{x},s)}{I_{\text{app}}(s)}\right|_{\tilde{x}=0,3} = \frac{1}{\bar{\sigma}\bar{\kappa}\bar{\psi}T\bar{Z}_{\tilde{\text{se}}}(s)} = \frac{\mu_1(s)}{\bar{\sigma}}, \tag{2.69}$$

where we have defined:

$$\mu_1^{\text{r}}(s) = \frac{1}{\bar{\psi}\bar{\kappa}^{\text{r}}T\bar{Z}^{\text{r}}_{\tilde{\text{se}}}(s)}. \tag{2.70}$$

At boundaries $\tilde{x} = 1, 2$ we have:

$$\bar{Z}_{\tilde{\text{se}}}(s)\left(-\bar{\psi}\bar{\kappa}T\frac{\partial^3\widetilde{\Theta}_{\text{e}}(\tilde{x},s)}{\partial\tilde{x}^3}+3600\bar{q}_{\text{e}}s\frac{\partial\widetilde{\Theta}_{\text{e}}(\tilde{x},s)}{\partial\tilde{x}}\right) = \frac{I_{\text{app}}(s)}{\bar{\kappa}}+\bar{\kappa}_D T\frac{\partial}{\partial\tilde{x}}\widetilde{\Theta}_{\text{e}}(\tilde{x},s).$$

Rearranging,

$$\begin{aligned}\left.\frac{\partial^3}{\partial\tilde{x}^3}\frac{\widetilde{\Theta}_{\text{e}}(\tilde{x},s)}{I_{\text{app}}(s)}\right|_{\tilde{x}=1,2} &= -\frac{1}{\bar{\psi}\bar{\kappa}^2T\bar{Z}_{\tilde{\text{se}}}(s)} - \frac{\bar{\kappa}_D T - 3600\bar{q}_{\text{e}}s\bar{Z}_{\tilde{\text{se}}}}{\bar{\psi}\bar{\kappa}T\bar{Z}_{\tilde{\text{se}}}(s)}\frac{\partial}{\partial\tilde{x}}\frac{\widetilde{\Theta}_{\text{e}}(\tilde{x},s)}{I_{\text{app}}(s)} \\ &= -\frac{\mu_1(s)}{\bar{\kappa}} - \mu_2(s)\left.\frac{\partial}{\partial\tilde{x}}\frac{\widetilde{\Theta}_{\text{e}}(\tilde{x},s)}{I_{\text{app}}(s)}\right|_{\tilde{x}=1,2},\end{aligned} \tag{2.71}$$

where we have defined:

$$\mu_2^{\text{r}}(s) = \left(\bar{\kappa}_D T - 3600\bar{q}_{\text{e}}^{\text{r}}s\bar{Z}^{\text{r}}_{\tilde{\text{se}}}(s)\right)\mu_1^{\text{r}}(s). \tag{2.72}$$

We know the form of the electrolyte-concentration equation for electrode regions from Eq. (2.64), so we can compute the derivatives:

$$\frac{\partial}{\partial z}\frac{\widetilde{\Theta}_{\text{e}}(z,s)}{I_{\text{app}}(s)} = c_1\Lambda_1 e^{\Lambda_1(z-1)} - c_2\Lambda_1 e^{-\Lambda_1 z} + c_3\Lambda_2 e^{\Lambda_2(z-1)} - c_4\Lambda_2 e^{-\Lambda_2 z} \tag{2.73}$$

$$\frac{\partial^3}{\partial z^3}\frac{\widetilde{\Theta}_{\text{e}}(z,s)}{I_{\text{app}}(s)} = c_1\Lambda_1^3 e^{\Lambda_1(z-1)} - c_2\Lambda_1^3 e^{-\Lambda_1 z} + c_3\Lambda_2^3 e^{\Lambda_2(z-1)} - c_4\Lambda_2^3 e^{-\Lambda_2 z}. \tag{2.74}$$

We can also compute the separator-region derivative from Eq. (2.67):

$$\frac{\partial}{\partial z}\frac{\widetilde{\Theta}_{\text{e}}(z,s)}{I_{\text{app}}(s)} = c_1^{\text{s}}\Lambda_1^{\text{s}} e^{\Lambda_1^{\text{s}}(z-1)} - c_2^{\text{s}}\Lambda_1^{\text{s}} e^{-\Lambda_1^{\text{s}} z}. \tag{2.75}$$

With this background, we can develop ten equations summarizing the ten BCs in terms of the ten unknown parameter values.[39]

- At $\tilde{x} = 0$ ($z^{\text{n}} = 0$), we have two BC equations:
  - Combine Eqs. (2.41) and (2.73):

$$c_1^{\text{n}}\Lambda_1^{\text{n}} e^{-\Lambda_1^{\text{n}}} - c_2^{\text{n}}\Lambda_1^{\text{n}} + c_3^{\text{n}}\Lambda_2^{\text{n}} e^{\Lambda_2^{\text{n}}} - c_4^{\text{n}}\Lambda_2^{\text{n}} = 0.$$

[39] The purpose of the following analysis is to show that the ten BCs yield a system of equations that is linear in the ten unknown parameter values, as summarized by Eq. (2.81). One way to find a solution to the unknown parameter values is to solve Eq. (2.81), which involves the inversion of a 10×10 matrix (or equivalent operations using back-substitution, etc.).

There are other ways to simplify the problem. For example, Eqs. (2.76) and (2.77) can be combined to solve for c_1^{s} and c_2^{s} in terms of c_1^{n}, c_2^{n}, c_3^{n}, and c_4^{n}. This solution can be substituted into Eqs. (2.79) and (2.80) to eliminate c_1^{s} and c_2^{s} from the system of equations to be solved, allowing us to reformulate the eight remaining unknown parameter values in an 8×8 framework similar to Eq. (2.81). The 8×8 framework can be solved for c_1^{n}, c_2^{n}, c_3^{n}, c_4^{n}, c_1^{p}, c_2^{p}, c_3^{p}, and c_4^{p}. Then we can find c_1^{s} and c_2^{s} from the values of these eight parameters. This approach requires less computation than using the 10×10 framework presented in Eq. (2.81).

Through a lot of manual effort, it is possible to combine equations manually to find a completely closed-form result: this solution is given in Sect. 2.C, and is the method we prefer for solving for the parameter values in practice.

- Combine Eqs. (2.69) and (2.74):

$$c_1^{\rm n}(\Lambda_1^{\rm n})^3 e^{-\Lambda_1^{\rm n}} - c_2^{\rm n}(\Lambda_1^{\rm n})^3 + c_3^{\rm n}(\Lambda_2^{\rm n})^3 e^{-\Lambda_2^{\rm n}} - c_4^{\rm n}(\Lambda_2^{\rm n})^3 = \frac{\mu_1^{\rm n}}{\bar{\sigma}^{\rm n}}.$$

- At $\tilde{x} = 1$ ($z^{\rm n} = 1$ and $z^{\rm s} = 0$), we have three BC equations:
  - Combine Eqs. (2.42), (2.73), and (2.75):

$$\bar{\kappa}^{\rm n}\left(\Lambda_1^{\rm n}\left(c_1^{\rm n} - c_2^{\rm n} e^{-\Lambda_1^{\rm n}}\right) + \Lambda_2^{\rm n}\left(c_3^{\rm n} - c_4^{\rm n} e^{-\Lambda_2^{\rm n}}\right)\right) = \bar{\kappa}^{\rm s}\Lambda_1^{\rm s}\left(c_1^{\rm s} e^{-\Lambda_1^{\rm s}} - c_2^{\rm s}\right). \tag{2.76}$$

  - Combine Eqs. (2.43), (2.64), and (2.67):

$$c_1^{\rm n} + c_2^{\rm n} e^{-\Lambda_1^{\rm n}} + c_3^{\rm n} + c_4^{\rm n} e^{-\Lambda_2^{\rm n}} = c_1^{\rm s} e^{-\Lambda_1^{\rm s}} + c_2^{\rm s}. \tag{2.77}$$

  - Combine Eqs. (2.71), (2.73), and (2.74) at $\tilde{x} = 1^-$:

$$-\frac{\mu_1^{\rm n}}{\bar{\kappa}^{\rm n}} = c_1^{\rm n}\lambda_1^{\rm n} - c_2^{\rm n}\lambda_1^{\rm n} e^{-\Lambda_1^{\rm n}} + c_3^{\rm n}\lambda_2^{\rm n} - c_4^{\rm n}\lambda_2^{\rm n} e^{-\Lambda_2^{\rm n}},$$

  where we have defined:

$$\lambda_k^{\rm r} = (\Lambda_k^{\rm r})^3 + \mu_2^{\rm r}\Lambda_k^{\rm r}. \tag{2.78}$$

- At $\tilde{x} = 2$ ($z^{\rm s} = z^{\rm p} = 1$), we have three BC equations (notice sign changes in positive-electrode derivatives because $dz/d\tilde{x} = -1$):
  - Combine Eqs. (2.42), (2.73), and (2.75):

$$\bar{\kappa}^{\rm s}\Lambda_1^{\rm s}\left(c_1^{\rm s} - c_2^{\rm s} e^{-\Lambda_1^{\rm s}}\right) = -\bar{\kappa}^{\rm p}\left(\Lambda_1^{\rm p}\left(c_1^{\rm p} - c_2^{\rm p} e^{-\Lambda_1^{\rm p}}\right) + \Lambda_2^{\rm p}\left(c_3^{\rm p} - c_4^{\rm p} e^{-\Lambda_2^{\rm p}}\right)\right). \tag{2.79}$$

  - Combine Eqs. (2.43), (2.64), and (2.67):

$$c_1^{\rm s} + c_2^{\rm s} e^{-\Lambda_1^{\rm s}} = c_1^{\rm p} + c_2^{\rm p} e^{-\Lambda_1^{\rm p}} + c_3^{\rm p} + c_4^{\rm p} e^{-\Lambda_2^{\rm p}}. \tag{2.80}$$

  - Combine Eqs. (2.71), (2.73), and (2.74) at $\tilde{x} = 2^+$:

$$\frac{\mu_1^{\rm p}}{\bar{\kappa}^{\rm p}} = c_1^{\rm p}\lambda_1^{\rm p} - c_2^{\rm p}\lambda_1^{\rm p} e^{-\Lambda_1^{\rm p}} + c_3^{\rm p}\lambda_2^{\rm p} - c_4^{\rm p}\lambda_2^{\rm p} e^{-\Lambda_2^{\rm p}}.$$

- At $\tilde{x} = 3$ ($z^{\rm p} = 0$), we have two BC equations (notice sign changes in positive-electrode derivatives because $dz/d\tilde{x} = -1$):
  - Combine Eqs. (2.41) and (2.73):

$$c_1^{\rm p}\Lambda_1^{\rm p} e^{-\Lambda_1^{\rm p}} - c_2^{\rm p}\Lambda_1^{\rm p} + c_3^{\rm p}\Lambda_2^{\rm p} e^{-\Lambda_2^{\rm p}} - c_4^{\rm p}\Lambda_2^{\rm p} = 0.$$

  - Combine Eqs. (2.69) and (2.74):

$$c_1^{\rm p}(\Lambda_1^{\rm p})^3 e^{-\Lambda_1^{\rm p}} - c_2^{\rm p}(\Lambda_1^{\rm p})^3 + c_3^{\rm p}(\Lambda_2^{\rm p})^3 e^{-\Lambda_2^{\rm p}} - c_4^{\rm p}(\Lambda_2^{\rm p})^3 = -\frac{\mu_1^{\rm p}}{\bar{\sigma}^{\rm p}}.$$

All of these BCs may be assembled in matrix form $\mathbf{A}\mathbf{c} = \mathbf{b}$:

$$\begin{bmatrix} \Lambda_1^n e^{-\Lambda_1^n} & -\Lambda_1^n & \Lambda_2^n e^{-\Lambda_2^n} & -\Lambda_2^n & 0 & 0 & 0 & 0 & 0 & 0 \\ (\Lambda_1^n)^3 e^{-\Lambda_1^n} & -(\Lambda_1^n)^3 & (\Lambda_2^n)^3 e^{-\Lambda_2^n} & -(\Lambda_2^n)^3 & 0 & 0 & 0 & 0 & 0 & 0 \\ \bar{\kappa}^n \Lambda_1^n & -\bar{\kappa}^n \Lambda_1^n e^{-\Lambda_1^n} & \bar{\kappa}^n \Lambda_2^n & -\bar{\kappa}^n \Lambda_2^n e^{-\Lambda_2^n} & -\bar{\kappa}^s \Lambda_1^s e^{-\Lambda_1^s} & \bar{\kappa}^s \Lambda_1^s & 0 & 0 & 0 & 0 \\ 1 & e^{-\Lambda_1^n} & 1 & e^{-\Lambda_2^n} & -e^{-\Lambda_1^s} & -1 & 0 & 0 & 0 & 0 \\ \lambda_1^n & -\lambda_1^n e^{-\Lambda_1^n} & \lambda_2^n & -\lambda_2^n e^{-\Lambda_2^n} & 0 & 0 & 0 & 0 & 0 & 0 \\ 0 & 0 & 0 & 0 & \bar{\kappa}^s \Lambda_1^s & -\bar{\kappa}^s \Lambda_1^s e^{-\Lambda_1^s} & \bar{\kappa}^p \Lambda_1^p & -\bar{\kappa}^p \Lambda_1^p e^{-\Lambda_1^p} & \bar{\kappa}^p \Lambda_2^p & -\bar{\kappa}^p \Lambda_2^p e^{-\Lambda_2^p} \\ 0 & 0 & 0 & 0 & 1 & e^{-\Lambda_1^s} & -1 & -e^{-\Lambda_1^p} & -1 & -e^{-\Lambda_2^p} \\ 0 & 0 & 0 & 0 & 0 & 0 & \lambda_1^p & -\lambda_1^p e^{-\Lambda_1^p} & \lambda_2^p & -\lambda_2^p e^{-\Lambda_2^p} \\ 0 & 0 & 0 & 0 & 0 & 0 & \Lambda_1^p e^{-\Lambda_1^p} & -\Lambda_1^p & \Lambda_2^p e^{-\Lambda_2^p} & -\Lambda_2^p \\ 0 & 0 & 0 & 0 & 0 & 0 & (\Lambda_1^p)^3 e^{-\Lambda_1^p} & -(\Lambda_1^p)^3 & (\Lambda_2^p)^3 e^{-\Lambda_2^p} & -(\Lambda_2^p)^3 \end{bmatrix} \begin{bmatrix} c_1^n \\ c_2^n \\ c_3^n \\ c_4^n \\ c_1^s \\ c_2^s \\ c_1^p \\ c_2^p \\ c_3^p \\ c_4^p \end{bmatrix} = \begin{bmatrix} 0 \\ \mu_1^n/\bar{\sigma}^n \\ 0 \\ 0 \\ -\mu_1^n/\bar{\kappa}^n \\ 0 \\ 0 \\ \mu_1^p/\bar{\kappa}^p \\ 0 \\ -\mu_1^p/\bar{\sigma}^p \end{bmatrix}. \tag{2.81}$$

As long as $\mathbf{A}$ is nonsingular and well-conditioned, we can determine the constants numerically as $\mathbf{c} = \mathbf{A}^{-1}\mathbf{b}$ at every frequency of interest. Alternately, we can solve these ten equations to find a closed-form result: this solution is given in Sect. 2.C.

2.9.3 *Interfacial lithium-flux TF*

The derivation of $\widetilde{\Theta}_e(\tilde{x},s)/I_{\text{app}}(s)$ unlocks all remaining electrochemical-variable TFs as they can all be written in terms of $\widetilde{\Theta}_e(\tilde{x},s)$. We begin by looking at the interfacial lithium flux, which is expressed as a function of the electrolyte concentration in Eq. (2.53). In the negative electrode, we substitute Eq. (2.65):

$$\begin{aligned} \frac{I_{\text{f+dl}}^n(z,s)}{I_{\text{app}}(s)} &= -\bar{\psi}\bar{\kappa}^n T\Big(c_1^n\,(\Lambda_1^n)^2\,e^{\Lambda_1^n(z-1)} + c_2^n\,(\Lambda_1^n)^2\,e^{-\Lambda_1^n z} \\ &\quad + c_3^n\,(\Lambda_2^n)^2\,e^{\Lambda_2^n(z-1)} + c_4^n\,(\Lambda_2^n)^2\,e^{-\Lambda_2^n z}\Big) \\ &\quad + 3600\bar{q}_e^n s\left(c_1^n e^{\Lambda_1^n(z-1)} + c_2^n e^{-\Lambda_1^n z} + c_3^n e^{\Lambda_2^n(z-1)} + c_4^n e^{-\Lambda_2^n z}\right) \\ &= j_1^n e^{\Lambda_1^n(z-1)} + j_2^n e^{-\Lambda_1^n z} + j_3^n e^{\Lambda_2^n(z-1)} + j_4^n e^{-\Lambda_2^n z}, \end{aligned} \tag{2.82}$$

where the equation subfunctions $j_1^n(s)$ through $j_4^n(s)$ are given by:

$$j_1^n = c_1^n\left(-\bar{\psi}\bar{\kappa}^n T\,(\Lambda_1^n)^2 + 3600\bar{q}_e^n s\right),\quad j_2^n = c_2^n\left(-\bar{\psi}\bar{\kappa}^n T\,(\Lambda_1^n)^2 + 3600\bar{q}_e^n s\right),$$
$$j_3^n = c_3^n\left(-\bar{\psi}\bar{\kappa}^n T\,(\Lambda_2^n)^2 + 3600\bar{q}_e^n s\right),\quad j_4^n = c_4^n\left(-\bar{\psi}\bar{\kappa}^n T\,(\Lambda_2^n)^2 + 3600\bar{q}_e^n s\right).$$

There is no reaction in the separator region because of the absence of solid material, so $I_{\text{f+dl}}^s(z,s) = 0$. In the positive electrode, we substitute Eq. (2.66) to find:

$$\frac{I_{\text{f+dl}}^p(z,s)}{I_{\text{app}}(s)} = j_1^p e^{\Lambda_1^p(z-1)} + j_2^p e^{-\Lambda_1^p z} + j_3^p e^{\Lambda_2^p(z-1)} + j_4^p e^{-\Lambda_2^p z}, \qquad (2.83)$$

where the equation subfunctions $j_1^p(s)$ through $j_4^p(s)$ are given by:

$$j_1^p = c_1^p\left(-\bar{\psi}\bar{\kappa}^p T\,(\Lambda_1^p)^2 + 3600\bar{q}_e^p s\right),\quad j_2^p = c_2^p\left(-\bar{\psi}\bar{\kappa}^p T\,(\Lambda_1^p)^2 + 3600\bar{q}_e^p s\right),$$
$$j_3^p = c_3^p\left(-\bar{\psi}\bar{\kappa}^p T\,(\Lambda_2^p)^2 + 3600\bar{q}_e^p s\right),\quad j_4^p = c_4^p\left(-\bar{\psi}\bar{\kappa}^p T\,(\Lambda_2^p)^2 + 3600\bar{q}_e^p s\right).$$

2.9.4 *Faradaic interfacial lithium-flux TF*

We will need to know the faradaic lithium flux i_f^r to compute overpotential and cell voltage; however, its calculation depends on the exact interfacial model used. For the default simplified model of Fig. 2.15,

$$\frac{I_f^r(\tilde{x},s)}{I_{\text{app}}(s)} = \frac{\bar{Z}_{\widetilde{\text{dl}}}^r(s)}{\bar{R}_{\text{ct}}^r + \bar{Z}_{\tilde{s}}^r(s) + \bar{Z}_{\widetilde{\text{dl}}}^r(s)}\,\frac{I_{\text{f+dl}}^r(\tilde{x},s)}{I_{\text{app}}(s)}, \qquad (2.84)$$

where

$$\bar{Z}_{\widetilde{\text{dl}}}(s) = \bar{R}_{\text{dl}} + \frac{(s\bar{C}_{\text{dl}} + \omega_{\text{dl}})^{(1-n_{\text{dl}})}}{s(\bar{C}_{\text{dl}})^{2-n_{\text{dl}}}}.$$

2.9.5 *Nonfaradaic interfacial lithium-flux TF*

By similar arguments, the double-layer flux TF for the default simplified model of Fig. 2.15 is:

$$\frac{I_{\text{dl}}^r(\tilde{x},s)}{I_{\text{app}}(s)} = \frac{\bar{R}_{\text{ct}}^r + \bar{Z}_{\tilde{s}}^r(s)}{\bar{R}_{\text{ct}}^r + \bar{Z}_{\tilde{s}}^r(s) + \bar{Z}_{\widetilde{\text{dl}}}^r(s)}\,\frac{I_{\text{f+dl}}^r(\tilde{x},s)}{I_{\text{app}}(s)}. \qquad (2.85)$$

2.9.6 *Phase-potential-difference TF*

The phase-potential-difference TF can be found by using Eq. (2.55). The negative-electrode TF can be written as:

$$\begin{aligned}\frac{\widetilde{\Phi}_{\text{s-e}}^n(z,s)}{I_{\text{app}}(s)} &= \bar{Z}_{\widetilde{\text{se}}}^n(s)\,\frac{I_{\text{f+dl}}^n(z,s)}{I_{\text{app}}(s)}\\ &= \bar{Z}_{\widetilde{\text{se}}}^n(s)\left(j_1^n e^{\Lambda_1^n(z-1)} + j_2^n e^{-\Lambda_1^n z} + j_3^n e^{\Lambda_2^n(z-1)} + j_4^n e^{-\Lambda_2^n z}\right). \qquad (2.86)\end{aligned}$$

There is no phase potential difference in the separator since $\widetilde{\Phi}_s$ is undefined in that region. The positive-electrode TF can be written in a similar way to the negative-electrode TF:

$$\frac{\widetilde{\Phi}^{\mathrm{p}}_{\text{s-e}}(z,s)}{I_{\text{app}}(s)} = \bar{Z}^{\mathrm{p}}_{\tilde{\text{se}}}(s)\left(\mathrm{j}^{\mathrm{p}}_1 e^{\Lambda^{\mathrm{p}}_1(z-1)} + \mathrm{j}^{\mathrm{p}}_2 e^{-\Lambda^{\mathrm{p}}_1 z} + \mathrm{j}^{\mathrm{p}}_3 e^{\Lambda^{\mathrm{p}}_2(z-1)} + \mathrm{j}^{\mathrm{p}}_4 e^{-\Lambda^{\mathrm{p}}_2 z}\right). \tag{2.87}$$

2.9.7 *Solid-surface-concentration TF*

We determine the solid-surface-concentration TF in a stepwise fashion by recognizing:

$$\frac{\widetilde{\Theta}_{\text{ss}}(\tilde{x},s)}{I_{\text{app}}(s)} = \frac{\widetilde{\Theta}_{\text{ss}}(\tilde{x},s)}{I_{\text{f}}(\tilde{x},s)}\frac{I_{\text{f}}(\tilde{x},s)}{I_{\text{app}}(s)}.$$

The first term depends on particle impedance and the OCP function:

$$\frac{\widetilde{\Theta}_{\text{ss}}(\tilde{x},s)}{I_{\text{f}}(\tilde{x},s)} = \frac{\bar{Z}_{\tilde{\text{s}}}(s)}{[U_{\text{ocp}}]'}.$$

Then, we can use Eq. (2.84) to write (for $\mathrm{r} \in \{\mathrm{n},\mathrm{p}\}$):

$$\frac{\widetilde{\Theta}^{\mathrm{r}}_{\text{ss}}(\tilde{x},s)}{I_{\text{app}}(s)} = \frac{\bar{Z}^{\mathrm{r}}_{\tilde{\text{s}}}(s)}{[U^{\mathrm{r}}_{\text{ocp}}]'}\frac{\bar{Z}^{\mathrm{r}}_{\tilde{\text{dl}}}(s)}{\bar{R}^{\mathrm{r}}_{\text{ct}} + \bar{Z}^{\mathrm{r}}_{\tilde{\text{s}}}(s) + \bar{Z}^{\mathrm{r}}_{\tilde{\text{dl}}}(s)}\frac{I^{\mathrm{r}}_{\text{f+dl}}(\tilde{x},s)}{I_{\text{app}}(s)}. \tag{2.88}$$

2.9.8 *Solid-potential TF*

To find a TF for the solid potential, we must integrate Eq. (2.31) twice with respect to $\tilde{x}$. First, we write the electronic current that flows through the solid at any location $\tilde{x}$ as $i_s(\tilde{x},t)$, where:[40]

$$-\varepsilon_s A\frac{\partial}{\partial \tilde{x}}i_s(\tilde{x},t) = i_{\text{f+dl}}(\tilde{x},t), \tag{2.89}$$

which has limiting cases $\varepsilon_s A i_s(0,t) = i_{\text{app}}(t)$ and $\varepsilon_s A i_s(1,t) = 0$.

Integrating both sides of Eq. (2.89) in the negative electrode gives:

$$\begin{aligned}
-\varepsilon_s A\int_0^{\tilde{x}}\frac{\partial I^{\mathrm{n}}_s(\xi,s)}{\partial \xi}\,\mathrm{d}\xi &= \int_0^{\tilde{x}} I^{\mathrm{n}}_{\text{f+dl}}(\xi,s)\,\mathrm{d}\xi \\
-\varepsilon_s A\left[I^{\mathrm{n}}_s(\tilde{x},s) - I^{\mathrm{n}}_s(0,s)\right] &= \int_0^{\tilde{x}} I^{\mathrm{n}}_{\text{f+dl}}(\xi,s)\,\mathrm{d}\xi \\
-\varepsilon_s A I^{\mathrm{n}}_s(\tilde{x},s) + I_{\text{app}}(s) &= \int_0^{\tilde{x}} I^{\mathrm{n}}_{\text{f+dl}}(\xi,s)\,\mathrm{d}\xi \\
\varepsilon_s A\frac{I^{\mathrm{n}}_s(\tilde{x},s)}{I_{\text{app}}(s)} &= 1 - \int_0^{\tilde{x}}\frac{I^{\mathrm{n}}_{\text{f+dl}}(\xi,s)}{I_{\text{app}}(s)}\,\mathrm{d}\xi.
\end{aligned}$$

The TF for $I_s(z,s)$ with respect to the input current for the negative electrode can then be found (substituting $z = \tilde{x}$, $\zeta = \xi$):

$$\varepsilon^{\mathrm{n}}_s A\frac{I^{\mathrm{n}}_s(z,s)}{I_{\text{app}}(s)} = 1 - \int_0^{z}\frac{I^{\mathrm{n}}_{\text{f+dl}}(\zeta,s)}{I_{\text{app}}(s)}\,\mathrm{d}\zeta$$

[40] We temporarily reintroduce standard nonlumped parameters ε_s and A from Vol. I. In the LPM, we do not know either ε_s or A, but these quantities will disappear from the equations by the time we reach the final result.

$$= 1 - \int_0^z j_1^n e^{\Lambda_1^n(\zeta-1)} + j_2^n e^{-\Lambda_1^n \zeta} + j_3^n e^{\Lambda_2^n(\zeta-1)} + j_4^n e^{-\Lambda_2^n \zeta}\, d\zeta$$

$$= 1 - j_1^n e^{-\Lambda_1^n} \frac{e^{\Lambda_1^n z}-1}{\Lambda_1^n} + j_2^n \frac{e^{-\Lambda_1^n z}-1}{\Lambda_1^n} - j_3^n e^{-\Lambda_2^n} \frac{e^{\Lambda_2^n z}-1}{\Lambda_2^n} + j_4^n \frac{e^{-\Lambda_2^n z}-1}{\Lambda_2^n}.$$

We can now solve for solid potential by defining debiased variable $\tilde{\phi}_s(\tilde{x},t) = \phi_s(\tilde{x},t) - \phi_s(0,t)$ and integrating again, giving:

$$\begin{aligned}
\frac{\widetilde{\Phi}_s^n(z,s)}{I_{\text{app}}(s)} &= -\frac{1}{\bar{\sigma}^n} \int_0^z \varepsilon_s^n A \frac{I_s^n(\zeta,s)}{I_{\text{app}}(s)}\, d\zeta \\
&= -\frac{z}{\bar{\sigma}^n} + \frac{j_1^n \left(e^{\Lambda_1^n(z-1)} - (1+z\Lambda_1^n)e^{-\Lambda_1^n}\right)}{\bar{\sigma}^n \left(\Lambda_1^n\right)^2} + \frac{j_2^n \left(e^{-\Lambda_1^n z} - 1 + z\Lambda_1^n\right)}{\bar{\sigma}^n \left(\Lambda_1^n\right)^2} \\
&\quad + \frac{j_3^n \left(e^{\Lambda_2^n(z-1)} - (1+z\Lambda_2^n)e^{-\Lambda_2^n}\right)}{\bar{\sigma}^n \left(\Lambda_2^n\right)^2} + \frac{j_4^n \left(e^{-\Lambda_2^n z} - 1 + z\Lambda_2^n\right)}{\bar{\sigma}^n \left(\Lambda_2^n\right)^2}.
\end{aligned} \tag{2.90}$$

We find the TF for the positive-electrode solid potential in a similar way but the result is slightly different because of the sign change in the BC at $z = 0$:[41]

$$\begin{aligned}
\frac{\widetilde{\Phi}_s^p(z,s)}{I_{\text{app}}(s)} &= -\frac{1}{\bar{\sigma}^p} \int_0^z \varepsilon_s^p A \frac{I_s^p(\zeta,s)}{I_{\text{app}}(s)}\, d\zeta \\
&= \frac{z}{\bar{\sigma}^p} + \frac{j_1^p \left(e^{\Lambda_1^p(z-1)} - (1+z\Lambda_1^p)e^{-\Lambda_1^p}\right)}{\bar{\sigma}^p \left(\Lambda_1^p\right)^2} + \frac{j_2^p \left(e^{-\Lambda_1^p z} - 1 + z\Lambda_1^p\right)}{\bar{\sigma}^p \left(\Lambda_1^p\right)^2} \\
&\quad + \frac{j_3^p \left(e^{\Lambda_2^p(z-1)} - (1+z\Lambda_2^p)e^{-\Lambda_2^p}\right)}{\bar{\sigma}^p \left(\Lambda_2^p\right)^2} + \frac{j_4^p \left(e^{-\Lambda_2^p z} - 1 + z\Lambda_2^p\right)}{\bar{\sigma}^p \left(\Lambda_2^p\right)^2}.
\end{aligned} \tag{2.91}$$

[41] As a reminder, recall that we debias $\Phi_s^p(z,s)$ relative to $\Phi_s^p(z=0,s) = \Phi_s^p(\tilde{x}=3,s)$.

2.9.9 *Electrolyte potential TF*

We obtain a TF for the electrolyte potential by integrating Eq. (2.36) twice with respect to $\tilde{x}$. The ionic current flowing through the electrolyte, $i_e(\tilde{x},t)$, is defined as:[42]

$$\varepsilon_e A i_e(\tilde{x},t) = -\bar{\kappa} \frac{\partial \phi_e(\tilde{x},t)}{\partial \tilde{x}} - \bar{\kappa}\bar{\kappa}_D T \frac{\partial \ln \theta_e(\tilde{x},t)}{\partial \tilde{x}}.$$

[42] We again temporarily reintroduce standard nonlumped parameters ε_e and A from Vol. I. In the LPM, we do not know either ε_e or A, but these quantities will disappear from the equations by the time we reach the final result.

Hence, we can rewrite the electrolyte-potential PDE as:

$$\varepsilon_e A \frac{\partial i_e(\tilde{x},t)}{\partial \tilde{x}} = i_{\text{f+dl}}(\tilde{x},t).$$

NEGATIVE ELECTRODE: The TF for $i_e(z,t)$ in the negative electrode is:

$$\varepsilon_e A \frac{I_e^n(z,s)}{I_{\text{app}}(s)} = \int_0^z \frac{I_{\text{f+dl}}^r(\zeta,s)}{I_{\text{app}}(s)}\, d\zeta$$

$$= j_1^n e^{-\Lambda_1^n} \frac{e^{\Lambda_1^n z}-1}{\Lambda_1^n} - j_2^n \frac{e^{-\Lambda_1^n z}-1}{\Lambda_1^n} + j_3^n e^{-\Lambda_2^n} \frac{e^{\Lambda_2^n z}-1}{\Lambda_2^n} - j_4^n \frac{e^{-\Lambda_2^n z}-1}{\Lambda_2^n}.$$

If we define debiased electrolyte potential as $\tilde{\phi}_e(\tilde{x},t) = \phi_e(\tilde{x},t) - \phi_e^n(0,t)$ and integrate again:

$$\begin{aligned}\tilde{\phi}_e^n(z,t) &= -\frac{1}{\bar{\kappa}^n}\int_0^z \varepsilon_e^n A i_e^n(\zeta,t)\,d\zeta - \bar{\kappa}_D T \int_0^z \frac{\partial \ln \theta_e^n(\zeta,t)}{\partial \zeta}\,d\zeta \\ &\approx \underbrace{-\frac{1}{\bar{\kappa}^n}\int_0^z \varepsilon_e^n A i_e^n(\zeta,t)\,d\zeta}_{[\tilde{\phi}_e^n(z,t)]_1} \underbrace{-\bar{\kappa}_D T \int_0^z \frac{\partial \tilde{\theta}_e^n(\zeta,t)}{\partial \zeta}\,d\zeta}_{[\tilde{\phi}_e^n(z,t)]_2},\end{aligned} \tag{2.92}$$

where the natural logarithm has been linearized by a Taylor series.

The TF for $[\tilde{\phi}_e^n(z,t)]_1$ will be:

$$\begin{aligned}\frac{\left[\tilde{\Phi}_e^n(z,s)\right]_1}{I_{app}(s)} &= -\frac{1}{\bar{\kappa}^n}\int_0^z \varepsilon_e A \frac{I_e^n(\zeta,s)}{I_{app}(s)}\,d\zeta \\ &= -\left(\frac{j_1^n\left(e^{\Lambda_1^n(z-1)} - (1+z\Lambda_1^n)e^{-\Lambda_1^n}\right)}{\bar{\kappa}^n\left(\Lambda_1^n\right)^2} + \frac{j_2^n\left(e^{-\Lambda_1^n z} - 1 + z\Lambda_1^n\right)}{\bar{\kappa}^n\left(\Lambda_1^n\right)^2}\right. \\ &\quad \left. + \frac{j_3^n\left(e^{\Lambda_2^n(z-1)} - (1+z\Lambda_2^n)e^{-\Lambda_2^n}\right)}{\bar{\kappa}^n\left(\Lambda_2^n\right)^2} + \frac{j_4^n\left(e^{-\Lambda_2^n z} - 1 + z\Lambda_2^n\right)}{\bar{\kappa}^n\left(\Lambda_2^n\right)^2}\right).\end{aligned} \tag{2.93}$$

The TF for the second term is:

$$\begin{aligned}\frac{\left[\tilde{\Phi}_e^n(z,s)\right]_2}{I_{app}(s)} &= -\bar{\kappa}_D T \int_0^z \frac{\partial}{\partial \zeta}\frac{\tilde{\Theta}_e^n(\zeta,s)}{I_{app}(s)}\,d\zeta \\ &= -\bar{\kappa}_D T \int_0^z \Lambda_1^n\left(c_1^n e^{\Lambda_1^n(\zeta-1)} - c_2^n e^{-\Lambda_1^n \zeta}\right) + \Lambda_2^n\left(c_3^n e^{\Lambda_2^n(\zeta-1)} - c_4^n e^{-\Lambda_2^n \zeta}\right)\,d\zeta \\ &= -\bar{\kappa}_D T\left(c_1^n\left(e^{\Lambda_1^n(z-1)} - e^{-\Lambda_1^n}\right) + c_2^n\left(e^{-\Lambda_1^n z} - 1\right)\right. \\ &\quad \left. + c_3^n\left(e^{\Lambda_2^n(z-1)} - e^{-\Lambda_2^n}\right) + c_4^n\left(e^{-\Lambda_2^n z} - 1\right)\right).\end{aligned} \tag{2.94}$$

Overall, the negative-electrode-region electrolyte-potential TF is:

$$\frac{\tilde{\Phi}_e^n(z,s)}{I_{app}(s)} = \frac{\left[\tilde{\Phi}_e^n(z,s)\right]_1}{I_{app}(s)} + \frac{\left[\tilde{\Phi}_e^n(z,s)\right]_2}{I_{app}(s)}. \tag{2.95}$$

SEPARATOR REGION: The TF for $i_e(z,t)$ in the separator region is:

$$\varepsilon_e A \frac{I_e^s(z,s)}{I_{app}(s)} = 1.$$

We integrate again (over the separator region only, for now) and use the same nomenclature as in Eq. (2.92). The TF for the first term is:

$$\frac{\left[\widetilde{\Phi}_e^s(z,s)\right]_1}{I_{app}(s)} = -\frac{1}{\bar{\kappa}^s}\int_0^z \varepsilon_e^s A \frac{I_e^s(\zeta,s)}{I_{app}(s)}\,d\zeta = -\frac{z}{\bar{\kappa}^s}. \tag{2.96}$$

The TF for the second term is:

$$\begin{aligned}\frac{\left[\widetilde{\Phi}_e^s(z,s)\right]_2}{I_{app}(s)} &= -\bar{\kappa}_D T\int_0^z \frac{\partial}{\partial\zeta}\frac{\widetilde{\Theta}_e^s(\zeta,s)}{I_{app}(s)}\,d\zeta \\ &= -\bar{\kappa}_D T\int_0^z (c_1^s\Lambda_1^s e^{\Lambda_1^s(\zeta-1)} - c_2^s\Lambda_1^s e^{-\Lambda_1^s\zeta})\,d\zeta \\ &= -\bar{\kappa}_D T\left(c_1^s\left(e^{\Lambda_1^s(z-1)} - e^{-\Lambda_1^s}\right) + c_2^s\left(e^{-\Lambda_1^s z} - 1\right)\right).\end{aligned} \tag{2.97}$$

The electrolyte-potential TF in the separator region is therefore:

$$\frac{\widetilde{\Phi}_e^s(z,s)}{I_{app}(s)} = \frac{\widetilde{\Phi}_e^n(1,s)}{I_{app}(s)} + \frac{\left[\widetilde{\Phi}_e^s(z,s)\right]_1}{I_{app}(s)} + \frac{\left[\widetilde{\Phi}_e^s(z,s)\right]_2}{I_{app}(s)}. \tag{2.98}$$

POSITIVE ELECTRODE: In the positive electrode, we write:

$$\varepsilon_e^p A\frac{\partial I_e^p(\tilde{x},s)}{\partial\tilde{x}} = I_{f+dl}^p(\tilde{x},s).$$

Integrating,

$$\begin{aligned}\varepsilon_e^p A\int_2^{\tilde{x}}\frac{\partial I_e^p(\xi,s)}{\partial\xi}\,d\xi &= \int_2^{\tilde{x}} I_{f+dl}^p(\xi,s)\,d\xi \\ \varepsilon_e^p A\left[I_e^p(\tilde{x},s) - I_e^p(2,s)\right] &= \int_2^{\tilde{x}} I_{f+dl}^p(\xi,s)\,d\xi \\ \varepsilon_e^p A I_e^p(\tilde{x},s) - I_{app}(s) &= \int_2^{\tilde{x}} I_{f+dl}^p(\xi,s)\,d\xi \\ \varepsilon_e^p A\frac{I_e^p(\tilde{x},s)}{I_{app}(s)} &= 1 + \int_2^{\tilde{x}}\frac{I_{f+dl}^p(\xi,s)}{I_{app}(s)}\,d\xi.\end{aligned}$$

Changing variables, $z = 3 - \tilde{x}$, $\zeta = 3 - \xi$,

$$\varepsilon_e^p A\frac{I_e^p(z,s)}{I_{app}(s)} = 1 - \int_1^z \frac{I_{f+dl}^p(\zeta,s)}{I_{app}(s)}\,d\zeta.$$

So, the TF for $i_e(z,t)$ in the positive electrode is:

$$\begin{aligned}\varepsilon_e^p A\frac{I_e^p(z,t)}{I_{app}(s)} &= 1 - \int_1^z \frac{I_{f+dl}^p(\zeta,s)}{I_{app}(s)}\,d\zeta \\ &= 1 + \left(\frac{j_1^p(e^{\Lambda_1^p(z-1)} - 1)}{\Lambda_1^p} - \frac{j_2^p(e^{-\Lambda_1^p z} - e^{-\Lambda_1^p})}{\Lambda_1^p}\right. \\ &\quad \left. + \frac{j_3^p(e^{\Lambda_2^p(z-1)} - 1)}{\Lambda_2^p} - \frac{j_4^p(e^{-\Lambda_2^p z} - e^{-\Lambda_2^p})}{\Lambda_2^p}\right).\end{aligned}$$

We integrate again (over the positive-electrode region only) and use the same nomenclature as in Eq. (2.92). The TF for the first term is:

$$\begin{aligned}
\frac{\left[\widetilde{\Phi}_{\mathrm{e}}^{\mathrm{p}}(z,s)\right]_1}{I_{\mathrm{app}}(s)} &= \frac{1}{\bar{\kappa}^{\mathrm{p}}}\int_1^z \varepsilon_{\mathrm{e}}^{\mathrm{p}} A \frac{I_{\mathrm{e}}^{\mathrm{p}}(\zeta,s)}{I_{\mathrm{app}}(s)}\,\mathrm{d}\zeta \\
&= \frac{(z-1)}{\bar{\kappa}^{\mathrm{p}}} - \left(\frac{\mathrm{j}_1^{\mathrm{p}}\left(e^{\Lambda_1^{\mathrm{p}}(z-1)} + (1-z)\Lambda_1^{\mathrm{p}} - 1\right)}{\bar{\kappa}^{\mathrm{p}}\left(\Lambda_1^{\mathrm{p}}\right)^2} \right. \\
&\quad + \frac{\mathrm{j}_2^{\mathrm{p}}\left(e^{-\Lambda_1^{\mathrm{p}}z} + (z-1)\Lambda_1^{\mathrm{p}}e^{-\Lambda_1^{\mathrm{p}}} - e^{-\Lambda_1^{\mathrm{p}}}\right)}{\bar{\kappa}^{\mathrm{p}}\left(\Lambda_1^{\mathrm{p}}\right)^2} \\
&\quad + \frac{\mathrm{j}_3^{\mathrm{p}}\left(e^{\Lambda_2^{\mathrm{p}}(z-1)} + (1-z)\Lambda_2^{\mathrm{p}} - 1\right)}{\bar{\kappa}^{\mathrm{p}}\left(\Lambda_2^{\mathrm{p}}\right)^2} \\
&\quad \left. + \frac{\mathrm{j}_4^{\mathrm{p}}\left(e^{-\Lambda_2^{\mathrm{p}}z} + (z-1)\Lambda_2^{\mathrm{p}}e^{-\Lambda_2^{\mathrm{p}}} - e^{-\Lambda_2^{\mathrm{p}}}\right)}{\bar{\kappa}^{\mathrm{p}}\left(\Lambda_2^{\mathrm{p}}\right)^2} \right). \qquad (2.99)
\end{aligned}$$

The TF for the second term is:

$$\begin{aligned}
\frac{\left[\widetilde{\Phi}_{\mathrm{e}}^{\mathrm{p}}(z,s)\right]_2}{I_{\mathrm{app}}(s)} &= -\bar{\kappa}_D T \int_1^z \frac{\partial}{\partial\zeta}\frac{\widetilde{\Theta}_{\mathrm{e}}^{\mathrm{p}}(\zeta,s)}{I_{\mathrm{app}}(s)}\,\mathrm{d}\zeta \\
&= -\bar{\kappa}_D T \int_1^z \Lambda_1^{\mathrm{p}}\left(c_1^{\mathrm{p}}e^{\Lambda_1^{\mathrm{p}}(\zeta-1)} - c_2^{\mathrm{p}}e^{-\Lambda_1^{\mathrm{p}}(s)\zeta}\right) + \Lambda_2^{\mathrm{p}}\left(c_3^{\mathrm{p}}e^{\Lambda_2^{\mathrm{p}}(\zeta-1)} - c_4^{\mathrm{p}}e^{-\Lambda_2^{\mathrm{p}}\zeta}\right)\mathrm{d}\zeta \\
&= -\bar{\kappa}_D T \left(c_1^{\mathrm{p}}\left(e^{\Lambda_1^{\mathrm{p}}(z-1)} - 1\right) + c_2^{\mathrm{p}}\left(e^{-\Lambda_1^{\mathrm{p}}z} - e^{-\Lambda_1^{\mathrm{p}}}\right) \right. \\
&\quad \left. + c_3^{\mathrm{p}}\left(e^{\Lambda_2^{\mathrm{p}}(z-1)} - 1\right) + c_4^{\mathrm{p}}\left(e^{-\Lambda_2^{\mathrm{p}}z} - e^{-\Lambda_2^{\mathrm{p}}}\right) \right). \qquad (2.100)
\end{aligned}$$

The electrolyte-potential TF in the positive electrode is therefore:

$$\frac{\widetilde{\Phi}_{\mathrm{e}}^{\mathrm{p}}(z,s)}{I_{\mathrm{app}}(s)} = \frac{\widetilde{\Phi}_{\mathrm{e}}^{\mathrm{s}}(1,s)}{I_{\mathrm{app}}(s)} + \frac{\left[\widetilde{\Phi}_{\mathrm{e}}^{\mathrm{p}}(z,s)\right]_1}{I_{\mathrm{app}}(s)} + \frac{\left[\widetilde{\Phi}_{\mathrm{e}}^{\mathrm{p}}(z,s)\right]_2}{I_{\mathrm{app}}(s)}. \qquad (2.101)$$

2.10 *Frequency-response example*

In this section, we present some summary frequency-response plots of different electrochemical variables, computed using an example set of parameter values.[43] The purpose is to help develop intuitive insight regarding the frequency-dependent behaviors of each variable. For simplicity, the model being presented does *not* contain a double layer in either electrode: this would add some dynamics at high frequencies. Instead, we wish to compare and contrast the frequency

[43] The parameter values for the "NMC30" cell used in this example are listed in Sect. 2.B. As mentioned in the main text, we temporarily delete the double layer from the NMC30 cell by setting $\bar{C}_{\mathrm{dl}}^{\mathrm{r}} = 0$ and $\bar{R}_{\mathrm{dl}}^{\mathrm{r}} = \infty$ in this example. The frequency responses are shown for a cell SOC of 80 %.

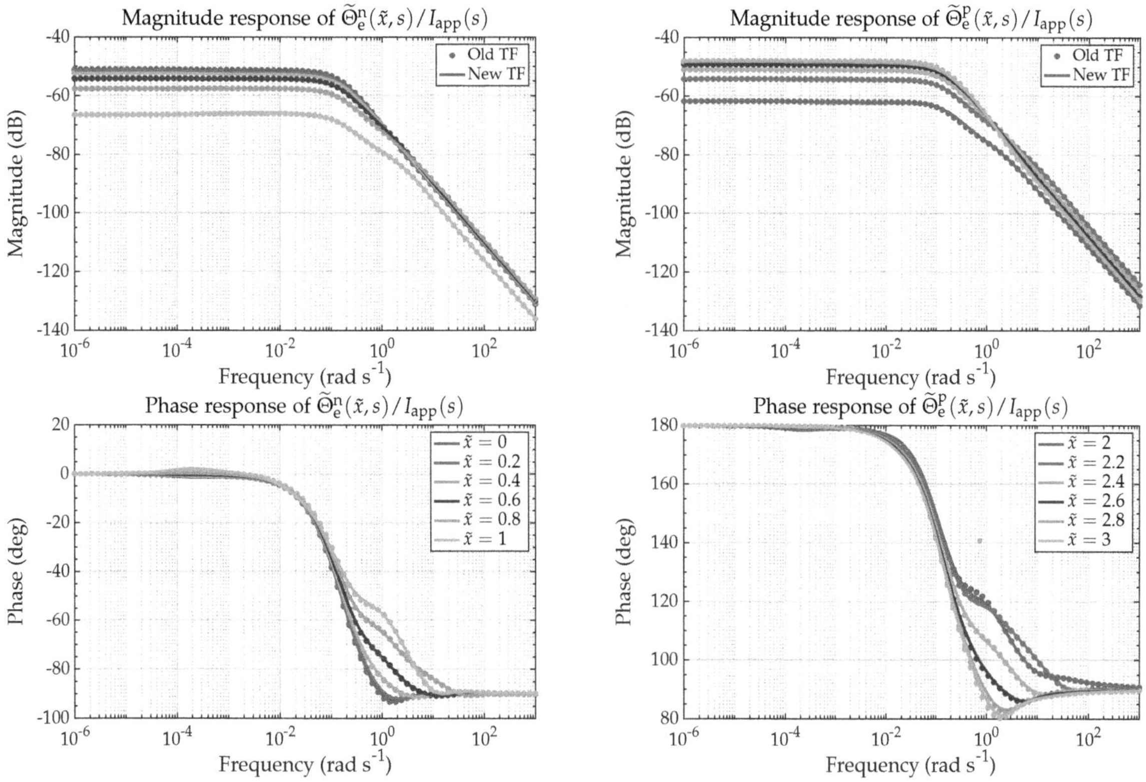

Figure 2.17: Frequency responses of debiased electrolyte-concentration.

responses of the cell as computed using the "old" approximate TFs developed in Vol. I and the "new" exact TFs just presented.

Fig. 2.17 presents the magnitude and phase responses for the debiased electrolyte stoichiometry, $\widetilde{\Theta}_e^r(\tilde{x},s)/I_{app}(s)$.[44] We show this result first because the new derivation develops all other TFs from it. The figure shows that the frequency responses for this variable are almost identical for old and new TFs.[45] This explains, in part, why the old TFs worked well for the applications presented in Vol. I. But, curiously, TFs of other cell variables will show greater differences even though all are computed directly from $\widetilde{\Theta}_e^r(\tilde{x},s)/I_{app}(s)$.

Fig. 2.18 presents the magnitude and phase responses for the interfacial current TF, $I_{f+dl}^r(\tilde{x},s)/I_{app}(s)$. Recall that for this example there is no double layer, and so $i_{f+dl}^r = i_f^r$ and also $i_{dl}^r = 0$. So, all differences between the old-TF and new-TF results are because we no longer make Assumption A2 in the derivation. We notice that the frequency responses found using the old and new methods have the same dc and infinite-frequency values. However, responses at intermediate frequencies strongly depend on the effect of the electrolyte concentration gradient on the electrolyte potential, which was previously ignored by Assumption A2.[46]

The top two rows of Fig. 2.19 display the phase-potential-difference

[44] Notice that we return to using $\tilde{x}$ as the spatial coordinate. The variable z was used temporarily in Sect. 2.9 to simplify some of the mathematical steps. In an implementation, $\tilde{x}$ can be converted to z to compute intermediate results, but the convention of this book is to present final results in terms of $\tilde{x}$.

[45] Fig. 2.17 plots results in terms of the new normalized electrolyte stoichiometry θ_e rather than the standard-model variable c_e. However, if we happen to know $c_{e,0}$, then we have:

$$\frac{\widetilde{C}_e(\tilde{x},s)}{I_{app}(s)} = c_{e,0}\frac{\widetilde{\Theta}_e(\tilde{x},s)}{I_{app}(s)}.$$

This operation shifts the magnitude response by the gain factor $c_{e,0}$ but leaves the phase response unchanged since the phase of a complex number is unchanged when multiplying by a positive real constant.

[46] Also note that if we know $\bar{j}^r$,

$$\frac{J^r(\tilde{x},s)}{I_{app}(s)} = \frac{\bar{j}^r}{F}\frac{I_{f+dl}^r(\tilde{x},s)}{I_{app}(s)}.$$

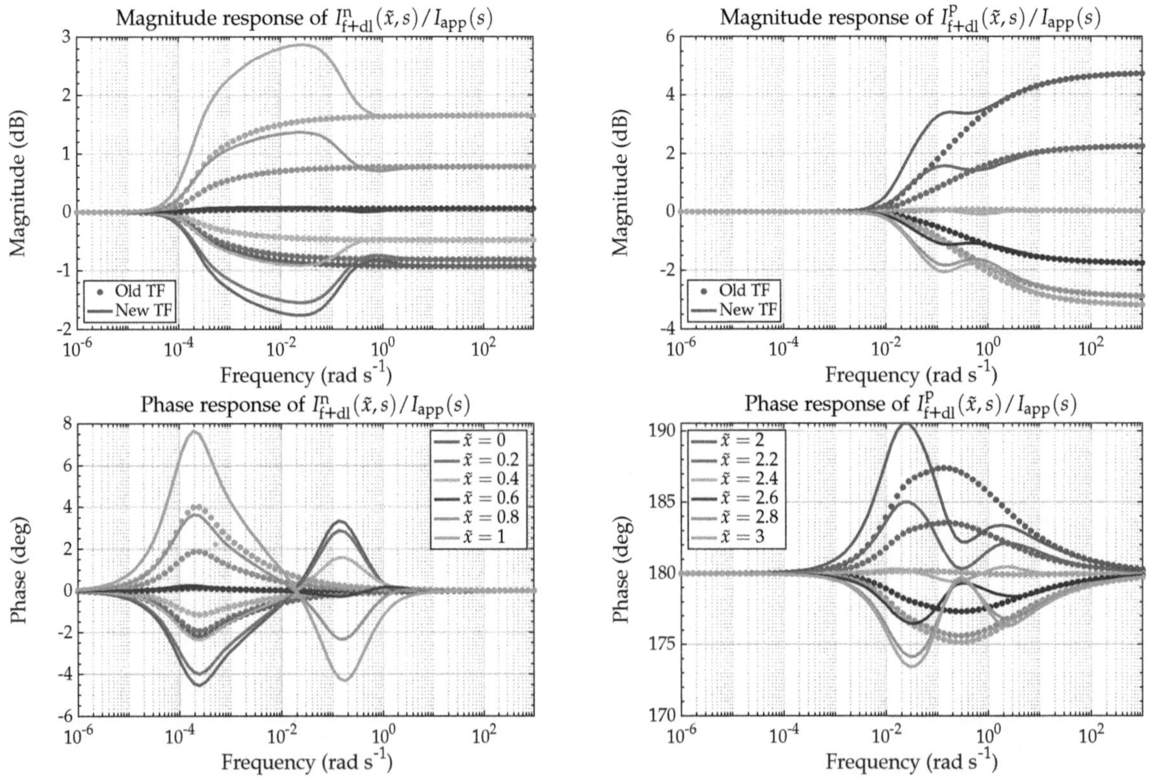

Figure 2.18: Frequency responses of total interfacial current.

frequency responses for different locations within the cell. Again, we see that low- and high-frequency responses appear unchanged, but intermediate frequencies have different values for the new TFs. The magnitude responses at low frequency have slope of -20 dB per decade, indicating the presence of integration dynamics. When using these TFs to enable time-domain simulation in Chap. 4, we will need to compute "integrator-removed" TFs. The bottom two rows of the figure show frequency responses for the integrator-removed phase-potential-difference TF.[47] In this case, the dc gain is affected by the electrolyte concentration gradient, which was previously ignored by Assumption A2. Furthermore, removing the integrator has made it more apparent visually that there are differences between the new and old TFs at intermediate frequencies, but the values at high frequency are identical.

[47] Recall from Vol. I that the $[\cdot]^*$ notation indicates that we have removed the integrator term from a TF.

The top two rows of Fig. 2.20 display the frequency responses for solid surface stoichiometry. We notice that the magnitude responses at low frequency have a slope of -20 dB per decade, indicating the presence of integration dynamics; the magnitude responses at high frequency have a slope of -10 dB per decade, indicating the presence of spherical solid diffusion. The new and old TFs give similar results, except for some differences in the phase response in the mid-

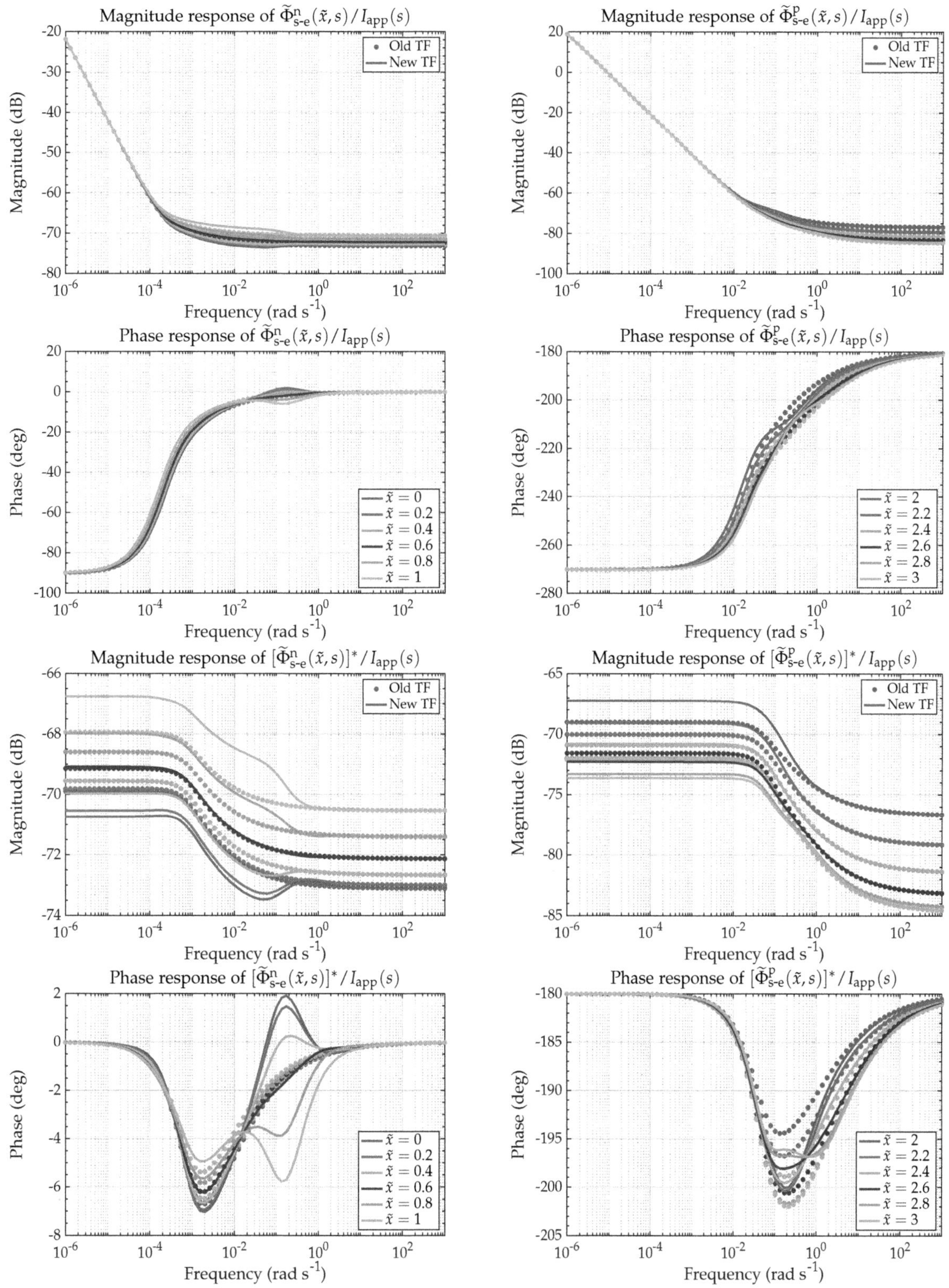

Figure 2.19: Frequency responses of phase-potential-difference.

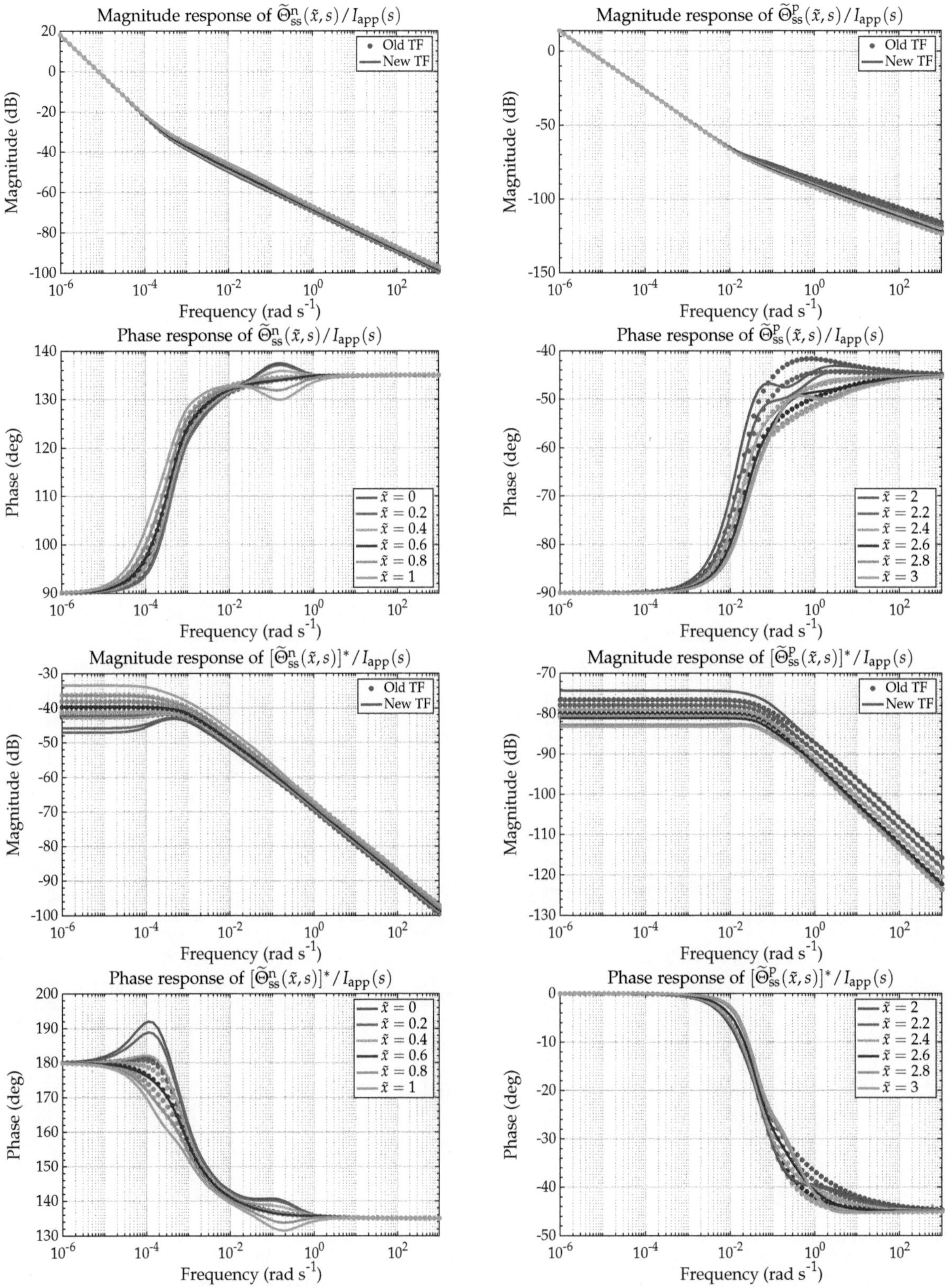

Figure 2.20: Frequency responses of solid-surface stoichiometry.

frequency range. When creating time-domain models, we will again need to remove the integration dynamics from the TFs. When we do so, we obtain the frequency responses displayed in the bottom two rows of the figure. At this scale, we can see that the dc gains of the new and old integrator-removed TFs are different: dc gain is a function of θ_e in a way that was ignored by Assumption A2.[48]

The solid-potential frequency responses are displayed in the top two rows of Fig. 2.21. In this case, dc and infinite-frequency gains have the same values for the old and new approach but differ at intermediate frequencies. The bottom two rows of the figure show frequency responses for electrolyte potential. The old TFs include only the $[\widetilde{\Phi}_e(\tilde{x},s)]_1/I_{app}(s)$ term but the new TFs also include $[\widetilde{\Phi}_e(\tilde{x},s)]_2/I_{app}(s)$ in these plots, which explains some of the differences at low frequency. Overall, we see that the dc gain strongly depends on the contribution of the electrolyte concentration (previously ignored by Assumption A2). Importantly, with this new approach we account for the dynamics that Assumption A2 was ignoring.

[48] Also note that if we know $c^r_{s,max}$, we can convert from the solid-surface stoichiometry to the solid-surface concentration via:

$$\frac{[\widetilde{C}^r_{ss}(\tilde{x},s)]^*}{I_{app}(s)} = c^r_{s,max}\frac{[\widetilde{\Theta}^r_{ss}(\tilde{x},s)]^*}{I_{app}(s)}.$$

2.11 *Full-cell impedance*

We are now ready to build a full-cell impedance model. In a laboratory, impedance is measured using electrochemical impedance spectroscopy (EIS). A small-magnitude sinusoidal input current (or voltage) is applied to a cell, and the voltage (or current) response is monitored. For example, we might apply:

$$i_{app}(t) = M_1(\omega)\cos(\omega t + \phi_1(\omega)).$$

The procedure assumes that the cell has a linear perturbation response around its present operating condition, and uses a small-magnitude input signal to ensure this. Therefore, (steady-state) output will also be sinusoidal at the same frequency, but having different magnitude and phase from the input. For example,

$$v(t) = \text{OCV} + M_2(\omega)\cos(\omega t + \phi_2(\omega)).$$

We compute the magnitude response of the cell as the ratio of output-perturbation to input-signal magnitude; we compute phase response as the difference between output and input phases:

$$M(\omega) = M_2(\omega)/M_1(\omega), \quad \text{and} \quad \phi(\omega) = \phi_2(\omega) - \phi_1(\omega).$$

We will see that cell impedance is related to cell frequency response, which is $H(j\omega) = M(\omega)\exp(j\phi(\omega))$.

The cell impedance can also be expressed using our TF models. To see how to do this, we start with the voltage equation (using $\tilde{x}$ coordinates in the electrodes): $v(t) = \phi^p_s(3,t) - \phi^n_s(0,t)$.

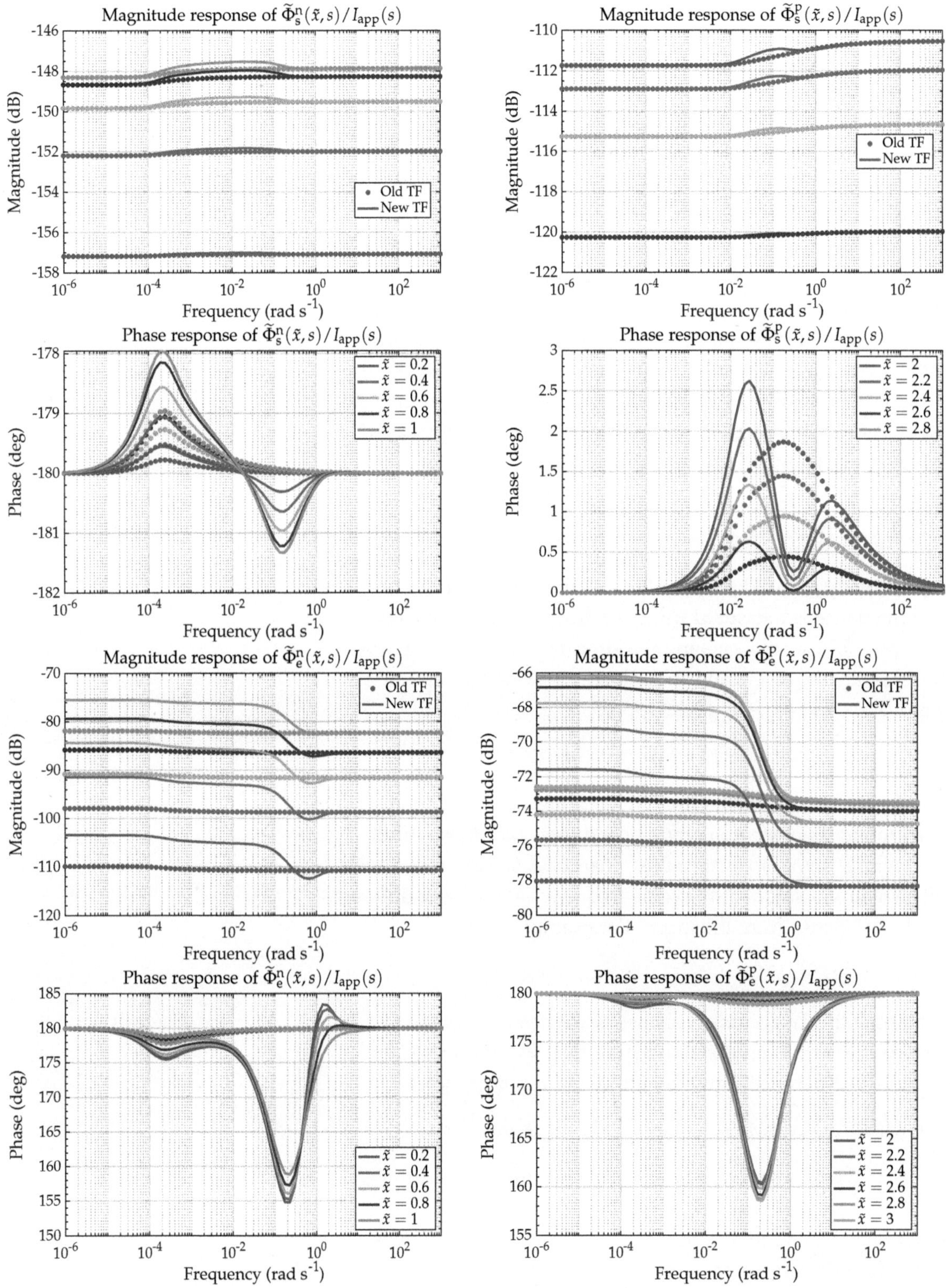

Figure 2.21: Frequency responses of solid and electrolyte potential.

Recall that our TFs compute debiased solid potential $\tilde{\phi}_s^r$ and not absolute solid potential ϕ_s^r directly. The biasing term is cell voltage itself in the positive electrode, so we cannot use the TFs for $\tilde{\phi}_s^r$ to find cell voltage, since we would need to know cell voltage in order to find cell voltage.

Instead, we recall $\tilde{\phi}_e^r(3,t) = \phi_e^r(3,t) - \phi_e^n(0,t)$ and write $\phi_s^r = \phi_{s\text{-}e}^r + \phi_e^r$:

$$\begin{aligned} v(t) &= \left(\phi_{s\text{-}e}^p(3,t) + \phi_e^p(3,t)\right) - \left(\phi_{s\text{-}e}^n(0,t) + \phi_e^n(0,t)\right) \\ &= \left(\tilde{\phi}_{s\text{-}e}^p(3,t) + U_{ocp}^p(\theta_{s,0}^p) + \tilde{\phi}_e^p(3,t)\right) - \left(\tilde{\phi}_{s\text{-}e}^n(0,t) + U_{ocp}^n(\theta_{s,0}^n)\right). \end{aligned}$$

We define debiased (linearized) cell voltage:

$$\begin{aligned} \tilde{v}(t) &= v(t) - \left(U_{ocp}^p(\theta_{s,0}^p) - U_{ocp}^n(\theta_{s,0}^n)\right) \\ &= \tilde{\phi}_{s\text{-}e}^p(3,t) + \tilde{\phi}_e^p(3,t) - \tilde{\phi}_{s\text{-}e}^n(0,t). \end{aligned}$$

Then, we can take Laplace transforms of $\tilde{v}(t)$ to find the TF:

$$\frac{\widetilde{V}(s)}{I_{app}(s)} = \frac{\widetilde{\Phi}_{s\text{-}e}^p(3,s)}{I_{app}(s)} + \frac{\widetilde{\Phi}_e^p(3,s)}{I_{app}(s)} - \frac{\widetilde{\Phi}_{s\text{-}e}^n(0,s)}{I_{app}(s)}. \tag{2.102}$$

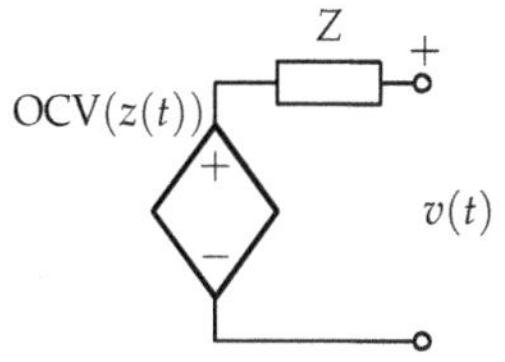

Figure 2.22: Circuit describing cell when seeking to determine impedance.

This voltage variation is related to cell impedance. To see this, consider the simplified cell model depicted in the circuit of Fig. 2.22. In the circuit, cell voltage is equal to open-circuit voltage minus the voltage drop over the impedance Z:

$$\begin{aligned} v(t) &= \text{OCV}(z(t)) - v_Z(t) \\ \tilde{v}(t) &= -v_Z(t). \end{aligned}$$

In the frequency domain, $V_Z(s) = Z(s)I_{app}(s)$. Therefore,

$$\frac{\widetilde{V}(s)}{I_{app}(s)} = -Z(s).$$

In summary, we can write overall cell impedance using TFs as:

$$Z(s) = -\left[\frac{\widetilde{\Phi}_{s\text{-}e}^p(3,s)}{I_{app}(s)} + \frac{\widetilde{\Phi}_e^p(3,s)}{I_{app}(s)} - \frac{\widetilde{\Phi}_{s\text{-}e}^n(0,s)}{I_{app}(s)}\right], \tag{2.103}$$

where the TFs for $\widetilde{\Phi}_{s\text{-}e}^r(\tilde{x},s)/I_{app}(s)$ include the integrator dynamics. And, of course, $Z(j\omega) = Z(s)|_{s=j\omega}$.[49]

Using this result, we can compute the impedance of a cell using either new or old TFs. Fig. 2.23 shows example frequency-response plots. In this case, the results for "Old TF" do not include the electrical double layer but the results for "New TF" do include the double layer. This modifies results for frequencies above about 1 rad sec^{-1}.

[49] Some cell models include a contact or current-collector resistance R_c that describes a resistance external to the main electrode stack. In this case, Eq. (2.103) is modified by adding to it the value of R_c.

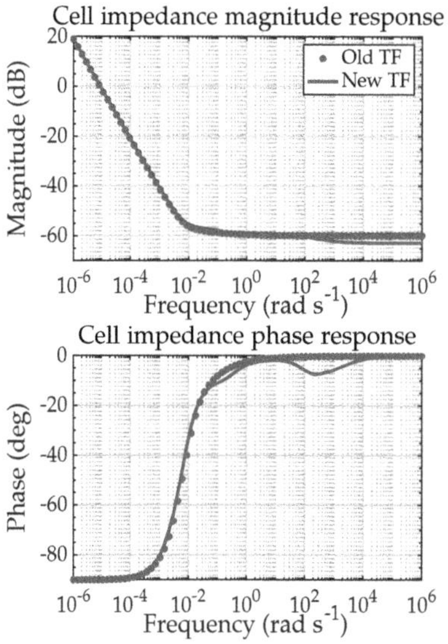

Figure 2.23: Full-cell frequency responses.

2.12 *Nyquist (Cole–Cole) plots*

Bode frequency-response plots such as shown in Fig. 2.23 are not very helpful for visualizing an impedance response. That is, it is not very easy to see features of interest plotted this way.

We will find it more helpful to display Nyquist-style plots.[50] A standard Nyquist plot is a parametric plot (over frequency) that displays the real value of impedance on the horizontal axis and the imaginary value on the vertical axis. If we were to plot the impedance of a battery cell directly in this format, it would result in diagrams where the detail of interest is in the fourth quadrant, which is a little awkward. So, the electrochemical literature instead uses the convention of plotting the real value of impedance versus the *negative* of the imaginary value, producing diagrams where the detail of interest is in first quadrant. Even though the procedure of negating the imaginary values is not standard when producing Nyquist plots, we still refer to the result simply as a Nyquist plot (but are careful to label the vertical axis as "$-\text{Imag}(Z)$").

[50] These are also sometimes referred to as Cole–Cole plots.

This is very simple to do in engineering software environments such as MATLAB. If `Z` is a vector of complex impedances, we produce the Nyquist plot using the command:

```
plot(real(Z),-imag(Z));
```

Nyquist plots can appear very strange at first and take a little getting used to! Let's start by looking at Nyquist plots of some simple circuits. Fig. 2.24 is our first example. Recall that the impedance of a capacitor is $Z_c = (j\omega C)^{-1}$. This result is purely imaginary and its magnitude approaches infinity at low frequencies and zero at high frequencies. The impedance of a series resistor–capacitor (RC) circuit is simply $Z = R + (j\omega C)^{-1}$. This is now complex-valued with real part equal to R and imaginary part equal to $-(\omega C)^{-1}$. When we plot this complex number on a Nyquist plot for all frequencies $0 \le \omega \le \infty$, the result is a vertical line shifted such that the real part is R and the imaginary part forms a vertical line between zero and infinity. Notice that the change in impedance as frequency increases is indicated on the diagram by an arrow.

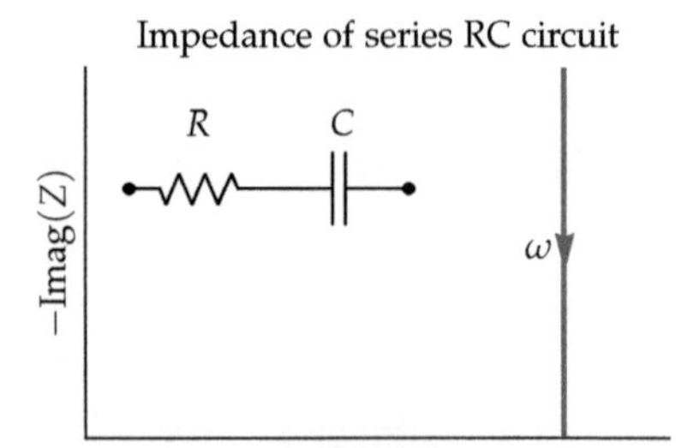

Figure 2.24: Nyquist impedance of a series RC circuit.

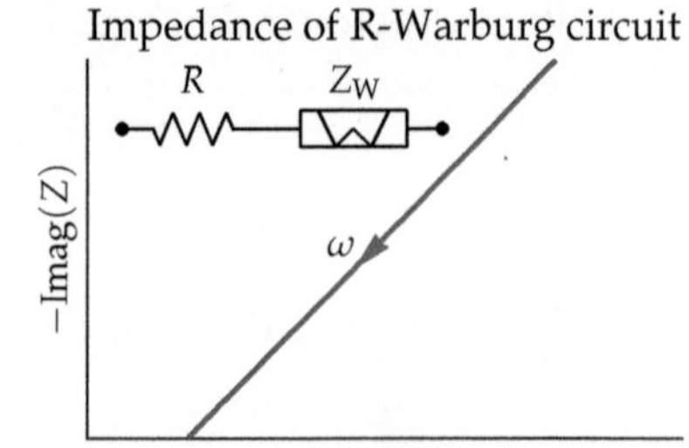

Figure 2.25: Nyquist impedance of a series R–Warburg circuit.

The second example shows the Nyquist plot of a series resistor–Warburg circuit in Fig. 2.25. Recall that a Warburg element has impedance $Z_W = W/\sqrt{j\omega}$, which by itself is a 45° line ending at the origin.[51] The resistance shifts this line by R since the impedance of the series resistor–Warburg circuit is $Z = R + W/\sqrt{j\omega}$.

[51] Note that the Warburg impedance is a CPE with phase of $-45°$.

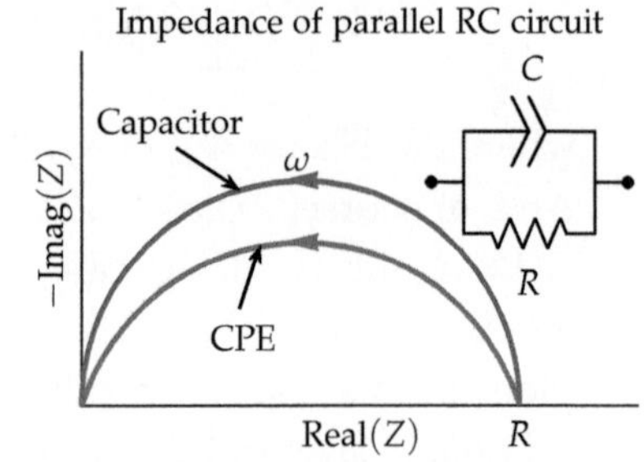

Figure 2.26: Nyquist impedance of parallel RC and ZARC circuits.

The third example, in Fig. 2.26, shows Nyquist plots for parallel resistor–capacitor and ZARC circuits. The Nyquist drawing of the resistor–capacitor impedance is an exact semicircle; the impedance of the ZARC circuit is a flattened semicircle. Again, this arc shape of the impedance "Z" gives rise to the "ZARC" name of the parallel resistor–CPE element circuit.

Fig. 2.27 shows one final example: the Nyquist diagram of the impedance of a series resistor–inductor circuit. The impedance of the inductance is $j\omega L$, which is completely imaginary. The Nyquist plot of the impedance of an inductor by itself is a vertical line that approaches zero at low frequencies and negative infinity at high frequencies. When combined with the resistance, the line is shifted over to have real part R.

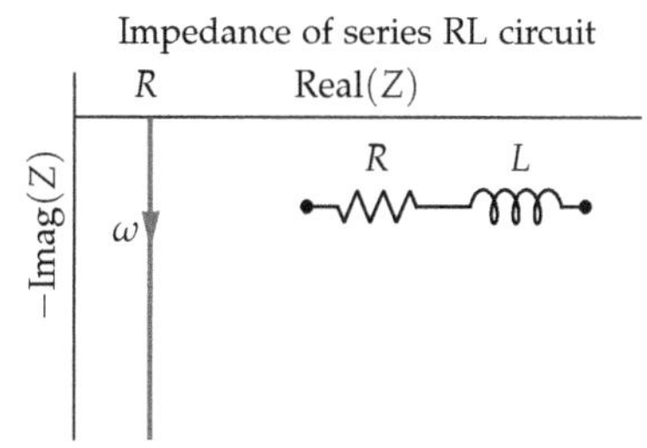

Figure 2.27: Nyquist impedance of series RL circuit.

The purpose of these examples is to build a visual dictionary of Nyquist impedances of simple circuits. Since impedances in series add, we can visualize Nyquist diagrams for simple circuits having nonoverlapping frequency ranges of interest by adding together simple responses. For example, a commonly seen equivalent-circuit model—comprising several subcircuits wired in series—has the Nyquist diagram shown in Fig. 2.28. The resistance R shifts the entire diagram to the right; the inductance L has a vertical section below the real axis; both ZARC circuits form flattened semicircles; and the Warburg element contributes a 45° line. The (shifted) responses of the ZARC elements by themselves are drawn with dotted red and purple lines; the (shifted) response of the Warburg element is drawn as a dotted green line. The overall response is drawn as the blue line.

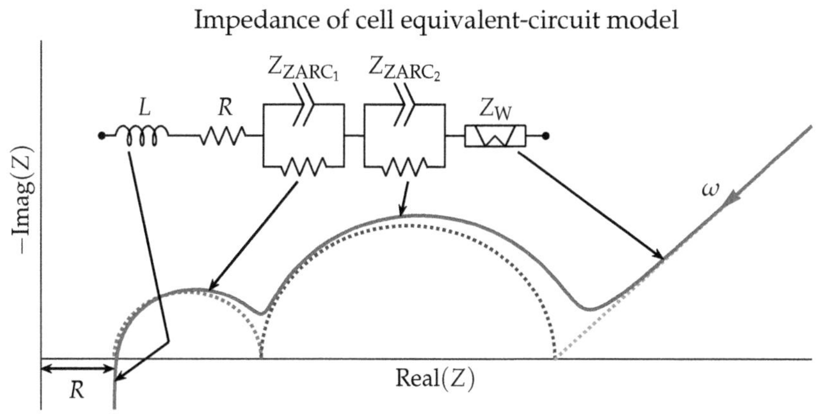

Figure 2.28: Understanding a Nyquist diagram by breaking it down into segments that can be modeled by simple circuits.

Hopefully, this exercise has helped show how to interpret Nyquist diagrams. You should now be able to look at such a diagram and visualize what kind of circuit might have led to that impedance response. In particular, it can be helpful to understand the impedance response of a battery by showing how it can be decomposed by frequency range into responses that we can understand as typical of resistances or inductances or resistor–capacitor circuits, and so forth.

With this in mind, we consider Fig. 2.29, which uses the TFs we have developed and the impedance model of Eq. (2.103) to compute the Nyquist diagram of the NMC30 cell from Sect. 2.B, but deleting its double-layer dynamics. We notice that this Nyquist diagram does not look very much like the diagram of Fig. 2.28. The reason is the omission of the double layer in this example.

Fig. 2.30 shows how the Nyquist diagram changes when we use the unmodified NMC30-cell parameters. Now, we clearly see a double-layer bump in the Nyquist plot that is also observed in the measured impedance of physical cells. We also notice that impedance is a strong function of the SOC of the cell around which the impedance data are collected.

The parameter values used in TFs can be highly temperature-dependent. Most often, we use Arrhenius relationships (as discussed in more detail in Chap. 3). Fig. 2.31 shows how the Nyquist diagram of impedance changes as a function of temperature (impedance is higher at cold temperatures) when cell SOC is set to 80 %.

One of the principal applications of the TFs in this book is in conjunction with estimating the parameter values of a physics-based model for a commercial cell. We have now seen that we can write the TFs as closed-form expressions of the parameter values and that we can use those TFs to produce Nyquist diagrams of cell impedance. We can also use laboratory instruments such as potentiostats to measure the impedance response of commercial cells. We will see that we can then also regress the parameter values of the TFs to attempt to match the computed impedance with the measured impedance. The fact that the impedance shows a richness of shape as a function of frequency, SOC, and temperature gives hope that we can estimate many parameter values from this kind of data. We will study this more in Chap. 3.

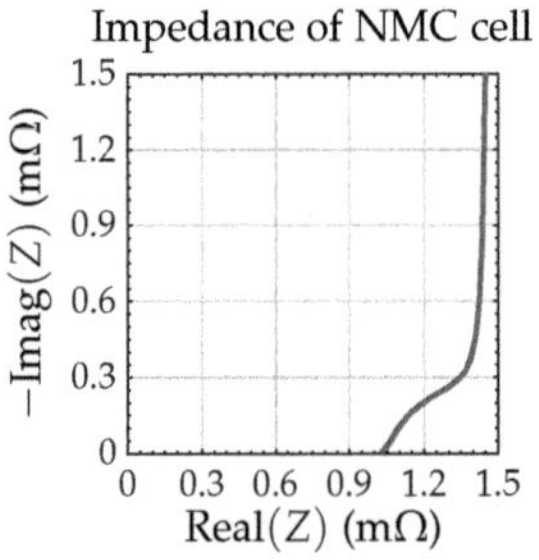

Figure 2.29: Nyquist diagram of the NMC30 cell at 80 % SOC and 25 °C, without a double layer.

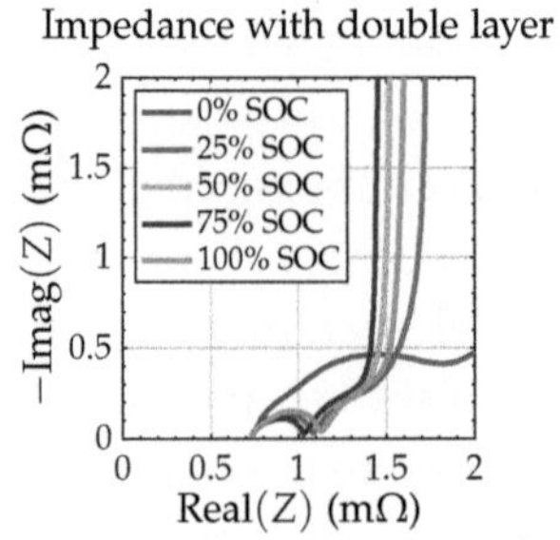

Figure 2.30: Nyquist diagram of the NMC30 cell at 25 °C, with a double layer, around different setpoint SOCs.

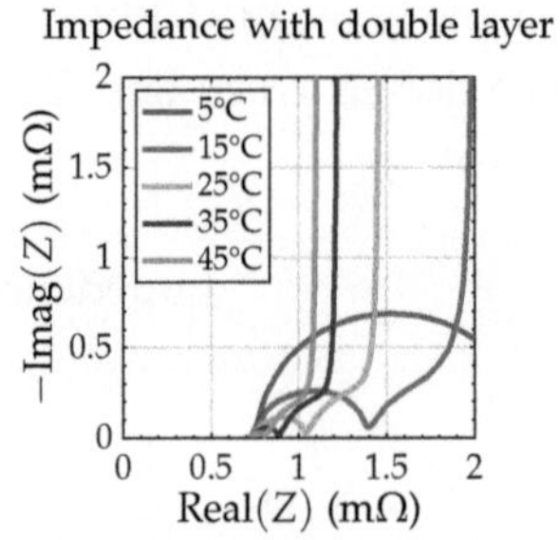

Figure 2.31: Nyquist diagram of the NMC30 cell at 80 % SOC, with a double layer, around different setpoint temperatures.

2.13 *MATLAB toolbox*

Before ending the chapter, we would like to spend a few pages introducing a MATLAB-based toolbox of routines that accompanies this book. Based on our own experience, we recognize that it is very difficult to implement the TFs correctly in an engineering programming language since there are many, many details. It is easy to enter the equations incorrectly by misplacing parentheses or by omitting negative signs, and so forth. So, to assist researchers wishing to use these methods in their own work, we are providing a MATLAB toolbox of verified functions as a companion to this book.[52]

[52] See: `http://mocha-java.uccs.edu/BMS3/`.

Fig. 2.32 illustrates the parameter-estimation and model-simulation components of the toolbox. Most of this functionality has not been discussed yet, but you will learn about it as you progress through

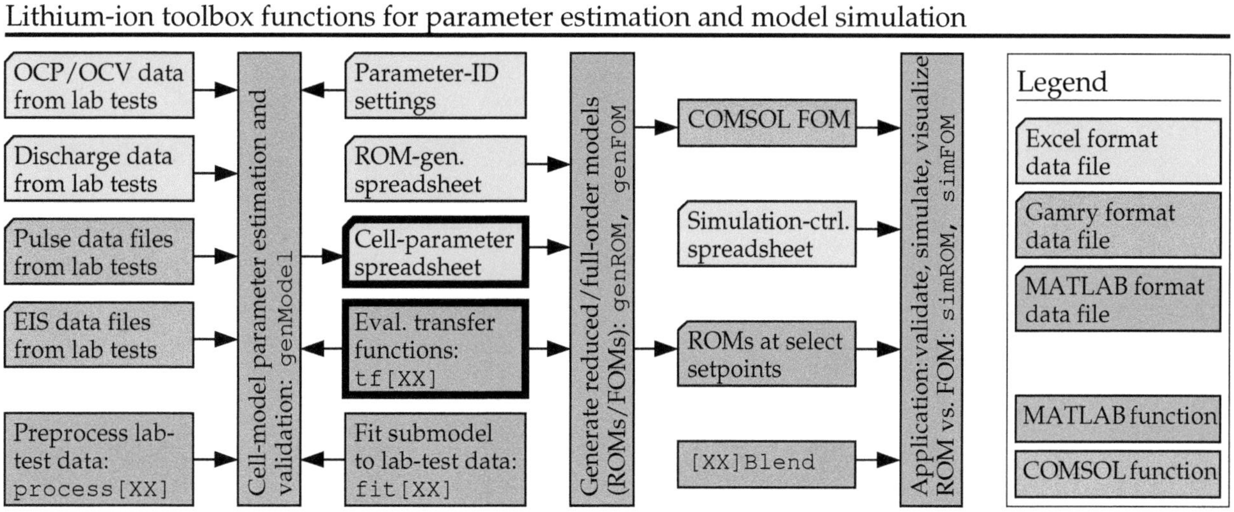

Figure 2.32: MATLAB-toolbox functionality, highlighting the focus of Chap. 2.

the book. The most critical elements to notice at this point are highlighted as "Cell-parameter spreadsheet" and "TF" tf[XX] components in the figure.

The toolbox stores cell parameter values in easy-to-use custom-format Excel spreadsheets. Fig. 2.33 shows an example of what this looks like to the user. The spreadsheet has sections for cell generic information, as well as for parameter values that apply over the whole cell, or only to negative-electrode, separator, or positive-electrode regions. The toolbox works both for cells defined using standard parameter names and values (e.g., from Vol. I) and for cells defined using lumped-parameter names and values.

See *Instructions* tab for more information				
Cell Information				
Parameter	**Code Name**	**Value**		**Unit**
Cell name	name	NMC30		n/a
Standard variables (0) or lumped variables (1)	lumped	1		n/a
Standard kinetics (0) or MSMR kinetics (1)	MSMR	0		n/a
Cell Parameters				
Parameter	**Code Name**	**Value**	**Eact [kJ mol^{-1}]**	**Unit**
Cell total capacity	Q	29.8587	0	Ah
Electrolyte diffusivity / electrolyte conductivity	psi	0.000116451	0	$V\,K^{-1}$
Contact (current collector) resistance	Rc	0.000534	0	Ω
kappa D	kD	-0.000127537	0	$V\,K^{-1}$
Negative Electrode Parameters				
Parameter	**Code Name**	**Value**	**Eact [kJ mol^{-1}]**	**Unit**
OCP as function of Stoichiometry	Uocp	#UocpNeg	0	V
dOCP as function of Stoichiometry	dUocp	#dUocpNeg	0	V
Electrode Stoichiometry at 0% SOC	theta0	0.014	0	unitless

Figure 2.33: Example of cell parameter spreadsheet.

The toolbox function `loadCellParams` loads the spreadsheet into MATLAB's workspace as a structure having fields that store cell constants, functions, whether the cell is defined using lumped parameters and MSMR values, and even an optional cell name.

```
>> cellParams = loadCellParams('cellNMC30.xlsx')

cellParams =

  struct with fields:

       const: [1x1 struct]
        name: 'NMC30'
      lumped: 1
        MSMR: 0
    function: [1x1 struct]
```

Every parameter value in the spreadsheet may either be a constant, or a function, or a lookup table. Details for how to use each type of input are written in the "Instructions" tab of the spreadsheet. After being loaded into MATLAB, parameters are stored internally as functions of electrode stoichiometry `x` and temperature `T` in MATLAB:

```
>> cellParams.function.neg

ans =

  struct with fields:

        Uocp: [function_handle]
       dUocp: [function_handle]
      theta0: @(x,T)(0.014)
    theta100: @(x,T)(0.884)
       sigma: @(x,T)(1.29981e+07)
       kappa: @(x,T)(6198.99)
          nF: @(x,T)(1)
       Dsref: @(x,T)(0.000666742).*exp(28800*(1/298.15-1/T)/8.31446)
          qe: @(x,T)(0.876003)
          k0: @(x,T)(288.672).*exp(48900*(1/298.15-1/T)/8.31446)
       alpha: @(x,T)(0.5)
          Rf: @(x,T)(5e-5)
         Cdl: @(x,T)(28.5562)
         nDL: @(x,T)(1)
         wDL: @(x,T)(6.2832e-05)
         Rdl: @(x,T)(1e-5)
         soc: @(x,T)(0.014+x*(0.87))
```

To convert the generic functions of SOC and temperature to a specific operational setpoint (and precompute common TF terms for every frequency in complex-frequency vector `s`), use `evalSetpoint`:

```
>> cellData = evalSetpoint(cellParams,s,0.5,25)

cellData =

  struct with fields:
       const: [1x1 struct]
        name: 'NMC30'
      lumped: 1
        MSMR: 0
    function: [1x1 struct]
         neg: [1x1 struct]
```

```
        sep: [1x1 struct]
        pos: [1x1 struct]
     common: [1x1 struct]
```

Functions are also provided that implement the ten TFs:

- `tfIfdl.m` implements $I^{\mathrm{r}}_{\mathrm{f+dl}}(\tilde{x},s)/I_{\mathrm{app}}(s)$ for $\mathrm{r} \in \{\mathrm{n},\mathrm{p}\}$.
- `tfIf.m` implements $I^{\mathrm{r}}_{\mathrm{f}}(\tilde{x},s)/I_{\mathrm{app}}(s)$ for $\mathrm{r} \in \{\mathrm{n},\mathrm{p}\}$.
- `tfIdl.m` implements $I^{\mathrm{r}}_{\mathrm{dl}}(\tilde{x},s)/I_{\mathrm{app}}(s)$ for $\mathrm{r} \in \{\mathrm{n},\mathrm{p}\}$.
- `tfPhie.m` implements $\widetilde{\Phi}^{\mathrm{r}}_{\mathrm{e}}(\tilde{x},s)/I_{\mathrm{app}}(s)$ for $\mathrm{r} \in \{\mathrm{n},\mathrm{s},\mathrm{p}\}$.
- `tfPhis.m` implements $\widetilde{\Phi}^{\mathrm{r}}_{\mathrm{s}}(\tilde{x},s)/I_{\mathrm{app}}(s)$ for $\mathrm{r} \in \{\mathrm{n},\mathrm{p}\}$.
- `tfPhiseInt.m` implements $\widetilde{\Phi}^{\mathrm{r}}_{\mathrm{s\text{-}e}}(\tilde{x},s)/I_{\mathrm{app}}(s)$ for $\mathrm{r} \in \{\mathrm{n},\mathrm{p}\}$.
- `tfPhise.m` implements $[\widetilde{\Phi}^{\mathrm{r}}_{\mathrm{s\text{-}e}}(\tilde{x},s)]^*/I_{\mathrm{app}}(s)$ for $\mathrm{r} \in \{\mathrm{n},\mathrm{p}\}$.
- `tfThetae.m` implements $\widetilde{\Theta}^{\mathrm{r}}_{\mathrm{e}}(\tilde{x},s)/I_{\mathrm{app}}(s)$ for $\mathrm{r} \in \{\mathrm{n},\mathrm{s},\mathrm{p}\}$.
- `tfThetassInt.m` implements $\widetilde{\Theta}^{\mathrm{r}}_{\mathrm{ss}}(\tilde{x},s)/I_{\mathrm{app}}(s)$ for $\mathrm{r} \in \{\mathrm{n},\mathrm{p}\}$.
- `tfThetass.m` implements $[\widetilde{\Theta}^{\mathrm{r}}_{\mathrm{ss}}(\tilde{x},s)]^*/I_{\mathrm{app}}(s)$ for $\mathrm{r} \in \{\mathrm{n},\mathrm{p}\}$.

All of these TFs have the same application programming interface (API) format. For example,

```
function [tf,aux] = tfIfdl(s,locs,cellData)
```

Outputs from the TFs are: frequency response `tf`, the high-frequency gains, the low-frequency (dc) gains, the integrator residue `res0`, and other auxiliary data that will become more important in Chap. 4.

The TFs are invoked with a vector of complex frequencies s and a set of $\tilde{x}$ locations. For example, this code creates a Bode plot:

```
clearvars; close all; clc;
cellData = loadCellParams('cellNMC30.xlsx');
omega = [0, logspace(-6,6,200)]; s = 1j*omega;
SOC = 0.8;              % Cell state of charge
T = 273.15 + 25;        % Temperature in K for 25 degC
cellData = evalSetpoint(cellData,s,SOC,T);

[ifdl_tf,~] = tfIfdl(s,0:0.2:3,cellData);

subplot(2,1,1); semilogx(omega,20*log10(abs(ifdl_tf)));
title('Magnitude response for i_{f+dl}');
ylabel('Magnitude (dB)'); xlabel('Frequency (rad/s)');
subplot(2,1,2); semilogx(omega,unwrap(angle(ifdl_tf),[],2)*180/pi);
title('Phase response for i_{f+dl}');
ylabel('Phase (deg)'); xlabel('Frequency (rad/s)');
```

As well, this sample code creates a Nyquist plot of cell impedance:

```
clearvars; close all; clc;
cellData = loadCellParams('cellNMC30.xlsx');
omega = [0, logspace(-6,6,200)]; s = 1j*omega;
SOC = 0.8;              % Cell state of charge
T = 273.15 + 25;        % Temperature in K for 25 degC
cellData = evalSetpoint(cellData,s,SOC,T);

[phise_tf,~] = tfPhiseInt(s,[0,3],cellData);
[phie_tf,~]  = tfPhie(s,3,cellData);
Z = -(phise_tf(2,:) - phise_tf(1,:) + phie_tf) + cellData.const.Rc;
```

```
plot(real(1000*Z),-imag(1000*Z));
xlabel('Real'); h = ylabel('-Imag'); title('Impedance (milliohms)');
xlim([0 2]); ylim([0 2]); grid on; axis square;
```

These two examples are included in the toolbox "`CH2`" folder.

From this point onward, most chapters will have a section that highlights portions of the toolbox that support the content of that chapter. By the time you reach the end of this book, you should have a good understanding of the functionality of the components of the toolbox and how to use them together. Examples provided with the toolbox will also help aid understanding. Furthermore, examination of the code in the toolbox functions will give even more clarity regarding how the theory from this volume can be converted to practical functioning programs.

2.14 *Where to from here?*

This brings us to the conclusion of the primary content of this chapter. We have now derived TFs of all electrochemical variables of a cell and have seen how to combine them to compute cell impedance. Implementing these TFs for a physical cell requires knowledge of the cell's model parameter values. The focus of Chap. 3 is to describe how to use the known model PDEs and known structure of the TFs along with data collected from cell tests in the laboratory to estimate parameter values of physical cells.

2.A *Summary of variables*

The following list itemizes all parameters and variables introduced in this chapter.

- $\beta_{\bar{s}}(s)$ [unitless], CPE factor for solid-diffusion, per Eq. (2.19).
- $c_k^{\mathrm{r}}(s)$ [A^{-1}], frequency-dependent functions out of which all TFs are constructed, per Eqs. (2.64), and (2.67).
- $\bar{C}_{\mathrm{dl}}$ [F], lumped double-layer capacitance. In terms of standard-parameter definitions, $\bar{C}_{\mathrm{dl}} = a_s A L C_{\mathrm{dl}}$ where C_{dl} has units [$\mathrm{F\,m^{-2}}$].
- $\bar{C}_{\mathrm{dl,eff}} = (\bar{C}_{\mathrm{dl}})^{2-n_{\mathrm{dl}}^{\mathrm{r}}}(\omega_{\mathrm{dl}})^{n_{\mathrm{dl}}^{\mathrm{r}}-1}$ [F] is an effective double-layer capacitance that appears in some equations because of how we modified the low-frequency portion of the CPE in Eq. (2.18).
- $\bar{D}_{\mathrm{s,ref}}$ [s^{-1}] is the constant in a description of SOC-dependent solid diffusivity, per Eq. (2.21).
- $f = F/(RT)$ [V^{-1}], is a short form used in many equations to condense the notation.
- $\bar{i}_0$ [A], linearized total exchange-current density, per Eq. (2.11).
- $\bar{i}_{0,j}$ [A], linearized exchange-current density for gallery j, per Eq. (2.12).

- i_{dl} [A], double-layer lithium flux. In terms of standard-parameter definitions, $i_{\mathrm{dl}} = a_s AL j_{\mathrm{dl}}/F$ where j_{dl} has units [$\mathrm{mol\,m^{-2}\,s^{-1}}$].
- i_{f} [A], flux of lithium exiting particle. In terms of standard-parameter definitions, $i_{\mathrm{f}} = a_s AL j_{\mathrm{f}}/F$ where j_{f} has units [$\mathrm{mol\,m^{-2}\,s^{-1}}$].
- $i_{\mathrm{f+dl}}$ [A], total lithium flux from double layer and particle into the electrolyte. In terms of standard-parameter definitions, $i_{\mathrm{f+dl}} = a_s AL j/F$ where j has units [$\mathrm{mol\,m^{-2}\,s^{-1}}$].
- $\mathrm{j}_k^{\mathrm{r}}(s)$ [unitless], coefficients needed to compute $I_{\mathrm{f+dl}}(\tilde{x},s)/I_{\mathrm{app}}(s)$, as used in Eqs. (2.82) and (2.83).
- $\lambda_k^{\mathrm{r}}(s)$ [unitless], temporary variable in TF derivations per Eq. (2.78).
- $\Lambda_k^{\mathrm{r}}(s)$ [unitless], TF frequency-dependent terms per Eqs. (2.61), (2.62), and (2.68).
- $\mu_1^{\mathrm{r}}(s)$ [$\mathrm{V^{-1}}$] and $\mu_2^{\mathrm{r}}(s)$ [unitless], temporary variables in TF derivations per Eqs. (2.70) and (2.72).
- ω_{dl} [$\mathrm{rad\,s^{-1}}$], breakpoint frequency for low-frequency-modified double-layer CPE as used in Eq. (2.18).
- R_{c} [Ω], contact or current-collector resistance.
- $\bar{R}_{\mathrm{ct}}$ [Ω], charge-transfer resistance modeling overpotential η versus faradaic current i_{f}, Eq. (2.22).
- $\bar{R}_{\mathrm{dl}}$ [Ω], lumped double-layer resistance. In terms of the standard-parameter definition, $\bar{R}_{\mathrm{dl}} = R_{\mathrm{dl}}/(a_s AL)$ where R_{dl} has units [$\Omega\,\mathrm{m}^2$].
- $\tau_k^{\mathrm{r}}(s)$ [unitless], TF frequency-dependent terms per Eqs. (2.58) and (2.59).
- $[U_{\mathrm{ocp}}^{\mathrm{r}}]'$ [$\mathrm{V^{-1}}$], derivative of the OCP function, per Eq. (2.13).
- z [unitless], alternate cell-region spatial variable, Eq. (2.63).
- $\bar{Z}_{\mathrm{s}}(s)$ [Ω], ideal particle diffusion impedance, Eq. (2.15).
- $\bar{Z}_{\tilde{\mathrm{s}}}(s)$ [Ω], particle diffusion impedance with CPE on $\bar{D}_{\mathrm{s}}$, Eq. (2.20).
- $\bar{Z}_{\mathrm{se}}(s)$ [Ω], ideal overall solid/electrolyte interface impedance, including double-layer and film resistance.
- $\bar{Z}_{\tilde{\mathrm{se}}}(s)$ [Ω], overall solid/electrolyte interface impedance, including double-layer and film resistance, with CPE on double layer (and possibly on $\bar{D}_{\mathrm{s}}$).

2.B NMC30-cell parameters

MSMR-format parameters for graphite and NMC111 are listed in Table 2.B.1. Cell parameter values in standard notation for the NMC30 cell (so-called because it has capacity of approximately 30 Ah) used in examples in this chapter are listed in Table 2.B.2. Cell parameter values converted to lumped-parameter notation for the NMC30 cell are listed in Table 2.C.1.

Table 2.B.1: MSMR parameter values for graphite and NMC111 (optimized to match data from: Johannes Schmalstieg, "Physikalisch-elektrochemische simulation von lithium-ionen-batterien: Implementierung, parametrierung und anwendung," PhD dissertation, RWTH Aachen University, 2017).

Graphite		
U_j^0 [V]	X_j	ω_j
0.0887	0.1672	0.0631
0.0913	0.2076	0.0140
0.1305	0.1979	0.0314
0.1330	0.3157	1.1177
0.2164	0.0591	0.0728
0.3673	0.0395	3.2005
1.2093	0.0130	10.7190

NMC111		
U_j^0 [V]	X_j	ω_j
3.7467	0.3559	1.8350
3.9215	0.0330	1.3443
4.0459	0.0734	2.6936
4.4178	0.3711	8.2197
7.0845	0.1666	3.3176

Symbol	Units	Negative electrode	Separator	Positive electrode
L	μm	46.6	18.7	43
R	μm	6.3	—	2.13
σ	$\mathrm{S\,m^{-1}}$	1000	—	10
ε_s	$\mathrm{m^3m^{-3}}$	0.4925	—	0.5724
ε_e	$\mathrm{m^3m^{-3}}$	0.292	0.3949	0.209
brug	—	1.52	1.62	1.44
$c_{s,\max}$	$\mathrm{mol\,m^{-3}}$	31 390	—	48 390
θ_0	—	0.014	—	0.9615
θ_{100}	—	0.884	—	0.4352
$D_{s,\mathrm{ref}}$	$\mathrm{m^2\,s^{-1}}$	2.6463×10^{-14}	—	7.6885×10^{-16}
k_{norm}	$\mathrm{mol\,m^2\,s^{-1}}$	1.5401×10^{-4}	—	1.0426×10^{-4}
α	—	0.5	—	0.5
R_{dl}	$\Omega\,\mathrm{m^2}$	1.9426×10^{-4}	—	6.162×10^{-5}
C_{dl}	$\mathrm{F\,m^{-2}}$	1.47	—	0.198
R_f	$\Omega\,\mathrm{m^2}$	9.713×10^{-4}	—	6.162×10^{-4}
A	$\mathrm{m^2}$		1.7775	
$c_{e,0}$	$\mathrm{mol\,m^{-3}}$		1000	
D_e	$\mathrm{m^2\,s^{-1}}$		2.7945×10^{-10}	
t_+^0	—		0.26	
R_c	$\mathrm{m}\Omega$		0.534	
$\mathrm{d}\ln f_\pm/\mathrm{d}\ln c_e$	—		0	

We compute $\sigma_{\mathrm{eff}} = \sigma\varepsilon_s^{\mathrm{brug}}$, $\kappa_{\mathrm{eff}} = \kappa\varepsilon_e^{\mathrm{brug}}$, $D_{e,\mathrm{eff}} = D_e\varepsilon_e^{\mathrm{brug}}$. In the electrolyte, conductivity is a function of concentration:

$$\kappa(c_e) = c_e(0.0420 - 1.0972 \times 10^{-5}c_e + 0.0132\exp(-0.0022c_e))^2.$$

The electrode OCP functions are evaluated using the MSMR-model parameters in Table 2.B.1; $n_f^r = n_{\mathrm{dl}}^r = 1$; $\omega_{\mathrm{dl}}^r = 2\pi \times 10^{-5}$.

Table 2.B.2: Standard values for the NMC30 cell used in examples (cf. Eq. (2.21) for the distinction between $\bar{D}_s$ and $\bar{D}_{s,\mathrm{ref}}$, which is analogous to the distinction here between D_s and $D_{s,\mathrm{ref}}$; parameter values from: Johannes Schmalstieg, Christiane Rahe, Madeleine Ecker, and Dirk Uwe Sauer, "Full cell parameterization of a high-power lithium-ion battery for a physico-chemical model: Part I. Physical and electrochemical parameters," *Journal of The Electrochemical Society*, 165(16):A3799, 2018, and Johannes Schmalstieg and Dirk Uwe Sauer, "Full cell parameterization of a high-power lithium-ion battery for a physico-chemical model: Part II. Thermal parameters and validation," *Journal of The Electrochemical Society*, 165(16):A3811, 2018).

2.C *Closed-form solution for TF $c_k^r(s)$ functions*

The coefficients $c_k^r(s)$ of the electrolyte-concentration TF follow the relationship shown in the matrix form in Eq. (2.81). When we are implementing the TFs, we might solve this equation numerically for every frequency of interest. Alternately, it is possible to determine a closed-form solution for all coefficients.[53] This appendix section gives the closed-form expressions, but omits the (very tedious, done by manual back-substitution and computer symbolic manipulation) algebra.

To express the solution, we first define (for r $\in \{\mathrm{n},\mathrm{p}\}$):

$$\omega_+^r = \bar{\kappa}^s\Lambda_1^s\,(\lambda_2^r\coth(\Lambda_1^r) - \lambda_1^r\coth(\Lambda_2^r)) + \bar{\kappa}^r\,(\lambda_2^r\Lambda_1^r - \lambda_1^r\Lambda_2^r)$$

[53] Closed-form impedance models of lithium-ion battery processes have been recognized as valuable by the research community for a long time. For historical perspective, the reader may be interested to confer the following reference:

- G. Paasch, K. Micka, and P. Gersdorf, "Theory of the electrochemical impedance of macrohomogeneous porous electrodes," *Electrochimica Acta*, 38(18):2653–2662, 1993.

Note that this article proposes impedance models that describe individual *processes* in a cell and not the overall impedances describing each variable in the DFN model, or cell voltage predicted by the DFN model. Indeed, the publication date of this paper precedes that of the original DFN model.

In this chapter, we have presented an exact linearized impedance model of the full DFN model, with enhancements to include MSMR kinetics, constant-phase elements, and the electric double layer. This appendix gives the closed-form equations for this more comprehensive impedance model.

Table 2.C.1: Lumped parameters for the NMC30 cell used in this chapter, converted from Table 2.B.2.

Symbol	Units	Negative electrode	Separator	Positive electrode
$\bar{\sigma}$	S	1.3×10^7	—	185109
$\bar{\kappa}$	S	6199	22274	4580
θ_0	—	0.014	—	0.9615
θ_{100}	—	0.884	—	0.4352
$\bar{D}_{\mathrm{s,ref}}$	s^{-1}	6.6674×10^{-4}	—	1.695×10^{-4}
$\bar{q}_{\mathrm{e}}$	Ah	0.876	0.4754	0.5786
$\bar{k}_0$	A	288.67	—	619.9
α	—	0.5	—	0.5
$\bar{R}_{\mathrm{dl}}$	Ω	1×10^{-5}	—	1×10^{-6}
$\bar{C}_{\mathrm{dl}}$	F	28.556	—	12.2
$\bar{R}_{\mathrm{f}}$	Ω	5×10^{-5}	—	1×10^{-5}
Q	Ah		29.86	
$\bar{\psi}$	$\mathrm{V\,K}^{-1}$		1.1577×10^{-4}	
$\bar{\kappa}_D$	$\mathrm{V\,K}^{-1}$		-1.275×10^{-4}	
R_{c}	mΩ		0.534	

The electrode OCP functions are evaluated using the MSMR-model parameters in Table 2.B.1; $n_{\mathrm{f}}^{\mathrm{r}} = n_{\mathrm{dl}}^{\mathrm{r}} = 1$; $\omega_{\mathrm{dl}}^{\mathrm{r}} = 2\pi \times 10^{-5}$.

$$\omega_-^{\mathrm{r}} = \bar{\kappa}^{\mathrm{s}}\Lambda_1^{\mathrm{s}}\left(\lambda_2^{\mathrm{r}}\coth(\Lambda_1^{\mathrm{r}}) - \lambda_1^{\mathrm{r}}\coth(\Lambda_2^{\mathrm{r}})\right) - \bar{\kappa}^{\mathrm{r}}\left(\lambda_2^{\mathrm{r}}\Lambda_1^{\mathrm{r}} - \lambda_1^{\mathrm{r}}\Lambda_2^{\mathrm{r}}\right)$$

$$\mathrm{den}^{\mathrm{r}} = \bar{\sigma}^{\mathrm{r}}\bar{\kappa}^{\mathrm{r}}\left(\omega_-^{\mathrm{n}}\omega_-^{\mathrm{p}}e^{-2\Lambda_1^{\mathrm{s}}} - \omega_+^{\mathrm{n}}\omega_+^{\mathrm{p}}\right).$$

Then, separator-region electrolyte-concentration TF coefficients are:

$$\begin{aligned} c_1^{\mathrm{s}} = &-\frac{\bar{\kappa}^{\mathrm{n}}\mu_1^{\mathrm{n}}\omega_-^{\mathrm{p}}e^{-\Lambda_1^{\mathrm{s}}}}{\mathrm{den}^{\mathrm{n}}}\Big[\bar{\sigma}^{\mathrm{n}}\left\{\Lambda_2^{\mathrm{n}}\coth(\Lambda_1^{\mathrm{n}}) - \Lambda_1^{\mathrm{n}}\coth(\Lambda_2^{\mathrm{n}})\right\} \\ &+ \bar{\kappa}^{\mathrm{n}}\left\{\Lambda_2^{\mathrm{n}}\mathrm{csch}(\Lambda_1^{\mathrm{n}}) - \Lambda_1^{\mathrm{n}}\mathrm{csch}(\Lambda_2^{\mathrm{n}})\right\}\Big] \\ &+ \frac{\bar{\kappa}^{\mathrm{p}}\mu_1^{\mathrm{p}}\omega_+^{\mathrm{n}}}{\mathrm{den}^{\mathrm{p}}}\Big[\bar{\sigma}^{\mathrm{p}}\left\{\Lambda_2^{\mathrm{p}}\coth(\Lambda_1^{\mathrm{p}}) - \Lambda_1^{\mathrm{p}}\coth(\Lambda_2^{\mathrm{p}})\right\} \\ &+ \bar{\kappa}^{\mathrm{p}}\left\{\Lambda_2^{\mathrm{p}}\mathrm{csch}(\Lambda_1^{\mathrm{p}}) - \Lambda_1^{\mathrm{p}}\mathrm{csch}(\Lambda_2^{\mathrm{p}})\right\}\Big] \end{aligned}$$

$$\begin{aligned} c_2^{\mathrm{s}} = &-\frac{\bar{\kappa}^{\mathrm{n}}\mu_1^{\mathrm{n}}\omega_+^{\mathrm{p}}}{\mathrm{den}^{\mathrm{n}}}\Big[\bar{\sigma}^{\mathrm{n}}\left\{\Lambda_2^{\mathrm{n}}\coth(\Lambda_1^{\mathrm{n}}) - \Lambda_1^{\mathrm{n}}\coth(\Lambda_2^{\mathrm{n}})\right\} \\ &+ \bar{\kappa}^{\mathrm{n}}\left\{\Lambda_2^{\mathrm{n}}\mathrm{csch}(\Lambda_1^{\mathrm{n}}) - \Lambda_1^{\mathrm{n}}\mathrm{csch}(\Lambda_2^{\mathrm{n}})\right\}\Big] \\ &+ \frac{\bar{\kappa}^{\mathrm{p}}\mu_1^{\mathrm{p}}\omega_-^{\mathrm{n}}e^{-\Lambda_1^{\mathrm{s}}}}{\mathrm{den}^{\mathrm{p}}}\Big[\bar{\sigma}^{\mathrm{p}}\left\{\Lambda_2^{\mathrm{p}}\coth(\Lambda_1^{\mathrm{p}}) - \Lambda_1^{\mathrm{p}}\coth(\Lambda_2^{\mathrm{p}})\right\} \\ &+ \bar{\kappa}^{\mathrm{p}}\left\{\Lambda_2^{\mathrm{p}}\mathrm{csch}(\Lambda_1^{\mathrm{p}}) - \Lambda_1^{\mathrm{p}}\mathrm{csch}(\Lambda_2^{\mathrm{p}})\right\}\Big]. \end{aligned}$$

Coefficients for electrode regions are expressed in terms of c_1^{s} and c_2^{s}. Negative-electrode region TF coefficients are:

$$
\begin{aligned}
c_1^{\mathrm{n}} &= \frac{(1-e^{-2\Lambda_1^{\mathrm{n}}})^{-1}}{\bar{\kappa}^{\mathrm{n}}\left(\lambda_2^{\mathrm{n}}\Lambda_1^{\mathrm{n}}-\lambda_1^{\mathrm{n}}\Lambda_2^{\mathrm{n}}\right)}\left[\mu_1^{\mathrm{n}}\Lambda_2^{\mathrm{n}}\left(1+\frac{\bar{\kappa}^{\mathrm{n}}e^{-\Lambda_1^{\mathrm{n}}}}{\bar{\sigma}^{\mathrm{n}}}\right)+\bar{\kappa}^{\mathrm{s}}\Lambda_1^{\mathrm{s}}\lambda_2^{\mathrm{n}}\left(c_1^{\mathrm{s}}e^{-\Lambda_1^{\mathrm{s}}}-c_2^{\mathrm{s}}\right)\right]\\
c_2^{\mathrm{n}} &= \frac{\operatorname{csch}(\Lambda_1^{\mathrm{n}})/2}{\bar{\kappa}^{\mathrm{n}}\left(\lambda_2^{\mathrm{n}}\Lambda_1^{\mathrm{n}}-\lambda_1^{\mathrm{n}}\Lambda_2^{\mathrm{n}}\right)}\left[\mu_1^{\mathrm{n}}\Lambda_2^{\mathrm{n}}\left(1+\frac{\bar{\kappa}^{\mathrm{n}}e^{\Lambda_1^{\mathrm{n}}}}{\bar{\sigma}^{\mathrm{n}}}\right)+\bar{\kappa}^{\mathrm{s}}\Lambda_1^{\mathrm{s}}\lambda_2^{\mathrm{n}}\left(c_1^{\mathrm{s}}e^{-\Lambda_1^{\mathrm{s}}}-c_2^{\mathrm{s}}\right)\right]\\
c_3^{\mathrm{n}} &= \frac{-(1-e^{-2\Lambda_2^{\mathrm{n}}})^{-1}}{\bar{\kappa}^{\mathrm{n}}\left(\lambda_2^{\mathrm{n}}\Lambda_1^{\mathrm{n}}-\lambda_1^{\mathrm{n}}\Lambda_2^{\mathrm{n}}\right)}\left[\mu_1^{\mathrm{n}}\Lambda_1^{\mathrm{n}}\left(1+\frac{\bar{\kappa}^{\mathrm{n}}e^{-\Lambda_2^{\mathrm{n}}}}{\bar{\sigma}^{\mathrm{n}}}\right)+\bar{\kappa}^{\mathrm{s}}\Lambda_1^{\mathrm{s}}\lambda_1^{\mathrm{n}}\left(c_1^{\mathrm{s}}e^{-\Lambda_1^{\mathrm{s}}}-c_2^{\mathrm{s}}\right)\right]\\
c_4^{\mathrm{n}} &= \frac{-\operatorname{csch}(\Lambda_2^{\mathrm{n}})/2}{\bar{\kappa}^{\mathrm{n}}\left(\lambda_2^{\mathrm{n}}\Lambda_1^{\mathrm{n}}-\lambda_1^{\mathrm{n}}\Lambda_2^{\mathrm{n}}\right)}\left[\mu_1^{\mathrm{n}}\Lambda_1^{\mathrm{n}}\left(1+\frac{\bar{\kappa}^{\mathrm{n}}e^{\Lambda_2^{\mathrm{n}}}}{\bar{\sigma}^{\mathrm{n}}}\right)+\bar{\kappa}^{\mathrm{s}}\Lambda_1^{\mathrm{s}}\lambda_1^{\mathrm{n}}\left(c_1^{\mathrm{s}}e^{-\Lambda_1^{\mathrm{s}}}-c_2^{\mathrm{s}}\right)\right].
\end{aligned}
$$

Positive-electrode region electrolyte-concentration TF coefficients are:

$$
\begin{aligned}
c_1^{\mathrm{p}} &= \frac{-(1-e^{-2\Lambda_1^{\mathrm{p}}})^{-1}}{\bar{\kappa}^{\mathrm{p}}\left(\lambda_2^{\mathrm{p}}\Lambda_1^{\mathrm{p}}-\lambda_1^{\mathrm{p}}\Lambda_2^{\mathrm{p}}\right)}\left[\mu_1^{\mathrm{p}}\Lambda_2^{\mathrm{p}}\left(1+\frac{\bar{\kappa}^{\mathrm{p}}e^{-\Lambda_1^{\mathrm{p}}}}{\bar{\sigma}^{\mathrm{p}}}\right)+\bar{\kappa}^{\mathrm{s}}\Lambda_1^{\mathrm{s}}\lambda_2^{\mathrm{p}}\left(c_1^{\mathrm{s}}-c_2^{\mathrm{s}}e^{-\Lambda_1^{\mathrm{s}}}\right)\right]\\
c_2^{\mathrm{p}} &= \frac{-\operatorname{csch}(\Lambda_1^{\mathrm{p}})/2}{\bar{\kappa}^{\mathrm{p}}\left(\lambda_2^{\mathrm{p}}\Lambda_1^{\mathrm{p}}-\lambda_1^{\mathrm{p}}\Lambda_2^{\mathrm{p}}\right)}\left[\mu_1^{\mathrm{p}}\Lambda_2^{\mathrm{p}}\left(1+\frac{\bar{\kappa}^{\mathrm{p}}e^{\Lambda_1^{\mathrm{p}}}}{\bar{\sigma}^{\mathrm{p}}}\right)+\bar{\kappa}^{\mathrm{s}}\Lambda_1^{\mathrm{s}}\lambda_2^{\mathrm{p}}\left(c_1^{\mathrm{s}}-c_2^{\mathrm{s}}e^{-\Lambda_1^{\mathrm{s}}}\right)\right]\\
c_3^{\mathrm{p}} &= \frac{(1-e^{-2\Lambda_2^{\mathrm{p}}})^{-1}}{\bar{\kappa}^{\mathrm{p}}\left(\lambda_2^{\mathrm{p}}\Lambda_1^{\mathrm{p}}-\lambda_1^{\mathrm{p}}\Lambda_2^{\mathrm{p}}\right)}\left[\mu_1^{\mathrm{p}}\Lambda_1^{\mathrm{p}}\left(1+\frac{\bar{\kappa}^{\mathrm{p}}e^{-\Lambda_2^{\mathrm{p}}}}{\bar{\sigma}^{\mathrm{p}}}\right)+\bar{\kappa}^{\mathrm{s}}\Lambda_1^{\mathrm{s}}\lambda_1^{\mathrm{p}}\left(c_1^{\mathrm{s}}-c_2^{\mathrm{s}}e^{-\Lambda_1^{\mathrm{s}}}\right)\right]\\
c_4^{\mathrm{p}} &= \frac{\operatorname{csch}(\Lambda_2^{\mathrm{p}})/2}{\bar{\kappa}^{\mathrm{p}}\left(\lambda_2^{\mathrm{p}}\Lambda_1^{\mathrm{p}}-\lambda_1^{\mathrm{p}}\Lambda_2^{\mathrm{p}}\right)}\left[\mu_1^{\mathrm{p}}\Lambda_1^{\mathrm{p}}\left(1+\frac{\bar{\kappa}^{\mathrm{p}}e^{\Lambda_2^{\mathrm{p}}}}{\bar{\sigma}^{\mathrm{p}}}\right)+\bar{\kappa}^{\mathrm{s}}\Lambda_1^{\mathrm{s}}\lambda_1^{\mathrm{p}}\left(c_1^{\mathrm{s}}-c_2^{\mathrm{s}}e^{-\Lambda_1^{\mathrm{s}}}\right)\right].
\end{aligned}
$$

2.D *Transfer-function limits*

At various times in this book, we will need closed-form solutions for low- ($s \to 0$) and high-frequency ($s \to \infty$) responses of all TFs, as well as integrator residues. This section summarizes those results.[54]

[54] Most of these results were solved in Mathematica by taking limits of expressions as $s \to 0$ and as $s \to \infty$.

2.D.1 *Integrator residues*

During one step of the ROM-generation process described in Chap. 4, we must remove integrator dynamics from all TFs. The integrator residue of TF $G^{\mathrm{r}}(s)$ is defined as $\mathrm{res}_0^{\mathrm{r}} = \lim_{s\to 0} sG^{\mathrm{r}}(s)$. Integrator-removed TFs are denoted with square brackets and an asterisk as:

$$\frac{[\mathrm{Variable}(\tilde{x},s)]^*}{I_{\mathrm{app}}(s)} = \frac{\mathrm{Variable}(\tilde{x},s)}{I_{\mathrm{app}}(s)} - \frac{\mathrm{res}_0^{\mathrm{r}}}{s}.$$

Many of the TFs do not have integration dynamics. In those cases, we simply state that the residues are zero.

ELECTROLYTE CONCENTRATION: $\mathrm{res}_0^{\mathrm{r}} = 0$ for $\widetilde{\Theta}_{\mathrm{e}}^{\mathrm{r}}(\tilde{x},s)/I_{\mathrm{app}}(s)$.

SOLID SURFACE CONCENTRATION: The negative- and positive-electrode residues for $\widetilde{\Theta}^{\mathrm{r}}_{\mathrm{ss}}(\tilde{x},s)/I_{\mathrm{app}}(s)$ are:

$$\mathrm{res}^{\mathrm{n}}_0 \text{ for } \frac{\widetilde{\Theta}^{\mathrm{n}}_{\mathrm{ss}}(\tilde{x},s)}{I_{\mathrm{app}}(s)} = -\frac{\left|\theta^{\mathrm{n}}_{100}-\theta^{\mathrm{n}}_{0}\right|}{3600Q-\bar{C}^{\mathrm{n}}_{\mathrm{dl,eff}}\left|\theta^{\mathrm{n}}_{100}-\theta^{\mathrm{n}}_{0}\right|[U^{\mathrm{n}}_{\mathrm{ocp}}]'}$$
$$\mathrm{res}^{\mathrm{p}}_0 \text{ for } \frac{\widetilde{\Theta}^{\mathrm{p}}_{\mathrm{ss}}(\tilde{x},s)}{I_{\mathrm{app}}(s)} = \frac{\left|\theta^{\mathrm{p}}_{100}-\theta^{\mathrm{p}}_{0}\right|}{3600Q-\bar{C}^{\mathrm{p}}_{\mathrm{dl,eff}}\left|\theta^{\mathrm{p}}_{100}-\theta^{\mathrm{p}}_{0}\right|[U^{\mathrm{p}}_{\mathrm{ocp}}]'}.$$

ELECTROLYTE POTENTIAL: $\mathrm{res}^{\mathrm{r}}_0 = 0$ for $\widetilde{\Phi}^{\mathrm{r}}_{\mathrm{e}}(\tilde{x},s)/I_{\mathrm{app}}(s)$.
SOLID POTENTIAL: $\mathrm{res}^{\mathrm{r}}_0 = 0$ for $\widetilde{\Phi}^{\mathrm{r}}_{\mathrm{s}}(\tilde{x},s)/I_{\mathrm{app}}(s)$.
PHASE POTENTIAL DIFFERENCE: The negative- and positive-electrode residues for $\widetilde{\Phi}^{\mathrm{r}}_{\mathrm{s\text{-}e}}(\tilde{x},s)/I_{\mathrm{app}}(s)$ are equal to the residues for $\widetilde{\Theta}^{\mathrm{r}}_{\mathrm{ss}}(\tilde{x},s)/I_{\mathrm{app}}(s)$ but multiplied by $[U^{\mathrm{r}}_{\mathrm{ocp}}]'$.

$$\mathrm{res}^{\mathrm{r}}_0 \text{ for } \frac{\widetilde{\Phi}^{\mathrm{r}}_{\mathrm{s\text{-}e}}(\tilde{x},s)}{I_{\mathrm{app}}(s)} = [U^{\mathrm{r}}_{\mathrm{ocp}}]' \times \left[\mathrm{res}^{\mathrm{r}}_0 \text{ for } \frac{\widetilde{\Theta}^{\mathrm{r}}_{\mathrm{ss}}(\tilde{x},s)}{I_{\mathrm{app}}(s)}\right].$$

INTERFACIAL LITHIUM FLUX: $\mathrm{res}^{\mathrm{r}}_0 = 0$ for $I^{\mathrm{r}}_{\mathrm{f+dl}}(\tilde{x},s)/I_{\mathrm{app}}(s)$.
INTERFACIAL FARADAIC LITHIUM FLUX: $\mathrm{res}^{\mathrm{r}}_0 = 0$ for $I^{\mathrm{r}}_{\mathrm{f}}(\tilde{x},s)/I_{\mathrm{app}}(s)$.
INTERFACIAL DOUBLE-LAYER FLUX: $\mathrm{res}^{\mathrm{r}}_0 = 0$ for $I^{\mathrm{r}}_{\mathrm{dl}}(\tilde{x},s)/I_{\mathrm{app}}(s)$.

2.D.2 *Low-frequency gains*

ELECTROLYTE CONCENTRATION: The low-frequency gains are:

$$\frac{\widetilde{\Theta}^{\mathrm{n}}_{\mathrm{e}}(\tilde{x},0)}{I_{\mathrm{app}}(0)} = \frac{\bar{q}^{\mathrm{n}}_{\mathrm{e}}\bar{\kappa}^{\mathrm{s}}\bar{\kappa}^{\mathrm{p}}+3\bar{q}^{\mathrm{s}}_{\mathrm{e}}(\bar{\kappa}^{\mathrm{s}}\bar{\kappa}^{\mathrm{p}}+\bar{\kappa}^{\mathrm{n}}\bar{\kappa}^{\mathrm{p}})+\bar{q}^{\mathrm{p}}_{\mathrm{e}}(2\bar{\kappa}^{\mathrm{n}}\bar{\kappa}^{\mathrm{s}}+3\bar{\kappa}^{\mathrm{s}}\bar{\kappa}^{\mathrm{p}}+6\bar{\kappa}^{\mathrm{n}}\bar{\kappa}^{\mathrm{p}})}{6\bar{\psi}\bar{\kappa}^{\mathrm{n}}\bar{\kappa}^{\mathrm{s}}\bar{\kappa}^{\mathrm{p}}T(\bar{q}^{\mathrm{n}}_{\mathrm{e}}+\bar{q}^{\mathrm{s}}_{\mathrm{e}}+\bar{q}^{\mathrm{p}}_{\mathrm{e}})} - \frac{\tilde{x}^2}{2\bar{\psi}\bar{\kappa}^{\mathrm{n}}T} \tag{2.104}$$

$$\frac{\widetilde{\Theta}^{\mathrm{s}}_{\mathrm{e}}(\tilde{x},0)}{I_{\mathrm{app}}(0)} = \frac{-2\bar{q}^{\mathrm{n}}_{\mathrm{e}}\bar{\kappa}^{\mathrm{s}}\bar{\kappa}^{\mathrm{p}}+3\bar{q}^{\mathrm{s}}_{\mathrm{e}}\bar{\kappa}^{\mathrm{n}}\bar{\kappa}^{\mathrm{p}}+2\bar{q}^{\mathrm{p}}_{\mathrm{e}}(\bar{\kappa}^{\mathrm{n}}\bar{\kappa}^{\mathrm{s}}+3\bar{\kappa}^{\mathrm{n}}\bar{\kappa}^{\mathrm{p}})}{6\bar{\psi}\bar{\kappa}^{\mathrm{n}}\bar{\kappa}^{\mathrm{s}}\bar{\kappa}^{\mathrm{p}}T(\bar{q}^{\mathrm{n}}_{\mathrm{e}}+\bar{q}^{\mathrm{s}}_{\mathrm{e}}+\bar{q}^{\mathrm{p}}_{\mathrm{e}})} - \frac{\tilde{x}-1}{\bar{\psi}\bar{\kappa}^{\mathrm{s}}T}$$

$$\frac{\widetilde{\Theta}^{\mathrm{p}}_{\mathrm{e}}(\tilde{x},0)}{I_{\mathrm{app}}(0)} = \frac{-\bar{q}^{\mathrm{n}}_{\mathrm{e}}(2\bar{\kappa}^{\mathrm{s}}\bar{\kappa}^{\mathrm{p}}+3\bar{\kappa}^{\mathrm{n}}\bar{\kappa}^{\mathrm{s}}+6\bar{\kappa}^{\mathrm{n}}\bar{\kappa}^{\mathrm{p}})-3\bar{q}^{\mathrm{s}}_{\mathrm{e}}(\bar{\kappa}^{\mathrm{n}}\bar{\kappa}^{\mathrm{s}}+\bar{\kappa}^{\mathrm{n}}\bar{\kappa}^{\mathrm{p}})-\bar{q}^{\mathrm{p}}_{\mathrm{e}}\bar{\kappa}^{\mathrm{n}}\bar{\kappa}^{\mathrm{s}}}{6\bar{\psi}\bar{\kappa}^{\mathrm{n}}\bar{\kappa}^{\mathrm{s}}\bar{\kappa}^{\mathrm{p}}T(\bar{q}^{\mathrm{n}}_{\mathrm{e}}+\bar{q}^{\mathrm{s}}_{\mathrm{e}}+\bar{q}^{\mathrm{p}}_{\mathrm{e}})} + \frac{(3-\tilde{x})^2}{2\bar{\psi}\bar{\kappa}^{\mathrm{p}}T}. \tag{2.105}$$

SOLID SURFACE CONCENTRATION: The low-frequency gain of the $\widetilde{\Theta}^{\mathrm{r}}_{\mathrm{ss}}(\tilde{x},s)/I_{\mathrm{app}}(s)$ TF is infinite. The low-frequency gain of integrator-removed $[\widetilde{\Theta}^{\mathrm{r}}_{\mathrm{ss}}(\tilde{x},s)]^*/I_{\mathrm{app}}(s)$ can be expressed as:

$$\frac{[\widetilde{\Theta}^{\mathrm{r}}_{\mathrm{ss}}(\tilde{x},0)]^*}{I_{\mathrm{app}}(0)} = z^{\mathrm{r}}_0 - \mathrm{res}^{\mathrm{r}}_0\bar{\psi}\bar{\kappa}^{\mathrm{r}}T\frac{[\widetilde{\Theta}''_1(\tilde{x},0)]^{\mathrm{r}}}{I_{\mathrm{app}}(0)} + 3600\mathrm{res}^{\mathrm{r}}_0\bar{q}^{\mathrm{r}}_{\mathrm{e}}\frac{\widetilde{\Theta}^{\mathrm{r}}_{\mathrm{e}}(\tilde{x},0)}{I_{\mathrm{app}}(0)},$$

where $\widetilde{\Theta}^{\mathrm{r}}_{\mathrm{e}}(\tilde{x},0)/I_{\mathrm{app}}(0)$ are the low-frequency gains of the electrolyte concentration, from above,

$$z_0^{\mathrm{r}} = \frac{\left|\theta_{100}^{\mathrm{r}} - \theta_0^{\mathrm{r}}\right| \bar{C}_{\mathrm{dl,eff}}^{\mathrm{r}} \left(\left|\theta_{100}^{\mathrm{r}} - \theta_0^{\mathrm{r}}\right| [U_{\mathrm{ocp}}^{\mathrm{r}}]' \bar{R}_{\mathrm{dl}}^{\mathrm{r}} \bar{C}_{\mathrm{dl,eff}}^{\mathrm{r}} + 3600 Q \bar{R}_{\mathrm{ct}}^{\mathrm{r}}\right)}{\left(\bar{C}_{\mathrm{dl,eff}}^{\mathrm{r}} \left|\theta_{100}^{\mathrm{r}} - \theta_0^{\mathrm{r}}\right| [U_{\mathrm{ocp}}^{\mathrm{r}}]' - 3600Q\right)^2} + \frac{(1 - n_{\mathrm{dl}}^{\mathrm{r}}) \bar{C}_{\mathrm{dl,eff}}^{\mathrm{r}} \left|\theta_{100}^{\mathrm{r}} - \theta_0^{\mathrm{r}}\right|^2 [U_{\mathrm{ocp}}^{\mathrm{r}}]' \left(\bar{C}_{\mathrm{dl}}^{\mathrm{r}} / \omega_{\mathrm{dl}}^{\mathrm{r}}\right)}{\left(\bar{C}_{\mathrm{dl,eff}}^{\mathrm{r}} \left|\theta_{100}^{\mathrm{r}} - \theta_0^{\mathrm{r}}\right| [U_{\mathrm{ocp}}^{\mathrm{r}}]' - 3600Q\right)^2}.$$

and $[\widetilde{\Theta}_1''(\tilde{x},0)]^{\mathrm{r}}/I_{\mathrm{app}}(0)$ are coefficients multiplying s in the Taylor-series expansion of the second spatial derivative of the electrolyte-concentration TF, which are (written in terms of z):

$$[\widetilde{\Theta}_1''(z)]^{\mathrm{n}} = \frac{3600 \times \mathrm{num}^{\mathrm{n}}}{18 \bar{D}_{\mathrm{s}}^{\mathrm{n}} [U_{\mathrm{ocp}}^{\mathrm{n}}]' (\bar{q}_{\mathrm{e}}^{\mathrm{n}} + \bar{q}_{\mathrm{e}}^{\mathrm{s}} + \bar{q}_{\mathrm{e}}^{\mathrm{p}}) (\bar{\kappa}^{\mathrm{n}})^2 \bar{\kappa}^{\mathrm{s}} \bar{\kappa}^{\mathrm{p}} \bar{\sigma}^{\mathrm{n}} (\bar{\psi} T)^2}$$

$$\begin{aligned} \mathrm{num}^{\mathrm{n}} = -\Bigg[& -3\bar{D}_{\mathrm{s}}^{\mathrm{n}} [U_{\mathrm{ocp}}^{\mathrm{n}}]' \bar{q}_{\mathrm{e}}^{\mathrm{n}} \Big(2\bar{q}_{\mathrm{e}}^{\mathrm{p}} \bar{\kappa}^{\mathrm{n}} \bar{\kappa}^{\mathrm{s}} + 3\bar{q}_{\mathrm{e}}^{\mathrm{s}} \bar{\kappa}^{\mathrm{p}} (\bar{\kappa}^{\mathrm{n}} + \bar{\kappa}^{\mathrm{s}} - z^2 \bar{\kappa}^{\mathrm{s}}) \\ & + \bar{q}_{\mathrm{e}}^{\mathrm{n}} (1 - 3z^2) \bar{\kappa}^{\mathrm{s}} \bar{\kappa}^{\mathrm{p}} + 3\bar{q}_{\mathrm{e}}^{\mathrm{p}} \bar{\kappa}^{\mathrm{p}} (2\bar{\kappa}^{\mathrm{n}} + \bar{\kappa}^{\mathrm{s}} - z^2 \bar{\kappa}^{\mathrm{s}}) \Big) \bar{\sigma}^{\mathrm{n}} \\ & + \frac{3Q\bar{D}_{\mathrm{s}}^{\mathrm{n}}}{\left|\theta_{100}^{\mathrm{n}} - \theta_0^{\mathrm{n}}\right|} (\bar{q}_{\mathrm{e}}^{\mathrm{n}} + \bar{q}_{\mathrm{e}}^{\mathrm{s}} + \bar{q}_{\mathrm{e}}^{\mathrm{p}}) \bar{\kappa}^{\mathrm{s}} \bar{\kappa}^{\mathrm{p}} \Big(T(3z^2 - 1) \bar{\kappa}_D \bar{\sigma}^{\mathrm{n}} \\ & + \bar{\psi} T((-2 + 6z - 3z^2) \bar{\kappa}^{\mathrm{n}} + \bar{\sigma}^{\mathrm{n}} - 3z^2 \bar{\sigma}^{\mathrm{n}}) \Big) \Bigg] \end{aligned}$$

$$[\widetilde{\Theta}_1''(z)]^{\mathrm{p}} = \frac{3600 \times \mathrm{num}^{\mathrm{p}}}{18 \bar{D}_{\mathrm{s}}^{\mathrm{p}} [U_{\mathrm{ocp}}^{\mathrm{p}}]' (\bar{q}_{\mathrm{e}}^{\mathrm{n}} + \bar{q}_{\mathrm{e}}^{\mathrm{s}} + \bar{q}_{\mathrm{e}}^{\mathrm{p}}) \bar{\kappa}^{\mathrm{n}} \bar{\kappa}^{\mathrm{s}} (\bar{\kappa}^{\mathrm{p}})^2 \bar{\sigma}^{\mathrm{p}} (\bar{\psi} T)^2}$$

$$\begin{aligned} \mathrm{num}^{\mathrm{p}} = \Bigg[& -3\bar{D}_{\mathrm{s}}^{\mathrm{p}} [U_{\mathrm{ocp}}^{\mathrm{p}}]' \bar{q}_{\mathrm{e}}^{\mathrm{p}} \Big(2\bar{q}_{\mathrm{e}}^{\mathrm{n}} \bar{\kappa}^{\mathrm{s}} \bar{\kappa}^{\mathrm{p}} + 3\bar{q}_{\mathrm{e}}^{\mathrm{s}} \bar{\kappa}^{\mathrm{n}} (\bar{\kappa}^{\mathrm{p}} + \bar{\kappa}^{\mathrm{s}} - z^2 \bar{\kappa}^{\mathrm{s}}) \\ & + \bar{q}_{\mathrm{e}}^{\mathrm{p}} (1 - 3z^2) \bar{\kappa}^{\mathrm{n}} \bar{\kappa}^{\mathrm{s}} + 3\bar{q}_{\mathrm{e}}^{\mathrm{n}} \bar{\kappa}^{\mathrm{n}} (2\bar{\kappa}^{\mathrm{p}} + \bar{\kappa}^{\mathrm{s}} - z^2 \bar{\kappa}^{\mathrm{s}}) \Big) \bar{\sigma}^{\mathrm{p}} \\ & + \frac{3Q\bar{D}_{\mathrm{s}}^{\mathrm{p}}}{\left|\theta_{100}^{\mathrm{p}} - \theta_0^{\mathrm{p}}\right|} (\bar{q}_{\mathrm{e}}^{\mathrm{n}} + \bar{q}_{\mathrm{e}}^{\mathrm{s}} + \bar{q}_{\mathrm{e}}^{\mathrm{p}}) \bar{\kappa}^{\mathrm{n}} \bar{\kappa}^{\mathrm{s}} \Big(T(3z^2 - 1) \bar{\kappa}_D \bar{\sigma}^{\mathrm{p}} \\ & + \bar{\psi} T((-2 + 6z - 3z^2) \bar{\kappa}^{\mathrm{p}} + \bar{\sigma}^{\mathrm{p}} - 3z^2 \bar{\sigma}^{\mathrm{p}}) \Big) \Bigg]. \end{aligned}$$

ELECTROLYTE POTENTIAL: The low-frequency gains are:

$$\begin{aligned} \frac{\widetilde{\Phi}_{\mathrm{e}}^{\mathrm{n}}(\tilde{x},0)}{I_{\mathrm{app}}(0)} &= -\frac{\tilde{x}^2}{2\bar{\kappa}^{\mathrm{n}}} + \frac{\bar{\kappa}_D \tilde{x}^2}{2\bar{\psi}\bar{\kappa}^{\mathrm{n}}} \\ \frac{\widetilde{\Phi}_{\mathrm{e}}^{\mathrm{s}}(\tilde{x},0)}{I_{\mathrm{app}}(0)} &= \frac{\widetilde{\Phi}_{\mathrm{e}}^{\mathrm{n}}(1,0)}{I_{\mathrm{app}}(0)} - \frac{\tilde{x}-1}{\bar{\kappa}^{\mathrm{s}}} + \frac{\bar{\kappa}_D(\tilde{x}-1)}{\bar{\psi}\bar{\kappa}^{\mathrm{s}}} \\ \frac{\widetilde{\Phi}_{\mathrm{e}}^{\mathrm{p}}(\tilde{x},0)}{I_{\mathrm{app}}(0)} &= \frac{\widetilde{\Phi}_{\mathrm{e}}^{\mathrm{s}}(2,0)}{I_{\mathrm{app}}(0)} + \frac{(3-\tilde{x})^2 - 1}{2\bar{\kappa}^{\mathrm{p}}} - \frac{\bar{\kappa}_D\left((3-\tilde{x})^2 - 1\right)}{2\bar{\psi}\bar{\kappa}^{\mathrm{p}}}. \end{aligned} \tag{2.106}$$

SOLID POTENTIAL: The low-frequency gains are:

$$\frac{\widetilde{\Phi}_{\mathrm{s}}^{\mathrm{n}}(\tilde{x},0)}{I_{\mathrm{app}}(0)} = \frac{\tilde{x}(\tilde{x}-2)}{2\bar{\sigma}^{\mathrm{n}}} \quad \text{and} \quad \frac{\widetilde{\Phi}_{\mathrm{s}}^{\mathrm{p}}(\tilde{x},0)}{I_{\mathrm{app}}(0)} = \frac{-(3-\tilde{x})(1-\tilde{x})}{2\bar{\sigma}^{\mathrm{p}}}.$$

PHASE POTENTIAL DIFFERENCE: The low-frequency gain of the $\widetilde{\Phi}^{\mathrm{r}}_{\text{s-e}}(\tilde{x},s)/I_{\mathrm{app}}(s)$ TF is infinite. The low-frequency gain of integrator-removed $[\widetilde{\Phi}^{\mathrm{r}}_{\text{s-e}}(\tilde{x},s)]^*/I_{\mathrm{app}}(s)$ can be expressed as:

$$\frac{[\widetilde{\Phi}^{\mathrm{r}}_{\text{s-e}}(\tilde{x},0)]^*}{I_{\mathrm{app}}(0)} = z_0^{\mathrm{r}} - \mathrm{res}_0^{\mathrm{r}}\bar{\psi}\bar{\kappa}^{\mathrm{r}}T\frac{[\widetilde{\Theta}_1''(\tilde{x},0)]^{\mathrm{r}}}{I_{\mathrm{app}}(0)} + 3600\mathrm{res}_0^{\mathrm{r}}\bar{q}_{\mathrm{e}}^{\mathrm{r}}\frac{\widetilde{\Theta}^{\mathrm{r}}_{\mathrm{e}}(\tilde{x},0)}{I_{\mathrm{app}}(0)},$$

where the $\widetilde{\Theta}^{\mathrm{r}}_{\mathrm{e}}(\tilde{x},0)/I_{\mathrm{app}}(0)$ and $[\widetilde{\Theta}_1''(\tilde{x},0)]^{\mathrm{r}}/I_{\mathrm{app}}(0)$ terms are the same as above, but where we now have:

$$\begin{aligned}
z_0^{\mathrm{r}} &= \lim_{s\to 0}\bar{Z}^{\mathrm{r}}_{\mathrm{se}}(s) - \frac{\mathrm{res}_0^{\mathrm{r}}}{s} \\
&= \frac{\left(\bar{C}^{\mathrm{r}}_{\mathrm{dl,eff}}\left|\theta^{\mathrm{r}}_{100}-\theta^{\mathrm{r}}_{0}\right|[U^{\mathrm{r}}_{\mathrm{ocp}}]'\right)^2\bar{R}^{\mathrm{r}}_{\mathrm{dl}} + (3600Q)^2\bar{R}^{\mathrm{r}}_{\mathrm{ct}}}{\left(\bar{C}^{\mathrm{r}}_{\mathrm{dl,eff}}\left|\theta^{\mathrm{r}}_{100}-\theta^{\mathrm{r}}_{0}\right|[U^{\mathrm{r}}_{\mathrm{ocp}}]' - 3600Q\right)^2} \\
&\quad - \frac{3600Q\left|\theta^{\mathrm{r}}_{100}-\theta^{\mathrm{r}}_{0}\right|[U^{\mathrm{r}}_{\mathrm{ocp}}]'}{15\bar{D}^{\mathrm{r}}_{\mathrm{s}}\left(\bar{C}^{\mathrm{r}}_{\mathrm{dl,eff}}\left|\theta^{\mathrm{r}}_{100}-\theta^{\mathrm{r}}_{0}\right|[U^{\mathrm{r}}_{\mathrm{ocp}}]' - 3600Q\right)^2} \\
&\quad - \frac{(1-n^{\mathrm{r}}_{\mathrm{dl}})\left(\bar{C}^{\mathrm{r}}_{\mathrm{dl}}/\omega^{\mathrm{r}}_{\mathrm{dl}}\right)\bar{C}^{\mathrm{r}}_{\mathrm{dl,eff}}\left(\left|\theta^{\mathrm{r}}_{100}-\theta^{\mathrm{r}}_{0}\right|[U^{\mathrm{r}}_{\mathrm{ocp}}]'\right)^2}{\left(\bar{C}^{\mathrm{r}}_{\mathrm{dl,eff}}\left|\theta^{\mathrm{r}}_{100}-\theta^{\mathrm{r}}_{0}\right|[U^{\mathrm{r}}_{\mathrm{ocp}}]' - 3600Q\right)^2} + \bar{R}^{\mathrm{r}}_{\mathrm{f}}.
\end{aligned}$$

INTERFACIAL LITHIUM FLUX: The low-frequency gains are:

$$\frac{I^{\mathrm{n}}_{\mathrm{f+dl}}(\tilde{x},0)}{I_{\mathrm{app}}(0)} = 1 \quad\text{and}\quad \frac{I^{\mathrm{p}}_{\mathrm{f+dl}}(\tilde{x},0)}{I_{\mathrm{app}}(0)} = -1.$$

INTERFACIAL FARADAIC LITHIUM FLUX: The gains are:

$$\frac{I^{\mathrm{r}}_{\mathrm{f}}(\tilde{x},0)}{I_{\mathrm{app}}(0)} = \frac{3600Q}{3600Q - \bar{C}^{\mathrm{r}}_{\mathrm{dl,eff}}\left|\theta^{\mathrm{r}}_{100}-\theta^{\mathrm{r}}_{0}\right|[U^{\mathrm{r}}_{\mathrm{ocp}}]'}\frac{I^{\mathrm{r}}_{\mathrm{f+dl}}(\tilde{x},0)}{I_{\mathrm{app}}(0)}. \tag{2.107}$$

INTERFACIAL DOUBLE-LAYER FLUX: The gains are:

$$\frac{I^{\mathrm{r}}_{\mathrm{dl}}(\tilde{x},0)}{I_{\mathrm{app}}(0)} = \frac{-\bar{C}^{\mathrm{r}}_{\mathrm{dl,eff}}\left|\theta^{\mathrm{r}}_{100}-\theta^{\mathrm{r}}_{0}\right|[U^{\mathrm{r}}_{\mathrm{ocp}}]'}{3600Q - \bar{C}^{\mathrm{r}}_{\mathrm{dl,eff}}\left|\theta^{\mathrm{r}}_{100}-\theta^{\mathrm{r}}_{0}\right|[U^{\mathrm{r}}_{\mathrm{ocp}}]'}\frac{I^{\mathrm{r}}_{\mathrm{f+dl}}(\tilde{x},0)}{I_{\mathrm{app}}(0)}.$$

2.D.3 *High-frequency gains*

To find high-frequency gains, we define:

$$\nu^{\mathrm{r}}_{\infty} = \lim_{s\to\infty}\nu^{\mathrm{r}}(s) = \sqrt{\frac{1}{\bar{\sigma}^{\mathrm{r}}} + \frac{1}{\bar{\kappa}^{\mathrm{r}}}}\Bigg/\sqrt{\bar{R}^{\mathrm{r}}_{\mathrm{f}} + \left(\frac{1}{\bar{R}^{\mathrm{r}}_{\mathrm{ct}}} + \frac{1}{\bar{R}^{\mathrm{r}}_{\mathrm{dl}}}\right)^{-1}}.$$

Then the high-frequency gains of the TFs of interest are:

ELECTROLYTE CONCENTRATION: The high-frequency gains of $\widetilde{\Theta}^{\mathrm{r}}_{\mathrm{e}}(\tilde{x},s)/I_{\mathrm{app}}(s)$ are zero.

SOLID SURFACE CONCENTRATION: High-frequency gains of $\widetilde{\Theta}^{\mathrm{r}}_{\mathrm{ss}}(\tilde{x},s)/I_{\mathrm{app}}(s)$ are zero.

ELECTROLYTE POTENTIAL: High-frequency gains are:

$$\begin{aligned}
\frac{\widetilde{\Phi}^{\mathrm{n}}_{\mathrm{e}}(\tilde{x},\infty)}{I_{\mathrm{app}}(\infty)} &= \frac{\cosh(\nu^{\mathrm{n}}_{\infty}) - \cosh(\nu^{\mathrm{n}}_{\infty}(1-\tilde{x}))}{(\bar{\sigma}^{\mathrm{n}}+\bar{\kappa}^{\mathrm{n}})\sinh(\nu^{\mathrm{n}}_{\infty})\nu^{\mathrm{n}}_{\infty}} \\
&\quad + \frac{\bar{\sigma}^{\mathrm{n}}\left(1-\cosh(\nu^{\mathrm{n}}_{\infty}\tilde{x})\right)}{\bar{\kappa}^{\mathrm{n}}(\bar{\sigma}^{\mathrm{n}}+\bar{\kappa}^{\mathrm{n}})\sinh(\nu^{\mathrm{n}}_{\infty})\nu^{\mathrm{n}}_{\infty}} - \frac{\tilde{x}}{\bar{\sigma}^{\mathrm{n}}+\bar{\kappa}^{\mathrm{n}}} \\
\frac{\widetilde{\Phi}^{\mathrm{s}}_{\mathrm{e}}(\tilde{x},\infty)}{I_{\mathrm{app}}(\infty)} &= \frac{\widetilde{\Phi}^{\mathrm{n}}_{\mathrm{e}}(1,\infty)}{I_{\mathrm{app}}(\infty)} - \frac{\tilde{x}-1}{\bar{\kappa}^{\mathrm{s}}} \\
\frac{\widetilde{\Phi}^{\mathrm{p}}_{\mathrm{e}}(\tilde{x},\infty)}{I_{\mathrm{app}}(\infty)} &= \frac{\widetilde{\Phi}^{\mathrm{s}}_{\mathrm{e}}(2,\infty)}{I_{\mathrm{app}}(\infty)} - \frac{1-\cosh(\nu^{\mathrm{p}}_{\infty}(\tilde{x}-2))}{(\bar{\sigma}^{\mathrm{p}}+\bar{\kappa}^{\mathrm{p}})\sinh(\nu^{\mathrm{p}}_{\infty})\nu^{\mathrm{p}}_{\infty}} \\
&\quad - \frac{\bar{\sigma}^{\mathrm{p}}\left(\cosh(\nu^{\mathrm{p}}_{\infty}) - \cosh(\nu^{\mathrm{p}}_{\infty}(3-\tilde{x}))\right)}{\bar{\kappa}^{\mathrm{p}}(\bar{\sigma}^{\mathrm{p}}+\bar{\kappa}^{\mathrm{p}})\sinh(\nu^{\mathrm{p}}_{\infty})\nu^{\mathrm{p}}_{\infty}} - \frac{\tilde{x}-2}{\bar{\sigma}^{\mathrm{p}}+\bar{\kappa}^{\mathrm{p}}}.
\end{aligned}$$

SOLID POTENTIAL: High-frequency gains are:

$$\begin{aligned}
\frac{\widetilde{\Phi}^{\mathrm{n}}_{\mathrm{s}}(\tilde{x},\infty)}{I_{\mathrm{app}}(\infty)} &= \frac{-\bar{\kappa}^{\mathrm{n}}(\cosh(\nu^{\mathrm{n}}_{\infty}) - \cosh(\nu^{\mathrm{n}}_{\infty}(\tilde{x}-1))) - \bar{\sigma}^{\mathrm{n}}(1-\cosh(\nu^{\mathrm{n}}_{\infty}\tilde{x}))}{\bar{\sigma}^{\mathrm{n}}\left(\bar{\kappa}^{\mathrm{n}}+\bar{\sigma}^{\mathrm{n}}\right)\sinh(\nu^{\mathrm{n}}_{\infty})\nu^{\mathrm{n}}_{\infty}} - \frac{\tilde{x}}{\bar{\kappa}^{\mathrm{n}}+\bar{\sigma}^{\mathrm{n}}} \\
\frac{\widetilde{\Phi}^{\mathrm{p}}_{\mathrm{s}}(\tilde{x},\infty)}{I_{\mathrm{app}}(\infty)} &= \frac{\bar{\kappa}^{\mathrm{p}}(\cosh(\nu^{\mathrm{p}}_{\infty}) - \cosh(\nu^{\mathrm{p}}_{\infty}(2-\tilde{x}))) + \bar{\sigma}^{\mathrm{p}}(1-\cosh(\nu^{\mathrm{p}}_{\infty}(3-\tilde{x})))}{\bar{\sigma}^{\mathrm{p}}\left(\bar{\kappa}^{\mathrm{p}}+\bar{\sigma}^{\mathrm{p}}\right)\sinh(\nu^{\mathrm{p}}_{\infty})\nu^{\mathrm{p}}_{\infty}} + \frac{3-\tilde{x}}{\bar{\kappa}^{\mathrm{p}}+\bar{\sigma}^{\mathrm{p}}}.
\end{aligned}$$

PHASE POTENTIAL DIFFERENCE: High-frequency gains of both $\widetilde{\Phi}^{\mathrm{r}}_{\mathrm{s\text{-}e}}(\tilde{x},s)/I_{\mathrm{app}}(s)$ and $[\widetilde{\Phi}^{\mathrm{r}}_{\mathrm{s\text{-}e}}(\tilde{x},s)]^{*}/I_{\mathrm{app}}(s)$ are (even when $n_{\mathrm{sf}} \neq 1$):

$$\begin{aligned}
\frac{\widetilde{\Phi}^{\mathrm{n}}_{\mathrm{s\text{-}e}}(\tilde{x},\infty)}{I_{\mathrm{app}}(\infty)} &= \frac{\bar{\sigma}^{\mathrm{n}}\cosh(\nu^{\mathrm{n}}_{\infty}\tilde{x}) + \bar{\kappa}^{\mathrm{n}}\cosh(\nu^{\mathrm{n}}_{\infty}(\tilde{x}-1))}{\bar{\sigma}^{\mathrm{n}}\bar{\kappa}^{\mathrm{n}}\sinh(\nu^{\mathrm{n}}_{\infty})\nu^{\mathrm{n}}_{\infty}} \\
\frac{\widetilde{\Phi}^{\mathrm{p}}_{\mathrm{s\text{-}e}}(\tilde{x},\infty)}{I_{\mathrm{app}}(\infty)} &= \frac{-\bar{\sigma}^{\mathrm{p}}\cosh(\nu^{\mathrm{p}}_{\infty}(3-\tilde{x})) - \bar{\kappa}^{\mathrm{p}}\cosh(\nu^{\mathrm{p}}_{\infty}(2-\tilde{x}))}{\bar{\sigma}^{\mathrm{p}}\bar{\kappa}^{\mathrm{p}}\sinh(\nu^{\mathrm{p}}_{\infty})\nu^{\mathrm{p}}_{\infty}}.
\end{aligned}$$

INTERFACIAL LITHIUM FLUX: The high-frequency gains are:

$$\begin{aligned}
\frac{I^{\mathrm{n}}_{\mathrm{f+dl}}(\tilde{x},\infty)}{I_{\mathrm{app}}(\infty)} &= \frac{\nu^{\mathrm{n}}_{\infty}[\bar{\sigma}^{\mathrm{n}}\cosh(\nu^{\mathrm{n}}_{\infty}\tilde{x}) + \bar{\kappa}^{\mathrm{n}}\cosh(\nu^{\mathrm{n}}_{\infty}(\tilde{x}-1))]}{(\bar{\kappa}^{\mathrm{n}}+\bar{\sigma}^{\mathrm{n}})\sinh(\nu^{\mathrm{n}}_{\infty})} \\
\frac{I^{\mathrm{p}}_{\mathrm{f+dl}}(\tilde{x},\infty)}{I_{\mathrm{app}}(\infty)} &= -\frac{\nu^{\mathrm{p}}_{\infty}[\bar{\sigma}^{\mathrm{p}}\cosh(\nu^{\mathrm{p}}_{\infty}(3-\tilde{x})) + \bar{\kappa}^{\mathrm{p}}\cosh(\nu^{\mathrm{p}}_{\infty}(2-\tilde{x}))]}{(\bar{\kappa}^{\mathrm{p}}+\bar{\sigma}^{\mathrm{p}})\sinh(\nu^{\mathrm{p}}_{\infty})}.
\end{aligned}$$

INTERFACIAL FARADAIC LITHIUM FLUX: The gains are:

$$\frac{I^{\mathrm{r}}_{\mathrm{f}}(\tilde{x},\infty)}{I_{\mathrm{app}}(\infty)} = \frac{\bar{R}^{\mathrm{r}}_{\mathrm{dl}}}{\bar{R}^{\mathrm{r}}_{\mathrm{ct}}+\bar{R}^{\mathrm{r}}_{\mathrm{dl}}}\frac{I^{\mathrm{r}}_{\mathrm{f+dl}}(\tilde{x},\infty)}{I_{\mathrm{app}}(\infty)}.$$

INTERFACIAL DOUBLE-LAYER FLUX: The gains are:

$$\frac{I^{\mathrm{r}}_{\mathrm{dl}}(\tilde{x},\infty)}{I_{\mathrm{app}}(\infty)} = \frac{\bar{R}^{\mathrm{r}}_{\mathrm{ct}}}{\bar{R}^{\mathrm{r}}_{\mathrm{dl}}+\bar{R}^{\mathrm{r}}_{\mathrm{ct}}}\frac{I^{\mathrm{r}}_{\mathrm{f+dl}}(\tilde{x},\infty)}{I_{\mathrm{app}}(\infty)}.$$

3
Model Parameter Estimation

To use a PBM in BMS algorithms, we must estimate the values of its parameters so that the model's predictions closely match the dynamics of the physical cells that will be managed by the BMS. Two distinct approaches can be taken to find these estimates:

1. We may disassemble a cell and directly measure some values. We can also design experiments that use specialized laboratory equipment applied to subcomponents of a disassembled cell to estimate other values. Any such parameter-estimation methods that require cell disassembly are called teardown approaches.
2. We may subject an intact cell to carefully crafted excitation and regress the cell's response to the model's parameters to estimate their values. For this second approach even to be mathematically sensible, it is necessary to eliminate redundant parameters from the model equations as we did in Chap. 1 when we created the LPM. It is also helpful to develop alternate forms of the model equations against which to regress experimental data as we began to do in Chap. 2 by deriving an impedance model of the cell.

Considering the roadmap of Fig. 3.1, we are now well prepared to proceed to the third step—model parameter estimation—which is the focus of this chapter. We begin by discussing the relative advantages and disadvantages of teardown versus nonteardown approaches to parameter estimation. We then explore how well we might expect to

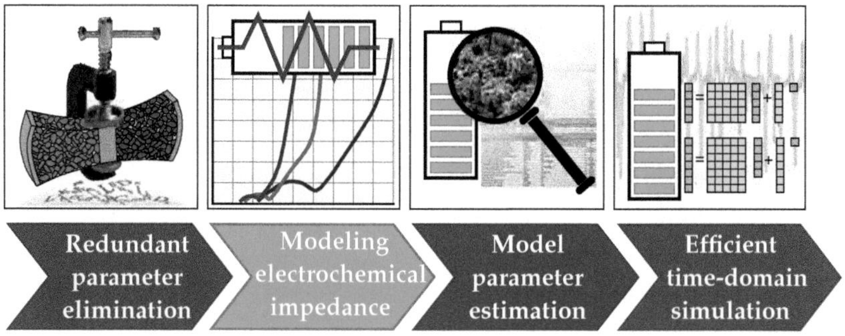

Figure 3.1: Topics in lithium-ion cell modeling that we cover in this volume.

estimate parameter values without cell disassembly, since there are benefits to avoiding teardown.

Once a PBM is fully parameterized it may then be used for many purposes. In this book, Chap. 4 shows how to convert the fully parameterized FOM to ROMs for efficient time-domain simulation. Chaps. 5–8 then show how to use these ROMs in BMS algorithms.

3.1 *Teardown versus nonteardown approaches*

3.1.1 *Parameter estimation via cell teardown*

The teardown approach to parameter estimation begins by disassembling the cell. This must be done in a glovebox in an inert environment (e.g., argon gas) to avoid unintended reactions between internal cell components and elements in the atmosphere and to avoid exposing the scientist performing the teardown to potentially toxic chemicals in the cell. Fig. 3.2 shows this being done for a prismatic cell, where the operator is unwrapping the negative electrode.

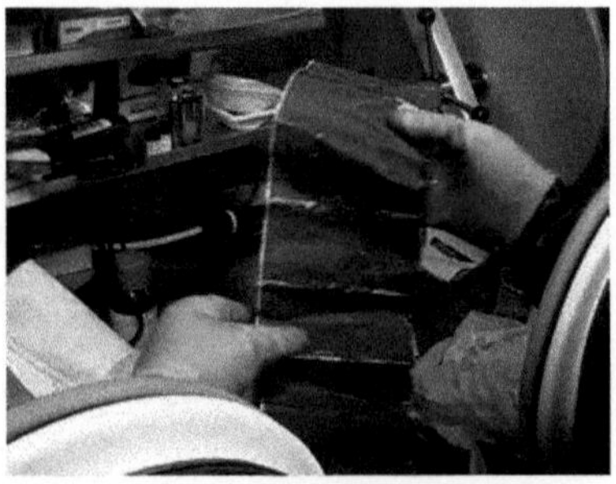

Figure 3.2: Opening a commercial prismatic cell in a glovebox, preparing to make coin cells of electrode material versus lithium. The negative electrode is shown. (Thanks to Drs. Kandler Smith, Ying Shi, Lei Cao, and Shriram Santhanagopalan of the National Renewable Energy Laboratory for this photo and for fabricating the coin cells.)

Certain measurements can be made directly. Current-collector area A can be found by measuring the current collector's length and width using a ruler and then computing their product. A micrometer can be used to determine electrode and separator thicknesses L^{r}. Samples of the electrodes may be imaged using a scanning electron microscope (SEM) to estimate particle sizes $R_{\mathrm{s}}^{\mathrm{r}}$. A focused ion beam (FIB) can be used to slice through the electrodes to make cross-sectional images from which porosity $\varepsilon_{\mathrm{e}}^{\mathrm{r}}$ can be estimated.

Other parameters require much more elaborate procedures.[1] Electrode porosity can also be estimated using mercury porosimetry. Electrode diffusivity can be estimated by forming half cells of electrode material versus lithium and conducting galvanostatic intermittent titration technique (GITT) experiments, which subject the half cells to pulses of current and measure the voltage response versus time. These voltage responses are regressed against a simplified model, producing estimates of $D_{\mathrm{s}}^{\mathrm{r}}$.

Electrode composition can be estimated using inductively coupled plasma–optical emission spectrometry (ICP–OES), which determines the relative fractions of different atomic elements present in a sample. For example, this can discover whether the positive-electrode material is NCA or NMC (or others), and what ratios of nickel, cobalt, manganese, and aluminum are present. Values of $c_{\mathrm{s,max}}^{\mathrm{r}}$ can be found experimentally but are usually adopted from the literature for the electrode composition that is detected. The volume fraction of the active material $\varepsilon_{\mathrm{s}}^{\mathrm{r}}$ can be found by comparing the measured cell capacity to the maximum capacity of the electrode, which itself is

[1] This is a very active research field. Some modern references include:

- Laura Oca, Eduardo Miguel, Eneko Agirrezabala, Alvaro Herran, Emanuele Gucciardi, Laida Otaegui, Émilie Bekaert, Aitor Villaverde, and Unai Iraola, "Physico-chemical parameter measurement and model response evaluation for a pseudo-two-dimensional model of a commercial lithium-ion battery," *Electrochimica Acta*, 382:138287, 2021.
- Johannes Schmalstieg, Christiane Rahe, Madeleine Ecker, and Dirk Uwe Sauer, "Full cell parameterization of a high-power lithium-ion battery for a physico-chemical model: Part I. Physical and electrochemical parameters," *Journal of The Electrochemical Society*, 165(16):A3799, 2018.
- Johannes Schmalstieg and Dirk Uwe Sauer, "Full cell parameterization of a high-power lithium-ion battery for a physico-chemical model: Part II. Thermal parameters and validation," *Journal of The Electrochemical Society*, 165(16):A3811, 2018.

computed assuming that the active material occupies the entire volume. Interfacial surface area can be estimated if we assume that all particles are spherical via $a_s^r = 3\varepsilon_s^r / R_s^r$.

Electrode OCP as a function of lithiation levels can be estimated using methods such as those presented in Sect. 3.2, where the absolute lithiation levels for any particular OCP values can also be found by determining the ratio of Li to other elements using ICP–OES or X-ray methods applied to electrode materials evaluated at different resting potentials. By comparing electrode OCP to OCV, estimates of θ_0^r and θ_{100}^r can also be found, much like as discussed in Sect. 3.5.

Exchange-current density and double-layer capacitance are difficult to measure directly. Instead, half cells or full cells are tested using electrochemical impedance spectroscopy (EIS). Parameter values for a simplified electrode model are fit to the EIS data. Effective electrode conductivity σ_{eff}^r is also difficult to measure since making half cells of electrode material substantially modifies contact resistance, which is part of the measured electrode resistance. Since electrode conductivity is generally much higher than electrolyte conductivity, it is often claimed that accurate values of σ_{eff}^r are not critical to be able to model a lithium-ion cell reasonably well, and so values of σ_{eff}^r that are representative of the electrode active materials discovered to be present in the cell are then adopted from the literature.

Lithium-ion cells tend to contain very little electrolyte, so it is extremely difficult to identify electrolyte properties accurately. Centrifuge processes can be successful in extracting a sufficient volume to perform some analysis. Gas chromatography can be applied to analyze the components of the extracted electrolyte; then, additional electrolytes having the same solvent composition and varying salt concentration can be mixed for more detailed analysis. Conductivity meters can be used to measure conductivity κ as a function of temperature and lithium-ion concentration and the Nernst–Einstein relationship can be used to estimate diffusivity via:

$$D_e = \frac{\kappa RT}{F^2 c_e}.$$

Note that conductivity and diffusivity measured this way are bulk values and must be converted to estimates of the effective values, for example by the Bruggeman relationship (see Chap. 4 of Vol. I):

$$\kappa_{\text{eff}} \approx \kappa \varepsilon_e^{\text{brug}} \quad \text{and} \quad D_{e,\text{eff}} \approx D_e \varepsilon_e^{\text{brug}}.$$

The reconstituted electrolytes can also be used in experiments to determine transference number t_0^+ and the electrolyte mean molar activity term $\partial \ln f_\pm / \partial c_e$, but these values are also often simply adopted from the literature for an electrolyte deemed to be sufficiently similar to the one found in the cell.

Other methods are outlined in the literature. A summary observation is that specialized (often very expensive) equipment and highly trained scientists are needed. Example experiments used to determine model parameters via direct measurement include FTIR, GC–MS, NMR, XRD, ICP–OES, TGA, DLS, PEIS, SSPP, PITT, SEM–EDS, pycnometry, and Hg-porosimetry.[2] The equipment to perform these experiments is out of reach of many engineering teams desiring to work with PBMs. Further, the DFN model often used as a basis for PBMs is a homogenization of a microscale model, which is itself a homogenization of an atomistic-scale model; therefore, true values do not necessarily exist for all parameters in the P2D model (e.g., electrode particles are not perfect spheres of uniform radius, current density is not uniform across all 1D cross sections of the 3D electrode, and so forth). Experimental errors on the individual parameter values tend to add together when the entire PBM is evaluated. For these reasons, the final model still requires tuning of the parameter values to fit observed behaviors. Also, in many cases cell-supplier legal agreements will prohibit cell teardown, and so alternate means to determine the PBM parameter values have appeal. Nonteardown methods provide such alternate means.

[2] See Oca et al. in sidenote 1 for definitions of these terms and how they apply to the parameter-estimation problem.

3.1.2 *Parameter estimation without cell teardown*

The nonteardown approach to parameter estimation stimulates an intact physical cell electrically and/or thermally and measures its response. It then applies the same input stimulus to the cell model and computes the model's predicted response. The output measurements from the physical cell are compared to the model predictions and the parameters of the model are adjusted to make the model predictions match the physical measurements as closely as possible.

With this background, we can say that the nonteardown approach has three main components, as illustrated in Fig. 3.3:

1. A *model* that describes how the cell is expected to respond to an experiment,
2. *Measurements* from the experiment on a physical cell, and
3. An *optimization method* that adjusts the model's parameter values to maximize agreement between the model and the measurements.

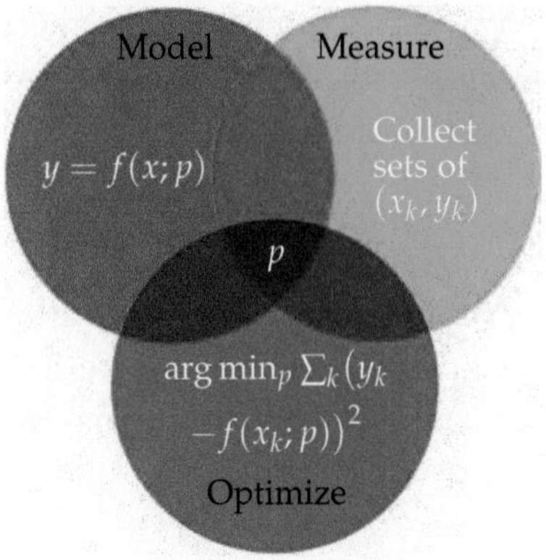

Figure 3.3: Three components comprising generic parameter estimation.

We illustrate the nonteardown parameter-estimation process with a simple example, not directly related to the PBM parameter-estimation problem. In this example, we assume that some system has input x and output y, which are related via $y = mx + b$. We wish to use this equation along with measurements of input/output pairs (x, y) to estimate the parameters m and b of the model.

Considering the three components of Fig. 3.3, the model is simply $y = mx + b$. The experiment applies different input conditions x and measures the corresponding outputs y. The optimization adjusts values of the parameter set $p = \{m, b\}$ such that the model predictions fit the data as well as possible.

For an affine model—such as the one considered in this example—and a least-squares cost function, there are clever ways to compute the optimal value of the model's parameters in closed form.[3] This is not possible for most parameter-estimation problems, so we present a more general optimization-based strategy here. As an example, our objective might be to determine the values for the parameters m and b that minimize the root-mean-squared (rms) difference between the model predictions and the measured data.

[3] This is known as the linear regression problem and is heavily studied in many disciplines.

The example MATLAB code presented below illustrates this process. Noisy "measured" x- and y-coordinate data are first generated synthetically using a random-number generator. A model equation is fit to the data using MATLAB's `fminsearch` method from its optimization toolbox to minimize a cost function that is defined as the rms difference between model and data. The result is a 2-vector `pOpt` where the first component of `pOpt` is the estimated slope m and the second component is the estimated y-intercept b.

```
% First, generate some data to use in the identification
m = 0.1; b = 0.2; % true slope and y-intercept
x = rand(100,1);  % 100 random x locations
y = m*x + b + 0.01*randn(size(x)); % noisy "measured" data
plot([0 1],[b m+b],x,y,'.','markersize',15); hold on

% Now, regress the data to a line "the hard way" (instead of using known
% optimal ways for fitting a line, assume we need to do nonlinear
% optimization)
cost = @(p) rms(p(1)*x+p(2) - y); % "cost" to minimize
pOpt = fminsearch(cost,[0 0]); % init mhat=bhat=0; then seek best fit

plot([0 1],[pOpt(2),pOpt(1)+pOpt(2)]); grid on
xlabel('x coordinate'); ylabel('y coordinate');
title('Parameter-estimation example');
legend('True line','"Measured" data','Estimated line')
```

A sample result from this process is illustrated in Fig. 3.4. The blue line plots $y = 0.1x + 0.2$, which is the relationship that we are seeking to find; the red markers show the noisy (x, y) measured data pairs; and the yellow line is the relationship found by this optimization. We observe that the line fit is quite good, but that noise on the data in this example makes it impossible to guarantee a perfect fit. The estimate of m is 0.1008 rather than the true value of 0.1; the estimate of b is 0.2002 rather than the true value of 0.2. In the more general problem of optimizing parameter values for a PBM, we will also converge to imperfect estimates due to measurement noise.

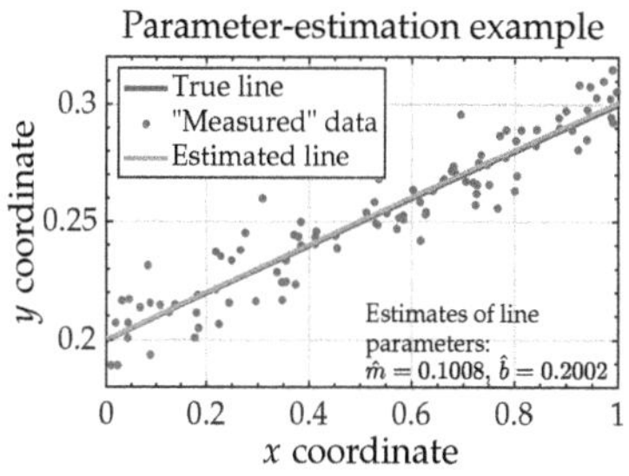

Figure 3.4: Optimizing the fit of a line to noisy data.

The primary purpose of this example is to clarify that nonteardown parameter estimation is merely an application of nonlinear optimization methods and theory. The concept is fundamentally simple but there are hazards to applying the theory without giving sufficient consideration to known risk factors relating to nonlinear optimization. For example, a model is said to be nonidentifiable if there exist two different sets of model parameter values that give the same input/output function; we showed in Chap. 1 that the DFN model is nonidentifiable in this sense.[4,5] This is why we created the LPM for which it is mathematically possible, in principle, to estimate all parameter values from input/output data.

[4] This fact is not widely understood and we still see many papers where authors attempt to estimate the entire set of DFN parameter values from nondestructive tests. It simply is not mathematically possible to do so.

[5] In other words, for a model to be identifiable, there should be a unique globally optimal solution when regressing measured input/output data to the parameter set. If the DFN model were identifiable, then an identified set of parameter values would lead to a unique cell design since the DFN parameters include dimensional quantities. In reality, this is clearly not true because DFN models for cells having different sizes can still produce identical input/output responses as long as some material properties are scaled by cell dimensions. The LPM eliminates the dimensional considerations and the resulting lumped parameters simply describe the scaled electrochemical properties of certain sets of cell materials.

It is also important to give thought to the cost function we are seeking to minimize. In the example just considered, the cost function was the rms error between the model of a line and measured data points. It turns out that this combination of model and cost function has a property known as convexity, which implies that there is a single minimum to the cost that also corresponds to the optimal solution. The cost function for this problem is illustrated in Fig. 3.5 as a function of estimates of m and b. It is not difficult for an optimization method to find the best solution; even simple methods such as gradient descent will converge to the correct values.

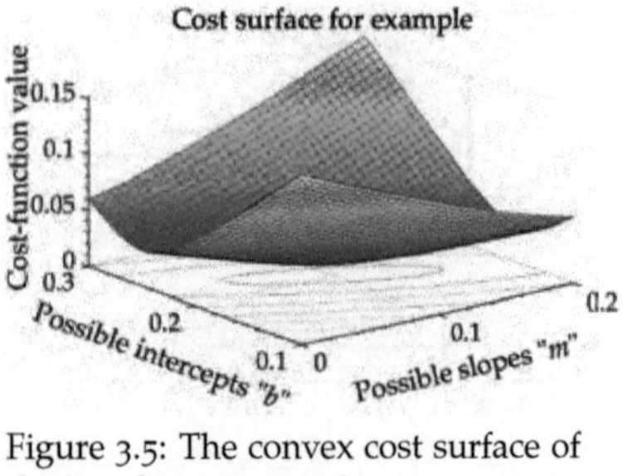

Figure 3.5: The convex cost surface of the line-fitting example.

More complicated models—such as the LPMs we are seeking to parameterize—lead to nonconvex cost functions which have multiple local minima, as illustrated for a generic example in Fig. 3.6. No known optimizing method (except exhaustive search, which is impractical) is guaranteed to find the globally optimal solution for a nonconvex cost function. Instead, all optimization algorithms will converge to solutions that are only locally optimal but globally suboptimal (and hence not strictly correct). This is a factor of which we must be aware when using optimization to estimate parameter values: the solutions will certainly be at least somewhat wrong; however, they may be close enough to predict cell internal electrochemical variables and cell voltage quite well. In order to find a good solution to the optimization, it is helpful:

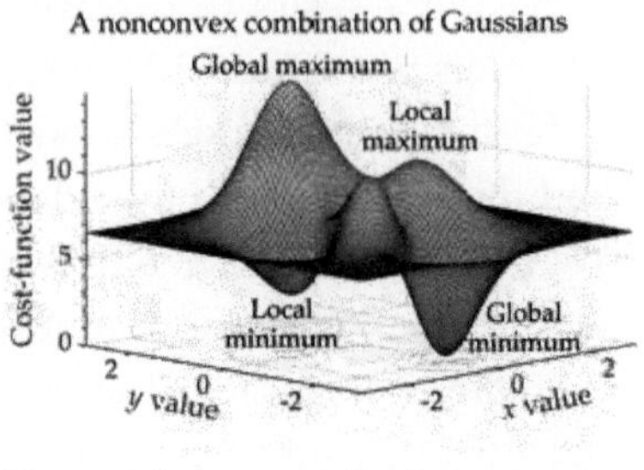

Figure 3.6: An example of a nonconvex cost surface, which is more typical of the problems we seek to solve in this chapter.

- To have initial guesses for the parameter values that are close to their true values (so that the optimization is more likely to converge to the correct minima);
- To have relatively tight bounds on possible values for the parameters (so the optimization does not have to search over a large number of local minima);
- To select a good optimization algorithm that is able to escape poor local minima while seeking better near-global minima;
- To use an optimization cost function that gives weight to measured data samples according to their quality and importance.

A lot can be learned by studying the field of numeric optimization, but much research remains to be done on this topic.

Another factor that we will need to keep in mind when seeking to estimate LPM parameter values is the possibility of model mismatch. That is, it is possible (and indeed essentially certain) that our LPMs do not perfectly describe the dynamics of a physical lithium-ion cell. Because of this, there is no set of parameter estimates that will describe the cell exactly. Various assumptions were made in Vol. I when deriving the microscale model and additional assumptions were made when applying volume averaging to homogenize the microscale to develop a continuum-scale model.[6] Also, we can be certain that some (hopefully minor) physical phenomena were overlooked when developing the LPM, and that some of our model equations are simply not quite right.

Fig. 3.7 illustrates model mismatch applied to the data of the prior example. The true model remains $y = 0.1x + 0.2$. However, we attempt to fit a quadratic function to the same data. The yellow line shows a fit of the form $\hat{y} = p_1 x^2 + p_2 x + p_3$. We see that the optimized solution does not produce $p_1 = 0$, nor does it find $p_2 = 0.1$ or $p_3 = 0.2$. Therefore, we can state that the solution is *wrong* because it does not match the truth. However, despite being wrong, it is still *useful* since the quality of its predictions are quite good as can be seen by comparing the yellow and blue lines visually. The same will be true with our parameter-estimation results. We will never find exactly the correct parameter values for the LPM; however, the model that we do find may still have the ability to estimate cell internal electrochemical variables and cell voltage very well.[7] That is, the model will certainly be wrong, but it may still be very useful.

Finally, we must also consider that our laboratory-test setup will always be imperfect. Measured data will include biases, nonlinear errors, random noises, and even errors that are correlated over time.[8] Our parameter-estimation methods must carefully calibrate raw test results to clean up measured data as much as possible before optimizing parameter values.

3.1.3 *A balanced approach*

It should be evident by now that there are advantages and disadvantages to both the teardown and nonteardown approaches to estimate LPM parameter values. Teardown methods are very good for measuring dimensional quantities; nonteardown methods tend to be better for estimating effective lumped transport parameters.

So, perhaps we should do both. However, notice that the process for converting the original DFN model to the LPM in Chap. 1 elimi-

[6] For example, the 1D DFN model assumed that all electrode particles were spherical; it assumed that all parameters were uniform across a given cell region; it assumed that the current distribution throughout the cell's current collectors is uniform (which allows us to assume that a pseudo-two-dimensional model can adequately capture cell behaviors). None of these assumptions are true in practice.

We believe that with careful cell design, uniform current densities across a cell's current collector can be approached and that the LPMs we have developed are quite good. However, they are not perfect.

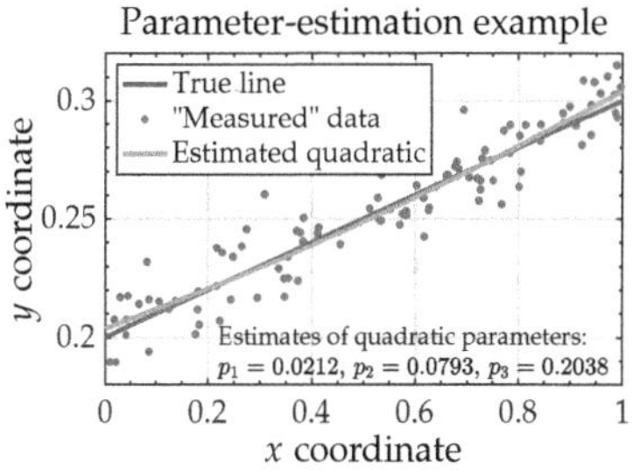

Figure 3.7: An example of model/data mismatch.

[7] Note the distinction between cell internal electrochemical *variables* and cell-model *parameters*. Variables are physical quantities that can change rapidly as a cell operates. Parameters are values that quantify a cell model, and are either constant or evolve very slowly as a cell ages. Example variables include ϕ_s, ϕ_e, θ_s, θ_e, and i_{f+dl}. Example parameters include $\bar{D}_s^r$, $\bar{R}_f$, and Q.

[8] For example, due to mutual inductance in measurement cables that connect the test equipment to the cell under test.

nated all dimensional parameters. Therefore, it might seem like cell teardown can be avoided entirely. But later in this chapter we will encounter fundamental obstacles to determining unique electrode OCP functions without performing at least a minimal cell teardown.

Our observation is that the research topic of parameter estimation for LPMs is developing very rapidly. At this point in time, our recommendation is to perform teardown—if permitted by cell-manufacturer agreements—to find the values most easily determined by those methods; that is, to find dimensional quantities (which are needed if you desire to convert LPM parameters to standard parameters) and to determine the electrode OCP functions. We recommend using nonteardown methods to estimate all remaining parameter values. More complex experiments based on teardown could be performed to augment the nonteardown methods to validate estimates made without teardown. Also, if cell disassembly is not permitted by the cell manufacturer and so must be avoided entirely, we present some approaches in this chapter to estimating all LPM parameters, including electrode OCP functions, that do not require teardown.

So, since teardown can largely be avoided and since we assume that most readers of this book will have a stronger background in engineering topics than in materials-science measurement techniques, the focus of the remainder of this chapter is on presenting our best understanding at this point in time regarding how to estimate LPM parameter values with minimal or no reliance on cell teardown.

Fig. 3.8 illustrates the strategy that we will present. Six different laboratory tests are used to collect data that are regressed to specific subsets of the overall set of parameters in the LPM. By optimizing several smaller groups of parameter values instead of attempting to optimize all parameter values at the same time, we reduce the dimensionality of the individual optimization problems. This improves our chances of finding a good local minimum in the optimization cost function, resulting in a good parameterization of the overall model. The OCP test, presented in Sect. 3.2, estimates the negative-electrode OCP function and produces a preliminary estimate of the positive-electrode OCP function. The OCV test, in Sect. 3.5, determines cell capacity, OCV function, and the negative-electrode operating boundaries. It also produces a preliminary estimate of the positive-electrode operating boundaries. The discharge test, in Sect. 3.9, calibrates the positive-electrode OCP function and boundaries and produces an initial estimate of the solid diffusivities of each electrode. The pulse test, in Sect. 3.12 determines the values of all parameters used in the ϕ_s^r, ϕ_e^r, and i_{f+dl}^r equations. The EIS test, in Sect. 3.15, estimates all remaining parameter values except $\bar{\psi}$. Finally, the pseudo-steady-state (PSS) test, in Sect. 3.17, estimates $\bar{\psi}$.

Laboratory test   Final parameter estimate   Reference

Order of data processing

OCP test $U_{ocp}^n(\theta_s^n),\ \{U_j^0, X_j, \omega_j\}^n$ $\hat{U}_{ocp}^p(\hat{\theta}_s^p)$

OCV test $Q,\ U_{ocv}(z),\ \theta_0^n,\ \theta_{100}^n$ $\hat{\theta}_0^p,\ \hat{\theta}_{100}^p$

Discharge test $U_{ocp}^p(\theta_s^p),\ \{U_j^0, X_j, \omega_j\}^p,\ \theta_0^p,\ \theta_{100}^p$ $\hat{D}_{s,ref}^r$

Pulse test $\alpha^r,\ \bar{\kappa}^r,\ \bar{\sigma}^r,\ \bar{R}_f^r,\ \bar{R}_{dl}^r$ $\hat{k}_0^r$

EIS test $\bar{D}_{s,ref}^r,\ n_f^r,\ \bar{C}_{dl}^r,\ n_{dl}^r,\ \bar{k}_0^r$ $\bar{q}_e^r/\bar{\psi},\ \bar{\kappa}_D/\bar{\psi}$

PSS test $\bar{q}_e^r,\ \bar{\kappa}_D,\ \bar{\psi}$

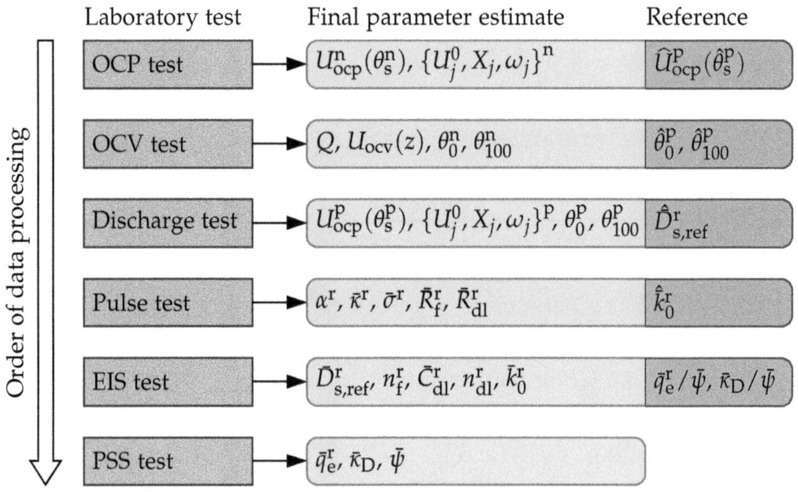

Figure 3.8: Parameter-estimation strategy, indicating which final parameter values and which preliminary reference values are estimated using data from each test.

3.2 OCP testing

The first step in the strategy of Fig. 3.8 is to estimate the electrode OCP relationships $U_{ocp}^r(\theta_s^r)$ as functions of local stoichiometry θ_s^r.[9] These relationships are important for developing good SOC estimators, for diagnosing and modeling some kinds of cell aging mechanisms, and for creating an accurate dynamic model of the lithium-ion cell that is not biased by deterministic dc errors.

The most direct approach to finding $U_{ocp}^r(\theta_s^r)$ for both electrodes requires cell teardown: we discuss this first. In some cases it is possible to avoid teardown: we will discuss that scenario in Sect. 3.7.

Although we have seen that applying teardown to estimate cell parameter values in general can be very complicated, it turns out that estimating electrode OCP using teardown is fairly straightforward. The cell is first opened in a glovebox in an inert environment, as illustrated in Fig. 3.2. Usually, we find that the current collectors are coated on both sides with electrode materials: these materials must be removed (scraped off and/or dissolved) from one side of the current collector to make a bare copper or aluminum surface for electrical contact. A circular punch is used to make an electrode disk which is assembled into a coin cell versus lithium metal foil, as illustrated in Fig. 3.9. Once electrolyte is added, the coin cell is sealed and can be removed from the glovebox.

The total capacity of such a coin cell will be small, on the order of a few mAh. Not all cell-test equipment can reliably produce currents small enough to meet the OCP-test low-C-rate criteria to be presented shortly, and so we often wire several coin cells in parallel to increase the total capacity of the grouping, as illustrated in Fig. 3.10. This also

[9] The content of this section has been adapted from: Dongliang Lu, M. Scott Trimboli, Guodong Fan, Ruigang Zhang, and Gregory L. Plett, "Implementation of a physics-based model for half-cell open-circuit potential and full-cell open-circuit voltage estimates: part I. Processing half-cell data," *Journal of The Electrochemical Society*, 168(7):070532, 2021.

Coin-cell lid

Spring

Spacer

Lithium foil

Separator

Electrode material (coated on one side)

Coin-cell base

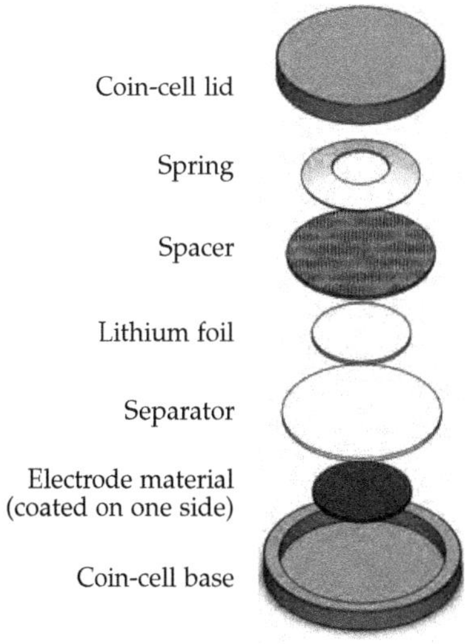

Figure 3.9: Components of a lithium metal versus electrode material coin cell used for OCP testing.

electrically averages the behaviors of several samples together, which may produce more consistent results (especially considering that these half-cell electrodes were harvested by hand).

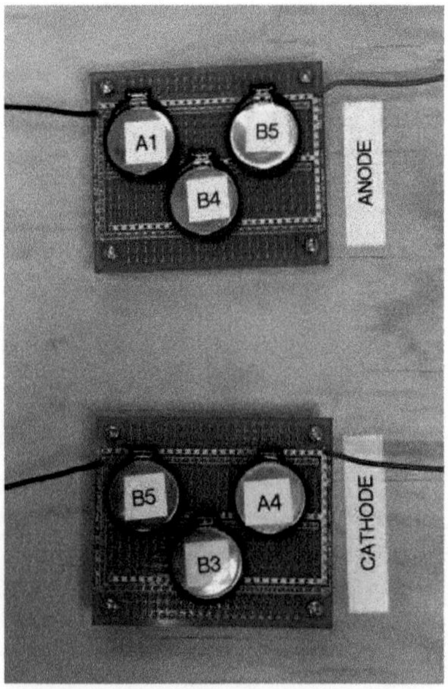

Figure 3.10: Wiring coin cells in parallel for higher overall capacity.

3.2.1 *Half-cell testing to determine electrode OCP relationship*

A coin cell constructed in this way has a single insertion electrode that cycles versus metallic lithium and not two insertion electrodes that cycle versus one other. Since only one of the electrodes has an intercalation property, we refer to the coin cell as a half cell.[10] Lithium metal has resting potential of 0 V versus Li/Li^+, so the resting half-cell voltage equals its insertion electrode's OCP at its present θ_s versus Li/Li^+. By estimating the OCV of the half cell, we actually estimate the OCP of the insertion-electrode material in the half cell.

The OCV of the half cell is measured in the same way as cell OCV was measured in Vol. I of this series. The following sections review the process and describe some improvements that have been made to the data-processing methods since that volume was published. We use four lab-test scripts, executed sequentially, to collect a half-cell's voltage across as much of its stoichiometry range as possible. The half cell is very slowly discharged, then very slowly charged while measuring its terminal voltage and accumulated ampere hours. Since the half cell has electrical and electrochemical resistance, its OCP must lie above the discharge and below the charge voltages for every value of θ_s. Roughly speaking, we will average the discharge and charge curves to find OCP.[11] Tests are run at several temperatures to determine the temperature dependence of the U_{ocp} function.

Specifically, the following laboratory procedure is conducted for each temperature T to be modeled (where v_{min} and v_{max} are minimum and maximum cutoff voltages). Four test scripts are executed.

OCP test script 1 (at test temperature)

The first OCP test script determines low-rate constant-current discharge voltage as a function of ampere-hours discharged:

1. Soak the fully charged half cell at the test temperature T for at least two hours to ensure a uniform temperature throughout.
2. Discharge the half cell at a constant rate until its voltage equals v_{min} (e.g., the constant-current rate might be C/30 or C/100).

Note that half-cell terminal and resting voltages are not equal; if the half cell were allowed to rest following step 2, voltage would rebound above v_{min}. Therefore the OCP of the half cell at the test temperature T at the final moment of step 2 is known to be greater than v_{min}.

[10] That is, the term "coin cell" refers to the physical enclosure (e.g., in a 2032 format) whereas "half cell" refers to its generic schematic form as electrode versus lithium metal. The coin-cell form factor is not required for OCP testing; for example, it is also possible to build pouch-format half cells. So, we will use the more general "half cell" term in this chapter when not taking about a specific physical example.

[11] An alternate approach is to discharge and/or charge the half cell to a set of θ_s values, then allow it to rest until its voltage stabilizes. Voltages measured after resting are recorded as the OCP of the cell. An example of applying this approach to estimate full-cell OCV is reported in: Anup Barai, W. Dhammika Widanage, James Marco, Andrew McGordon, and Paul Jennings, "A study of the open circuit voltage characterization technique and hysteresis assessment of lithium-ion cells," *Journal of Power Sources*, 295:99–107, 2015.

An advantage of the dis/charge and rest (DR) approach versus the constant-current dis/charge (CC) approach we present here is that it eliminates linear polarization from the voltage, so the measurement is closer to the true OCP. However, it does not eliminate hysteresis, so there is still a difference between measured voltage and OCP. This can be overcome, in part, by

(continued on next page)

OCP test script 2 (at 25 °C)

Electrode OCP is temperature-dependent. Therefore, for the final OCP relationship at all temperatures to be consistent, we must calibrate the tests at a fixed temperature. We choose to do so at 25 °C since this is the default calibration temperature for cell-level OCV testing. The goal of test script 2 is to move the half cell's (resting) OCP to v_{min} at this calibration temperature. We perform these steps:

3. Soak the half cell at 25 °C for at least two hours to ensure a uniform temperature throughout the cell.
4. Dis/charge (i.e., either discharge or charge, as appropriate) the half cell to v_{min} at the same rate as used in step 2. Then, hold the half cell at that voltage for several hours.[12]

At the end of step 4, the OCP of the electrode under test in the half cell is as close to v_{min} as we are able to achieve at the calibration temperature in a lab setting. We will assume that the half cell's OCP is equal to v_{min} at this point although we realize that there will be some testing error introduced by this assumption.

OCP test script 3 (at test temperature)

Next, we determine the charge voltage as a function of ampere hours charged:

5. Soak the half cell at the test temperature T for at least two hours to ensure a uniform temperature throughout.
6. Charge the half cell at the same constant-current rate used in step 2 until the half-cell terminal voltage is equal to v_{max}.

Note again that terminal and resting voltages are unequal; if the half cell were allowed to rest following step 6, its voltage would relax below v_{max}. Therefore the half cell's OCP at the test temperature T at the final moment of step 6 is known to be lower than v_{max}.

OCP test script 4 (at 25 °C)

The final test script is implemented to calibrate the tests once again. The goal of test script 4 is to move the half cell (resting) OCP to v_{max} at this calibration temperature. We now:

7. Soak the half cell at 25 °C for at least two hours to ensure a uniform temperature throughout.
8. Dis/charge the half cell to v_{max} at the same rate as used in step 2. Then hold the half cell at that voltage for several hours.[13]

At this point, we assume that the electrode OCP is equal to v_{max}.

11(*cont.*) making measurements at the same θ_s values after charging and discharging, and averaging both of those measurements together.

We prefer the CC approach for three reasons: (1) we find that the rest duration required by the DR method before a stable voltage is attained can make the test slower than the CC approach (e.g., if a cell has very slow solid diffusion); (2) the CC approach provides voltage data over the entire SOC range instead of only at a few selected points, which ensures that no features of the OCP curve are missed; and (3) the CC approach enables direct computation of electrode differential capacity, which can be helpful to know.

12 We can also choose to perform one or more empirically motivated dither steps, much like we did when OCV testing in Vol. I. This helps to reduce the effect of hysteresis on the calibration of results. A voltage dither step at the end of step 4 for a positive-electrode half cell might look like the figure below:

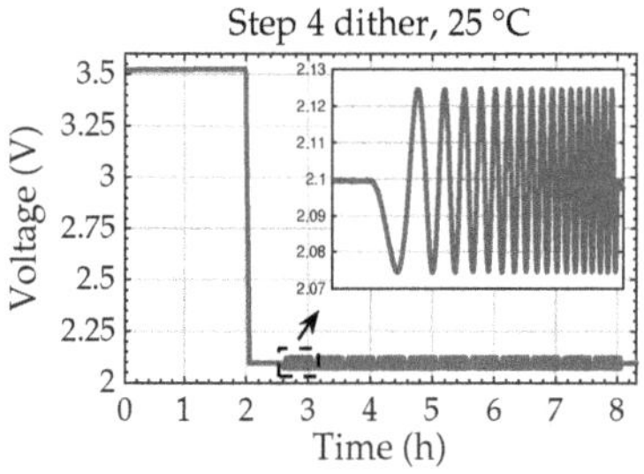

13 We can choose to perform one or more dither steps here as well. A voltage dither step at the end of step 8 for a positive-electrode half cell might look like the figure below:

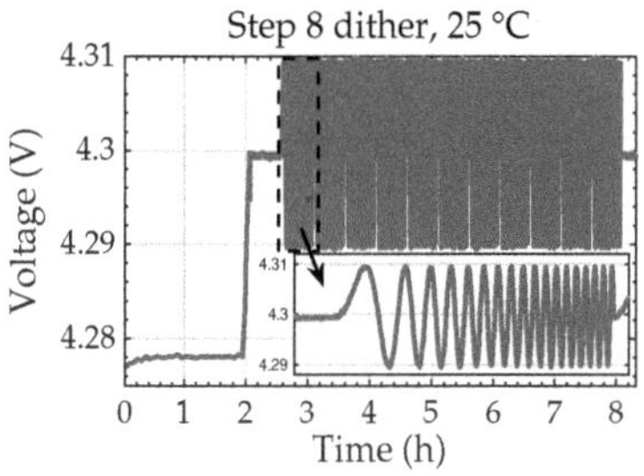

Script	Temp.	Step number and details
1	T	1. Rest ($\geq 2\,\text{h}$). 2. Discharge at a constant rate to v_{min}.
2	25 °C	3. Rest ($\geq 2\,\text{h}$). 4. Dis/charge to v_{min}; hold at (and/or dither around) v_{min}.
3	T	5. Rest ($\geq 2\,\text{h}$). 6. Charge at constant rate to v_{max}.
4	25 °C	7. Rest ($\geq 2\,\text{h}$). 8. Dis/charge to v_{max}; hold at (and/or dither around) v_{max}.

Table 3.1: Summary of the OCP test scripts. The constant-current rate used in steps 2 and 6 should be the same (e.g., C/30 or C/100).

The test steps are summarized in Table 3.1. During every step of the test, voltage, accumulated ampere-hours discharged, and accumulated ampere-hours charged are recorded periodically (e.g., once per second). Script 4 is run first to "fully charge" the half cell; then scripts 1–4 are run in sequence for a particular test temperature. Because a very low current rate is used, there is negligible heat generation in the cell, and we can consider all data points to be collected at the ambient test temperature for every script (but temperature data can be measured as well to verify this assumption).

3.2.2 *Initial data processing*

Measured data from the OCP test scripts must undergo initial processing to create calibrated discharge voltage and calibrated charge voltage versus some measure of electrode SOC. Ideally, this measure of SOC would be the *absolute stoichiometry* θ_s of the electrode. However, we do not yet know the absolute level of lithiation $\theta_s = \theta_{\text{min}}$ at the beginning of the test when $U_{\text{ocp}} = v_{\text{max}}$, nor do we know $\theta_s = \theta_{\text{max}}$ at the most lithiated point in the test when $U_{\text{ocp}} = v_{\text{min}}$. So, we temporarily define a *relative stoichiometry* $\tilde{\theta}_s = 0$ at the beginning of the test and $\tilde{\theta}_s = 1$ at the most lithiated point in the test and will later seek a relationship to convert between $\tilde{\theta}_s$ and θ_s.[14] Relative stoichiometry $\tilde{\theta}_s$ will be our measure of electrode SOC for now.

[14] Amplifying: θ_s (without the tilde symbol) is absolute stoichiometry and $\tilde{\theta}_s$ (with the tilde symbol) is relative stoichiometry.

There are some subtle points we must keep in mind when processing lab data if $T \neq 25\,°\text{C}$, since OCP is temperature-dependent:

- Prior to step 2, the half cell is resting with $\tilde{\theta}_s = 0$ but its voltage is no longer v_{max} since $T \neq 25\,°\text{C}$.
- At the end of step 2, $\tilde{\theta}_s \neq 1$ because the cell is not resting and $T \neq 25\,°\text{C}$, and the v_{min} specification that defines $\tilde{\theta}_s = 1$ applies only when the cell is resting at 25 °C. The actual value of $\tilde{\theta}_s$ may be above or below 1.0 at the end of step 2.
- Similarly, at the end of step 6, the half cell will have $\tilde{\theta}_s \neq 0$ because the v_{max} specification that defines $\tilde{\theta}_s = 0$ applies only when the

cell is resting at 25 °C. The actual value of $\tilde{\theta}_s$ may be above or below zero at the end of step 6.

- This is why we must execute test scripts 2 and 4, to ensure that the half cell has reached a calibrated point before starting script 3 and then script 1, respectively (for the next temperature to be tested).

These considerations require some careful processing of the data. We discuss processing data for test temperature $T = 25\,°\text{C}$ first. This is the easiest case because all four scripts are then executed at 25 °C—no other temperatures are involved. Also, since calibration voltages v_{max} and v_{min} define $\tilde{\theta}_s = 0$ and $\tilde{\theta}_s = 1$ for 25 °C (only), the net ampere-hours discharged over all steps equals the relative capacity $\widetilde{Q}_h^r$ of the half cell over that voltage range.[15,16]

The net ampere-hours charged over all steps will be slightly higher than $\widetilde{Q}_h^r$ since the coulombic efficiency of a physical half cell is not ideal. We compute the coulombic efficiency at 25 °C as:

$$\eta(25\,°\text{C}) = \frac{\text{total ampere-hours discharged in all steps at } 25\,°\text{C}}{\text{total ampere-hours charged in all steps at } 25\,°\text{C}}.$$

At a test temperature $T \neq 25\,°\text{C}$, we must follow a different approach. We don't know the value of $\tilde{\theta}_s$ a priori at the end of steps 2 and 6; but, we assume that $\tilde{\theta}_s = 1$ at the end of step 4 and that $\tilde{\theta}_s = 0$ at the end of step 8. Knowing this, we compute the half cell's coulombic efficiency at temperature T:

$$\begin{aligned}\eta(T) = &\frac{\text{total ampere-hours discharged}}{\text{total ampere-hours charged at temperature } T}\\ &- \eta(25\,°\text{C})\frac{\text{total ampere-hours charged at } 25\,°\text{C}}{\text{total ampere-hours charged at temperature } T}.\end{aligned}$$

We can now evaluate $\tilde{\theta}_s$ for every data sample in the test. The depth-of-discharge (in ampere hours) at every point in time is calculated as:[17]

$$\begin{aligned}\text{DOD}(t) = &\text{ total Ah discharged until time } t\\ &- \eta(25\,°\text{C}) \times \text{total Ah charged at } 25\,°\text{C until } t\\ &- \eta(T) \times \text{total Ah charged at temperature } T \text{ until } t.\end{aligned}$$

Using this metric, the half-cell relative capacity $\widetilde{Q}_h^r$ over the OCP range spanning v_{min} to v_{max} (measured at temperature T) is equal to the depth-of-discharge at the end of step 4. Likewise, the value of $\tilde{\theta}_s$ corresponding to every data sample is then:

$$\tilde{\theta}_s(t) = \text{DOD}(t)/\widetilde{Q}_h^r.$$

As a check, the value of $\tilde{\theta}_s$ at the end of step 4 must be 1, and the value of $\tilde{\theta}_s$ at the end of step 8 must be zero.

[15] We use the term "relative capacity" to refer to the capacity of the half cell between $\tilde{\theta}_s = 0$ and $\tilde{\theta}_s = 1$.

[16] In the new notation, subscript "h" identifies this parameter as the *half-cell* capacity as opposed to a *full-cell* capacity. The superscript $r \in \{n, p\}$ refers to the electrode of the full cell that is being tested versus lithium metal in the half cell.

[17] Both the ampere-hour counter and time t are set to zero before executing step 1.

In what follows, we use the calibrated lithiation (discharge) voltage versus $\tilde{\theta}_s$ from step 2 (always plotted as a blue line) and the calibrated delithiation (charge) voltage versus $\tilde{\theta}_s$ from step 6 (always plotted as a red line) in our calculations to determine electrode OCP. Fig. 3.11 presents lab results that show the outcome of the process as described so far for graphite and NMC materials. The insets in the plots highlight regions of the calibrated discharge and charge curves that illustrate the "missing-data problem," described next.

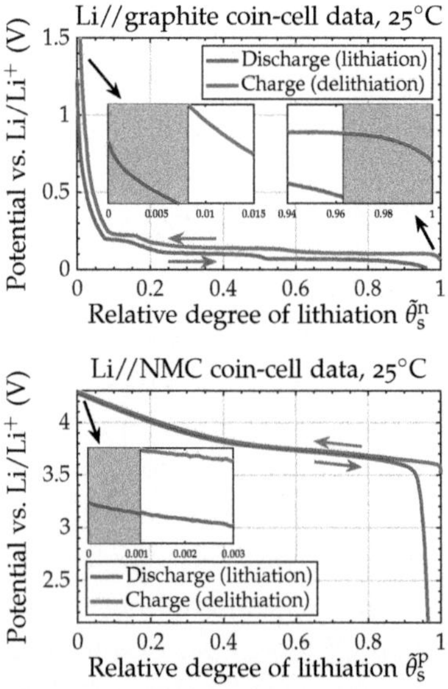

Figure 3.11: Calibrated discharge and charge voltages of two coin cells. Three of the four missing-data regions are shaded in gray in the plot insets.

3.2.3 *Overcoming three data-processing problems*

Data collected using this procedure can be processed further to produce good estimates of electrode OCP. However, there are several problems this processing must overcome.

The missing-data problem

First, we recognize that a half cell's $U_{\text{ocp}}(\theta_s)$ is an electrode equilibrium property as a function of stoichiometry θ_s (and temperature). But, voltages measured in steps 2 and 6 are not collected in equilibrium. At any point in time, we can model measured voltage as:

$$v(t) = U_{\text{ocp}}(\theta_s(t)) + \underbrace{\text{hysteresis} - \text{impedance} \times \text{current}}_{\text{polarization}}. \tag{3.1}$$

The difference between $v(t)$ and $U_{\text{ocp}}(\theta_s(t))$ depends on half-cell polarization—hysteresis and voltage drop due to linear impedance—which itself is dependent on the magnitude of the input current. Impedance is further a function of electrode stoichiometry $\theta_s(t)$. We recommend using a low rate for the constant-current discharge and charge steps to minimize polarization so that the system is in a quasi-equilibrium state at all times.[18] Even so, $v(t) \neq U_{\text{ocp}}(\theta_s(t))$ during constant-current discharge and charge steps 2 and 6.

As a result, constant-current-discharge step 2 never lithiates the half cell all the way to $U_{\text{ocp}} = v_{\text{min}}$ even though measured voltage reaches v_{min}. Therefore $\tilde{\theta}_s$ does not reach 1.0 in step 2. This is evident by examining the blue lines in Fig. 3.11. Similarly, constant-current-charge step 6 never delithiates the half cell all the way to $U_{\text{ocp}} = v_{\text{max}}$ even though measured voltage reaches v_{max}. Therefore, $\tilde{\theta}_s$ does not reach 0.0 in step 6. This can be seen in the red lines in Fig. 3.11.

The horizontal gaps between the blue and red lines near v_{min} and v_{max} illustrate the missing-data problem. For $\tilde{\theta}_s$ close to 1.0, we do not have discharge data; for $\tilde{\theta}_s$ close to 0.0, we do not have charge data. So, we cannot perform a direct average of discharge and charge data to find OCP as there are some values of $\tilde{\theta}_s$ for which we do not have both discharge and charge data to do so.[19]

[18] The voltage drop over a linear impedance is equal to the current magnitude multiplied by the impedance. Ideally, we would perform low-current tests to minimize the difference between $v(t)$ and $U_{\text{ocp}}(\theta_s(t))$ in Eq. (3.1). However, even for vanishing current, nonlinear hysteresis still exists in the measurements and so we will never achieve $v(t) = U_{\text{ocp}}(\theta_s(t))$ exactly, even after resting a cell until it achieves equilibrium.

The choice of using a C/30 rate in this chapter is a compromise between a desire to minimize the current magnitude and practical test-duration considerations. With this selection, we will later observe nonrandom patterns in the residuals due to unmodeled polarization. We often use rates as low as C/100 when seeking very-high-fidelity models.

[19] Some researchers stretch the discharge and charge curves so that both span from $\tilde{\theta}_s = 0$ to $\tilde{\theta}_s = 1$. However, this is not correct and will degrade the OCP estimates. It is even possible when doing so for the stretched discharge and charge curves to intersect, such that the values of the discharge curve become greater than the values of the charge curve in some places!

The inaccessible-lithium problem

To understand the second data-processing problem, notice that we must select physically achievable values for $v_{\min}$ and $v_{\max}$. This is usually done with some knowledge of the materials involved.

For example, we don't choose $v_{\min} < 0$ since this will cause lithium plating. Also, low values of $v_{\min}$ may cause very high stresses and cracking for positive-electrode materials. So, we always constrain $v_{\min} \geq 0$ and for positive-electrode materials we use higher values (often around $v_{\min} \approx 2\,\text{V}$) based on experience.[20]

[20] When the material approaches full lithiation, its potential drops extremely quickly—we can use a voltage just under the dropoff point for $v_{\min}$.

Similarly, we know that negative-electrode materials are generally affixed to a copper current-collector foil. Copper corrodes in most electrolytes at potentials above about 3 V. Therefore, we should not charge a negative-electrode half cell above about $v_{\max} \approx 3\,\text{V}$ (lower voltages are more practical). Positive-electrode materials can have structural breakdown at high voltages, and most electrolytes presently in use decompose above about 4.3 V. Therefore, we select $v_{\max} \approx 1.5\,\text{V}$ for negative-electrode materials and $v_{\max} \approx 4.3\,\text{V}$ for positive-electrode materials.

Whatever value is chosen for $v_{\min}$, the electrode is never completely lithiated even when the half cell is discharged to $v_{\min}$. That is, absolute stoichiometry $\theta_s \neq 1$ at the end of step 4, even though we define relative stoichiometry $\tilde{\theta}_s = 1$ at that point. Likewise, not all lithium is removed from the active material when the half cell is charged to $v_{\max}$. That is, absolute stoichiometry $\theta_s \neq 0$ at the end of step 8, even though we define relative stoichiometry $\tilde{\theta}_s = 0$ there.

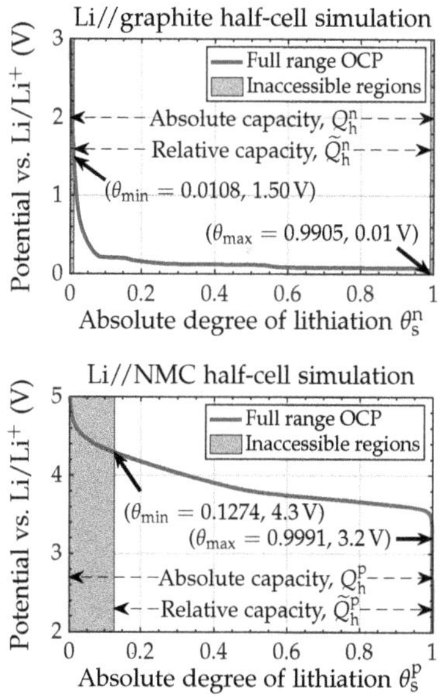

Figure 3.12: Illustrating the inaccessible-lithium problem.

As illustrated by the gray shaded regions in Fig. 3.12, there are portions of the active materials that are never accessed by the OCP tests since we cannot achieve either high-enough or low-enough voltages in a practical laboratory test. We refer to these unaccessed portions of the electrode as "inaccessible regions" and to this overall phenomenon as the "inaccessible-lithium problem."

This problem is why we must make a distinction between $\tilde{\theta}_s$ and θ_s and also why we make a distinction between relative and absolute capacities $\tilde{Q}_h^r$ and Q_h^r.[21] If we were somehow able to use the entire range of lithiation and delithiation, we could achieve absolute capacity Q_h^r. But we cannot, and so the quantity contained between cutoff voltages $v_{\min}$ and $v_{\max}$ is then a relative capacity $\tilde{Q}_h^r$.

[21] We use the term "absolute capacity" to refer to the capacity of the half cell between $\theta_s = 0$ and $\theta_s = 1$. Again, subscript "h" reminds us that this is *half-cell* capacity and not *full-cell* capacity. The superscript "$r \in \{n, p\}$" refers to the electrode of the full cell that is being tested versus lithium metal in the half cell.

The lab test measures only $\tilde{Q}_h^r$, not Q_h^r; so, the stoichiometric values computed from experimental data are also relative values, denoted as $\tilde{\theta}_s$. The corresponding voltage versus SOC relationships for the "absolute" and "relative" cases are written as $U_{\text{ocp}}(\theta_s)$ and $\tilde{U}_{\text{ocp}}(\tilde{\theta}_s)$, respectively.

Note that there is a one-to-one transformation between $\tilde{\theta}_s$ and θ_s, and hence $U_{\text{ocp}}(\theta_{\text{s}}) = \widetilde{U}_{\text{ocp}}(\tilde{\theta}_{\text{s}})$. Knowing that an OCP test allows us to find only $\widetilde{U}(\tilde{\theta}_{\text{s}})$, we must find a way to convert between $\tilde{\theta}_{\text{s}}$ and θ_{s}. The mapping between relative and absolute stoichiometries is:

$$\theta_{\text{s}} = \theta_{\text{min}} + \tilde{\theta}_{\text{s}}(\theta_{\text{max}} - \theta_{\text{min}}), \tag{3.2}$$

where θ_{min} and θ_{max} are the absolute stoichiometries when $U_{\text{ocp}} = v_{\text{max}}$ and $U_{\text{ocp}} = v_{\text{min}}$, respectively. Moreover, we will later find it helpful to be able to compute the electrode's differential capacity. The relationship between differential capacities under the two coordinate systems is (recalling that $U_{\text{ocp}}(\theta_{\text{s}}) = \widetilde{U}_{\text{ocp}}(\tilde{\theta}_{\text{s}})$):

$$\frac{\mathrm{d}\tilde{\theta}_{\text{s}}}{\mathrm{d}\widetilde{U}_{\text{ocp}}(\tilde{\theta}_{\text{s}})} = \frac{\mathrm{d}\theta_{\text{s}}}{\mathrm{d}U_{\text{ocp}}(\theta_{\text{s}})}\frac{\mathrm{d}\tilde{\theta}_{\text{s}}}{\mathrm{d}\theta_{\text{s}}} = \frac{1}{\theta_{\text{max}} - \theta_{\text{min}}}\frac{\mathrm{d}\theta_{\text{s}}}{\mathrm{d}U_{\text{ocp}}(\theta_{\text{s}})}.$$

Similarly, we can write:

$$\frac{\mathrm{d}\theta_{\text{s}}}{\mathrm{d}U_{\text{ocp}}(\theta_{\text{s}})} = (\theta_{\text{max}} - \theta_{\text{min}})\frac{\mathrm{d}\tilde{\theta}_{\text{s}}}{\mathrm{d}\widetilde{U}_{\text{ocp}}(\tilde{\theta}_{\text{s}})}.$$

Practical computation of differential capacity

In addition to the fundamental missing-data and inaccessible-lithium problems, we also encounter a practical data-quality problem. We wish to use differential-capacity estimates to identify equilibrium voltages and other electrochemical-reaction characteristics. However, as Fig. 3.13 illustrates, lab data will always contain measurement noise and analog-to-digital quantization errors:

- Quantization errors cause flat zero-change regions and result in divide-by-zero problems when evaluating $\mathrm{d}\tilde{\theta}_{\text{s}}/\mathrm{d}\widetilde{U}_{\text{ocp}}$ directly.
- Further, we know that the true OCP function and differential capacity should be monotonically decreasing ($\mathrm{d}\widetilde{U}_{\text{ocp}}/\mathrm{d}\tilde{\theta}_{\text{s}} < 0$ and $\mathrm{d}\tilde{\theta}_{\text{s}}/\mathrm{d}\widetilde{U}_{\text{ocp}} < 0$).[22] However, measurement noise can cause OCP to appear to increase with $\tilde{\theta}_{\text{s}}$ at some points.

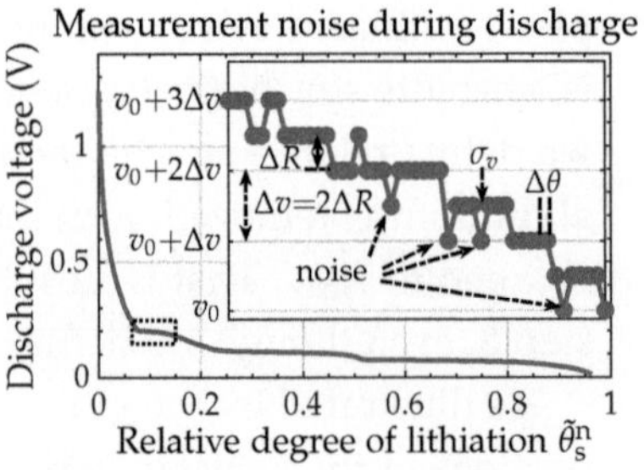

Figure 3.13: Zoom of discharge (lithiation) voltage curve showing measurement noise and analog-to-digital quantization errors.

[22] Most differential-capacity figures in this chapter are drawn with positive magnitudes and are denoted as "absolute differential capacity" for clearer presentation.

Smoothing/filtering algorithms can be used to preprocess lab-test data to mitigate these problems partially before evaluating the differentiation. However, established smoothing methods tend to cause under-fitting or over-fitting problems, thereby distorting the information of the actual cell behaviors.

In this section, we propose a simple histogram counting method, which can be programmed in MATLAB with only three lines of code. The general idea is to partition the entire voltage range into constant-width intervals and count the number of data samples within each interval to determine the relative capacity per window. A similar idea was proposed by Feng et al. and called the level evaluation analysis

(LEAN) method.[23] The histogram counting method is similar to the LEAN method but simpler to implement in MATLAB.

[23] Xuning Feng, Yu Merla, Caihao Weng, Minggao Ouyang, Xiangming He, Bor Yann Liaw, Shriram Santhanagopalan, Xuemin Li, Ping Liu, Languang Lu *et al.*, "A reliable approach of differentiating discrete sampled-data for battery diagnosis," *eTransportation*, 3:100051, 2020.

There are three steps. First, we select a voltage interval Δv, which determines the resolution of peaks in the differential-capacity result. As suggested by Feng et al., Δv should be k times the data sampling resolution ΔR, where k is an integer. To minimize fluctuation, Δv should also be greater than the magnitude of data-sampling error σ_v, yielding $(k-1)\,\Delta R \le \sigma_v < k \cdot \Delta R = \Delta v$.

After selecting the voltage window, the second step is to count the number of occurrences n_v of measured data points within each window. This can be achieved simply in MATLAB via `histcounts`.

The last step is to compute the stoichiometric interval spanned by each voltage window and from that the differential capacity. The change in relative stoichiometry $\Delta\tilde{\theta}_s$ across an interval Δv (for constant current) is equal to the number of data samples in that interval multiplied by the stoichiometric "width" of each data sample.[24] This "width" is $\max(\tilde{\theta}_s) - \min(\tilde{\theta}_s)$ for the discharge or charge segment being analyzed divided by the number of samples N across that entire segment. Therefore, for some interval centered at voltage v_0, we can write:

[24] Note that the histogram method assumes that the data being analyzed result from either constant-current charge or discharge, and so we expect that true voltage will be monotonically increasing or decreasing, respectively.

$$\begin{bmatrix} \Delta\tilde{\theta}_s \text{ across} \\ \text{an interval} \end{bmatrix} = \begin{bmatrix} \text{count of samples} \\ n_v \text{ in the interval} \end{bmatrix} \times \frac{\max(\tilde{\theta}_s) - \min(\tilde{\theta}_s)}{N}.$$

Then, (where $\max(\tilde{\theta}_s) - \min(\tilde{\theta}_s) < 1$ because of "missing-data"):

$$\left.\frac{d\tilde{\theta}_s}{d\tilde{U}_{ocp}}\right|_{v_0} \approx \begin{bmatrix} \text{count of samples } n_v(v_0) \\ \text{in the interval for } v_0 \end{bmatrix} \times \frac{\max(\tilde{\theta}_s) - \min(\tilde{\theta}_s)}{N\Delta v}.$$

In summary, we compute smoothed differential capacity by:

1. Specifying a constant voltage interval Δv (e.g., $\Delta v = 2\,\text{mV}$).
2. Determining the count of data samples, n_v, within each Δv (e.g., via `histcounts`).[25]
3. Computing the differential capacity via:

$$\left.\frac{d\tilde{\theta}_s}{d\tilde{U}_{ocp}}\right|_{v_0} \approx n_v(v_0)\frac{\max(\tilde{\theta}_s) - \min(\tilde{\theta}_s)}{N\Delta v}.$$

[25] A smoothed $\tilde{U}_{ocp}(\tilde{\theta}_s)$ relationship can be also be extracted from this step by integrating the histogram.

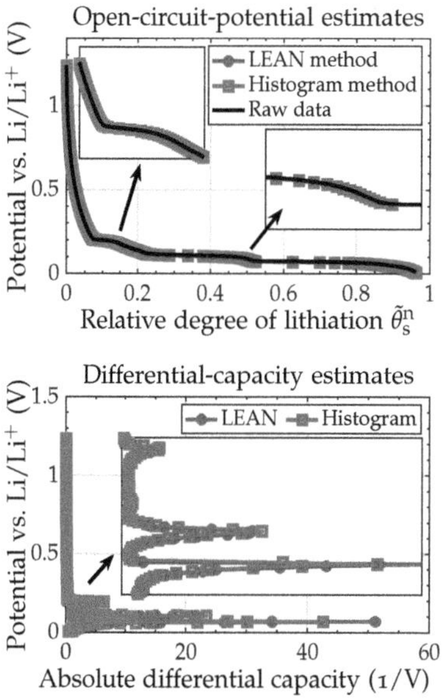

Figure 3.14: Illustrating the histogram-counting method.

Sample results are shown in Fig. 3.14. The output of the histogram-counting method is nearly indistinguishable from that of the LEAN method, at much lower computational cost.

3.3 *Estimating OCP*

We are now ready to present methods to combine the calibrated discharge and charge relationships to estimate the electrode's OCP function. Whatever method we use, it must address the missing-data and inaccessible-lithium problems. Vol. I of this series presented one approach (applied there to full-cell OCV estimation), which we will call the resistance-blending method. We have investigated several other strategies since the publication of that volume, and find them generally to be preferable to resistance blending. In this volume, we will look at four new methods.

Three of these new methods comprise two steps:

1. Estimate a relative $\widetilde{U}_{\text{ocp}}(\tilde{\theta}_s)$ from the OCP-test data.
2. Then, estimate absolute $U_{\text{ocp}}(\theta_s)$ by fitting an MSMR model to $\widetilde{U}_{\text{ocp}}(\tilde{\theta}_s)$.

The fourth method is a little different, addressing both of these steps in a more direct way.

3.3.1 *Voltage averaging (method 1)*

Fig. 3.15 illustrates the principle of operation of the first three OCP-estimation methods. We begin with the calibrated relationships of constant-current discharge voltage versus electrode relative stoichiometry (the blue line) and of constant-current charge voltage versus relative stoichiometry (the red line). We know that the true OCP must lie somewhere between these two curves.

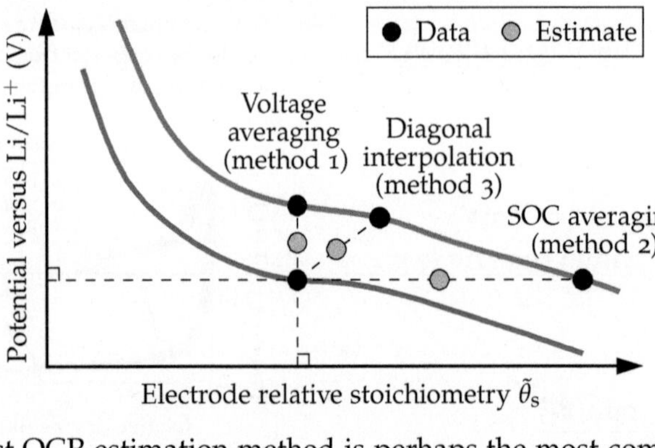

Figure 3.15: Illustrating the principle behind OCP-estimation methods 1–3.

The first OCP-estimation method is perhaps the most commonly seen approach in the literature. It assumes that the deviations between true OCP and the discharge and charge voltages at any given relative stoichiometry have equal magnitude.[26] OCP is then estimated by averaging the charge and discharge voltages at every calibrated relative stoichiometry value $\tilde{\theta}_s$.

Missing data are approximated by linearly extrapolating lithiation (discharge) voltage to $\tilde{\theta}_s = 1$ and delithiation (charge) voltage to $\tilde{\theta}_s = 0$ before averaging. The direct output of Method 1 is an estimate of OCP versus relative stoichiometry, which we denote as $\widehat{\widetilde{U}}_{\text{ocp}}(\tilde{\theta}_s)$.

[26] We will see that this assumption causes problems, especially in regions where the SOC-dependent impedance is changing or where there are missing data. This is because average stoichiometry lags behind surface stoichiometry in a constant-current test and therefore the measured voltage corresponds to an OCP value at a stoichiometry that leads the average stoichiometry. The discharge curve is shifted left compared with OCP and the charge curve is shifted right. Performing a vertical average amounts to averaging voltages from different SOCs, which is not what is intended. These problems are mitigated but not completely overcome by using low input-current amplitudes.

An additional step 2 must be taken later to convert this to the desired estimate versus absolute stoichiometry $\widehat{U}_{\text{ocp}}(\theta_{\text{s}})$.

3.3.2 *SOC averaging (method 2)*

It is also possible to average $\tilde{\theta}_{\text{s}}$ (electrode relative SOC) of the lithiation and delithiation curves at points where their measured terminal voltages are the same. Method 2 assumes that the deviation between true $\tilde{\theta}_{\text{s}}$ and its experimental values for the discharge and charge curves is equal in magnitude. Missing-data regions are linearly extrapolated to v_{min} and v_{max} before averaging.

Again, the direct output of Method 2 is $\widetilde{U}_{\text{ocp}}(\tilde{\theta}_{\text{s}})$. "Step 2" must be taken later to convert this to the desired $\widehat{U}_{\text{ocp}}(\theta_{\text{s}})$.

3.3.3 *Diagonal interpolation (method 3)*

Method 3 seeks to interpolate between the discharge and charge curves in a diagonal sense to form an estimate of OCP. This is motivated by recognizing that data samples of discharge and charge voltage versus electrode SOC have x and y coordinates. Here, we assume that there are errors in both x and in y due to polarization. The source of the errors in y are obvious when considering Eq. (3.1): the measured voltage at any point in time is different from OCP. The errors in x are caused by the fact that measured half-cell voltage depends on electrode *surface* stoichiometry and not *average* stoichiometry. Since average stoichiometry lags behind surface stoichiometry in a constant-current test, the measured voltage corresponds to an OCP value at a stoichiometry that leads the average stoichiometry.

Generally speaking, Method 3 shifts the discharge curve up and to the right; it shifts the charge curve down and to the left; it then either selects one of these results or averages the voltages of the two shifted results. The magnitude of the displacement used in both x and y directions is determined by first finding the differential-capacity curves of the discharge and charge data, which tends to magnify or accentuate the locations of important features versus voltage and relative stoichiometry. We determine the shift values by computing the cross correlations between the discharge and charge differential-capacity curves. This cross correlation of discharge and charge differential capacity versus voltage will be greatest for a shift value τ_U. The discharge and charge curves will be most similar if the they are both shifted toward each other by an amount $\Delta U = \tau_U/2$. Similarly, the cross correlation of discharge and charge differential capacity versus stoichiometry is maximized at some shift $\tau_{\tilde{\theta}_{\text{s}}}$. The discharge and charge curves will be most similar if they are both shifted toward each other by an amount $\Delta\tilde{\theta}_{\text{s}} = \tau_{\tilde{\theta}_{\text{s}}}/2$.

To visualize this data-processing approach, Fig. 3.16 shows simulated discharge and charge $\mathrm{d}\tilde{\theta}_s/\mathrm{d}U_{\mathrm{ocp}}$ versus both voltage and $\tilde{\theta}_s$ for a graphite material in the top two plots; it also plots the cross correlations between the discharge and charge differential capacities in the bottom two plots. The locations of the peaks in the cross correlations are identified. The peak in the voltage plot occurs at a shift of $\tau_U = 38\,\mathrm{mV}$ and so $\Delta U = 19\,\mathrm{mV}$; the peak in the stoichiometry plot is at a shift of $\tau_{\tilde{\theta}_s} = 0.017$ and so $\Delta\tilde{\theta}_s = 0.0085$.

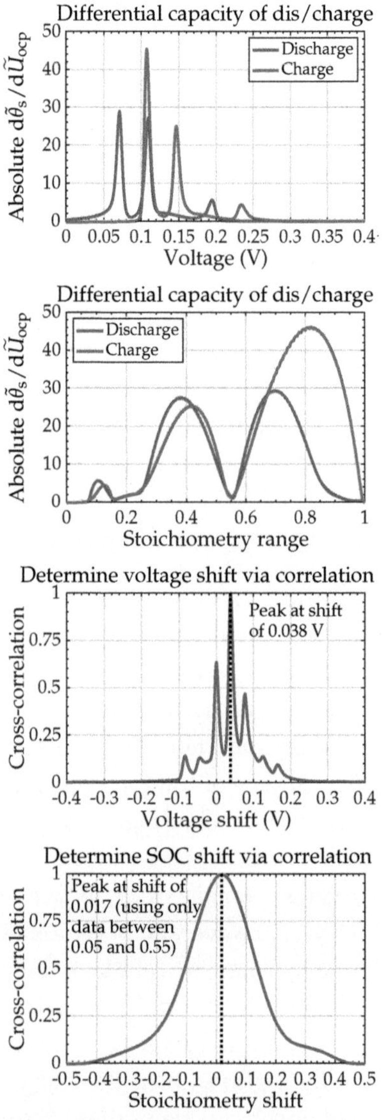

Figure 3.16: Visualizing the diagonal-interpolation method.

We have a choice in how to apply ΔU and $\Delta\tilde{\theta}_s$ to make an OCP estimate. We can shift only the discharge curve (producing $\widehat{\widetilde{U}}^{(\mathrm{d})}_{\mathrm{ocp}}(\tilde{\theta}_s)$) and use the resulting curve as the OCP estimate. Or we can shift only the charge curve (producing $\widehat{\widetilde{U}}^{(\mathrm{c})}_{\mathrm{ocp}}(\tilde{\theta}_s)$) or we can shift both curves and average the result (producing $\widehat{\widetilde{U}}^{(\mathrm{a})}_{\mathrm{ocp}}(\tilde{\theta}_s)$):

$$
\begin{aligned}
\widehat{\widetilde{U}}^{(\mathrm{d})}_{\mathrm{ocp}}(\tilde{\theta}_s) &= U_{\mathrm{dis}}\left(\tilde{\theta}_s^{\mathrm{dis}} - \Delta\tilde{\theta}_s\right) + \Delta U \\
\widehat{\widetilde{U}}^{(\mathrm{c})}_{\mathrm{ocp}}(\tilde{\theta}_s) &= U_{\mathrm{chg}}\left(\tilde{\theta}_s^{\mathrm{chg}} + \Delta\tilde{\theta}_s\right) - \Delta U \\
\widehat{\widetilde{U}}^{(\mathrm{a})}_{\mathrm{ocp}}(\tilde{\theta}_s) &= \frac{\widehat{\widetilde{U}}^{(\mathrm{d})}_{\mathrm{ocp}}(\tilde{\theta}_s) + \widehat{\widetilde{U}}^{(\mathrm{c})}_{\mathrm{ocp}}(\tilde{\theta}_s)}{2}.
\end{aligned}
$$

We have no general guidance regarding which of these three variants to use. We usually choose to plot all three overlaid on the original discharge and charge curves and select the most reasonable-looking candidate as the OCP estimate. We find that $\widehat{\widetilde{U}}^{(\mathrm{a})}_{\mathrm{ocp}}(\tilde{\theta}_s)$ is most often the best-looking estimate.

As with the methods discussed earlier, the direct output of Method 3 is $\widehat{\widetilde{U}}_{\mathrm{ocp}}(\tilde{\theta}_s)$. "Step 2" (described next) must be taken later to convert this to the desired $\widehat{U}_{\mathrm{ocp}}(\theta_s)$.

3.3.4 *Step 2: Convert relative to absolute relationships for Methods 1–3*

Methods 1–3 address the missing-data problem by combining the extrapolated discharge and curves in different ways. But none of these approaches determine OCP as a function of absolute stoichiometry to overcome the inaccessible-lithium problem. To do so, we must extrapolate the estimates $\widehat{\widetilde{U}}_{\mathrm{ocp}}(\tilde{\theta}_s)$ output by Methods 1–3 beyond the measured data; then, find $\theta_{\min}$ and $\theta_{\max}$ in Eq. (3.2).

We rely on the MSMR model to extrapolate $\widehat{\widetilde{U}}_{\mathrm{ocp}}(\tilde{\theta}_s)$ in a physically meaningful way, and then use nonlinear optimization to determine $\{\theta_{\min}, \theta_{\max}, U_j^0, X_j, \omega_j\}$ such that a MSMR model $\widehat{U}_{\mathrm{ocp}}(\theta_s)$ agrees with $\widehat{\widetilde{U}}_{\mathrm{ocp}}(\tilde{\theta}_s)$ as closely as possible. Inside the optimization, $\widehat{U}^{\mathrm{data}}_{\mathrm{ocp}}(\theta_s)$ is computed from $\widehat{\widetilde{U}}_{\mathrm{ocp}}(\tilde{\theta}_s)$ by finding θ_s for every $\tilde{\theta}_s$ for every data point using $\theta_{\min}$ and $\theta_{\max}$ via Eq. (3.2). Then U_j^0, X_j, and ω_j are used to create an MSMR estimate $\widehat{U}^{\mathrm{mdl}}_{\mathrm{ocp}}(\theta_s)$: a nonlinear-optimization routine adjusts $\{\theta_{\min}, \theta_{\max}, U_j^0, X_j, \omega_j\}_{j=1}^J$ until $\widehat{U}_{\text{ocp}}^{\text{mdl}}(\theta_s)$ agrees with $\widehat{U}_{\text{ocp}}^{\text{data}}(\theta_s)$ as closely as possible.

Specifically, OCP-estimation "step 2" seeks parameter values p^*:[27]

$$p^* = \arg\min_p \sum_{n=1}^{N} \Big[\theta_{s,n}(\widehat{U}_{\text{ocp}}^{\text{data}}) - \theta_{s,n}(\widehat{U}_{\text{ocp}}^{\text{mdl}})\Big]^2 + w\left[\frac{d\theta_s}{\widehat{U}_{\text{ocp}}^{\text{data}}(\theta_{s,n})} - \frac{d\theta_s}{\widehat{U}_{\text{ocp}}^{\text{mdl}}(\theta_{s,n})}\right]^2,$$

where w is a weighting factor that blends modeling errors corresponding to raw OCP estimates and differential-capacity estimates. We use MATLAB's `fmincon` function to perform the regression.

The final result is a dependable conversion between relative and absolute stoichiometries as long as $\widehat{\widehat{U}}_{\text{ocp}}(\tilde{\theta}_s)$ covers a wide enough voltage range to include measurements spanning every physically existing gallery in the MSMR model. The method is not clever enough to invent correct data for any physical gallery whose occupancy is not excited by the lab test. That is, if a gallery is always essentially full or empty for all the data collected by the OCP test, we will not be able to model the MSMR parameter values for that gallery.

We can verify whether the model will compute absolute stoichiometry faithfully by checking the slope of $\widehat{\widehat{U}}_{\text{ocp}}(\tilde{\theta}_s)$ near $\tilde{\theta}_s = 0$ and $\tilde{\theta}_s = 1$. If the slope near $\tilde{\theta}_s = 0$ is steep, then the MSMR model tells us that we are approaching true $\theta_s = 0$ and so we will model all galleries well in their empty state. If the slope near $\tilde{\theta}_s = 1$ is steep then we know that we are approaching the true $\theta_s = 1$ and so we will model all galleries well in their fully occupied state. Both of these conditions need to be true for us to have confidence that the absolute value of θ_s computed by this step 2 is accurate.

In practice, we find these conditions to be satisfied when estimating the OCP of graphite materials. However, for many cathode materials (e.g., NMC) the slope of $\widehat{\widehat{U}}_{\text{ocp}}(\tilde{\theta}_s)$ is not steep near $\tilde{\theta}_s = 0$ and so we have less confidence in the calibration of absolute capacity produced by this method. Nonetheless, we believe that step 2 still provides an improvement over performing only step 1. If the equipment is available, perhaps a better approach would be to perform ICP–OES of the electrode material at a few known resting OCP values to determine the relative fraction of lithium in the electrode sample and therefore the absolute stoichiometry of the electrode experimentally; this can then be a part of the optimization that calibrates the OCP estimate.[28] Another calibration approach is presented in Sect. 3.9.

The output of this step 2 is an MSMR model that fits measured data as closely as possible using absolute stoichiometries. The overall OCP-estimation procedures use Methods 1–3 to overcome the

[27] The total number of galleries J of the MSMR model $\widehat{U}_{\text{ocp}}^{\text{mdl}}$ is a user-selected parameter, not optimized automatically.

[28] For example, this was done in:

- Johannes Schmalstieg, Christiane Rahe, Madeleine Ecker, and Dirk Uwe Sauer, "Full cell parameterization of a high-power lithium-ion battery for a physico-chemical model: Part I. Physical and electrochemical parameters," *Journal of The Electrochemical Society*, 165(16):A3799, 2018.

Another option uses X-ray diffraction methods:

- Franziska Friedrich, Benjamin Strehle, Anna T. S. Freiberg, Karin Kleiner, Sarah J. Day, Christoph Erk, Michele Piana, and Hubert A. Gasteiger, "Capacity fading mechanisms of NCM-811 cathodes in lithium-ion batteries studied by X-ray diffraction and other diagnostics," *Journal of The Electrochemical Society*, 166(15):A3760, 2019.

missing-data problem and then fit an MSMR model to address the inaccessible-lithium problem.

3.3.5 *Direct MSMR model fit (method 4)*

The final OCP-estimation method we present fits separate MSMR models directly to the discharge and charge curves. $\widehat{U}_{\text{ocp}}(\theta_s)$ is then estimated by averaging parameters from both models (i.e., $\{\theta_{\min}, \theta_{\max}, U_j^0, X_j, \omega_j\}_{\text{dis}}$ and $\{\theta_{\min}, \theta_{\max}, U_j^0, X_j, \omega_j\}_{\text{chg}}$). That is, we assume that the true OCP MSMR parameters have equal distance to the discharge and charge MSMR-model parameters.

Example fits to calibrated discharge and charge lab data are shown in Fig. 3.17. The data are closely fit by the model, especially at warm temperatures where voltage polarization is smaller.

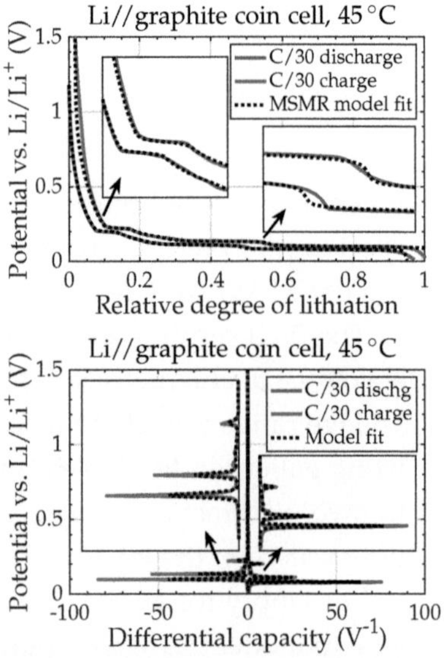

Figure 3.17: Method 4 fits to discharge and charge voltage and differential capacity.

3.4 *Validating OCP*

3.4.1 *Validation of methods using simulation*

It is impossible to judge the accuracy of physical half-cell OCP estimates since there is strictly no known truth against which to compare the estimate. The only fact we can verify at this point is that the $\widehat{\widehat{U}}_{\text{ocp}}(\tilde{\theta}_s)$ estimates should reside between the discharge and charge data curves of the slow constant-current tests.

We can, however, validate the methods in simulation: We simulate the OCP test scripts from Sect. 3.2; we then apply the different methods to the simulated discharge and charge voltages and compare the different estimates of $\widehat{U}_{\text{ocp}}(\theta_s)$ to the true $U_{\text{ocp}}(\theta_s)$ used by the simulator as part of producing simulated voltages. The best method is the one that produces the best match.

In this section, a physics-based half-cell model was implemented in COMSOL and output data were processed by MATLAB scripts. A Li//graphite half cell was lithiated from θ_s=0.0108 to $v_{\min}$=0.01 V and delithiated from θ_s=0.9905 to $v_{\max}$=1.5 V. A Li//NMC half cell was lithiated from θ_s=0.1274 to $v_{\min}$=3.0 V and delithiated from θ_s=0.9992 to $v_{\max}$=4.3 V. Constant-current steps used a C/30 rate.

Fig. 3.18 displays the simulation-voltage outputs at 25 °C ("C/30 data") along with the relative-stoichiometry OCP estimates. Method 2 struggles with the flat plateaus of graphite and Method 3 appears to work best overall.

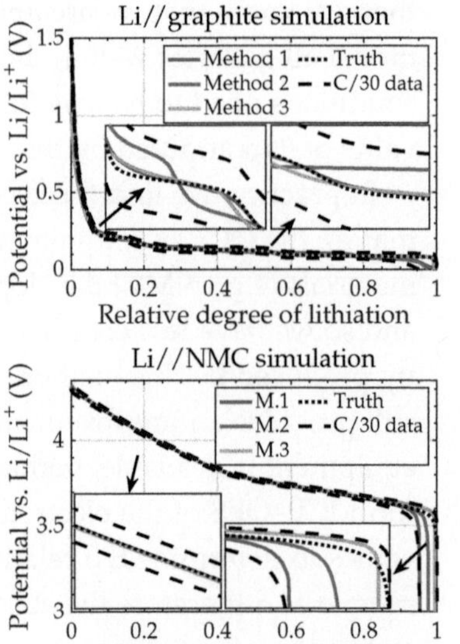

Figure 3.18: Step 1 outputs from Methods 1–3 applied to simulation data. Note that "M.1" is an abbreviation for "Method 1" (and so forth for "M.2," "M.3," and "M.4") in various places in this chapter.

Fig. 3.19 shows the conversion to absolute-stoichiometry OCP estimates, along with root-mean-squared error (RMSE): RMSE was calculated over the range of $0.07 < \theta_s^n < 0.99$ for the Li//graphite half cell and over the range of $0.1 < \theta_s^p < 1$ for the Li//NMC half cell, which should extend beyond the typical operating ranges used in commercial cells. Method 3 works well for the Li//graphite half cell, which exhibits wide voltage plateaus; Method 4 also produces quite good results. Method 2 produces the best results for the Li//NMC half cell, and Methods 3 and 4 also produce good results.

We hesitate to recommend Method 2 in general as it does not work well for materials having wide, flat voltage plateaus (e.g., Li//graphite). Both Methods 1 and 2 tend to break down in high-curvature regions of OCP, especially at very high and very low levels of lithiation; while they are simple to implement, they do not predict OCV accurately when there is a significant level of missing-data.

For further insight, Fig. 3.20 shows the residuals (errors) for each method, which we desire to be small and without SOC dependence. Methods 3 and 4 are the most accurate, but also the slowest. But even Method 4 (the slowest) executes within about 3 s in MATLAB, which is not excessive for a one-time data-processing method.

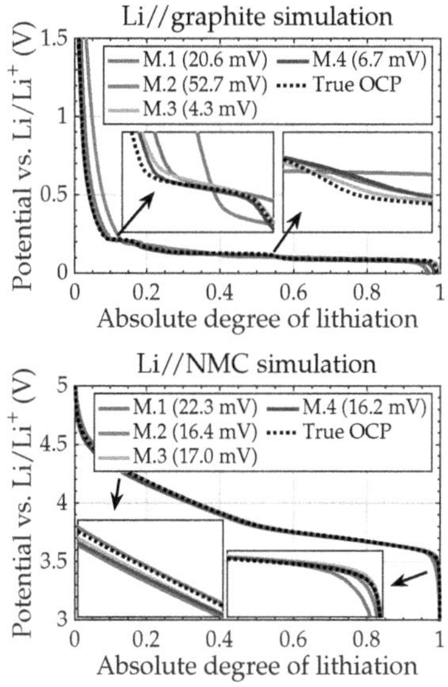

Figure 3.19: Step 2 outputs from all methods applied to simulation data. (Values in parentheses are RMS errors.)

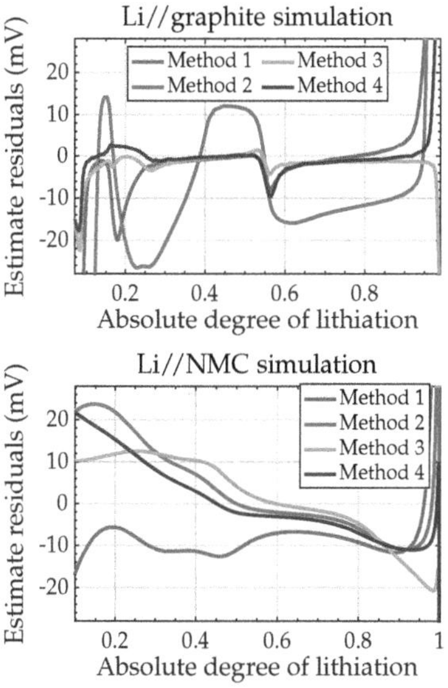

Figure 3.20: OCP-estimation residuals (true OCP minus estimate) when using simulation data.

3.4.2 *Application of methods to physical half cells*

We are interested in this chapter to understand how well all of the proposed parameter-estimation methods work for a commercial cell. To do so, we selected several 25 Ah graphite//NMC prismatic PHEV2-form-factor Panasonic cells from a Ford C-MAX Eneregi PHEV battery pack (see Fig. 3.21). Colleagues from the National Renewable Energy Laboratory disassembled one of these cells in a glovebox (see Fig. 3.2) and made coin cells of electrode material versus lithium (see Fig. 3.10). Figs. 3.11, 3.13, 3.14, and 3.17 have already presented data collected from these coin cells, as will all figures from this point on that are labeled as coin-cell data.

We applied the four OCP-estimation methods to data collected from these physical coin cells where the test temperature was 25 °C. The Li//graphite coin cell was cycled between v_{min}=0.01 V and v_{max}=1.5 V and the Li//NMC coin cell was cycled between v_{min}=2.1 V and v_{max}=4.3 V.

Fig. 3.22 presents the final results in terms of relative and absolute lithiation. We cannot validate the correctness of these results beyond noticing whether the relative OCP estimates lie between the discharge and charge curves, which they all do. (We cannot plot residuals since the true OCP is unknown.)

We observe that the missing-data problem is more prominent at low levels of lithiation in the Li//graphite coin cell than it was in the simulations; in all other respects, simulation and lab results appear to be similar. We find less variation in the Method-4 MSMR parameter values versus temperature (presented next) than those produced by Method 3, so we believe it to produce the better results.

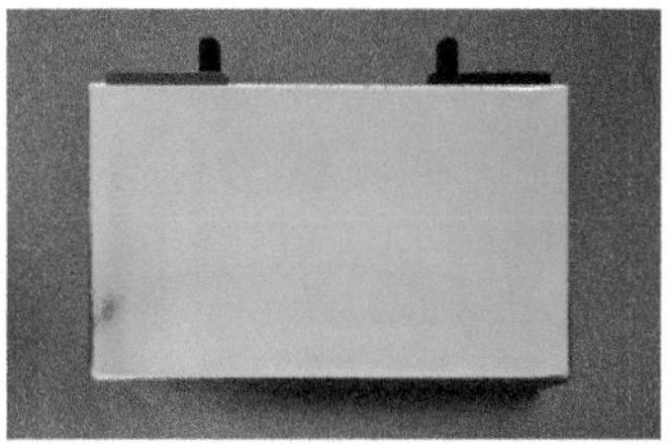

Figure 3.21: The Panasonic cell used to demonstrate the parameter-estimation methods presented in this chapter.

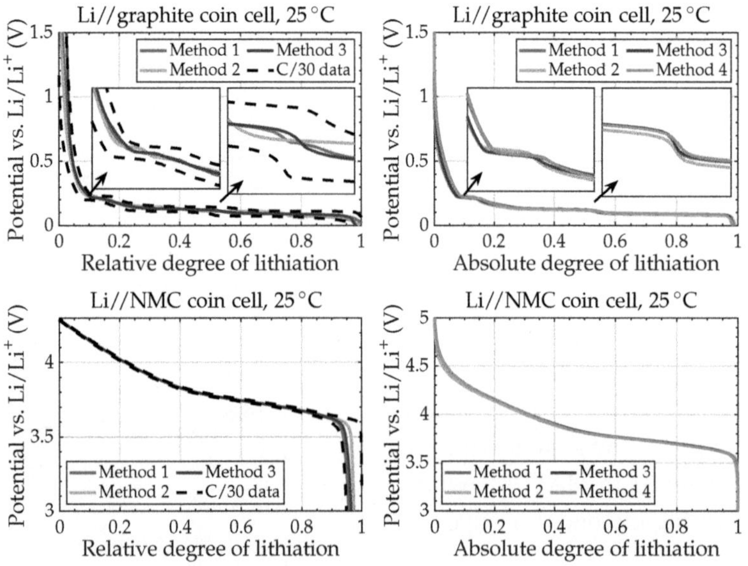

Figure 3.22: The OCP-estimation methods applied to physical coin cells using electrodes harvested from the Panasonic cell.

We also note that the step-1 results for Method 2 (and to a lesser extent for Method 1) are clearly biased, as shown in the zoom plots, but that fitting the MSMR model to these biased results in step 2 greatly improves the quality and reasonableness of the final OCP model produced by these methods.

3.4.3 *Temperature dependence of OCP*

Our LPMs will need to be able to model the temperature-dependence of electrode OCP. But how should we do so? To begin, we note that the MSMR model has built-in explicit dependence on temperature in the term $f = F/(RT)$, even if U_j^0, X_j, and ω_j remain constants:

$$U_j = U_j^0 + \frac{\omega_j}{f} \ln\left(\frac{X_j - x_j}{x_j}\right).$$

Evaluating this model in simulation with fixed U_j^0, X_j, and ω_j produces the OCP and differential-capacity plots of Fig. 3.23, which show some variation with temperature. The largest differences between OCP at different temperatures are in the high-curvature regions; we also note sharper and higher peaks in differential capacity at cold temperatures.

So there is some temperature dependence built in to the MSMR model itself. But is it sufficient to describe all the temperature variation observed in a physical electrode? That is, perhaps the MSMR parameter values $\{U_j^0, X_j, \omega_j\}$ must also be functions of temperature. Fig. 3.24 plots the experimental charge and discharge differential

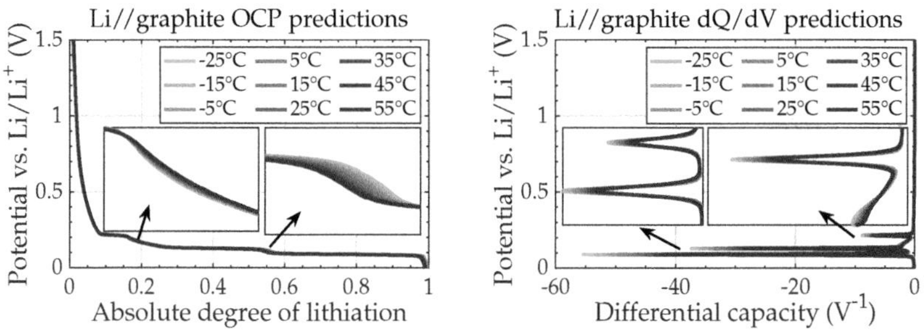

Figure 3.23: Built-in MSMR-model temperature dependence, illustrated for a simulated graphite electrode.

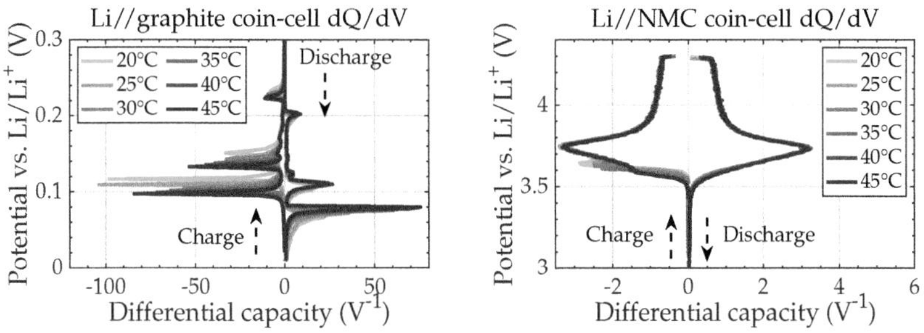

Figure 3.24: Experimental differential capacity of discharge and charge of coin cells at different temperatures.

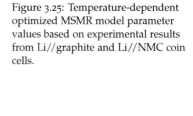

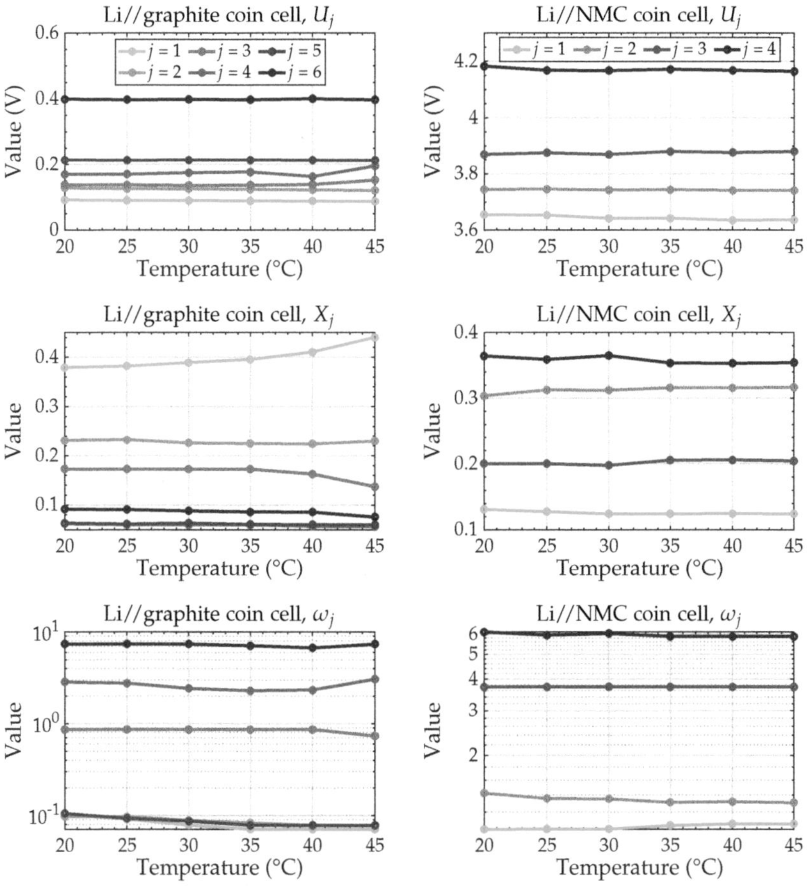

Figure 3.25: Temperature-dependent optimized MSMR model parameter values based on experimental results from Li//graphite and Li//NMC coin cells.

capacity for the Li//graphite and Li//NMC coin cells. There is noticeable variation in the displacement of peaks for data collected at different temperatures, especially for the graphite material.[29] These displacements might be due to temperature-dependent polarization or to material thermodynamics.

It remains to see whether shifts in experimental differential capacity are due to temperature-dependent OCP or polarization. Fig. 3.25 displays Model-4 estimates of MSMR parameters individually optimized from experimental coin-cell data at different temperatures. Different colors represent different galleries in the MSMR model. We see very little fluctuation in the optimized U_j^0, X_j, and ω_j values. It seems reasonable to assume that all temperature dependence is captured by "f" and so fixed values may be used for U_j^0, X_j, and ω_j. We will proceed with this assumption.

[29] Displacement is still present in the NMC material, but it is less obvious visually in comparison with the overall cycling voltage range. We do notice an interesting feature: one of the peaks from the Li//NMC charge curve (at around 3.65 V) depends strongly on temperature. The phenomenon is observed in a number of publications, but as far as we know its origin still remains an open question to investigate.

3.5 *OCV testing*

We now have estimates of the electrode OCP functions $\widehat{U}_{\text{ocp}}^{\text{r}}(\theta_{\text{s}}^{\text{r}})$, but we still must determine operating boundaries θ_0^{n}, θ_{100}^{n}, θ_0^{p}, and θ_{100}^{p} between which the electrodes cycle as the full cell is cycled between 0 % and 100 % SOC.[30] To assist with this task, we recognize that cell OCV at any specific SOC equals positive-electrode OCP at the value of $\theta_{\text{s}}^{\text{p}}$ corresponding to that cell SOC minus negative-electrode OCP at the value of $\theta_{\text{s}}^{\text{n}}$ corresponding to that SOC. That is, if cell SOC is denoted by z, then we have:

$$\theta_{\text{s}}^{\text{r}}(z) = \theta_0^{\text{r}} + z\,(\theta_{100}^{\text{r}} - \theta_0^{\text{r}}) \tag{3.3}$$

$$U_{\text{ocv}}^{\text{ocp}}(z) = U_{\text{ocp}}^{\text{p}}(\theta_{\text{s}}^{\text{p}}(z)) - U_{\text{ocp}}^{\text{n}}(\theta_{\text{s}}^{\text{n}}(z)). \tag{3.4}$$

This is illustrated in Fig. 3.26, where the green "o" and dark red "×" markers show absolute degrees of electrode lithiation when the cell is resting at 100 % and 0 % SOC, respectively. Dashed lines connect matching operating points in the electrodes and the cell.

To determine θ_0^{r} and θ_{100}^{r}, we must first compute an estimate of cell OCV versus cell SOC. We then optimize the operating boundaries θ_0^{n}, θ_{100}^{n}, θ_0^{p}, and θ_{100}^{p} to get the best match between electrode OCP regions and cell OCV via Eqs. (3.3) and (3.4).

[30] The content of this section has been adapted from: Dongliang Lu, M. Scott Trimboli, Guodong Fan, Ruigang Zhang, and Gregory L. Plett, "Implementation of a physics-based model for half-cell open-circuit potential and full-cell open-circuit voltage estimates: part II. Processing full-cell data," *Journal of The Electrochemical Society*, 168(7):070533, 2021.

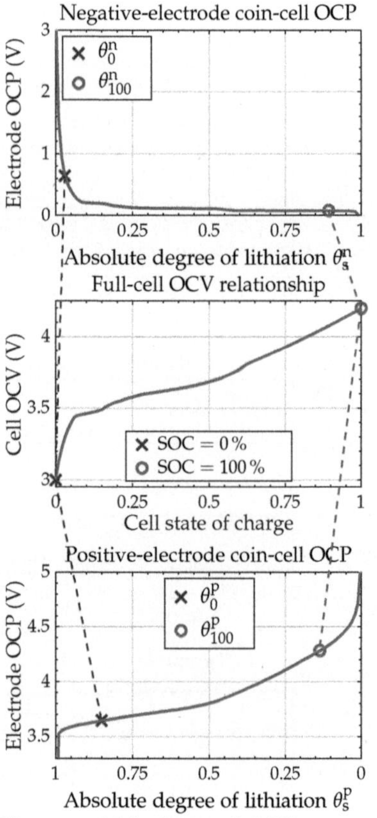

Figure 3.26: Matching cell OCV to operating windows in electrode OCP functions. (Note that the horizontal axis on the positive-electrode OCP function is reversed.)

3.5.1 *Determining cell OCV versus SOC*

To find the cell's OCV versus SOC relationship, we use exactly the same procedure (on a cell level) that we used to find electrode OCP (on a half-cell level). We need to perform the lab test only at 25 °C since all temperature dependence is included in the OCP MSMR models; we simply need to correlate the 25 °C cell OCV with the

Table 3.2: Summary of OCV test script.

Step number and details (all steps execute with cell soaked at 25 °C)
1. Discharge at a constant rate (e.g., C/30 or C/100) to v_{min}.
2. Dis/charge to v_{min}; hold at (and/or dither around) v_{min}.
3. Charge at a constant rate to v_{max} (same rate as step 1).
4. Dis/charge to v_{max}; hold at (and/or dither around) v_{max}.

25 °C electrode OCPs. The lab-test procedure is summarized in Table 3.2. We can use the same four discharge and charge voltage-processing methods (i.e., Methods 1–4 from Sect. 3.3) to convert cell discharge and charge data to an OCV versus SOC estimate.

We can also postprocess an OCV model found this way to refine it for BMS application. We know that $U_{\mathrm{ocv}}(0) = v_{\mathrm{min}}$ and $U_{\mathrm{ocv}}(1) = v_{\mathrm{max}}$ at the reference temperature. However, the output of the four OCV-estimation methods will not generally meet these constraints exactly and so we might choose to adjust the OCV model to enforce these limits.

Note that the modification to be proposed is contrary to the caution of sidenote 19 because it involves stretching the already-estimated OCV relationship. However, we find that the amount of stretching required by this postprocessing tends to be small relative to the stretching warned against in sidenote 19; therefore, it will not distort the physical relationship being estimated to the same degree.

The overall modified OCV-estimation process is then:

1. *Initial estimation*: One of the four OCV-estimation methods is used to provide an initial estimate $\widehat{U}_{\mathrm{ocv}}(\hat{z})$ of true $U_{\mathrm{ocv}}(z)$.
2. *Extrapolation*: We linearly extrapolate $\widehat{U}_{\mathrm{ocv}}(\hat{z})$ over a standard voltage vector U_{std}, assuring that $\widehat{U}_{\mathrm{ocv}}(\hat{z})$ spans v_{min} to v_{max}.
3. *Shift*: We shift the curve horizontally (adding a constant Δz to $\hat{z}$) to enforce $U_{\mathrm{ocv}} = v_{\mathrm{min}}$ at $z = 0\,\%$.
4. *Stretch*: We stretch the curve horizontally (multiplying $\hat{z}$ by a constant k) to enforce $U_{\mathrm{ocv}} = v_{\mathrm{max}}$ at $z = 100\,\%$.

This can be summarized by stating that we replace $\hat{z}$ with $k(\hat{z} + \Delta z)$. Fig. 3.27 illustrates the steps.

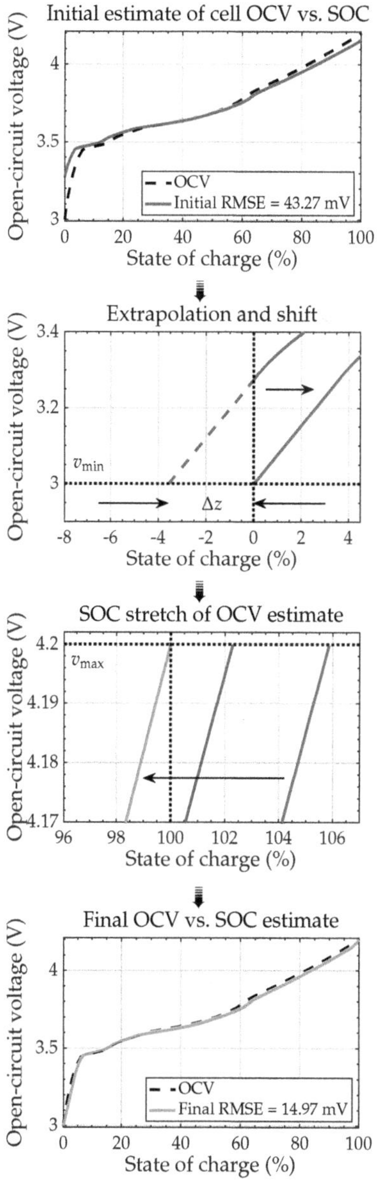

Figure 3.27: Postprocessing OCV to enforce constraints. (Initial OCV in the top frame was made using resistance-blending. See Vol. I.)

Fig. 3.28 shows results from the OCV-estimation methods. Three of the subfigures apply the methods to C/30 discharge and charge data for a simulated graphite//NMC cell; the final subfigure shows results from application to the Panasonic cell. The simulation results are helpful since the true OCV relationship is known, against which we can compare the estimated OCV.

The top-left subplot shows overall OCV estimates versus truth. Method 3 gives the closest match and Method 1 also appears quite good in this case. The top-right subplot shows differential capacity estimated using the four methods, compared against the truth. Here

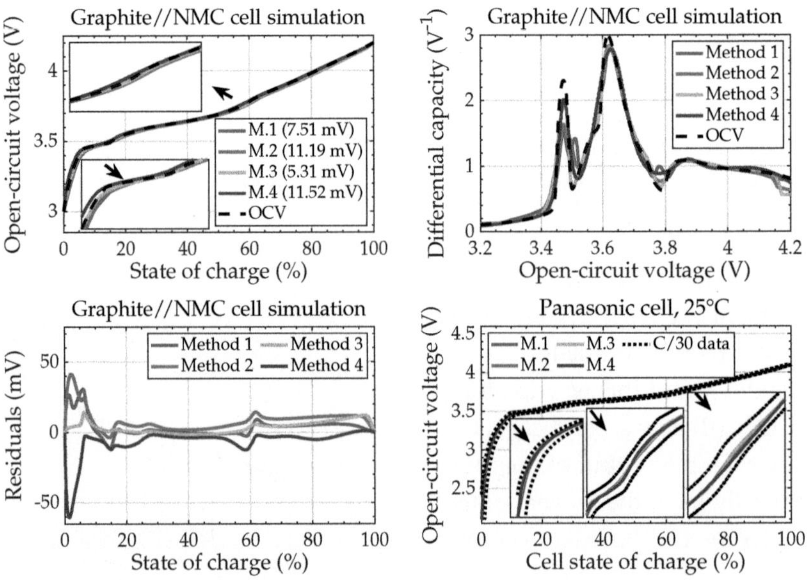

Figure 3.28: Example illustrating OCV-estimation methods.

we see that Method 1 produced artifacts in the differential-capacity plot, which decreases our level of trust in this approach. The bottom-left subplot shows modeling error as a function of SOC. Method 3 is able to capture the low-SOC behaviors better than the others. Finally, the bottom-right subplot shows data from a physical cell. We don't know the truth in this case, but we do know that the OCV estimate must lie between the charge and discharge C/30 data. All of the methods produce similar results, and visually we might determine that Method 3 produces the most appealing (best centered) result.

3.5.2 *Estimating boundaries*

Now, knowing cell-level OCV versus SOC, we seek to estimate electrode operating boundaries $\{\theta_0^{\rm n}, \theta_{100}^{\rm n}, \theta_0^{\rm p}, \theta_{100}^{\rm p}\}$ that produce the best match between a cell's OCV $\widehat{U}_{\rm ocv}^{\rm cell}(z_k)$ and its electrode OCP functions $\widehat{U}_{\rm ocp}^{\rm r}(\theta_{\rm s}^{\rm r})$. Boundaries can be estimated by minimizing the cost:

$$J = \sum_{k=1}^{N} \left[\left(\widehat{U}_{\rm ocv}^{\rm cell}(z_k) - \widehat{U}_{\rm ocv}^{\rm ocp}(z_k) \right)^2 + w \left(\frac{dz}{d\widehat{U}_{\rm ocv}^{\rm cell}(z_k)} - \frac{dz}{d\widehat{U}_{\rm ocv}^{\rm ocp}(z_k)} \right)^2 \right],$$

where w balances between optimizing OCV fit and differential-capacity fit, and:

$$\hat{\theta}_{\rm s}^{\rm r}(z) = \hat{\theta}_0^{\rm r} + z\,(\hat{\theta}_{100}^{\rm r} - \hat{\theta}_0^{\rm r}) \tag{3.5}$$

$$\widehat{U}_{\rm ocv}^{\rm ocp}(z) = \widehat{U}_{\rm ocp}^{\rm p}(\hat{\theta}_{\rm s}^{\rm p}(z)) - \widehat{U}_{\rm ocp}^{\rm n}(\hat{\theta}_{\rm s}^{\rm n}(z)), \tag{3.6}$$

$$\frac{dz}{d\widehat{U}_{\rm ocv}^{\rm ocp}(z)} = \frac{1}{(\hat{\theta}_{100}^{\rm p} - \hat{\theta}_0^{\rm p})\frac{d\widehat{U}_{\rm ocp}^{\rm p}(\hat{\theta}_{\rm s}^{\rm p})}{d\hat{\theta}_{\rm s}^{\rm p}} - (\hat{\theta}_{100}^{\rm n} - \hat{\theta}_0^{\rm n})\frac{d\widehat{U}_{\rm ocp}^{\rm n}(\hat{\theta}_{\rm s}^{\rm n})}{d\hat{\theta}_{\rm s}^{\rm n}}}. \tag{3.7}$$

	Method	RMSE (mV) U_{ocv}	Absolute error (%) θ_0^n	θ_{100}^n	θ_0^p	θ_{100}^p
(a)	1	3.70	0.98	0.27	2.14	0.64
	2	9.26	3.69	5.51	4.21	1.98
	3	3.30	0.05	0.87	0.63	0.42
	4	6.17	0.22	4.98	0.24	2.29
(b)	1	5.55	0.14	1.05	0.54	0.11
	2	5.16	0.17	2.52	1.17	0.43
	3	2.82	0.03	0.68	0.11	0.48
	4	5.36	0.30	3.93	0.65	0.00

Table 3.3: Estimating boundaries using simulation data. Absolute estimation error (in percent) is computed as $|\theta_{\text{true}} - \theta_{\text{estimate}}| \times 100$ and RMSE is the root-mean-squared error between cell OCV $\widehat{U}_{\text{ocv}}^{\text{cell}}$ estimated from full-cell data and cell OCV $\widehat{U}_{\text{ocv}}^{\text{ocp}}$ estimated from half-cell data using Eqs. (3.5) and (3.6).

During optimization, we can use the domain knowledge that $0 \leq \theta_0^n < \theta_{100}^n \leq 1$ and $0 \leq \theta_{100}^p < \theta_0^p \leq 1$ to constrain the solution.

3.6 *Validating boundaries*

3.6.1 *Estimating operating boundaries using simulation data*

We now seek to evaluate the effectiveness of this approach to estimating electrode operating boundaries. We first use simulation to validate results in two steps. The first step assumes that we have exact knowledge of the electrode OCP functions but that we must estimate the cell OCV function from discharge and charge data using one of the four methods. Results from this step are listed in the rows labeled "(a)" in Table 3.3.

The second step assumes that we must estimate electrode OCP functions from half-cell discharge and charge data in addition to estimating cell OCV from full-cell discharge and charge data. The same data-processing method is applied to half-cell and full-cell data (i.e., if method n is applied to the negative-electrode data then the same method n is also applied to the positive-electrode data and to the full-cell data). Results from this step are listed in the rows labeled "(b)" in the table.

It is difficult to state in general which method will give the best results, but in these simulations Method 3 works best under both scenarios. The most valuable conclusion from this study is: Results having lowest RMSE for the cell model also have lowest absolute boundary-estimation error. For a physical cell, we can compute RMSE but we cannot compute boundary-value estimation error as the true boundaries are unknown. The observation that low RMSE correlates well with low absolute error in the boundary estimates gives a way to establish confidence in how well each method performs when using data from a physical cell.

3.6.2 *Estimating operating boundaries using laboratory data*

We now seek estimates of $\{\theta_0^{\mathrm{n}}, \theta_{100}^{\mathrm{n}}, \theta_0^{\mathrm{p}}, \theta_{100}^{\mathrm{p}}\}$ using experimental data collected from the Panasonic cell. Table 3.4 displays values for the optimized operating boundaries and the voltage RMSE from fitting the electrode OCP functions estimated from half-cell data to full-cell estimated OCV.

Method	RMSE (mV)	θ_0^{n}	$\theta_{100}^{\mathrm{n}}$	θ_0^{p}	$\theta_{100}^{\mathrm{p}}$
1	30.47	0.0010	0.8960	0.9990	0.1653
2	41.78	0.0009	0.8026	0.9990	0.1510
3	25.36	0.0008	0.8255	0.9284	0.1691
4	19.54	0.0007	0.8279	0.9089	0.1769

Table 3.4: Estimating boundaries using data collected from the Panasonic cell.

Estimates having lower RMSE are expected to be better in LPMs because closer agreement between cell-level OCV and electrode-level OCP functions can yield better model performance. It is also important to realize that the RMS errors calibrate only the voltage estimates within the operating windows. We are unable to evaluate the OCP models outside the operating boundaries (i.e., models where $\theta_{\mathrm{s}}^{\mathrm{n}} \notin [\theta_0^{\mathrm{n}}, \theta_{100}^{\mathrm{n}}]$ and/or $\theta_{\mathrm{s}}^{\mathrm{p}} \notin [\theta_{100}^{\mathrm{p}}, \theta_0^{\mathrm{p}}]$). But MSMR-model constraints provide some confidence regarding reasonableness of OCP estimates somewhat outside normal operating regions.

We note that Methods 1–2 saturated θ_0^{p} to its maximum permitted value during optimization. This seems unreasonable for a physical cell as some positive-electrode lithium inventory is permanently consumed by SEI-layer growth in the negative electrode during cell formation, even for a new cell. Thus, we believe that the results computed by Methods 3 and 4 are more physically meaningful.

3.7 *Estimating OCP without requiring cell teardown*

We have now estimated electrode OCP and operating boundaries using a teardown approach. However, if possible, we would prefer to find all LPM parameter values without cell teardown. So, we are compelled to ask: Can we determine both electrode OCP functions uniquely from the cell OCV function? The answer turns out to be no: OCV information alone is insufficient to determine electrode OCP functions uniquely. We refer to this fact as the observability problem, which can be demonstrated fairly quickly.

Consider an arbitrary function $f(z)$, where z is once again cell SOC. We can index $f(\cdot)$ using either $\theta_{\mathrm{s}}^{\mathrm{n}}$ or $\theta_{\mathrm{s}}^{\mathrm{p}}$ via:

$$z = \frac{\theta_{\mathrm{s}}^{\mathrm{n}} - \theta_0^{\mathrm{n}}}{\theta_{100}^{\mathrm{n}} - \theta_0^{\mathrm{n}}} = \frac{\theta_{\mathrm{s}}^{\mathrm{p}} - \theta_0^{\mathrm{p}}}{\theta_{100}^{\mathrm{p}} - \theta_0^{\mathrm{p}}}.$$

Suppose we define biased estimates $\widehat{U}_{\text{ocp}}^{\text{r,bias}}$ of the true OCP by adding $f(z)$ to the true OCP function:

$$\widehat{U}_{\text{ocp}}^{\text{n,bias}}(\theta_{\text{s}}^{\text{n}}) = U_{\text{ocp}}^{\text{n}}(\theta_{\text{s}}^{\text{n}}) + f((\theta_{\text{s}}^{\text{n}} - \theta_{0}^{\text{n}})/(\theta_{100}^{\text{n}} - \theta_{0}^{\text{n}}))$$
$$\widehat{U}_{\text{ocp}}^{\text{p,bias}}(\theta_{\text{s}}^{\text{p}}) = U_{\text{ocp}}^{\text{p}}(\theta_{\text{s}}^{\text{p}}) + f((\theta_{\text{s}}^{\text{p}} - \theta_{0}^{\text{p}})/(\theta_{100}^{\text{p}} - \theta_{0}^{\text{p}})).$$

Because $f(z)$ is arbitrary, $\widehat{U}_{\text{ocp}}^{\text{r,bias}} \neq U_{\text{ocp}}^{\text{r}}$ in general. But we cannot observe the bias in OCP estimates from an OCV measurement. If we attempted to do so, we would use the estimates of $\widehat{U}_{\text{ocp}}^{\text{r,bias}}$ to compute an estimate of OCV, $\widehat{U}_{\text{ocv}}$:

$$\begin{aligned}\widehat{U}_{\text{ocv}}(z) &= \widehat{U}_{\text{ocp}}^{\text{p,bias}}(\theta_{\text{s}}^{\text{p}}(z)) - \widehat{U}_{\text{ocp}}^{\text{n,bias}}(\theta_{\text{s}}^{\text{n}}(z)) \\ &= \Big[U_{\text{ocp}}^{\text{p}}(\theta_{\text{s}}^{\text{p}}) + f((\theta_{\text{s}}^{\text{p}} - \theta_{0}^{\text{p}})/(\theta_{100}^{\text{p}} - \theta_{0}^{\text{p}}))\Big] \\ &\quad - \Big[U_{\text{ocp}}^{\text{n}}(\theta_{\text{s}}^{\text{n}}) + f((\theta_{\text{s}}^{\text{n}} - \theta_{0}^{\text{n}})/(\theta_{100}^{\text{n}} - \theta_{0}^{\text{n}}))\Big] \\ &= U_{\text{ocp}}^{\text{p}}(\theta_{\text{s}}^{\text{p}}) - U_{\text{ocp}}^{\text{n}}(\theta_{\text{s}}^{\text{n}}) = U_{\text{ocv}}(z).\end{aligned}$$

That is, the biased estimates of OCP combine in such a way that the biases cancel and the estimated OCV matches the true OCV exactly. The error in electrode OCP estimates is not observable from a cell-level OCV measurement.[31]

[31] We might also ask whether differential capacity might improve observability of the OCP functions from an OCV relationship. Again, the answer is no (math omitted here).

In retrospect, this result is not surprising. We write OCV as the difference of two OCP functions. We have two unknowns (the OCP functions) and a single equation. We know that it is not possible to solve one equation uniquely for two unknowns.

In sum, the observability problem is that cell-level OCV alone is insufficient to determine electrode-level OCP functions uniquely. Instead, we proceed by making one of two assumptions:

A1: We are able to fabricate half cells from electrode materials and measure OCP directly, as has been our assumption to this point.
A2: Or, we will use a database approach, where we correlate cell-level OCV data with a database of known OCP functions for a variety of electrode materials.

This section proposes two different strategies for how we might proceed if we must make Assumption A2:

A2.1: We assume that the negative electrode has a known composition and identify positive-electrode OCP from OCV tests.
A2.2: Or, we assume that the composition of neither electrode is known but that we have a database of models of known materials against which we can compare cell measured OCV.

These two approaches are presented as follows.

A2.1: Assume that negative-electrode OCP is known

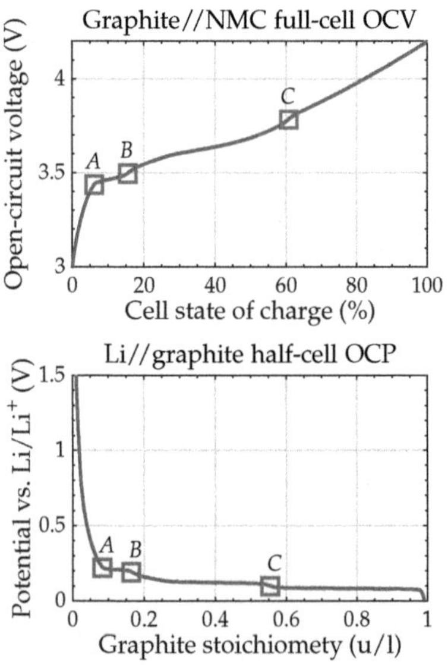

Figure 3.29: Distinct features in graphite OCP and cell OCV for a cell having a graphite negative electrode.

We can combine a known negative-electrode OCP with a cell-level OCV estimate to develop a model of positive-electrode OCP. We restrict our discussion here to cells having graphite negative electrodes; however, we believe this approach can generalize to other materials. Graphite's OCP function has distinct features located at points "A," "B," and "C" shown in Fig. 3.29 that also appear quite visibly in the cell's OCV function. These points can be found automatically:

- A is determined from a peak in $\mathrm{d}^2U_{\mathrm{ocv}}/\mathrm{d}z^2$ and $\mathrm{d}^2U^{\mathrm{n}}_{\mathrm{ocp}}/\mathrm{d}\theta_{\mathrm{s}}^2$.
- B and C are located at peaks in $\mathrm{d}U_{\mathrm{ocv}}/\mathrm{d}z$ and $\mathrm{d}U^{\mathrm{n}}_{\mathrm{ocp}}/\mathrm{d}\theta_{\mathrm{s}}$.

Given the above information, the following four steps are then implemented to estimate cell- and electrode-level quantities:

- Step 0: Use any of the four methods to produce an estimate of cell OCV versus SOC, $\widehat{U}_{\mathrm{ocv}}(z)$, from cell discharge and charge data.
- Step 1: Correlate $\{A, B, C\}$ between cell OCV and graphite OCP.
  - Denote the cell-level SOCs of these points as z_A, z_B, and z_C, and the negative-electrode stoichiometries as $\theta^{\mathrm{n}}_{\mathrm{s},A}, \theta^{\mathrm{n}}_{\mathrm{s},B}$, and $\theta^{\mathrm{n}}_{\mathrm{s},C}$.
  - Use $\theta^{\mathrm{n}}_{\mathrm{s}}(z) = \theta^{\mathrm{n}}_0 + z\left(\theta^{\mathrm{n}}_{100} - \theta^{\mathrm{n}}_0\right)$ to solve for θ^{n}_0 and $\theta^{\mathrm{n}}_{100}$ via:

$$\begin{bmatrix} \theta^{\mathrm{n}}_{\mathrm{s},A} \\ \theta^{\mathrm{n}}_{\mathrm{s},B} \\ \theta^{\mathrm{n}}_{\mathrm{s},C} \end{bmatrix} = \begin{bmatrix} 1 - z_A & z_A \\ 1 - z_B & z_B \\ 1 - z_C & z_C \end{bmatrix} \begin{bmatrix} \theta^{\mathrm{n}}_0 \\ \theta^{\mathrm{n}}_{100} \end{bmatrix}$$

$$\begin{bmatrix} \theta^{\mathrm{n}}_0 \\ \theta^{\mathrm{n}}_{100} \end{bmatrix} = \begin{bmatrix} 1 - z_A & z_A \\ 1 - z_B & z_B \\ 1 - z_C & z_C \end{bmatrix}^{\dagger} \begin{bmatrix} \theta^{\mathrm{n}}_{\mathrm{s},A} \\ \theta^{\mathrm{n}}_{\mathrm{s},B} \\ \theta^{\mathrm{n}}_{\mathrm{s},C} \end{bmatrix},$$

  where the dagger (†) represents a matrix pseudoinverse.[32]
- Step 2: Using the computed negative-electrode operating boundaries, find an MSMR model of $\widehat{U}^{\mathrm{p}}_{\mathrm{ocp}}(\theta^{\mathrm{p}}_{\mathrm{s}})$ and optimize positive-electrode operating boundaries θ^{p}_0 and θ^{p}_0 by fitting to the cell SOC versus OCV estimate.
  - To do so, we minimize the cost function:

$$J = \sum_k \left(z_k^{\mathrm{cell}} - z_k^{\mathrm{ocp}}\right)^2 + w\left(\frac{\mathrm{d}z_k}{\mathrm{d}\widehat{U}^{\mathrm{cell}}_{\mathrm{ocv}}} - \frac{\mathrm{d}z_k}{\mathrm{d}\widehat{U}^{\mathrm{ocp}}_{\mathrm{ocv}}}\right)^2, \qquad (3.8)$$

  where $\mathrm{d}z_k/\mathrm{d}\widehat{U}^{\mathrm{cell}}_{\mathrm{ocv}}$ is cell differential capacity computed using OCV data, $\mathrm{d}z_k/\mathrm{d}\widehat{U}^{\mathrm{ocp}}_{\mathrm{ocv}}$ is cell differential capacity using electrode OCP data, and w is a weighting factor.
  - The SOC estimates in the above cost function can be expressed using the MSMR model:

$$z_k^{\mathrm{ocp}} = \frac{\theta^{\mathrm{p}}_{\mathrm{s}}(z_k^{\mathrm{cell}}) - \theta^{\mathrm{p}}_0}{\theta^{\mathrm{p}}_{100} - \theta^{\mathrm{p}}_0}$$

[32] This is an over-determined set of equations: The pseudoinverse solution gives the least-squares estimate of the unknown boundary values θ^{n}_0 and $\theta^{\mathrm{n}}_{100}$.

$$\theta_s^p(z_k^{\text{cell}}) = \sum_{j=1}^{J^p} \frac{X_j}{1 + \exp[f(\widehat{U}_{\text{ocp}}^p(\theta_s^p(z_k^{\text{cell}})) - U_j^0)/\omega_j]}$$
$$\widehat{U}_{\text{ocp}}^p(\theta_s^p(z_k^{\text{cell}})) = \widehat{U}_{\text{ocv}}(z_k^{\text{cell}}) + U_{\text{ocp}}^n(\theta_s^n(z_k^{\text{cell}})).$$

- Step 3: Since the cost value J in Eq. (3.8) will not be exactly zero, we recompute the OCV estimates:

$$\widehat{U}_{\text{ocv}}(z_k^{\text{cell}}) = \widehat{U}_{\text{ocp}}^p(\theta_s^p(z_k^{\text{cell}})) - U_{\text{ocp}}^n(\theta_s^n(z_k^{\text{cell}})).$$

When we implement this procedure, we encounter a variation of the observability problem discussed earlier. We find that we can make a good model of positive-electrode OCP over the stoichiometric range covered by the OCV test, but we cannot guarantee that the boundaries $\{\theta_0^p, \theta_{100}^p\}$ are physical. The estimate $\widehat{U}_{\text{ocp}}^p(\theta_s^p(z_k^{\text{cell}}))$ may be shifted and/or stretched with respect to the true function.

To see this, consider that both boundaries might be offset by the same constant $\Delta\theta$:

$$z_k^{\text{ocp}} = \frac{\theta_s^p(z_k^{\text{cell}}) - \theta_0^p}{\theta_{100}^p - \theta_0^p} = \frac{\left(\theta_s^p(z_k^{\text{cell}}) + \Delta\theta\right) - \left(\theta_0^p + \Delta\theta\right)}{\left(\theta_{100}^p + \Delta\theta\right) - \left(\theta_0^p + \Delta\theta\right)},$$

where:

$$\theta_s^p(z_k^{\text{cell}}) + \Delta\theta = \sum_j \left[\Delta\theta_j + \frac{X_j}{1 + \exp\left[f\left(\widehat{U}_{\text{ocp}}^p(\theta_s^p(z_k^{\text{cell}})) - U_j^0\right)/\omega_j\right]}\right].$$

There is no change in solution, regardless of the value of $\Delta\theta$.

This infers it is possible for the MSMR model to compute a result having a constant offset (to the truth) by biasing both operating boundaries by the same amount. $\Delta\theta$ is also eliminated by differentiation, so differential-capacity estimates do not overcome this problem.

Based on these observations, in order to make a good MSMR model using this approach we must start our optimizations with initial values for $\{U_j^0, X_j, \omega_j\}^p$ that are close to the true values and put narrow constraints on these values during optimization. This is not always feasible with physical cells; we rely on some insight regarding the material composition of the positive electrode and a literature MSMR model to use as an initialization for the optimizations.[33]

[33] See also Sect. 3.9 where we investigate a way to calibrate the OCP relationships.

To summarize, in the process of estimating OCV and OCP, we have discovered two observability problems.

1. Attempts to find both electrode OCP functions from cell-level OCV fail because the solution will be biased by a voltage offset.
2. Attempts to find positive-electrode OCP from cell-level OCV and negative-electrode OCP will be biased by a stoichiometric offset.

These observability problems appear whenever the electrode data are missing, which adds emphasis to our recommendation that collecting data from both electrodes is always preferred when estimating the OCP functions and boundaries.

A2.2: Assume that we have a database of known materials

We quickly present a second alternative that might be used to determine electrode OCP functions and operating boundaries. We assume that we possess a database of MSMR models of $N^{\mathrm{n}} \geq 1$ negative-electrode and $N^{\mathrm{p}} \geq 1$ positive-electrode materials. We attempt to fit cell OCV using exhaustive combinations of these materials to see which fits cell OCV best. In principle, this requires $N^{\mathrm{n}} \times N^{\mathrm{p}}$ optimizations, but in practice some of these combinations will not make sense and can be skipped. For example, cell chemistries having LFP positive electrodes have OCV relationships that are visually distinct from those having other electrodes; therefore, not every alternative from the database must to be checked if we can narrow down the reasonable candidates based on experience and simple observation.

The procedure to follow is:

1. Select one $U^{\mathrm{n}}_{\mathrm{ocp},j}(\theta^{\mathrm{n}}_{\mathrm{s}})$ and one $U^{\mathrm{p}}_{\mathrm{ocp},k}(\theta^{\mathrm{p}}_{\mathrm{s}})$ from the database of OCP functions, where $1 \leq j \leq N^{\mathrm{n}}$ and $1 \leq k \leq N^{\mathrm{p}}$.
2. Optimize $\{\theta^{\mathrm{n}}_{0}, \theta^{\mathrm{n}}_{100}, \theta^{\mathrm{p}}_{0}, \theta^{\mathrm{p}}_{100}\}$ to fit OCV for this combination.
3. Store the value of the optimization cost function as $J_{j,k} = J$ for this combination of j and k; similarly, store the optimized boundaries as $\Theta_{j,k} = \{\theta^{\mathrm{n}}_{0}, \theta^{\mathrm{n}}_{100}, \theta^{\mathrm{p}}_{0}, \theta^{\mathrm{p}}_{100}\}$.
4. Repeat steps 1–3 for all distinct combinations (that make sense) of $1 \leq j \leq N^{\mathrm{n}}$ and $1 \leq k \leq N^{\mathrm{p}}$.
5. Retain $\left\{\theta^{\mathrm{n}}_{0}, \theta^{\mathrm{n}}_{100}, \theta^{\mathrm{p}}_{0}, \theta^{\mathrm{p}}_{100}\right\} = \Theta_{j^*,k^*}$, $U^{\mathrm{n}}_{\mathrm{ocp}}(\theta^{\mathrm{n}}_{\mathrm{s}}) = U^{\mathrm{n}}_{\mathrm{ocp},j^*}(\theta^{\mathrm{n}}_{\mathrm{s}})$, and $U^{\mathrm{p}}_{\mathrm{ocp}}(\theta^{\mathrm{p}}_{\mathrm{s}}) = U^{\mathrm{p}}_{\mathrm{ocp},k^*}(\theta^{\mathrm{p}}_{\mathrm{s}})$ for the j^* and k^* that minimize $J_{j,k}$.

3.8 *Validating nonteardown OCP estimation*

3.8.1 *Nonteardown application to simulation data*

In order to demonstrate the nonteardown approach, we make Assumption A2.1 and seek to find positive-electrode OCP and boundaries $\{\theta^{\mathrm{n}}_{0}, \theta^{\mathrm{n}}_{100}, \theta^{\mathrm{p}}_{0}, \theta^{\mathrm{p}}_{100}\}$ that produce the best match between the cell's OCV and its OCP functions. We again begin with a simulation study. The performance of using each of the four OCV-estimation methods is shown in Table 3.5; results using the true (not estimated) OCV are also listed in the top row for reference.

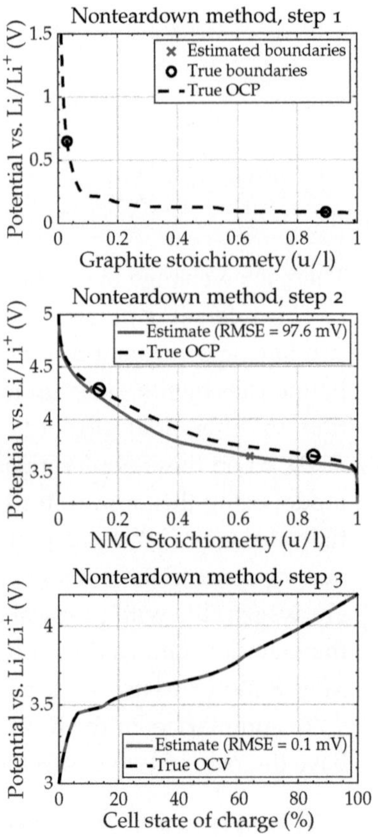

Figure 3.30: Illustrating the nonteardown method to find operating boundaries in simulation when the cell OCV and negative-electrode OCP are exact.

Despite the observability problem, constraints on positive-electrode MSMR parameter values help to give reasonable estimates. The re-

Step 0	Absolute Error (%)				RMSE (mV)	
	Step 1		Step 2		Step 2	Step 3
	$\theta_0^{\rm n}$	$\theta_{100}^{\rm n}$	$\theta_0^{\rm p}$	$\theta_{100}^{\rm p}$	$\widehat{U}_{\rm ocp}^{\rm p}(\theta_{\rm s}^{\rm p}(z))$	$\widehat{U}_{\rm ocv}(z)$
True OCV	0.03	0.00	6.71	0.17	22.17	0.07
Method 1	2.06	4.17	17.07	5.36	88.72	85.07
Method 2	0.39	0.14	12.22	7.55	92.52	14.00
Method 3	0.52	0.48	13.57	8.23	41.97	5.34
Method 4	0.07	3.86	13.25	0.52	39.90	6.60

Table 3.5: Results when making Assumption A2 in simulation.

sults were generated using the following constraints on the positive-electrode MSMR parameters:

$$3.5\,\mathrm{V} \le U_j^0 \le 4.3\,\mathrm{V}, \quad 0.1 \le X_j \le 0.45, \quad 0.5 \le \omega_j \le 6,$$

and the optimization used `fmincon` in MATLAB.

Overall, $\theta_0^{\rm p}$ proves to be the most difficult boundary to identify. The first table row shows that even when the true OCV is known, small errors in $\theta_0^{\rm n}$ can cause significant bias to the $\theta_0^{\rm p}$ estimate. When true OCV is unknown and must be estimated, positive-electrode boundary estimates are worse because of the second observability problem, resulting in large RMSE in the step-2 results. However, when the estimated OCP is stretched and/or shifted to span the cell SOC operating range (step 3), the RMSE is greatly reduced. This implies that the MSMR model between the operating boundaries is reasonable but that the operating boundaries themselves are biased.

Considering the final results, OCV-estimation Methods 3 and 4 lead to the best positive-electrode OCP models. The performance of this approach can also be graphically depicted: Fig. 3.30 shows boundary estimates and the positive-electrode OCP estimate when cell OCV and negative-electrode OCP are exact; Fig. 3.31 shows results when the negative-electrode OCP is exact but cell OCV is estimated from simulated discharge and charge data. In both cases, the step-2 outputs are clearly distorted compared with the truth. However, the shape of $\widehat{U}_{\rm ocp}^{\rm p}$ between the estimated boundaries $\hat{\theta}_0^{\rm p}$ and $\hat{\theta}_{100}^{\rm p}$ is good, as seen by the output of step 3. Therefore, we conclude that the method is producing inaccurate estimates of the boundaries. In Sect. 3.9, we propose a method to calibrate these boundary estimates.

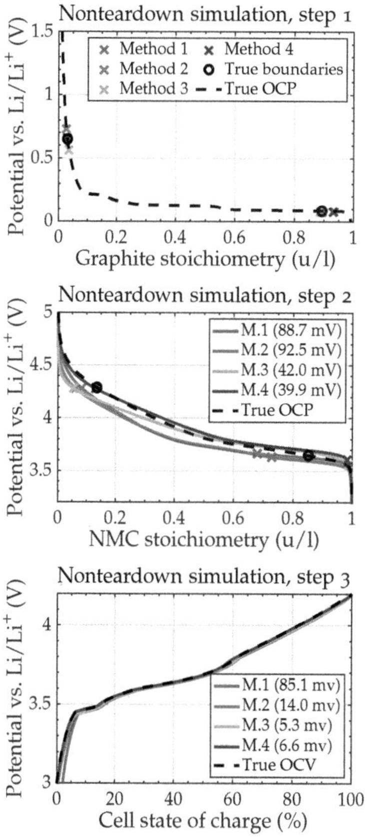

Figure 3.31: Illustrating the nonteardown method to find operating boundaries in simulation when electrode OCP is exact but cell OCV is estimated.

3.8.2 *Nonteardown application to laboratory cell data*

We again make Assumption A2.1 and seek to find positive-electrode OCP and estimates of boundaries $\{\theta_0^{\rm n}, \theta_{100}^{\rm n}, \theta_0^{\rm p}, \theta_{100}^{\rm p}\}$ corresponding to the Panasonic cell using lab data. We assume that the negative-electrode OCP function can be modeled using the MSMR model of

Step 0	Step 1		Step 2	
	θ_0^{n}	$\theta_{100}^{\mathrm{n}}$	θ_0^{p}	$\theta_{100}^{\mathrm{p}}$
Method 1 OCV	−0.0035	0.8499	0.6846	0.0943
Method 2 OCV	−0.0040	0.8323	0.6692	0.0876
Method 3 OCV	−0.0023	0.8436	0.6285	0.2115
Method 4 OCV	0.0007	0.8279	0.6478	0.2118

Table 3.6: Estimated operating boundaries for the Panasonic cell.

graphite parameterized with the values in Table 2.B.1. Optimizations to find the positive-electrode OCP model and boundaries were initialized using the MSMR model of NMC from the same table, and MSMR parameter values were constrained to vary from those in the table by no more than ±5 % during optimization.

Table 3.6 lists operating-boundary estimation results when cell OCV is estimated using the four methods. Note that step 1 produces θ_0^{n} estimates that are negative in most cases. This is a nonphysical result, which gives evidence that the Panasonic cell has a negative electrode that does not have an OCP that is exactly identical to that of the assumed MSMR model.

The estimates of $\theta_{100}^{\mathrm{n}}$ are all fairly similar, and are plotted in the top frame of Fig. 3.32. These results appear to be reasonable, and largely agree with those presented in the simulation table.

As expected, the positive-electrode boundaries do not agree well with those presented in the simulation table, again illustrating the second observability problem. This is seen clearly in the middle frame of Fig. 3.32. But when we combine the assumed negative-electrode OCP and the computed estimates of positive-electrode OCP to make overall OCV estimates, the results agree quite well, as shown in the lower frame of the figure. As with the simulation study, we believe that $\widehat{U}_{\mathrm{ocp}}^{\mathrm{p}}$ has an accurate shape but inaccurate boundary estimates, which we address in the next section. In the meantime, we might decide qualitatively that Methods 3 and 4 give the best results to this point since their OCV estimates fall closest to the center of the region spanned by the C/30 discharge and charge data.

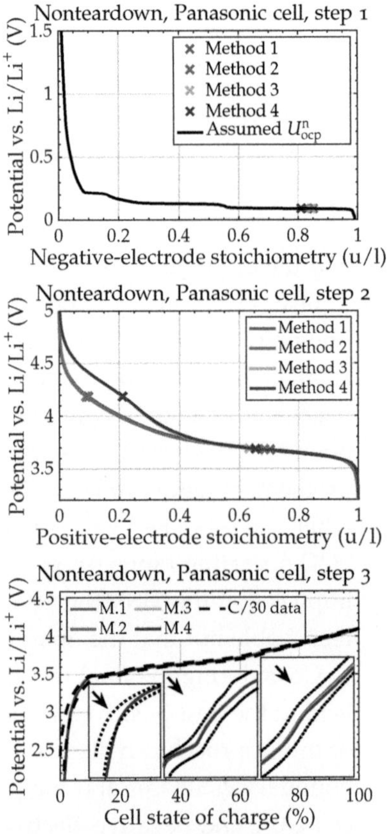

Figure 3.32: Illustrating the nonteardown method to find operating boundaries for the Panasonic cell.

3.9 *Discharge testing*

Reviewing the parameter-estimation roadmap of Fig. 3.8 to get a better sense of our progress to date, we observe that we have now studied OCP and OCV testing.[34] As we will review in this section, we believe that the OCP test is successful in producing a usable OCP model for graphite-dominant negative-electrode materials. That is, we believe that both the $\widehat{U}_{\mathrm{ocp}}^{\mathrm{n}}(\hat{\theta}_{\mathrm{s}}^{\mathrm{n}})$ function and the estimates $\hat{\theta}_0^{\mathrm{n}}$ and

[34] The content of this section has been adapted from: Dongliang Lu, M. Scott Trimboli, and Gregory L. Plett, "Cell discharge testing to calibrate a positive-electrode open-circuit-potential model for lithium-ion cells," *Journal of The Electrochemical Society*, 169(7):070524, 2022.

$\hat{\theta}^{\mathrm{n}}_{100}$ of the electrode's operating boundaries are accurate. However, there remains some uncertainty regarding the quality of the OCP model that we have made for positive-electrode materials.

The OCV test confirms that the shape of $\widehat{U}^{\mathrm{p}}_{\mathrm{ocp}}(\hat{\theta}^{\mathrm{p}}_{\mathrm{s}})$ is correct between the estimated positive-electrode operating boundaries $\hat{\theta}^{\mathrm{p}}_0$ and $\hat{\theta}^{\mathrm{p}}_{100}$; however, there have been some indications that the estimates of the operating boundaries themselves are biased. If this turns out to be the case, estimated electrode stoichiometry $\hat{\theta}^{\mathrm{p}}_{\mathrm{s}}$ at any point in time will possibly be quite different from the true $\theta^{\mathrm{p}}_{\mathrm{s}}$ since:

$$\begin{aligned}\theta^{\mathrm{p}}_{\mathrm{s}} &= \theta^{\mathrm{p}}_0 + z(\theta^{\mathrm{p}}_{100} - \theta^{\mathrm{p}}_0)\\ &\neq \hat{\theta}^{\mathrm{p}}_0 + z(\hat{\theta}^{\mathrm{p}}_{100} - \hat{\theta}^{\mathrm{p}}_0) = \hat{\theta}_{\mathrm{s}}.\end{aligned}$$

A plot of $\widehat{U}^{\mathrm{p}}_{\mathrm{ocp}}(\hat{\theta}^{\mathrm{p}}_{\mathrm{s}})$ versus $\hat{\theta}_{\mathrm{s}}$ then would be shifted and/or stretched with respect to a plot of $U^{\mathrm{p}}_{\mathrm{ocp}}(\theta^{\mathrm{p}}_{\mathrm{s}})$.

If teardown is possible, we might be able to calibrate the operating boundaries of the positive-electrode OCP model by using ICP–OES to determine the ratios of lithium and electrode elements at different resting voltages;[35] however, we believe that results obtained from ICP–OES from teardown of a commercial cell might overestimate the actual lithium content in the electrode due to the presence of lithium salts in the electrolyte, which cannot be completely washed from the electrode materials. Calibration might also be done using X-ray diffraction methods,[36] but the equipment and processing to do so is not readily available.

[35] This was done in Johannes Schmalstieg, Christiane Rahe, Madeleine Ecker, and Dirk Uwe Sauer, "Full cell parameterization of a high-power lithium-ion battery for a physico-chemical model: Part I. Physical and electrochemical parameters," *Journal of The Electrochemical Society*, 165(16):A3799, 2018.

[36] This was done in Franziska Friedrich, Benjamin Strehle, Anna T. S. Freiberg, Karin Kleiner, Sarah J. Day, Christoph Erk, Michele Piana, and Hubert A. Gasteiger, "Capacity fading mechanisms of NCM-811 cathodes in lithium-ion batteries studied by X-ray diffraction and other diagnostics," *Journal of The Electrochemical Society*, 166(15):A3760, 2019.

If the shape of $\widehat{U}^{\mathrm{p}}_{\mathrm{ocp}}(\hat{\theta}^{\mathrm{p}}_{\mathrm{s}})$ is correct, as we assume, we might wonder if it is necessary to calibrate its inputs $\hat{\theta}^{\mathrm{p}}_{\mathrm{s}}$. Perhaps we can use $\widehat{U}^{\mathrm{p}}_{\mathrm{ocp}}(\hat{\theta}^{\mathrm{p}}_{\mathrm{s}})$ and the biased operating boundaries $\hat{\theta}^{\mathrm{p}}_0$ and $\hat{\theta}^{\mathrm{p}}_{100}$ (and therefore biased operating points $\hat{\theta}^{\mathrm{p}}_{\mathrm{s}}$) without penalty in our LPM. Specifically, we wonder if we can substitute all instances of $U^{\mathrm{p}}_{\mathrm{ocp}}(\theta_{\mathrm{s}})$ and θ_{s} with the biased $\widehat{U}^{\mathrm{p}}_{\mathrm{ocp}}(\hat{\theta}^{\mathrm{p}}_{\mathrm{s}})$ and $\hat{\theta}^{\mathrm{p}}_{\mathrm{s}}$ and have an equivalent overall mathematical model, and hence compute the same result for all LPM internal electrochemical variables and voltage.

We investigate this idea here, starting by searching the LPM equations to discover where $U^{\mathrm{p}}_{\mathrm{ocp}}(\theta^{\mathrm{p}}_{\mathrm{s}})$ and $\theta^{\mathrm{p}}_{\mathrm{s}}$ appear. We find several references. One is when computing the overpotential η for the kinetics equation:

$$\eta^{\mathrm{r}} = \phi^{\mathrm{r}}_{\mathrm{s}} - \phi^{\mathrm{r}}_{\mathrm{e}} - U^{\mathrm{r}}_{\mathrm{ocp}}(\theta^{\mathrm{r}}_{\mathrm{ss}}),$$

and another is when initializing $\phi^{\mathrm{p}}_{\mathrm{s}}$ and $\phi^{\mathrm{r}}_{\mathrm{e}}$. We must also know the boundaries θ^{r}_0 and $\theta^{\mathrm{r}}_{100}$ of the operating range of $\theta^{\mathrm{r}}_{\mathrm{s}}$ during normal cell operation to be able to scale $\theta^{\mathrm{r}}_{\mathrm{s}}$ properly when accessing $U^{\mathrm{r}}_{\mathrm{ocp}}(\theta^{\mathrm{r}}_{\mathrm{s}})$ in the above applications. Additionally, $\theta^{\mathrm{r}}_{\mathrm{s}}$ is needed when computing the Butler–Volmer prefactor, which in our LPM is:[37]

$$i^{\mathrm{r}}_0 = \bar{k}^{\mathrm{r}}_0(\theta^{\mathrm{r}}_{\mathrm{e}})^{1-\alpha^{\mathrm{r}}}(1-\theta^{\mathrm{r}}_{\mathrm{ss}})^{1-\alpha^{\mathrm{r}}}(\theta^{\mathrm{r}}_{\mathrm{ss}})^{\alpha^{\mathrm{r}}},$$

[37] The same basic principle applies to the MSMR kinetics-model prefactor.

and if we are modeling SOC-dependent solid diffusivity as:[38]

$$\bar{D}_{\rm s}^{\rm r}(\theta_{\rm s}^{\rm r}) = \bar{D}_{\rm s,ref}^{\rm r} \frac{F}{RT} \theta_{\rm s}^{\rm r}(\theta_{\rm s}^{\rm r} - 1) \left. \frac{{\rm d}U_{\rm ocp}^{\rm r}(\theta^{\rm r})}{{\rm d}\theta^{\rm r}} \right|_{\theta^{\rm r} = \theta_{\rm s}^{\rm r}} . \tag{3.9}$$

[38] Daniel R. Baker and Mark W. Verbrugge, "Intercalate diffusion in multiphase electrode materials and application to lithiated graphite," *Journal of The Electrochemical Society*, 159(8):A1341, 2012; and Mark Verbrugge, Daniel Baker, and Xingcheng Xiao, "Formulation for the treatment of multiple electrochemical reactions and associated speciation for the lithium-silicon electrode," *Journal of The Electrochemical Society*, 163(2):A262–A271, 2015.

Having searched the LPM for references to $U_{\rm ocp}^{\rm p}(\theta_{\rm s}^{\rm p})$ and $\theta_{\rm s}^{\rm p}$, we must now consider the following questions: Are biased OCP and lithiation sufficient to evaluate a LPM correctly? Or must we know unbiased OCP and lithiation? Some simple analysis will show that the biased OCP relationship is not a suitable replacement for the unbiased OCP relationship in all places in a LPM. That is, we require accurate knowledge of $\theta_{\rm s}$ and not merely biased $\hat{\theta}_{\rm s}$.

Consider first the overpotential equation:

$$\begin{aligned} \eta^{\rm r} &= \phi_{\rm s}^{\rm r} - \phi_{\rm e}^{\rm r} - U_{\rm ocp}^{\rm r}(\theta_{\rm ss}^{\rm r}) \\ &= \phi_{\rm s}^{\rm r} - \phi_{\rm e}^{\rm r} - \widehat{U}_{\rm ocp}^{\rm r}(\hat{\theta}_{\rm ss}^{\rm r}), \end{aligned}$$

since we have assumed that the shape of $\widehat{U}_{\rm ocp}(\hat{\theta}_{\rm s})$ is correct. This result shows that we can compute the same overpotential regardless of whether our LPM uses $\theta_{\rm s}$ or $\hat{\theta}_{\rm s}$. We find that the same is true when initializing $\phi_{\rm s}^{\rm p}$ and $\phi_{\rm e}^{\rm r}$. But we run into problems when we examine the prefactor to the Butler–Volmer equation:

$$i_0^{\rm r} = \bar{k}_0^{\rm r}(\theta_{\rm e}^{\rm r})^{1-\alpha^{\rm r}}(\theta_{\rm ss}^{\rm r})^{\alpha^{\rm r}}(1 - \theta_{\rm ss}^{\rm r})^{1-\alpha^{\rm r}}.$$

We define an affine relationship between $\theta_{\rm s}^{\rm p}$ and $\hat{\theta}_{\rm s}^{\rm p}$ to be:[39]

[39] The interpretation of scale factor m and shift factor b will be discussed shortly.

$$(1 - \theta_{\rm s}^{\rm p}) = m(1 - \hat{\theta}_{\rm s}^{\rm p}/b). \tag{3.10}$$

Then we can write:

$$i_0^{\rm r} = \bar{k}_0^{\rm r}(\theta_{\rm e}^{\rm r})^{1-\alpha^{\rm r}}(1 - m(1 - \hat{\theta}_{\rm ss}/b))^{\alpha^{\rm r}}(m(1 - \hat{\theta}_{\rm ss}/b))^{1-\alpha^{\rm r}}.$$

It is not possible to evaluate this expression based only on $\hat{\theta}_{\rm ss}$ without knowledge of m and b. Using a similar analysis, we find that there is no way to evaluate Eq. (3.9) having knowledge of only $\hat{\theta}_{\rm s}$.

So, it is important to calibrate our OCP relationships. Sect. 3.3 proposed that relative OCP relationships might be calibrated by regressing them to an MSMR model. The reasoning behind this idea was that the MSMR model has the ability to extrapolate in a physically meaningful way somewhat beyond measured data into lithiation regions that are inaccessible by the OCP tests (e.g., if lithiating above $\theta_{\rm max}$ or delithiating below $\theta_{\rm min}$ would cause material failure). In so doing, we hoped to overcome the inaccessible-lithium problem, which was earlier illustrated in Fig. 3.12.

Referring once again to that figure, we observe that the relative OCP estimate for graphite (in the nonshaded regions in the top frame) has steep slopes at both low and high lithiation levels. According to the MSMR-model framework, this implies that we have most likely collected data spanning all galleries of the material, and so regressing the relative OCP estimate to an MSMR model does not change it much; only the narrow gray shaded regions must be added to calibrate the values of θ_{min} and θ_{max}. We conclude that the calibrated absolute OCP relationship is likely to be accurate because nearly all of the OCP relationship was accessible between v_{min} and v_{max}.

The situation is different for the NMC622 electrode in the lower frame of the figure. We do observe a steep slope in the measured OCP near v_{min}, so we have high confidence that we have lithiated the half cell to a level very close to $\theta_s = 1$. However, we do not observe a steep slope in the measured OCP near v_{max}; therefore, we can be certain that the absolute level of lithiation of the half cell at that point is not close to zero. By regressing an MSMR model to the measured data, we extrapolate beyond the measured range into the gray inaccessible range. However, we cannot be certain that we have achieved the correct result. There may be one or more MSMR galleries in the inaccessible range that have not been sufficiently excited by the OCP test, and so their contribution is either underestimated or completely omitted from the MSMR model that we regress. That is, the inaccessible region may be wider than what is shown, in an absolute sense.

This problem becomes even more evident when we consider how it might be possible to determine electrode OCP without teardown as was proposed in Sect. 3.7. If we are willing to assume that the negative electrode is dominated by graphite, as is presently common, then we can compute a segment of the positive-electrode relative OCP by adding graphite's OCP relationship to the cell's OCV. By regressing an MSMR model to this segment, we can also compute an estimate of the electrode operating boundaries. Fig. 3.32 illustrated a problem with this method, as applied to a physical graphite//NMC cell. The top frame shows the assumed negative-electrode OCP function as well as the operating boundaries found by four different methods. The estimated boundaries are not identical but are quite consistent. The problem appears in the middle frame, which shows the estimates that were made of the positive-electrode OCP function and its operating boundaries. Different OCV-processing methods produced results that were very different from one another in terms of their calibration to an absolute stoichiometry. Several of the methods regressed MSMR models that had very wide inaccessible regions (by adding a high-voltage gallery to the MSMR model) and several of the methods regressed MSMR models that had narrower inaccessible regions.

However, the OCP models produced by all methods led to essentially equivalent OCV estimates, so OCV itself cannot be used to judge the quality of the absolute MSMR model.[40]

[40] The problem is due in part to the narrow voltage range in the positive electrode spanned by the nonteardown OCV test; therefore the regression of the extracted portion of the positive-electrode OCP to an MSMR model fails to calibrate well. The teardown OCP test executed on a half cell is able to span a wider voltage range, enabling better calibration, especially close to $\theta_s = 1$.

The point of the example is that while regressing an MSMR model can improve calibration of a relative OCP model, unless the measured slopes at both ends of the collected data are steep it is not guaranteed to produce a valid calibration. We have seen that an accurate absolute OCP relationship is needed by the LPM; therefore, we desire to find a better way to calibrate a biased OCP model.

3.10 *Calibrating OCP*

Before discussing how to calibrate the positive-electrode OCP model, we review the lab-test and parameter-estimation steps we must take to reach this point:

- Estimate absolute OCP $U_{\text{ocp}}^{\text{n}}(\theta_{\text{s}}^{\text{n}})$ for the negative electrode and its corresponding MSMR model $\{U_j^0, X_j, \omega_j\}^{\text{n}}$ from the OCP test.
- Create a preliminary biased OCP estimate $\widehat{U}_{\text{ocp}}^{\text{p}}(\hat{\theta}_{\text{s}}^{\text{p}})$ for the positive electrode from the OCP test. For the purposes of this section, this biased OCP estimate may be equal either to the relative OCP $\widetilde{U}_{\text{ocp}}^{\text{p}}(\tilde{\theta}_{\text{s}}^{\text{p}})$ produced by the first step of Methods 1–3 or to the final estimate of absolute OCP produced by any of the four methods.
- Estimate cell total capacity Q, cell OCV $U_{\text{ocv}}(z)$, and the negative-electrode operating boundaries θ_0^{n}, and θ_{100}^{n} from the OCV test.
- Determine preliminary biased estimates $\hat{\theta}_0^{\text{p}}$ and $\hat{\theta}_{100}^{\text{p}}$ of the positive-electrode boundaries from the OCV test.

The form of the MSMR model leads us to expect that all OCP relationships will have steep slopes near both $\theta_{\text{s}} = 0$ and $\theta_{\text{s}} = 1$. This was true for $\widetilde{U}_{\text{ocp}}^{\text{n}}(\tilde{\theta}_{\text{s}}^{\text{n}})$ with graphite negative electrodes, which is why we believe that the MSMR model $U_{\text{ocp}}^{\text{n}}(\theta_{\text{s}}^{\text{n}})$ and estimates of θ_0^{n} and θ_{100}^{n} are accurate. However, for many positive-electrode chemistries such as LCO, NCA, and NMC, we will not observe a steep slope in the voltage at low levels of lithiation because we are not able to delithiate the materials sufficiently without causing damage to the electrodes and/or the electrolyte. This means that simply regressing $\widetilde{U}_{\text{ocp}}^{\text{p}}(\tilde{\theta}_{\text{s}}^{\text{p}})$ to an MSMR model will not yield a reliably calibrated $U_{\text{ocp}}^{\text{p}}(\theta_{\text{s}}^{\text{p}})$. This is why we do not yet have final $U_{\text{ocp}}^{\text{p}}(\theta_{\text{s}}^{\text{p}})$, θ_0^{p}, or θ_{100}^{p}.

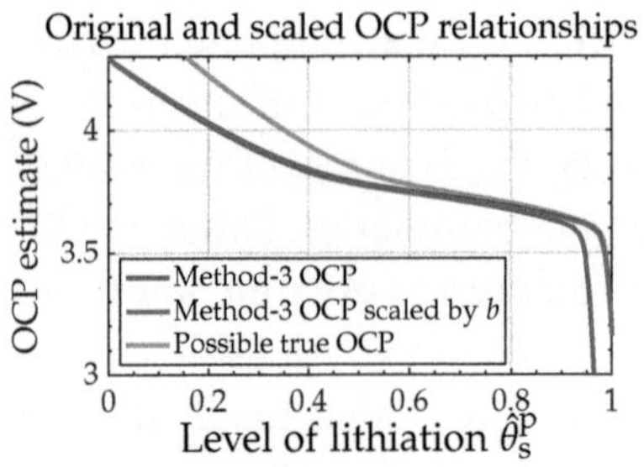

Figure 3.33: Illustrating the two steps of the calibration process.

Fig. 3.33 illustrates the type of transformation that will be required to calibrate $\widehat{U}_{\text{ocp}}^{\text{p}}(\hat{\theta}_{\text{s}}^{\text{p}})$. The blue line shows the positive-electrode relative OCP relationship $\widetilde{U}_{\text{ocp}}^{\text{p}}(\tilde{\theta}_{\text{s}}^{\text{p}})$ that was output by OCP Method 3, which as input to the calibration process we now rename to be $\widehat{U}_{\text{ocp}}^{\text{p}}(\hat{\theta}_{\text{s}}^{\text{p}})$. We notice that it has a steep slope near $\hat{\theta}_{\text{s}}^{\text{p}} = 1$ but not exactly at $\hat{\theta}_{\text{s}}^{\text{p}} = 1$. We will need to adjust the blue curve manually using a shift factor we denote as "b" to move this slope to $\hat{\theta}_s^p = 1$, as illustrated by the red line. Then, we will somehow need to determine how much compression is required to calibrate the red line to achieve the calibrated OCP curve, drawn as the green line. We will implement this using a scale factor "m," which will be found automatically.

The fact that solid diffusivity $\bar{D}_s^r$ is SOC-dependent gives us an opportunity to calibrate θ_s^p, and therefore $U_{\text{ocp}}^p(\theta_s^p)$. Here, we rely on literature that shows that single-particle models (SPMs) can predict cell voltage well for low-rate constant-current profiles.[41] So, we collect constant-current data from the cell; we then optimize the parameter values of the SPM—including scale factor m—to fit these data. Upon doing so, the optimized m calibrates our OCP relationship.

In the SPM, we assume that our existing model of $U_{\text{ocp}}^n(\theta_s^n)$ and estimates of θ_0^n and θ_{100}^n are accurate and that biased $\widehat{U}_{\text{ocp}}^p(\hat{\theta}_s^p)$, $\hat{\theta}_0^p$, and $\hat{\theta}_{100}^p$ can be corrected by the conversion of Eq. (3.10). The shift factor "b" adjusts the $\widehat{U}_{\text{ocp}}^p(\hat{\theta}_s^p)$ relationship so that the steep slope moves from $\hat{\theta}_s^p \approx b$ to $\hat{\theta}_s^p \approx 1$ (see Fig. 3.33). For now, we select b manually to achieve this goal.

The $\widehat{U}_{\text{ocp}}^p(\hat{\theta}_s^p/b)$ relationship then reflects a fully lithiated positive electrode when $\hat{\theta}_s^p/b = 1$. The implications of scale factor m then are: (1) when $\hat{\theta}_s^p/b = 1$, then $\theta_s^p = 1$ also; and (2) when $\hat{\theta}_s^p/b = 0$, then $\theta_s^p = 1 - m$. So, the m scale factor stretches or compresses $\widehat{U}_{\text{ocp}}^p(\hat{\theta}_s^p/b)$ to determine $U_{\text{ocp}}^p(\theta_s^p)$.

3.10.1 *Rederiving SPM equations to calibrate OCP*

There are multiple SPM variations in the literature.[42] Here, we adopt a very simple polynomial-based SPM, where the stoichiometry profile inside a particle is modeled as a fourth-order polynomial in r.[43] Using lumped parameter values, this polynomial-based SPM can be expressed as:[44]

$$\begin{bmatrix} \dot{\theta}_{\text{avg}}^r(t) \\ \nabla\dot{\theta}_{\text{avg}}^r(t) \end{bmatrix} = \begin{bmatrix} 0 & 0 \\ 0 & -30\bar{D}_s^r(\theta_{\text{ss}}^r(t)) \end{bmatrix} \begin{bmatrix} \theta_{\text{avg}}^r(t) \\ \nabla\theta_{\text{avg}}^r(t) \end{bmatrix} + \begin{bmatrix} \frac{-1}{3600Q^r} \\ \frac{-1}{480Q^r} \end{bmatrix} i_f^r(t)$$
$$[\theta_{\text{ss}}^r(t)] = \begin{bmatrix} 1 & \frac{8}{35} \end{bmatrix} \begin{bmatrix} \theta_{\text{avg}}^r(t) \\ \nabla\theta_{\text{avg}}^r(t) \end{bmatrix} + \begin{bmatrix} \frac{-(1/378\,000)}{Q^r\bar{D}_s^r(\theta_{\text{ss}}^r(t))} \end{bmatrix} i_f^r(t), \tag{3.11}$$

where we set $i_f^n(t) = i_{\text{app}}(t)$ and $i_f^p(t) = -i_{\text{app}}(t)$.

The state variables of this SPM model are: $\theta_{\text{avg}}(t)$, which is the average stoichiometry of the particle, and $\nabla\theta_{\text{avg}}(t)$, which is the average gradient of the stoichiometry within the particle. The output variable of the SPM is surface stoichiometry $\theta_{\text{ss}}(t)$. In the equations, we define electrode absolute capacity as a function of cell total capacity Q to be: $Q^r = Q/\left|\theta_{100}^r - \theta_0^r\right|$. Cell voltage is approximated

[41] For some SPM variations found in the literature, confer:

- Dong Zhang, Satadru Dey, Luis D. Couto, and Scott J. Moura, "Battery adaptive observer for a single-particle model with intercalation-induced stress," *IEEE Transactions on Control Systems Technology*, 28(4):1363–1377, 2019.
- Dong Zhang, Branko N. Popov, and Ralph E. White, "Modeling lithium intercalation of a single spinel particle under potentiodynamic control," *Journal of the Electrochemical Society*, 147(3):831, 2000.
- Venkat R. Subramanian, James A. Ritter, and Ralph E. White, "Approximate solutions for galvanostatic discharge of spherical particles I. Constant diffusion coefficient," *Journal of the Electrochemical Society*, 148(11):E444, 2001.
- Meng Guo, Godfrey Sikha, and Ralph E. White, "Single-particle model for a lithium-ion cell: Thermal behavior," *Journal of The Electrochemical Society*, 158(2):A122, 2010.
- Tanvir R. Tanim, Christopher D. Rahn, and Chao-Yang Wang, "A temperature dependent, single particle, lithium ion cell model including electrolyte diffusion," *Journal of Dynamic Systems, Measurement, and Control*, 137(1), 2015.
- Krishnakumar Gopalakrishnan and Gregory J. Offer, "A composite single particle lithium-ion battery model through system identification," *IEEE Transactions on Control Systems Technology*, 30(1):1–13, 2021.

[42] For example, confer those listed in the previous sidenote.

[43] Venkat R. Subramanian, Vinten D. Diwakar, and Deepak Tapriyal, "Efficient macro-micro scale coupled modeling of batteries," *Journal of The Electrochemical Society*, 152(10):A2002, 2005.

[44] The derivation details are presented in App. 3.B.

as:

$$v(t) = U_{\text{ocp}}^{\text{p}}(\theta_{\text{ss}}^{\text{p}}(t)) - U_{\text{ocp}}^{\text{n}}(\theta_{\text{ss}}^{\text{n}}(t)) - R_0 i_{\text{app}}(t). \tag{3.12}$$

The benefit of using this very simple polynomial form of SPM over more complex SPMs is that we do not require knowledge of many parameter values to apply it to predict cell voltage. This is important because we will not have estimates of any cell parameter values (besides estimates of biased OCP and electrode operating boundaries) at this point in the process of parameter estimation.[45] We can simulate the negative electrode using this SPM model, where the only unknown parameter value is $\bar{D}_{\text{s,ref}}^{\text{n}}$. However, when we seek to simulate the positive electrode, we know neither $\bar{D}_{\text{s,ref}}^{\text{p}}$, nor $U_{\text{ocp}}^{\text{p}}$, nor θ_0^{p}, nor θ_{100}^{p}. To use the SPM to model the positive electrode, we must convert its expressions to be in terms of $\hat{\theta}_{\text{s}}^{\text{p}}$, $\widehat{U}_{\text{ocp}}^{\text{p}}(\hat{\theta}_{\text{s}}^{\text{p}})$, m, and b. Then we can optimize the parameter values $\bar{D}_{\text{s,ref}}^{\text{n}}$, $\bar{D}_{\text{s,ref}}^{\text{p}}$, and m until the SPM model matches the measured data as closely as possible.[46] We will also need to optimize R_0, but since it is "linear in the parameters" this turns out to be a linear-optimization substep of the overall nonlinear optimization to find the three other values, as we will see shortly.

[45] Two significant limitations to Eq. (3.12) are: (1) it ignores the spatial dependence of $\theta_{\text{ss}}^{\text{r}}(\tilde{x}, t)$ in each electrode, and (2) it assumes that R_0 is not a function of SOC, which in fact it is. Both of these limitations are mitigated to a large extent by using a constant-current input having low magnitude. Therefore, Eq. (3.12) "works" reasonably well for the purpose of this section—calibrating a positive-electrode OCP function—but is inadequate as a voltage model suitable for general application.
Once the positive-electrode OCP function is calibrated, Sects. 3.12 through 3.17 will focus on estimating LPM parameter values necessary to predict voltage more accurately under general conditions.

[46] Note that in the approach as we have presented it so far we have already chosen b manually, so we do not need to use optimization to find its value.

The desired output of this overall procedure is the set of factors b and m, which calibrate the positive-electrode OCP function. However, a further benefit of this process is that it gives rough estimates of $\bar{D}_{\text{s,ref}}^{\text{n}}$ and $\bar{D}_{\text{s,ref}}^{\text{p}}$ as well, which will be used as initial values to the follow-on frequency-response parameter-estimation methods.

3.10.2 *Converting the SPM to biased stoichiometry*

We seek to convert the positive-electrode SPM equations, which originate as functions of $\theta_{\text{s}}^{\text{p}}$, to be in terms of $\hat{\theta}_{\text{s}}^{\text{p}}$, m, and b. We will substitute the following identities, which are consequences of Eq. (3.10):

$$\begin{aligned}
\theta_{\text{s}}^{\text{p}} &= 1 - \frac{m}{b}\left(b - \hat{\theta}_{\text{s}}^{\text{p}}\right) \\
\theta_{0}^{\text{p}} &= 1 - \frac{m}{b}\left(b - \hat{\theta}_{0}^{\text{p}}\right) \\
\theta_{100}^{\text{p}} &= 1 - \frac{m}{b}\left(b - \hat{\theta}_{100}^{\text{p}}\right).
\end{aligned}$$

We start the conversion process by considering the positive-electrode absolute capacity Q^{p}:

$$Q^{\text{p}} = \frac{Q}{\left|\theta_{100}^{\text{p}} - \theta_0^{\text{p}}\right|} = \frac{b}{m}\frac{Q}{\left|\hat{\theta}_{100}^{\text{p}} - \hat{\theta}_0^{\text{p}}\right|} = \frac{b}{m}\widetilde{Q}^{\text{p}},$$

where $\widetilde{Q}^{\text{p}} = Q/\left|\hat{\theta}_{100}^{\text{p}} - \hat{\theta}_0^{\text{p}}\right|$. Therefore, in the SPM we replace positive-electrode absolute capacity, which we do not know, a scaled version of the relative capacity, which we do know.

We next seek to convert the SPM equation for the electrode-average stoichiometry, which is:

$$\frac{\mathrm{d}\theta^{\mathrm{p}}_{\mathrm{avg}}(t)}{\mathrm{d}t} = -\frac{1}{3600Q^{\mathrm{p}}} i^{\mathrm{p}}_{\mathrm{f}}(t). \tag{3.13}$$

Substituting the relationships between $\theta^{\mathrm{p}}_{\mathrm{s}}$ and $\hat{\theta}^{\mathrm{p}}_{\mathrm{s}}$ per Eq. (3.10):

$$\begin{aligned}
\frac{m}{b}\frac{\mathrm{d}\hat{\theta}^{\mathrm{p}}_{\mathrm{avg}}(t)}{\mathrm{d}t} &= -\frac{m}{b}\frac{1}{3600\widetilde{Q}^{\mathrm{p}}} i^{\mathrm{p}}_{\mathrm{f}}(t) \\
\frac{\mathrm{d}\hat{\theta}^{\mathrm{p}}_{\mathrm{avg}}(t)}{\mathrm{d}t} &= -\frac{1}{3600\widetilde{Q}^{\mathrm{p}}} i^{\mathrm{p}}_{\mathrm{f}}(t),
\end{aligned} \tag{3.14}$$

which has exactly the same form as Eq. (3.13), which we cannot evaluate; however, we can evaluate Eq. (3.14) since we know all terms.

Next, we consider the SPM equation for electrode average gradient of stoichiometry, which is:

$$\frac{\mathrm{d}\nabla\theta^{\mathrm{p}}_{\mathrm{avg}}(t)}{\mathrm{d}t} = -30\bar{D}^{\mathrm{p}}_{\mathrm{s}}(\theta^{\mathrm{r}}_{\mathrm{ss}}(t))\nabla\theta^{\mathrm{p}}_{\mathrm{avg}}(t) - \frac{1}{480Q^{\mathrm{p}}} i^{\mathrm{p}}_{\mathrm{f}}(t).$$

Substituting the relationships between $\theta^{\mathrm{p}}_{\mathrm{s}}$ and $\hat{\theta}^{\mathrm{p}}_{\mathrm{s}}$ per Eq. (3.10):[47]

$$\begin{aligned}
\frac{m}{b}\frac{\mathrm{d}\nabla\hat{\theta}^{\mathrm{p}}_{\mathrm{avg}}(t)}{\mathrm{d}t} &= -30\frac{m}{b}\bar{D}^{\mathrm{p}}_{\mathrm{s}}\left(1 - \frac{m}{b}(b - \hat{\theta}^{\mathrm{r}}_{\mathrm{ss}}(t^-))\right)\nabla\hat{\theta}^{\mathrm{p}}_{\mathrm{avg}}(t) - \frac{m}{b}\frac{1}{480\widetilde{Q}^{\mathrm{p}}} i^{\mathrm{p}}_{\mathrm{f}}(t) \\
\frac{\mathrm{d}\nabla\hat{\theta}^{\mathrm{p}}_{\mathrm{avg}}(t)}{\mathrm{d}t} &= -30\,\bar{D}^{\mathrm{p}}_{\mathrm{s}}\left(1 - \frac{m}{b}(b - \hat{\theta}^{\mathrm{r}}_{\mathrm{ss}}(t^-))\right)\nabla\hat{\theta}^{\mathrm{p}}_{\mathrm{avg}}(t) - \frac{1}{480\widetilde{Q}^{\mathrm{p}}} i^{\mathrm{p}}_{\mathrm{f}}(t).
\end{aligned}$$

[47] To do so, we recognize that $\nabla\tilde{\theta}^{\mathrm{p}}_{\mathrm{avg}}$ is itself a gradient, which eliminates constant terms in the transformation.

This also has exactly the same form as before, except for the argument to the $\bar{D}^{\mathrm{p}}_{\mathrm{s}}$ function, which we need to examine closely. To condense notation, we create a temporary definition:

$$\theta^*_{\mathrm{s}} = 1 - \frac{m}{b}(b - \hat{\theta}^{\mathrm{r}}_{\mathrm{ss}}).$$

Substituting this into Eq. (3.9), we have:

$$\bar{D}^{\mathrm{p}}_{\mathrm{s}}(\theta^*_{\mathrm{s}}) = \frac{m}{b}\bar{D}^{\mathrm{p}}_{\mathrm{s,ref}}\frac{F}{RT}\left(1 - \frac{m}{b}(b - \hat{\theta}^{\mathrm{p}}_{\mathrm{s}})\right)\left(\hat{\theta}^{\mathrm{p}}_{\mathrm{s}} - b\right)\left.\frac{\mathrm{d}U^{\mathrm{p}}_{\mathrm{ocp}}(\theta^{\mathrm{p}})}{\mathrm{d}\theta^{\mathrm{p}}}\right|_{\theta^{\mathrm{p}}=\theta^*_{\mathrm{s}}}.$$

However, $U^{\mathrm{p}}_{\mathrm{ocp}}(\theta^{\mathrm{p}}) = \widehat{U}^{\mathrm{p}}_{\mathrm{ocp}}\left(b - \frac{b}{m}(1 - \theta^{\mathrm{p}})\right)$, so:

$$\begin{aligned}
\left.\frac{\mathrm{d}U^{\mathrm{p}}_{\mathrm{ocp}}(\theta^{\mathrm{p}})}{\mathrm{d}\theta^{\mathrm{p}}}\right|_{\theta^{\mathrm{p}}=\theta^*_{\mathrm{s}}} &= \left.\frac{\mathrm{d}\widehat{U}^{\mathrm{p}}_{\mathrm{ocp}}\left(b - \frac{b}{m}(1 - \theta^{\mathrm{p}})\right)}{\mathrm{d}\left(b - \frac{b}{m}(1 - \theta^{\mathrm{p}})\right)}\frac{\mathrm{d}\left(b - \frac{b}{m}(1 - \theta^{\mathrm{p}})\right)}{\mathrm{d}\theta^{\mathrm{p}}}\right|_{\theta^{\mathrm{p}}=\theta^*_{\mathrm{s}}} \\
&= \frac{b}{m}\left.\frac{\mathrm{d}\widehat{U}^{\mathrm{p}}_{\mathrm{ocp}}(\hat{\theta}^{\mathrm{p}})}{\mathrm{d}\hat{\theta}^{\mathrm{p}}}\right|_{\hat{\theta}^{\mathrm{p}}=\hat{\theta}^{\mathrm{p}}_{\mathrm{s}}}.
\end{aligned}$$

So, finally, redefining $\widetilde{\bar{D}}_s^p(\cdot)$ to have input argument $\hat{\theta}_s^p$:

$$\widetilde{\bar{D}}_s^p(\hat{\theta}_s^p) = \bar{D}_{s,\mathrm{ref}}^p \frac{F}{RT}\left(1 - \frac{m}{b}(b - \hat{\theta}_s^p)\right)\left(\hat{\theta}_s^p - b\right)\left.\frac{\mathrm{d}\widehat{U}_{\mathrm{ocp}}^p(\hat{\theta}^p)}{\mathrm{d}\hat{\theta}^p}\right|_{\hat{\theta}^p = \hat{\theta}_s^p}.$$

Putting everything together, we can simulate the true SPM using biased inputs via the following definitions in the positive electrode:

$$\widetilde{Q}^p = \frac{Q}{\left|\hat{\theta}_{100}^p - \hat{\theta}_0^p\right|}$$

$$\begin{bmatrix} \dot{\hat{\theta}}_{\mathrm{avg}}^p(t) \\ \nabla\dot{\hat{\theta}}_{\mathrm{avg}}^p(t) \end{bmatrix} = \begin{bmatrix} 0 & 0 \\ 0 & -30\widetilde{\bar{D}}_s^p(\hat{\theta}_{ss}^p(t)) \end{bmatrix}\begin{bmatrix} \hat{\theta}_{\mathrm{avg}}^p(t) \\ \nabla\hat{\theta}_{\mathrm{avg}}^p(t) \end{bmatrix} + \begin{bmatrix} \frac{-1}{3600\widetilde{Q}^p} \\ \frac{-1}{480\widetilde{Q}^p} \end{bmatrix} i_f^r(t)$$

$$\left[\hat{\theta}_{ss}^p(t)\right] = \begin{bmatrix} 1 & \frac{8}{35} \end{bmatrix}\begin{bmatrix} \hat{\theta}_{\mathrm{avg}}^p(t) \\ \nabla\hat{\theta}_{\mathrm{avg}}^p(t) \end{bmatrix} + \begin{bmatrix} \dfrac{-(1/378\,000)}{\widetilde{Q}^p\widetilde{\bar{D}}_s^p(\hat{\theta}_{ss}^p(t))} \end{bmatrix} i_f^p(t)$$

$$\widetilde{\bar{D}}_s^p(\hat{\theta}_s^p) = \bar{D}_{s,\mathrm{ref}}^p \frac{F}{RT}\left(1 - \frac{m}{b}(b - \hat{\theta}_s^p)\right)\left(\hat{\theta}_s^p - b\right)\left.\frac{\mathrm{d}\widehat{U}_{\mathrm{ocp}}^p(\hat{\theta}^p)}{\mathrm{d}\hat{\theta}^p}\right|_{\hat{\theta}^p = \hat{\theta}_s^p}.$$

Optimization produces an estimate of m which we use to convert:[48]

$$U_{\mathrm{ocp}}^p(\theta_s^p) = \widehat{U}_{\mathrm{ocp}}^p\left(b - \frac{b}{m}\left(1 - \theta_s^p\right)\right)$$

$$\theta_0^p = 1 - \frac{m}{b}\left(b - \hat{\theta}_0^p\right)$$

$$\theta_{100}^p = 1 - \frac{m}{b}\left(b - \hat{\theta}_{100}^p\right).$$

Finally, we create an MSMR model $\{U_j^0, X_j, \omega_j\}^p$ of $U_{\mathrm{ocp}}^p(\theta_s^p)$.

[48] When we use the SPM inside an optimization code, it is helpful to have bounds for m. Suppose that we "know" that the true θ_s^p for $U_{\mathrm{ocp}}^p = 4.3\,\mathrm{V}$ is between 0.05 and 0.5. Also, suppose that from the OCP test we have that $\widehat{U}_{\mathrm{ocp}}^p = 4.3\,\mathrm{V}$ when $\hat{\theta}_s^p = \hat{\theta}_{4.3}$. Then optimization bounds on m are:

$$\frac{1 - 0.5}{b - \hat{\theta}_{4.3}^p} < \frac{m}{b} < \frac{1 - 0.05}{b - \hat{\theta}_{4.3}^p}.$$

3.10.3 Setting up the SPM regression

To use the SPM in an optimization process, we need to design a lab test to collect current/voltage data to compare with the output of the SPM to estimate $\{\bar{D}_{s,\mathrm{ref}}^r, m\}$. In doing so, we note the following:

- Polynomial SPMs work best for constant-current tests.
- To get the best estimates of $\bar{D}_{s,\mathrm{ref}}^r$, we need to test a cell over as wide a range of SOC as possible.
- If we use too low a C-rate, the modeling error will be dominated by estimation errors in $U_{\mathrm{ocp}}^n(\theta_s^n)$ and $\widehat{U}_{\mathrm{ocp}}^p(\hat{\theta}_s^p)$ since cell polarization will be very low and measurements will be dominated by OCV.
- If we use too high a C-rate, the SPM will not model cell voltage very well (the literature suggests that SPMs of the nature we are proposing work reasonably well up to about a 1C rate, or perhaps a little higher, depending on the actual cell design).

We choose to implement a variation of the OCV test presented in Table 3.2, with all steps executed at 25 °C, and where the constant-current rate is chosen to be in the neighborhood of 1C. The discharge voltage and charge voltage versus SOC are calibrated in the same way as we calibrated OCV data in Sect. 3.5. Discharge and/or charge voltage are regressed against voltage predictions made by the SPM to find optimized estimates of $\bar{D}^{\mathrm{n}}_{\mathrm{s,ref}}$, $\bar{D}^{\mathrm{p}}_{\mathrm{s,ref}}$, and m.

3.10.4 *Resolving R_0 when regressing parameter values*

When evaluating the SPM inside of an optimization routine, the $U^{\mathrm{r}}_{\mathrm{ocp}}(\theta^{\mathrm{r}}_{\mathrm{ss}}(t))$ terms can be computed if $\bar{D}^{\mathrm{r}}_{\mathrm{s,ref}}$ and m are given. But to predict voltage, we also need to know R_0, per Eq. (3.12).

When optimizing a set of parameter values, one possibility would be: (1) construct a cost function in terms of: $p_4 = \{\bar{D}^{\mathrm{n}}_{\mathrm{s,ref}}, \bar{D}^{\mathrm{p}}_{\mathrm{s,ref}}, m, R_0\}$; then, (2) find the p_4^* that minimizes the difference between predicted cell voltage and true $v(t)$. However, this is a 4D nonlinear optimization; we can reduce it to a 3D nonlinear optimization plus a 1D linear optimization.

To do so, we define values found by nonlinear optimization as: $p_3 = \{\bar{D}^{\mathrm{n}}_{\mathrm{s,ref}}, \bar{D}^{\mathrm{p}}_{\mathrm{s,ref}}, m\}$. Then, every time the cost function is evaluated for a specific p_3, we:

1. Compute $U^{\mathrm{p}}_{\mathrm{ocp}}(\theta^{\mathrm{p}}_{\mathrm{ss}}(t))$ and $U^{\mathrm{n}}_{\mathrm{ocp}}(\theta^{\mathrm{n}}_{\mathrm{ss}}(t))$ using the SPM and p_3.
2. Construct the vector (which ideally equals $-R_0 i_{\mathrm{app}}(t)$):

$$e(t) = v(t) - \left(U^{\mathrm{p}}_{\mathrm{ocp}}(\theta^{\mathrm{p}}_{\mathrm{ss}}(t)) - U^{\mathrm{n}}_{\mathrm{ocp}}(\theta^{\mathrm{n}}_{\mathrm{ss}}(t)) \right).$$

3. Solve the least-squares problem $\widehat{R}_0 = -\left[i_{\mathrm{app}}(t)\right]^{\dagger} e(t)$, where the † symbol denotes the pseudoinverse operation.
4. Define cost as:

$$\mathrm{rms}\left(v(t) - \left(U^{\mathrm{p}}_{\mathrm{ocp}}(\theta^{\mathrm{p}}_{\mathrm{ss}}(t)) - U^{\mathrm{n}}_{\mathrm{ocp}}(\theta^{\mathrm{n}}_{\mathrm{ss}}(t)) - \widehat{R}_0 i_{\mathrm{app}}(t) \right) \right).$$

Minimizing this cost function over the three variables of p_3 will automatically choose the best value of $\widehat{R}_0$ as well. This speeds up optimization and improves the likelihood of converging to a good local-minimum solution for p_3.

3.10.5 *Evaluating the SPM with ideal parameter values*

In the next section, we will use simulation to gain insight into the effectiveness of this method to calibrate $U^{\mathrm{p}}_{\mathrm{ocp}}(\theta^{\mathrm{p}}_{\mathrm{s}})$. Before we do so, we would like to understand how well the SPM can predict voltage when the parameter values $\{\bar{D}^{\mathrm{n}}_{\mathrm{s,ref}}, \bar{D}^{\mathrm{p}}_{\mathrm{s,ref}}, m\}$ have exactly the correct values. We use the parameter values for the NMC30 cell, which were listed in Table 2.C.1.

Fig. 3.34 shows discharge and charge data simulated for the NMC30 cell using COMSOL as solid lines (for C/2, 1C, and 2C rates); dashed lines show the SPM predictions. We see that the SPM and full-order LPM are in good agreement over much of the SOC range for the C-rates chosen for this demonstration. In order to visualize the modeling fidelity more easily, Fig. 3.35 shows the error between the full-order LPM and the SPM. We see that the SPM gives accurate predictions for most SOCs and that errors are largest for discharge when SOC $< 10\,\%$ for rates of 1C or greater. Based on this observation, when optimizing parameter values using discharge data we crop the dataset to use only voltages measured between $5\,\% < \text{SOC} < 100\,\%$ for C-rates greater than or equal to 1C. We believe that this will help to avoid biasing parameter estimates by the error contributed by the approximations of the SPM itself.

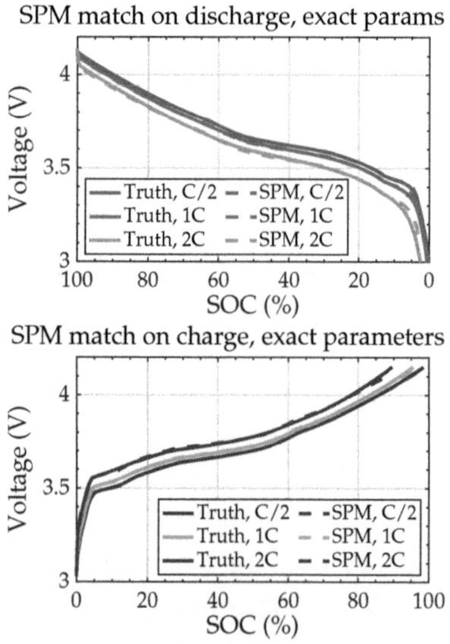

Figure 3.34: Evaluating the SPM match using a simulated LPM having known parameter values for different discharge and charge rates.

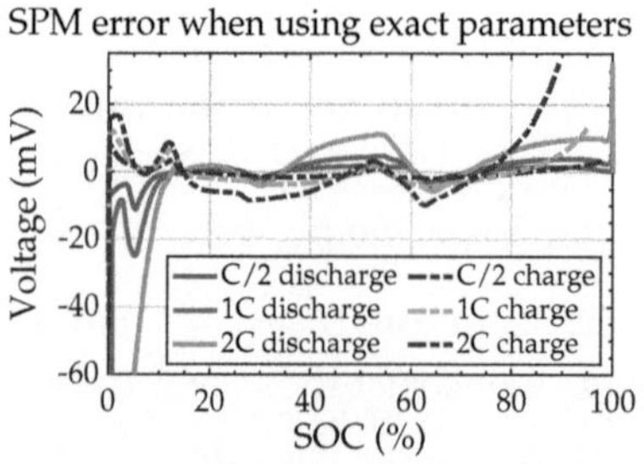

Figure 3.35: SPM modeling error for the simulations in Fig. 3.34.

3.11 *Validating calibration*

3.11.1 *Simulation results and strategy*

We now wish to validate the approach to estimating m. Our first step is to apply it to synthetic data computed by simulating a full-order LPM. This will allow us to evaluate whether we can find m reliably, since its value is known. We will then show the method applied to data collected from the Panasonic cell in our laboratory.

The simulation study was run using the parameter values for the NMC30 cell. We first applied the optimization to data collected at a 1C rate, regressing parameters for three different true values of m. For each of the three datasets (one for each value of m), we ran MATLAB's `particleswarm` function from its optimization toolbox 100 times with different initial randomized particle distributions and evaluated the mean and relative errors of the estimates. Table 3.7 summarizes the results, where boldface entries are the ones that we consider to be the best for each parameter being estimated.

We notice sizable variation in the results, especially for $\bar{D}^{\text{n}}_{\text{s,ref}}$. Estimates of $\bar{D}^{\text{n}}_{\text{s,ref}}$ are always best using charge data. Estimates of $\bar{D}^{\text{p}}_{\text{s,ref}}$ are best when the optimization uses both charge and discharge data, but are very similar when using only discharge data or only charge data. Estimates of m are best using discharge data.

Tables 3.8 and 3.9 list results when the constant-current rates were C/2 and 2C instead. Estimates of $\bar{D}^{\text{n}}_{\text{s,ref}}$ are greatly improved versus the 1C case when using discharge data for these cases. Estimation errors for $\bar{D}^{\text{p}}_{\text{s,ref}}$ are generally similar to the 1C case. Once again, estimates of m are uniformly best when using discharge data only.

Since our primary goal is to calibrate the positive electrode's OCP

True	Estimated Values and Percent Relative Error					
m	$\bar{D}^{n}_{s,ref}$ [s^{-1}]	Error	$\bar{D}^{p}_{s,ref}$ [s^{-1}]	Error	m	Error
	Estimates made using discharge data only					
0.9	3.01×10^{-3}	−351.2 %	1.06×10^{-4}	37.5 %	**0.90**	**0.4** %
1.0	3.11×10^{-3}	−367.1 %	1.06×10^{-4}	37.6 %	**0.99**	**0.9** %
1.1	3.07×10^{-3}	−359.8 %	1.06×10^{-4}	37.4 %	**1.09**	**0.5** %
	Estimates made using charge data only					
0.9	$\mathbf{3.70\times10^{-4}}$	**44.5** %	1.02×10^{-4}	40.0 %	1.33	−48.0 %
1.0	$\mathbf{3.70\times10^{-4}}$	**44.5** %	1.02×10^{-4}	40.0 %	1.48	−48.0 %
1.1	$\mathbf{3.70\times10^{-4}}$	**44.5** %	1.02×10^{-4}	40.0 %	1.63	−48.0 %
	Estimates made using both discharge and charge data					
0.9	3.46×10^{-4}	48.1 %	$\mathbf{1.19\times10^{-4}}$	**29.9** %	1.25	−38.8 %
1.0	3.44×10^{-4}	48.3 %	$\mathbf{1.19\times10^{-4}}$	**29.8** %	1.39	−38.8 %
1.1	3.44×10^{-4}	48.4 %	$\mathbf{1.19\times10^{-4}}$	**29.8** %	1.53	−38.9 %

Table 3.7: 1C simulation results. Note that the true reference solid diffusivities were: $\bar{D}^{n}_{s,ref} = 6.67\times10^{-4}$ [s^{-1}] and $\bar{D}^{p}_{s,ref} = 1.70\times10^{-4}$ [s^{-1}].

True	Estimated Values and Percent Relative Error					
m	$\bar{D}^{n}_{s,ref}$ [s^{-1}]	Error	$\bar{D}^{p}_{s,ref}$ [s^{-1}]	Error	m	Error
	Estimates made using discharge data only					
0.9	$\mathbf{7.57\times10^{-4}}$	**−13.6** %	$\mathbf{1.20\times10^{-4}}$	**29.4** %	**0.87**	**3.0** %
1.0	$\mathbf{7.59\times10^{-4}}$	**−13.8** %	$\mathbf{1.20\times10^{-4}}$	**29.4** %	**0.96**	**3.8** %
1.1	$\mathbf{7.57\times10^{-4}}$	**−13.5** %	$\mathbf{1.20\times10^{-4}}$	**29.3** %	**1.06**	**3.9** %
	Estimates made using charge data only					
0.9	3.74×10^{-4}	44.0 %	1.16×10^{-4}	31.6 %	1.22	−36.0 %
1.0	3.73×10^{-4}	44.0 %	1.21×10^{-4}	28.4 %	1.36	−36.1 %
1.1	3.74×10^{-4}	44.0 %	1.16×10^{-4}	31.6 %	1.50	−36.0 %
	Estimates made using both discharge and charge data					
0.9	3.74×10^{-4}	43.9 %	1.20×10^{-4}	29.4 %	1.16	−29.4 %
1.0	3.74×10^{-4}	43.9 %	1.20×10^{-4}	29.4 %	1.29	−29.4 %
1.1	3.74×10^{-4}	43.9 %	1.20×10^{-4}	29.4 %	1.42	−29.4 %

Table 3.8: C/2 simulation results. Note that the true reference solid diffusivities were: $\bar{D}^{n}_{s,ref} = 6.67\times10^{-4}$ [s^{-1}] and $\bar{D}^{p}_{s,ref} = 1.70\times10^{-4}$ [s^{-1}].

True	Estimated Values and Percent Relative Error					
m	$\bar{D}^{n}_{s,ref}$ [s^{-1}]	Error	$\bar{D}^{p}_{s,ref}$ [s^{-1}]	Error	m	Error
	Estimates made using discharge data only					
0.9	9.94×10^{-4}	−49.1 %	1.30×10^{-4}	23.2 %	**0.90**	**0.4** %
1.0	9.68×10^{-4}	−45.2 %	1.30×10^{-4}	23.3 %	**0.99**	**0.5** %
1.1	9.69×10^{-4}	−45.4 %	1.31×10^{-4}	22.8 %	**1.10**	**0.0** %
	Estimates made using charge data only					
0.9	4.19×10^{-4}	37.2 %	$\mathbf{1.52\times10^{-4}}$	**10.2** %	1.40	−55.2 %
1.0	4.16×10^{-4}	37.5 %	$\mathbf{1.54\times10^{-4}}$	**9.4** %	1.56	−55.7 %
1.1	4.15×10^{-4}	37.7 %	$\mathbf{1.55\times10^{-4}}$	**8.7** %	1.69	−54.0 %
	Estimates made using both discharge and charge data					
0.9	$\mathbf{4.67\times10^{-4}}$	**29.9** %	1.39×10^{-4}	18.1 %	1.14	−26.6 %
1.0	$\mathbf{4.59\times10^{-4}}$	**31.1** %	1.36×10^{-4}	19.5 %	1.27	−26.6 %
1.1	$\mathbf{4.70\times10^{-4}}$	**29.5** %	1.38×10^{-4}	18.7 %	1.39	−26.3 %

Table 3.9: 2C simulation results. Note that the true reference solid diffusivities were: $\bar{D}^{n}_{s,ref} = 6.67\times10^{-4}$ [s^{-1}] and $\bar{D}^{p}_{s,ref} = 1.70\times10^{-4}$ [s^{-1}].

function, we are most interested in assessing the quality of the estimate of m. And since the simulation study showed that estimates of m were always best using discharge data, we will restrict ourselves to using only discharge data from this point forward (and so we call this overall test the discharge test). There is a further benefit to this choice: it is generally possible to perform constant-current discharge of a physical cell at a variety of different C-rates; however, there are usually restrictions to the maximum C-rate that is advised when charging the cell (to avoid lithium plating, and so forth). By using discharge data only, we have greater flexibility in choosing the C-rates we wish to use. And while computing estimates of $\bar{D}^{\mathrm{r}}_{\mathrm{s,ref}}$ is not the focus of this effort, the proposed procedure does produce estimates that are almost always within a factor of 2 of the true value, and so the output estimates of $\bar{D}^{\mathrm{r}}_{\mathrm{s,ref}}$ will be used to initialize follow-on optimizations of these parameters using frequency-response data and parameter-estimation methods.

3.11.2 *Application to physical cell*

Having gained some insight into the performance of the proposed approach using simulation, we now seek to apply it to the challenge of calibrating the positive-electrode OCP function for the Panasonic cell. The precise NMC structure (i.e., x, y, and z in $\mathrm{Li}\left[\mathrm{Ni_xMn_yCo_z}\right]\mathrm{O_2}$) for this cell is unknown, and since different compositions can have distinct properties,[49] we cannot rely on literature results to calibrate our OCP estimate. Instead, we will use the discharge test to do so.

[49] Hyung-Joo Noh, Sungjune Youn, Chong Seung Yoon, and Yang-Kook Sun, "Comparison of the structural and electrochemical properties of layered $\mathrm{Li[Ni_xCo_yMn_z]O_2}$ ($x = 1/3, 0.5, 0.6, 0.7, 0.8$ and 0.85) cathode material for lithium-ion batteries," *Journal of Power Sources*, 233:121–130, 2013.

When implementing the discharge test, we must select one or more C-rates to use. We choose to do so in a way that produces similar voltage traces to the ones encountered during the simulation study. The top frame of Fig. 3.36 shows voltage-simulation output for the discharge and charge tests. We see that the discharge curves span a wider range of cell SOC (and therefore electrode stoichiometry) than the charge curves. This may be one reason that the discharge data produced better estimates of m than did the charge data.

In the simulation studies, we observe voltage spreads between the discharge and charge curves of 100–250 mV for the cases where the discharge test led to good estimates of m. Therefore, we will seek to use discharge rates that give voltage differences in this range.

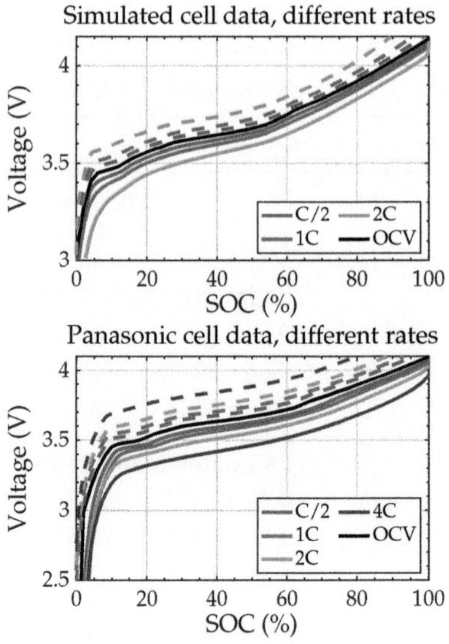

Figure 3.36: Comparing simulation data to measured data. Solid lines are for discharge; dashed lines are for charge.

3.11.3 *Application to Panasonic-cell data using teardown OCP*

We are now ready to apply the procedure to the Panasonic cell. We performed the modified OCV tests where scripts 1 and 3 used rates between C/2 to 4C. The bottom frame of Fig. 3.36 plots the measured voltages versus cell SOC, which are qualitatively similar to those

Teardown OCP method	Rate	$\bar{D}^{\mathrm{n}}_{\mathrm{s,ref}}$ [s^{-1}]	$\bar{D}^{\mathrm{p}}_{\mathrm{s,ref}}$ [s^{-1}]	m
Method 3, cell 41	C/2	5.61×10^{-5}	1.52×10^{-3}	0.82
Method 3, cell 41	1C	1.00×10^{-4}	6.79×10^{-5}	0.83
Method 3, cell 41	2C	1.79×10^{-4}	1.86×10^{-3}	0.82
Method 3, cell 41	4C	3.47×10^{-4}	1.63×10^{-3}	0.81
Method 3, cell 42	1C	9.99×10^{-5}	6.47×10^{-5}	0.82
Method 4, cell 41	C/2	5.69×10^{-5}	1.74×10^{-3}	0.94
Method 4, cell 41	1C	1.03×10^{-4}	6.99×10^{-5}	0.96
Method 4, cell 41	2C	1.79×10^{-4}	1.97×10^{-3}	0.91
Method 4, cell 41	4C	3.48×10^{-4}	1.77×10^{-3}	0.91
Method 4, cell 42	1C	1.05×10^{-4}	5.55×10^{-5}	0.98

Table 3.10: Estimation parameter values using discharge data from commercial cells and teardown-based $\widetilde{U}^{\mathrm{p}}_{\mathrm{ocp}}(\tilde{\theta}^{\mathrm{p}}_{\mathrm{s}})$.

used in simulations in the top frame, and so we anticipate that some of the general principles observed using simulation will also apply to the physical cell. We notice that the voltage data collected from the Panasonic cell for rates up to 2C give the desired spread of about 100–250 mV, and so we anticipate that estimates of m using these data will produce the most reliable estimates.

When optimizing m using an SPM and the lab data, we have an option regarding which $\widehat{U}^{\mathrm{p}}_{\mathrm{ocp}}(\hat{\theta}^{\mathrm{p}}_{\mathrm{s}})$ estimate to use.[50] In Sect. 3.3, when using teardown to build coin cells of the positive-electrode material versus lithium, we concluded that the Method-3 and Method-4 estimates were best. Therefore, we made estimates of m starting with two different $\widehat{U}^{\mathrm{p}}_{\mathrm{ocp}}(\hat{\theta}^{\mathrm{p}}_{\mathrm{s}})$ functions, to see whether calibration would result in the same final relationship. The relative $\widetilde{U}^{\mathrm{p}}_{\mathrm{ocp}}(\tilde{\theta}^{\mathrm{p}}_{\mathrm{s}})$ computed by Method 3 is plotted in Fig. 3.33 as the blue line, where we observe that the steep slope at high lithiation occurs near $\tilde{\theta}_{\mathrm{s}} \approx 0.96$. Therefore, we manually chose $b = 0.96$ when using this relative OCP as input to the SPM. The MSMR-model $\widehat{U}^{\mathrm{p}}_{\mathrm{ocp}}(\hat{\theta}^{\mathrm{p}}_{\mathrm{s}})$ estimate produced by Method 4 had the steep slope located at $\hat{\theta}_{\mathrm{s}} = 1$ and so we used $b = 1$ when using this absolute OCP estimate as input to the SPM.

We ran `particleswarm` 100 times on each discharge dataset using these two OCP models. Four of the datasets were collected from one physical cell (serial number 41) and one dataset was collected from a different cell (serial number 42) to see whether the results would be consistent. Estimates are listed in Table 3.10.

We see very consistent estimates of m for the C/2, 1C, 2C, and even the 4C data fit to each OCP method (we do not expect consistency between the two OCP methods, as we will illustrate). Using these data, we average all values of m for each method to arrive at a final estimate $m = 0.82$ for the Method-3 OCP function and $m = 0.94$ for the Method-4 OCP function.

The top frame of Fig. 3.37 shows the uncalibrated Method-3 relative $\widetilde{U}^{\mathrm{p}}_{\mathrm{ocp}}(\tilde{\theta}^{\mathrm{p}}_{\mathrm{s}})$ scaled by b as the blue dashed line and the uncalibrated

[50] For the negative electrode, we always use the Method-4 OCP function $U^{\mathrm{n}}_{\mathrm{ocp}}(\theta_{\mathrm{s}})$ from earlier, with boundaries $\theta^{\mathrm{n}}_{0} = 0.0007$ and $\theta^{\mathrm{n}}_{100} = 0.8279$ and MSMR parameter values:

j	U^0_j [V]	X_j	ω_j
1	0.09024	0.38223	0.09141
2	0.12612	0.23217	0.09667
3	0.13748	0.17290	0.86186
4	0.17046	0.06024	2.76685
5	0.21309	0.06152	0.09307
6	0.39732	0.09095	7.41078

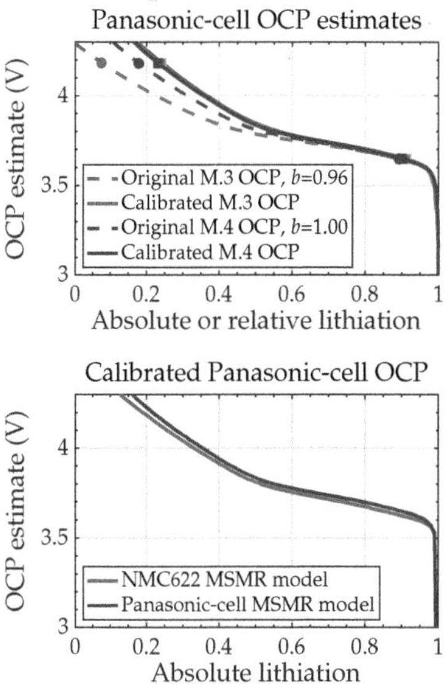

Figure 3.37: Calibrated Panasonic-cell positive-electrode OCP versus uncalibrated OCP and a literature NMC622 model.

Method-4 MSMR-model absolute $\widehat{U}^{\mathrm{p}}_{\mathrm{ocp}}(\hat{\theta}^{\mathrm{p}}_{\mathrm{s}})$ as the purple dashed line. Circle markers show uncalibrated estimates of θ^{p}_{0} and $\theta^{\mathrm{p}}_{100}$. The figure also shows the calibrated versions—scaled by both b and m—as solid lines having the same colors. Square markers show the calibrated θ^{p}_{0} and $\theta^{\mathrm{p}}_{100}$. We see that the calibration process has taken two quite different initial estimates of OCP and has scaled them to be nearly identical to one another. The agreement between both calibrated estimates of $U^{\mathrm{p}}_{\mathrm{ocp}}(\theta^{\mathrm{p}}_{\mathrm{s}})$ leads us to believe that either one of these calibrated relationships would serve as a good estimate of absolute OCP in our final model of the Panasonic cell.

For comparison purposes, we fit a single MSMR model to the average of these two calibrated curves. The optimized MSMR-model parameter values are listed in Table 3.11, and its OCP relationship is plotted in the bottom frame of Fig. 3.37. We notice by comparing the OCP curves and the values listed in Tables 3.11 and 3.12 that the calibrated model is similar to but not identical to the model of NMC622 presented by Verbrugge et al.[51,52] Since the Ford C-MAX Energi from which these Panasonic cells were harvested was launched in October 2012, its positive electrode likely has an older formulation than NMC622; we cannot be certain, but we suspect that it is NMC111. Therefore, it is not surprising to us that the OCP relationships in the figure are somewhat different from one another.

Table 3.11: Calibrated MSMR model parameter values for the positive electrode of the Panasonic cell. For these values, $\theta^{\mathrm{p}}_{0} = 0.9144$ and $\theta^{\mathrm{p}}_{100} = 0.2263$.

j	U^0_j [V]	X_j	ω_j
1	3.6536	0.1196	1.0272
2	3.7456	0.2936	1.3558
3	3.8772	0.2061	3.8248
4	4.1677	0.2870	5.5140
5	4.7569	0.0937	10.9713

Table 3.12: NMC622 MSMR model parameter values from Verbrugge et al.

j	U^0_j [V]	X_j	ω_j
1	3.62274	0.13442	0.96710
2	3.72645	0.32460	1.39712
3	3.90575	0.21118	3.50500
4	4.22955	0.32980	5.52757

3.11.4 *Application to Panasonic-cell data using nonteardown OCP*

The previous section demonstrated that the discharge test is adept at calibrating teardown-based positive-electrode OCP. Now, we consider whether it is capable of providing a good calibration to nonteardown OCP estimates made using Assumption A2.1. The nonteardown approach has two main limitations:

1. We cannot know the negative-electrode OCP for certain—even if we are confident that the physical cell has a negative electrode with a dominant chemistry—since most manufacturers will use some kind of blend. Some amount of hard carbon, natural graphite, artificial graphite, and even silicon are likely to be blended together in a negative-electrode chemistry that is nominally graphite. Differences between the assumed OCP and the true OCP for the negative electrode will bias the estimates of positive-electrode OCP.
2. The full cell can be cycled over only a relatively narrow voltage range compared with the voltage range over which we cycled the coin cells to find electrode OCP functions using the teardown approach. Therefore, the positive electrode is unlikely to reach an average lithiation near $\theta_{\mathrm{s}} = 1$ during the OCV test. Since the

[51] Mark Verbrugge, Daniel Baker, Brian Koch, Xingcheng Xiao, and Wentian Gu, "Thermodynamic model for substitutional materials: Application to lithiated graphite, spinel manganese oxide, iron phosphate, and layered nickel-manganese-cobalt oxide," *Journal of The Electrochemical Society*, 164(11):E3243, 2017.

[52] We added a fifth gallery to our MSMR model to give a slightly better match to our high-voltage OCP data. The most important difference between the Panasonic-cell and NMC622 models, however, is the small voltage shift seen in the bottom frame of Fig. 3.37.

Nonteardown Method	Rate	$\bar{D}^{\mathrm{n}}_{\mathrm{s,ref}}$ [s^{-1}]	$\bar{D}^{\mathrm{p}}_{\mathrm{s,ref}}$ [s^{-1}]	m	b
OCV method 2, cell 41	C/2	6.19×10^{-5}	1.12×10^{-4}	0.79	0.80
OCV method 2, cell 41	1C	1.07×10^{-4}	1.32×10^{-4}	0.77	0.86
OCV method 2, cell 41	2C	2.04×10^{-4}	1.87×10^{-4}	0.82	0.86
OCV method 2, cell 41	4C	3.96×10^{-4}	1.36×10^{-4}	0.89	0.93
OCV method 2, cell 42	1C	1.52×10^{-4}	8.76×10^{-5}	0.75	0.88
OCV method 4, cell 41	C/2	6.12×10^{-5}	5.90×10^{-5}	1.02	0.86
OCV method 4, cell 41	1C	1.01×10^{-4}	9.16×10^{-5}	0.90	0.83
OCV method 4, cell 41	2C	1.99×10^{-4}	7.72×10^{-5}	1.02	0.87
OCV method 4, cell 41	4C	3.71×10^{-4}	8.60×10^{-5}	1.05	0.93
OCV method 4, cell 42	1C	9.78×10^{-5}	9.60×10^{-5}	0.88	0.83

Table 3.13: Estimation parameter values using discharge data from commercial cells and nonteardown-based $\widetilde{U}^{\mathrm{p}}_{\mathrm{ocp}}(\tilde{\theta}^{\mathrm{p}}_{\mathrm{s}})$.

"knee" in the positive-electrode OCP is not crossed by the OCV test, the MSMR model fit to the nonteardown OCP estimate will likely be biased.

As a reminder, Fig. 3.32 shows estimates of positive-electrode OCP made using the nonteardown methods. We see that there are considerable differences between the estimates made by the different methods, depending on which local minimum was chosen by the optimization routines.

Here, we adopt one version of each of the two main variations—the Method-2 and Method-4 OCP functions—as input to the SPM optimization routine.[53] For this case, we choose to calibrate both b and m during the optimization rather than choosing b manually as we did when using the Method-3 teardown OCP estimate.

Table 3.13 lists the results. We see more variation in the estimates of m than we did for the nonteardown case, likely due to the two main error sources just enumerated. Fig. 3.38 shows results graphically, where we averaged the estimates of m and b optimized for all the data sets for each OCV method to produce the final calibration shown in the figure. In the top frame, the dashed lines show the two uncalibrated OCP curves and the solid lines show the calibrated versions. An MSMR model was fit to the average of the calibrated versions between the θ_0 and θ_{100} points, which was then evaluated over the entire stoichiometry range as shown by the dotted line in the lower figure. For comparison purposes, the lower frame of the figure shows the NMC622 model, the calibrated teardown MSMR model from Fig. 3.37, and this new calibrated nonteardown MSMR model superimposed. There is good agreement over much of the stoichiometry range between the nonteardown and teardown MSMR calibrated models, indicating that the calibration approach is working as desired. However, at low levels of absolute lithiation there are noticeable differences between the two relationships. The calibra-

[53] To do so, we adjusted the Method-2 values of θ^{n}_0 and $\theta^{\mathrm{n}}_{100}$ from Table 3.6 manually so that $\theta^{\mathrm{n}}_{\mathrm{s}}$ would not be negative: we added 0.005 to each of these boundaries.

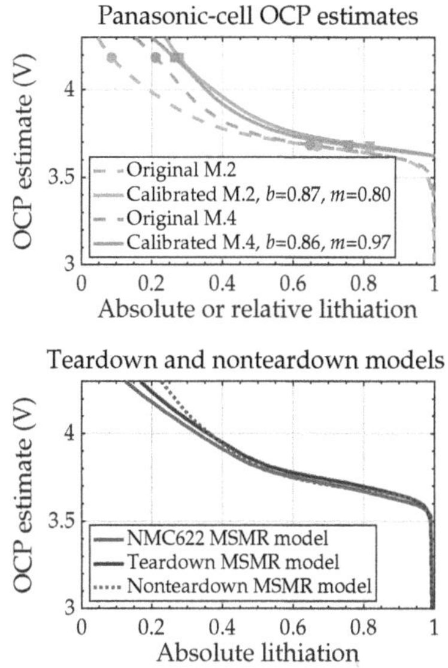

Figure 3.38: Calibrated Panasonic-cell positive-electrode OCP using nonteardown data versus uncalibrated OCP and a literature NMC622 model.

tion approach can correct the incorrect shift and stretch factors of the original relative OCP relationship, but it cannot modify errors to the shape of the relationship. Therefore, while the calibrated nonteardown version will represent the true cell better than the uncalibrated nonteardown version over most of its SOC range, we believe that the calibrated teardown OCP is the better estimate to use, if available.

3.12 *Pulse-resistance testing*

By this point in the parameter-estimation process, we have determined cell total capacity Q, OCPs $U^{\mathrm{r}}_{\mathrm{ocp}}(\theta^{\mathrm{r}}_{\mathrm{s}})$, and boundaries θ^{r}_{0} and $\theta^{\mathrm{r}}_{100}$ for both electrodes.[54] It remains to find all LPM parameter values related to cell kinetics (dynamic effects).

[54] The content of this section has been adapted from: Dongliang Lu, M. Scott Trimboli, Guodong Fan, Ruigang Zhang, and Gregory L. Plett, "Nondestructive pulse testing to estimate a subset of physics-based-model parameter values for lithium-ion cells," *Journal of The Electrochemical Society*, 168(8):080533, 2021.

As illustrated in Fig. 3.8, we implement three final laboratory tests to do so. The first test applies very-short-duration pulses to the cell to enable isolating equations describing nonlinear portions of the LPM and estimating their parameter values. The second test—described in Sect. 3.15—applies small-signal sinusoidal inputs to the cell to enable estimating values for all but one of the remaining parameters that describe primarily linear portions of the LPM. The third test—in Sect. 3.17—reuses the constant-current discharge data to find the final remaining LPM parameter value.

In this section, we introduce the first test, which eliminates the lithium-concentration (θ_{e} and θ_{s}) equations from consideration when computing cell voltage. To understand how this is possible, notice that these two PDEs involve time derivatives and so the values of θ_{e} and θ_{s} cannot change instantly (in zero time) with finite inputs. The lab test applies a *very* short-duration (a few μs or ms) current pulse to the cell: θ_{e} and θ_{s} remain essentially unchanged from their original values but ϕ_{s}, ϕ_{e}, and $i_{\mathrm{f+dl}}$ change instantly. If the cell is in equilibrium prior to applying the pulse, we can compute the initial at-rest variable values: we know that $\theta_{\mathrm{e}} = \theta_{\mathrm{e},0} = 1$ and $\theta_{\mathrm{s}} = \theta_{\mathrm{s},0}$; also, from the equations in Chap. 1, we can compute initial ϕ_{s} and ϕ_{e} from the $U^{\mathrm{r}}_{\mathrm{ocp}}(\theta^{\mathrm{r}}_{\mathrm{s},0})$ functions. Therefore, initial values of ϕ_{s} and ϕ_{e} prior to applying the pulse are known and θ_{e} and θ_{s} remain constant over the duration of the test. Any changes to cell voltage must arise due to changes in ϕ_{s}, ϕ_{e}, and $i_{\mathrm{f+dl}}$.

We will see shortly that the change in cell voltage over the short-duration pulse is a nonlinear function of about 13 model parameter values and a further function of cell SOC and pulse-current magnitude (and temperature). By conducting at least 13 experiments at different SOCs and using a variety of pulse-current magnitudes, we can in principle optimize a fit between the nonlinear function and the changes in cell voltages to solve for the unknown parameter values.

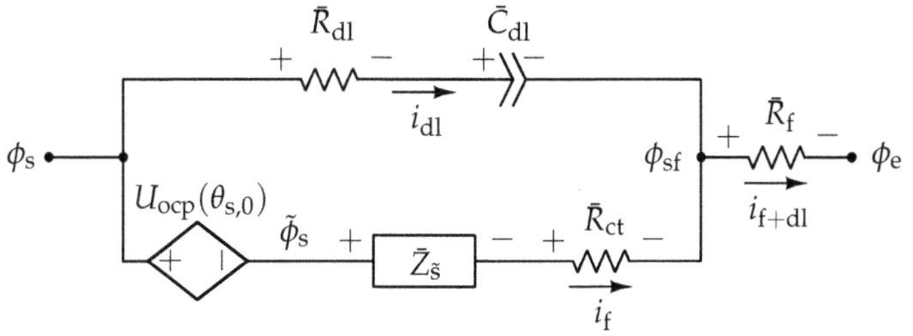

Figure 3.39: Simplified solid/electrolyte interphase model, reproduced from Fig. 2.15 in this book.

To begin analyzing the pulse-resistance test, recall that cell voltage $v(t) = \phi_{\mathrm{s}}^{\mathrm{p}}(3,t) - \phi_{\mathrm{s}}^{\mathrm{n}}(0,t)$. To model the change in cell voltage caused by a pulse input, we must develop equations that describe changes in $\phi_{\mathrm{s}}^{\mathrm{r}}$. However, since we are relying on the nonlinearity of changes to cell voltage to estimate LPM parameter values, we cannot use the linear TFs from Chap. 2 for this purpose. Instead, we consider again the simplified solid–electrolyte interphase model of Fig. 3.39, which applies to every spatial location $\tilde{x}$ in both electrodes.[55]

For very-short-duration (infinite-frequency) current pulses, we can make the following observations relating to this figure:[56]

- The voltage drop over $\bar{Z}_{\tilde{\mathrm{s}}}$ is zero since lithium concentrations in the solid cannot change instantly.
- The voltage drop over $\bar{C}_{\mathrm{dl}}$ remains unchanged from its initial value as lithium concentration in the electrolyte cannot change instantly.
- $U_{\mathrm{ocp}}(\theta_{\mathrm{s},0})$ is a constant since $\theta_{\mathrm{s},0}$ is a constant.
- We treat $\bar{R}_{\mathrm{dl}}$ and $\bar{R}_{\mathrm{f}}$ as linear resistances.
- Finally, $\bar{R}_{\mathrm{ct}}$ is a nonlinear resistance, whose voltage drop η is governed by the Butler–Volmer equation:[57]

$$
\begin{aligned}
i_{\mathrm{f}} &= \bar{i}_0 \left\{ \exp\left(\frac{(1-\alpha)F}{RT}\eta \right) - \exp\left(\frac{-\alpha F}{RT}\eta \right) \right\} \\
\bar{i}_0 &= \bar{k}_0 \left(\theta_{\mathrm{e}}\right)^{1-\alpha} \left(1-\theta_{\mathrm{s},0}\right)^{1-\alpha} \left(\theta_{\mathrm{s},0}\right)^{\alpha} \quad (\text{and } \theta_{\mathrm{e}} = 1) \\
\eta &= \tilde{\phi}_{\text{s-e}} - \bar{R}_{\mathrm{f}} i_{\mathrm{f+dl}} \\
i_{\mathrm{f+dl}} &= i_{\mathrm{f}} + \frac{\eta}{\bar{R}_{\mathrm{dl}}}.
\end{aligned} \tag{3.15}
$$

To seek connections between LPM parameter values and pulse-resistance-test measurements, we expand the cell's voltage equation:

$$
\begin{aligned}
v(t) &= \phi_{\mathrm{s}}^{\mathrm{p}}(3,t) - \phi_{\mathrm{s}}^{\mathrm{n}}(0,t) \\
&= \left[\phi_{\text{s-e}}^{\mathrm{p}}(3,t) + \phi_{\mathrm{e}}^{\mathrm{p}}(3,t)\right] - \left[\phi_{\text{s-e}}^{\mathrm{n}}(0,t) + \phi_{\mathrm{e}}^{\mathrm{n}}(0,t)\right] \\
&= \left[\phi_{\text{s-e}}^{\mathrm{p}}(3,t) - \phi_{\text{s-e}}^{\mathrm{n}}(0,t)\right] + \phi_{\mathrm{e}}^{\mathrm{p}}(3,t) - \phi_{\mathrm{e}}^{\mathrm{n}}(0,t).
\end{aligned} \tag{3.16}
$$

Since the cell is in equilibrium before the current pulse is applied (i.e., at time $t = 0^-$),

$$
\phi_{\mathrm{e}}^{\mathrm{p}}(3,0^-) = \phi_{\mathrm{e}}^{\mathrm{n}}(0,0^-) = \phi_{\mathrm{e},0} = -U_{\mathrm{ocp}}^{\mathrm{n}}(\theta_{\mathrm{s},0}^{\mathrm{n}})
$$

[55] To amplify, this circuit does not describe the composite behavior of an entire porous electrode but instead describes the solid–electrolyte potential difference at every spatial location in an electrode. That is, the values of $\phi_{\mathrm{s}}(\tilde{x},t)$ and $\phi_{\mathrm{e}}(\tilde{x},t)$ may be different at every location $\tilde{x}$ but the impedance model describing the difference between these potentials is uniform for every $\tilde{x}$. The subsequent development takes into consideration the spatial variation of $\phi_{\mathrm{s}}(\tilde{x},t)$ and $\phi_{\mathrm{e}}(\tilde{x},t)$ to derive a model that applies specifically (only) to the pulse-resistance test, which is summarized in Fig. 3.41 where the impact of solid and electrolyte conductivities across the thickness of the electrodes becomes more apparent.

[56] Note that this analysis shows that $\bar{R}_{\mathrm{dl}}^{\mathrm{r}}$ and $\bar{R}_{\mathrm{f}}^{\mathrm{r}}$ cannot both be zero. If they were both zero and a current pulse were applied when the cell was in equilibrium, then the instantaneous voltage drops over $\bar{R}_{\mathrm{dl}}^{\mathrm{r}}$ and $\bar{R}_{\mathrm{f}}^{\mathrm{r}}$ and $\bar{C}_{\mathrm{dl}}^{\mathrm{r}}$ must all be zero, forcing $\tilde{\phi}_{\text{s-e}}^{\mathrm{r}}(\tilde{x},0^+) = 0$ for all $\tilde{x}$ and hence also $\partial\tilde{\phi}_{\text{s-e}}^{\mathrm{r}}(\tilde{x},0^+)/\partial\tilde{x} = 0$. However, this violates the boundary conditions for the $\tilde{\phi}_{\text{s-e}}^{\mathrm{r}}$ PDE that will be presented in Eq. (3.21). Therefore, either $\bar{R}_{\mathrm{dl}}^{\mathrm{r}}$ or $\bar{R}_{\mathrm{f}}^{\mathrm{r}}$ or both must be nonzero for the interphase model to be physically realistic.

[57] We use the standard DFN Butler–Volmer kinetics model for now; however, lab data presented later in the chapter will motivate replacing the Butler–Volmer model with an MSMR model.

$$\begin{aligned}
\phi^{\mathrm{p}}_{\text{s-e}}(3,0^-) &= \phi^{\mathrm{p}}_{\mathrm{s}}(3,0^-) - \phi^{\mathrm{p}}_{\mathrm{e}}(3,0^-)\\
&= \left[U^{\mathrm{p}}_{\mathrm{ocp}}(\theta^{\mathrm{p}}_{\mathrm{s},0}) - U^{\mathrm{n}}_{\mathrm{ocp}}(\theta^{\mathrm{n}}_{\mathrm{s},0})\right] - \left[-U^{\mathrm{n}}_{\mathrm{ocp}}(\theta^{\mathrm{n}}_{\mathrm{s},0})\right] = U^{\mathrm{p}}_{\mathrm{ocp}}(\theta^{\mathrm{p}}_{\mathrm{s},0})\\
\phi^{\mathrm{n}}_{\text{s-e}}(0,0^-) &= \phi^{\mathrm{n}}_{\mathrm{s}}(0,0^-) - \phi^{\mathrm{n}}_{\mathrm{e}}(0,0^-)\\
&= 0 + U^{\mathrm{n}}_{\mathrm{ocp}}(\theta^{\mathrm{n}}_{\mathrm{s},0}) = U^{\mathrm{n}}_{\mathrm{ocp}}(\theta^{\mathrm{n}}_{\mathrm{s},0}),
\end{aligned}$$

due to initial conditions. Therefore, cell voltage immediately prior to applying the pulse is $v(0^-) = U^{\mathrm{p}}_{\mathrm{ocp}}(\theta^{\mathrm{p}}_{\mathrm{s},0}) - U^{\mathrm{n}}_{\mathrm{ocp}}(\theta^{\mathrm{n}}_{\mathrm{s},0})$. Then, the moment after a current pulse is applied (i.e., at time 0^+), it is:

$$v(0^+) = \left[\phi^{\mathrm{p}}_{\text{s-e}}(3,0^+) - \phi^{\mathrm{n}}_{\text{s-e}}(0,0^+)\right] + \tilde{\phi}^{\mathrm{p}}_{\mathrm{e}}(3,0^+),$$

where $\tilde{\phi}^{\mathrm{r}}_{\mathrm{e}}(\tilde{x},t) = \phi^{\mathrm{r}}_{\mathrm{e}}(\tilde{x},t) - \phi^{\mathrm{n}}_{\mathrm{e}}(0,t)$. So, the cell-voltage change is:

$$\begin{aligned}
\Delta v &= v(0^+) - v(0^-)\\
&= \lim_{t\to 0^+}\left[\tilde{\phi}^{\mathrm{p}}_{\text{s-e}}(3,t) - \tilde{\phi}^{\mathrm{n}}_{\text{s-e}}(0,t) + \tilde{\phi}^{\mathrm{p}}_{\mathrm{e}}(3,t)\right],
\end{aligned} \tag{3.17}$$

where $\tilde{\phi}^{\mathrm{r}}_{\text{s-e}} = \phi^{\mathrm{r}}_{\text{s-e}} - U^{\mathrm{r}}_{\mathrm{ocp}}$.

The equation for Δv can be written more explicitly in terms of LPM parameter values. Recall that the PDEs for ϕ_{s} and ϕ_{e} are:

$$\bar{\sigma}\frac{\partial^2 \phi_{\mathrm{s}}}{\partial \tilde{x}^2} = i_{\mathrm{f+dl}} \tag{3.18}$$

$$\frac{\partial}{\partial \tilde{x}}\left(\bar{\kappa}\left(\frac{\partial}{\partial \tilde{x}}\phi_{\mathrm{e}} + \bar{\kappa}_D T\frac{\partial \ln(\theta_{\mathrm{e}})}{\partial \tilde{x}}\right)\right) = -i_{\mathrm{f+dl}}, \tag{3.19}$$

having nonzero boundary conditions:

$$\begin{aligned}
\bar{\sigma}^{\mathrm{n}}\frac{\partial}{\partial \tilde{x}}\phi_{\mathrm{s}}\bigg|_{\tilde{x}=0} = \bar{\sigma}^{\mathrm{p}}\frac{\partial}{\partial \tilde{x}}\phi_{\mathrm{s}}\bigg|_{\tilde{x}=3} &= -i_{\mathrm{app}}\\
-\bar{\kappa}\left[\frac{\partial}{\partial \tilde{x}}\phi_{\mathrm{e}} + \bar{\kappa}_D T\frac{\partial \ln(\theta_{\mathrm{e}})}{\partial \tilde{x}}\right]_{\tilde{x}=1,2} &= i_{\mathrm{app}},
\end{aligned}$$

and initial values $\phi^{\mathrm{n}}_{\mathrm{s},0} = 0$, $\phi^{\mathrm{p}}_{\mathrm{s},0} = U^{\mathrm{p}}_{\mathrm{ocp}}(\theta^{\mathrm{p}}_{\mathrm{s},0}) - U^{\mathrm{n}}_{\mathrm{ocp}}(\theta^{\mathrm{n}}_{\mathrm{s},0})$, and $\phi_{\mathrm{e},0} = -U^{\mathrm{n}}_{\mathrm{ocp}}(\theta^{\mathrm{n}}_{\mathrm{s},0})$.

For the conditions of the pulse-resistance test, θ_{e} remains uniform and so the gradient of $\ln(\theta_{\mathrm{e}})$ is zero. Combining Eqs. (3.18) and (3.19) gives a debiased solid–electrolyte potential-difference expression:

$$\frac{\partial^2 \tilde{\phi}_{\text{s-e}}(\tilde{x},t)}{\partial \tilde{x}^2} = \left(\frac{1}{\bar{\sigma}} + \frac{1}{\bar{\kappa}}\right) i_{\mathrm{f+dl}}(\tilde{x},t). \tag{3.20}$$

Boundary conditions on the potential equations are also combined:

$$\bar{\sigma}^{\mathrm{n}}\frac{\partial \tilde{\phi}_{\text{s-e}}}{\partial \tilde{x}}\bigg|_{\tilde{x}=0} = -\bar{\kappa}^{\mathrm{n}}\frac{\partial \tilde{\phi}_{\text{s-e}}}{\partial \tilde{x}}\bigg|_{\tilde{x}=1} = -\bar{\kappa}^{\mathrm{p}}\frac{\partial \tilde{\phi}_{\text{s-e}}}{\partial \tilde{x}}\bigg|_{\tilde{x}=2} = \bar{\sigma}^{\mathrm{p}}\frac{\partial \tilde{\phi}_{\text{s-e}}}{\partial \tilde{x}}\bigg|_{\tilde{x}=3} = -i_{\mathrm{app}}. \tag{3.21}$$

The initial conditions for the debiased potentials are:

$$\tilde{\phi}^{\mathrm{n}}_{\text{s-e},0} = \phi^{\mathrm{n}}_{\mathrm{s},0} - \phi^{\mathrm{n}}_{\mathrm{e},0} - U^{\mathrm{n}}_{\mathrm{ocp}}\left(\theta^{\mathrm{n}}_{\mathrm{s},0}\right) = 0$$
$$\tilde{\phi}^{\mathrm{p}}_{\text{s-e},0} = \phi^{\mathrm{p}}_{\mathrm{s},0} - \phi^{\mathrm{p}}_{\mathrm{e},0} - U^{\mathrm{p}}_{\mathrm{ocp}}\left(\theta^{\mathrm{p}}_{\mathrm{s},0}\right) = 0. \tag{3.22}$$

For the pulse-resistance test assumptions, the PDE for $\tilde{\phi}_{\text{s-e}}$ is reduced to an ODE, Eq. (3.20), together with the interfacial flux Eq. (3.15), and subject to boundary conditions Eq. (3.21) and initial conditions Eq. (3.22). If all parameter values are fixed, we can solve for $\tilde{\phi}_{\text{s-e}}$ in both electrodes directly in MATLAB by using boundary-value ODE solvers (e.g., `bvp5c`). Then, we combine values of $\tilde{\phi}_{\text{s-e}}$ (as described in detail below) to predict pulse resistance.

In terms of the parameter-estimation approach illustrated in Fig. 3.3, the model that we will use to describe the pulse-resistance test is this boundary-value ODE. The measurements will come from the pulse-resistance test, and optimizations will regress a subset of the LPM parameter values to the model. The ODE model describing the pulse-resistance test is orders of magnitude simpler computationally than the overall LPM PDE model. Therefore, regressing parameters to the ODE model can be done more quickly and robustly than if we were to attempt to regress them to the full PDE model instead.

3.12.1 *Determine $\tilde{\phi}^{\mathrm{n}}_{\text{s-e}}(\tilde{x} = 0, t = 0^+)$ and $\tilde{\phi}^{\mathrm{p}}_{\text{s-e}}(\tilde{x} = 3, t = 0^+)$*

To use one of MATLAB's boundary-value ODE solvers, we must provide the ODE to be solved, boundary conditions, and an initial guess for the solution, all in a prescribed format. The ODE being solved is assumed to be in the standard first-order form:

$$\frac{\partial y}{\partial x} = y' = f(x, y),$$

where y is permitted to be a vector function of x. To convert our problem to this form, we let $x = \tilde{x}$ and we substitute $y_1(\tilde{x}) = \tilde{\phi}_{\text{s-e}}(\tilde{x}, 0^+)$ and $y_2 = \partial\tilde{\phi}_{\text{s-e}}(\tilde{x}, 0^+)/\partial\tilde{x}$ to rewrite the equations as a vector of first-order equations. That is,

$$y_1'(\tilde{x}) = \frac{\partial\tilde{\phi}_{\text{s-e}}(\tilde{x}, 0^+)}{\partial\tilde{x}} = y_2(\tilde{x}), \qquad \text{and} \qquad y_2'(\tilde{x}) = \frac{\partial^2\tilde{\phi}_{\text{s-e}}(\tilde{x}, 0^+)}{\partial\tilde{x}^2}.$$

The boundary conditions must be coded in the form $g(\cdot) = 0$. Equation (3.21) is translated to:

$$\frac{\partial\tilde{\phi}_{\text{s-e}}(\tilde{x} = 0, t = 0^-)}{\partial\tilde{x}} + \frac{i_{\mathrm{app}}}{\bar{\sigma}^{\mathrm{n}}} = 0, \qquad \frac{\partial\tilde{\phi}_{\text{s-e}}(\tilde{x} = 2, t = 0^-)}{\partial\tilde{x}} - \frac{i_{\mathrm{app}}}{\bar{\kappa}^{\mathrm{p}}} = 0,$$
$$\frac{\partial\tilde{\phi}_{\text{s-e}}(\tilde{x} = 1, t = 0^-)}{\partial\tilde{x}} - \frac{i_{\mathrm{app}}}{\bar{\kappa}^{\mathrm{n}}} = 0, \qquad \frac{\partial\tilde{\phi}_{\text{s-e}}(\tilde{x} = 3, t = 0^-)}{\partial\tilde{x}} + \frac{i_{\mathrm{app}}}{\bar{\sigma}^{\mathrm{p}}} = 0.$$

Finally, we use the MATLAB function "`bvpinit`" to create an initial guess for the solution to the equations at time $t = 0^+$. The solver is robust to errors in this guess, and works even with a zero initialization. However, we will be calling the boundary-value ODE solver within an optimization loop, so it is beneficial to make the solver as fast as possible. We find that `bvp5c` executes more quickly if the initial guess for $\tilde{\phi}_{\text{s-e}}$ is assigned to a function that is closer to its final solution than simply initializing it to zero.

Our choice for an initial guess of the profile of $\tilde{\phi}_{\text{s-e}}(\tilde{x}, 0^+)$ in the negative electrode is to assume that it has a quadratic form:

$$\tilde{\phi}^{\text{n}}_{\text{s-e}}(\tilde{x}, 0^+) \approx a\tilde{x}^2 + b\tilde{x} + c, \tag{3.23}$$

for $0 \le \tilde{x} \le 1$. The derivative of Eq. (3.23) at $\tilde{x} = 0$ is $\partial\tilde{\phi}_{\text{s-e}}/\partial\tilde{x}|_{\tilde{x}=0} = b$. So, via Eq. (3.21) we have:

$$b = -\frac{i_{\text{app}}}{\bar{\sigma}^{\text{n}}}.$$

Similarly, $\partial\tilde{\phi}_{\text{s-e}}/\partial\tilde{x}|_{\tilde{x}=1} = 2a + b$. So, again via Eq. (3.21):

$$a = \frac{1}{2}\left(\frac{i_{\text{app}}}{\bar{\kappa}^{\text{n}}} - b\right).$$

The mean across $0 \le \tilde{x} \le 1$ is $\bar{\tilde{\phi}}_{\text{s-e}} = a/3 + b/2 + c$. Rearranging:

$$c = \bar{\tilde{\phi}}_{\text{s-e}} - \frac{a}{3} - \frac{b}{2}.$$

To approximate the true mean of $\tilde{\phi}_{\text{s-e}}$, we assume that $i_{\text{f+dl}}(\tilde{x}, 0^+)$ is uniform across the entire electrode and equal to i_{app}. We solve nonlinear equations to estimate the mean of $\tilde{\phi}_{\text{s-e}}$ using `fzero`:

```
K1n = (1-alphan)*F/(R*T); K2n = -alphan*F/(R*T);
i0n = k0n*((1-thetan)^(1-alphan))*(thetan^alphan);
negMean = @(x,i) i0n.*(exp(K1n.*x)-exp(K2n.*x))+x/Rdln-i; % x = etan
if Rdln == 0
  phisemean = Rfn*i;
else
  phisemean = fzero(negMean,0,[],i) + Rfn*i;
end
```

We now have values for a, b, and c and can provide `bvp5c` with an initial estimate via Eq. (3.23).

The approach to initializing the solution in the positive electrode is similar, except that we approximate $\tilde{\phi}^{\text{p}}_{\text{s-e}}(\tilde{x}, 0^+) \approx a(3 - \tilde{x})^2 + b(3 - \tilde{x}) + c$ for $2 \le \tilde{x} \le 3$. The equations for finding a, b, and c for this initialization are similar to those used in the negative electrode. With this initialization, we find that computation time is $\lesssim 5\,\text{ms}$ per (SOC, rate) pair in MATLAB on an ordinary laptop computer.

An example solution for $\tilde{\phi}_{\text{s-e}}(\tilde{x})$ and $\partial\tilde{\phi}_{\text{s-e}}(\tilde{x})/\partial\tilde{x}$ (evaluated using `deval` based on structure returned by `bvp5c`) when the cell is initially resting at 80 % SOC and then subjected to a 4C-rate short-duration discharge pulse current is displayed in Fig. 3.40.

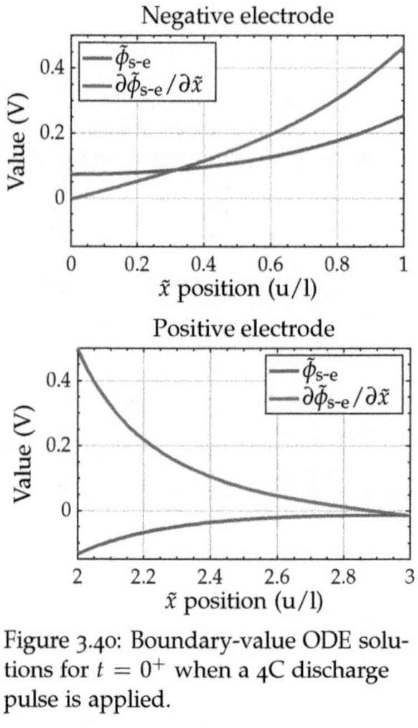

Figure 3.40: Boundary-value ODE solutions for $t = 0^+$ when a 4C discharge pulse is applied.

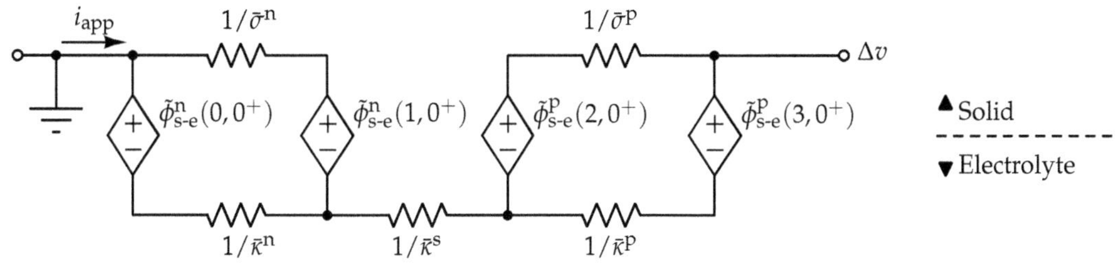

Figure 3.41: Equivalent circuit describing a cell's response to a very-short-duration current pulse.

3.12.2 *Determine $\tilde{\phi}^p_e(\tilde{x} = 3, t = 0^+)$ and Δv*

Referring back to Eq. (3.17), we notice that we also must be able to find $\tilde{\phi}^p_e(3,0^+)$ to compute $\Delta v(t)$. App. 3.C shows (after a lengthy but straightforward derivation) that:

$$\begin{aligned}\tilde{\phi}^p_e(3,0^+) = &-\frac{i_{app}}{\bar{\kappa}^n + \bar{\sigma}^n} - \frac{i_{app}}{\bar{\kappa}^s} - \frac{i_{app}}{\bar{\kappa}^p + \bar{\sigma}^p} \\ &- \frac{\bar{\sigma}^n}{\bar{\kappa}^n + \bar{\sigma}^n}\left(\tilde{\phi}^n_{s\text{-}e}(1,0^+) - \tilde{\phi}^n_{s\text{-}e}(0,0^+)\right) \\ &- \frac{\bar{\sigma}^p}{\bar{\kappa}^p + \bar{\sigma}^p}\left(\tilde{\phi}^p_{s\text{-}e}(3,0^+) - \tilde{\phi}^p_{s\text{-}e}(2,0^+)\right).\end{aligned}$$

This can be readily calculated given the values of $\tilde{\phi}^r_{s\text{-}e}(\tilde{x},0^+)$ produced by the `bvp5c` solution to the boundary-value ODE.

So, we are now able to compute the change in cell voltage due to the application of a short-duration current pulse:

$$\begin{aligned}\Delta v &= \tilde{\phi}^p_{s\text{-}e}(3,0^+) - \tilde{\phi}^n_{s\text{-}e}(0,0^+) + \tilde{\phi}^p_e(3,0^+) \\ &= -\frac{i_{app}}{\bar{\kappa}^n + \bar{\sigma}^n} - \frac{i_{app}}{\bar{\kappa}^s} - \frac{i_{app}}{\bar{\kappa}^p + \bar{\sigma}^p} - \frac{\bar{\sigma}^n}{\bar{\kappa}^n + \bar{\sigma}^n}\tilde{\phi}^n_{s\text{-}e}(1,0^+) \\ &\quad - \frac{\bar{\kappa}^n}{\bar{\kappa}^n + \bar{\sigma}^n}\tilde{\phi}^n_{s\text{-}e}(0,0^+) + \frac{\bar{\kappa}^p}{\bar{\kappa}^p + \bar{\sigma}^p}\tilde{\phi}^p_{s\text{-}e}(3,0^+) + \frac{\bar{\sigma}^p}{\bar{\kappa}^p + \bar{\sigma}^p}\tilde{\phi}^p_{s\text{-}e}(2,0^+).\end{aligned} \tag{3.24}$$

If we normalize this voltage change by the value of applied current, we define a nonlinear pulse resistance R_0:

$$\begin{aligned}R_0 &= -\frac{\Delta v}{i_{app}} \\ &= \frac{1}{\bar{\kappa}^n + \bar{\sigma}^n} + \frac{1}{\bar{\kappa}^s} + \frac{1}{\bar{\kappa}^p + \bar{\sigma}^p} + \frac{\bar{\kappa}^n}{\bar{\kappa}^n + \bar{\sigma}^n}\frac{\tilde{\phi}^n_{s\text{-}e}(0,0^+)}{i_{app}} \\ &\quad + \frac{\bar{\sigma}^n}{\bar{\kappa}^n + \bar{\sigma}^n}\frac{\tilde{\phi}^n_{s\text{-}e}(1,0^+)}{i_{app}} - \frac{\bar{\sigma}^p}{\bar{\kappa}^p + \bar{\sigma}^p}\frac{\tilde{\phi}^p_{s\text{-}e}(2,0^+)}{i_{app}} - \frac{\bar{\kappa}^p}{\bar{\kappa}^p + \bar{\sigma}^p}\frac{\tilde{\phi}^p_{s\text{-}e}(3,0^+)}{i_{app}}.\end{aligned}$$

It is interesting to observe that we can construct a circuit diagram from Eq. (3.24) that helps to illustrate how a cell responds to a short-duration current pulse. This is drawn in Fig. 3.41. Note that this circuit is derived directly from Eq. (3.24)—which is itself derived from the LPM equations and the assumptions of the pulse-resistance test—and not vice versa. The circuit diagram gives insight into the terms of the model—a current split between solid and electrolyte in the two electrodes, nonlinear voltage drops due to $\tilde{\phi}^{\rm r}_{\rm s\text{-}e}(\tilde{x},0^+)$ at four locations in the cell, and linear voltage drop over the electrolyte in the separator region.

3.12.3 *Summary of the pulse-resistance calculation*

Fig. 3.42 summarizes the ODE-simulation method for computing R_0. For every (SOC, rate) input, we initialize the `bvp5c` solution using the steps labeled "[1]." The FOM equations have been combined to make the boundary-value ODE that we seek to solve, labeled "[2]." The forcing function to this ODE comprises the coupled equations labeled "[3]," solved in practice using `fzero`. The boundary conditions for the ODE that are supplied to `bvp5c` are listed in [4]. The output of `bvp5c` comprises $\tilde{\phi}^{\rm r}_{\rm s\text{-}e}(\tilde{x},0^+)$ for locations $\tilde{x} \in \{0,1,2,3\}$. These are combined together as shown in the rightmost box in the figure to compute R_0 for this combination of SOC and rate.

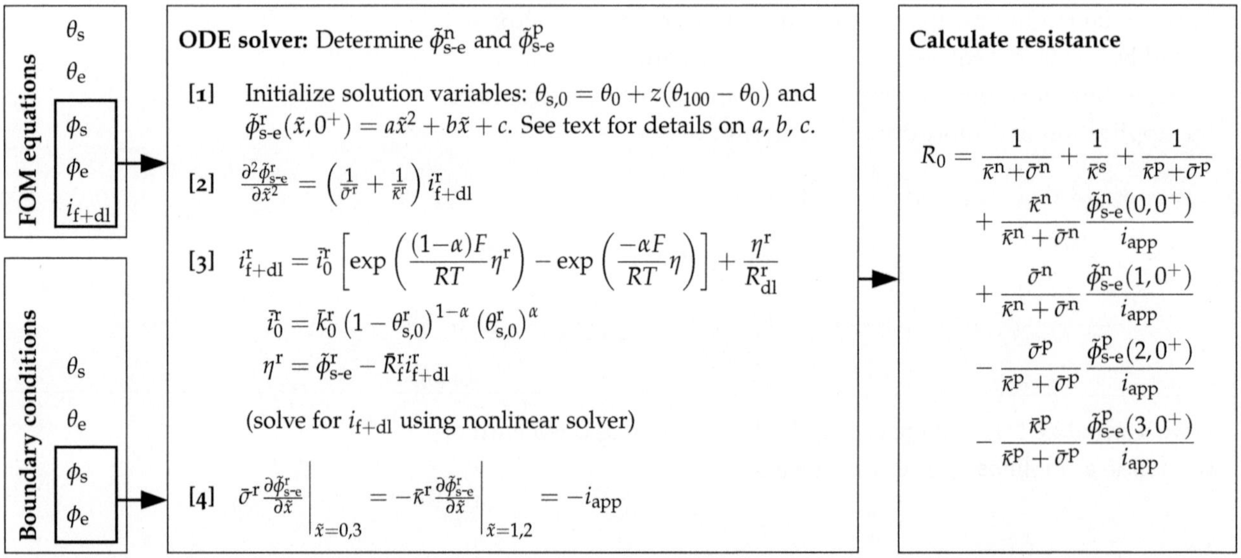

Figure 3.42: Summarizing the ODE pulse-resistance simulation method.

Some representative examples are shown in Fig. 3.43. The top-left plot shows R_0 as a function of SOC and C-rate for the NMC30 cell. There is strong dependence of R_0 on SOC, especially at low values of SOC; there is weak dependence on the input-current rate. The top-right plot shows R_0 for the same cell, but with the double-layer removed from the model. We see somewhat stronger dependence of R_0 on the input-current rate in this case and conclude that the double layer flattens the R_0 relationship. The bottom-left figure shows R_0 for

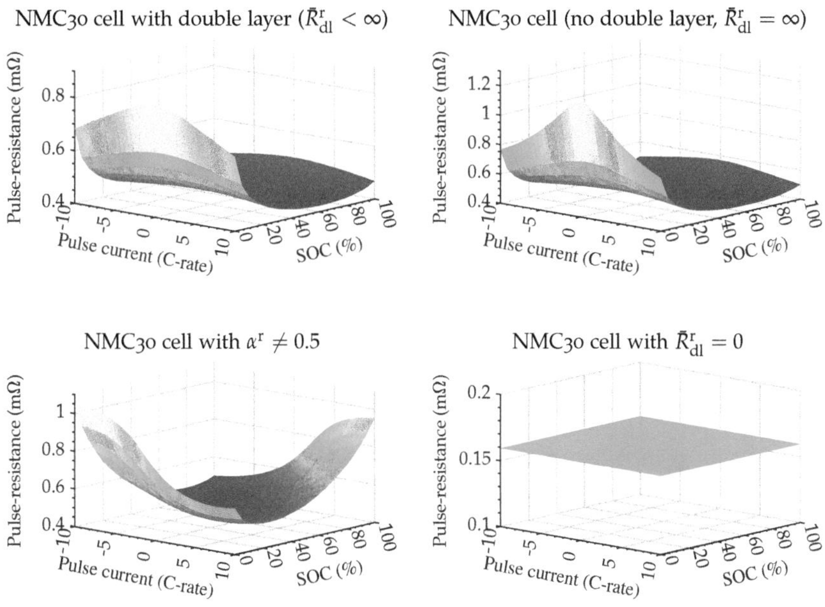

Figure 3.43: Representative pulse resistances.

the NCM30 cell with α^{r} artificially set to values different from 0.5. We see that this distorts the R_0 figure and indicates that we would expect to be able to estimate α^{r} well from R_0 if we observe asymmetric R_0. The bottom-right figure shows R_0 if the double-layer resistance is set to zero (which is the case with most or all models of the double layer in the literature). If $\bar{R}^{r}_{dl} = 0$, then R_0 has no dependence on rate or SOC; however, if we observe any dependence of R_0 on rate or SOC in measured lab-test data, then we must conclude that $\bar{R}^{r}_{dl} \neq 0$.

3.13 *Estimating pulse-resistance-test parameters*

3.13.1 *Challenges when estimating R_0*

The LPM parameters required to compute a cell's nonlinear pulse resistance R_0 comprise: α^{r}, $\bar{\kappa}^{r}$, $\bar{\sigma}^{r}$, $\bar{R}^{r}_{f}$, $\bar{R}^{r}_{dl}$, and $\bar{k}^{r}_{0}$.[58] The procedure we implement to estimate the values of these parameters must first determine a cell's R_0 at different initial SOC values, current amplitudes, and temperatures. Then it regresses these values of R_0 to the boundary-value ODE model to determine estimates of the set of parameter values.

[58] Electrolyte intrinsic conductivity κ is a function of electrolyte concentration. However, we are estimating lumped parameters $\bar{\kappa}^{n}$, $\bar{\kappa}^{s}$, and $\bar{\kappa}^{p}$ (based on κ) around the cell's equilibrium electrolyte lithium concentration, $\theta_e = \theta_{e,0} = 1$. Therefore, κ is a constant for all of the pulse-resistance tests and $\bar{\kappa}^{n}$, $\bar{\kappa}^{s}$, and $\bar{\kappa}^{p}$ are also constants.

We find that we must overcome three challenges to implement the pulse-resistance parameter-estimation method on a physical cell:

1. We cannot instantly change applied current from zero to i_{app}. The electronics driving the cell will have nonzero rise time, overshoot, and settling time as the test equipment's controller stabilizes its

output current. Also, the cabling connecting the test equipment to a cell will have inductance that limits the achievable slew rate.

2. We cannot apply a pulse having exactly zero duration. Finite-duration pulses will cause concentrations and double-layer charge to begin to change (somewhat) over this interval, and this will introduce error into our results.
3. Measurements are not noise-free. We will need to apply multiple pulses and perform some kind of averaging in an attempt to reduce the impact of noise.

We now look at a method to convert realistic time-domain data to an estimate of R_0 that considers these three factors.

3.13.2 *Development of a practical pulse-resistance-test model*

A short-duration real-world pulse process can be modeled by:

$$v_{\text{meas}}(t) = \text{OCV}(0^-) - R_0 i(t) + \text{dynamic terms}(t) + \eta_v(t),$$

where voltage and current are now assumed to evolve over time in nonideal ways and "meas" refers to measurements that are noise-corrupted. We no longer assume a perfect pulse of current having value i_{app}, but instead consider a more realistic approximate pulse $i(t)$; we no longer assume that no dynamics are excited over duration 0^+, but consider some very-high-frequency "dynamic terms(t)"; we also model noise $\eta_v(t)$ which is additive to the voltage measurement.

The dynamic terms are modeled to include inductive coupling between current-carrying and voltage-sensing cables as well as a capacitance due to connections and perhaps excitation of the cell's electrical double layer. We describe measured voltage as the response of a series resistor–inductor–capacitor (RLC) circuit, where we must identify the values of inductance L and capacitance C in order to obtain an unbiased estimate of R_0.[59]

The overall model of the change in voltage away from the initial equilibrium value is expressed as follows:

$$\Delta v_{\text{meas}}(t) = -\left[R_0 i(t) + L\frac{\text{d}i(t)}{\text{d}t} + C\int_0^t i(\tau)\,\text{d}\tau\right] + \eta_v(t)$$

$$i_{\text{meas}}(t) = i(t) + \eta_i(t),$$

where $\eta_v(t)$ and $\eta_i(t)$ are independent white noises, $i(t)$ is the true current, and $i_{\text{meas}}(t)$ is the measured current.

3.13.3 *Implementation of a pulse-resistance-test data-processing method*

We desire to use this model of voltage change to determine an estimate of R_0 from lab-test data $\{i_{\text{meas}}(t), \Delta v_{\text{meas}}(t)\}$. We begin by esti-

[59] The purpose of this model of the experimental setup is to enable fitting the experimentally measured voltage response $v_{\text{meas}}(t)$ to the experimentally measured approximate pulse current $i(t)$ as closely as possible to isolate an unbiased estimate of R_0. After fitting the measurements, the values of L and C are discarded.

Alternate models may produce better fits to the measured data for some experimental setups, and therefore better estimates of R_0. We illustrate the general approach with an RLC circuit, but have found that augmenting the model with parallel resistor/capacitor Voigt elements has improved results for some tests we have done, especially when the pulse duration is long enough that some double-layer dynamics are activated and must be estimated and removed from the voltage to produce a good estimate of R_0.

For the Panasonic cell and our experimental setup, Fig. 3.48 shows the fit of the RLC circuit to measured data; while the match is not perfect, it appears to produce reasonable estimates of R_0, which can be used to estimate the pulse-resistance-test parameter values.

mating the capacitance in the model. To do so, we define OCV_1 to be the rest voltage prior to applying the pulse and OCV_2 to be the rest voltage after removing the pulse. Any difference $\Delta v = \mathrm{OCV}_2 - \mathrm{OCV}_1$ must be due to charge accumulated during the test by the RLC model's capacitance. We estimate the effective capacitance over the measurement interval, where $t_{\max}$ is the experiment duration, as:

$$\widehat{C} = \frac{1}{\Delta v} \int_0^{t_{\max}} i_{\mathrm{meas}}(t)\,\mathrm{d}t.$$

We then calibrate the measured voltage by removing the effect of this capacitance:

$$\begin{aligned}
\Delta v_{\mathrm{cal}}(t) &= v_{\mathrm{meas}}(t) - \mathrm{OCV}_1 - \frac{1}{\widehat{C}} \int_0^t i_{\mathrm{meas}}(\tau)\,\mathrm{d}\tau \\
&= v_{\mathrm{meas}}(t) - \mathrm{OCV}_1 - \Delta v \frac{\int_0^t i_{\mathrm{meas}}(\tau)\,\mathrm{d}\tau}{\int_0^{t_{\max}} i_{\mathrm{meas}}(\tau)\,\mathrm{d}\tau}.
\end{aligned}$$

After calibration, the beginning and ending values of Δv_{cal} are zero and the intermediate values result from the voltage-change response of a resistor–inductor circuit to $i_{\mathrm{meas}}(t)$.

Next, we fit resistance $\widehat{R}_0$ and inductance $\widehat{L}$ estimates to the $\Delta v_{\mathrm{cal}}(t)$ and $i_{\mathrm{meas}}(t)$ data. This involves a regression of data to these two remaining unknown parameters. We seek to minimize the integral of squared errors of the voltage response for a certain nonideal pulse-current input:

$$\left\{\widehat{L}, \widehat{R}_0\right\} = \arg \min_{\{L,R_0\}} \int_0^{t_{\max}} \left(\Delta v_{\mathrm{cal}}(t) - \Delta v_{\mathrm{model}}(t)\right)^2 \mathrm{d}t$$

where,

$$\Delta v_{\mathrm{model}}(t) = -\left[R_0 i_{\mathrm{meas}}(t) + L \frac{\mathrm{d} i_{\mathrm{meas}}(t)}{\mathrm{d}t}\right].$$

Note that evaluating this cost function requires differentiating the measured electrical current, which amplifies the measurement noise in the signal and tends to produce erratic results. So, instead, we redefine the cost function to seek to minimize the integral of squared errors of the *integrated* voltage responses:

$$\left\{\widehat{L}, \widehat{R}_0\right\} = \arg \min_{\{L,R_0\}} \int_0^{t_{\max}} \left(\int_0^t \Delta v_{\mathrm{cal}}(\tau) - \Delta v_{\mathrm{model}}(\tau)\,\mathrm{d}\tau\right)^2 \mathrm{d}t$$

where,

$$\int_0^t \Delta v_{\mathrm{model}}(\tau)\,\mathrm{d}\tau = -\left[R_0 \int_0^t i_{\mathrm{meas}}(\tau)\,\mathrm{d}\tau + L i_{\mathrm{meas}}(t)\right].$$

The trapezoid rule is used when approximating the integrals using measured data. Note that this regression is linear in the data so linear-regression tools can be used to find $\widehat{L}$ and $\widehat{R}_0$ very quickly.

Example simulation results are shown in Fig. 3.44. Estimates $\widehat{R}_0$ of true R_0 for tests having small current magnitudes are poorest: The small pulse magnitude excites a small voltage change; this small voltage change is similar in size to the magnitude of the voltage-measurement noise; and the measurement noise corrupts the estimate of R_0. This result suggests that we may not be able to obtain accurate R_0 estimates from experimental data at low input C-rates. However, we do see that tests having larger current magnitudes excite a larger voltage change that is easier to detect since the measurement-noise power is maintained constant for all C-rates and SOCs. Estimates $\widehat{R}_0$ are visually indistinguishable from the true values R_0 for tests having larger-magnitude input currents.

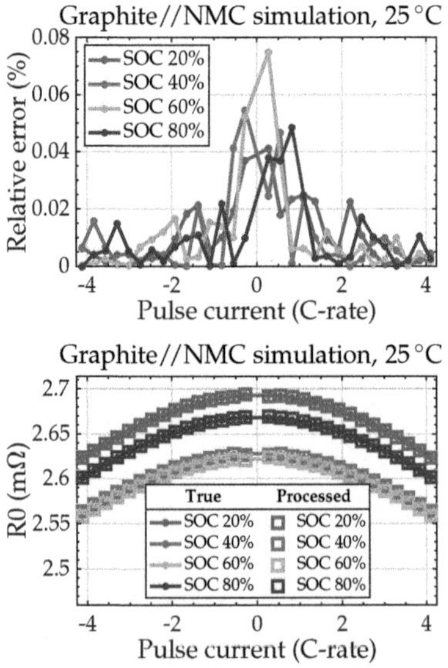

Figure 3.44: Example results for estimating pulse resistance from realistic simulated pulse-resistance-test data. "True" values depict R_0 and "processed" values depict $\widehat{R}_0$.

3.14 *Validating pulse-resistance-test parameters*

3.14.1 *Estimating cell-model parameter values using synthetic data*

As we have already noted, pulse resistance R_0 is a nonlinear function of about 13 model parameter values, and a further function of cell SOC and pulse magnitude (and also temperature). Therefore, by collecting at least 13 data points at different SOCs and pulse magnitudes, we can fit the nonlinear model function to our estimates $\widehat{R}_0$ and optimize the estimates of the parameter values.

The underlying optimization task seeks to find a set of parameter values that can make the model predictions agree as closely as possible to the processed $\widehat{R}_0$ at all setpoints. If we define a vector of M input magnitudes and N SOC points, we can state the objective as seeking to minimize the following RMSE cost function:

$$p^* = \arg\min_p \sqrt{\frac{1}{MN}\sum_{m=1}^{M}\sum_{n=1}^{N}\left[\widehat{R}_0(i_m, z_n) - R_0(i_m, z_n; p)\right]^2},$$

where p is the set of model parameters to identify, $\widehat{R}_0$ is processed estimate of measured resistance, and R_0 is the resistance simulated by the ODE when using parameters p.

In this section, we seek to estimate cell-model parameter values using $\widehat{R}_0$ estimates found by time-domain simulation of a noisy LPM for $M = 30$ and $N = 4$. Optimizations were initialized with parameter estimates chosen randomly around the true values.

Graphical results are plotted in Fig. 3.45, evaluated at many more SOC and magnitude points than were included in the data used to estimate the model parameter values. Overall, the physics-based pulse-resistance model has excellent agreement to the processed $\widehat{R}_0$. The model generalizes well. Numeric results are shown in Table 3.14.

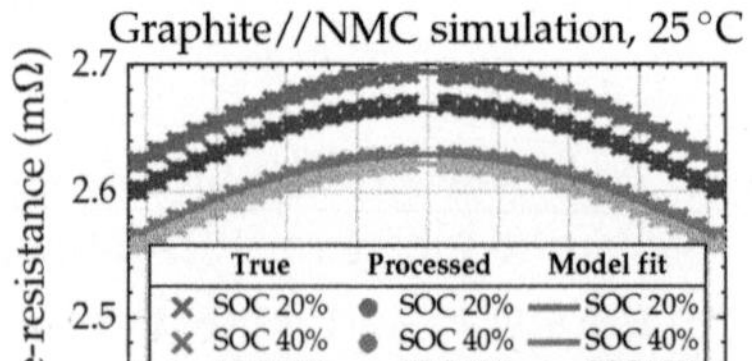

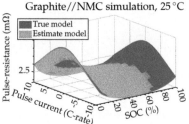

Figure 3.45: Fitting a model to simulated pulse-resistance-test data.

Table 3.14: Parameter-estimation results from simulated pulse-resistance-test data. Note that [u/l] stands for "unitless."

Lumped Parameter	Lower Limit	Upper Limit	Initial Value	Optimized Estimate	True Value	Relative Error
α^{n} [u/l]	0.37	0.68	0.53	0.52	0.50	5 %
α^{p} [u/l]	0.41	0.77	0.59	0.52	0.50	4 %
$\bar{k}_0^{\mathrm{n}}$ [A]	41.9	77.8	59.8	64.6	77.3	16 %
$\bar{k}_0^{\mathrm{p}}$ [A]	19.9	37.0	28.4	25.3	26.1	3 %
$\bar{\sigma}^{\mathrm{n}}$ [kS]	316	948	632	626	612	2 %
$\bar{\sigma}^{\mathrm{p}}$ [kS]	12.6	37.8	25.2	21.9	29.0	24 %
$\bar{\kappa}^{\mathrm{n}}$ [kS]	0.65	1.94	1.29	1.84	1.35	36 %
$\bar{\kappa}^{\mathrm{s}}$ [kS]	2.98	8.93	5.96	5.56	5.76	3 %
$\bar{\kappa}^{\mathrm{p}}$ [kS]	1.66	4.97	3.31	3.12	2.44	28 %
$\bar{R}_{\mathrm{dl}}^{\mathrm{n}}$ [mΩ]	0.894	1.67	1.28	0.959	1	4 %
$\bar{R}_{\mathrm{dl}}^{\mathrm{p}}$ [mΩ]	0.606	1.12	0.865	0.983	1	2 %
$\bar{R}_{\mathrm{f}}^{\mathrm{n}}$ [mΩ]	0.912	1.69	1.30	1.04	1	4 %
$\bar{R}_{\mathrm{f}}^{\mathrm{p}}$ [μΩ]	0.751	1.39	1.07	1.07	1	7 %

Traces of parameter values versus algorithm iteration during the `particleswarm` optimization process are shown in Fig. 3.46. We see that parameters converge to the correct neighborhoods. Further, the optimization is not confusing corresponding parameters in negative and positive-electrode regions.[60]

[60] This is due to the $\theta_{\mathrm{ss}}^{\mathrm{r}}$ influence in the Butler–Volmer equations of both electrodes varying in opposition to one another as a function of cell SOC.

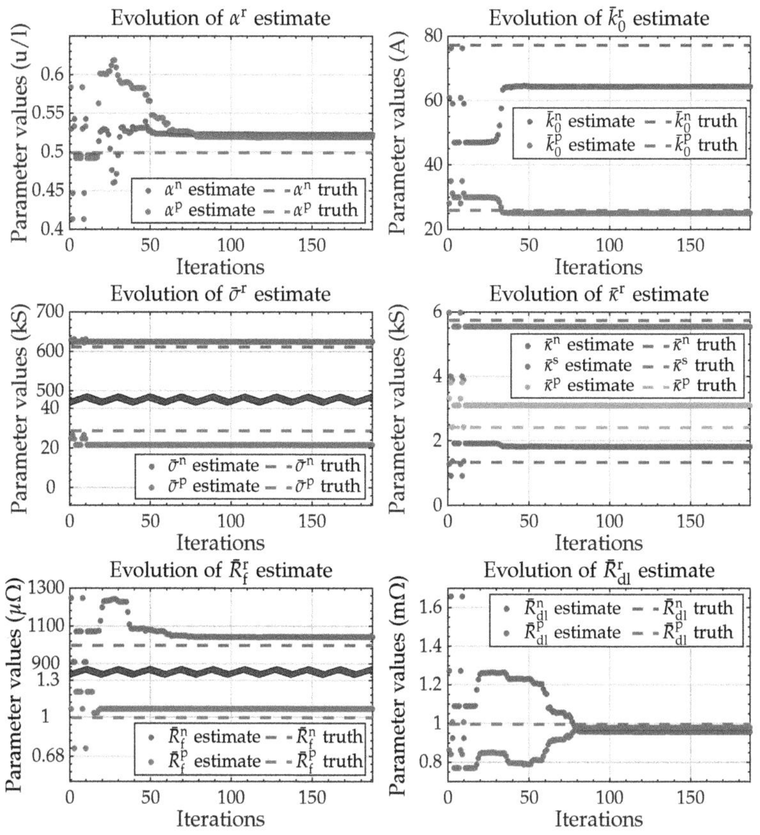

Figure 3.46: Traces of the evolution of the pulse-resistance-test parameter estimates versus iteration during the optimization process. (The $\bar{\sigma}^{\mathrm{r}}$ and $\bar{R}_{\mathrm{f}}^{\mathrm{r}}$ plots have a split vertical scale due to the large difference in values for these parameters in the negative and positive electrodes.)

We make the following summary observations: Estimates of charge-transfer coefficients α^r have good accuracies as they exhibit the highest nonlinear influences on R_0 responses. $\bar{k}_0^p$ is estimated well, while $\bar{k}_0^n$ still has some offset to its true value. This may be due to the processed $\widehat{R}_0$ being noisy at low-input-magnitude setpoints.

R_0 is a very weak function of $\bar{\sigma}^n$, so it is difficult to identify a high-accuracy estimate for $\bar{\sigma}^n$. However, even large-magnitude absolute error produces small relative estimation error due to the large magnitude of $\bar{\sigma}^n$. The value of $\bar{\sigma}^p$ is reasonably well estimated in the study.

Electrolyte conductances $\bar{\kappa}^r$ have large relative errors because they are coupled with large-valued $\bar{\sigma}^r$ in the model: small relative errors in identifying $\bar{\sigma}^r$ will require large errors in $\bar{\kappa}^r$ to compensate. Estimates of lumped film resistance $\bar{R}_f^r$ and lumped double-layer resistance $\bar{R}_{dl}^r$ are highly accurate.

3.14.2 *Application to a physical cell*

We now discuss results obtained by applying the pulse-resistance-test method to the Panasonic cell. The test procedure is summarized in Table 3.15. We used three Gamry Reference 3000 potentiostats plus three Reference 30k power boosters, connected in parallel, to create the current pulses: this enabled us to generate pulses with amplitude up to ± 90 A. Pulse-response data were gathered at four SOC setpoints (nominally 20 %, 40 %, 60 %, and 80 % but actual values were calibrated post-test). At every SOC setpoint, current pulses between -90 A to 90 A, in steps of 6 A, were applied (skipping 0 A). For every pulse magnitude, ten charge pulses were applied, followed by ten discharge pulses, with 5 min of rest in between each pulse. Applying multiple pulses allowed us to average computed resistance estimates $\widehat{R}_0$ to reduce the influence of measurement noises and also enabled determination of the consistency of the measured data.

Script	Temp.	Step Number and Details
1	25 °C	1. Charge until $U_{ocv} = v_{max}$ (100 % SOC). 2. Rest ($\geq$ 2 h).
2 (loop)	T	3. Slowly discharge to desired SOC setpoint. 4. Rest ($\geq$ 2 h). 5. Apply a series of pulses (rest 5 min in between).

Table 3.15: Laboratory procedure used for implementing the pulse-resistance test.

Each pulse experiment first rested the cell for 200 μs to collect the voltage data that enable computing the initial OCV estimate OCV_1, then the pulse was applied for 200 μs, then the cell was rested for 200 μs to collect voltage data that enable computing the final OCV estimate OCV_2. We found that it took about 50 μs to reach a stable

90 A pulse; this motivated the 200 μs duration for the applied pulse. Current and voltages were sampled at a 300 kHz rate ($T_s = 3.3\,\mu$s).

Sample current-pulse and voltage-response profiles are displayed in Fig. 3.47. The color of each line in the figures is a function of current magnitude and helps somewhat to distinguish the responses. The inset zoom axes on the voltage plot show the presence of significant measurement noise. The presence of this noise motivated repeating measurements of pulse responses for every (SOC, rate) setpoint multiple times. We also see large voltage overshoot, especially on discharge, indicating the presence of nonnegligible inductance. This is why we must fit an RLC circuit to the data to find a minimally biased estimate $\widehat{R}_0$ of pulse resistance.

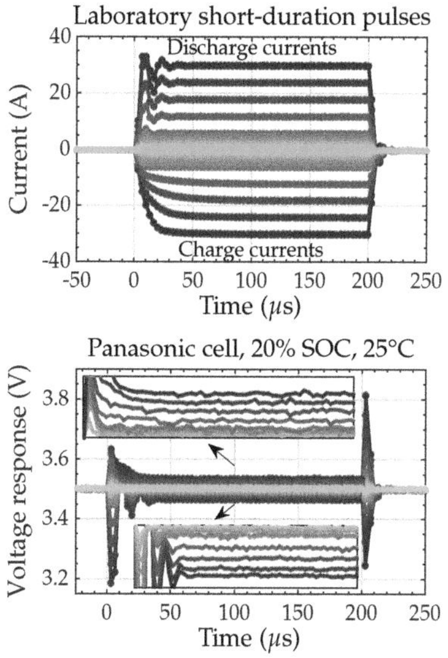

Figure 3.47: Current and voltage samples recorded during pulse-resistance-test experiments. Recall that per the convention we use in our models, discharge currents have a positive sign and charge currents have a negative sign.

3.14.3 *Implementation of data-processing method: Panasonic cell*

Lab-data processing is complicated by our use of up to three instruments to deliver current pulses and measure voltage responses. For current magnitudes where $|i_{\text{app}}| \leq 30$ A, a single Gamry Reference 3000 potentiostat plus Reference 30k power-booster "unit" was used and the prior methodology was followed. When higher current magnitudes were required, we used either two units ($30 < |i_{\text{app}}| \leq 60$ A) or three units ($|i_{\text{app}}| > 60$ A).

When using more than one unit, we must modify the data-processing procedure. First, we must sometimes time-shift voltage measurements by one sample to synchronize responses among the three units. Second, we make individual estimates $\widehat{R}_0$, $\widehat{L}$, and $\widehat{C}$ for data measured by each unit (up to three estimates per variable per setpoint rather than one estimate per variable per setpoint). Finally, there is an artifact that we cannot explain at this point whereby $\widehat{R}_0$ computed for one-unit tests are shifted vertically versus those from two-unit tests or three-unit tests. We manually shift two-unit and three-unit resistances in an attempt to correct this anomaly.

Example time-domain fits of RLC models to measured data are displayed in Fig. 3.48. Voltage traces for 30 A discharge and charge pulses (requiring a single measurement unit) are presented in the top row; those for 60 A pulses (requiring two measurement units in parallel) are shown in the second row; and those for 90 A pulses (using three measurement units in parallel) are displayed in the bottom row. The RLC-model fits are quite good in the steady-state regions, and are moderately good in the transition regions. The large-magnitude overshoot on current transitions illustrates the need for an inductance in the model of the experiment. However, even with optimized values of $\widehat{L}$ in the model, we don't predict measured voltage perfectly.

Results of fitting $\widehat{R}_0$ and $\widehat{L}$ values to the data collected by the pulse

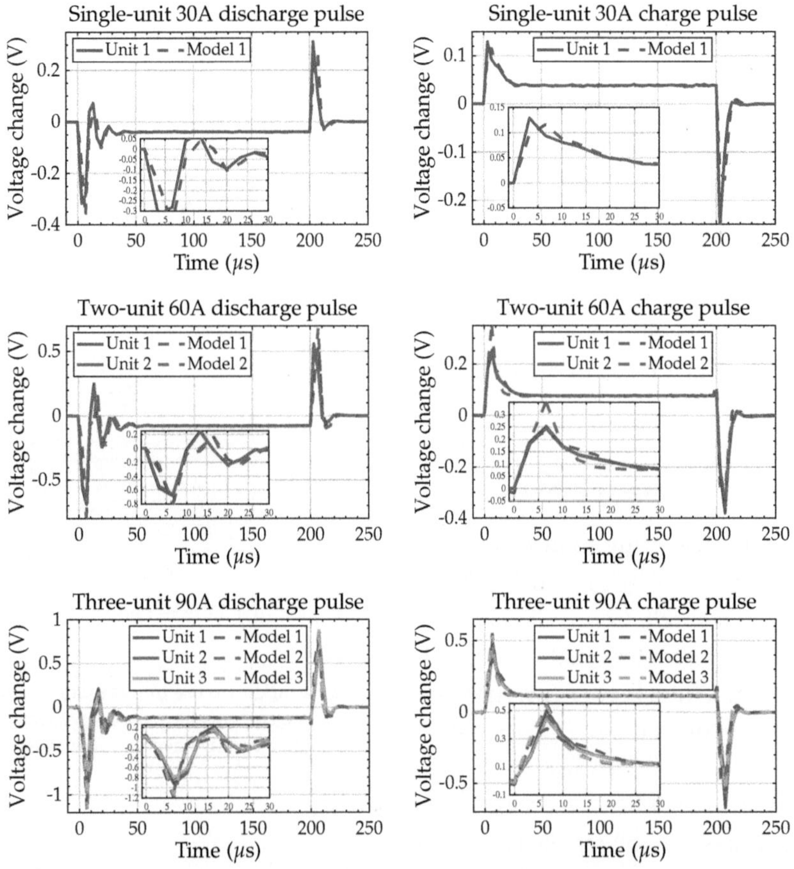

Figure 3.48: Time-domain voltage responses of the Panasonic cell to different short-duration current pulses.

experiments are depicted graphically in Fig. 3.49. The vertical black dotted lines denote the boundaries between tests conducted using one, two, and three units. Each plotted value of $\widehat{R}_0$ and $\widehat{L}$ is the average of the ten values computed from the ten pulses conducted at each setpoint. The error bars indicate the 99 % confidence interval based on fitting a normal distribution to the ten values computed for each point: narrow error bars indicate that the tests were very consistent; wide error bars indicate that the test results were inconsistent. Error bars are widest for $\widehat{R}_0$ estimates at low input-current levels. We believe this to be due to the relatively high level of voltage-measurement noise we see in the experiments, which is of similar order to the voltage change due to $i_{\text{app}} \times R_0$. In retrospect, this motivates applying more pulses at low levels of input current so that there is a greater opportunity for averaging the computed $\widehat{R}_0$ to obtain more confidence in the results.

We observe very high consistency between estimates of inductance at all SOC setpoints and all temperatures. This leads us to believe that the inductance is largely external to the cell—otherwise we would expect it to be a function of temperature and most likely SOC.

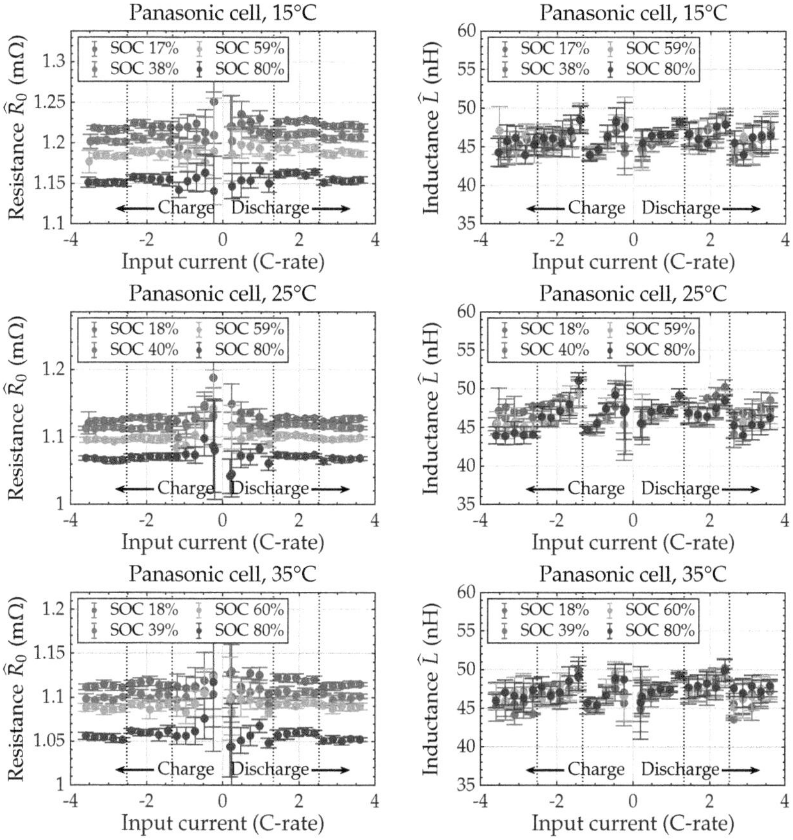

Figure 3.49: Fitting resistance estimates $\widehat{R}_0$ and inductance estimates $\widehat{L}$ to Panasonic-cell pulse-resistance-test data. The error bars indicate the relative confidence of the parameter estimates.

We also see very similar variation in estimated inductance versus applied-current magnitude. This leads us to believe that this effect is largely a function of the mutual inductance between the current-source and voltage-measurement wiring.

We notice a distinct variation in pulse resistance versus cell SOC that is consistent across temperatures. These results give physical support for including a double-layer resistance $\bar{R}_{\text{dl}}$ in the solid–electrolyte interphase model. If $\bar{R}_{\text{dl}}$ were zero, then the cell's R_0 would not be a function of SOC (or rate). We also notice a variation in pulse resistance versus rate that generally supports the model presented earlier. However, the greatest apparent variation in R_0 versus rate is in the range of $|i_{\text{app}}| \leq 30\,\text{A}$, where uncertainty in the $\widehat{R}_0$ estimates is also largest due to the small true voltage change and comparably large voltage-measurement noise. Therefore, we have less confidence than we would like in interpreting these $\widehat{R}_0$ values.

The results show that $\widehat{R}_0$ decreases as temperature increases, as expected. We see very little asymmetry between charge and discharge resistances, which allows us to assume that $\alpha^{\text{r}} \approx 0.5$. If we were to enforce this assumption (which we do not do in the following analysis), we could average charge and discharge resistances at each input C-rate, somewhat reducing the effective noise level.

3.14.4 *Parameter estimation using lab $\widehat{R}_0$ data*

We use MATLAB's `particleswarm` to optimize LPM parameter values to fit $\widehat{R}_0$ estimates.[61] Fits of the pulse-resistance model to the data are plotted in Fig. 3.50. We make two key observations:

[61] The ODE model predictions were made using $U^{\mathrm{n}}_{\mathrm{ocp}}(\theta_{\mathrm{s}})$ from sidenote 50 and $U^{\mathrm{p}}_{\mathrm{ocp}}(\theta_{\mathrm{s}})$ from Table 3.11.

1. The $\widehat{R}_0$ values are too noisy to observe clear patterns in the ± 1C region. The pulse-resistance-test model fits these data, but less-noisy $\widehat{R}_0$ might enable fine-tuning model parameters better. This motivates making more pulse-resistance measurements between ± 1C and averaging them to reduce noise.[62]
2. Modeled average resistance at different SOC levels does not agree with the actual average level of resistance as a function of SOC. The collected data show monotonically decreasing resistance as SOC increases; however, the model predicts that resistances for 80 % SOC must be higher than for 60 % SOC.

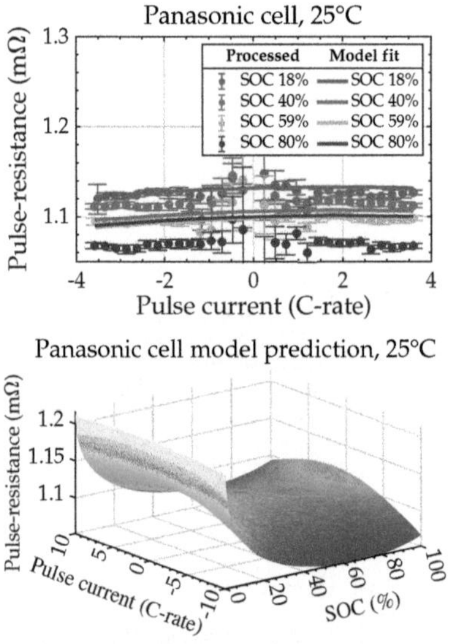

Figure 3.50: Pulse fitting results when using the standard Butler–Volmer interface model.

[62] It also motivates developing additional different lab tests that actuate the nonlinearities of the cell to estimate parameter values of the nonlinear portions of the LPM, particularly the kinetics.

Further analysis exposes the likely source of this second problem. Fundamentally, the mismatch demonstrates that the model we have used to this point does not have sufficient SOC degrees of freedom to describe the observed changes in resistance as a function of SOC. The only occurrence of SOC in the model is in the prefactor to the Butler–Volmer equation. Therefore, this prefactor is somehow too simple to describe the data we have measured. We believe that the solution to this problem is to replace the standard DFN (Butler–Volmer) kinetics model with the MSMR kinetics model.

Fig. 3.51 shows results of a simulation study where we replaced the DFN kinetics with the MSMR model. We notice that more SOC degrees of freedom are enabled, which produces pulse resistances R_0 having similar features to those observed experimentally.

MSMR-model $\widehat{R}_0$ fits for the Panasonic cell are shown in Fig. 3.52. At this point, we don't believe that we have collected data at sufficient SOC setpoints to determine all $\bar{k}^{\mathrm{n}}_{0,j}$ and $\bar{k}^{\mathrm{p}}_{0,j}$ without ambiguity, which we suppose to be a reason for the bumpiness of the 3D surface. However, the 2D fit to all $\widehat{R}_0$ data that have been measured to this point is greatly improved over the non-MSMR model fit.

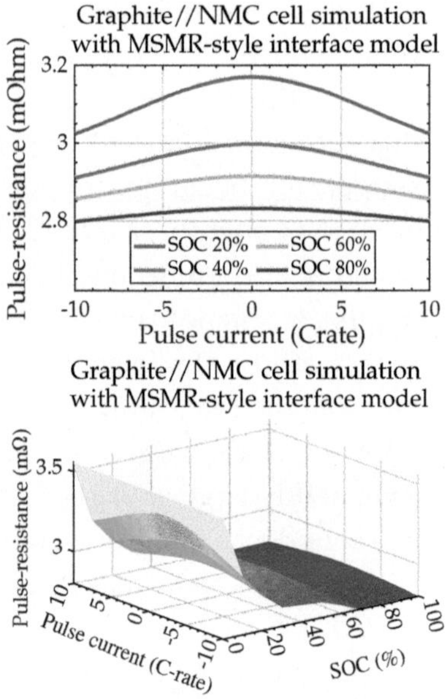

Figure 3.51: Simulation of R_0 when an MSMR model replaces the standard Butler–Volmer interface model.

3.15 *Frequency-response (EIS) testing*

The pulse-resistance test enables us to estimate all parameters of the LPM kinetics and potential equations. However, by its design, lithium concentrations do not change during the laboratory experiment and so the pulse-resistance test does not enable us to estimate the values of the parameters that are unique to the concentration equations. To be able to estimate the values of these parameters, we leverage data collected using a common lab experiment known as electrochemical impedance spectroscopy (EIS).[63]

EIS was introduced in Sect. 2.11. In summary, the technique applies a small-signal sinusoidal current (or voltage) to the terminals of the cell and measures the sinusoidal perturbation in the cell's voltage (or current). In steady state, the output perturbation will have the same frequency as the input sinusoid but will have a different magnitude and phase with respect to the input. The collection of these magnitude gains and phase shifts versus frequency is known as the frequency response of the cell. We have already seen how to model this frequency response using the TFs from Chap. 2. Specifically, we evaluate the cell's impedance response $Z(s)$ via Eq. (2.103), which can be computed using closed-form equations. Then, the frequency response is $Z(j\omega) = Z(s)|_{s=j\omega}$. Therefore, referring back to Fig. 3.3, we have an available lab experiment with corresponding model. It remains to regress lab-test EIS data against the TF-based cell impedance model to estimate all (but one) of the remaining model parameter values:[64]

- We collect EIS data at multiple SOC setpoints. These are the measurement data from Fig. 3.3.
- The regression model is the cell impedance from Eq. (2.103).
- We use nonlinear optimization to adjust the parameter values of the impedance model until its impedance predictions agree with the measured EIS data as closely as possible.

While the premise of using EIS data to estimate parameter values is straightforward, a robust implementation can be quite complicated.

There are many parameter values that we hope to estimate using the EIS data, and we must overcome some inherent obstacles associated with nonlinear optimization to do so well. Nonlinear optimization can be very sensitive to initial guesses of the parameter values and the limits or boundaries placed on the possible values that may be assumed by the parameters during optimization. This is especially true when jointly optimizing a set comprising many parameters. In cases where there is no a priori knowledge that might assist with making these initial estimates, a guess of zero might be applied by default to all parameter values. However, the majority of LPM parameters have no physical meaning at zero values. The cost function we aim to minimize has numerous local minima corresponding to suboptimal solutions, and so bad initial values will sometimes push

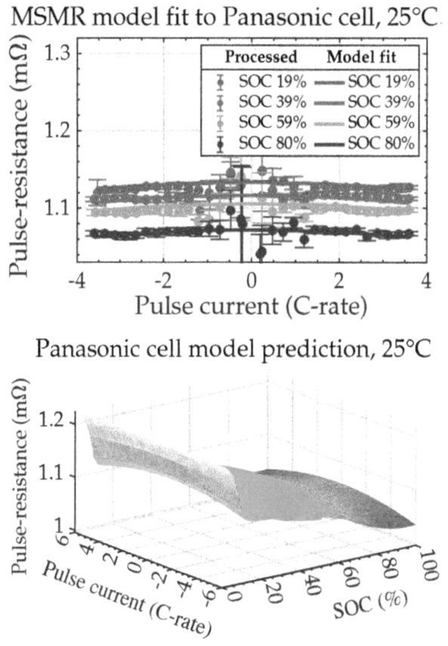

Figure 3.52: Fits of MSMR-model-based R_0 to lab-test data.

[63] The content of this section has been adapted from: Dongliang Lu, M. Scott Trimboli, Guodong Fan, Yujun Wang, and Gregory L. Plett, "Nondestructive EIS testing to estimate a subset of physics-based-model parameter values for lithium-ion cells," *Journal of The Electrochemical Society*, 169(8):080504, 2022.

[64] The final parameter, $\bar{\psi}$, is found via the pseudo-steady-state test in Sect. 3.17.

the model to extreme or unrealistic parameter-space regions that will result in the identification of a bad local minimum. Consequently, the physical meaning of each parameter and its search boundaries should be carefully considered prior to optimization.

In the next section, we present a tool that will help initialize optimizations with reasonable initial guesses for the parameter values to boost the optimization speed and avoid over/under tuning. Whenever possible, we prefer to compute initial guesses based on the available cell data rather than on an exhaustive search of the literature database. Modern battery chemistries often blend multiple materials into the electrode, which can drastically change the physical interpretation of many lumped-parameter values even if a dominant chemistry is assumed. The initialization ideas developed in the following work are based on making appropriate assumptions and analyzing the available data. The major advantages are that the proposed approach requires minimum or even no manual tuning, and it is also applicable to a variety of cell chemistries.[65]

3.15.1 *Distribution of relaxation times (DRT)*

A tool that we will use when initializing parameter values is a method that converts impedance data into a distribution of relaxation times (DRT). This technique is widely used in the literature to analyze impedance spectra.[66] It recognizes that the bump in the Nyquist plot of a lithium-ion cell arises from processes having different time constants: one can imagine that the particle surfaces or porous materials are providing many different circuit pathways for charge transfer and storage in parallel, each with a different relaxation time τ_i. Nontrivial electrochemical systems exhibit a continuous distribution of an essentially infinite number of relaxation times over a given range of input frequencies. In this section, we will briefly discuss the properties of the DRT method and how we will apply it to help initialize parameter estimates prior to optimization.

For illustration, Fig. 3.53 demonstrates the ideal and practical DRT functions for an example electrochemical process that is represented by two parallel resistor–capacitor (RC) networks, themselves wired in series. The combined impedance shown in the Nyquist plot obeys:

$$
\begin{aligned}
Z(f) &= \frac{R_1}{1+j2\pi f R_1 C_1} + \frac{R_2}{1+j2\pi f R_2 C_2} \\
&= \frac{R_1}{1+j2\pi f \tau_1} + \frac{R_2}{1+j2\pi f \tau_2},
\end{aligned}
\tag{3.25}
$$

where $\tau_i = R_i C_i = (2\pi f_i)^{-1}$. Note that f is the frequency at which the impedance is being evaluated while f_i is an inherent property

[65] One challenge with using EIS results to parameterize a cell model is that of separating the impedance correctly into the parts contributed by each electrode and the separator region. How do we know we are not confusing the impedances of the negative and positive electrodes?

Our experience is that by using impedance data measured at a multiplicity of SOC setpoints, the nonlinear dependence of impedance on SOC correctly identifies the negative-electrode versus positive-electrode contributions. Since the focus of this chapter is on nonteardown methods, we satisfy ourselves with these results. However, if teardown facilities are available, three-terminal impedance measurements can be made that enable direct attribution of portions of the impedance to the two electrodes. Two papers that consider this possibility are:

- Daniel Juarez-Robles, Chien-Fan Chen, Yevgen Barsukov, and Partha P. Mukherjee, "Impedance evolution characteristics in lithium-ion batteries," *Journal of The Electrochemical Society*, 164(4):A837, 2017.
- Zhengyu Chu, Ryan Jobman, Albert Rodriguez, Gregory L. Plett, M. Scott Trimboli, Xuning Feng, and Minggao Ouyang, "A control-oriented electrochemical model for lithium-ion battery. Part II: Parameter identification based on reference electrode," *Journal of Energy Storage*, 27:101101, 2020.

[66] For example:

- Ole Christensen and Khadija L. Christensen, *Approximation Theory*. Birkhäuser Boston, 2005.
- Francesco Ciucci and Chi Chen, "Analysis of electrochemical impedance spectroscopy data using the distribution of relaxation times: A Bayesian and hierarchical Bayesian approach," *Electrochimica Acta*, 167:439–454, 2015.
- Mohammed B. Effat and Francesco Ciucci, "Bayesian and hierarchical Bayesian based regularization for deconvolving the distribution of relaxation times from electrochemical impedance spectroscopy data," *Electrochimica Acta*, 247:1117–1129, 2017.

(continued on next page)

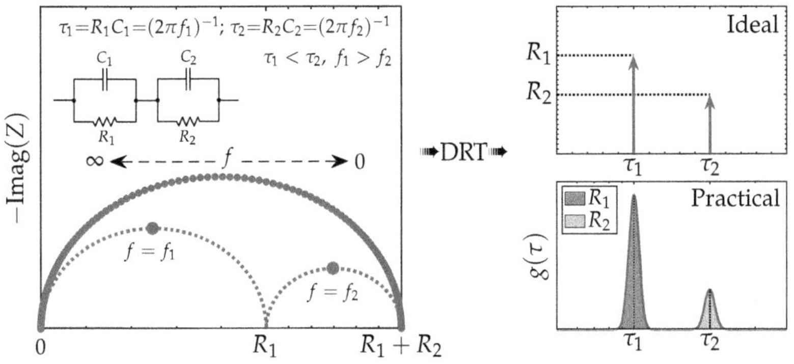

Figure 3.53: An illustration of Nyquist versus DRT estimates for a simple double resistor-capacitor-pair circuit. The figure was inspired by the work reported in: J. Illig, M. Ender, T. Chrobak, J. P. Schmidt, D. Klotz, and E. Ivers-Tiffée, "Separation of charge transfer and contact resistance in $LiFePO_4$ cathodes by impedance modeling," *Journal of The Electrochemical Society*, 159(7):A952–A960, 2012.

of the underlying process (it is the reciprocal of the RC time constant, converted from radians per second to hertz). When $f = f_i$, the impedance of a single RC network can be written as:

$$Z_i(f = f_i) = \frac{R_i}{1+j} = \left(\frac{1}{2} - j\frac{1}{2}\right) R_i,$$

where f_1 and f_2 are annotated under their corresponding Nyquist bumps in Fig. 3.53, representing the frequencies at which the impedance of that subcircuit has the maximum imaginary magnitude. The combination of these two smaller bumps is what produces the observed larger bump: that is, we do not measure the impedance spectra of the two individual processes separately but rather the overall impedance of the combined process, from which f_1 and f_2 are not directly observable. The DRT method seeks to deconvolve the overall impedance into its representative components so that we can observe f_1 and f_2 (or, equivalently, τ_1 and τ_2).

In Eq. (3.25), two types of parameters, R and τ, characterize the underlying physical processes. Specifically, R (measured in Ω) gives the polarization and τ (measured in s) represents the relaxation time of the corresponding process. An exact (ideal) DRT solution can be drawn if we are given values for R and τ as illustrated by the Dirac delta impulse arrows in the top-right frame of the figure. However, we do not see this sharp distinction of time constants produced by a practical DRT method for nontrivial electrochemical systems in which many polarization processes are blended together. Typical-looking results for a practical system are shown in the bottom-right frame of Fig. 3.53. Each polarization process is now characterized in the DRT by the center frequency of its peak which is the corresponding relaxation frequency and by the shape surrounding the peak. In this case, the DRT function is denoted by $g(\tau)$, and the area under the peak surrounding the peak frequency corresponds to the polarization of the loss process (i.e., the area is equal to R_i).[67] The shape of each peak depends on the nature of the underlying process.

66(*cont.*)

- Markus Hahn, Stefan Schindler, Lisa-Charlotte Triebs, and Michael A. Danzer, "Optimized process parameters for a reproducible distribution of relaxation times analysis of electrochemical systems," *Batteries*, 5(2):43, 2019.
- J. Illig, M. Ender, T. Chrobak, J. P. Schmidt, D. Klotz, and E. Ivers-Tiffée, "Separation of charge transfer and contact resistance in $LiFePO_4$ cathodes by impedance modeling," *Journal of The Electrochemical Society*, 159(7):A952–A960, 2012.
- Mattia Saccoccio, Ting Hei Wan, Chi Chen, and Francesco Ciucci, "Optimal regularization in distribution of relaxation times applied to electrochemical impedance spectroscopy: Ridge and lasso regression methods—a theoretical and experimental study," *Electrochimica Acta*, 147:470–482, 2014.
- Pouyan Shafiei Sabet, Gereon Stahl, and Dirk Uwe Sauer, "Non-invasive investigation of predominant processes in the impedance spectra of high energy lithium-ion batteries with Nickel-Cobalt-Aluminum cathodes," *Journal of Power Sources*, 406:185–193, 2018.

[67] J. Illig, M. Ender, T. Chrobak, J. P. Schmidt, D. Klotz, and E. Ivers-Tiffée, "Separation of charge transfer and contact resistance in $LiFePO_4$ cathodes by impedance modeling," *Journal of The Electrochemical Society*, 159(7):A952–A960, 2012.

The main goal of the DRT method is to process full-cell impedance data to identify the characteristic distribution of its underlying timescales, especially those that are hidden together in the Nyquist bump. In order to do that, the experimental data (e.g., the large blue dots in the Nyquist plot of Fig. 3.53) measured at a given frequency, $Z_{\text{meas}}(f)$, are "fitted" against the following model:

$$Z_{\text{DRT}}(f) = \int_0^{\infty} \frac{g(\tau)}{1+j2\pi f\tau}\,\mathrm{d}\tau = \int_0^{\infty} \frac{\gamma(\ln(\tau))}{1+j2\pi f\tau}\,\mathrm{d}\ln\tau, \qquad (3.26)$$

where $g(\tau)$ is the DRT function of the underlying system. Eq. (3.26) can be understood as the sum of an infinite series of RC elements where each has its own separate time constant. Since the impedance data are often collected on a logarithmic frequency scale with a given number of frequencies per decade; that is, $[\log f_1, \log f_2, , \cdots]$, then $g(\tau)$ can be written as $\gamma(\ln(\tau)) = \tau g(\tau)$. Note, $g(\tau)$ is measured in $\Omega\,\mathrm{s}^{-1}$ but $\gamma(\ln(\tau))$ is measured in Ω.

A challenge when conducting DRT analysis is to find an accurate $\gamma(\ln(\tau))$ from the measured data. Solving for $\gamma(\ln(\tau))$ using Eq. (3.26) is known to be an ill-posed problem, and a proper approach involves two complicated steps: discretization and regularization. In this work, we follow the approach developed by Wan et al.[68] and use a modified version of their open-source DRT toolbox. Our modification allows automatic processing of the cell impedance data.

[68] Ting Hei Wan, Mattia Saccoccio, Chi Chen, and Francesco Ciucci, "Influence of the discretization methods on the distribution of relaxation times deconvolution: Implementing radial basis functions with DRTtools," *Electrochimica Acta*, 184:483–499, 2015; and Francesco Ciucci and Chi Chen, "Analysis of electrochemical impedance spectroscopy data using the distribution of relaxation times: A Bayesian and hierarchical Bayesian approach," *Electrochimica Acta*, 167:439–454, 2015.

3.15.2 *Decomposing full-cell impedance*

Fig. 3.53 illustrated that a single Nyquist bump can arise from the combination of two (or more) time constants whose impedances overlap in the frequency domain. The DRT can help uncover the individual time constants, but then we wish to understand how those time constants relate to the TF components that describe cell impedance.

The full-cell impedance is described by Eq. (2.103). Examining this equation, we notice that it comprises three components: two are due to the solid/electrolyte potential differences at the current collectors, and one is due to the electrolyte potential difference spanning the thickness of the cell between the current collectors. The two necessary solid/electrolyte potential differences can be expressed using Eqs. (2.86) and (2.87) as:

$$+\frac{\widetilde{\Phi}^{\mathrm{n}}_{\text{s-e}}(0,s)}{I_{\text{app}}(s)} = +\bar{Z}^{\mathrm{n}}_{\widetilde{\mathrm{se}}}(s)\cdot\underbrace{\left(\mathrm{j}^{\mathrm{n}}_1 e^{-\Lambda^{\mathrm{n}}_1} + \mathrm{j}^{\mathrm{n}}_2 + \mathrm{j}^{\mathrm{n}}_3 e^{-\Lambda^{\mathrm{n}}_2} + \mathrm{j}^{\mathrm{n}}_4\right)}_{I^{\mathrm{n}}_{\mathrm{f+dl}}(0,s)/I_{\text{app}}(s)}$$

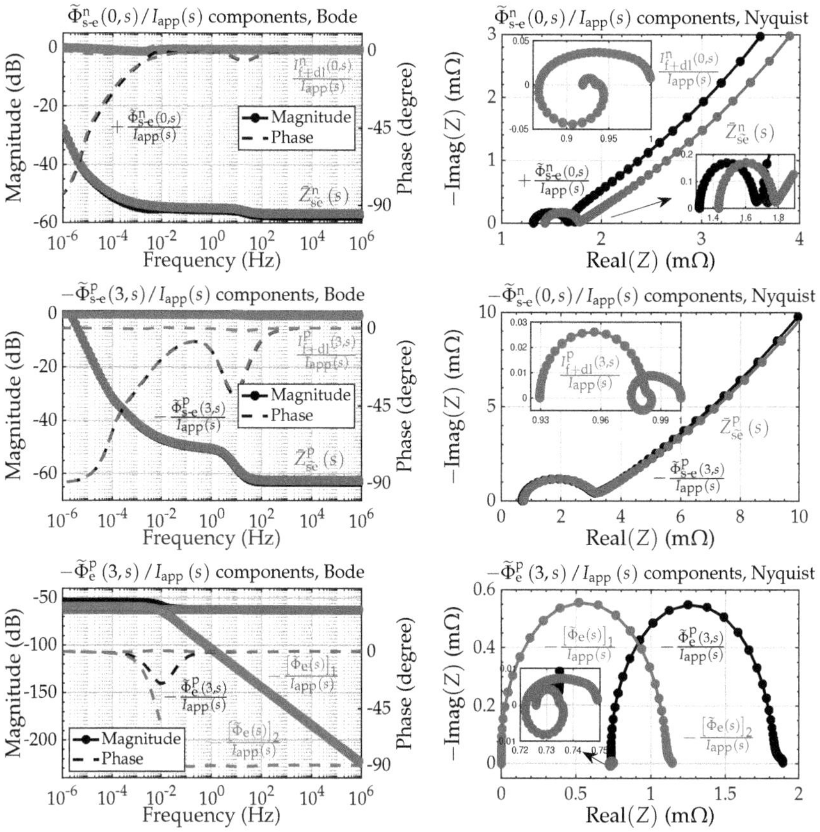

Figure 3.54: Examining simulated Bode and Nyquist components of full-cell impedance (plotted at 80 % SOC).

$$-\frac{\widetilde{\Phi}^{p}_{s\text{-}e}(3,s)}{I_{app}(s)} = -\bar{Z}^{p}_{\widetilde{se}}(s)\cdot \underbrace{\left(j^{p}_{1}e^{-\Lambda^{p}_{1}} + j^{p}_{2} + j^{p}_{3}e^{-\Lambda^{p}_{2}} + j^{p}_{4}\right)}_{I^{p}_{f+dl}(3,s)/I_{app}(s)}.$$

The Bode and Nyquist diagrams for these two transfer functions are displayed in the top two rows of Fig. 3.54, where the diagrams show the impact of the components of the TF on the overall frequency response. The results suggest that both $I^{n}_{f+dl}(0,s)/I_{app}(s)$ and $I^{p}_{f+dl}(3,s)/I_{app}(s)$ terms have magnitudes close to one; therefore, the electrochemical features we observe from a full-cell Nyquist plot (e.g., the charge-transfer semicircle and solid-diffusion impedance) are dominated by the nonspatially varying solid-electrolyte interface impedances $\bar{Z}^{r}_{\widetilde{se}}(s)$ of both electrodes. If the cell impedance data can be properly decoded (as will be discussed in conjunction with Fig. 3.56), we can perhaps find good approximations for the interface model parameters without needing to find all parameter values describing j^{r}_{k} of the $I^{r}_{f+dl}(\tilde{x},s)/I_{app}(s)$ TFs.

The contribution of the electrolyte TF to the full-cell impedance is given in Eq. (2.101). At the current collector, it can be divided into two parts as:

$$-\frac{\widetilde{\Phi}_e^p(3,s)}{I_{app}(s)} = -\frac{[\widetilde{\Phi}_e(3,s)]_1}{I_{app}(s)} - \frac{[\widetilde{\Phi}_e(3,s)]_2}{I_{app}(s)},$$

where:

$$\begin{aligned}
-\frac{[\widetilde{\Phi}_e(3,s)]_1}{I_{app}(s)} &= \frac{1}{\bar{\kappa}^s} + \frac{1}{\bar{\kappa}^p} \\
&+ \frac{1}{\bar{\kappa}^n}\left(\frac{\left(1-e^{-\Lambda_1^n}\right)(j_1^n - j_2^n) - \Lambda_1^n\left(j_1^n e^{-\Lambda_1^n} - j_2^n\right)}{(\Lambda_1^n)^2}\right. \\
&\left. + \frac{\left(1-e^{-\Lambda_2^n}\right)(j_3^n - j_4^n) - \Lambda_2^n\left(j_3^n e^{-\Lambda_2^n} - j_4^n\right)}{(\Lambda_2^n)^2}\right) \\
&+ \frac{1}{\bar{\kappa}^p}\left(\frac{\left(1-e^{-\Lambda_1^p}\right)\left(j_2^p - j_1^p\right) - \Lambda_1^p\left(j_2^p e^{-\Lambda_1^p} - j_1^p\right)}{\left(\Lambda_1^p\right)^2}\right. \\
&\left. + \frac{\left(1-e^{-\Lambda_2^p}\right)\left(j_4^p - j_3^p\right) - \Lambda_2^p\left(j_4^p e^{-\Lambda_2^p} - j_3^p\right)}{\left(\Lambda_2^p\right)^2}\right) \\
-\frac{[\widetilde{\Phi}_e(3,s)]_2}{I_{app}(s)} &= \bar{\kappa}_D T\left(\left(1-e^{-\Lambda_1^n}\right)(c_1^n - c_2^n) + \left(1-e^{-\Lambda_2^n}\right)(c_3^n - c_4^n)\right. \\
&+ \left(1-e^{-\Lambda_1^s}\right)(c_1^s - c_2^s) \\
&\left. + \left(1-e^{-\Lambda_1^p}\right)\left(c_2^p - c_1^p\right) + \left(1-e^{-\Lambda_2^p}\right)\left(c_4^p - c_3^p\right)\right).
\end{aligned}$$

The $[\cdot]_1$ terms are associated with the ionic current flow in the electrolyte and the $[\cdot]_2$ terms are associated with the lithium concentration gradient in the electrolyte. The bottom row of Fig. 3.54 shows that $-\widetilde{\Phi}_e^p(3,s)/I_{app}(s)$ looks almost like a copy of $-[\widetilde{\Phi}_e(s)]_2/I_{app}(s)$ except shifted along the real axis, indicating that the impedance due to ionic current flow might be approximated by a real constant value (i.e., an ohmic resistance). The DRT technique can perhaps be used here to approximate $-\widetilde{\Phi}_e^p(3,s)/I_{app}(s)$ since it has a semicircular shape in a Nyquist plot that is associated with an RC model.

3.15.3 *Understanding the interface model components*

Ultimately, we will regress the measured EIS data to the full-cell impedance model of Eq. (2.103), but here we are building an approximate model solely for the purpose of initializing some of the parameter values prior to optimization. The previous section demonstrated that full-cell impedance is dominated by the nonspatially varying interfacial impedance models of both electrodes plus the electrolyte impedance due to a lithium concentration gradient. The frequencies over which the interface impedances $\bar{Z}^{\mathrm{r}}_{\widetilde{\mathrm{se}}}(s)$ dominate are those that appear in the Nyquist bump of the full-cell impedance whereas the electrolyte impedance usually has its dominant contribution at somewhat lower frequencies. Therefore, we conclude that the full-cell Nyquist bump contains information primarily determined by $\bar{Z}^{\mathrm{r}}_{\widetilde{\mathrm{se}}}(s)$, and hence that we can initialize components of $\bar{Z}^{\mathrm{r}}_{\widetilde{\mathrm{se}}}(s)$ using EIS data from the bump.

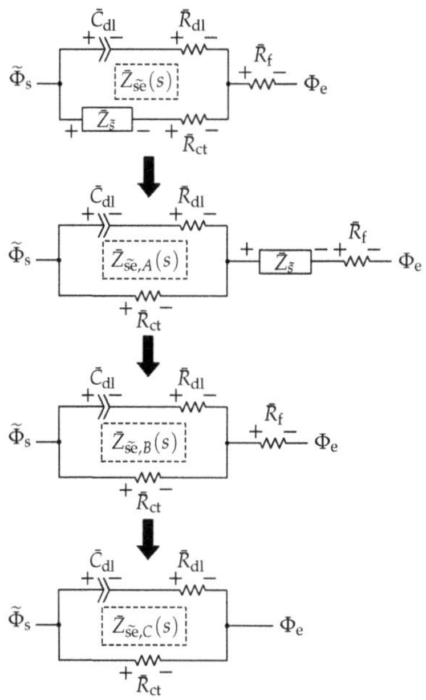

Figure 3.55: Approximations to the interface model $\bar{Z}_{\widetilde{\mathrm{se}}}(s)$ (also shown in Fig. 2.15). $\bar{Z}_{\widetilde{\mathrm{se}},A}(s)$ relocates the ideal solid diffusion impedance $\bar{Z}_{\tilde{\mathrm{s}}}(s)$ outside of the parallel loop; $\bar{Z}_{\widetilde{\mathrm{se}},B}(s)$ eliminates the $\bar{Z}_{\tilde{\mathrm{s}}}(s)$ component; $\bar{Z}_{\widetilde{\mathrm{se}},C}(s)$ removed the series $\bar{R}_{\mathrm{f}}$.

The interphase circuit in Fig. 2.15 illustrates the exact expression of $\bar{Z}^{\mathrm{r}}_{\widetilde{\mathrm{se}}}(s)$ using five impedance terms: the double-layer resistance and capacitance $\bar{R}_{\mathrm{dl}}$ and $\bar{C}_{\mathrm{dl}}$, the film resistance $\bar{R}_{\mathrm{f}}$, the solid-diffusion impedance $\bar{Z}_{\tilde{\mathrm{s}}}(s)$, and the charge-transfer resistance $\bar{R}_{\mathrm{ct}}$. Of these, $\bar{Z}_{\tilde{\mathrm{s}}}(s)$ and $\bar{R}_{\mathrm{ct}}$ are SOC-dependent and the rest are constant. We would like to use the DRT to gain insight into how these components affect the overall interface impedance, and hence also cell impedance.

We perform several transformations to the overall interphase impedance $\bar{Z}_{\widetilde{\mathrm{se}}}(s)$ as presented in Fig. 3.55. These transformations are not exact equivalents to the original circuit; they only approximate the original circuit in simplified forms to help gain intuition regarding the impact of the subcomponents. In $\bar{Z}_{\widetilde{\mathrm{se}},A}(s)$, we modify $\bar{Z}_{\widetilde{\mathrm{se}}}(s)$ by moving $\bar{Z}_{\tilde{\mathrm{s}}}(s)$ out of the parallel RC branch; in $\bar{Z}_{\widetilde{\mathrm{se}},B}(s)$, we delete the $\bar{Z}_{\tilde{\mathrm{s}}}(s)$ term; in $\bar{Z}_{\widetilde{\mathrm{se}},C}(s)$, we further delete the series resistance $\bar{R}_{\mathrm{f}}$.

Interface model comparisons, 80 % SOC

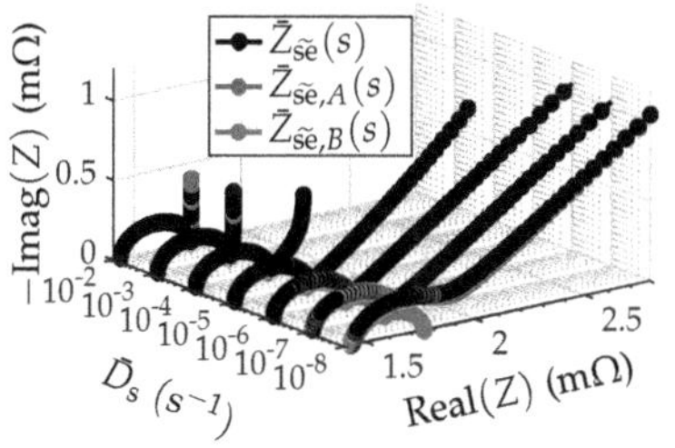

Figure 3.56: Analyzing the interface models shown in Fig. 3.55: Comparing Nyquist plots between $\bar{Z}_{\tilde{\mathrm{s}}}(s)$, $\bar{Z}_{\widetilde{\mathrm{se}},A}(s)$ and $\bar{Z}_{\widetilde{\mathrm{se}},B}(s)$ for different $\bar{D}_{\mathrm{s}}$ values.

Comparisons among $\bar{Z}_{\widetilde{\mathrm{se}}}(s)$, $\bar{Z}_{\widetilde{\mathrm{se}},A}(s)$, and $\bar{Z}_{\widetilde{\mathrm{se}},B}(s)$ for various $\bar{D}_{\mathrm{s}}$ values are computed and presented in Fig. 3.56. We notice that $\bar{Z}_{\widetilde{\mathrm{se}}}(s)$ and $\bar{Z}_{\widetilde{\mathrm{se}},A}(s)$ are very similar when $\bar{D}_{\mathrm{s}}$ is large and start to diverge only when $\bar{D}_{\mathrm{s}}$ is small ($\leq 10^{-6}\,\mathrm{s}^{-1}$). This is because $\bar{Z}_{\tilde{\mathrm{s}}}(s)$ dominates interface impedance only at low frequencies and the impedances that form the Nyquist bump dominate at high frequencies. When there is sufficient frequency separation between the impedance of solid diffusion and the impedance of the double layer, the approximation made by $\bar{Z}_{\widetilde{\mathrm{se}},A}(s)$ does not lose much fidelity when describing the Nyquist bump. The $\bar{Z}_{\widetilde{\mathrm{se}},B}(s)$ approximation amplifies this observation by deleting $\bar{Z}_{\tilde{\mathrm{s}}}(s)$ entirely, showing that its contribution impacts the charge-transfer region (the bump) only when $\bar{D}_{\mathrm{s}}$ is small. In practice, we have found $\bar{D}^{\mathrm{n}}_{\mathrm{s}} \approx 1 \times 10^{-4}\,\mathrm{s}^{-1}$ and $\bar{D}^{\mathrm{p}}_{\mathrm{s}} \approx 7 \times 10^{-5}\,\mathrm{s}^{-1}$ for the physical Panasonic cell we are using as the demonstration article in this chapter. Therefore, if the bump impedance can be isolated from $\bar{Z}_{\widetilde{\mathrm{se}}}(s)$, then we can train initial guesses to component values by fitting the Nyquist-bump impedances to the $\bar{Z}_{\widetilde{\mathrm{se}},B}(s)$ model. Keep in mind that the only data that are available to us are the full-cell impedances rather than the individual interfacial impedances. As the full-cell impedance comprises many components that have a real (ohmic) part, $\bar{R}_{\mathrm{f}}$ is difficult to expose directly from the bump analysis, and therefore we ignore it in the $\bar{Z}_{\widetilde{\mathrm{se}},C}(s)$ configuration.

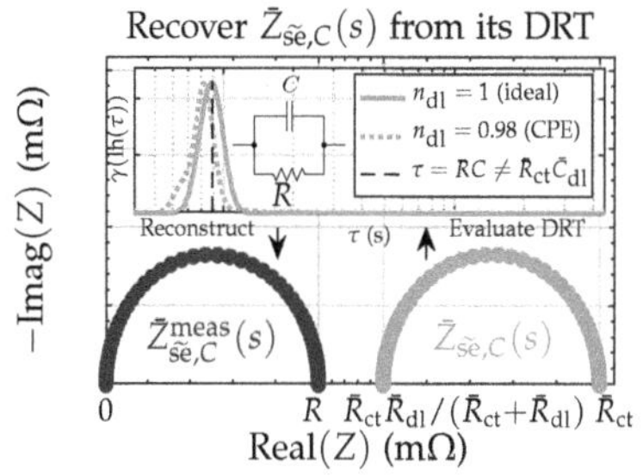

Figure 3.57: Illustrating the idea of using the DRT for $\bar{Z}_{\widetilde{\mathrm{se}},C}(s)$ to reconstruct its impedance spectrum with shifted real axis.

Fig. 3.57 illustrates how we estimate initial guesses for the interfacial model component values from the full-cell Nyquist bump. If we perform the DRT on the impedances from the bump region, we will observe a spectrum of $\gamma(\ln(\tau))$; the figure shows only a single time constant, but there will be (at least) two because the bump comprises impedances from (at least) the double layers of both electrodes. The centers of the peaks in the distribution correspond to the RC time constants of that feature; in principle, the area under each peak corresponds to its resistance, but this can be difficult to evaluate due to the discrete nature of a practical DRT implementation and because peaks are not always clearly separated. Instead, we essentially invert the DRT function inside an optimization to find the $\bar{R}_{\text{ct}}$ values that reproduce the original bump shape. This inversion or reconstruction of the bump from a segment of the DRT loses the horizontal shift of the original contribution to the bump, but retains its width. Since the DRT describes RC behaviors, it recognizes $\tilde{Z}^{\text{meas}}_{\tilde{\text{se}},C}(s)$ as $R/(1+sRC)$, which is theoretically identical to $\tilde{Z}_{\tilde{\text{se}},C}(s)$ but shifted to the origin along the real axis (the absolute resistance shift is lost in this operation). The amount of the real-part shift is mathematically known for one electrode to be $\bar{R}_{\text{ct}}\bar{R}_{\text{dl}}/(\bar{R}_{\text{ct}}+\bar{R}_{\text{dl}})$, but the real-part shift for the entire cell also includes electrolyte terms. Mathematically, we can derive the following relationship:[69]

[69] Here, we treat $\bar{C}_{\text{dl}}$ as ideal for simplicity rather than as a CPE term.

$$\begin{aligned}
\tilde{Z}^{\text{meas}}_{\tilde{\text{se}},C}(s) &= \tilde{Z}_{\tilde{\text{se}},C}(s) - \frac{\bar{R}_{\text{ct}}\bar{R}_{\text{dl}}}{\bar{R}_{\text{ct}}+\bar{R}_{\text{dl}}} \\
&= \frac{\bar{R}_{\text{ct}}+s\bar{C}_{\text{dl}}\bar{R}_{\text{ct}}\bar{R}_{\text{dl}}}{1+s\bar{C}_{\text{dl}}(\bar{R}_{\text{ct}}+\bar{R}_{\text{dl}})} - \frac{\bar{R}_{\text{ct}}\bar{R}_{\text{dl}}}{\bar{R}_{\text{ct}}+\bar{R}_{\text{dl}}} \\
&= \frac{\bar{R}^2_{\text{ct}}/(\bar{R}_{\text{ct}}+\bar{R}_{\text{dl}})}{1+s\bar{C}_{\text{dl}}(\bar{R}_{\text{ct}}+\bar{R}_{\text{dl}})} = \frac{R}{1+sRC},
\end{aligned} \tag{3.27}$$

where:

$$R = \frac{\bar{R}^2_{\text{ct}}}{\bar{R}_{\text{ct}}+\bar{R}_{\text{dl}}} \quad \text{and} \quad C = \bar{C}_{\text{dl}}\frac{(\bar{R}_{\text{ct}}+\bar{R}_{\text{dl}})^2}{\bar{R}^2_{\text{ct}}}.$$

We then use an optimization routine to minimize the RMSE between the DRT function and the reconstructed impedance data, and regress the values for $\bar{R}_{\text{dl}}$, $\bar{C}_{\text{dl}}$, and $\bar{R}_{\text{ct}}$ (also n_{dl}) at a specific cell SOC level. The cost function is defined as seeking to minimize the root-mean-squared error between the reconstructed impedance spectrum (violet data points in Fig. 3.57) and the model Eq. (3.27):

$$\begin{aligned}
p^* = \arg\min_p \sum_{m=1}^{M} &\left[\frac{\mathbb{R}\left(\tilde{Z}^{\text{meas}}_{\tilde{\text{se}},C}(s_m,z) - \hat{\tilde{Z}}^{\text{meas}}_{\tilde{\text{se}},C}(s_m,z)\right)}{\left|\tilde{Z}^{\text{meas}}_{\tilde{\text{se}},C}(s_m,z)\right|}\right]^2 \\
+ \sum_{m=1}^{M} &\left[\frac{\mathbb{I}\left(\tilde{Z}^{\text{meas}}_{\tilde{\text{se}},C}(s_m,z) - \hat{\tilde{Z}}^{\text{meas}}_{\tilde{\text{se}},C}(s_m,z)\right)}{\left|\tilde{Z}^{\text{meas}}_{\tilde{\text{se}},C}(s_m,z)\right|}\right]^2,
\end{aligned}$$

where $p = \{\bar{R}^{\mathrm{r}}_{\mathrm{ct}}, \bar{C}^{\mathrm{r}}_{\mathrm{dl}}, n^{\mathrm{r}}_{\mathrm{dl}}\}$.

Fig. 3.58 demonstrates optimization results for different assumed $\bar{R}_{\mathrm{dl}}$ values, which are known approximately from the pulse-resistance test. In both optimization scenarios, we set the constraints to be $R \leq \bar{R}^{\mathrm{r}}_{\mathrm{ct}} \leq 0.01\,\Omega$, $0.01\,\mathrm{F} \leq \bar{C}^{\mathrm{r}}_{\mathrm{dl}} \leq 200\,\mathrm{F}$, and $0.7 \leq n^{\mathrm{r}}_{\mathrm{dl}} \leq 1$. If $\bar{R}_{\mathrm{dl}}$ is known exactly, the parameters of the interphase model converge to their true values; if $\bar{R}_{\mathrm{dl}}$ is known only approximately, the results still indicate that the final estimates are always within one order of magnitude from the their true values, which we consider to be good initial approximations.

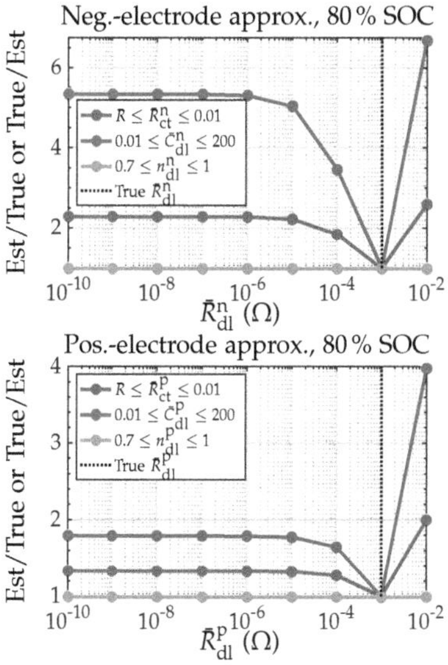

Figure 3.58: Estimating $\{\bar{R}_{\mathrm{ct}}, \bar{C}_{\mathrm{dl}}, n_{\mathrm{dl}}\}$ from $\tilde{Z}^{\mathrm{meas}}_{\tilde{\mathrm{se}},\mathrm{C}}(s)$ with given $\bar{R}_{\mathrm{dl}}$ values. Parameter optimization boundaries are stated in the legend.

Fig. 3.59 illustrates this process further by showing the simulated full-cell DRT. From this DRT, we can associate the two highest-frequency DRT peaks to the electrode-interface RC responses, and recover their impedance spectra $\tilde{Z}^{\mathrm{r,meas}}_{\tilde{\mathrm{se}},\mathrm{C}}(s)$, shown in Fig. 3.60.[70] We estimate the $\bar{C}^{\mathrm{r}}_{\mathrm{dl}}$ and $\bar{R}^{\mathrm{r}}_{\mathrm{ct}}$ values and compare the reconstructed impedances of each interface (the colored lines) to the true values (the black lines). In practice, we are unable to distinguish the ordering of time constants from the two electrodes. Presently, we simply assume that the smaller time constant (at higher frequency) is associated with the negative electrode and that the larger one (at lower frequency) is associated with the positive electrode.

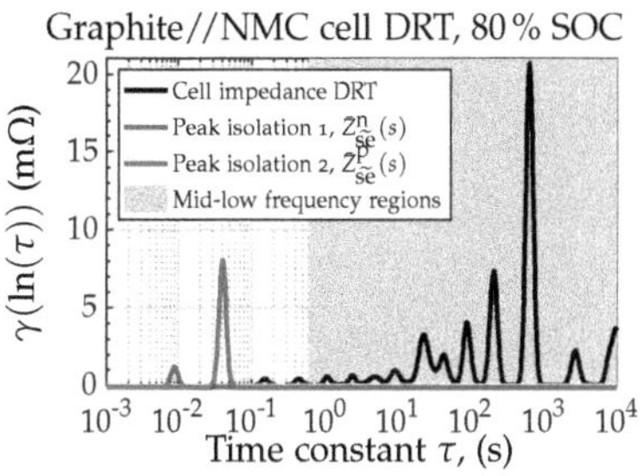

Figure 3.59: Isolating the DRT peaks that are associated with charge-transfer from the cell data.

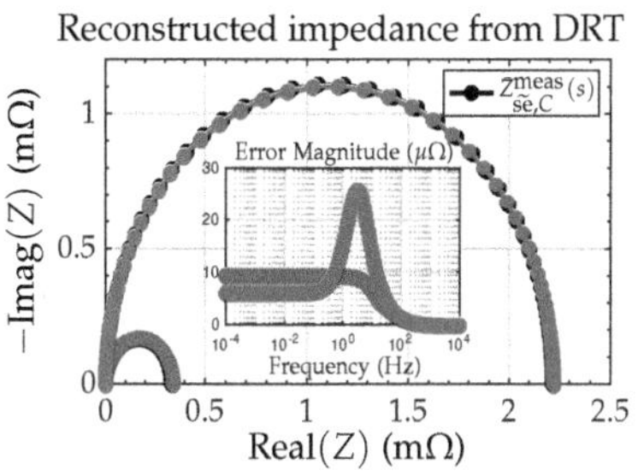

Figure 3.60: Recovering the impedance spectra that are associated to $\bar{R}_{\mathrm{dl}}$, $\bar{C}_{\mathrm{dl}}$, and $\bar{R}_{\mathrm{ct}}$.

3.15.4 *Initializing specific model parameter estimates*

With this background, we recognize that the DRT can be used to isolate time constants from the Nyquist bump region. Recreating these time constants with an RC circuit provides initial estimates of $\bar{C}^{\mathrm{r}}_{\mathrm{dl}}$ as well as $\bar{R}^{\mathrm{r}}_{\mathrm{ct}}$. The latter can be used to fit initial values to $\bar{k}^{\mathrm{r}}_{0,j}$. In the next section, we will also see how the DRT can be used to preprocess lab data to remove the effects of mutual inductance arising from the experimental wiring harness from the measured frequency response.

In order to maintain the momentum of the chapter, details regarding how we might initialize other model parameter estimates prior to invoking nonlinear optimization are summarized in App. 3.D. We simply mention here that some of the initializations use the DRT method, as just discussed; others are ad hoc and there remains room for improvement. However, we find that even these crude approximations are helpful to provide reasonable initial guesses. Also note that some of the parameter values we discuss can be identified from the pulse-resistance test. The initializations we present in the appendix can be applied prior to the optimizations for that test; and then the optimized values from the pulse-resistance test can be used to ini-

[70] Peaks at lower frequencies in Fig. 3.59 arise due to electrolyte and solid-diffusion impedance.

tialize the optimizations that use EIS data. We use the latter method when presenting results for the Panasonic cell later in this chapter.

3.15.5 Preprocessing laboratory EIS data

When applying the parameter-estimation method to measured EIS data, we encounter features in those data that are not represented by the LPM. Specifically, the imaginary part of $Z(s)$ computed using the TF models is always strictly nonpositive. However, we commonly observe that EIS impedances measured for high-frequency inputs are reported as having positive imaginary parts.

These positive imaginary impedance measurements give evidence of inductance in the lab setup, which we believe to be dominated by the mutual inductance of the wires connecting the potentiostat to the cell. In a lab procedure, even small movements to the current-carrying and measurement wires will cause unpredictable changes in the observed inductance and so the effect of mutual inductance is very difficult to eliminate and even to calibrate unless the positions of the wires are strictly controlled and reproducible;[71] even so, deterministic calibration may not succeed in all cases. Therefore, we desire to develop methods to quantify the inductance directly from the measured impedance spectra rather than physically measuring its value. The effect of the inductance will be subtracted from the measured impedance to calibrate the measurements.

[71] Manuel Kasper, Arnd Leike, Johannes Thielmann, Christian Winkler, Nawfal Al-Zubaidi R-Smith, and Ferry Kienberger, "Electrochemical impedance spectroscopy error analysis and round robin on dummy cells and lithium-ion-batteries," *Journal of Power Sources*, 536:231407, 2022; and Ryan R. Jobman, "Identification of lithium-ion-cell physics-model parameter values," PhD dissertation, University of Colorado Colorado Springs, (`http://hdl.handle.net/10976/166641`), 2016.

We should also not simply ignore the inductive impact by cropping and deleting all impedance measurements having positive imaginary parts; doing so may distort some important information in the measurements. The blue lines of the top frame of Fig. 3.61 show the Nyquist plot of two RC networks, without inductance. The lines plotted with circle markers are for a small RC time constant and the lines plotted with the square markers are for a large RC time constant. The red lines plot the impedances of the same RC networks with an added series inductance. We see that the impedance of the RC circuit having large time constant is not greatly affected by the inductance, and so simply cropping the measured impedance to frequencies having a negative imaginary part might be acceptable to provide input to characterize the circuit. However, the Nyquist bump for the RC circuit having small time constant is greatly distorted by the impact of the inductance. If we were simply to crop the impedance measurements, this distorted bump would lead to incorrect parameter estimates. Instead, we must estimate the inductance and subtract its effect from the overall estimated impedance to recreate the original impedance drawn by the blue line. The bottom frame of Fig. 3.61 presents another view of the damage caused by simply truncating

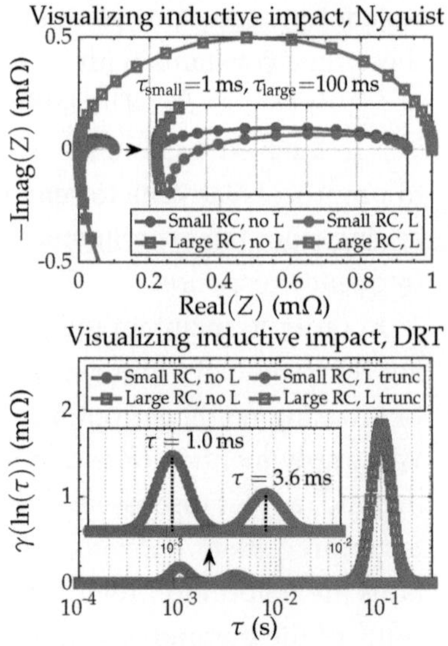

Figure 3.61: Impact of ignoring inductance on systems with large and small time constants: impedance responses (top) and their computed DRT after trimming the impedance data with positive imaginary parts (bottom).

the EIS data. The blue lines and circle and square markers show the true DRT for the small and large time constants, respectively. The red lines and markers show the DRT of the truncated EIS data. We see again that truncation does not distort the DRT of the data from the large-time-constant RC circuit, but that it causes a significant misidentification of the time constant (by 3.6 times) of the small-time-constant RC circuit.

Approach #1: Discrete model fit

How should we estimate the effect of this inductance, in order to remove it? One possibility is to notice that we can fit the high-frequency data to various hand-selected inductance models. Fig 3.62 demonstrates two model candidates, where the top frame shows the Nyquist plot of an ideal inductor connected in series with a parallel resistor–inductor (RL) network and the bottom frame shows the Nyquist plot of an ideal inductance having a CPE exponent. Models such as these can be manually chosen and adjusted to fit measured high-frequency data (e.g., 10 kHz to 100 kHz) in which the impedance of the cell itself is assumed to be essentially ohmic. To be able to do this well, we recommend collecting as many high-frequency data points as possible since they can be measured quickly and they can assist in determining an appropriate inductance model.

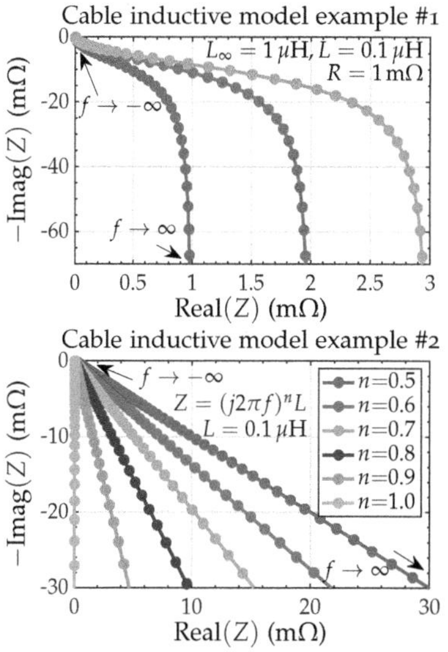

Figure 3.62: Nyquist plots for two different inductance models: an ideal inductor connected in series with parallel RL components (top), and ideal inductance with constant-phase-element exponent (bottom).

The advantage of this discrete model fit approach is that one can manually choose the model to fit. However, this process requires user intervention. We find that a model that is selected and regressed for one set of data may not be applicable to other sets of data that are collected using different cable configurations. Whenever the test cables are moved, the model should also be adjusted.

Approach #2: Generalized DRT (GDRT)

Another approach to estimating the inductance is to use a generalized DRT (GDRT). The DRT can be extended to estimate the distribution of RL time constants in addition to RC time constants. The GDRT analysis presented here applies to complex superimposed impedance spectra that include ohmic, inductive, capacitive, resistive–inductive, and resistive–capacitive effects.

The GDRT was first proposed by Danzer,[72] and we implement his structure in this work. Using the GDRT approach to estimating the inductance simplifies the workflow because it does not require that the user manually select models to fit the high-frequency impedance, and it also leads to a fast, reliable, and reproducible DRT analysis.

[72] Michael A. Danzer, "Generalized distribution of relaxation times analysis for the characterization of impedance spectra," *Batteries*, 5(3):53, 2019.

To proceed, the DRT model in Eq. (3.26) is extended to include a series resistor, a series inductor, series inductor–resistor distributed time constants, and a series capacitor to model the impedance as:

$$\begin{aligned} Z_{\text{GDRT}}(s) &= R_\infty + sL_\infty + \frac{1}{sC_0} \\ &+ \int_0^\infty \frac{s\tau_{\text{RL}}}{1+s\tau_{\text{RL}}}\gamma_{\text{RL}}(\ln(\tau_{\text{RL}}))\,\mathrm{d}\ln\tau_{\text{RL}} \\ &+ \int_0^\infty \frac{1}{1+s\tau_{\text{RC}}}\gamma_{\text{RC}}(\ln(\tau_{\text{RC}}))\,\mathrm{d}\ln\tau_{\text{RC}}, \end{aligned} \tag{3.28}$$

where R_∞ is an ohmic resistance, L_∞ is an ideal inductance, C_0 is an ideal capacitor, $\tau_{\text{RL}} = L/R$, and $\tau_{\text{RC}} = RC$. In such a GDRT framework, the calculation of $Z_{\text{DRT}}(s)$ is extended. The subscripts on R, L, and C denote the frequencies (i.e., very low or very high) that they dominate. Additionally, the standard impedance of the distribution function for the resistive–capacitor behavior, an additional impedance of a distribution function for the resistive–inductive behavior is calculated. The added distribution function is also a series connection of parallel circuits with individual time constants of the form $\tau_{\text{RL}} = L/R$. Notice that the numerator of the RL impedance is different from the numerator of the RC impedance within the integrals. Identification of the two distribution functions, $\gamma_{\text{RL}}(\ln(\tau_{\text{RL}}))$ and $\gamma_{\text{RC}}(\ln(\tau_{\text{RC}}))$, and the three parameters, R_∞, L_∞, and C_0 in the GDRT analysis can also follow the computational procedures as described by Wan et al.[73] Mathematical details will be briefly illustrated in App. 3.E, where we adopt the GDRT framework proposed by Danzer and implement in the computational framework proposed by Wan et al. Per Danzer, a cost function is defined as:

[73] Ting Hei Wan, Mattia Saccoccio, Chi Chen, and Francesco Ciucci, "Influence of the discretization methods on the distribution of relaxation times deconvolution: Implementing radial basis functions with DRTtools," *Electrochimica Acta*, 184:483–499, 2015.

$$J = \underbrace{\left\|\mathbf{A}'\mathrm{x} - \mathbf{Z}'_{\text{meas}}\right\|^2 + \left\|\mathbf{A}''\mathrm{x} - \mathbf{Z}''_{\text{meas}}\right\|^2}_{\text{least squares}} + \underbrace{\lambda \mathrm{x}^T\mathbf{M}\mathrm{x}}_{\text{penalty}}, \tag{3.29}$$

where λ is a user-selected regularization parameter, $\mathbf{Z}'_{\text{meas}}$ is a vector of real–impedance data, $\mathbf{Z}''_{\text{meas}}$ is a vector of imaginary–impedance data, $\mathbf{A}'$, $\mathbf{A}''$, and $\mathbf{M}$ are matrices built based on the input/output data and their structures are listed in the appendix. The vector x is targeted to estimate:

$$\mathrm{x} = \left[R_\infty, L_\infty, C_0, x_1^{\text{RC}}, x_2^{\text{RC}}, \cdots, x_M^{\text{RC}}, x_1^{\text{RL}}, x_2^{\text{RL}}, \cdots, x_M^{\text{RL}}\right], \tag{3.30}$$

where x are unknown parameters to structure the distribution functions. In summary, by implementing the GDRT method, modeled by Eqs. (3.28)–(3.29), we can now automatically separate the inductive impedance from the actual cell impedance, and isolate double-layer contributions of each electrode by examining the RC GDRT peaks. We will implement this method when processing any further synthetic and laboratory data.[74]

[74] It is important to emphasize that the GDRT model Eq. (3.28) is very sensitive to the highest frequency $f_{\max}$ at which the data are collected. It is required to have $f_{\max} \gg 1/(2\pi\tau_{\text{RL}})$; otherwise we will face an under-fitting problem (the optimization task in Eq. (3.29) will not give satisfying fits to the data). To avoid this, we can simply adopt the discrete model-fit approach to fit the high-frequency data (e.g., 1 kHz to 100 kHz), then manually extrapolate the measured data to even higher frequency regions using the inductive model.

	R_∞ (mΩ)	L_∞ (μH)	C_0 (F)	τ_{RL1} (μs)	τ_{RL2} (μs)	τ_{RC1} (ms)	τ_{RC2} (ms)
True	2	1	400	25	75	15	52
Estimate	1.98	1	400	26.26	78.80	15.30	52.85

Table 3.16: Estimates of time constants using the GDRT framework. Graphical results are shown in the top frame of Fig. 3.63.

To validate the performance of the GDRT method, the top frame of Fig. 3.63 likewise displays the results when applying the GDRT analysis to a circuit model having known electrochemical components. The cell and inductive models are designed as:

$$
\begin{aligned}
Z_{\text{cell}}(s) &= R_\infty + \frac{R_1}{1+s\tau_{\text{RC1}}} + \frac{R_2}{1+s\tau_{\text{RC2}}} + \bar{Z}_{\tilde{s}}^{\text{n}*}(s) + \bar{Z}_{\tilde{s}}^{\text{p}*}(s) + \frac{1}{sC_0} \\
Z_L(s) &= sL_0 + \frac{sR_3\tau_{\text{RL1}}}{1+s\tau_{\text{RL1}}} + \frac{sR_4\tau_{\text{RL2}}}{1+s\tau_{\text{RL2}}},
\end{aligned}
\tag{3.31}
$$

where $\bar{Z}_{\tilde{s}}^{\text{n}*}(s)$ and $\bar{Z}_{\tilde{s}}^{\text{p}*}(s)$ are the electrode interface diffusion impedances incorporating constant-phase elements but after removing integrators. The overall impedance is the sum of the two components, $Z(s) = Z_{\text{cell}}(s) + Z_L(s)$. The impedance data are simulated at a sampling rate of 10 ppd within $1\,\text{mHz} \le f \le 100\,\text{kHz}$ ($N = 81$). The model parameter values used to simulate the data are listed in Table 3.16 along with the GDRT estimates. The two diffusion-impedance values, $\bar{Z}_{\tilde{s}}^{\text{n}*}(s)$ and $\bar{Z}_{\tilde{s}}^{\text{p}*}(s)$, can also be characterized using series RC components which describe the charge-transfer behaviors from one pathway to another inside the electrode. The estimated impedance in the top frame of the figure has almost no visual difference from the simulated data. The estimated distribution functions are also displayed, in which two peaks at high frequencies are caused by the two RL elements, the middle two peaks are associated with the two RC elements, and the remaining peaks at low frequencies correlate to the solid-diffusion impedance.

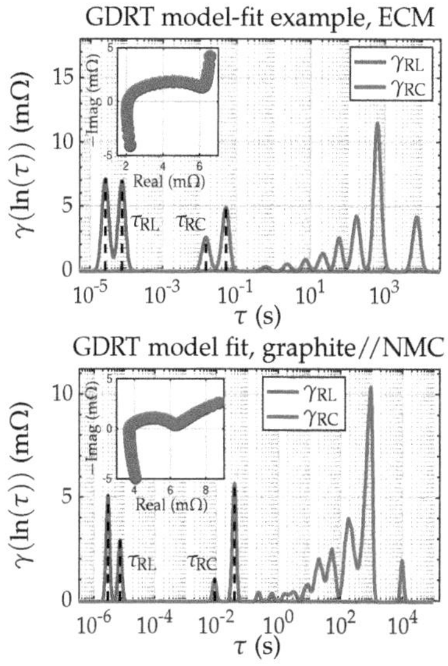

Figure 3.63: Demonstration of using the GDRT method to fit an equivalent-circuit model simulation (top), and the synthetic graphite//NMC cell impedance (bottom) when measured data are corrupted by noise.

Numeric results are listed in Table 3.16. Note that since the DRT provides its solution only for discrete values of τ, the peak locations and heights are not resolved exactly. It is likely that improved estimates of the listed parameter values could be found simply by performing an additional optimization of the overall impedance using the values in the table as initializations.

The lower frame of Fig. 3.63 demonstrates GDRT model fits to simulated graphite//NMC cell data contaminated by synthetic measurement noise. We model cell impedance using the system transfer-functions as written in Eq. (2.103), and we also describe inductance using Eq. (3.31). In addition, we include stochastic measurement noise having magnitude proportional to the absolute value of the impedance.[75] An overall computed noisy synthetic impedance for the

[75] Mattia Saccoccio, Ting Hei Wan, Chi Chen, and Francesco Ciucci, "Optimal regularization in distribution of relaxation times applied to electrochemical impedance spectroscopy: Ridge and lasso regression methods—a theoretical and experimental study," *Electrochimica Acta*, 147:470–482, 2014; and Ting Hei Wan, Mattia Saccoccio, Chi Chen, and Francesco Ciucci, "Influence of the discretization methods on the distribution of relaxation times deconvolution: Implementing radial basis functions with DRTtools," *Electrochimica Acta*, 184:483–499, 2015.

virtual graphite//NMC cell is written as:

$$Z_{\text{meas}}(s) = |Z_{\text{cell}}(s) + Z_L(s)| + \varepsilon\,|Z_{\text{cell}}(s) + Z_L(s)|\,(\eta' + j\eta''), \quad (3.32)$$

where $\varepsilon = 0.5\,\%$ is the noise level, and η' and η'' are two independently distributed random variables where both have standard normal distribution $\mathcal{N}(0, 0.2)\ \Omega$. The figure for this example also shows excellent agreement between simulated data and GDRT estimates.

3.16 *Validating frequency-response-test parameters*

3.16.1 *Application to a simulated cell*

We now proceed to estimate the full-cell model parameter values from simulated EIS-test data. The synthetic full-cell impedance data were computed by Eq. (3.32) and the impact of the cable inductance was carefully estimated and removed using the GDRT method. For reference, Fig. 3.64 displays sample processed data at 95 % and 30 % SOC setpoints. It might be hard to see the inductive impact on the simulated cell due to the large cell impedance, but we will see it more clearly in the Panasonic-cell results later. Table 3.17 lists the configurations for the impedance simulations. The assignments for SOC setpoints and input frequencies are identical to those we use for the experiment we perform in the laboratory on the Panasonic cell.

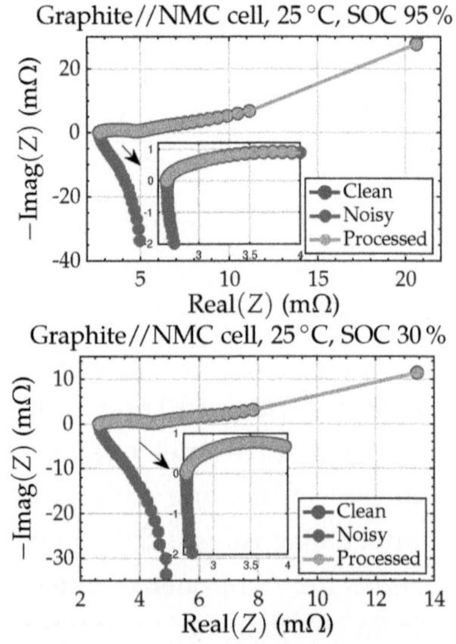

Figure 3.64: Results of preprocessing synthetic EIS-test data using the GDRT method.

Table 3.17: EIS-test configurations for simulation (ppd = points per decade).

SOC	100% : −5% : 5%
Freq.	100 kHz − 0.1 Hz, 20 ppd 0.1 Hz − 1 mHz, 10 ppd 0.1 mHz
Temp.	25 °C
Noise	$L_0 = 52\,\text{nH}$ $R_1 = 1.2\,\text{m}\Omega,\ \tau_1 = 7.1\,\mu\text{s}$ $R_2 = 1.5\,\text{m}\Omega,\ \tau_2 = 2.7\,\mu\text{s}$ $\varepsilon = 0.5\,\%$ $\eta' = \mathcal{N}(0, 0.2)\ \Omega$ $\eta'' = \mathcal{N}(0, 0.2)\ \Omega$

In the following two subsections, we will use the processed simulation data to estimate the LPM model parameter values. In addition, we also add an ideal series ohmic resistance to the processed simulation data emulating the contact resistance from the testing cables and the two current collectors,[76] namely R_c, which is an extra parameter we aim to identify in the work. Therefore, the overall cell impedance model is then $Z(s) + R_c$. Knowing that we have also emulated the measurement noise (with standard normal distributions) that cannot be removed from the simulated cell data, parameter-estimation results are not expected to be exact for any parameter. The primary purpose of studying and analyzing the synthetic data is to validate the underlying parameter-estimation methods since no truth is known for evaluating results from a laboratory process.

Ideally, we seek to estimate the parameter values:

$$p_{\text{EIS}} = \{\bar{D}^{\text{r}}_{\text{s,ref}}, n^{\text{r}}_{\text{f}}, \bar{C}^{\text{r}}_{\text{dl}}, n^{\text{r}}_{\text{dl}}, \bar{\psi}, \bar{q}^{\text{r}}_{\text{e}}, \bar{\kappa}_{\text{D}}, R_{\text{c}}\}$$

from the EIS data. We begin by assuming that the other model kinetics parameter values:

$$p_{\text{pulse}} = \{\alpha^{\text{r}}, \bar{k}^{\text{r}}_{0,j}, \bar{\kappa}^{\text{r}}, \bar{\sigma}^{\text{r}}, \bar{R}^{\text{r}}_{\text{f}}, \bar{R}^{\text{r}}_{\text{dl}}\}$$

have already been estimated by the pulse-resistance test. However, we encounter an identifiability problem when doing so. We know

[76] Note that the LPM describes only the dynamics between the two current collectors, but we also need to consider the resistance arising from causes such as particle binders, current collectors, and cable clamps when modeling the measured impedance of a physical cell.

from Chap. 1 that all the parameters of the PDE LPM model are identifiable; however, not all parameters of the linearized TF model turn out to be identifiable. We describe this phenomenon by saying that the LPM is "nonlinearly identifiable" but "not linearly identifiable."

A quick example will help to illustrate the problem. Consider a simple model, $y(x) = ax + b\sin(x)$. If we collect (x, y) data from this model for a variety of x, we can estimate both parameters a and b accurately. We say that this model is nonlinearly identifiable. However, when we linearize the model by performing a Taylor-series expansion at $x = 0$ and deleting the nonlinear terms, we find that the linearized model is $y(x) = (a + b)x$. In the linearized model, we cannot distinguish between the effects of a and b, as illustrated in Fig. 3.65. We can identify the sum, but not the individual values. We say that this model is not linearly identifiable.

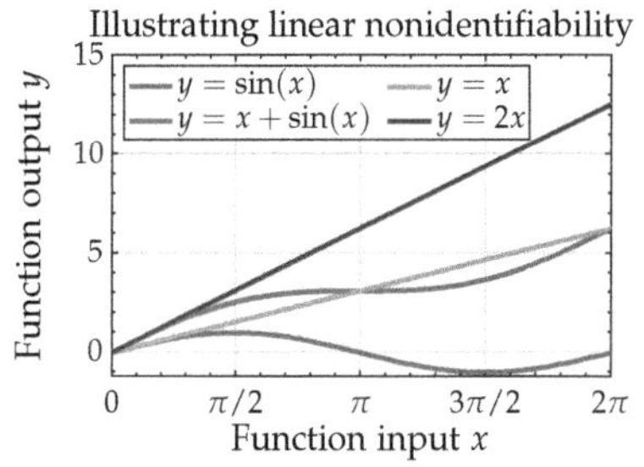

Figure 3.65: Illustrating how $y = x + \sin(x)$ is indistinguishable from $y = 2x$ for small values of x.

It is not at all obvious from a casual glance at the TF equations that there is a redundancy such as this in the TFs. However, some detailed analysis shows that the model never has $\bar{\psi}$, $\bar{q}_e^r$, and $\bar{\kappa}_D$ alone; they always occur as ratios of $\bar{q}_e^r/\bar{\psi}$ and $\bar{\kappa}_D/\bar{\psi}$. So, the EIS test cannot identify all parameters in the desired set; rather, it can identify:

$$p_{\text{EIS}}^{\text{lin}} = \{\bar{D}_{\text{s,ref}}^r, n_f^r, \bar{C}_{\text{dl}}^r, n_{\text{dl}}^r, \bar{q}_e^r/\bar{\psi}, \bar{\kappa}_D/\bar{\psi}, R_c\}.$$

We rely on the pseudo-steady-state test of Sect. 3.17 to identify $\bar{\psi}$, which allows us to recover $\bar{q}_e^r$ and $\bar{\kappa}_D$ also.

Estimating subset of LPM parameters using simulated data

In the first simulation study we perform, we analyze the case where the parameters estimated using the pulse-resistance test p_{pulse} are assumed to be exact; then, we identify only the remaining subset of model parameters $p_{\text{EIS}}^{\text{lin}}$ using the EIS-test data. Fig. 3.66 visualizes the parameter trajectories during the progress of the optimization. The underlying optimization task aims to find the set of parameter values that makes the model predictions agree as closely as possible to the processed cell impedance $Z(s) + R_c$ at all SOC and frequency setpoints. If we define a vector of M input frequencies and N SOC points, we can state the objective as seeking to minimize the following sum-of-squared error (SSE) cost function:

$$p^* = \arg\min_p \sum_{m=1}^{M}\sum_{n=1}^{N}\left[\frac{\mathbb{R}\left(Z(s_m, z_n) - \hat{Z}(s_m, z_n)\right)}{|Z(s_m, z_n)|}\right]^2 + \sum_{m=1}^{M}\sum_{n=1}^{N}\left[\frac{\mathbb{I}\left(Z(s_m, z_n) - \hat{Z}(s_m, z_n)\right)}{|Z(s_m, z_n)|}\right]^2, \quad (3.33)$$

where p is the set of model parameters to identify, and $\mathbb{R}$ and $\mathbb{I}$ represent real and imaginary parts, respectively. Since cell impedance

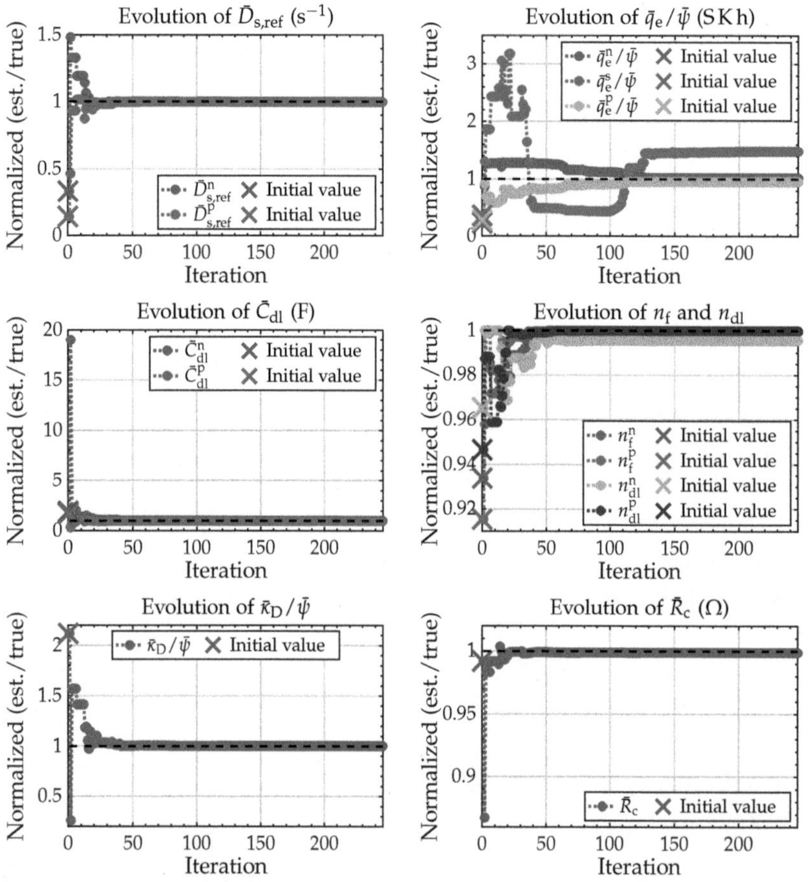

Figure 3.66: Convergence trajectories of 13 normalized parameter estimates during optimization; pulse-resistance-test parameters are set to their true values.

varies with frequency, the real- and imaginary-part errors are normalized by the magnitudes of the underlying data samples. Parameter initial values are also indicated in the figures and the optimization results are reproducible since the random number generator in MATLAB is set to be the default seed. The MATLAB function `particleswarm` from the Global Optimization Toolbox was used in hybrid mode along with `fmincon` to perform the optimization.

In Fig. 3.66, all parameter values are plotted in a normalized sense such that we desire for the trajectories to converge to "1." We notice that this is nearly universally the case. There is some deviation in $\bar{q}_e^s/\bar{\psi}$ and in n_{dl}^n. A sensitivity study shows that full-cell impedance is extremely insensitive to these parameter values, which makes them difficult to estimate perfectly from EIS data. The converse is also true: we do not require highly accurate values in order for the LPM to make excellent predictions of internal variables and cell voltage. We consider the quality of the estimates that we observe in the figure to be more than adequate for parameterizing an LPM.

Estimating full set of LPM parameters using simulated data

In the previous section, we assumed that the parameter values estimated using the pulse-resistance test were exact. This, of course, will not be true in practice, and so we would also like to see whether they might be estimated (or at least refined) using EIS data.

In this section, we again use simulated data but now attempt to identify the full set of model parameter values that are needed to compute the cell impedance response. Pulse-resistance-test parameters were initialized using the "optimized estimate" results reported in Table 3.14, while its MSMR parameters $\bar{k}^{\mathrm{r}}_{0,j}$ were initialized using the results reported in Table 3.D.1. Alternatively, one can consider the initialization strategies in App. 3.D if the pulse-resistance-test results are not available. Since the system TFs are linearized models, a pulse-resistance test is always recommended in order to expose and understand cell nonlinearities.

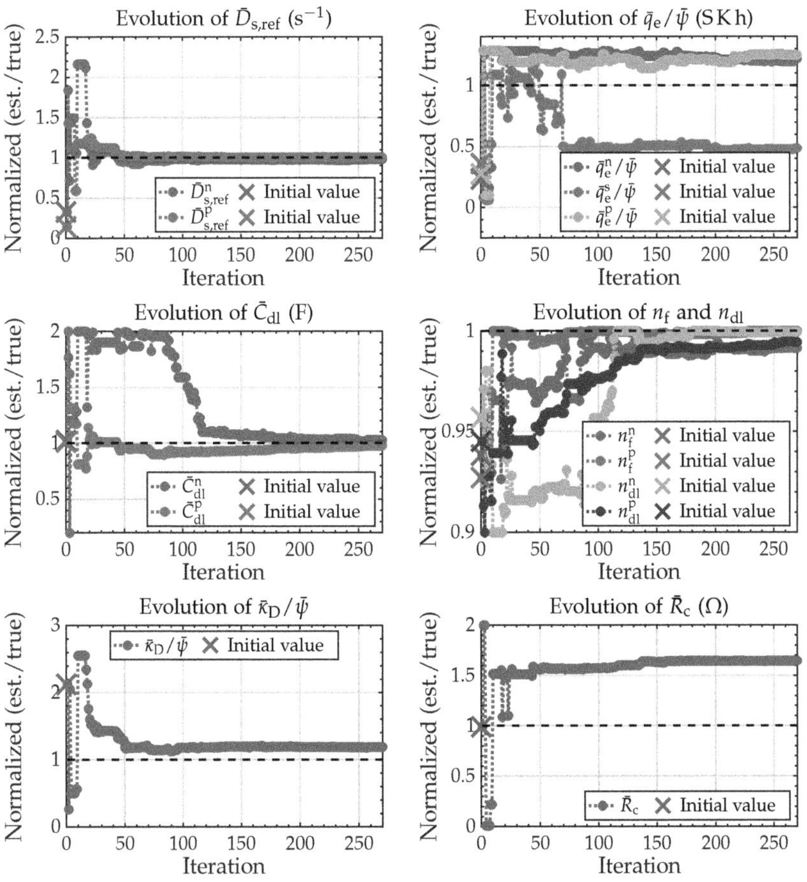

Figure 3.67: Convergence trajectories of each model parameter value during the optimization process. Pulse-resistance-test parameters are initialized to the values from that test but are allowed to adjust. The vertical axes are the normalized by their true values.

Figs. 3.67–3.68 display the parameter trajectories during optimization, where Fig. 3.67 reports the same parameters as previously shown in Fig. 3.66 and Fig. 3.68 reports the results for pulse-

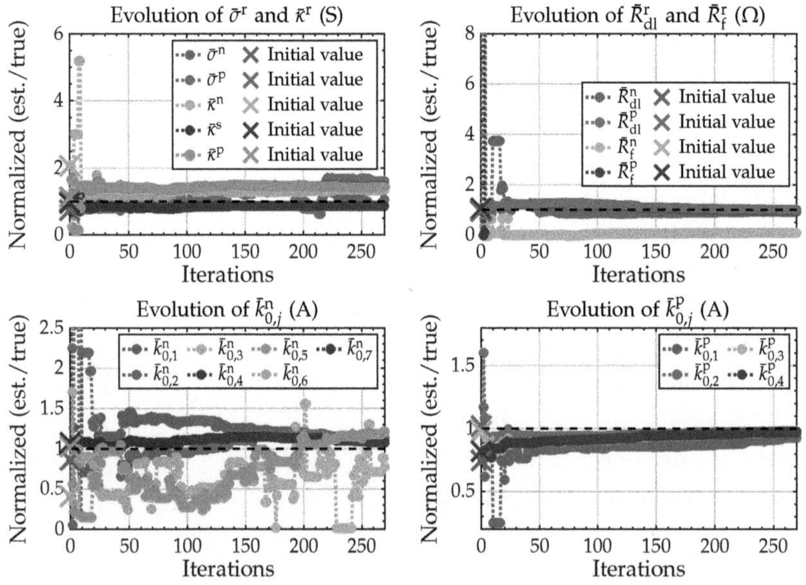

Figure 3.68: Convergence trajectories of pulse-resistance-test parameters (that are being estimated here using EIS data instead) during the optimization process. The vertical axes are the normalized by their true values.

resistance parameters. We see some degradation in the quality of parameter estimates for p_{EIS} in Fig. 3.67. We attribute some of this to the fact that the optimization must take place over a much larger parameter space, and it is more likely that the optimizer will select a local minimum that is different from the global minimum. However, another factor is that not all of the combined sets of parameters $\{p_{\text{EIS}}^{\text{lin}}, p_{\text{pulse}}\}$ are linearly identifiable. For example, the impacts of R_c and $1/\bar{\kappa}^s$ on the full-cell impedance are identical. The terms cannot be separated in any linear way. This is one reason that we see a lack of convergence of R_c in Fig. 3.67.

Fig. 3.68 shows optimization trajectories for the parameters in p_{pulse}, augmented with $\bar{k}_{0,j}^{r}$. We notice that the EIS data allow some refinement of the initial parameter guesses in most cases. However, as observed in the discussion of the pulse-resistance test and in App. 3.D, impedance is insensitive to the value of $\bar{k}_{0,j}^{n}$ for some galleries, which makes finding good estimates of those values challenging. (Conversely, we do not require precise estimates to get good predictions from the LPM.)

For reference, Fig. 3.69 visualizes the performance of impedance-model fit to the simulated data, from which we observe larger estimation errors at lower frequencies. Although the final estimates of many parameters are not exact, it is still difficult to visualize differences in the Nyquist plots, showing that we are modeling the impedance well.

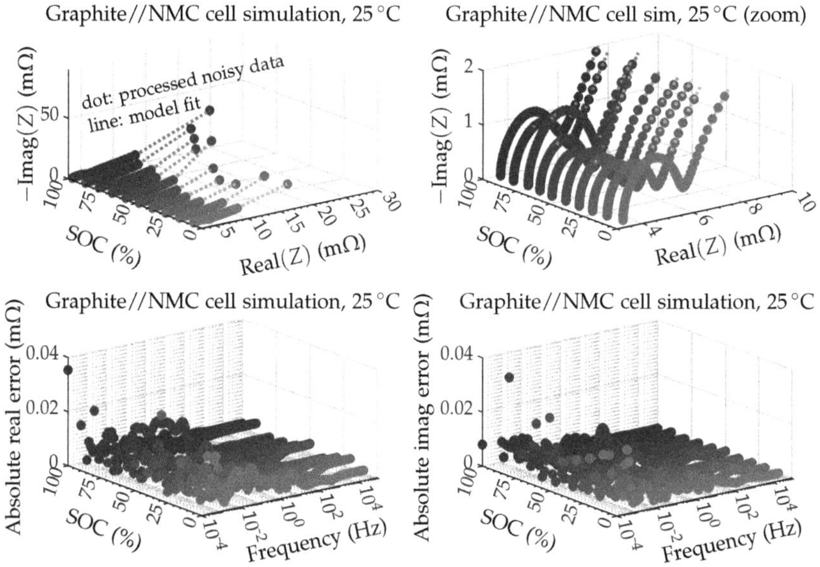

Figure 3.69: Results when estimating LPM parameter values from emulated cell impedance data. Top row shows Nyquist fits, and bottom row shows absolute real-part and imaginary-part errors.

3.16.2 *Application to a physical cell*

Data collection and processing

We now apply the methods to frequency-response data collected from the Panasonic cell. These data were measured using a Gamry Reference 3000 potentiostat. Since the Panasonic cell has very low pulse resistance ($< 2\,\mathrm{m}\Omega$), we use galvanostatic control to perform the EIS test. We recommend using a fixed cable configuration through the entire EIS process since any movement during the test will change the mutual inductance of the test setup and will lead to significant measurement errors.[77]

When selecting the sinusoidal current amplitude, we considered several factors. The value should be as small as possible to avoid moving the cell's SOC, but large enough to observe stable output with minimum influence of the measurement noise. Although Gamry Instruments suggests a current amplitude of (at most) 5 % of the standard charge current as reported by the manufacturer, it is always advantageous to conduct quick measurements to verify the choices. We performed some test runs at 0 % and 100 % SOC in galvanostatic mode with various input amplitudes, from 1 mHz to 100 kHz, sampled at 5 points-per-decade (ppd). One of the approaches to validate the impedance data quality is to perform a Kramers–Kronig (K–K) Transform on the experimental data,[78] which calculates the spectrum of a linear, stable, and causal circuit. If the K–K residuals of the experimental data exceed a certain threshold, then the samples may exhibit nonlinearity, instability, or the influence of unknown variables (i.e., the input/output relationship is noncausal). For our

[77] We also recommend using a 4-wire Kelvin connection between the test equipment and the cell, where the two sense leads connect directly to the cell's terminals. This eliminates the effects of the resistance of the two power leads and their contact with the cell terminals. In the results presented in this chapter, we used Kelvin connections for all tests, which is especially important when the cell impedance is small as is true for the Panasonic cell.

[78] Craig F. Bohren, "What did Kramers and Kronig do and how did they do it?" *European Journal of Physics*, 31(3):573–577, 2010; Bernard A. Boukamp, "A linear Kronig-Kramers transform test for immittance data validation," *Journal of The Electrochemical Society*, 142(6):1885, 1995; and M. Schönleber, D. Klotz, and E. Ivers-Tiffée, "A method for improving the robustness of linear Kramers–Kronig validity tests," *Electrochimica Acta*, 131:20–27, 2014.

Table 3.18: Laboratory EIS-test configuration used for the Panasonic cell.

Sequence	Temp.	Details
#1	25 °C	1. Slowly charge to $v_{\max}$ (100 % SOC) 2. Rest (≥ 2 h)
#2 (Loop)	T	1. Rest (≥ 6 h) 2. Galvanostatic EIS (100 kHz − 0.1 Hz, 20 ppd) 3. Galvanostatic EIS (0.1 Hz − 1 mHz, 10 ppd) 4. Galvanostatic EIS (0.1 mHz) 5. Slow discharge to desired SOC setpoint 6. Repeat Steps 1–5
SOC setpoints, input frequencies are identical to Table 3.17		

experiments, we observed that an input rate of C/50 produced results having the minimal overall K–K residuals; hence we used that rate (≈ 0.5 A) as the input amplitude in this work.

Table 3.18 summarizes the EIS-test steps. The cell is initially charged to a calibrated 100 % SOC point, then slowly discharged to the desired SOC setpoints. The underlying sampling-rate designs compromise to balance an acceptable test duration while maintaining proper resolution. The impedance measurements at each SOC setpoint take about one day to finish. The Gamry potentiostat automatically selects and runs multiple cycles of input current at each setpoint to reach a steady-state cell response, and we are unable to adjust the number of input cycles manually. In total, impedance data for nineteen distinct SOC setpoints were measured. At low cell SOCs, the impedance data are extremely hard to collect because of the significant voltage drift; hence, we use Gamry's hybrid (combined potentiostatic and galvanostatic) mode to measure the data for SOCs at or below 10 %.

Sample processed Panasonic-cell Nyquist plots are shown in the top-left frame of Fig. 3.70, and the computed GDRT functions are displayed in the top-right frame. To visualize the data-processing performance, the middle-left frame zooms impedance data samples at a few SOC setpoints, graphed along with the raw impedance data with cable inductance. This shows how important it is to remove the inductance rather than truncating the impedance—removing the inductance recovers the expected Nyquist bump, but truncating the impedance would have deleted the bump. The middle-right frame demonstrates the peak-analysis process to isolate $\bar{C}_{\text{dl}}$ and $\bar{R}_{\text{ct}}$ from the two electrodes. We assume that peak A is associated with the negative electrode, peak B is associated with the positive electrode, and peak C is associated with the solid-diffusion impedance $\bar{Z}_{\text{s}}$.

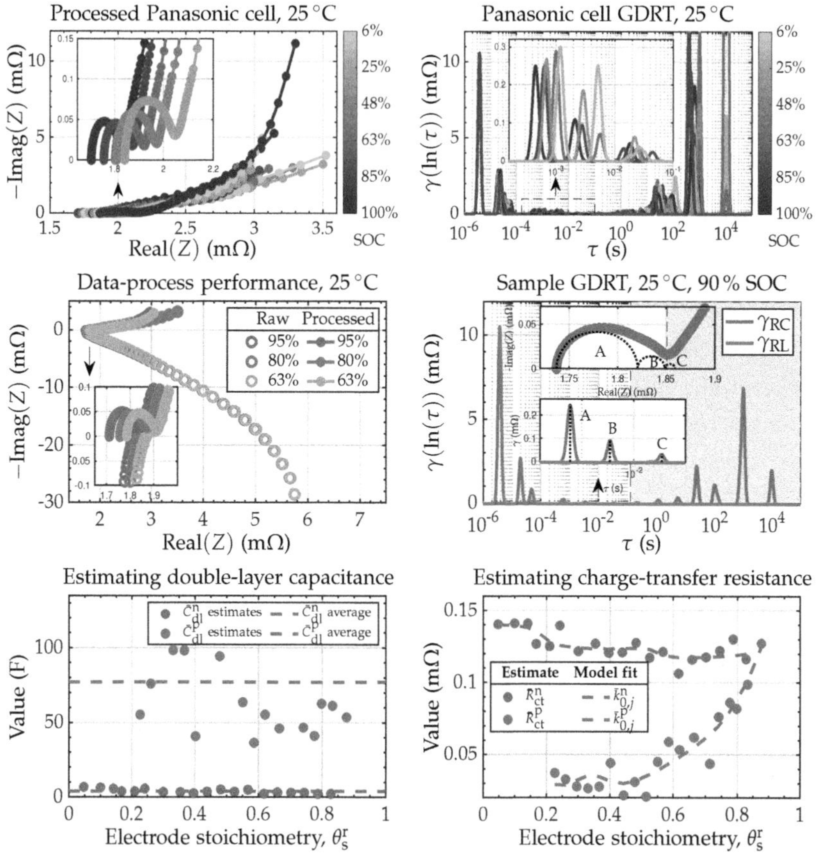

Figure 3.70: The processed Panasonic-cell impedance data at 25 °C (top-left); the computed GDRT functions (top-right); visualizing the data-processing performance at three distinct SOC setpoints (middle-left); demonstrating the process for peak analysis to isolate $\bar{C}_{\rm dl}$ and $\bar{R}_{\rm ct}$ from the two electrodes (middle-right); approximating $\bar{C}^{\rm r}_{\rm dl}$, $\bar{R}^{\rm r}_{\rm ct}$, and $\bar{k}^{\rm r}_{0,j}$ values from DRT peak analysis (bottom row).

Further, the approximations of $\bar{C}^{\rm r}_{\rm dl}$, $\bar{R}^{\rm r}_{\rm ct}$, and $\bar{k}^{\rm r}_{0,j}$ from the GDRT peak analysis are shown in the bottom row.

Different parameter-optimization strategies

In simulation studies, the true values of all parameters are known and the impedance data SOC setpoints are exact—the cell is always in an equilibrium state. However, a laboratory process will introduce SOC drift and its impact becomes significant especially at low states of charge. Unmodeled hysteresis will also contaminate EIS measurements, especially at low frequencies.[79] Therefore, it will be extremely hard to achieve a high degree of fit among all SOC setpoints, and the accuracy of model parameter estimates can be distorted.

In addition to the optimization strategy that we studied previously in Eq. (3.33), there is another approach that we can consider.[80] For clarification, we designate these as Strategy 1 and Strategy 2, where:

- Strategy 1: Fits a single model to cell data collected at all SOC setpoints; in this case, the optimization routines will be executed only once using the full data set.
- Strategy 2: Fits individual models to cell data collected at each

[79] M. Oldenburger, B. Bedürftig, E. Richter, R. Findeisen, A. Hintennach, and A. Gruhle, "Analysis of low frequency impedance hysteresis of li-ion cells by time-and frequency domain measurements and its relation to the ocv hysteresis," *Journal of Energy Storage*, 26:101000, 2019.

[80] Qi Zhang, Dafang Wang, Bowen Yang, Haosong Dong, Cheng Zhu, and Ziwei Hao, "An electrochemical impedance model of lithium-ion battery for electric vehicle application," *Journal of Energy Storage*, 50:104182, 2022.

SOC setpoint, and eventually averages the parameter values (except for $\bar{D}_s$). Note that in this case we identify $\bar{D}_s$ rather than $\bar{D}_{s,ref}$; we then use the estimates of $\bar{D}_s$ at a variety of SOC points to optimize a single $\bar{D}_{s,ref}$ value.

Ideally, Strategy 2 can yield better Nyquist fits to EIS data at each SOC setpoint (the number of data samples we gather at each SOC is sufficient for this approach) but perhaps less-accurate parameter values; while Strategy 1 can find more accurate sets of model parameter values but the model frequency-response fits to EIS data collected at individual SOC setpoints are not generally as close. Strategy 2 is also more difficult to implement well at temperatures other than 25 °C because the initializations and parameter boundaries could be very different for the data at different temperatures; otherwise, wide search-ranges should be used for temperature-dependent parameters. The cost function for Strategy 1 was reported in Eq. (3.33); the one for Strategy 2 is written as:

$$p_z^* = \arg\min_{p_z} \sum_{m=1}^{M} \left[\frac{\mathbb{R}\left(Z(s_m, z) - \hat{Z}(s_m, z)\right)}{|Z(s_m, z)|} \right]^2 + \sum_{m=1}^{M} \left[\frac{\mathbb{I}\left(Z(s_m, z) - \hat{Z}(s_m, z)\right)}{|Z(s_m, z)|} \right]^2,$$

where p_z is the set of model parameters to identify at given cell SOC z, and $\mathbb{R}$ and $\mathbb{I}$ represent real and imaginary parts, respectively. In the following studies, we will explore both strategies.

3.16.3 *Estimating LPM parameter values using laboratory data*

We now present results from seeking to estimate parameter values for an LPM of the Panasonic cell using its impedance data. We have discovered from the simulation studies that it is difficult to achieve good accuracies for some model parameters because impedance is insensitive to their values; however, we might still be able to yield excellent agreement between the cell impedance model and EIS data. The parameter-estimation results in this section are found using MATLAB's `particleswarm` function with the same algorithm settings as configured in the simulation studies.

Fig. 3.71 presents the impedance fit for a variety of SOC setpoints at which frequency-response data were collected (at 25 °C), where the top row shows results using Strategy 1, and the remaining plots present results using Strategy 2. The frequency-match agreement is excellent for Strategy 2 results, although the low-frequency fit is slightly skewed due to the optimization method trying to find insensitive electrolyte components such as $\bar{q}_e^r/\bar{\psi}$. Our hypotheses for

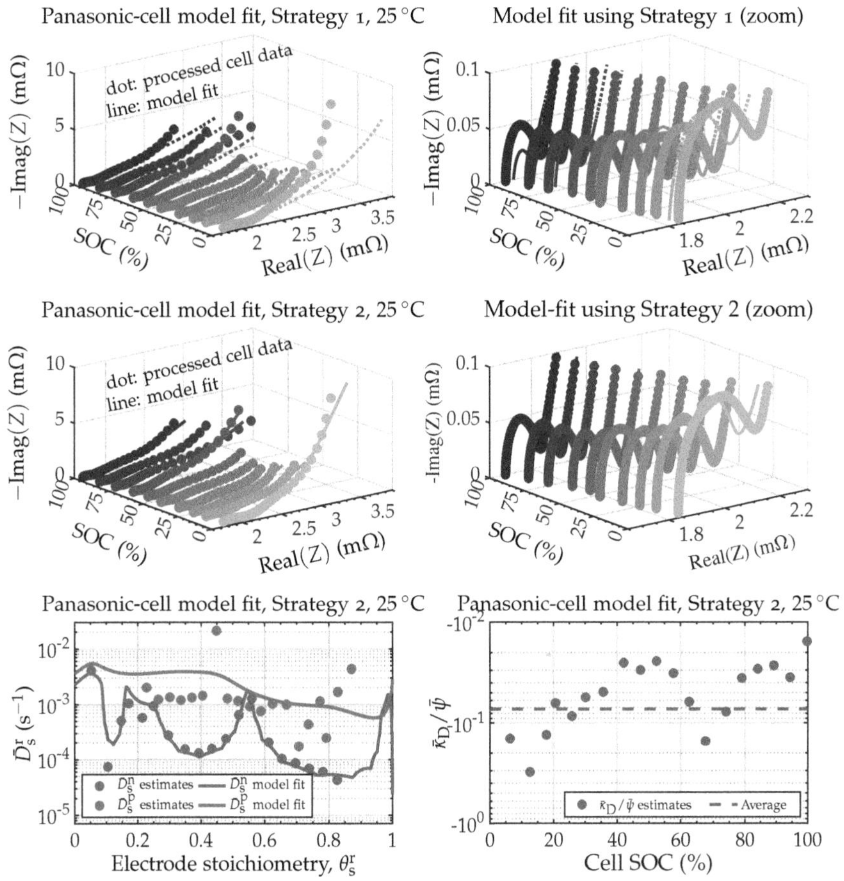

Figure 3.71: Fitting the EIS-test data via Strategy 1 (top row) and Strategy 2 (middle row). Parameter estimates for $\bar{D}_s$ and $\bar{\kappa}_D/\bar{\psi}$ are graphed in the bottom row.

the mismatch using Strategy 1 are that (i) the cell experiences slight SOC drift during the frequency-response measurement (especially at low frequencies), and (ii) the cell SOC setpoints we determine from the electrode operating boundaries are not exact.[81] The bottom row of plots display sample parameter estimates from Strategy 2. A more comprehensive comparison among the two optimization strategies (such as parameter estimates implemented in time-domain simulations) is a topic for future research. For reference, the final set of model parameter values found using optimization Strategy 1 is listed in Table 3.19.[82]

3.17 *Pseudo-steady-state (PSS) testing to find $\bar{\psi}$*

Referring back to Fig. 3.8, we are now at the final step of the parameter-estimation process. By this point, we have found values for all parameters except $\bar{\psi}$. To estimate $\bar{\psi}$, we will reuse the constant-current discharge data collected for the discharge test of Sect. 3.9 and will regress those data against an enhanced single-particle model that includes electrolyte effects and is optimized to predict pseudo-steady-

[81] This is important because we are using the SOC-dependent model of solid diffusivity of Eq. (3.9) with Strategy 1. If electrode SOC is modeled incorrectly, then the estimates of $\bar{D}^r_{s,\text{ref}}$ found by the optimizations will be biased. We observe from the lower-left figure that Eq. (3.9) matches $\bar{D}^n_s$ very well but that the fit is not as good for $\bar{D}^p_s$. It is possible that errors in the estimates of θ^p_0 and θ^p_{100} are biasing estimates of kinetics parameters, which will lead to biases in the estimates of $\bar{D}^p_{s,\text{ref}}$ also.

[82] The optimizations returned either the maximum or minimum permitted values for some parameters. This leads us to think that these values may not be as accurate as we might like. It is possible that parameter-estimation inaccuracy from earlier stages of optimizations have caused a bias in their estimates. More work remains to be done to refine the parameter-estimation methods and to make them more robust.

	Unit	Negative electrode	Separator	Positive electrode
$\bar{\sigma}$	kS	100	—	10000
$\bar{\kappa}$	kS	0.707	8	8
$\bar{R}_{\mathrm{f}}$	mΩ	1.03	—	1×10^{-7}
$\bar{R}_{\mathrm{dl}}$	mΩ	1×10^{-7}	—	3.98×10^{-1}
$\bar{D}_{\mathrm{s,ref}}$	s^{-1}	2.75×10^{-3}	—	1.17×10^{-4}
$\bar{q}_{\mathrm{e}}/\bar{\psi}$	SKh	229773	459544	33613
$\bar{C}_{\mathrm{dl}}$	F	10.03	—	5.38
n_{f}	u/l	0.90	—	0.9
n_{dl}	u/l	1	—	0.9
$\bar{k}_{0,1}$	A	1.66×10^{-5}	—	6.23×10^{2}
$\bar{k}_{0,2}$	A	9.75×10^{2}	—	2.97×10^{2}
$\bar{k}_{0,3}$	A	2.35×10^{-4}	—	5.91×10^{5}
$\bar{k}_{0,4}$	A	7.08×10^{-1}	—	1.17×10^{7}
$\bar{k}_{0,5}$	A	4.41×10^{4}	—	6.02×10^{9}
$\bar{k}_{0,6}$	A	4.73×10^{7}	—	
$\bar{\kappa}_{\mathrm{D}}/\bar{\psi}$	u/l	−11.83 (spans all regions)		
R_{c}	Ω	1×10^{-5} (spans all regions)		

Table 3.19: Identified Panasonic cell parameter values at 25 °C.

state behaviors of a cell. We call this model the pseudo-steady-state or PSS model.[83]

The PSS approximation relies on the observation that when a constant current is applied to a lithium-ion cell, two things will happen. After some initial transient:

- All cell electrochemical variables described by stable TFs will approach steady-state values.
- Cell electrochemical variables whose TFs have integration dynamics will continue to evolve over time, but their integrator-removed TFs will approach steady-state values.

Therefore, while some variables continuously change, there is still an aspect of steady-state behavior that applies to all variables of the cell.[84] The PSS model takes advantage of this observation to build an enhanced SPM of the cell that can be used inside an optimization routine to determine $\bar{\psi}$.

Fig. 3.72 illustrates the interactions between the different components of the pseudo-steady-state reduced-order model (PSS-ROM) that we will use. It calculates voltage derived from Eq. (3.16):

$$\begin{aligned} v(t) &= \left[\phi^{\mathrm{p}}_{\text{s-e}}(3,t) - \phi^{\mathrm{n}}_{\text{s-e}}(0,t)\right] + \phi^{\mathrm{p}}_{\mathrm{e}}(3,t) - \phi^{\mathrm{n}}_{\mathrm{e}}(0,t) \\ &= \left[\phi^{\mathrm{p}}_{\text{s-e}}(3,t) - \phi^{\mathrm{n}}_{\text{s-e}}(0,t)\right] + \tilde{\phi}^{\mathrm{p}}_{\mathrm{e}}(3,t). \end{aligned}$$

To find $\phi^{\mathrm{r}}_{\text{s-e}}(\tilde{x},t)$, we recognize that this variable exists in the overpo-

[83] This section is simplified from: Brandon Guest, M. Scott Trimboli, and Gregory L. Plett, "Pseudo-steady-state reduced-order-model approximation for constant-current parameter identification in lithium-ion cells," *Journal of The Electrochemical Society*, 167(16):160546, 2020. It is also adapted to use LPM notation and associated steady-state results from Chap. 2.

[84] Since not all variables achieve steady state, we refer to the overall model as the *pseudo*-steady-state model.

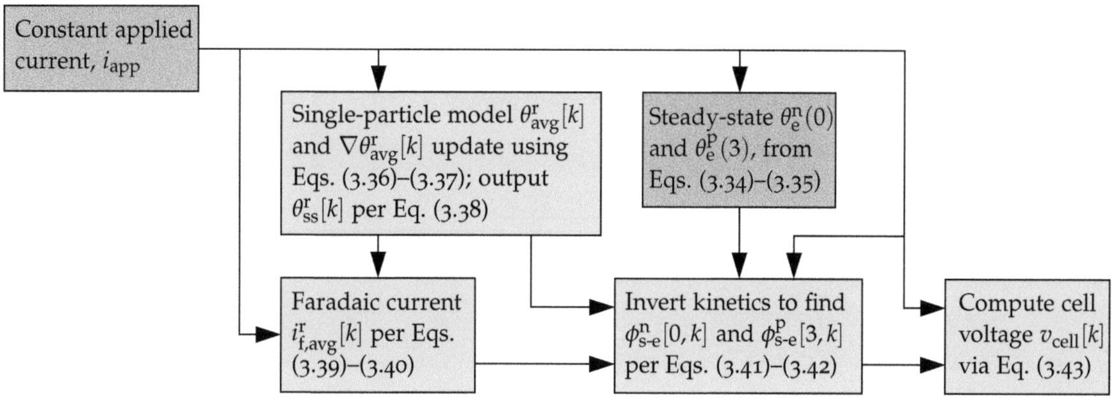

Figure 3.72: A flowchart showing each step of the approximation. Boxes shaded blue represent time-invariant quantities; boxes shaded green compute time-varying values.

tential term:

$$\eta^r(\tilde{x},t) = \phi^r_{s\text{-}e}(\tilde{x},t) - U^r_{ocp}(\theta^r_{ss}(\tilde{x},t)).$$

We solve for its values by inverting the kinetics equation, which further requires knowledge of $\theta^r_e(\tilde{x},t)$ and $\theta^r_{ss}(\tilde{x},t)$ at $\tilde{x} \in \{0,3\}$.

The electrolyte profiles are assumed to attain steady state quickly during the discharge test, and so we solve for their values using TF limits. The time-varying average (across electrode thickness) solid surface stoichiometries are calculated using an SPM for each electrode. The SPM that we use has the same form as the one presented in Sect. 3.9, but here we adjust the model so that it executes in discrete-time. The solid/electrolyte potential differences are found by calculating the cell overpotential, which is accomplished by inverting the kinetics equations. Finally, the electrolyte potential and solid/electrolyte potential difference are summed at the positive current collector to form the cell voltage. The components of this process are described in the following subsections.

3.17.1 *PSS electrolyte approximation*

The PSS model assumes that the electrolyte variables can be adequately described by their steady-state values. These, in turn, are computed by evaluating their corresponding TFs in the limit as $s \to 0$. These low-frequency limits have already been calculated, and are listed in App. 2.D. For reference, we will need:

- The low-frequency gain for debiased potential of the electrolyte, $\widetilde{\Phi}^r_e(\tilde{x},0)/I_{app}(0)$ at $\tilde{x}=3$, which is computed from Eq. (2.106). For convenient reference,

$$\frac{\widetilde{\Phi}^r_e(3,0)}{I_{app}(0)} = \left(\frac{\bar{\kappa}_D}{\bar{\psi}} - 1\right)\left(\frac{1}{2\bar{\kappa}^n} + \frac{1}{2\bar{\kappa}^p} + \frac{1}{\bar{\kappa}^s}\right).$$

Notice that the $\bar{\kappa}_D/\bar{\psi}$ term is not decoupled in this expression.

Therefore, its contribution to cell voltage will not assist us in distinguishing $\bar{\kappa}_\mathrm{D}$ versus $\bar{\psi}$.

- The low-frequency gains for debiased concentration of lithium in the electrolyte, $\widetilde{\Theta}_\mathrm{e}^\mathrm{r}(\tilde{x},0)/I_\mathrm{app}(0)$ at $\tilde{x} \in \{0,3\}$, which are computed from Eqs. (2.104) and (2.105). At the points of interest:

$$\frac{\widetilde{\Theta}_\mathrm{e}^\mathrm{n}(0,0)}{I_\mathrm{app}(0)} = \frac{\bar{q}_\mathrm{e}^\mathrm{n}\bar{\kappa}^\mathrm{s}\bar{\kappa}^\mathrm{p} + 3\bar{q}_\mathrm{e}^\mathrm{s}(\bar{\kappa}^\mathrm{s}\bar{\kappa}^\mathrm{p} + \bar{\kappa}^\mathrm{n}\bar{\kappa}^\mathrm{p}) + \bar{q}_\mathrm{e}^\mathrm{p}(2\bar{\kappa}^\mathrm{n}\bar{\kappa}^\mathrm{s} + 3\bar{\kappa}^\mathrm{s}\bar{\kappa}^\mathrm{p} + 6\bar{\kappa}^\mathrm{n}\bar{\kappa}^\mathrm{p})}{6\bar{\kappa}^\mathrm{n}\bar{\kappa}^\mathrm{s}\bar{\kappa}^\mathrm{p}\bar{\psi}T(\bar{q}_\mathrm{e}^\mathrm{n} + \bar{q}_\mathrm{e}^\mathrm{s} + \bar{q}_\mathrm{e}^\mathrm{p})} \tag{3.34}$$

$$\frac{\widetilde{\Theta}_\mathrm{e}^\mathrm{p}(3,0)}{I_\mathrm{app}(0)} = \frac{-\bar{q}_\mathrm{e}^\mathrm{n}(2\bar{\kappa}^\mathrm{s}\bar{\kappa}^\mathrm{p} + 3\bar{\kappa}^\mathrm{n}\bar{\kappa}^\mathrm{s} + 6\bar{\kappa}^\mathrm{n}\bar{\kappa}^\mathrm{p}) - 3\bar{q}_\mathrm{e}^\mathrm{s}(\bar{\kappa}^\mathrm{n}\bar{\kappa}^\mathrm{s} + \bar{\kappa}^\mathrm{n}\bar{\kappa}^\mathrm{p}) - \bar{q}_\mathrm{e}^\mathrm{p}\bar{\kappa}^\mathrm{n}\bar{\kappa}^\mathrm{s}}{6\bar{\kappa}^\mathrm{n}\bar{\kappa}^\mathrm{s}\bar{\kappa}^\mathrm{p}\bar{\psi}T(\bar{q}_\mathrm{e}^\mathrm{n} + \bar{q}_\mathrm{e}^\mathrm{s} + \bar{q}_\mathrm{e}^\mathrm{p})}. \tag{3.35}$$

In these equations, note that we can multiply numerator and denominator by $\bar{\psi}$ to achieve:

$$\frac{\widetilde{\Theta}_\mathrm{e}^\mathrm{n}(0,0)}{I_\mathrm{app}(0)} = \frac{\frac{\bar{q}_\mathrm{e}^\mathrm{n}}{\bar{\psi}}\bar{\kappa}^\mathrm{s}\bar{\kappa}^\mathrm{p} + 3\frac{\bar{q}_\mathrm{e}^\mathrm{s}}{\bar{\psi}}(\bar{\kappa}^\mathrm{s}\bar{\kappa}^\mathrm{p} + \bar{\kappa}^\mathrm{n}\bar{\kappa}^\mathrm{p}) + \frac{\bar{q}_\mathrm{e}^\mathrm{p}}{\bar{\psi}}(2\bar{\kappa}^\mathrm{n}\bar{\kappa}^\mathrm{s} + 3\bar{\kappa}^\mathrm{s}\bar{\kappa}^\mathrm{p} + 6\bar{\kappa}^\mathrm{n}\bar{\kappa}^\mathrm{p})}{6\bar{\kappa}^\mathrm{n}\bar{\kappa}^\mathrm{s}\bar{\kappa}^\mathrm{p}[\boldsymbol{\bar{\psi}}]T\frac{\bar{q}_\mathrm{e}^\mathrm{n} + \bar{q}_\mathrm{e}^\mathrm{s} + \bar{q}_\mathrm{e}^\mathrm{p}}{\bar{\psi}}}$$

$$\frac{\widetilde{\Theta}_\mathrm{e}^\mathrm{p}(3,0)}{I_\mathrm{app}(0)} = \frac{-\frac{\bar{q}_\mathrm{e}^\mathrm{n}}{\bar{\psi}}(2\bar{\kappa}^\mathrm{s}\bar{\kappa}^\mathrm{p} + 3\bar{\kappa}^\mathrm{n}\bar{\kappa}^\mathrm{s} + 6\bar{\kappa}^\mathrm{n}\bar{\kappa}^\mathrm{p}) - 3\frac{\bar{q}_\mathrm{e}^\mathrm{s}}{\bar{\psi}}(\bar{\kappa}^\mathrm{n}\bar{\kappa}^\mathrm{s} + \bar{\kappa}^\mathrm{n}\bar{\kappa}^\mathrm{p}) - \frac{\bar{q}_\mathrm{e}^\mathrm{p}}{\bar{\psi}}\bar{\kappa}^\mathrm{n}\bar{\kappa}^\mathrm{s}}{6\bar{\kappa}^\mathrm{n}\bar{\kappa}^\mathrm{s}\bar{\kappa}^\mathrm{p}[\boldsymbol{\bar{\psi}}]T\frac{\bar{q}_\mathrm{e}^\mathrm{n} + \bar{q}_\mathrm{e}^\mathrm{s} + \bar{q}_\mathrm{e}^\mathrm{p}}{\bar{\psi}}}.$$

In both expressions, we retain coupled $\bar{q}_\mathrm{e}^\mathrm{r}/\bar{\psi}$ terms. However, they also have a decoupled $\bar{\psi}$ term in the denominators (typeset in boldface with brackets as $[\boldsymbol{\bar{\psi}}]$ for emphasis). It is the presence of this decoupled $\bar{\psi}$ term that allows us to use the PSS test to estimate $\bar{\psi}$ independently of $\bar{\kappa}_\mathrm{D}$ and $\bar{q}_\mathrm{e}^\mathrm{r}$.

3.17.2 *PSS average solid-surface approximation*

To compute the time-varying average solid-surface stoichiometry, we adopt the SPM of Eq. (3.11). However, this equation is expressed as a continuous-time state-space model and we desire to convert it to a discrete-time model for efficient computation. That is, we wish to convert from a generic single-variable state-space form:

$$\begin{aligned} \dot{x}(t) &= ax(t) + bu(t) \\ y(t) &= cx(t) + du(t), \end{aligned}$$

to a model that is equivalent at the sample points:

$$\begin{aligned} x[k] &= a_\mathrm{d}x[k-1] + b_\mathrm{d}u[k-1] \\ y[k] &= cx[k] + du[k]. \end{aligned}$$

We employed a zero-order-hold approach to perform this conversion in Chap. 2 of Vol. I of this series, where the result was found to be:[85]

[85] This applies only when $a \neq 0$. When $a = 0$, we simply integrate over one time step, so:

$$x[k] = x[k-1] + bT_s u[k-1].$$

$$x[k] = \underbrace{e^{aT_s}}_{a_d} x[k-1] + \frac{1}{a}\left(e^{aT_s} - 1\right) bu[k-1],$$

where T_s is the sampling period.

Applying this general form to Eq. (3.11), we achieve discrete-time recurrences for the two SPM states:

$$\theta^r_{avg}[k] = \theta^r_{avg}[k-1] - \frac{T_s}{3600Q^r} i^r_{f,avg}[k-1] \quad (3.36)$$

$$\nabla\theta^r_{avg}[k] = a^r_d \nabla\theta^r_{avg}[k-1] + \frac{a^r_d - 1}{14400Q^r\bar{D}^r_s(\theta^r_{ss}[k-1])} i^r_{f,avg}[k-1], \quad (3.37)$$

where: $a^r_d = \exp\left(-30T_s\bar{D}^r_s(\theta^r_{ss}[k-1])\right)$. The output average surface concentration is computed as:

$$\theta^r_{ss}[k] = \theta^r_{avg}[k] + \frac{8}{35}\nabla\theta^r_{avg}[k] + \frac{-1/378\,000}{Q^r\bar{D}^r_s(\theta^r_{ss}[k-1])} i^r_{f,avg}[k]. \quad (3.38)$$

In these equations, the average faradaic molar flux is computed via the dc gain of Eqs. (3.39)–(3.40):

$$i^n_{f,avg}[k] = \frac{3600Q}{3600Q - \bar{C}^n_{dl,eff}\left|\theta^n_{100} - \theta^n_0\right|[U^n_{ocp}]'} i_{app} \quad (3.39)$$

$$i^p_{f,avg}[k] = \frac{-3600Q}{3600Q - \bar{C}^p_{dl,eff}\left|\theta^p_{100} - \theta^p_0\right|[U^p_{ocp}]'} i_{app}. \quad (3.40)$$

Note that these values are time-varying, since the input to the OCP derivative function is θ^r_{ss}, which is also time-varying.

3.17.3 *Kinetics inversion*

The interfacial solid/electrolyte potential difference is found by inverting the kinetics equations. In general, this can be done using a nonlinear optimization. For simplicity, we consider a simplified MSMR model where $\alpha^r_j = 0.5$. In this case, we can solve in closed form. We start with the faradaic current, which can be written as:

$$i_f = 2\left(\sum_{j=1}^{J} i_{0,j}\right)\sinh\left(\frac{f}{2}\eta\right),$$

where $\eta = \phi_{s\text{-}e} - U_{ocp}(\theta_{ss}) - \bar{R}_f i_{f+dl}$, and:

$$i_{0,j} = \bar{k}_{0,j}\sqrt{(\theta_e)(X_j - x_j)^{\omega_j}(x_j)^{\omega_j}}$$

$$x_j = \frac{X_j}{1 + \exp(f(U_{ocp}(\theta_{ss}) - U^0_j)/\omega_j)}.$$

The interfacial potential difference is then approximated as:

$$\phi_{s\text{-}e}[\tilde{x},k] = \frac{2}{f}\operatorname{asinh}\left(\frac{i_{f,avg}[\tilde{x},k]}{2\sum_{j=1}^{J} i_{0,j}}\right) + U_{ocp}(\theta_{ss}[\tilde{x},k]) + \bar{R}_f i_{f+dl,avg}.$$

The interfacial potential differences at each current collector are required to compute the cell's voltage. They are:

$$\phi^{\mathrm{n}}_{\text{s-e}}[0,k] = \frac{2}{f}\operatorname{asinh}\left(\frac{i^{\mathrm{n}}_{\mathrm{f,avg}}[k]}{2\sum_{j=1}^{J}\bar{k}^{\mathrm{n}}_{0,j}\sqrt{\theta^{\mathrm{n}}_{\mathrm{e}}(0)(X^{\mathrm{n}}_j - x^{\mathrm{n}}_j[k])^{\omega^{\mathrm{n}}_j}(x^{\mathrm{n}}_j[k])^{\omega^{\mathrm{n}}_j}}}\right) + U^{\mathrm{n}}_{\mathrm{ocp}}(\theta^{\mathrm{n}}_{\mathrm{ss}}[k]) + \bar{R}^{\mathrm{n}}_{\mathrm{f}} i_{\mathrm{app}}, \tag{3.41}$$

which appears in the expression for $\phi^{\mathrm{p}}_{\mathrm{e}}[3,k]$, and,

$$\phi^{\mathrm{p}}_{\text{s-e}}[3,k] = \frac{2}{f}\operatorname{asinh}\left(-\frac{i^{\mathrm{p}}_{\mathrm{f,avg}}[k]}{2\sum_{j=1}^{J}\bar{k}^{\mathrm{p}}_{0,j}\sqrt{\theta^{\mathrm{p}}_{\mathrm{e}}(3)(X^{\mathrm{p}}_j - x^{\mathrm{p}}_j[k])^{\omega^{\mathrm{p}}_j}(x^{\mathrm{p}}_j[k])^{\omega^{\mathrm{p}}_j}}}\right) + U^{\mathrm{p}}_{\mathrm{ocp}}(\theta^{\mathrm{p}}_{\mathrm{ss}}[k]) - \bar{R}^{\mathrm{p}}_{\mathrm{f}} i_{\mathrm{app}}. \tag{3.42}$$

3.17.4 *Cell voltage*

The cell voltage is synthesized from the potential difference between the current collectors. Since the potential reference was set to zero at the negative current collector, the voltage can be expressed as the solid potential at the positive current collector. This is equivalent to: $v_{\mathrm{cell}}[k] = \phi^{\mathrm{p}}_{\text{s-e}}[3,k] + \phi^{\mathrm{p}}_{\mathrm{e}}[3,k]$. Using the profile calculated previously for the electrolyte potential, the expression becomes:

$$v_{\mathrm{cell}}[k] = \phi^{\mathrm{p}}_{\text{s-e}}[3,k] - \phi^{\mathrm{n}}_{\text{s-e}}[0,k] + \left(\frac{\bar{\kappa}_D}{\bar{\psi}} - 1\right)\left(\frac{1}{2\bar{\kappa}^{\mathrm{n}}} + \frac{1}{\bar{\kappa}^{\mathrm{s}}} + \frac{1}{2\bar{\kappa}^{\mathrm{p}}}\right) i_{\mathrm{app}}. \tag{3.43}$$

3.17.5 *Summary of the PSS-ROM*

The entire PSS-ROM has now been derived. Here is a quick summary of what is necessary to compute cell voltage for a constant-current discharge simulation:

- At the beginning of the simulation, we must initialize variables.
  - We compute $\theta^{\mathrm{n}}_{\mathrm{e}}(0)$ via Eq. (2.104) and $\theta^{\mathrm{p}}_{\mathrm{e}}(3)$ via Eq. (2.105).
  - We initialize $\theta^{\mathrm{r}}_{\mathrm{avg}}[0] = \theta^{\mathrm{r}}_{\mathrm{s},0}$ and $\nabla\theta^{\mathrm{r}}_{\mathrm{avg}}[0] = 0$.
- During the simulation, every time step:
  - We update $\theta^{\mathrm{r}}_{\mathrm{avg}}[k]$ using Eq. (3.36).
  - We also update $\nabla\theta^{\mathrm{r}}[k]$ using Eq. (3.37).
  - We then calculate $\theta^{\mathrm{r}}_{\mathrm{ss}}[k]$ using Eq. (3.38).
  - We solve for $\phi^{\mathrm{n}}_{\text{s-e}}[0,k]$ and $\phi^{\mathrm{p}}_{\text{s-e}}[3,k]$ using Eqs. (3.41)–(3.42).
  - Finally, we solve for cell voltage using Eq. (3.43).

3.18 *Validating the $\bar{\psi}$ estimate*

The PSS method was compared extensively to COMSOL full-order-model results in the original reference by Guest et al. Here, we quickly comment on some results when using it to estimate $\bar{\psi}$.

The first set of results sought to estimate $\bar{\psi}$ for the NMC30 cell when all other parameter values were considered to be known exactly. Table 3.20 shows optimization relative errors in estimating $\bar{\psi}$ when using simulated constant-current discharges. At both very low and high discharge rates, the voltage-prediction PSS-model approximation errors rival the contribution to voltage by $\bar{\psi}$ and so the results begin to degrade. However, at moderate rates in the neighborhood of C/5 for this simulation test case, estimates of $\bar{\psi}$ are very good.

Table 3.20: Estimating $\bar{\psi}$ in simulation using the PSS method. The true value was $8.15\times10^{-5}\,\mathrm{V\,K^{-1}}$.

Rate	Estimate	Rel. Error
C/10	6.55×10^{-5}	19.7 %
C/5	8.63×10^{-5}	−5.9 %
C/2	1.05×10^{-4}	−28.6 %
1C	1.10×10^{-4}	−34.7 %

We also applied the methods to the Panasonic cell using C/2 data. When doing so, we took the opportunity also to adjust the estimates of $\bar{\kappa}^{\mathrm{r}}$ when fitting the discharge voltage as we felt that the estimates made using EIS data were poor (the frequency response is relatively insensitive to $\bar{\kappa}^{\mathrm{r}}$). The optimized value of $\bar{\psi}$ compares favorably with the value for the NMC30 cell. The revised values for $\bar{\kappa}^{\mathrm{n}}$and computed values for $\bar{q}_{\mathrm{e}}^{\mathrm{r}}$ and $\bar{\kappa}_{\mathrm{D}}$ are listed in Table 3.21.

Table 3.21: Estimating $\bar{\psi}$ and $\bar{\kappa}^{\mathrm{r}}$ for the Panasonic cell using the PSS method. The corresponding values of $\bar{q}_{\mathrm{e}}^{\mathrm{r}}$ and $\bar{\kappa}_{\mathrm{D}}$ computed from $\bar{q}_{\mathrm{e}}^{\mathrm{r}}/\bar{\psi}$ and $\bar{\kappa}_{\mathrm{D}}/\bar{\psi}$ are also listed.

	Unit	Negative electrode	Separator	Positive electrode
$\bar{\kappa}$	kS	99.4	100	15.3
$\bar{q}_{\mathrm{e}}$	Ah	30.1	60.2	4.4
$\bar{\psi}$	$\mathrm{V\,K^{-1}}$	1.31×10^{-4} (spans all regions)		
$\bar{\kappa}_{\mathrm{D}}$	$\mathrm{V\,K^{-1}}$	-1.55×10^{-4} (spans all regions)		

3.19 *Temperature dependence*

The focus of this chapter has been on estimating the parameter values of a LPM at a single temperature. We have already commented that the MSMR model for electrode OCP has built-in temperature-dependence, but we have not discussed how other model parameter values vary with temperature.

We briefly consider how the preceding methods might be used to make a temperature-varying LPM. From Vol. I, we recall that the temperature dependence of many LPM parameters can be well described using the Arrhenius equation:

$$\bar{\Psi}(T) = \bar{\Psi}_{\mathrm{ref}} \underbrace{\exp\left(\frac{E_{\bar{\Psi}}}{R}\left(\frac{1}{T_{\mathrm{ref}}} - \frac{1}{T}\right)\right)}_{\text{temperature-dependent factor}}, \tag{3.44}$$

where $\bar{\Psi}$ is the value of a parameter at a reference temperature T_{ref} and $E_{\bar{\Psi}}$ is the activation energy [$\mathrm{J\,mol^{-1}}$] for that parameter.[86]

[86] All temperatures in this equation are in K; the reference temperature is frequently chosen to be $T_{\mathrm{ref}} = 25\,^{\circ}\mathrm{C} = 298.15\,\mathrm{K}$.

In the LPM, conductivities, diffusivities, and reaction-rate constants increase as temperature increases and so these parameters have positive activation energy. Resistances decrease as temperature increases; we can use the same equation, but activation energies must be negative. Several model parameter values are believed not to be temperature-varying (such as $\bar{\kappa}_D$ and $\bar{\psi}$). Activation energies that must be found are listed in Table 3.22.

Several strategies might be used to find these activation energies and we do not have a recommended practice at this point in time. For example, we might:

1. First, estimate all parameter values for a single model at a reference temperature. Then, use these values as $\bar{\Psi}_{\text{ref}}$ and fit E_{Ψ} so that the overall model fits the data collected at all temperatures as well as possible. Or, we might instead...
2. Estimate all parameter values for individual models at every setpoint temperature. Then, use these parameter values as noisy inputs to fitting Eq. (3.44) to the parameters of each model to yield an overall best $\bar{\Psi}_{\text{ref}}$ and E_{Ψ} for every temperature-varying parameter. Or, we might instead...
3. Estimate the reference values $\bar{\Psi}_{\text{ref}}$ and the activation energies E_{Ψ} at the same time, using a single (very large) optimization.

Table 3.22: Parameters that vary with temperature require estimation of the activation energies, tabulated here.

Negative electrode	Separator	Positive electrode
$E^{\text{n}}_{\bar{\sigma}}$		$E^{\text{p}}_{\bar{\sigma}}$
$E^{\text{n}}_{\bar{k}_{0,j}}$		$E^{\text{p}}_{\bar{k}_{0,j}}$
$E^{\text{n}}_{\bar{D}_{\text{s}}}$		$E^{\text{p}}_{\bar{D}_{\text{s}}}$
$E^{\text{n}}_{\bar{R}_{\text{f}}}$		$E^{\text{p}}_{\bar{R}_{\text{f}}}$
$E^{\text{n}}_{\bar{R}_{\text{dl}}}$		$E^{\text{p}}_{\bar{R}_{\text{dl}}}$
$E^{\text{n}}_{\bar{C}_{\text{dl}}}$		$E^{\text{p}}_{\bar{C}_{\text{dl}}}$
$E_{\bar{\kappa}}$ spans all regions		

3.20 *MATLAB toolbox*

As we reach the end of this chapter, we return to the overview illustration of the MATLAB toolbox in Fig. 3.73, now highlighting the portions of its functionality discussed in this chapter. Lab data from different tests are preprocessed by different routines denoted `process[XX]` for each test. Then, optimization routines `fit[XX]` are executed to fit subsets of model parameter values to the data from individual lab tests. The process can be automated via `genModel` or subfunctions can be executed manually.

Fig. 3.74 presents in more detail the functionality of the toolbox related to parameter estimation. In general, lab-test data collected from different equipment will originate in multiple formats. For example, data collected from Arbin cell-cycling equipment is stored in Excel spreadsheet format; data collected from Gamry Instruments potentiostats is stored in text "`dat`" file format. These data must be converted to a standard format to be used by the processing routines. The toolbox provides functions `makeMATfile[XX]` that convert the demonstration data provided with the toolbox to the standard format. If you happen to use different equipment, you will need to create your own `makeMATfile` codes to convert between the format

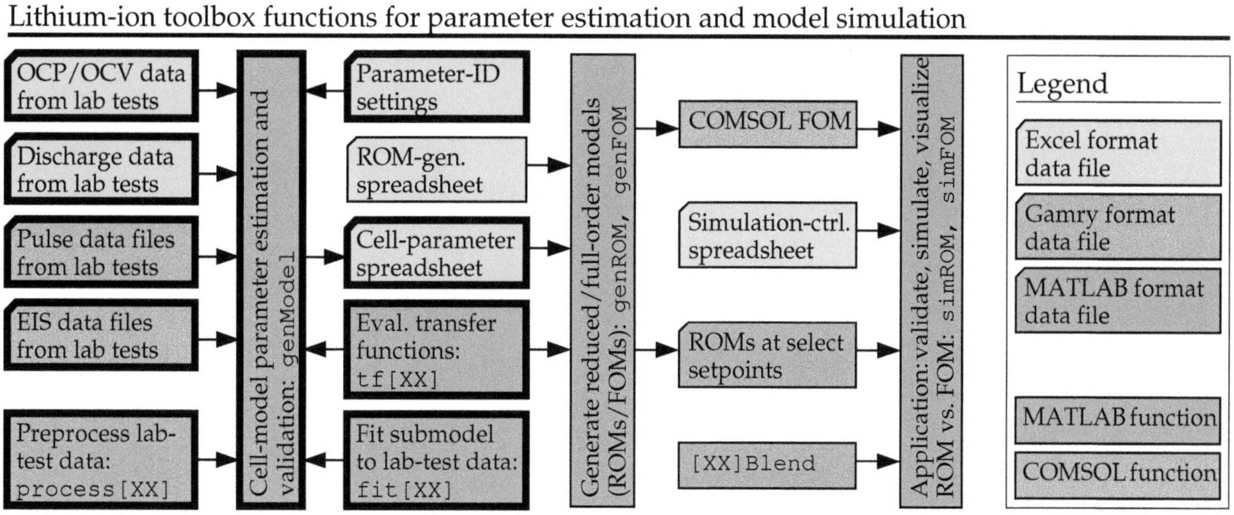

Figure 3.73: MATLAB-toolbox functionality, highlighting the focus of this chapter.

exported by that equipment and the required format for use by the toolbox. Examining the code in the existing `makeMATfile[XX]` functions should help to see what these required formats happen to be.

Once the data are in the desired format, different `process[XX]` functions perform some preliminary preprocessing steps on the data. For example, `processOCP` uses the histogram method to compute calibrated discharge voltage versus capacity and charge voltage versus capacity curves, along with differential capacity estimates. As another example, `processPulse` estimates the $\widehat{R}_0$ values for every testing setpoint. The `process[XX]` functions do not directly run optimizations to determine estimates of LPM parameter values; they simply preprocess data to make them ready for the optimizations.

The `fit[XX]` functions perform the parameter-estimation optimizations using the data collected at the present step in the process. The final parameter estimates produced at each step are indicated in the figure, as are reference results that are produced for use as initial guesses in later steps in the parameter-estimation process. These outputs are the same as those presented in Fig. 3.8, although shown here in a more detailed format.

Fig. 3.74 also shows the variables and files produced by each toolbox function. The most common variable is the `cellData` structure, which ultimately contains estimates of all model parameter values. These parameters can be written to a human-readable Excel spreadsheet format using the `saveCellParams` function (not shown in the figure). The spreadsheet can be loaded into MATLAB using `loadCellParams`, as discussed in Chap. 2.

For further reference, the reader is encouraged to examine the

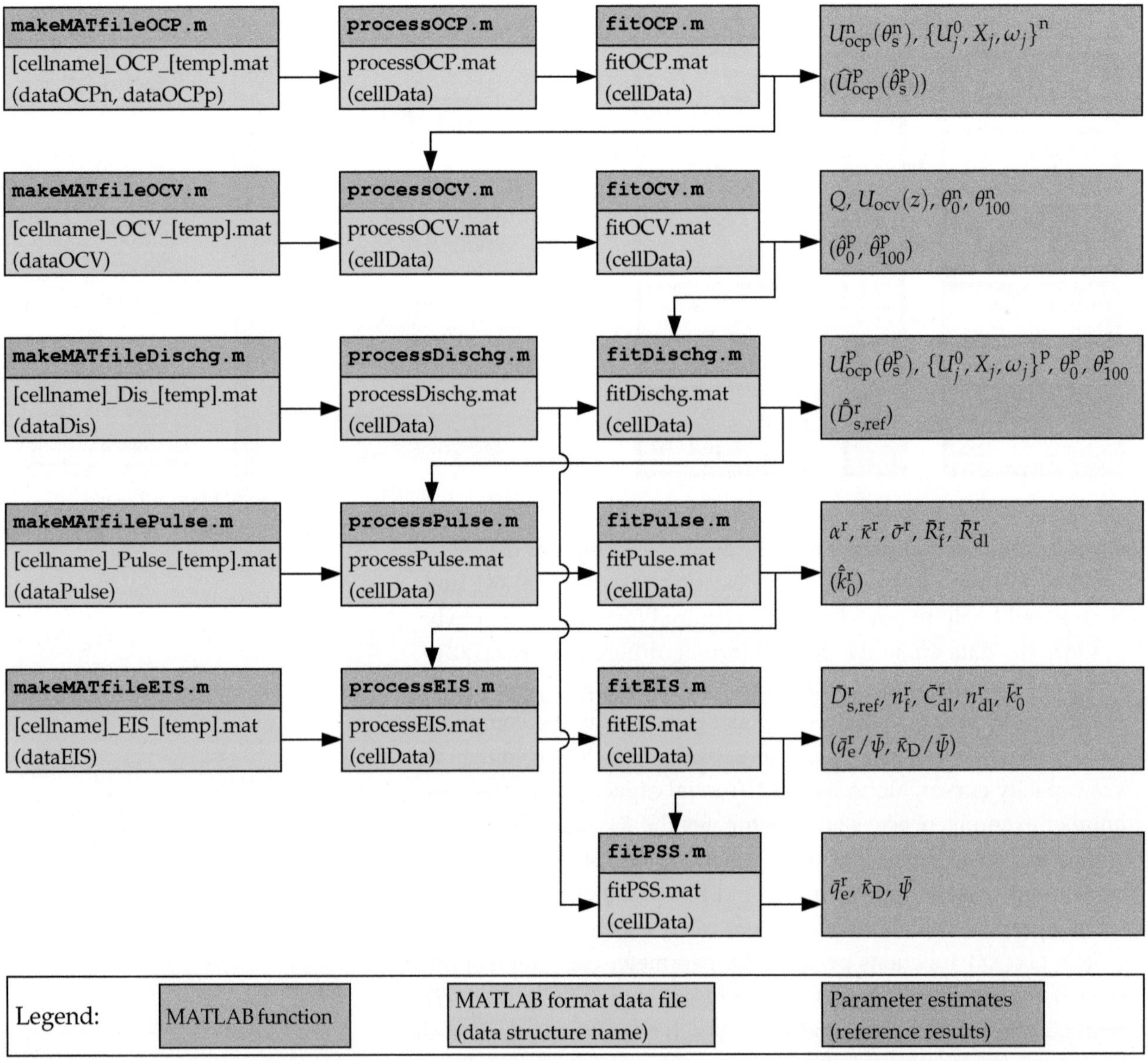

Figure 3.74: Details of the parameter-estimation portions of the MATLAB toolbox.

MATLAB example functions in the Chap. 3 folder of the toolbox. These will help to illustrate how the toolbox functions.

3.21 *Where to from here?*

In this chapter, we have considered some methods that might be used to estimate the parameter values of a LPM. This is still a rapidly evolving area of research and even though we have found the tests and data-processing methodology that we propose in this chapter to be useful, there remains room for refinement. We hope that our discussion of these techniques serves as inspiration to prompt readers of this volume to develop them further. We have focused on non-teardown approaches, but until these methods are robustly validated and in routine usage, some hybrid combination of teardown and nonteardown parameter-estimation may be the best approach.

The content of this chapter has focused on characterizing a fresh cell, before it ages. As there will be cell-to-cell variation even among new high-quality cells, and since parameter values change as a cell degrades, we will also need to consider how to combine fresh-cell parameters with real-time data to adjust the model to specific cells at specific points in life in a BMS. This is one topic of Chap. 6. Since teardown cannot be used in situ in a BMS, we must be able to accomplish this adaptation using only current/voltage/temperature data. For this reason, our research program continues to seek to understand and improve nonteardown parameter-estimation approaches.

But first, we consider direct applications of the fresh-cell LPM. In Chap. 4, we will see how to perform efficient time-domain simulations of cells and battery packs using reduced-order models. In Chap. 5, we will show how to perform real-time electrochemical state estimation for lithium-ion cells.

3.A *Summary of variables*

The following list itemizes all parameters and variables introduced in this chapter.

- b [unitless] is the shift factor used to calibrate the positive-electrode OCP in the discharge test (see Eq. (3.10)).
- m [unitless] is the scale factor used to calibrate the positive-electrode OCP in the discharge test (see Eq. (3.10)).
- $\widetilde{Q}_{\mathrm{h}}^{\mathrm{r}}$ [Ah] is the relative capacity of a half cell—constructed using the electrode from full-cell region r versus lithium metal—between voltages v_{min} and v_{max}.
- $Q_{\mathrm{h}}^{\mathrm{r}}$ [Ah] is the absolute capacity of a half cell—constructed using the electrode from full-cell region r versus lithium metal: $Q_{\mathrm{h}}^{\mathrm{r}} = \widetilde{Q}_{\mathrm{h}}^{\mathrm{r}} / |\theta_{\mathrm{max}}^{\mathrm{r}} - \theta_{\mathrm{min}}^{\mathrm{r}}|$.
- $\tilde{\theta}_{\mathrm{s}}^{\mathrm{r}}$ [unitless] is the relative stoichiometry of the half-cell electrode taken from full-cell region r. $\tilde{\theta}_{\mathrm{s}}^{\mathrm{r}} = 0$ when the electrode OCP is equal to v_{max}; $\tilde{\theta}_{\mathrm{s}}^{\mathrm{r}} = 1$ when the electrode OCP is equal to v_{min}.
- $\theta_{\mathrm{min}}^{\mathrm{r}}$ [unitless] is the value of absolute stoichiometry $\theta_{\mathrm{s}}^{\mathrm{r}}$ when the OCP of the electrode is equal to v_{max}.
- $\theta_{\mathrm{max}}^{\mathrm{r}}$ [unitless] is the value of absolute stoichiometry $\theta_{\mathrm{s}}^{\mathrm{r}}$ when the OCP of the electrode is equal to v_{min}.
- $\widetilde{U}_{\mathrm{ocp}}^{\mathrm{r}}(\tilde{\theta}_{\mathrm{s}}^{\mathrm{r}})$ [V] is electrode OCP evaluated at relative electrode stoichiometry point $\tilde{\theta}_{\mathrm{s}}^{\mathrm{r}}$.

3.B *SPM derivation*

In Sect. 3.9, we used a polynomial-type SPM to assist with calibrating the positive-electrode OCP estimate. The SPM we adopted was first proposed by Subramanian et al., where their original definition was presented in terms of standard DFN parameters.[87] Here, for convenient reference, we rederive their SPM using lumped parameters.

Recall that the LPM PDE describing lithium conservation in the solid is:[88]

$$\frac{\partial \theta_s}{\partial t} = \frac{1}{\tilde{r}^2}\frac{\partial}{\partial \tilde{r}}\left(\bar{D}_s \tilde{r}^2 \frac{\partial \theta_s}{\partial \tilde{r}}\right), \tag{3.45}$$

with $\theta_{s,0} = \theta_0 + z_0(\theta_{100} - \theta_0)$ at $t = 0$ for $0 < \tilde{r} < 1$, where its boundary conditions are:

$$\bar{D}_s \frac{\partial \theta_s}{\partial \tilde{r}}\bigg|_{\tilde{r}=0} = 0 \tag{3.46}$$

$$\bar{D}_s \frac{\partial \theta_s}{\partial \tilde{r}}\bigg|_{\tilde{r}=1} = -\frac{|\theta_{100} - \theta_0|}{10800Q} i_f. \tag{3.47}$$

[87] Venkat R. Subramanian, Vinten D. Diwakar, and Deepak Tapriyal, "Efficient macro-micro scale coupled modeling of batteries," *Journal of The Electrochemical Society*, 152(10):A2002, 2005.

[88] We treat $\bar{D}_s$ as uniform throughout the particle at every point in time, but as a time-varying function of the SOC at the particle surface.

Following Subramanian et al., we model θ_s using a fourth-order polynomial of the general form:

$$\theta_s(\tilde{r}, t) = \alpha(t) + \beta(t)\tilde{r}^2 + \gamma(t)\tilde{r}^4. \tag{3.48}$$

Due to the monotonic nature of the concentration profile under constant-current discharge or charge, we need to consider only even-ordered terms in this polynomial.

Substituting Eq. (3.48) into Eq. (3.45) gives:

$$\begin{aligned}\frac{\partial \theta_s(\tilde{r},t)}{\partial t} &= \frac{\partial \alpha(t)}{\partial t} + \frac{\partial \beta(t)}{\partial t}\tilde{r}^2 + \frac{\partial \gamma(t)}{\partial t}\tilde{r}^4 \\ &= \bar{D}_s(\theta_s)\frac{1}{\tilde{r}^2}\frac{\partial}{\partial \tilde{r}}\left(\tilde{r}^2 2\tilde{r}\beta(t) + \tilde{r}^2 4\tilde{r}^3\gamma(t)\right) \\ &= \bar{D}_s(\theta_s)\left(6\beta(t) + 20\tilde{r}^2\gamma(t)\right).\end{aligned}$$

Examining the boundary conditions, we have:

$$\frac{\partial \theta_s(\tilde{r},t)}{\partial \tilde{r}}\bigg|_{\tilde{r}=0} = 0$$

$$\frac{\partial \theta_s(\tilde{r},t)}{\partial \tilde{r}}\bigg|_{\tilde{r}=1} = 2\beta(t) + 4\gamma(t) = -\frac{|\theta_{100} - \theta_0|}{10800\bar{D}_s Q} i_f. \tag{3.49}$$

We proceed to find expressions for time-dependent functions $\alpha(t)$, $\beta(t)$, and $\gamma(t)$ in terms of the solid surface concentration $\theta_{ss}(t)$ and the average concentration θ_{avg}. The solid surface concentration is obtained by setting $\tilde{r} = 1$ in Eq. (3.48) giving:

$$\theta_{ss}(t) = \alpha(t) + \beta(t) + \gamma(t). \tag{3.50}$$

The volume-averaged concentration is again obtained by integrating the local concentration across the radial dimension and dividing by the total volume, giving:

$$\begin{aligned}\theta_{\text{avg}}(t) &= 3\int_{\tilde{r}=0}^{\tilde{r}=1} \tilde{r}^2\theta(\tilde{r},t)\,\mathrm{d}\tilde{r} \\ &= 3\int_{\tilde{r}=0}^{\tilde{r}=1} \tilde{r}^2\left[\alpha(t)+\beta(t)\tilde{r}^2+\gamma(t)\tilde{r}^4\right]\mathrm{d}\tilde{r} \\ &= \left[\frac{3\alpha(t)\tilde{r}^3}{3}+\frac{3\beta(t)\tilde{r}^5}{5}+\frac{3\gamma(t)\tilde{r}^7}{7}\right]_0^1 \\ &= \alpha(t)+\frac{3}{5}\beta(t)+\frac{3}{7}\gamma(t).\end{aligned} \tag{3.51}$$

We need a third equation as we have three time-dependent functions to identify. We introduce here the volume-averaged concentration flux, $\nabla\theta_{\text{avg}}(t)$, for this purpose. The process for generating the volume-averaged concentration flux proceeds in a similar way to that of the concentration; thus we write:

$$\begin{aligned}\nabla\theta_{\text{avg}}(t) &= 3\int_{\tilde{r}=0}^{\tilde{r}=1} \tilde{r}^2\frac{\partial\theta_s(\tilde{r},t)}{\partial\tilde{r}}\,\mathrm{d}\tilde{r} \\ &= \int_{\tilde{r}=0}^{\tilde{r}=1} \left[6\beta(t)\tilde{r}^3+12\gamma(t)\tilde{r}^5\right]\mathrm{d}\tilde{r} \\ &= \left[\frac{6\beta(t)\tilde{r}^4}{4}+\frac{12\gamma(t)\tilde{r}^6}{6}\right]_0^1 \\ &= \frac{3}{2}\beta(t)+2\gamma(t).\end{aligned} \tag{3.52}$$

Equations (3.50) through (3.52) can be combined to find:

$$\begin{aligned}\alpha(t) &= \frac{39}{4}\theta_{\text{ss}}(t)-\frac{35}{4}\theta_{\text{avg}}(t)-3\nabla\theta_{\text{avg}}(t) \\ \beta(t) &= -35\theta_{\text{ss}}(t)+35\theta_{\text{avg}}(t)+10\nabla\theta_{\text{avg}}(t) \\ \gamma(t) &= \frac{105}{4}\theta_{\text{ss}}(t)-\frac{105}{4}\theta_{\text{avg}}(t)-7\nabla\theta_{\text{avg}}(t).\end{aligned} \tag{3.53}$$

We can now write a single expression for the concentration profile in terms of the solid-surface and average concentrations and the concentration flux by substituting Eqs. (3.53) into Eq. (3.48) to write:

$$\begin{aligned}\theta_s(\tilde{r},t) &= \left(\frac{39}{4}\theta_{\text{ss}}(t)-\frac{35}{4}\theta_{\text{avg}}(t)-3\nabla\theta_{\text{avg}}(t)\right) \\ &\quad+\left(-35\theta_{\text{ss}}(t)+35\theta_{\text{avg}}(t)+10\nabla\theta_{\text{avg}}(t)\right)\tilde{r}^2 \\ &\quad+\left(\frac{105}{4}\theta_{\text{ss}}(t)-\frac{105}{4}\theta_{\text{avg}}(t)-7\nabla\theta_{\text{avg}}(t)\right)\tilde{r}^4 \\ &= \left(\frac{39}{4}-35\tilde{r}^2+\frac{105}{4}\tilde{r}^4\right)\theta_{\text{ss}}(t)-\left(\frac{1}{4}-\tilde{r}^2+\frac{3}{4}\tilde{r}^4\right)\theta_{\text{avg}}(t) \\ &\quad-\left(3-10\tilde{r}^2+7\tilde{r}^4\right)\nabla\theta_{\text{avg}}(t).\end{aligned}$$

In order to obtain a complete solution for the fourth-order case, we will need three additional equations. These will come from: (i) the volume-averaged concentration, (ii) the volume-averaged concentration flux, and (iii) the boundary condition at the particle surface.

The volume-averaged concentration flux is obtained via:

$$\frac{\int_0^1 \frac{\partial}{\partial \tilde{r}}\left(\frac{\partial \theta_s}{\partial t}\right) 4\pi\tilde{r}^2\,d\tilde{r}}{\int_0^1 \frac{4}{3}\pi\tilde{r}^3\,d\tilde{r}} = \frac{\bar{D}_s \int_0^1 \frac{\partial}{\partial \tilde{r}}\left[\frac{1}{\tilde{r}^2}\frac{\partial}{\partial \tilde{r}}\left(\tilde{r}^2\frac{\partial \theta_s}{\partial \tilde{r}}\right) 4\pi\tilde{r}^2\right]\,d\tilde{r}}{\int_0^1 \frac{4}{3}\pi\tilde{r}^3\,d\tilde{r}}.$$

The left-hand side evaluates to equal the time derivative of the concentration flux, $d\nabla\theta_{avg}(t)/dt$. The right-hand side evaluates to:

$$\frac{\bar{D}_s \int_0^1 \frac{\partial}{\partial \tilde{r}}\left[\frac{1}{\tilde{r}^2}\frac{\partial}{\partial \tilde{r}}\left(\tilde{r}^2\frac{\partial \theta_s}{\partial \tilde{r}}\right) 4\pi\tilde{r}^2\right]\,d\tilde{r}}{\int_0^1 \frac{4}{3}\pi\tilde{r}^3\,d\tilde{r}} = 3\bar{D}_s \int_0^1 \frac{\partial}{\partial \tilde{r}}\left[\frac{1}{\tilde{r}^2}\frac{\partial}{\partial \tilde{r}}\left(\tilde{r}^2\frac{\partial \theta_s}{\partial \tilde{r}}\right)\right]\tilde{r}^2\,d\tilde{r}.$$

Taking first the inner derivative,

$$\frac{\partial}{\partial \tilde{r}}\left(\tilde{r}^2\frac{\partial \theta_s}{\partial \tilde{r}}\right) = r^2\frac{\partial \theta_s^2}{\partial \tilde{r}^2} + 2\tilde{r}\frac{\partial \theta_s}{\partial \tilde{r}}.$$

Upon substituting,

$$\begin{aligned}\frac{\partial}{\partial \tilde{r}}\left[\frac{1}{\tilde{r}^2}\frac{\partial}{\partial \tilde{r}}\left(\tilde{r}^2\frac{\partial \theta_s}{\partial \tilde{r}}\right)\right] &= \frac{\partial}{\partial \tilde{r}}\left[\frac{1}{r^2}\left(r^2\frac{\partial \theta_s^2}{\partial \tilde{r}^2} + 2\tilde{r}\frac{\partial \theta_s}{\partial \tilde{r}}\right)\right] \\ &= \frac{\partial \theta_s^3}{\partial \tilde{r}^3} + 2\tilde{r}^{-1}\frac{\partial \theta_s^2}{\partial \tilde{r}^2} - 2\tilde{r}^{-2}\frac{\partial \theta_s}{\partial \tilde{r}}.\end{aligned}$$

Again, substituting,

$$\begin{aligned}&3\bar{D}_s \int_0^1 \frac{\partial}{\partial \tilde{r}}\left[\frac{1}{\tilde{r}^2}\frac{\partial}{\partial \tilde{r}}\left(\tilde{r}^2\frac{\partial \theta_s}{\partial \tilde{r}}\right)\right]\tilde{r}^2\,d\tilde{r} \\ &\quad = 3\bar{D}_s \int_0^1 \left[\frac{\partial \theta_s^3}{\partial \tilde{r}^3} + 2\tilde{r}^{-1}\frac{\partial \theta_s^2}{\partial \tilde{r}^2} - 2\tilde{r}^{-2}\frac{\partial \theta_s}{\partial \tilde{r}}\right]\tilde{r}^2\,d\tilde{r} && (3.54) \\ &\quad = 3\bar{D}_s \int_0^1 \left[\tilde{r}^2\frac{\partial \theta_s^3}{\partial \tilde{r}^3} + 2\tilde{r}\frac{\partial \theta_s^2}{\partial \tilde{r}^2} - 2\frac{\partial \theta_s}{\partial \tilde{r}}\right]\,d\tilde{r}. && (3.55)\end{aligned}$$

From the polynomial form, we compute the derivatives,

$$\begin{aligned}\frac{d\theta_s(t)}{d\tilde{r}} &= 2\beta(t)\tilde{r} + 4\gamma(t)\tilde{r}^3 \\ \frac{d^2\theta_s(t)}{d\tilde{r}^2} &= 2\beta(t) + 12\gamma(t)\tilde{r}^2 \\ \frac{d^3\theta_s(t)}{d\tilde{r}^3} &= 24\gamma(t)\tilde{r},\end{aligned}$$

We insert these into Eq. (3.55) to write:

$$\begin{aligned}\frac{d\nabla\theta_{avg}(t)}{dt} &= 3\bar{D}_s \int_0^1 \Big[\tilde{r}^2 24\gamma(t)\tilde{r} + 2\tilde{r}\left(2\beta(t) + 12\gamma(t)\tilde{r}^2\right) \\ &\qquad - 2\left(2\beta(t)\tilde{r} + 4\gamma(t)\tilde{r}^3\right)\Big]\,d\tilde{r} \\ &= 3\bar{D}_s \int_0^1 40\gamma(t)\tilde{r}^3 d\tilde{r} = 30\bar{D}_s\gamma(t). && (3.56)\end{aligned}$$

At this point, we apply the boundary conditions at the particle surface. From Eq. (3.49), solving for $\beta(t)$ gives:

$$\beta(t) = -2\gamma(t) - \frac{1}{2}\frac{|\theta_{100} - \theta_0|}{10800\bar{D}_s Q} i_f. \tag{3.57}$$

Substituting into Eq. (3.52) gives,

$$\begin{aligned}\nabla\theta_{avg}(t) &= -\frac{3}{2}\left[2\gamma(t) + \frac{|\theta_{100} - \theta_0|}{2\bar{D}_s 10800 Q} i_f\right] + 2\gamma(t) \\ &= -\gamma(t) - \frac{3}{4}\frac{|\theta_{100} - \theta_0|}{10800\bar{D}_s Q} i_f.\end{aligned}$$

Solving for $\gamma(t)$, we obtain:

$$\gamma(t) = -\nabla\theta_{avg}(t) - \frac{3}{4}\frac{|\theta_{100} - \theta_0|}{10800\bar{D}_s Q} i_f. \tag{3.58}$$

Substituting Eq. (3.58) into Eq. (3.56) gives a first-order ordinary differential equation for the volume-averaged flux density,

$$\frac{d\nabla\theta_{avg}(t)}{dt} + 30\bar{D}_s\nabla\theta_{avg}(t) + \frac{45}{2}\frac{|\theta_{100} - \theta_0|}{10800Q} i_f = 0. \tag{3.59}$$

The fourth-order polynomial approximation features two dynamic states: $\theta_{avg}(t)$ and $\nabla\theta_{avg}(t)$—both given by first-order ordinary differential equations—and a single output, $\theta_{ss}(t)$, obtained from an algebraic equation. We have already found a differential equation describing $\nabla\theta_{avg}(t)$. We find the differential equation for θ_{avg} as:

$$\begin{aligned}\frac{d\theta_{avg}}{dt} &= 3\bar{D}_s \int_0^1 \frac{\partial}{\partial\tilde{r}}\left(\tilde{r}^2\frac{\partial\theta_s}{\partial\tilde{r}}\right) d\tilde{r} \\ &= 3\bar{D}_s\left[\tilde{r}^r\frac{\partial\theta_s}{\partial\tilde{r}}\right]_0^1 \\ &= 3\bar{D}_s\left(-\frac{|\theta_{100} - \theta_0|}{\bar{D}_s 10800Q} i_f\right),\end{aligned} \tag{3.60}$$

where the final term is supplied from the boundary condition at $\tilde{r} = 1$. This gives the final form of the equation as:

$$\frac{d\theta_{avg}}{dt} = -\frac{3|\theta_{100} - \theta_0|}{10800Q} i_f, \tag{3.61}$$

which is now an ordinary differential equation in $\theta_{avg}(t)$.

To find the output equation, we combine Eqs. (3.50) and (3.51):

$$\theta_{ss}(t) = \theta_{avg}(t) + \frac{2}{5}\beta(t) + \frac{4}{7}\gamma(t).$$

We substitute $\beta(t)$ and $\gamma(t)$ from Eqs. (3.57) and (3.58):

$$\theta_{ss}(t) = \theta_{avg}(t) + \frac{8}{35}\nabla\theta_{avg}(t) - \frac{1}{35}\frac{|\theta_{100} - \theta_0|}{10800\bar{D}_s Q} i_f(t).$$

To simplify notation, we define $Q^{\mathrm{r}} = Q/\left|\theta^{\mathrm{r}}_{100} - \theta^{\mathrm{r}}_{0}\right|$. Finally, we condense the model equations into a time-varying state-space form, which is:

$$\begin{bmatrix} \dot{\theta}^{\mathrm{r}}_{\mathrm{avg}}(t) \\ \nabla\dot{\theta}^{\mathrm{r}}_{\mathrm{avg}}(t) \end{bmatrix} = \begin{bmatrix} 0 & 0 \\ 0 & -30\bar{D}^{\mathrm{r}}_{\mathrm{s}}(\theta^{\mathrm{r}}_{\mathrm{ss}}(t)) \end{bmatrix} \begin{bmatrix} \theta^{\mathrm{r}}_{\mathrm{avg}}(t) \\ \nabla\theta^{\mathrm{r}}_{\mathrm{avg}}(t) \end{bmatrix} + \begin{bmatrix} \frac{-1}{3600Q^{\mathrm{r}}} \\ \frac{-1}{480Q^{\mathrm{r}}} \end{bmatrix} i^{\mathrm{r}}_{\mathrm{f}}(t)$$

$$\left[\theta^{\mathrm{r}}_{\mathrm{ss}}(t)\right] = \begin{bmatrix} 1 & \frac{8}{35} \end{bmatrix} \begin{bmatrix} \theta^{\mathrm{r}}_{\mathrm{avg}}(t) \\ \nabla\theta^{\mathrm{r}}_{\mathrm{avg}}(t) \end{bmatrix} + \left[\frac{-(1/378\,000)}{Q^{\mathrm{r}}\bar{D}^{\mathrm{r}}_{\mathrm{s}}(\theta^{\mathrm{r}}_{\mathrm{ss}}(t))}\right] i^{\mathrm{r}}_{\mathrm{f}}(t).$$

This is the form of the SPM that is used in Sect. 3.9.

3.C *Finding* $\tilde{\phi}^{\mathrm{p}}_{\mathrm{e}}(3,0^+)$ *from* $\tilde{\phi}^{\mathrm{r}}_{\text{s-e}}(\tilde{x},0^+)$

In Sect. 3.12, we developed Eq. (3.17), which describes the instantaneous change in cell voltage due to a suddenly applied pulse in terms of $\tilde{\phi}^{\mathrm{p}}_{\text{s-e}}(3,0^+)$, $\tilde{\phi}^{\mathrm{n}}_{\text{s-e}}(0,0^+)$, and $\tilde{\phi}^{\mathrm{p}}_{\mathrm{e}}(3,0^+)$. The MATLAB `bvp5c` solver can find $\tilde{\phi}^{\mathrm{r}}_{\text{s-e}}(\tilde{x},0^+)$ directly, but we also need to compute $\tilde{\phi}^{\mathrm{p}}_{\mathrm{e}}(3,0^+)$ in order to evaluate the voltage change. Here, we see how to use the available information to do so.

3.C.1 *Negative electrode*

To begin, recall from Vol. I that the ionic current flowing through the electrolyte is defined by the ODE:[89]

$$\varepsilon^{\mathrm{r}}_{\mathrm{e}} A i^{\mathrm{r}}_{\mathrm{e}}(\tilde{x},t) = -\bar{\kappa}^{\mathrm{r}} \frac{\partial \phi^{\mathrm{r}}_{\mathrm{e}}(\tilde{x},t)}{\partial \tilde{x}}. \tag{3.62}$$

[89] We recognize that neither porosity $\varepsilon^{\mathrm{r}}_{\mathrm{e}}$ nor plate area A are lumped parameters. However, we will discover that these parameters disappear from the equations in the final result.

To find $\tilde{\phi}^{\mathrm{p}}_{\mathrm{e}}(3,0^+)$, we will need to integrate this equation across the entire width of the cell. We will do this in a stepwise fashion by integrating over each cell region separately. Beginning with the negative-electrode region, we also recognize:

$$\varepsilon^{\mathrm{n}}_{\mathrm{e}} A \frac{\partial i^{\mathrm{n}}_{\mathrm{e}}(\tilde{x},t)}{\partial \tilde{x}} = i^{\mathrm{n}}_{\mathrm{f+dl}}(\tilde{x},t). \tag{3.63}$$

Our first step is to integrate both sides of Eq. (3.63) using dummy variable of integration χ to arrive at a solution for $i^{\mathrm{n}}_{\mathrm{e}}(\tilde{x},t)$:

$$\begin{aligned} \int_0^{\tilde{x}} \varepsilon^{\mathrm{n}}_{\mathrm{e}} A \frac{\partial i^{\mathrm{n}}_{\mathrm{e}}(\chi,t)}{\partial \chi} \mathrm{d}\chi &= \int_0^{\tilde{x}} i^{\mathrm{n}}_{\mathrm{f+dl}}(\chi,t)\,\mathrm{d}\chi \\ \varepsilon^{\mathrm{n}}_{\mathrm{e}} A \left[i^{\mathrm{n}}_{\mathrm{e}}(\tilde{x},t) - i^{\mathrm{n}}_{\mathrm{e}}(0,t)\right] &= \int_0^{\tilde{x}} i^{\mathrm{n}}_{\mathrm{f+dl}}(\chi,t)\,\mathrm{d}\chi \\ \varepsilon^{\mathrm{n}}_{\mathrm{e}} A i^{\mathrm{n}}_{\mathrm{e}}(\tilde{x},t) &= \int_0^{\tilde{x}} i^{\mathrm{n}}_{\mathrm{f+dl}}(\chi,t)\,\mathrm{d}\chi. \end{aligned}$$

Combining this with Eq. (3.62) and integrating once again, we find ϕ_{e} in the negative electrode:

$$-\int_0^{\tilde{x}} \bar{\kappa}^{\mathrm{n}} \frac{\partial \phi^{\mathrm{n}}_{\mathrm{e}}(\chi,t)}{\partial \chi} \mathrm{d}\chi = \int_0^{\tilde{x}} \varepsilon^{\mathrm{n}}_{\mathrm{e}} A i^{\mathrm{n}}_{\mathrm{e}}(\chi_1,t)\,\mathrm{d}\chi_1$$

$$
\begin{aligned}
&= \int_0^{\tilde{x}} \int_0^{\chi_1} i^{\mathrm{n}}_{\mathrm{f+dl}}(\chi_2, t)\,\mathrm{d}\chi_2\,\mathrm{d}\chi_1 \\
-\bar{\kappa}^{\mathrm{n}}\left[\phi^{\mathrm{n}}_{\mathrm{e}}(\tilde{x}, t) - \phi^{\mathrm{n}}_{\mathrm{e}}(0, t)\right] &= \int_0^{\tilde{x}} \int_0^{\chi_1} i^{\mathrm{n}}_{\mathrm{f+dl}}(\chi_2, t)\,\mathrm{d}\chi_2\,\mathrm{d}\chi_1 \\
\tilde{\phi}^{\mathrm{n}}_{\mathrm{e}}(\tilde{x}, t) &= -\frac{1}{\bar{\kappa}^{\mathrm{n}}} \int_0^{\tilde{x}} \int_0^{\chi_1} i^{\mathrm{n}}_{\mathrm{f+dl}}(\chi_2, t)\,\mathrm{d}\chi_2\,\mathrm{d}\chi_1. \qquad (3.64)
\end{aligned}
$$

3.C.2 *Separator*

In the separator region, we have: $\varepsilon^{\mathrm{s}}_{\mathrm{e}} A i^{\mathrm{s}}_{\mathrm{e}}(\tilde{x}, t) = i_{\mathrm{app}}$. We use this information when integrating Eq. (3.62) over the separator region:

$$
-\int_1^{\tilde{x}} \bar{\kappa}^{\mathrm{s}} \frac{\partial \phi^{\mathrm{s}}_{\mathrm{e}}(\chi, t)}{\partial \chi}\,\mathrm{d}\chi = \int_1^{\tilde{x}} \varepsilon^{\mathrm{s}}_{\mathrm{e}} A i^{\mathrm{s}}_{\mathrm{e}}(\chi, t)\,\mathrm{d}\chi = \int_1^{\tilde{x}} i_{\mathrm{app}}\,\mathrm{d}\chi.
$$

Since i_{app} is a constant over the vanishing time duration of the pulse-resistance test,

$$
\begin{aligned}
-\bar{\kappa}^{\mathrm{s}}\left[\phi^{\mathrm{s}}_{\mathrm{e}}(\tilde{x}, t) - \phi_{\mathrm{e}}(1, t)\right] &= i_{\mathrm{app}}\,(\tilde{x} - 1) \\
\tilde{\phi}^{\mathrm{s}}_{\mathrm{e}}(\tilde{x}, t) - \tilde{\phi}^{\mathrm{n}}_{\mathrm{e}}(1, t) &= -\frac{i_{\mathrm{app}}}{\bar{\kappa}^{\mathrm{s}}}\,(\tilde{x} - 1) \\
\tilde{\phi}^{\mathrm{s}}_{\mathrm{e}}(\tilde{x}, t) &= \tilde{\phi}^{\mathrm{n}}_{\mathrm{e}}(1, t) - \frac{i_{\mathrm{app}}}{\bar{\kappa}^{\mathrm{s}}}\,(\tilde{x} - 1). \qquad (3.65)
\end{aligned}
$$

3.C.3 *Positive electrode*

In the positive-electrode region, we have:

$$
\varepsilon^{\mathrm{p}}_{\mathrm{e}} A \frac{\partial i^{\mathrm{p}}_{\mathrm{e}}(\tilde{x}, t)}{\partial \tilde{x}} = i^{\mathrm{p}}_{\mathrm{f+dl}}(\tilde{x}, t). \qquad (3.66)
$$

Integrating this equation over the positive-electrode region only:

$$
\begin{aligned}
\int_2^{\tilde{x}} \varepsilon^{\mathrm{p}}_{\mathrm{e}} A \frac{\partial i^{\mathrm{p}}_{\mathrm{e}}(\chi, t)}{\partial \chi}\,\mathrm{d}\chi &= \int_2^{\tilde{x}} i^{\mathrm{p}}_{\mathrm{f+dl}}(\chi, t)\,\mathrm{d}\chi \\
\varepsilon^{\mathrm{p}}_{\mathrm{e}} A \left[i^{\mathrm{p}}_{\mathrm{e}}(\tilde{x}, t) - i^{\mathrm{p}}_{\mathrm{e}}(2, t)\right] &= \int_2^{\tilde{x}} i^{\mathrm{p}}_{\mathrm{f+dl}}(\chi, t)\,\mathrm{d}\chi \\
\varepsilon^{\mathrm{p}}_{\mathrm{e}} A i^{\mathrm{p}}_{\mathrm{e}}(\tilde{x}, t) &= i_{\mathrm{app}} + \int_2^{\tilde{x}} i^{\mathrm{p}}_{\mathrm{f+dl}}(\chi, t)\,\mathrm{d}\chi.
\end{aligned}
$$

We substitute this result into Eq. (3.62) and integrate once again:

$$
\begin{aligned}
-\bar{\kappa}^{\mathrm{p}} \int_2^{\tilde{x}} \frac{\partial \phi^{\mathrm{p}}_{\mathrm{e}}(\chi, t)}{\partial \chi}\,\mathrm{d}\chi &= \int_2^{\tilde{x}} i_{\mathrm{app}}\,\mathrm{d}\chi + \int_2^{\tilde{x}} \int_2^{\chi_1} i^{\mathrm{p}}_{\mathrm{f+dl}}(\chi_2, t)\,\mathrm{d}\chi_2\,\mathrm{d}\chi_1 \\
-\bar{\kappa}^{\mathrm{p}}\left[\phi^{\mathrm{p}}_{\mathrm{e}}(\tilde{x}, t) - \phi^{\mathrm{s}}_{\mathrm{e}}(2, t)\right] &= i_{\mathrm{app}}\,(\tilde{x} - 2) + \int_2^{\tilde{x}} \int_2^{\chi_1} i^{\mathrm{p}}_{\mathrm{f+dl}}(\chi_2, t)\,\mathrm{d}\chi_2\,\mathrm{d}\chi_1 \\
\tilde{\phi}^{\mathrm{p}}_{\mathrm{e}}(\tilde{x}, t) &= \tilde{\phi}^{\mathrm{s}}_{\mathrm{e}}(2, t) - \frac{i_{\mathrm{app}}\,(\tilde{x} - 2)}{\bar{\kappa}^{\mathrm{p}}} \\
&\quad - \frac{1}{\bar{\kappa}^{\mathrm{p}}} \int_2^{\tilde{x}} \int_2^{\chi_1} i^{\mathrm{p}}_{\mathrm{f+dl}}(\chi_2, t)\,\mathrm{d}\chi_2\,\mathrm{d}\chi_1.
\end{aligned}
$$

We now substitute Eqs. (3.64) and (3.65) into the electrolyte-potential expression in the positive electrode:

$$\begin{aligned}\tilde{\phi}_e^p(\tilde{x},t) &= -\frac{1}{\bar{\kappa}^n}\int_0^1\int_0^{\chi_1} i_{\text{f+dl}}^n(\chi_2,t)\,d\chi_2\,d\chi_1 - \frac{i_{\text{app}}}{\bar{\kappa}^s}\\ &\quad - \frac{i_{\text{app}}(\tilde{x}-2)}{\bar{\kappa}^p} - \frac{1}{\bar{\kappa}^p}\int_2^{\tilde{x}}\int_2^{\chi_1} i_{\text{f+dl}}^p(\chi_2,t)\,d\chi_2\,d\chi_1. \end{aligned} \tag{3.67}$$

Cell-voltage computation requires only that we know $\tilde{\phi}_e^p(3,t)$. Evaluating the previous equation at $\tilde{x}=3$ gives:

$$\begin{aligned}\tilde{\phi}_e^p(3,t) &= -\frac{1}{\bar{\kappa}^n}\int_0^1\int_0^{\chi_1} i_{\text{f+dl}}^n(\chi_2,t)\,d\chi_2\,d\chi_1 - \frac{i_{\text{app}}}{\bar{\kappa}^s}\\ &\quad - \frac{i_{\text{app}}}{\bar{\kappa}^p} - \frac{1}{\bar{\kappa}^p}\int_2^{3}\int_2^{\chi_1} i_{\text{f+dl}}^p(\chi_2,t)\,d\chi_2\,d\chi_1. \end{aligned} \tag{3.68}$$

3.C.4 *The double integral of $i_{\text{f+dl}}^r(\tilde{x},t)$*

The double-integration of $i_{\text{f+dl}}^r$ in the negative and positive electrodes can be evaluated directly from ODE-solver solutions. In the negative electrode, we have found

$$\tilde{\phi}_e^n(1,t) = -\frac{1}{\bar{\kappa}^n}\int_0^1\int_0^{\chi_1} i_{\text{f+dl}}^n(\chi_2,t)\,d\chi_2\,d\chi_1.$$

We will find a simpler solution by double-integrating

$$\frac{\partial^2\tilde{\phi}_{\text{s-e}}^n(\tilde{x},t)}{\partial\tilde{x}^2} = \left(\frac{1}{\bar{\sigma}^n}+\frac{1}{\bar{\kappa}^n}\right) i_{\text{f+dl}}^n(\tilde{x},t).$$

We begin with

$$\begin{aligned}\int_0^{\chi_1}\frac{\partial^2\tilde{\phi}_{\text{s-e}}^n(\chi_2,t)}{\partial\chi_2^2}\,d\chi_2 &= \left(\frac{1}{\bar{\sigma}^n}+\frac{1}{\bar{\kappa}^n}\right)\int_0^{\chi_1} i_{\text{f+dl}}^n(\chi_2,t)\,d\chi_2\\ \left.\frac{\partial\tilde{\phi}_{\text{s-e}}^n(\chi_2,t)}{\partial\chi_2}\right|_{\chi_1} - \underbrace{\frac{\partial\tilde{\phi}_{\text{s-e}}^n(0,t)}{\partial\chi_2}}_{-i_{\text{app}}(t)/\bar{\sigma}^n} &= \left(\frac{\bar{\sigma}^n+\bar{\kappa}^n}{\bar{\sigma}^n\bar{\kappa}^n}\right)\int_0^{\chi_1} i_{\text{f+dl}}^n(\chi_2,t)\,d\chi_2\\ \frac{\partial\tilde{\phi}_{\text{s-e}}^n(\chi_1,t)}{\partial\chi_1} &= \left(\frac{\bar{\sigma}^n+\bar{\kappa}^n}{\bar{\sigma}^n\bar{\kappa}^n}\right)\int_0^{\chi_1} i_{\text{f+dl}}^n(\chi_2,t)\,d\chi_2 - \frac{i_{\text{app}}(t)}{\bar{\sigma}^n}.\end{aligned}$$

We integrate once again:

$$\begin{aligned}\int_0^{\tilde{x}}\frac{\partial\tilde{\phi}_{\text{s-e}}^n(\chi_1,t)}{\partial\chi_1}\,d\chi_1 &= \left(\frac{\bar{\sigma}^n+\bar{\kappa}^n}{\bar{\sigma}^n\bar{\kappa}^n}\right)\int_0^{\tilde{x}}\int_0^{\chi_1} i_{\text{f+dl}}^n(\chi_2,t)\,d\chi_2\,d\chi_1\\ &\quad - \int_0^{\tilde{x}}\frac{i_{\text{app}}(t)}{\bar{\sigma}^n}\,d\chi_1\\ \tilde{\phi}_{\text{s-e}}^n(\tilde{x},t)-\tilde{\phi}_{\text{s-e}}^n(0,t) &= \left(\frac{\bar{\sigma}^n+\bar{\kappa}^n}{\bar{\sigma}^n\bar{\kappa}^n}\right)\int_0^{\tilde{x}}\int_0^{\chi_1} i_{\text{f+dl}}^n(\chi_2,t)\,d\chi_2\,d\chi_1 - \frac{\tilde{x}i_{\text{app}}(t)}{\bar{\sigma}^n}.\end{aligned}$$

Solving for the term of interest:

$$\frac{\bar{\sigma}^n\bar{\kappa}^n}{\bar{\sigma}^n+\bar{\kappa}^n}\left(\tilde{\phi}_{\text{s-e}}^n(\tilde{x},t)-\tilde{\phi}_{\text{s-e}}^n(0,t)+\frac{\tilde{x}i_{\text{app}}(t)}{\bar{\sigma}^n}\right) = \int_0^{\tilde{x}}\int_0^{\chi_1} i_{\text{f+dl}}^n(\chi_2,t)\,d\chi_2\,d\chi_1$$

$$\tilde{\phi}_{\text{e}}^{\text{n}}(1,t) = -\frac{\bar{\sigma}^{\text{n}}}{\bar{\sigma}^{\text{n}}+\bar{\kappa}^{\text{n}}}\left(\tilde{\phi}_{\text{s-e}}^{\text{n}}(1,t) - \tilde{\phi}_{\text{s-e}}^{\text{n}}(0,t)\right) - \frac{i_{\text{app}}(t)}{\bar{\sigma}^{\text{n}}+\bar{\kappa}^{\text{n}}}.$$

In the positive electrode, we follow the same strategy to find:

$$-\frac{1}{\bar{\kappa}^{\text{p}}}\int_2^3\int_2^{\chi_1} i_{\text{f+dl}}^{\text{p}}(\chi_2,t)\,\mathrm{d}\chi_2\,\mathrm{d}\chi_1.$$

We simplify by double-integrating:

$$\frac{\partial^2\tilde{\phi}_{\text{s-e}}^{\text{p}}(\tilde{x},t)}{\partial\tilde{x}^2} = \left(\frac{1}{\bar{\sigma}^{\text{p}}}+\frac{1}{\bar{\kappa}^{\text{p}}}\right) i_{\text{f+dl}}^{\text{p}}(\tilde{x},t).$$

We begin with:

$$\int_2^{\chi_1}\frac{\partial^2\tilde{\phi}_{\text{s-e}}^{\text{p}}(\chi_2,t)}{\partial\chi_2^2}\,\mathrm{d}\chi_2 = \left(\frac{1}{\bar{\sigma}^{\text{p}}}+\frac{1}{\bar{\kappa}^{\text{p}}}\right)\int_2^{\chi_1} i_{\text{f+dl}}^{\text{p}}(\chi_2,t)\,\mathrm{d}\chi_2$$

$$\left.\frac{\partial\tilde{\phi}_{\text{s-e}}^{\text{p}}(\chi_2,t)}{\partial\chi_2}\right|_{\chi_1} - \underbrace{\frac{\partial\tilde{\phi}_{\text{s-e}}^{\text{p}}(2,t)}{\partial\chi_2}}_{i_{\text{app}}(t)/\bar{\kappa}^{\text{p}}} = \left(\frac{\bar{\sigma}^{\text{p}}+\bar{\kappa}^{\text{p}}}{\bar{\sigma}^{\text{p}}\bar{\kappa}^{\text{p}}}\right)\int_2^{\chi_1} i_{\text{f+dl}}^{\text{p}}(\chi_2,t)\,\mathrm{d}\chi_2$$

$$\frac{\partial\tilde{\phi}_{\text{s-e}}^{\text{p}}(\chi_1,t)}{\partial\chi_1} = \left(\frac{\bar{\sigma}^{\text{p}}+\bar{\kappa}^{\text{p}}}{\bar{\sigma}^{\text{p}}\bar{\kappa}^{\text{p}}}\right)\int_2^{\chi_1} i_{\text{f+dl}}^{\text{p}}(\chi_2,t)\,\mathrm{d}\chi_2 + \frac{i_{\text{app}}(t)}{\bar{\kappa}^{\text{p}}}.$$

We integrate once again:

$$\begin{aligned}\int_2^{\tilde{x}}\frac{\partial\tilde{\phi}_{\text{s-e}}^{\text{p}}(\chi_1,t)}{\partial\chi_1}\,\mathrm{d}\chi_1 &= \left(\frac{\bar{\sigma}^{\text{p}}+\bar{\kappa}^{\text{p}}}{\bar{\sigma}^{\text{p}}\bar{\kappa}^{\text{p}}}\right)\int_2^{\tilde{x}}\int_2^{\chi_1} i_{\text{f+dl}}^{\text{p}}(\chi_2,t)\,\mathrm{d}\chi_2\,\mathrm{d}\chi_1\\ &\quad+\int_0^{\tilde{x}}\frac{i_{\text{app}}(t)}{\bar{\kappa}^{\text{p}}}\,\mathrm{d}\chi_1\\ \tilde{\phi}_{\text{s-e}}^{\text{p}}(\tilde{x},t)-\tilde{\phi}_{\text{s-e}}^{\text{p}}(2,t) &= \left(\frac{\bar{\sigma}^{\text{p}}+\bar{\kappa}^{\text{p}}}{\bar{\sigma}^{\text{p}}\bar{\kappa}^{\text{p}}}\right)\int_2^{\tilde{x}}\int_2^{\chi_1} i_{\text{f+dl}}^{\text{p}}(\chi_2,t)\,\mathrm{d}\chi_2\,\mathrm{d}\chi_1\\ &\quad+\frac{(\tilde{x}-2)i_{\text{app}}(t)}{\bar{\kappa}^{\text{p}}}.\end{aligned}$$

Solving for the term of interest:

$$\frac{\bar{\sigma}^{\text{p}}\bar{\kappa}^{\text{p}}}{\bar{\sigma}^{\text{p}}+\bar{\kappa}^{\text{p}}}\left(\tilde{\phi}_{\text{s-e}}^{\text{p}}(\tilde{x},t)-\tilde{\phi}_{\text{s-e}}^{\text{p}}(2,t)-\frac{(\tilde{x}-2)i_{\text{app}}(t)}{\bar{\kappa}^{\text{p}}}\right) = \int_2^{\tilde{x}}\int_2^{\chi_1} i_{\text{f+dl}}^{\text{p}}(\chi_2,t)\,\mathrm{d}\chi_2\,\mathrm{d}\chi_1.$$

Overall, we can now solve for $\tilde{\phi}_{\text{e}}^{\text{p}}(3,0^+)$ using the solutions to $\tilde{\phi}_{\text{s-e}}^{\text{r}}(\tilde{x},0^+)$ found by `bvp5c`:

$$\begin{aligned}\tilde{\phi}_{\text{e}}^{\text{p}}(3,t) &= -\frac{1}{\bar{\kappa}^{\text{n}}}\int_0^1\int_0^{\chi_1} i_{\text{f+dl}}^{\text{n}}(\chi_2,t)\,\mathrm{d}\chi_2\mathrm{d}\chi_1 - \frac{i_{\text{app}}}{\bar{\kappa}^{\text{s}}}\\ &\quad-\frac{i_{\text{app}}}{\bar{\kappa}^{\text{p}}} - \frac{1}{\bar{\kappa}^{\text{p}}}\int_2^3\int_2^{\chi_1} i_{\text{f+dl}}^{\text{p}}(\chi_2,t)\,\mathrm{d}\chi_2\,\mathrm{d}\chi_1\\ &= -\frac{\bar{\sigma}^{\text{n}}}{\bar{\sigma}^{\text{n}}+\bar{\kappa}^{\text{n}}}\left(\tilde{\phi}_{\text{s-e}}^{\text{n}}(1,t)-\tilde{\phi}_{\text{s-e}}^{\text{n}}(0,t)\right) - \frac{i_{\text{app}}(t)}{\bar{\sigma}^{\text{n}}+\bar{\kappa}^{\text{n}}} - \frac{i_{\text{app}}(t)}{\bar{\kappa}^{\text{s}}}\\ &\quad-\frac{i_{\text{app}}}{\bar{\kappa}^{\text{p}}} - \frac{\bar{\sigma}^{\text{p}}}{\bar{\sigma}^{\text{p}}+\bar{\kappa}^{\text{p}}}\left(\tilde{\phi}_{\text{s-e}}^{\text{p}}(3,t)-\tilde{\phi}_{\text{s-e}}^{\text{p}}(2,t)-\frac{i_{\text{app}}(t)}{\bar{\kappa}^{\text{p}}}\right)\\ &= -\frac{i_{\text{app}}(t)}{\bar{\sigma}^{\text{n}}+\bar{\kappa}^{\text{n}}} - \frac{i_{\text{app}}(t)}{\bar{\kappa}^{\text{s}}} - \frac{i_{\text{app}}}{\bar{\sigma}^{\text{p}}+\bar{\kappa}^{\text{p}}} - \frac{\bar{\sigma}^{\text{n}}}{\bar{\sigma}^{\text{n}}+\bar{\kappa}^{\text{n}}}\left(\tilde{\phi}_{\text{s-e}}^{\text{n}}(1,t)-\tilde{\phi}_{\text{s-e}}^{\text{n}}(0,t)\right)\\ &\quad-\frac{\bar{\sigma}^{\text{p}}}{\bar{\sigma}^{\text{p}}+\bar{\kappa}^{\text{p}}}\left(\tilde{\phi}_{\text{s-e}}^{\text{p}}(3,t)-\tilde{\phi}_{\text{s-e}}^{\text{p}}(2,t)\right).\end{aligned}$$

3.D *Initializing estimates*

In this appendix section, we summarize how we initialize parameter values prior to nonlinear optimization.

3.D.1 *Lumped double-layer capacitance* $\bar{C}^{\mathrm{r}}_{\mathrm{dl}}$

Previously, Figs. 3.56 and 3.60 illustrated an approach to isolate DRT peaks and approximate some interphase components. We expect to observe two distinct peaks in a DRT diagram within the Nyquist bump region as there are two electrodes, and those distinct time constants are associated with $\bar{C}^{\mathrm{r}}_{\mathrm{dl}}$ and $\bar{R}^{\mathrm{r}}_{\mathrm{ct}}$. Figs. 3.D.1 and 3.D.2 show results of $\bar{C}^{\mathrm{r}}_{\mathrm{dl}}$ and $\bar{R}^{\mathrm{r}}_{\mathrm{ct}}$ approximations. That is, we use the cell impedance model Eq. (2.103) and model parameters listed in Table 2.C.1 to compute simulated cell impedance at various cell SOC setpoints; then we evaluate their DRT functions, isolate peaks, and analyze them to approximate $\bar{C}^{\mathrm{r}}_{\mathrm{dl}}$ and $\bar{R}^{\mathrm{r}}_{\mathrm{ct}}$ values. The results presented in the figures assume $\bar{R}^{\mathrm{r}}_{\mathrm{dl}}$ and $n^{\mathrm{r}}_{\mathrm{dl}}$ to be true values. Since the DRT is a discrete method, the estimates of $\bar{C}^{\mathrm{r}}_{\mathrm{dl}}$ and $\bar{R}^{\mathrm{r}}_{\mathrm{ct}}$ will not be exact. However, we see that the average over all SOCs of both are sufficiently close to their actual values to provide a good starting point for optimization.

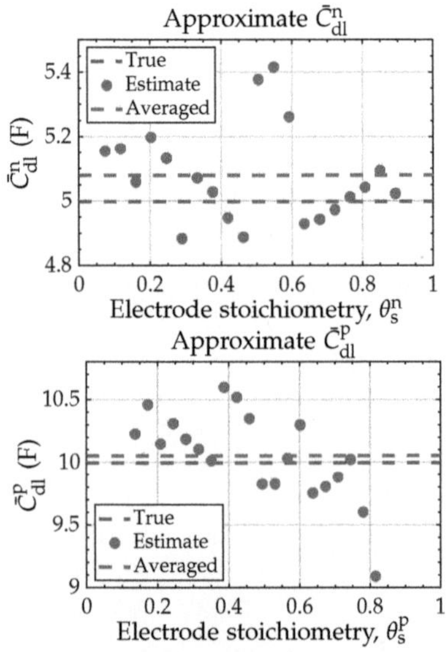

Figure 3.D.1: Approximate $\bar{C}^{\mathrm{r}}_{\mathrm{dl}}$ found from DRT peak analysis.

3.D.2 *Lumped resistances* $\bar{R}^{\mathrm{r}}_{\mathrm{dl}}$ *and* $\bar{R}^{\mathrm{r}}_{\mathrm{f}}$

We initialize the lumped double-layer and film resistances to arbitrary small nonzero values: $\bar{R}^{\mathrm{r}}_{\mathrm{dl},0} = 1 \times 10^{-4}\,\Omega$ and $\bar{R}^{\mathrm{r}}_{\mathrm{f},0} = 1 \times 10^{-6}\,\Omega$, where we use subscript "0" to indicate that this is an initial estimate only. Other small nonzero values should also work: while setting film resistance to zero admits a closed-form solution to the inverse Butler–Volmer equation, it is not physically realistic due to the presence of SEI and CEI films. We do not have general guidance for how to select the optimization boundaries/constraints for these variables. Since both resistances are expected to be small, we simply assume: $1 \times 10^{-6}\,[\Omega] < \bar{R}^{\mathrm{r}}_{\mathrm{dl}} < 1 \times 10^{-1}\,[\Omega]$ and $1 \times 10^{-9}\,[\Omega] < \bar{R}^{\mathrm{r}}_{\mathrm{f}} < 1 \times 10^{-1}\,[\Omega]$.

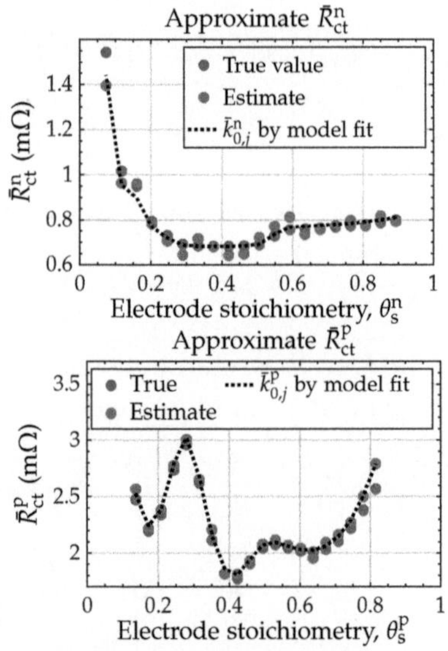

Figure 3.D.2: Approximate SOC-dependent $\bar{R}^{\mathrm{r}}_{\mathrm{ct}}$ found from DRT peak analysis.

3.D.3 *Lumped conductances* $\bar{\kappa}^{\mathrm{r}}$ *and* $\bar{\sigma}^{\mathrm{r}}$

Recall the lumped assignments for solid and electrolyte conductances:

$$\bar{\sigma}^{\mathrm{r}} = \frac{\sigma^{\mathrm{r}}_{\mathrm{eff}} A}{L^{\mathrm{r}}} = \sigma^{\mathrm{r}} \left(\varepsilon^{\mathrm{r}}_{\mathrm{s}}\right)^{\mathrm{brug}} \frac{A}{L^{\mathrm{r}}}$$

$$\bar{\kappa}^{\mathrm{r}} = \frac{\kappa^{\mathrm{r}}_{\mathrm{eff}} A}{L^{\mathrm{r}}} = \kappa \left(\varepsilon^{\mathrm{r}}_{\mathrm{e}}\right)^{\mathrm{brug}} \frac{A}{L^{\mathrm{r}}},$$

where ε^{r} are the phase volume fractions, brug is the Bruggeman constant (often ~ 1.5), A is the current-collector plate area, and L^{r} the electrode thickness. In principle, a cell's solid phase (electronic) conductance should be larger than its liquid phase (ionic) conductance, $\sigma^{\mathrm{r}} > \kappa^{\mathrm{r}}$. When initializing the optimization, we assume that $\bar{\sigma}^{\mathrm{r}}$ are an order of magnitude greater than the respective $\bar{\kappa}^{\mathrm{r}}$.

It is possible to perform a sensitivity study to gain intuition regarding the impact of each model parameter value on the full-cell impedance. We do not present detailed results here, but we have performed such studies and find that inaccurate values of $\bar{\sigma}^{\mathrm{r}}$ do not significantly impact the impedance-model predictions. We will initialize both parameters to constants prior to the optimization process. It might not be possible to directly predict the order of difference among $\bar{\sigma}^{\mathrm{n}}$ and $\bar{\sigma}^{\mathrm{p}}$. In terms of material properties, graphite has high electrical conductivity of $> 100\,\mathrm{S\,m^{-1}}$ at room temperature while NMC has slightly poorer conductivity of $\sim 3.8\,\mathrm{S\,m^{-1}}$.[90] Overall, using rough guesses for the dimensional quantities of the cell, we simply initialize $\bar{\sigma}^{\mathrm{n}} = 10^8\,[\mathrm{S}]$ and $\bar{\sigma}^{\mathrm{p}} = 10^6\,[\mathrm{S}]$.

We recognize that the DFN parameter κ is a function of electrolyte concentration and is often so modeled. However, since the EIS test is a small-signal experiment conducted near equilibrium, κ is treated as a constant that is evaluated at the equilibrium concentration $\theta_{\mathrm{e},0} = 1$. While κ is uniform among all regions, $\bar{\kappa}^{\mathrm{r}}$ are different among regions due to $\varepsilon_{\mathrm{e}}^{\mathrm{r}}$ and L^{r}. To proceed, we can address the following ratios:

$$\frac{\bar{\kappa}^{\mathrm{n}}}{\bar{\kappa}^{\mathrm{s}}} = \left(\frac{L^{\mathrm{s}}}{L^{\mathrm{n}}}\right)\left(\frac{\varepsilon_{\mathrm{e}}^{\mathrm{n}}}{\varepsilon_{\mathrm{e}}^{\mathrm{s}}}\right)^{\mathrm{brug}}, \quad \text{and} \quad \frac{\bar{\kappa}^{\mathrm{p}}}{\bar{\kappa}^{\mathrm{s}}} = \left(\frac{L^{\mathrm{s}}}{L^{\mathrm{p}}}\right)\left(\frac{\varepsilon_{\mathrm{e}}^{\mathrm{p}}}{\varepsilon_{\mathrm{e}}^{\mathrm{s}}}\right)^{\mathrm{brug}}.$$

It is expected that $L^{\mathrm{s}} < \{L^{\mathrm{n}}, L^{\mathrm{p}}\}$ and $\varepsilon_{\mathrm{e}}^{\mathrm{s}} > \{\varepsilon_{\mathrm{e}}^{\mathrm{n}}, \varepsilon_{\mathrm{e}}^{\mathrm{p}}\}$ for a commercial cell, which leads to $\bar{\kappa}^{\mathrm{s}} > \{\bar{\kappa}^{\mathrm{n}}, \bar{\kappa}^{\mathrm{p}}\}$. Electrolyte volume fraction in the separator is higher than those in the two electrodes but probably less than double, so we assume $0.5 < \varepsilon_{\mathrm{e}}^{\mathrm{n}}/\varepsilon_{\mathrm{e}}^{\mathrm{s}} < 1$ and $0.5 < \varepsilon_{\mathrm{e}}^{\mathrm{p}}/\varepsilon_{\mathrm{e}}^{\mathrm{s}} < 1$. The length of separator region is shorter than the electrodes but perhaps at least 1/4 of their length, so $1/4 < L^{\mathrm{s}}/L^{\mathrm{n}} < 1$ and $1/4 < L^{\mathrm{s}}/L^{\mathrm{p}} < 1$. Therefore, we assume $2^{-3.5} < \bar{\kappa}^{\mathrm{n}}/\bar{\kappa}^{\mathrm{s}} < 1$ and $2^{-3.5} < \bar{\kappa}^{\mathrm{p}}/\bar{\kappa}^{\mathrm{s}} < 1$.

In optimization studies, we prefer specifying constraints with deterministic values rather than depending on other model parameters. Here, we assign two unitless quantities, $K_{\bar{\kappa}}^{\mathrm{n}}$ and $K_{\bar{\kappa}}^{\mathrm{p}}$, where $\bar{\kappa}^{\mathrm{n}} = K_{\bar{\kappa}}^{\mathrm{n}} \cdot \bar{\kappa}^{\mathrm{s}}$ and $\bar{\kappa}^{\mathrm{p}} = K_{\bar{\kappa}}^{\mathrm{p}} \cdot \bar{\kappa}^{\mathrm{s}}$. This allows to write $2^{-3.5} < K_{\bar{\kappa}}^{\mathrm{n}} < 1$ and $2^{-3.5} < K_{\bar{\kappa}}^{\mathrm{p}} < 1$. Further, we can simply initialize them as $K_{\bar{\kappa},0}^{\mathrm{n}} = K_{\bar{\kappa},0}^{\mathrm{p}} = 0.5$. Note that both quantities should be identifiable because $\bar{\kappa}^{\mathrm{n}}$ and $\bar{\kappa}^{\mathrm{p}}$ are identifiable parameters. Overall, there are two benefits to defining $K_{\bar{\kappa}}^{\mathrm{n}}$ and $K_{\bar{\kappa}}^{\mathrm{p}}$: first, the parameter constraints are deterministic values rather than depending on other model parameters; second,

[90] Marc Doyle, John Newman, Antoni S. Gozdz, Caroline N. Schmutz, and Jean-Marie Tarascon, "Comparison of modeling predictions with experimental data from plastic lithium ion cells," *Journal of The Electrochemical Society*, 143(6):1890–1903, 1996; Apurba Sakti, Jeremy J. Michalek, Sang-Eun Chun, and Jay F. Whitacre, "A validation study of lithium-ion cell constant C-rate discharge simulation with Battery Design Studio®: Battery Design Studio® validation," *International Journal of Energy Research*, 37(12):1562–1568, 2013; Christian von Lüders, Jonas Keil, Markus Webersberger, and Andreas Jossen, "Modeling of lithium plating and lithium stripping in lithium-ion batteries," *Journal of Power Sources*, 414:41–47, 2019; and Jonas Keil and Andreas Jossen, "Electrochemical modeling of linear and nonlinear aging of lithium-ion cells," *Journal of The Electrochemical Society*, 167(11):110535, 2020.

this arrangement will ensure $\bar{\kappa}^{\rm s} > \{\bar{\kappa}^{\rm n}, \bar{\kappa}^{\rm p}\}$. To initialize $\bar{\kappa}^{\rm s}$, we simply assume $\bar{\kappa}^{\rm s}_0 = 5 \times 10^3$ [S] and 1×10^3 [S] $< \bar{\kappa}^{\rm s} < 1 \times 10^5$ [S].

3.D.4 *Lumped ionic conductivity coefficient* $\bar{\kappa}_{\rm D}$

The lumped parameter $\bar{\kappa}_{\rm D}$ is an abbreviation in the charge-conservation PDE, which is defined as:

$$\bar{\kappa}_{\rm D} = \frac{2R\left(t^0_+ - 1\right)}{F}\left(1 + \frac{\partial \ln f_\pm}{\partial \ln c_{\rm e}}\right),$$

where $f_\pm$ is the mean molar activity coefficient of the electrolyte, and $\partial \ln f_\pm / \partial \ln c_{\rm e}$ is often treated to be constant in literature. We choose somewhat typical values $t^0_+ = 0.363$ and $\partial \ln f_\pm / \partial \ln c_{\rm e} = 3$ to initialize $\bar{\kappa}_{\rm D}$:

$$\bar{\kappa}_{\rm D,0} = \frac{2R\,(0.363 - 1)}{F}(1+3) = -4.39 \times 10^{-4}\,[\mathrm{V\,K^{-1}}].$$

We also allow some variations for $0 < t^0_+ < 0.5$ and $0 < \partial \ln f_\pm / \partial \ln c_{\rm e} < 8$, from which:

$$-1.55 \times 10^{-3}\,[\mathrm{V\,K^{-1}}] < \bar{\kappa}_{\rm D} < -8.62 \times 10^{-5}\,[\mathrm{V\,K^{-1}}].$$

3.D.5 *Electrolyte transport ratio* $\bar{\psi}$

The lumped assignment for $\bar{\psi}$ is written as:

$$\bar{\psi}T = F\frac{\bar{D}^{\rm r}_{\rm e}}{\bar{\kappa}^{\rm r}} = F\frac{D^{\rm r}_{\rm e} c_{\rm e,0}}{\kappa^{\rm r}\left(1 - t^0_+\right)}.$$

Since it is difficult to assign reasonable estimates to the values for $D^{\rm r}_{\rm e}$, $c_{\rm e,0}$, and $\kappa^{\rm r}$ prior to optimization, we adopt an alternative expression. We begin with the Nernst–Einstein relationship:[91]

$$D^{\rm r}_{\rm e} = \frac{\kappa^{\rm r} RT}{c^{\rm r}_{\rm e} z^2 F^2},$$

where z is the charge number and $z = 1$ for Li^+. We then multiply both sides of the equation by $Fc_{\rm e,0}/\kappa^{\rm r} / \left(1 - t^0_+\right)$:

$$\underbrace{F\frac{D^{\rm r}_{\rm e} c_{\rm e,0}}{\kappa^{\rm r}\left(1 - t^0_+\right)}}_{\bar{\psi}T} = \frac{\kappa^{\rm r} RT}{c^{\rm r}_{\rm e} z^2 F}\cdot\frac{c_{\rm e,0}}{\kappa^{\rm r}\left(1 - t^0_+\right)} = \frac{c_{\rm e,0}}{c^{\rm r}_{\rm e}\left(1 - t^0_+\right)}\frac{RT}{F},$$

where $c^{\rm r}_{\rm e} \approx c_{\rm e,0}$ because we model a cell near equilibrium. Further, since we previously assumed $t^0_+ = 0.363$ for initialization,

$$\bar{\psi}_0 = \frac{1}{1 - 0.363}\cdot\frac{R}{F} = 1.353 \times 10^{-4}\left[\mathrm{V\,K^{-1}}\right].$$

[91] Madeleine Ecker, Thi Kim Dung Tran, Philipp Dechent, Stefan Käbitz, Alexander Warnecke, and Dirk Uwe Sauer, "Parameterization of a physico-chemical model of a lithium-ion battery: I. Determination of parameters," *Journal of the Electrochemical Society*, 162(9):A1836, 2015; and Johannes Schmalstieg, Christiane Rahe, Madeleine Ecker, and Dirk Uwe Sauer, "Full cell parameterization of a high-power lithium-ion battery for a physico-chemical model: Part I. Physical and electrochemical parameters," *Journal of The Electrochemical Society*, 165(16):A3799, 2018.

We have also previously set $0 < t_+^0 < 0.5$, which leads to $1 < (1 - t_+^0)^{-1} < 2$ and $R/F < \bar{\psi}_0 < 2R/F$. However, since the Nernst–Einstein relation is derived for highly diluted electrolyte and is only an approximation for the concentrated solutions that are used in commercial lithium-ion cells, we choose to expand the range for $\bar{\psi}$ slightly and let $R/\,(10F) < \bar{\psi} < 2.5R/F$. Numerically,

$$8.617 \times 10^{-6}\,[\mathrm{V\,K^{-1}}] < \bar{\psi} < 2.154 \times 10^{-4}\,[\mathrm{V\,K^{-1}}].$$

3.D.6 MSMR interface kinetics parameters α_j and $\bar{k}_{0,j}$

For simplicity, we initialize the charge-transfer coefficients in all galleries to be equal to one half, $\alpha_j = 0.5$. The lumped-reaction rate constants, $\bar{k}_{0,j}$, are blended in the cell TF models via the charge-transfer resistance as shown in Eq. (2.22). From what we observe, the MSMR galleries can have very different magnitudes of $\bar{k}_{0,j}$, and we cannot simply examine the reaction itself to determine $\bar{k}_{0,j}$. If there are six galleries in the (graphite) anode and four in the (NMC) cathode, there will be ten different $\bar{k}_{0,j}^{\mathrm{r}}$ to determine and we would need to gather cell impedance data for at least ten distinct SOC setpoints. In this section, we want to answer two questions:

- Are $\bar{k}_{0,j}$ in all galleries identifiable from the impedance data?
- Is impedance insensitive to any $\bar{k}_{0,j}$ so that we might make simplifying assumptions and minimize the degrees of freedom?

First, we will assume $\bar{k}_{0,j} = 1\,\mathrm{A}$ for all galleries j to isolate the SOC-dependent part in the exchange current rate equation $i_{0,j}$, and then examine its variations over the range of electrode stoichiometry, as displayed in Fig. 3.D.3. Recall that, at equilibrium, x_j and θ_s are related via Eq. (2.1). The set of MSMR model parameters $\{U_j^0, X_j^0, \omega_j\}$ for the Panasonic cell were identified earlier in this chapter, and their values are listed in sidenote 50 for $U_{\mathrm{ocp}}^{\mathrm{n}}(\theta_\mathrm{s})$ and in Table 3.11 for $U_{\mathrm{ocp}}^{\mathrm{p}}(\theta_\mathrm{s})$. The results indicate that $\bar{k}_{0,j}$ in both electrodes should be observable because each $i_{0,j}$ is a unique function of θ_s. If we collect enough data at various SOC setpoints, the function of $\bar{R}_{\mathrm{ct}}(\theta_\mathrm{s})$ can be exposed and $\bar{k}_{0,j}$ can be optimized.

In the top frame of Fig. 3.D.3, it appears that galleries 1–3 have orders-of-magnitude larger contribution to $i_{0,j}$ and hence $\bar{R}_{\mathrm{ct}}$ than galleries 4–6. In the bottom frame, galleries 1–2 have larger contribution than 3–4, also by orders of magnitude. The magnitude of $\bar{R}_{\mathrm{ct}}$ is dependent on the larger $i_{0,j}$ rather than the smaller ones. In the graphite electrode, if $\{\bar{k}_{0,j}\}_{j=4,5,6}$ are not orders of magnitudes larger than $\{\bar{k}_{0,j}\}_{j=1,2,3}$, then $\{\bar{k}_{0,j}\}_{j=4,5,6}$ will have almost no impact on cell impedance. In the NMC electrode, if $\{\bar{k}_{0,j}\}_{j=3,4}$ are not orders of magnitude larger than $\{\bar{k}_{0,j}\}_{j=1,2}$, then $\{\bar{k}_{0,j}\}_{j=3,4}$ will have almost

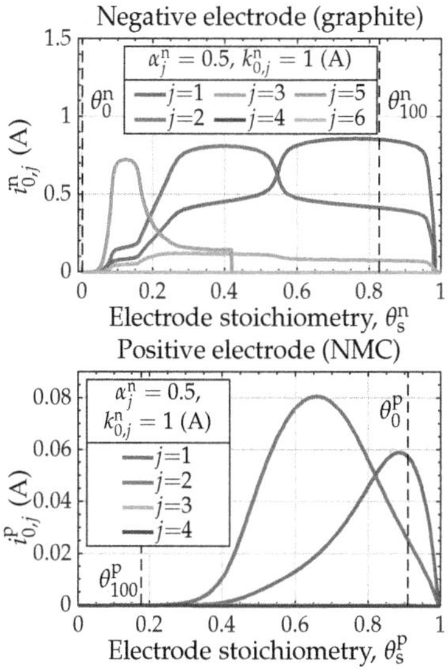

Figure 3.D.3: Examine the exchange current rate as $\bar{k}_{0,j} = 1\,\mathrm{A}$ and $\alpha_j = 0.5$ in both negative (top) and positive electrodes (bottom).

no impact on cell impedance. However, at this point (prior to performing experiments and regressing parameter values) we are still unable to guess their contributions to $\bar{R}_{\mathrm{ct}}$, for example, $\{\bar{k}_{0,j}\}_{j=4,5,6}$ in the graphite electrode could be significantly larger than $\{\bar{k}_{0,j}\}_{j=1,2,3}$, which leads to $\{i_{0,j}\}_{j=4,5,6} > \{i_{0,j}\}_{j=1,2,3}$.

Our present approach to initializing $\bar{k}_{0,j}$ uses the approximations of $\bar{R}_{\mathrm{ct}}^{\mathrm{r}}$ produced by the DRT method. We perform a side optimization to find $\bar{k}_{0,j}$ such that the computed $\bar{R}_{\mathrm{ct}}^{\mathrm{r}}$ agrees with these estimates as closely as possible. The fit to $\bar{R}_{\mathrm{ct}}$ data is shown by dashed lines in the lower row of Fig. 3.70, and the precise values compared to the truth are listed in Table 3.D.1. Despite the fact that $\bar{R}_{\mathrm{ct}}^{\mathrm{r}}$ is well approximated from the DRT analysis, we cannot find satisfying $\bar{k}_{0,j}$ estimates for all galleries, especially $\bar{k}_{0,6}^{\mathrm{n}}$ and $\bar{k}_{0,7}^{\mathrm{n}}$. This is because $\bar{R}_{\mathrm{ct}}^{\mathrm{r}}$ is extremely insensitive to $\bar{k}_{0,6}^{\mathrm{n}}$ and $\bar{k}_{0,7}^{\mathrm{n}}$. The boundaries for $\bar{k}_{0,j}$ are set to be one order of magnitude around their initial values; that is, $\bar{k}_{0,j,0}/10 < \bar{k}_{0,j,0} < 10\bar{k}_{0,j,0}$.

Table 3.D.1: Estimate $\bar{k}_{0,j}$ from $\bar{R}_{\mathrm{ct}}$ initial estimates. The fitting performance is also visualized in Fig. 3.D.2.

Electrode	j	True $\bar{k}_{0,j}$ [A]	Approximate $\bar{k}_{0,j}$ [A]
Negative (graphite)	1	1.6825×10^{1}	1.5543×10^{1}
	2	1.3654×10^{1}	1.2408×10^{1}
	3	4.8603×10^{-5}	1.6653×10^{-5}
	4	2.7755×10^{5}	3.0199×10^{5}
	5	7.7389×10^{-6}	8.3268×10^{-6}
	6	2.0028×10^{3}	2.5822×10^{7}
	7	1.6102×10^{3}	6.6166×10^{6}
Positive (NMC)	1	6.5792×10^{1}	4.9020×10^{1}
	2	1.0672×10^{2}	1.0953×10^{2}
	3	3.5123×10^{4}	3.4427×10^{4}
	4	2.4261×10^{5}	2.3800×10^{5}

3.D.7 *Electrolyte lithium content* $\bar{q}_{\mathrm{e}}$

The definitions of $\bar{q}_{\mathrm{e}}$ and cell capacity Q are:

$$\bar{q}_{\mathrm{e}}^{\mathrm{r}} = \frac{\varepsilon_{\mathrm{e}}^{\mathrm{r}} c_{\mathrm{e},0} A L^{\mathrm{r}} F}{3600(1-t_{+}^{0})} \,[\mathrm{Ah}]$$

$$Q = \varepsilon_{\mathrm{s}}^{\mathrm{r}} A L^{\mathrm{r}} F c_{\mathrm{s,max}}^{\mathrm{r}} \left|\theta_{100}^{\mathrm{r}} - \theta_{0}^{\mathrm{r}}\right| /3600\,[\mathrm{Ah}].$$

For the two electrodes ($\mathrm{r} \in \{\mathrm{n},\mathrm{p}\}$), we can write the following:

$$\begin{aligned}\frac{\bar{q}_{\mathrm{e}}^{\mathrm{r}}}{Q} &= \frac{\varepsilon_{\mathrm{e}}^{\mathrm{r}} c_{\mathrm{e},0}}{\varepsilon_{\mathrm{s}}^{\mathrm{r}} c_{\mathrm{s,max}}^{\mathrm{r}} \left(1-t_{+}^{0}\right)\left|\theta_{100}^{\mathrm{r}}-\theta_{0}^{\mathrm{r}}\right|} \\ &= \left(\frac{\varepsilon_{\mathrm{e}}^{\mathrm{r}}}{\varepsilon_{\mathrm{s}}^{\mathrm{r}}}\right)\left(\frac{c_{\mathrm{e},0}}{c_{\mathrm{s,max}}^{\mathrm{r}}}\right)\left(\frac{1}{1-t_{+}^{0}}\right)\left(\frac{1}{\left|\theta_{100}^{\mathrm{r}}-\theta_{0}^{\mathrm{r}}\right|}\right).\end{aligned}$$

We can use approximate knowledge from the literature regarding common values of these ratios to initialize $\bar{q}_e^r$. For example, we assume that the ratio between ε_e^r and ε_s^r is greater than 2/7 but less than 3/4; the ratio between $c_{e,0}$ and $c_{s,max}^r$ is greater than 1/50 but less than 1/10; we have previously decided $1 < (1 - t_+^0)^{-1} < 2$; the absolute difference of θ_{100}^r and θ_0^r should be anywhere between 0.7 to 1,[92] which leads to $1 < |\theta_{100}^r - \theta_0^r|^{-1} < 10/7$. Of course, other assumptions can be made, but with these, $Q/175 < \bar{q}_e^r < 3Q/14$ for $r \in \{n, p\}$. An initial value can be selected within the constraints; for example, $\bar{q}_{e,0}^r = Q/100$ for $r \in \{n, p\}$.

Further, in the separator ($r \in \{s\}$):

$$\frac{\bar{q}_e^s}{\bar{q}_e^n} = \left(\frac{\varepsilon_e^s}{\varepsilon_e^n}\right)\left(\frac{L^s}{L^n}\right), \quad \text{and} \quad \frac{\bar{q}_e^s}{\bar{q}_e^n} = \left(\frac{\varepsilon_e^s}{\varepsilon_e^p}\right)\left(\frac{L^s}{L^p}\right).$$

Previously from the $\bar{\kappa}$ initialization, we have assigned $1 < \varepsilon_e^s/\varepsilon_e^n < 2$, $1 < \varepsilon_e^p/\varepsilon_e^s < 2$, $1/4 < L^s/L^n < 1$, and $1/4 < L^s/L^p < 1$. Therefore, $\bar{q}_e^n/4 < \bar{q}_e^s < 2\bar{q}_e^n$ and $\bar{q}_e^p/4 < \bar{q}_e^s < 2\bar{q}_e^p$.

Once again, we can rewrite the model parameters by introducing a new constant $K_{\bar{q}_e}$, in which we assume $\bar{q}_e^s = K_{\bar{q}_e}\bar{q}_e^n$. Consequently, the constraints are transferred to $1/4 < K_{\bar{q}_e} < 2$. An initial value can be selected within the constraints (e.g., $K_{\bar{q}_e,0} = 1$).

[92] For a graphite electrode, it is most likely that $\theta_0 < 0.05$ while $0.8 < \theta_{100} < 0.95$. For a NMC electrode, θ_0 can be close to 1 while $0.1 < \theta_{100} < 0.35$.

3.D.8 *Lumped solid diffusivity $\bar{D}_{s,ref}$*

Initial guesses for solid diffusivity are set to the approximate values produced by the discharge test, which employs a generalized single-particle model to fit the slow-rate discharge data to calibrate positive-electrode OCP and to provide an initial estimate for $\bar{D}_{s,ref}$. These estimates are generally within a factor of two or so from the truth, so the bounds we use during optimization are set to a slightly looser range: $\bar{D}_{s,ref,0}/5 < \bar{D}_{s,ref,0} < 5\bar{D}_{s,ref,0}$.

3.D.9 *CPE exponents n_f and n_{dl}*

We simply initialize these parameters to 0.95 and bound them to be between 0.9 and 1.

3.E GDRT algorithm

In this chapter, we adopt the GDRT framework proposed by Danzer[93] and implement it in the computational framework proposed by Wan et al.[94] Estimating the DRT function is an optimization problem, where the cost function is expressed as Eq. (3.29), where λ is a user-selected regularization parameter, and the elements in $\mathbf{M}$ are proportional to the norm of the first derivative of the $\gamma(\ln \tau)$ function.

[93] Michael A. Danzer, "Generalized distribution of relaxation times analysis for the characterization of impedance spectra," *Batteries*, 5(3):53, 2019.

[94] Ting Hei Wan, Mattia Saccoccio, Chi Chen, and Francesco Ciucci, "Influence of the discretization methods on the distribution of relaxation times deconvolution: Implementing radial basis functions with DRTtools," *Electrochimica Acta*, 184:483–499, 2015.

Mathematical details for deriving $\mathbf{A}'$, $\mathbf{A}''$, and $\mathbf{M}$ are presented in the article by Wan et al. as well as Saccoccio et al.,[95] and we encourage the reader to confer their work if interested in the DRT implementation. In the following, we will briefly summarize how their expressions are implemented with RL components. We seek to estimate vector $\mathbf{x}$, presented in Eq. (3.30), where x^{RC} are the distribution functions for RC components, x^{RL} are the distribution functions for RL components, subscript M indicates the total number of elements.

[95] Mattia Saccoccio, Ting Hei Wan, Chi Chen, and Francesco Ciucci, "Optimal regularization in distribution of relaxation times applied to electrochemical impedance spectroscopy: Ridge and lasso regression methods—a theoretical and experimental study," *Electrochimica Acta*, 147:470–482, 2014.

System matrices $\mathbf{A}'$ and $\mathbf{A}''$ are reformed accordingly:

$$(\mathbf{A}')^{\mathrm{RC}}_{nm} = +\int_0^{\infty} \frac{1}{1+4\pi^2 e^{2(y^{\mathrm{RC}}+\ln f_n - \ln f_m)}} \phi_\mu(|y^{\mathrm{RC}}|)\, y^{\mathrm{RC}}$$

$$(\mathbf{A}')^{\mathrm{RL}}_{nm} = +\int_0^{\infty} \frac{\pi^2 e^{2(y^{\mathrm{RL}}+\ln f_n - \ln f_m)}}{1+4\pi^2 e^{2(y^{\mathrm{RL}}+\ln f_n - \ln f_m)}} \phi_\mu(|y^{\mathrm{RL}}|)\, \mathrm{d}y^{\mathrm{RL}}$$

$$(\mathbf{A}'')^{\mathrm{RC}}_{nm} = -\int_0^{\infty} \frac{2\pi e^{(y^{\mathrm{RC}}+\ln f_n - \ln f_m)}}{1+4\pi^2 e^{2(y^{\mathrm{RC}}+\ln f_n - \ln f_m)}} \phi_\mu(|y^{\mathrm{RC}}|)\, \mathrm{d}y^{\mathrm{RC}}$$

$$(\mathbf{A}'')^{\mathrm{RL}}_{nm} = +\int_0^{\infty} \frac{2\pi e^{(y^{\mathrm{RL}}+\ln f_n - \ln f_m)}}{1+4\pi^2 e^{2(y^{\mathrm{RL}}+\ln f_n - \ln f_m)}} \phi_\mu(|y^{\mathrm{RL}}|)\, \mathrm{d}y^{\mathrm{RL}}.$$

The elements in system matrices $\mathbf{M}$ remain unchanged from their original definitions, but its size is extended accordingly:

$$(\mathbf{M})_{lm} = \int_{-\infty}^{+\infty} \frac{\mathrm{d}\phi_\mu(|\ln\tau - \ln\tau_l|)}{\mathrm{d}\ln\tau} \cdot \frac{\mathrm{d}\phi_\mu(|\ln\tau - \ln\tau_m|)}{\mathrm{d}\ln\tau}\, \mathrm{d}\ln\tau.$$

Overall,

$$\mathbf{A}' = \begin{bmatrix} \mathbf{1} & \mathbf{0} & \mathbf{0} & (\mathbf{A}')^{\mathrm{RC}}_{nm} & (\mathbf{A}')^{\mathrm{RL}}_{nm} \end{bmatrix}$$
$$\mathbf{A}'' = \begin{bmatrix} \mathbf{0} & \mathbf{2\pi f} & (-\mathbf{2\pi f})^{-1} & (\mathbf{A}'')^{\mathrm{RC}}_{nm} & (\mathbf{A}'')^{\mathrm{RL}}_{nm} \end{bmatrix}$$
$$\mathbf{M} = \begin{bmatrix} [0] & [0] \\ [0] & \begin{bmatrix} (\mathbf{M})_{lm} & \mathbf{0} \\ \mathbf{0} & (\mathbf{M})_{lm} \end{bmatrix} \end{bmatrix},$$

where $\mathbf{1}$ and $\mathbf{0}$ are column vectors in $\mathbf{A}'$ and $\mathbf{A}''$. The $\mathbf{2\pi f}$ term represents the column vector with elements $2\pi f_n$,[96] and the $(-\mathbf{2\pi f})^{-1}$ term is analogous. For clarification, we let the number of τ_{RL} be M, number of τ_{RC} also be M, and assume the number of input frequency samples to be N. Therefore, vector $\mathbf{x}$ has a size of $2M+3$ by 1, both $\mathbf{A}'$ and $\mathbf{A}''$ have sizes of N by $2M+3$, and the matrix $\mathbf{M}$ has size of $2M+3$ by $2M+3$.

[96] That is,

$$\mathbf{2\pi f} = \begin{bmatrix} 2\pi f_1 \\ 2\pi f_2 \\ \vdots \\ 2\pi f_N \end{bmatrix}.$$

4
Efficient Time-Domain Simulation

Our progress to date can again be visualized using the roadmap of Fig. 4.1. We have eliminated unidentifiable parameters from the model equations, yielding a lumped-parameter version of the original DFN model. We have improved that model with an enhanced description of the solid/electrolyte interface and have derived TFs for all cell internal electrochemical variables and a full-cell impedance model. We have also developed laboratory and data-processing procedures to estimate all model parameter values to describe commercial cells. The outcome of these steps is a physics-based LPM. This is still a full-order model (FOM), comprising a coupled set of partial-differential, ordinary-differential, and algebraic equations.

Battery-cell models serve a range of purposes, and it is important to select an appropriate model for the intended use. FOMs can have prohibitively high computational complexity for direct application to controls in a battery-management system. So we turn our attention to exploring reduced order models (ROMs) that are computationally simple yet contain sufficient electrochemical information to instruct advanced control and estimation algorithms. There are a variety of methods that generate ROMs from FOMs; they are nontrivial and most are beyond the scope of this volume.[1] Broadly speaking (adopting the taxonomy of Li et al.), model-order reduction methods can

[1] For a survey of model-order-reduction techniques see: Yang Li, Dulmini Ralahamilage, Mahinda Vilathgamuwa, Yateendra Mishra, Troy Farrell, San Shing Choi, and Changfu Zou, "Model order reduction techniques for physics-based lithium-ion battery management: A survey," *IEEE Industrial Electronics Magazine*, 2021. Another good reference is: Guodong Fan, Ke Pan, and Marcello Canova, "A comparison of model order reduction techniques for electrochemical characterization of lithium-ion batteries," in *2015 54th IEEE Conference on Decision and Control (CDC)*. IEEE, 2015, pp. 3922–3931.

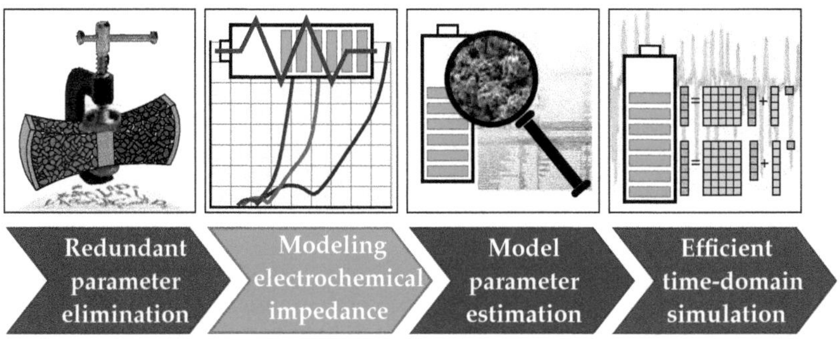

Figure 4.1: Topics in lithium-ion cell modeling that we cover in this volume.

be classified into four categories based on: (1) spatial discretization, (2) function approximation, (3) frequency-domain approximation, and (4) simplified physics. Each method exhibits its own set of advantages and disadvantages with respect to computational ease and the overall accuracy of the resultant ROM. For example, methods that rely on fine discretization of the PDEs can readily capture the nonlinear behavior of the underlying electrochemistry and thus deliver highly accurate results but become computationally heavy with closely spaced grid points. Conversely, ROMs based on simplified physics (e.g., single-particle models) are computationally very compact, but give up model accuracy to gain a speed advantage.

In this chapter, we will present a process for creating ROMs that employs a function approximation generated using subspace-projection methods. The technique spans several of the categories outlined above in that it begins with linearized frequency-domain TFs and effectively converts these into discrete-time state-space representations. The resulting state-space models can be used to compute all cell internal variables (and voltage) in the time domain. We make no claim here that this method is best by any measure, but our experience has shown that the resultant reduced-order state-space systems package important electrochemical dynamics economically into a very-low-order computationally light structure. One notable disadvantage, however, is that the ROMs are generated by a numeric procedure, not an analytic process. As such, they do not retain in their final form an explicit dependence on the values of specific physical parameters. This latter point becomes important when addressing model adaptivity due to cell aging effects—the ROMs must be regenerated from the TFs using a modified set of parameter values rather than simply adjusted based on how those parameter values may have changed over time.

The overall goal of this chapter is to show how to create and use these ROMs to simulate battery cells with either current, voltage, or power as the input signal, and also how to simulate battery packs comprising multiple cells connected in series and/or in parallel. Our first objective is to convert the TFs from Chap. 2, combined with the parameter values estimated in Chap. 3, into discrete-time state-space models of the form:

$$x[k+1] = Ax[k] + Bu[k] \tag{4.1}$$
$$y[k] = Cx[k] + Du[k], \tag{4.2}$$

where the state vector of the system is represented by $x[k] \in \mathbb{R}^{n\times 1}$, the input vector by $u[k] \in \mathbb{R}^{m\times 1}$, and the output vector by $y[k] \in \mathbb{R}^{q\times 1}$.[2,3]

When we apply this type of model to describe lithium-ion cells,

[2] Note that model matrices have dimensions: $A \in \mathbb{R}^{n\times n}$, $B \in \mathbb{R}^{n\times m}$, $C \in \mathbb{R}^{q\times n}$, and $D \in \mathbb{R}^{q\times m}$.

[3] By the end of this chapter, we augment the system with an integrator state, which has the overall effect of replacing n with $n+1$ in these dimensions.

$\boldsymbol{u}[k]$ corresponds to the electrical current applied to the cell (i.e., $\boldsymbol{u}[k] = i_{\text{app}}[k]$ and $m = 1$)[4] and $\boldsymbol{y}[k]$ corresponds to the set of electrochemical variables that we would like to predict using the model. The state vector $\boldsymbol{x}[k]$ will be defined numerically by the model-order reduction technique that we will describe.

[4] Note that cell temperature and other sensed quantities might also be considered as inputs to the model. However, in this chapter we address the temperature input in a different way, as described in Sect. 4.8.

Vol. I of this series proposed an approach to creating ROMs via the discrete-time realization algorithm (DRA). We have found that the DRA is effective, but that it has some practical limitations:

- For accurate steady-state (low-frequency) results, the DRA needed to be provided with long-duration unit-pulse responses as its input. These require substantial computer-memory resources when stored in the Hankel matrices used by the DRA.
- Computing the singular-value decomposition (SVD) of these Hankel matrices required significant CPU resources to process.
- The DRA also requires that all TFs be stable or stabilized.

More recent methods promise to reduce or remove these limitations.[5] Our research team first created the continuous-time realization algorithm (CRA), which processes the continuous-time TF outputs directly—without intermediate conversion to a discrete-time unit pulse response—to create a discrete-time state-space model. Since the CRA works in the frequency domain, it is able to model the high- and low-frequency dynamics of a system well without the long-duration time-domain unit-pulse responses required by the DRA. Because of this, the CRA needs far less computer memory and far less computation than the DRA. However, one step of the CRA requires inverting a Vandermonde matrix, which has very low numeric robustness. Double-precision floating-point rounding errors often cause the CRA to give poor solutions.

[5] For more detail regarding the continuous-time realization algorithm (CRA) and the Lagrange-interpolated realization algorithm (LRA), see: Albert Rodríguez, Gregory L. Plett, and M. Scott Trimboli, "Comparing four model-order reduction techniques, applied to lithium-ion battery-cell internal electrochemical transfer functions," *eTransportation*, 1:100009, 2019.

Our team then developed a method similar to the CRA, but which works with discrete-time frequency responses instead of continuous-time frequency responses. We call this method the hybrid realization algorithm (HRA) since it adopts a discrete-time component from the DRA and a continuous-time component from the CRA. The HRA converts a vector TF $\boldsymbol{G}(s)$ to discrete-time frequency response $\boldsymbol{G}(e^{j\omega T})$ and then uses a subspace system-identification method to compute a high- but finite-order discrete-time ROM. A standard balanced model-order reduction method is then used to arrive at a final low-order discrete-time ROM. As with the CRA, the HRA requires much less memory and computation than the DRA. This is our preferred method at this point, and the one we present in this chapter.

Both the CRA and HRA use the same ROM-generation framework as the DRA, illustrated in Fig. 4.2. The PDE models are converted to TFs, as derived in Chap. 2. These TFs are merged with parameter values estimated using methods presented in Chap. 3 to describe the dynamics of a commercial cell. The fully parameterized TFs are then input either to the DRA, CRA, or HRA (collectively, xRA). The output of this process is a linear discrete-time state-space model.

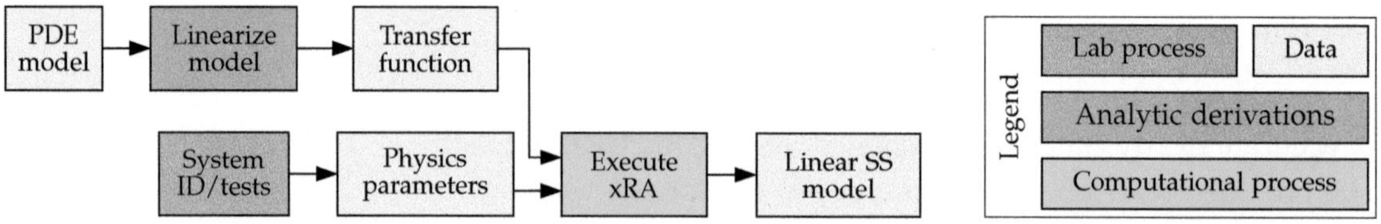

Figure 4.2: ROM-generation framework, based on xRA methods.

The models are then simulated to produce time-domain predictions, as illustrated in Fig. 4.3. Multiple linear state-space models—created around different SOC and temperature setpoints—are evaluated simultaneously to provide independent predictions of cell internal variables. The outputs of these models are blended and nonlinear corrections are applied. The final result is a set of predictions of internal cell electrochemical variables plus cell voltage for the present cell SOC and temperature.

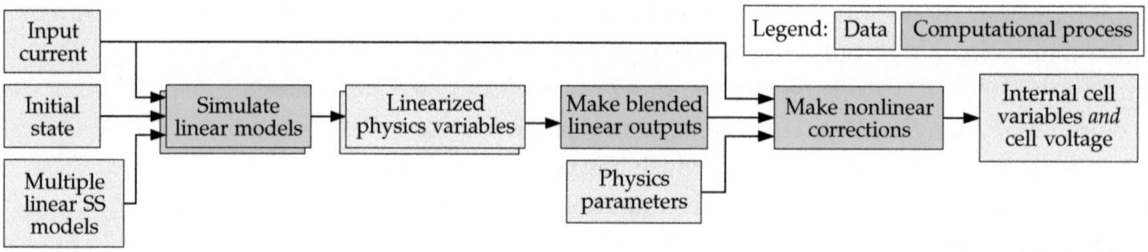

Figure 4.3: Simulating xRA models to produce time-domain predictions.

In this chapter, we first look at how to convert a generic frequency response from continuous-time to discrete-time. This is useful for validating the discrete-time ROMs produced by the xRAs. Next, we derive the HRA and show some results illustrating its ability to capture the frequency-domain dynamics of an electrochemical-variable TF. Then, we demonstrate how a model produced by the HRA can be used to simulate a cell in the time domain and validate the method against FOM simulations. Finally, we see how to implement a time-domain simulation of battery packs comprising multiple cells wired in parallel and/or in series.

4.1 *Convert continuous- to discrete-time frequency response*

The DRA from Vol. I is a method that converts continuous-time TFs into discrete-time state-space ROMs. When simulating a cell's response to a profile of input current versus time, the output of the DRA will not match the output of the FOM exactly, because:

- The DRA produces a ROM that is based on a linear approximation to the nonlinear dynamics of the cell;
- The DRA approximates the true infinite-order behavior of a cell with a model that has a small finite number of poles.[6]

When we wrote Vol. I, we did not have a good way to determine how much of any observed ROM error should be attributed to reducing the model order and how much to the linearization process.

Note that the TFs that are used as input to the DRA were derived by linearizing the FOM equations. If we were able to show that the ROM somehow has good agreement with these TFs, then we could infer that any time-domain prediction error must be due to the linearization process itself leading to the TFs.[7] However, if the DRA ROM does not match the TFs well, then we should focus on adjusting the tuning parameters of the DRA (e.g., unit-pulse-response signal length or number of poles) until it does.

The challenge is that there is no obvious way to compare the ROM to the TFs that we started with. We do know that continuous-time TFs may be converted readily into continuous-time frequency responses via $G(j\omega) = G(s)|_{s=j\omega}$. There are also well-established ways to find the discrete-time frequency response of the ROM. The issue is that continuous-time and discrete-time frequency responses of the same system are not themselves the same and cannot be compared directly.

This problem would be solved if there were a way to convert the continuous-time TFs into discrete-time frequency responses that we could compare directly to the discrete-time frequency responses of the ROM. Again, we run into a problem. There are well-established ways to convert continuous-time TFs that are of a rational-polynomial form in the Laplace variable s into discrete-time frequency responses, but there were no existing methods of which we were aware to convert the kind of transcendental TFs that describe lithium-ion cell models. We needed a new method.

Here, we present a numeric method to convert transcendental TFs (and indeed, rational-polynomial TFs also) into discrete-time frequency responses. The output is then directly comparable to the discrete-time frequency response of the ROM, so we can use this approach to determine whether the ROM is producing a good ap-

[6] Recall that a pole of a TF is any value of $s = \sigma + j\omega$ for which the magnitude of the TF approaches infinity. Practically speaking, the real part σ represents a time constant of the dynamics of the system it is modeling; the imaginary part represents a tendency of the system to oscillate at frequency ω.

[7] This linearization error is unavoidable using TF methods since TFs exist only for linear-time-invariant systems, although nonlinear corrections to the ROM outputs can improve time-domain predictions.

proximation to the already-linearized TFs. Even more important for this chapter, however, is that we can use the discrete-time frequency responses produced by this numeric method in a new xRA to generate ROMs more efficiently than by using the DRA.

The method depends on some background concepts relating to sampled-data systems. Fig. 4.4 illustrates components that are relevant to our discussion. In the figure, a discrete-time system (e.g., a computer) produces the discrete-time signal $u[k]$,[8] which we assume originated as continuous-time $u(t)$, sampled every Δt seconds. This sampled signal $u[k]$ is passed through a digital-to-analog converter incorporating a zero-order hold (ZOH) circuit to produce continuous-time $\bar{u}(t)$. This analog signal is input to a continuous-time system having TF $G(s)$, producing continuous-time output signal $y(t)$. Finally, $y(t)$ is sampled every Δt seconds to produce discrete-time $y[k]$.

[8] Square brackets denote a discrete-time signal and k is an (integer) discrete-time sampling index; parentheses denote a continuous-time signal. Therefore, $u[k] = u(k\Delta t)$.

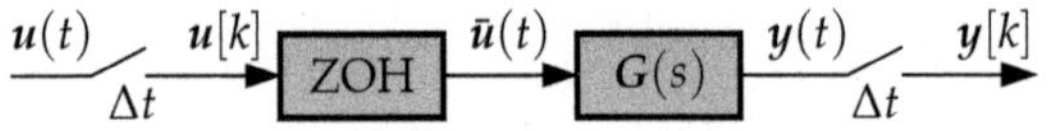

Figure 4.4: Signals relevant to sampled-data systems.

When we analyze this system, we can combine continuous-time elements $G(s)$ and the ZOH, which has known TF $(1-\exp(-s\Delta t))/s$, into a continuous-time subsystem having TF $\bar{G}(s)$:

$$\bar{G}(s) = \frac{1-\exp(-s\Delta t)}{s} G(s). \tag{4.3}$$

Then we find the continuous-time frequency response of this subsystem by substituting $s = j\omega$:

$$\bar{G}(j\omega) = \frac{1-\exp(-j\omega\Delta t)}{j\omega} G(j\omega). \tag{4.4}$$

The sampled-data system in the figure comprises both discrete-time and continuous-time elements. It is customary to use the discrete-time Fourier or the z-transform to analyze discrete-time systems and the continuous-time Fourier or Laplace transforms to analyze continuous-time systems. To analyze hybrid sampled-data systems (containing both continuous-time and discrete-time elements) using a common framework, we can use the starred Laplace transform.[9]

[9] Eliahu I. Jury, "Analysis and synthesis of sampled-data control systems," *Transactions of the American Institute of Electrical Engineers, Part I: Communication and Electronics*, 73(4):332–346, 1954.

To do so, note that a continuous-time representation of the sampled signal $u[k] = u(k\Delta t)$ is $u(t)\delta_{\Delta t}(t)$, where $\delta_{\Delta t}(t)$ is the Dirac comb function (also known as an impulse train) having period Δt:

$$\delta_{\Delta t}(t) = \sum_{k=0}^{\infty} \delta(t - k\Delta t),$$

where $\delta(t)$ is the Dirac impulse function.[10] The equivalent Laplace-

[10] The Dirac impulse function was introduced in Vol. I, Chap. 4. You may wish to review those details for deeper insight into how it is being used here.

domain representation of the sampled signal is then:

$$U^*(s) = \mathcal{L}\{u(t)\delta_{\Delta t}(t)\} = \sum_{k=0}^{\infty} u[k]e^{-sk\Delta t}.$$

The asterisk "*" symbol indicates that this is a starred Laplace transform. It works in much the same way as a standard Laplace transform, but describes sampled signals with a continuous-time transform. With this notation, it can be shown that $Y(s) = \bar{G}(s)U^*(s)$ and that $Y^*(s) = \bar{G}^*(s)U^*(s)$.

The starred Laplace transform is related to the standard Laplace transform via:

$$\bar{G}^*(s) = \frac{1}{\Delta t}\sum_{k=-\infty}^{\infty} \bar{G}\left(s + j\frac{2\pi}{\Delta t}k\right) + \frac{\bar{g}(0)}{2}. \tag{4.5}$$

The summand for $k = 0$ corresponds to the original frequency spectrum, which is copied directly into $\bar{G}^*(s)$. The summands for $k \neq 0$ comprise aliases of the original frequency spectrum that are introduced by the sampling operation. While the summation limits go from $-\infty$ to ∞ in principle, good approximations to $\bar{G}^*(s)$ are achieved using far fewer terms, for k between $-k_{\max}$ and $k_{\max}$, as we don't expect much aliasing if we have sampled rapidly enough.[11]

The value for $\bar{g}(0)$ in Eq. (4.5) can be found via the Laplace initial-value theorem as:

$$\bar{g}(0) = \lim_{s\to\infty} s\bar{G}(s) = \lim_{s\to\infty}(1 - e^{-s\Delta t})G(s) = \lim_{\omega\to\infty} G(j\omega).$$

This is also the value of the D matrix for our state-space models. Note that this term can be computed analytically in closed form from our TF models, and so it is possible for us to subtract it from $\bar{G}(j\omega)$ before converting from continuous-time to discrete-time and add it back in later. Therefore, for simplicity, we will assume that $\bar{g}(0) = 0$ in Eq. (4.5) without loss of generality.

The starred Laplace transform can also be related to the z-transform as $Y(z) = Y^*(s)|_{s=\ln(z)/\Delta t}$. Therefore, we have:

$$Y(z) = \underbrace{\left[\frac{1}{\Delta t}\sum_{k=-\infty}^{\infty} \bar{G}\left(\frac{\ln(z)}{\Delta t} + j\frac{2\pi}{\Delta t}k\right)\right]}_{G(z)} U(z).$$

The discrete-time frequency response that we desire to compute is $G(e^{j\omega\Delta t}) = G(z)|_{z=\exp(j\omega\Delta t)}$. Therefore,

$$G(e^{j\omega\Delta t}) = \frac{1}{\Delta t}\sum_{k=-k_{\max}}^{k_{\max}} \bar{G}\left(j\omega + j\frac{2\pi}{\Delta t}k\right). \tag{4.6}$$

[11] We can compute a finite value for $k_{\max}$ by looking at the range of the arguments to $\bar{G}(j\omega)$ that result in nonzero values. We assume that the frequency response $G(j\omega)$ is bandlimited to (has nonzero values only for) frequencies below $\omega = 2\pi f_{\max}$. The largest corresponding value of k occurs when:

$$2\pi f_{\max} = -2\pi f_{\max} + \frac{2\pi}{\Delta t}k_{\max}.$$

Rearranging gives:

$$k_{\max} = 2f_{\max}\Delta t.$$

While this analytic result is true, it can overestimate the practically necessary value of $k_{\max}$, depending on how strict we are in interpreting the term "bandlimited." We can often use smaller values of $k_{\max}$ with no observable degradation in performance.

Eq. (4.6) is our desired result. It shows how to compute a discrete-time frequency response from a continuous-time TF. To be clear, $G(e^{j\omega\Delta t})$ on the equation's left-hand side is the discrete-time frequency response of the sampled-data system that we wish to approximate; $\tilde{G}(j\omega)$ on the right-hand side is a known continuous-time frequency response calculated by Eq. (4.4). The result is valid for frequencies up to the Nyquist rate: $-\pi/\Delta t < \omega < \pi/\Delta t$.

4.2 *Illustrating frequency-response conversion*

The math for converting between a continuous-time TF and a discrete-time frequency response has now been derived. Summarizing, the following procedure implements Eq. (4.6):

1. Compute $\tilde{G}(s)$ from $G(s)$ using Eq. (4.3).
2. Compute $D = \tilde{g}(0) = \lim_{s\to\infty} G(s)$ and replace $\tilde{G}(s) := \tilde{G}(s) - D$.
3. Find continuous-time frequency response $\tilde{G}(j\omega)$ using Eq. (4.4).
4. For some $k_{\max}$, approximate discrete-time frequency response $\widehat{G}(e^{j\omega\Delta t})$ at N_ω output frequency points using Eq. (4.6).
5. Add back high-frequency gain, replacing $\widehat{G}(e^{j\omega\Delta t}) := \widehat{G}(e^{j\omega\Delta t}) + D$.

The new method can convert general continuous-time TFs to discrete-time frequency responses, unlike conventional methods that can be used only with finite-order rational-polynomial (or state-space) models. However, we choose to demonstrate the method first on a finite-order rational-polynomial TF, since we can then compare its solution to one obtained from these standard methods.

In particular, we choose (for sampling period $\Delta t = 0.1$ s):

$$G(s) = \frac{0.1s^2 + s + 1}{s^2 + 3s + 1}, \quad \text{so} \quad G(j\omega) = \frac{1 + j\omega - 0.1\omega^2}{1 + 3j\omega - \omega^2}.$$

For this simple case, we can find the exact equivalent discrete-time system using MATLAB's `c2d.m` function to be:

$$G(e^{j\omega\Delta t}) = \frac{0.1e^{j0.2\omega} - 0.1088e^{j0.1\omega} + 0.0174}{e^{j0.2\omega} - 1.732e^{j0.1\omega} + 0.7408}.$$

For illustration, we choose $k_{\max} = 5$ and a discrete-time frequency vector comprising $\omega = 0$ and 50 additional points spaced evenly on a logarithmic scale between 10^{-5} and 10^{0} times the Nyquist rate.

Fig. 4.5 compares the magnitude and phase responses of the proposed method to the exact result. A visually perfect match has been achieved, as hoped, showing that the method is accurate. We also note that the discrete-time and continuous-time frequency responses for this system are different from each other, as expected.

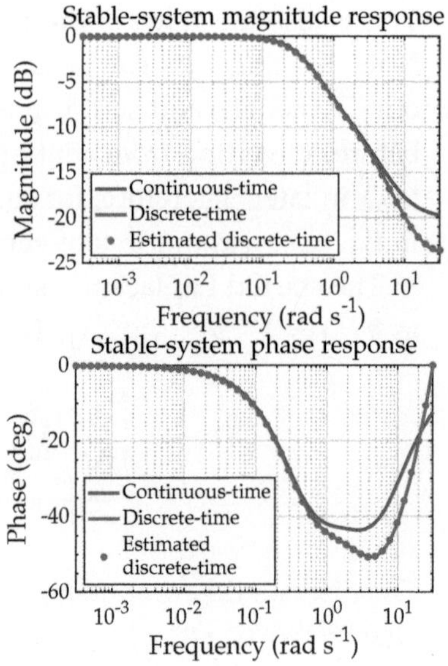

Figure 4.5: Illustrating the method to convert between continuous-time TFs and discrete-time frequency responses for a stable system.

This method also works for unstable systems. Consider:

$$G(s) = \frac{s^2 + 10s + 1}{10s^2 + 0.1s},$$

again for $\Delta t = 0.1$ s. Fig. 4.6 compares the magnitude and phase responses of the proposed method and the exact result. Again, we see visually perfect matches between the method's output and the desired result, giving confidence that it is working as hoped.

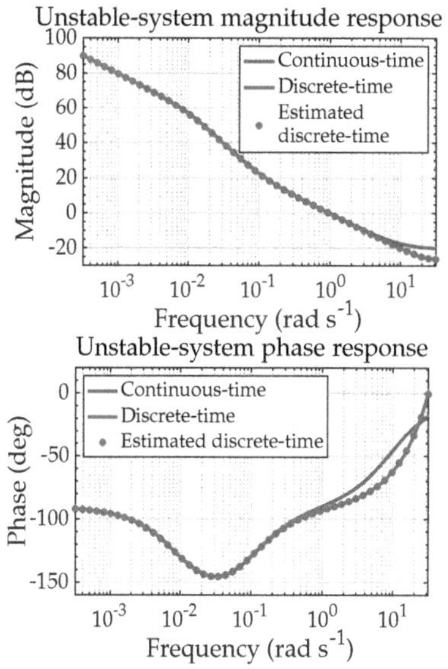

Figure 4.6: Illustrating the method to convert between continuous-time TFs and discrete-time frequency responses for an unstable system.

4.3 *The hybrid realization algorithm (HRA)*

We find that this new method is very helpful when seeking to understand how well the DRA (or any other xRA) approximates the TFs we derived in Chap. 2. However, it is also a critical step in the development of an alternate realization algorithm, the HRA.

Battery cells exhibit important features across frequencies spanning multiple decades. In order for the DRA to capture high-frequency dynamics well, its internal sampling period T_1 must be very short. To capture low-frequency dynamics well, the duration of the unit-pulse responses incorporated in its Hankel matrix must be long. This combination forces the Hankel matrix to become enormous, requiring that the computer running the DRA have more memory than is common; it also makes evaluating the SVD of this matrix very slow.[12] The eigensystem-realization algorithm (ERA) modification to the DRA proposed in Vol. I helps with both these concerns, but it remains an open question how to select its parameters well.

[12] Some clever ways for speeding up the DRA have been proposed in: Krishnakumar Gopalakrishnan, Teng Zhang, and Gregory J. Offer, "A fast, memory-efficient discrete-time realization algorithm for reduced-order Li-ion battery models," *Journal of Electrochemical Energy Conversion and Storage*, 14(1), 2017.

Here, we propose a different approach. We define a new realization algorithm that works directly in the frequency domain. This allows us to use frequency-response samples as its input instead of time-domain samples. We can choose a relatively small number of samples over a wide bandwidth, capturing both low- and high-frequency behaviors well. This overcomes the primary fundamental limitations of the DRA.

Our objective is to find a discrete-time state-space ROM having the form of Eqs. (4.1)–(4.2). In this section, we will show how the HRA determines the model's $\boldsymbol{A}$ matrix from the discrete-time frequency responses of the electrochemical variables, computed using method of Sect. 4.1. The derivation begins by taking z-transforms of the state-space model in Eqs. (4.1)–(4.2):[13]

[13] We assume that $x[0] = 0$, as is standard when computing TFs from difference equations. TFs describe the zero-state response of a system.

$$\begin{aligned} z\boldsymbol{X}(z) &= \boldsymbol{A}\boldsymbol{X}(z) + \boldsymbol{B}\boldsymbol{U}(z) \\ \boldsymbol{Y}(z) &= \boldsymbol{C}\boldsymbol{X}(z) + \boldsymbol{D}\boldsymbol{U}(z). \end{aligned}$$

Multiplying the output equation by z produces:

$$z\boldsymbol{Y}(z) = \boldsymbol{C}(z\boldsymbol{X}(z)) + \boldsymbol{D}(z\boldsymbol{U}(z))$$

$$= \boldsymbol{CAX}(z) + \boldsymbol{CBU}(z) + \boldsymbol{D}(z\boldsymbol{U}(z)).$$

Repeating this process $i-1$ times (where i is termed the oversizing parameter) and combining the i modified versions of the output equation into block matrices gives:

$$\begin{bmatrix} \boldsymbol{Y}(z) \\ z\boldsymbol{Y}(z) \\ \vdots \\ z^{i-1}\boldsymbol{Y}(z) \end{bmatrix} = \underbrace{\begin{bmatrix} \boldsymbol{C} \\ \boldsymbol{CA} \\ \vdots \\ \boldsymbol{CA}^{i-1} \end{bmatrix}}_{\mathcal{O}_i} \boldsymbol{X}(z) + \underbrace{\begin{bmatrix} \boldsymbol{D} & & & \\ \boldsymbol{CB} & \boldsymbol{D} & & \\ \vdots & \ddots & \ddots & \\ \boldsymbol{CA}^{i-2}\boldsymbol{B} & \cdots & \boldsymbol{CB} & \boldsymbol{D} \end{bmatrix}}_{\mathcal{T}_i} \begin{bmatrix} \boldsymbol{U}(z) \\ z\boldsymbol{U}(z) \\ \vdots \\ z^{i-1}\boldsymbol{U}(z) \end{bmatrix}, \tag{4.7}$$

where $\mathcal{T}_i$ has a block-Toeplitz form and $\mathcal{O}_i$ is called the extended observability matrix.

Now, suppose that a complex sinusoid of frequency ω_0 (where ω_0 is a constant) is applied to the jth system input:[14]

$$\boldsymbol{u}_j[k] = \mathbf{e}_j \exp(j\omega_0 k\Delta t).$$

We can write the z-transform of this input as:

$$\boldsymbol{U}_j(z) = \frac{z}{z - \exp(j\omega_0\Delta t)}\mathbf{e}_j.$$

The steady-state output corresponding to this input is, by definition, computed as the product between this input and the input/output frequency response $\boldsymbol{G}_j(e^{j\omega\Delta t})$ for frequency ω_0 and input j:

$$\boldsymbol{Y}(z) = \boldsymbol{G}_j(e^{j\omega_0\Delta t})\boldsymbol{U}_j(z).$$

Similarly, the steady-state version of the state vector can be related to the input/state frequency response $\boldsymbol{X}_j(e^{j\omega_0\Delta t})$ for frequency ω_0 and input j, which we write as:

$$\boldsymbol{X}(z) = \boldsymbol{X}_j(e^{j\omega_0\Delta t})\boldsymbol{U}_j(z).$$

That is, the steady-state state vector is computed as the input/state frequency response $\boldsymbol{X}_j(e^{j\omega_0\Delta t})$ multiplying the input.

We insert these steady-state relationships into Eq. (4.7):

$$\begin{bmatrix} \boldsymbol{G}_j(e^{j\omega_0 T_s})\boldsymbol{U}_j(z) \\ z\boldsymbol{G}_j(e^{j\omega_0 T_s})\boldsymbol{U}_j(z) \\ \vdots \\ z^{i-1}\boldsymbol{G}_j(e^{j\omega_0 T_s})\boldsymbol{U}_j(z) \end{bmatrix} = \mathcal{O}_i\boldsymbol{X}_j(e^{j\omega_0 T_s})\boldsymbol{U}_j(z) + \mathcal{T}_i \begin{bmatrix} \boldsymbol{U}_j(z) \\ z\boldsymbol{U}_j(z) \\ \vdots \\ z^{i-1}\boldsymbol{U}_j(z) \end{bmatrix}. \tag{4.8}$$

[14] In this notation, $\mathbf{e}_j$ is the standard-basis vector j. For example,

$$\mathbf{e}_1 = \begin{bmatrix} 1 \\ 0 \\ \vdots \\ 0 \end{bmatrix}, \quad \mathbf{e}_2 = \begin{bmatrix} 0 \\ 1 \\ \vdots \\ 0 \end{bmatrix},$$

and so forth.

Notice that $z/(z - \exp(j\omega_0 \Delta t))$ is common to all terms, and so we can cancel it from this final result. Also note that we can write similar equations for $1 \leq j \leq m$, and when we do so we can stack the results horizontally. That is, we define:

$$\begin{aligned} \boldsymbol{U}(e^{j\omega_0 T_s}) &= \left[\begin{array}{c:c:c} \boldsymbol{U}_1(e^{j\omega_0 T_s}) & \cdots & \boldsymbol{U}_m(e^{j\omega_0 T_s}) \end{array} \right] \\ \boldsymbol{G}(e^{j\omega_0 T_s}) &= \left[\begin{array}{c:c:c} \boldsymbol{G}_1(e^{j\omega_0 T_s}) & \cdots & \boldsymbol{G}_m(e^{j\omega_0 T_s}) \end{array} \right]. \end{aligned}$$

Then we note that $\boldsymbol{U}(e^{j\omega_0 \Delta t})$ divided by $z/(z - \exp(j\omega_0 \Delta t))$ is the identity matrix $\boldsymbol{I}_m$. With these definitions, we evaluate Eq. (4.8) at $z = \exp(j\omega_0 \Delta t)$, which gives:

$$\underbrace{\begin{bmatrix} \boldsymbol{G}(e^{j\omega_0 T_s}) \\ \hdashline e^{j\omega_0 T_s} \boldsymbol{G}(e^{j\omega_0 T_s}) \\ \hdashline \vdots \\ \hdashline e^{j\omega_0 (i-1) T_s} \boldsymbol{G}(e^{j\omega_0 T_s}) \end{bmatrix}}_{\mathcal{G}^c(\omega_0)} = \mathcal{O}_i \mathcal{X}^c(\omega_0) + \mathcal{T}_i \underbrace{\begin{bmatrix} \boldsymbol{I}_m \\ \hdashline e^{j\omega_0 T_s} \boldsymbol{I}_m \\ \hdashline \vdots \\ \hdashline e^{j\omega_0 (i-1) T_s} \boldsymbol{I}_m \end{bmatrix}}_{\mathcal{U}^c(\omega_0)}, \tag{4.9}$$

where

$$\mathcal{X}^c(\omega_0) = \left[\begin{array}{c:c:c:c} \boldsymbol{X}_1(e^{j\omega_0 T_s}) & \boldsymbol{X}_2(e^{j\omega_0 T_s}) & \cdots & \boldsymbol{X}_m(e^{j\omega_0 T_s}) \end{array} \right].$$

In this notation, we have also implicitly defined block matrices $\mathcal{G}^c(\omega_0)$ and $\mathcal{U}^c(\omega_0)$, where the superscript "c" indicates that these are complex-valued matrices.

We can further evaluate the terms of this equation at multiple frequency points $\omega_0 = \omega_1 \ldots \omega_N$, giving a combined representation:

$$\begin{aligned} \underbrace{\left[\begin{array}{c:c:c} \mathcal{G}^c(\omega_1) & \cdots & \mathcal{G}^c(\omega_N) \end{array} \right]}_{\mathcal{G}^c} = \mathcal{O}_i & \underbrace{\left[\begin{array}{c:c:c} \mathcal{X}^c(\omega_1) & \cdots & \mathcal{X}^c(\omega_N) \end{array} \right]}_{\mathcal{X}^c} \\ & + \mathcal{T}_i \underbrace{\left[\begin{array}{c:c:c} \mathcal{U}^c(\omega_1) & \cdots & \mathcal{U}^c(\omega_N) \end{array} \right]}_{\mathcal{U}^c}. \end{aligned} \tag{4.10}$$

A final step yields matrices containing only real values:

$$\begin{aligned} \underbrace{\left[\begin{array}{c:c} \mathrm{real}(\mathcal{G}^c) & \mathrm{imag}(\mathcal{G}^c) \end{array} \right]}_{\mathcal{G}^r} = \mathcal{O}_i & \underbrace{\left[\begin{array}{c:c} \mathrm{real}(\mathcal{X}^c) & \mathrm{imag}(\mathcal{X}^c) \end{array} \right]}_{\mathcal{X}^r} \\ & + \mathcal{T}_i \underbrace{\left[\begin{array}{c:c} \mathrm{real}(\mathcal{U}^c) & \mathrm{imag}(\mathcal{U}^c) \end{array} \right]}_{\mathcal{U}^r}. \end{aligned} \tag{4.11}$$

We use values computed via the method from Sect. 4.1 to populate $\mathcal{G}^r$. $\mathcal{U}^r$ can be computed directly from the vector of frequency points $\{\omega_1 \ldots \omega_N\}$ chosen to map between the TF and ROM. $\mathcal{X}^r$ is unknown but we will not require knowledge of its values. We then perform the LQ decomposition:[15]

$$\underbrace{\begin{bmatrix} \mathcal{U}^{\mathrm{r}} \\ \hdashline \mathcal{G}^{\mathrm{r}} \end{bmatrix}}_{\text{data matrix}} = LQ = \begin{bmatrix} L_{11} & 0 \\ \hdashline L_{21} & L_{22} \end{bmatrix} \begin{bmatrix} Q_1 \\ \hdashline Q_2 \end{bmatrix}, \tag{4.12}$$

where L has the same dimension as "data matrix," and where Q_1 and Q_2 are orthonormal.[16] Then,

$$\begin{aligned} \mathcal{U}^{\mathrm{r}} &= L_{11}Q_1 \\ \mathcal{G}^{\mathrm{r}} &= L_{21}Q_1 + L_{22}Q_2. \end{aligned}$$

Multiplying this last equation on the right by Q_2^T gives $\mathcal{G}^{\mathrm{r}}Q_2^T = L_{22}$. Multiplying Eq. (4.11) on the right by Q_2^T and substituting gives:

$$\mathcal{G}^{\mathrm{r}}Q_2^T = L_{22} = \mathcal{O}_i \mathcal{X}^{\mathrm{r}} Q_2^T.$$

The net effect of the LQ decomposition has been to project Eq. (4.11) onto a subspace orthogonal to the input $\mathcal{U}^{\mathrm{r}}$, thus removing the effect of $\mathcal{U}^{\mathrm{r}}$ from Eq. (4.11) and removing the need to know $\mathcal{T}_i$. The resulting equation states that L_{22} has the same column space as $\mathcal{O}_i$.

To recover $\mathcal{O}_i$ (up to an unknown but irrelevant similarity transformation) from L_{22}, we compute the SVD of L_{22}:

$$L_{22} = \begin{bmatrix} U_1 & \vdots & U_2 \end{bmatrix} \begin{bmatrix} \Sigma_1 & 0 \\ \hdashline 0 & \Sigma_2 \end{bmatrix} \begin{bmatrix} V_1^T \\ \hdashline V_2^T \end{bmatrix}.$$

L_{22} can be well approximated using the n_1 most significant singular values, those stored in Σ_1 (Σ is partitioned so $\Sigma_2 \approx 0$). We can then estimate the extended observability matrix as $\widehat{\mathcal{O}}_i = U_1\Sigma_1^{1/2}$. The shift property of the extended observability matrix allows computing A as the least-squares solution to:

$$A = (\widehat{\mathcal{O}}_i^{\downarrow})^{\dagger}\widehat{\mathcal{O}}_i^{\uparrow},$$

where $\dagger$ is the matrix pseudoinverse operation and the $\uparrow$ and $\downarrow$ symbols denote shifting the indicated matrix up or down.[17]

We find that the HRA is sensitive to model order n_1 and produces better models with large n_1 than with small n_1. So, we begin with order $n_1 > n$ to obtain an initial state-space model that closely approximates the infinite-order frequency response defined by the TFs. Then, we perform balanced model-order reduction (e.g., using MATLAB's `balred`) to obtain final state-space ROM having order n. We expect A to have real, stable poles, which is not guaranteed by the method as presented so far. So, we replace unstable poles in A by their reciprocals and complex poles by their magnitudes.[18]

[15] The LQ decomposition is one method to decompose a general matrix into factors. It is different from but related in concept to the QR and SVD matrix factorizing methods. Deriving the LQ decomposition is beyond the scope of our discussion here, but note that it is a standard function found in matrix toolboxes. It is built into MATLAB, for example.

[16] That is:

$$\begin{aligned} Q_1Q_1^T &= I, \\ Q_2Q_2^T &= I, \\ Q_1Q_2^T &= 0, \text{ and} \\ Q_2Q_1^T &= 0. \end{aligned}$$

[17] This effectively deletes its top or bottom block row:

$$\widehat{\mathcal{O}}_i^{\uparrow} \approx \mathcal{O}_i^{\uparrow} = \begin{bmatrix} CA \\ \hdashline CA^2 \\ \hdashline CA^3 \\ \hdashline \vdots \\ \hdashline CA^{i-1} \end{bmatrix}, \quad \text{and}$$

$$\widehat{\mathcal{O}}_i^{\downarrow} \approx \mathcal{O}_i^{\downarrow} = \begin{bmatrix} C \\ \hdashline CA \\ \hdashline CA^2 \\ \hdashline \vdots \\ \hdashline CA^{i-2} \end{bmatrix}.$$

[18] This substitution maintains the magnitude response exactly, but modifies the phase response of the ROM. We have attempted other more complicated modifications per: Daniel N. Miller and Raymond A. De Callafon, "Subspace identification with eigenvalue constraints," *Automatica*, 49(8):2468–2473, 2013. However, we find that this simple substitution works just about as well.

4.3.1 *Summarizing the method to determine* $\boldsymbol{A}$ *using the HRA*

We have now derived the steps needed to find the $\boldsymbol{A}$ matrix of the ROM. For convenience, we summarize those steps here.

1. Select a set of frequencies $\{\omega_1 \ldots \omega_N\}$ at which we want the final ROM to match the TFs well.
2. For each of these frequencies, compute $G(e^{j\omega_k \Delta t})$ from $G(j\omega_k)$ using the method of Sect. 4.1.
3. Extend each of these frequency responses to form $\mathcal{G}^{\text{c}}(\omega_k)$ and $\mathcal{U}^{\text{c}}(\omega_k)$ per Eq. (4.9).
4. Form the complex data matrices $\mathcal{G}^{\text{c}}$ and $\mathcal{U}^{\text{c}}$ per Eq. (4.10).
5. Form the real data matrices $\mathcal{G}^{\text{r}}$ and $\mathcal{U}^{\text{r}}$ per Eq. (4.11).
6. Find $\boldsymbol{L}_{22}$ via the LQ decomposition of Eq. (4.12).
7. Compute the SVD of $\boldsymbol{L}_{22}$ to give $\boldsymbol{U}_1$ and $\boldsymbol{\Sigma}_1$. At this point, the model order $n_1 > n$ corresponding to the size of $\boldsymbol{\Sigma}_1$ is selected such that we accurately approximate the TFs.
8. Compute $\widehat{\mathcal{O}}_i = \boldsymbol{U}_1 \boldsymbol{\Sigma}_1^{1/2}$.
9. Compute $\boldsymbol{A} = (\widehat{\mathcal{O}}_i^{\downarrow})^{\dagger} \widehat{\mathcal{O}}_i^{\uparrow}$.
10. Diagonalize $\boldsymbol{A}$ and replace unstable poles by their reciprocals and complex poles by their magnitudes.
11. Find $\boldsymbol{B}$ and $\boldsymbol{C}$ as described in the following section ($\boldsymbol{D}$ is found analytically).
12. Perform balanced model-order reduction to obtain a discrete-time state-space model having desired final model order n.
13. Recompute $\boldsymbol{C}$ as described in the following section for the nth-order model.

4.4 *Final form of* $\boldsymbol{A}$, $\boldsymbol{B}$, $\boldsymbol{C}$, *and* $\boldsymbol{D}$

We have now solved for the $\boldsymbol{A}$ matrix of the ROM; we must still compute $\boldsymbol{B}$, $\boldsymbol{C}$, and $\boldsymbol{D}$. We quickly look at how to do so here.

4.4.1 *Solving for the state-space* $\boldsymbol{B}$ *matrix*

State-space models have multiple degrees of freedom for choosing some parameter values, after which the remaining values will be fixed. For simplicity of the final implementation, we choose to assign $\boldsymbol{B} = \boldsymbol{1}_{n\times 1}$ without loss of generality, since the $\boldsymbol{C}$ matrix will be computed to scale outputs properly for this choice.

4.4.2 *Solving for the state-space* $\boldsymbol{C}$ *matrix*

When finding $\boldsymbol{C}$, we desire to minimize the 2-norm of errors between the actual and approximate frequency responses while forcing the dc gain of the ROM to match the dc gain of the TFs. To do so, we form a constrained-optimization scalar cost function to minimize:

$$J = \sum_{k=1}^{N} \|G(\omega_k) - H(\omega_k)\|_2^2 + \lambda^T \left(G(0) - H(0)\right),$$

where $G(\omega_k)$ is the actual discrete-time frequency response computed from the TF and $H(\omega_k)$ is its ROM approximation.[19]

For notational simplicity, we remove the D term (known analytically) and write $G(\omega_k) = G(e^{j\omega_k \Delta t}) - D$ and $M(\omega_k) = (e^{j\omega_k \Delta t} I - A)^{-1} B$. Then, $H(\omega_k) = CM(\omega_k)$ and:[20]

$$\begin{aligned} J &= \left\{ \sum_{k=1}^{N} \left[G(\omega_k) - H(\omega_k)\right]^* \left[G(\omega_k) - H(\omega_k)\right] \right\} + \lambda^T \left(G(0) - H(0)\right) \\ &= \left\{ \sum_{k=1}^{N} \operatorname{Tr}\left[G(\omega_k) G^*(\omega_k)\right] - \operatorname{Tr}\left[G(\omega_k) M^*(\omega_k) C^T\right] \right. \\ &\quad \left. - \operatorname{Tr}\left[CM(\omega_k) G^*(\omega_k)\right] + \operatorname{Tr}\left[CM(\omega_k) M^*(\omega_k) C^T\right] \right\} \\ &\quad + \operatorname{Tr}\left[\left(G(0) - CM(0)\right) \lambda^T\right]. \end{aligned}$$

Taking derivatives of J with respect to λ and C gives:

$$\frac{\partial J}{\partial \lambda} = G(0) - CM(0) = 0$$

$$\frac{\partial J}{\partial C} = -2\underbrace{\sum_{k=1}^{N} \mathbb{R}[G(\omega_k) M^*(\omega_k)]}_{M_1} + 2C\underbrace{\sum_{k=1}^{N} \mathbb{R}[M(\omega_k) M^*(\omega_k)]}_{M_2} - \lambda M^T(0) = 0.$$

Transposing and combining these two relationships into one matrix equation, we have:

$$\begin{bmatrix} 2M_2^T & -M(0) \\ M^T(0) & 0^T \end{bmatrix} \begin{bmatrix} C^T \\ \lambda^T \end{bmatrix} = \begin{bmatrix} 2M_1^T \\ G^T(0) \end{bmatrix}.$$

We solve this for C (and λ, which is not used) using least squares:

$$\begin{bmatrix} C^T \\ \lambda^T \end{bmatrix} = \begin{bmatrix} 2M_2^T & -M(0) \\ M^T(0) & 0^T \end{bmatrix}^{\dagger} \begin{bmatrix} 2M_1^T \\ G^T(0) \end{bmatrix}.$$

4.4.3 *Solving for the state-space **D** matrix*

The final model unknown is D, which describes instantaneous change in output $y[k]$ when the system is excited with input $u[k]$. For a PBM, the D matrix can be found in closed form since the electrochemical TFs are analytical (see Sect. 2.D). That is, we compute:

$$D = \left[\lim_{s\to\infty} G_1(s), \lim_{s\to\infty} G_2(s), \ldots, \lim_{s\to\infty} G_q(s)\right]^T.$$

[19] Note that for a discrete-time state-space model,

$$H(\omega_k) = C(e^{j\omega_k \Delta t} I - A)^{-1} B + D.$$

[20] We use the facts: (1) $m = \operatorname{Tr}[m]$ if m is a scalar and (2) $\operatorname{Tr}(ABC) = \operatorname{Tr}(BCA) = \operatorname{Tr}(CAB)$. This allows us to use theorems for the derivative of the trace of a quadratic matrix form with respect to a matrix to solve the optimization. Some steps are omitted here, but we make use of the following identities (if X is a real-valued matrix and A_1, A_2, and A_3 are complex-valued matrices of compatible size):

$$\frac{\partial \operatorname{Tr}[XA_1]}{\partial X} = A_1^T, \quad \frac{\partial \operatorname{Tr}[A_2 X^T]}{\partial X} = A_2,$$

$$\frac{\partial \operatorname{Tr}[XA_3X^T]}{\partial X} = X(A_3 + A_3^T).$$

We also make use of the fact that if A_4 is a square complex-valued matrix, then $A_4 + (A_4^*)^T = 2\mathbb{R}[A_4]$.

4.4.4 *Handling integration dynamics*

The DRA introduced in Vol. I cannot work directly with TFs having integration dynamics. In principle, the HRA *could* be made to do so, but the method just presented for calculating C would fail since it requires a finite dc gain. So, we choose to follow the same approach taken with the DRA:

- Compute the integrator residue as $\mathbf{res}^* = \lim_{s\to 0} sG(s)$.
- Remove the integrator by defining $[G(s)]^* = G(s) - \mathbf{res}^*/s$.
- Invoke the HRA to find a discrete-time state-space model of order n for integrator-removed $[G(s)]^*$.
- Finally, augment the identified model with the integration dynamics that were removed previously. That is, we write:

$$\mathfrak{A} = \left[\begin{array}{c|c} A & \mathbf{0} \\ \hline \mathbf{0}^T & 1 \end{array}\right], \quad \mathfrak{B} = \left[\begin{array}{c} B \\ \hline 1 \end{array}\right], \quad \mathfrak{C} = \left[\begin{array}{c|c} C & \mathbf{res}^* \end{array}\right], \tag{4.13}$$

  and $\mathfrak{D} = D$ is unchanged. The augmented model is of order $n+1$.
- Using this augmented system, we can write:[21]

$$\underbrace{\left[\begin{array}{c} x[k+1] \\ \hline x_0[k+1] \end{array}\right]}_{\mathfrak{X}[k+1]} = \underbrace{\left[\begin{array}{c|c} A & \mathbf{0} \\ \hline \mathbf{0}^T & 1 \end{array}\right]}_{\mathfrak{A}} \underbrace{\left[\begin{array}{c} x[k] \\ \hline x_0[k] \end{array}\right]}_{\mathfrak{X}[k]} + \underbrace{\left[\begin{array}{c} B \\ \hline 1 \end{array}\right]}_{\mathfrak{B}} u[k]$$

$$y[k] = \underbrace{\left[\begin{array}{c|c} C & \mathbf{res}^* \end{array}\right]}_{\mathfrak{C}} \underbrace{\left[\begin{array}{c} x[k] \\ \hline x_0[k] \end{array}\right]}_{\mathfrak{X}[k]} + \mathfrak{D}u[k].$$

  or

$$\mathfrak{X}[k+1] = \mathfrak{A}\mathfrak{X}[k] + \mathfrak{B}u[k] \tag{4.14}$$

$$y[k] = \mathfrak{C}\mathfrak{X}[k] + \mathfrak{D}u[k]. \tag{4.15}$$

Note that this first step of removing integration dynamics is common to all of the xRAs, but since integrator residues are given in closed form in Sect. 2.D, this is not a limitation.

[21] Note that we now make a notational distinction between state vectors for models having an integrator, denoted as $\mathfrak{X}[k]$, and models without integrator, denoted as $x[k]$. We do not make a notational distinction between outputs from these two types of systems, choosing to write both as $y[k]$ since they both define the same variables. However, they are technically distinct since $y[k]$ in Eq. (4.2) describes variables without integration dynamics and $y[k]$ in Eq. (4.15) describes the same variables but with integration dynamics. The context in which $y[k]$ is used should clarify how it should be interpreted.

4.5 *Sample HRA results*

We briefly examine some HRA-method results for the NMC30 cell with the double layer removed. In this example, we seek three discrete-time ROMs having 3...5 states to match $I^{\mathrm{p}}_{\mathrm{f+dl}}(\tilde{x}, s)/I_{\mathrm{app}}(s)$ over $\tilde{x} = 2 : 0.2 : 3$. The discrete-time sampling period was chosen to be $\Delta t = 1$.

The HRA was tuned by hand; better tuning might be possible, but these results are representative. The oversizing parameter was chosen to be $i = 10$. The method used $n_1 = 50$ initial poles. The discrete-time frequency vector for building $G(z)$ was chosen to comprise $\omega = 0$ and 99 additional points spaced evenly on a logarithmic scale between 10^{-4} and 10^0 times the Nyquist rate. The value of $k_{\max} = 50$ was chosen to estimate the discrete-time frequency response from the continuous-time infinite-order TFs.[22]

Actual TF and ROM discrete-time frequency-responses are shown in Fig. 4.7. We desire for the ROM discrete-time frequency responses (drawn with circle markers) to match the true discrete-time frequency responses (drawn with solid lines), computed directly from the TFs using the method from Sect. 4.1. The first row of the figure shows matching between the magnitude and phase responses when the ROM has size $n = 3$; the second row shows matching for $n = 4$; and the third row shows matching for $n = 5$.

The matches for $n = 3$ are perhaps surprisingly good considering that an infinite number of poles are required to match the true frequency response exactly. When model order is increased to $n = 4$, the differences between the ROM and TF become very small; when $n = 5$, it becomes nearly impossible to see the differences by eye. RMS frequency-response-matching errors for several cases are summarized in Table 4.1.

[22] The average time for MATLAB to execute the HRA for this example was 7.5 ms, which is much, much faster than the minutes or hours required by the DRA.

Table 4.1: RMS frequency-response errors between TF and ROM.

Poles	RMSE
$n = 3$	2.037×10^{-2}
$n = 4$	7.699×10^{-3}
$n = 5$	7.368×10^{-3}
$n = 6$	7.259×10^{-3}

4.6 *Simulating a cell in the time domain, near a setpoint*

We have now computed a discrete-time state-space ROM that approximates the TFs of a cell populated with parameter values corresponding to a specific SOC and temperature setpoint. Our next step is to use this model to simulate cell dynamics near that setpoint.

Simulation of the basic ROM form presented in Eqs. (4.14)–(4.15) is straightforward. We initialize the state $\mathfrak{X}[0] = 0$. Then, we iterate the equations in order to compute linearized variables $\boldsymbol{y}[k]$.

This can be implemented very simply in a "for" loop in MATLAB. Suppose that we are given the model $\mathfrak{A}$, $\mathfrak{B}$, $\mathfrak{C}$, and $\mathfrak{D}$ matrices, as well as a profile of current versus time stored in "ik." Then we can simulate the state update and output update:

```
x = zeros(n,1);             % Initialize x[0] to the correct size
for k = 0:length(ik)-1,     % MATLAB indexing: ik(1) = iapp[0]
  y = C*x + D*ik(k+1);      % The linearized outputs at this step, y[k]
  x = A*x + B*ik(k+1);      % The state for next time step, x[k+1]
end
```

Of course, we will want to do something with output "y," which holds linearized predictions of the internal variables chosen to simulate. To compute accurate predictions of the nonlinear variables we desire to model, we will also need to apply nonlinear corrections.

And to predict cell voltage, we will need to combine some variables in a nonlinear way. In the following pages, we see how to do both.

4.6.1 *Update cell state of charge $z[k]$*

First, we look at how to update electrode and cell states of charge. If the cell model has no double layer, then the electrode state of charge (i.e., average solid stoichiometry) can be computed as:

$$\theta^{\mathrm{r}}_{\mathrm{s,avg}}[k] = \theta^{\mathrm{r}}_{\mathrm{s,0}} + \tilde{\theta}^{\mathrm{res}^{\mathrm{r}}_0}_{\mathrm{ss}} x_0[k], \tag{4.16}$$

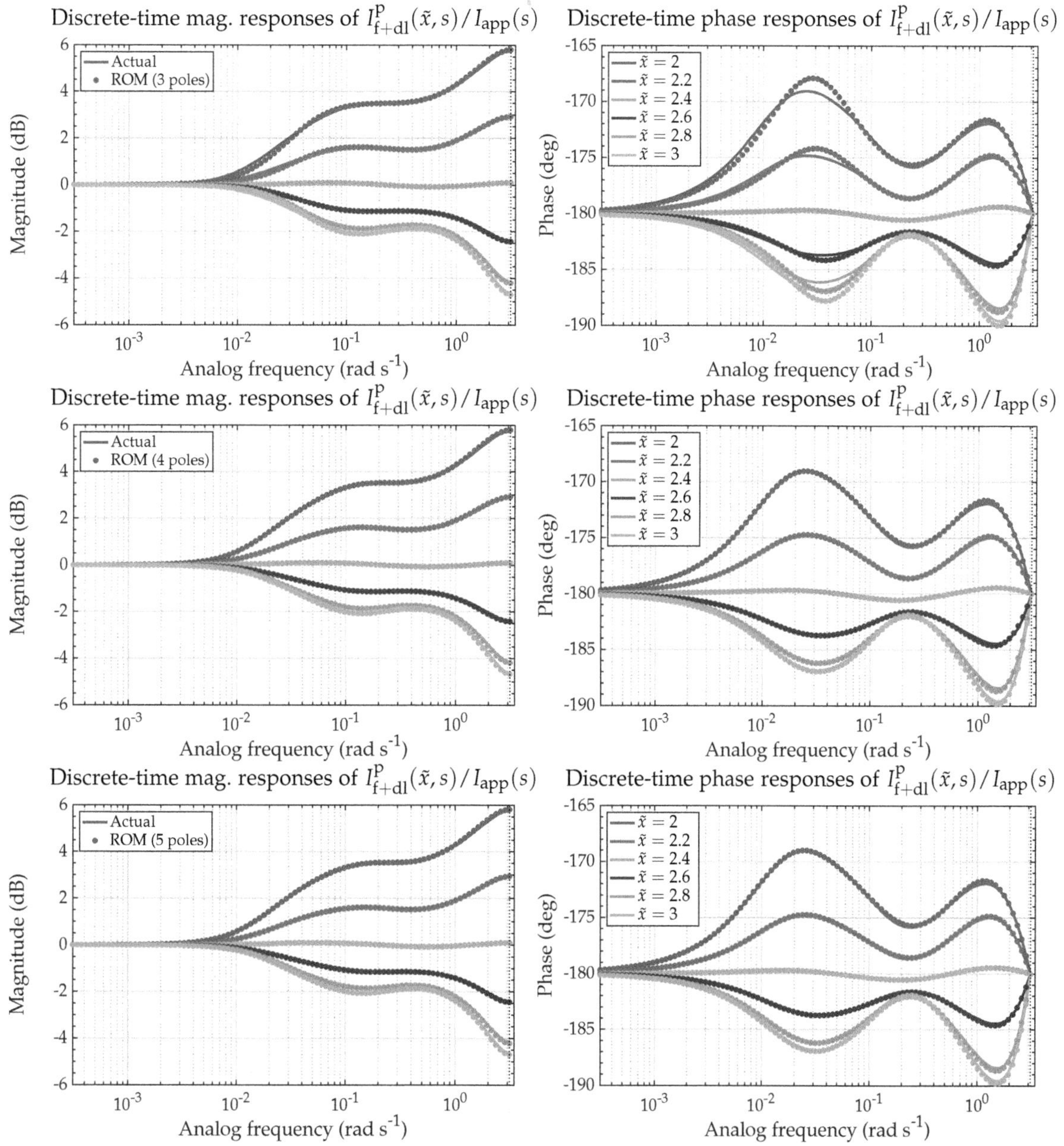

Figure 4.7: Frequency-response matching between actual TF (solid lines) and ROM (circle markers) for different model sizes.

where $\theta^{\mathrm{r}}_{\mathrm{s},0}$ is the initial stoichiometry, $x_0[k]$ represents the integrator state, and $\tilde{\theta}^{\mathrm{res}^{\mathrm{r}}_0}_{\mathrm{ss}}$ is the integrator residue for the $\widetilde{\Theta}^{\mathrm{r}}_{\mathrm{s}}(\tilde{x},s)/I_{\mathrm{app}}(s)$ TF. The initial stoichiometry is determined from cell SOC as:

$$\theta^{\mathrm{r}}_{\mathrm{s},0} = \theta^{\mathrm{r}}_0 + z_0\left(\theta^{\mathrm{r}}_{100} - \theta^{\mathrm{r}}_0\right). \tag{4.17}$$

The integrator residues are the following constants (see Sect. 2.D):

$$\tilde{\theta}^{\mathrm{res}^{\mathrm{n}}_0}_{\mathrm{ss}} = \frac{-\left|\theta^{\mathrm{n}}_{100} - \theta^{\mathrm{n}}_0\right|}{3600Q} \quad \text{and} \quad \tilde{\theta}^{\mathrm{res}^{\mathrm{p}}_0}_{\mathrm{ss}} = \frac{\left|\theta^{\mathrm{p}}_{100} - \theta^{\mathrm{p}}_0\right|}{3600Q}.$$

If the cell model has a double layer, integrator gains are a function of cell SOC; so, for the most accurate results we must consider a different approach. The integrator residues of $\widetilde{\Theta}^{\mathrm{r}}_{\mathrm{s}}(\tilde{x},s)/I_{\mathrm{app}}(s)$ are:

$$\begin{aligned}
\tilde{\theta}^{\mathrm{res}^{\mathrm{n}}_0}_{\mathrm{ss}} &= \frac{-\left|\theta^{\mathrm{n}}_{100} - \theta^{\mathrm{n}}_0\right|}{3600Q - \bar{C}^{\mathrm{n}}_{\mathrm{dl,eff}}\left|\theta^{\mathrm{n}}_{100} - \theta^{\mathrm{n}}_0\right|[U^{\mathrm{n}}_{\mathrm{ocp}}]'} \\
\tilde{\theta}^{\mathrm{res}^{\mathrm{p}}_0}_{\mathrm{ss}} &= \frac{\left|\theta^{\mathrm{p}}_{100} - \theta^{\mathrm{p}}_0\right|}{3600Q - \bar{C}^{\mathrm{p}}_{\mathrm{dl,eff}}\left|\theta^{\mathrm{p}}_{100} - \theta^{\mathrm{p}}_0\right|[U^{\mathrm{p}}_{\mathrm{ocp}}]'}.
\end{aligned}$$

Since $[U^{\mathrm{r}}_{\mathrm{ocp}}]'$ is a function of electrode SOC, we need to update electrode SOC using time-varying gains.

```
thetaavgn = thetas0n; % initialize stoichiometry in negative
thetaavgp = thetas0p; % initialize stoichiometry in positive
dQn = abs(theta100n - theta0n);
dQp = abs(theta100p - theta0p);

% Start main simulation loop
for k = 0:length(ik)-1, % MATLAB indexing: ik(1) = ik[0]
  res0n = -dQn/(F*Q-Cdleffn*dQn*dUocpn(thetaavgn));
  res0p =  dQp/(F*Q-Cdleffp*dQp*dUocpp(thetaavgp));
  thetaavgn = thetaavgn + res0n*ik(k+1);
  thetaavgp = thetaavgp + res0p*ik(k+1);
end
```

Next, cell-level SOC can now be predicted as:

$$z[k] = \frac{\theta^{\mathrm{n}}_{\mathrm{s,avg}}[k] - \theta^{\mathrm{n}}_0}{\theta^{\mathrm{n}}_{100} - \theta^{\mathrm{n}}_0} \quad \text{or} \quad z[k] = \frac{\theta^{\mathrm{p}}_{\mathrm{s,avg}}[k] - \theta^{\mathrm{p}}_0}{\theta^{\mathrm{p}}_{100} - \theta^{\mathrm{p}}_0}.$$

Or, maybe not.

The above code essentially replaces a single integrator with two—one for each electrode—making state estimation a challenge (Chap. 5). Also, the two equations for cell-level SOC may give different answers. Which $z[k]$ computation should be used and how important is it to use an integration residue that is SOC-varying rather than constant?

To address these questions, Fig. 4.8 plots SOC-varying and non-SOC-varying residue magnitudes for the NMC30 cell. Some differences are noticeable at very low negative-electrode stoichiometry,

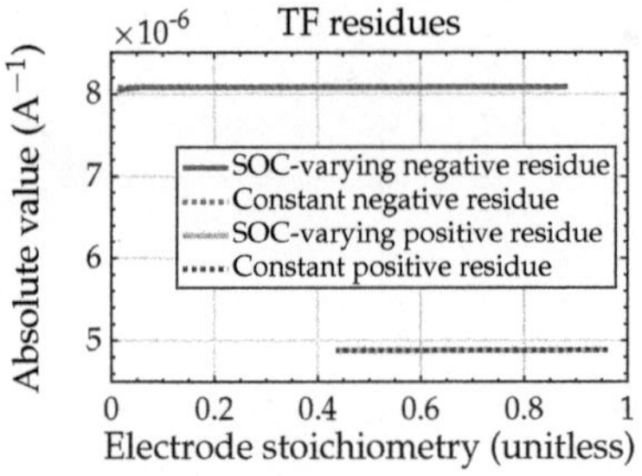

Figure 4.8: Comparing SOC-varying and constant residues.

indicating that SOC-varying residues will give better accuracy in open-loop predictions. But the differences are small. Notice that:

- $3600Q$ for many cells is large, frequently 100 000 or more;
- OCP changes are on the order of 1 V for the full SOC range so $[U^{\mathrm{r}}_{\mathrm{ocp}}]'$ is on the order of 1;
- $|\theta^{\mathrm{r}}_{100} - \theta^{\mathrm{r}}_{0}| < 1$ and $\bar{C}^{\mathrm{r}}_{\mathrm{dl,eff}}$ may be on the order of a few tens of farads (tends to be larger with larger-capacity cells);
- Altogether, $\bar{C}^{\mathrm{r}}_{\mathrm{dl,eff}}\,|\theta^{\mathrm{r}}_{100} - \theta^{\mathrm{r}}_{0}|\,[U^{\mathrm{r}}_{\mathrm{ocp}}]'$ has magnitude on the order of a few hundred, at most;
- So, the denominator term in $\tilde{\theta}^{\mathrm{res}_0,\mathrm{r}}_{\mathrm{ss}}$ is dominated by $3600Q$.

When using measurement feedback to perform state estimation, the adaptivity of the filter is sufficient to overcome errors introduced when using constant residues. So, in Chap. 5, we will use the simpler constant residues, even when the cell model has a double layer.

4.6.2 *Interfacial lithium fluxes*

We continue by considering how to convert ROM linear outputs to nonlinear predictions for every electrochemical variable that can be predicted by the TFs. In the case of the interfacial lithium fluxes, the linear outputs of the ROM for $i_{\mathrm{f+dl}}[\tilde{x},k]$, $i_{\mathrm{f}}[\tilde{x},k]$, and $i_{\mathrm{dl}}[\tilde{x},k]$ are unbiased predictions of the true fluxes. No corrections are needed.

4.6.3 *Solid surface stoichiometry*

The ROM can produce debiased outputs $\tilde{\theta}_{\mathrm{ss}}[\tilde{x},k]$, from which the nonlinear prediction is made via:

$$\theta_{\mathrm{ss}}[\tilde{x},k] = \tilde{\theta}_{\mathrm{ss}}[\tilde{x},k] + \theta^{\mathrm{r}}_{\mathrm{s},0}. \tag{4.18}$$

4.6.4 *Phase potential difference*

According to the original definition, the phase-potential difference variable can be computed as:

$$\phi^{\mathrm{r}}_{\mathrm{s\text{-}e}}[\tilde{x},k] = \tilde{\phi}^{\mathrm{r}}_{\mathrm{s\text{-}e}}[\tilde{x},k] + U^{\mathrm{r}}_{\mathrm{ocp}}(\theta^{\mathrm{r}}_{\mathrm{s},0}),$$

where $\tilde{\phi}^{\mathrm{r}}_{\mathrm{s\text{-}e}}[\tilde{x},k]$ is obtained directly from the system linear output, and the electrode OCP function is evaluated at initial local state of charge that is previously determined from Eq. (4.17).

However, performance can be improved by looking deeper at what is actually happening. The TF $\widetilde{\Phi}^{\mathrm{r}}_{\mathrm{s\text{-}e}}(\tilde{x},s)\,/\,I_{\mathrm{app}}(s)$ has a pole at the origin, which is removed prior to using the HRA to give $[\widetilde{\Phi}^{\mathrm{r}}_{\mathrm{s\text{-}e}}(\tilde{x},s)]^{*}/\,I_{\mathrm{app}}(s)$. If there is no double layer in the model, the integrator response could be added back manually as:

$$\tilde{\phi}^{\mathrm{r}}_{\mathrm{s\text{-}e}}[\tilde{x},k] = [\tilde{\phi}^{\mathrm{r}}_{\mathrm{s\text{-}e}}[\tilde{x},k]]^{*} + \tilde{\phi}^{\mathrm{res0,r}}_{\mathrm{s\text{-}e}} \times x_0[k],$$

where $x_0[k]$ is the integrator state of the state-space model.

However, recall the integrator residues of $\widetilde{\Phi}^{\mathrm{r}}_{\mathrm{s\text{-}e}}(\tilde{x},s)/I_{\mathrm{app}}(s)$,

$$\begin{aligned}\tilde{\phi}^{\mathrm{res}^{\mathrm{n}}_0}_{\mathrm{s\text{-}e}} &= \frac{-\left|\theta^{\mathrm{n}}_{100}-\theta^{\mathrm{n}}_0\right|\,[U^{\mathrm{n}}_{\mathrm{ocp}}]'}{3600Q-\bar{C}^{\mathrm{n}}_{\mathrm{dl,eff}}\left|\theta^{\mathrm{n}}_{100}-\theta^{\mathrm{n}}_0\right|\,[U^{\mathrm{n}}_{\mathrm{ocp}}]'}\\ \tilde{\phi}^{\mathrm{res}^{\mathrm{p}}_0}_{\mathrm{s\text{-}e}} &= \frac{\left|\theta^{\mathrm{p}}_{100}-\theta^{\mathrm{p}}_0\right|\,[U^{\mathrm{p}}_{\mathrm{ocp}}]'}{3600Q-\bar{C}^{\mathrm{p}}_{\mathrm{dl,eff}}\left|\theta^{\mathrm{p}}_{100}-\theta^{\mathrm{p}}_0\right|\,[U^{\mathrm{p}}_{\mathrm{ocp}}]'}.\end{aligned}$$

We notice that the residuals of the two unstable TFs are linked as:

$$\tilde{\phi}^{\mathrm{res}^{\mathrm{r}}_0}_{\mathrm{s\text{-}e}} = [U^{\mathrm{r}}_{\mathrm{ocp}}]' \times \tilde{\theta}^{\mathrm{res}^{\mathrm{r}}_0}_{\mathrm{ss}}.$$

Therefore, we can write:

$$\tilde{\phi}^{\mathrm{r}}_{\mathrm{s\text{-}e}}[\tilde{x},k] = [\tilde{\phi}^{\mathrm{r}}_{\mathrm{s\text{-}e}}[\tilde{x},k]]^* + [U^{\mathrm{r}}_{\mathrm{ocp}}]' \times \tilde{\theta}^{\mathrm{res}^{\mathrm{r}}_0}_{\mathrm{ss}} \times x_0[k],$$

Substituting the relationship from Eq. (4.16) yields:

$$\tilde{\phi}^{\mathrm{r}}_{\mathrm{s\text{-}e}}[\tilde{x},k] = [\tilde{\phi}^{\mathrm{r}}_{\mathrm{s\text{-}e}}[\tilde{x},k]]^* + [U^{\mathrm{r}}_{\mathrm{ocp}}]' \left(\theta^{\mathrm{r}}_{\mathrm{s,avg}}[k]-\theta^{\mathrm{r}}_{\mathrm{s,0}}\right).$$

We recognize the rightmost terms as a linearization of $U^{\mathrm{r}}_{\mathrm{ocp}}[\theta^{\mathrm{r}}_{\mathrm{s,avg}}[k]]$. Therefore, we achieve more accurate results if we compute:

$$\phi^{\mathrm{r}}_{\mathrm{s\text{-}e}}[\tilde{x},k] = [\tilde{\phi}^{\mathrm{r}}_{\mathrm{s\text{-}e}}[\tilde{x},k]]^* + U^{\mathrm{r}}_{\mathrm{ocp}}[\theta^{\mathrm{r}}_{\mathrm{s,avg}}[k]]. \tag{4.19}$$

4.6.5 *Potential in solid*

The ROM can directly compute debiased $\tilde{\phi}^{\mathrm{r}}_{\mathrm{s}}[\tilde{x},k]$. To use the nonlinear prediction, we must add back the term at the current collector:

$$\phi^{\mathrm{n}}_{\mathrm{s}}[\tilde{x},k] = \tilde{\phi}^{\mathrm{n}}_{\mathrm{s}}[\tilde{x},k] + \phi^{\mathrm{n}}_{\mathrm{s}}[0,k] = \tilde{\phi}^{\mathrm{n}}_{\mathrm{s}}[0,k] + 0 \tag{4.20}$$

$$\phi^{\mathrm{p}}_{\mathrm{s}}[\tilde{x},k] = \tilde{\phi}^{\mathrm{p}}_{\mathrm{s}}[\tilde{x},k] + \phi^{\mathrm{p}}_{\mathrm{s}}[3,k] = \tilde{\phi}^{\mathrm{p}}_{\mathrm{s}}[\tilde{x},k] + v_{\mathrm{cell}}[k], \tag{4.21}$$

where $v_{\mathrm{cell}}[k]$ will be computed from Eq. (4.25).

4.6.6 *Potential in electrolyte*

The ROM can directly compute debiased $\tilde{\phi}^{\mathrm{r}}_{\mathrm{e}}[\tilde{x},k]$. To compute the nonlinear prediction, notice:

$$\begin{aligned}\phi^{\mathrm{r}}_{\mathrm{e}}[\tilde{x},k] &= \tilde{\phi}^{\mathrm{r}}_{\mathrm{e}}[\tilde{x},k] + \phi^{\mathrm{n}}_{\mathrm{e}}[0,k] = \tilde{\phi}^{\mathrm{r}}_{\mathrm{e}}[\tilde{x},k] + \underbrace{\phi^{\mathrm{n}}_{\mathrm{s}}[0,k]}_{0} - \phi^{\mathrm{n}}_{\mathrm{s\text{-}e}}[0,k]\\ &= \tilde{\phi}^{\mathrm{r}}_{\mathrm{e}}[\tilde{x},k] - \phi^{\mathrm{n}}_{\mathrm{s\text{-}e}}[0,k],\end{aligned}$$

where $\phi^{\mathrm{n}}_{\mathrm{s\text{-}e}}[0,k]$ is found via Eq. (4.19). As a result,

$$\phi^{\mathrm{r}}_{\mathrm{e}}[\tilde{x},k] = \tilde{\phi}^{\mathrm{r}}_{\mathrm{e}}[\tilde{x},k] - [\tilde{\phi}^{\mathrm{n}}_{\mathrm{s\text{-}e}}[0,k]]^* - U^{\mathrm{n}}_{\mathrm{ocp}}(\theta^{\mathrm{n}}_{\mathrm{s,avg}}[k]). \tag{4.22}$$

4.6.7 *Lithium stoichiometry in electrolyte*

The ROM can directly compute debiased $\tilde{\theta}^{\mathrm{r}}_{\mathrm{e}}[\tilde{x},k]$. The actual electrolyte concentration ratio prediction is found by:

$$\theta^{\mathrm{r}}_{\mathrm{e}}[\tilde{x},k] = \tilde{\theta}^{\mathrm{r}}_{\mathrm{e}}[\tilde{x},k] + \theta_{\mathrm{e},0} = \tilde{\theta}^{\mathrm{r}}_{\mathrm{e}}[\tilde{x},k] + 1, \tag{4.23}$$

where we recall that $\theta_{\mathrm{e},0} = 1$.

4.6.8 *Cell voltage*

The voltage of the cell is computed by combining different electrochemical variables:

$$\begin{aligned} v_{\mathrm{cell}}[k] &= (\eta^{\mathrm{p}}[3,k] - \eta^{\mathrm{n}}[0,k]) + \left(U^{\mathrm{p}}_{\mathrm{ocp}}(\theta^{\mathrm{p}}_{\mathrm{ss}}[3,k]) - U^{\mathrm{n}}_{\mathrm{ocp}}(\theta^{\mathrm{n}}_{\mathrm{ss}}[0,k])\right) \\ &\quad + \left(\phi^{\mathrm{p}}_{\mathrm{e}}[3,k] - \phi^{\mathrm{n}}_{\mathrm{e}}[0,k]\right) + \left(\bar{R}^{\mathrm{p}}_{\mathrm{f}} i^{\mathrm{p}}_{\mathrm{f+dl}}[3,k] - \bar{R}^{\mathrm{n}}_{\mathrm{f}} i^{\mathrm{n}}_{\mathrm{f+dl}}[0,k]\right). \end{aligned}$$

If the MSMR charge-transfer coefficients $\alpha^{\mathrm{n}}_j = \alpha^{\mathrm{p}}_j = 0.5$, the overpotentials η^{r} are computed via inverting $i^{\mathrm{r}}_{\mathrm{f}}$:[23]

$$\begin{aligned} i^{\mathrm{r}}_{\mathrm{f}}[\tilde{x},k] &= \sum_{j=1}^{J^{\mathrm{r}}} i^{\mathrm{r}}_{0,j}[\tilde{x},k]\left\{\exp\left(\frac{f}{2}\eta^{\mathrm{r}}[\tilde{x},k]\right) - \exp\left(\frac{-f}{2}\eta^{\mathrm{r}}[\tilde{x},k]\right)\right\} \\ &= 2\sinh\left(\frac{f}{2}\eta^{\mathrm{r}}[\tilde{x},k]\right)\sum_{j=1}^{J^{\mathrm{r}}} i^{\mathrm{r}}_{0,j}[\tilde{x},k] \\ \eta^{\mathrm{r}}[\tilde{x},k] &= \frac{2}{f}\operatorname{asinh}\left(\frac{i^{\mathrm{r}}_{\mathrm{f}}[\tilde{x},k]}{2\sum_{j=1}^{J^{\mathrm{r}}}\bar{k}^{\mathrm{r}}_{0,j}\sqrt{(X^{\mathrm{r}}_j - x^{\mathrm{r}}_j[\tilde{x},k])^{\omega^{\mathrm{r}}_j}(x^{\mathrm{r}}_j[\tilde{x},k])^{\omega^{\mathrm{r}}_j}\theta^{\mathrm{r}}_{\mathrm{e}}[\tilde{x},k]}}\right). \end{aligned} \tag{4.24}$$

[23] The x^{r}_j are computed by first finding the electrode surface potential based on $\theta^{\mathrm{r}}_{\mathrm{ss}}$ via:

$$U[\tilde{x},k] = U^{\mathrm{r}}_{\mathrm{ocp}}(\theta^{\mathrm{r}}_{\mathrm{ss}}[\tilde{x},k]).$$

Then we compute:

$$x^{\mathrm{r}}_j[\tilde{x},k] = \frac{X^{\mathrm{r}}_j}{1+\exp\left[f(U[\tilde{x},k] - U^{0,\mathrm{r}}_j)/\omega^{\mathrm{r}}_j\right]}.$$

The electrolyte-potential difference between the two current collectors is known to be:

$$\phi^{\mathrm{p}}_{\mathrm{e}}[3,k] - \phi^{\mathrm{n}}_{\mathrm{e}}[0,k] = \tilde{\phi}^{\mathrm{p}}_{\mathrm{e}}[3,k],$$

where $\tilde{\phi}^{\mathrm{p}}_{\mathrm{e}}[3,k]$ is a direct linear output from the ROM.

With all previous definitions, we can finally write cell voltage as:

$$\begin{aligned} v_{\mathrm{cell}}[k] &= (\eta^{\mathrm{p}}[3,k] - \eta^{\mathrm{n}}[0,k]) + \left(U^{\mathrm{p}}_{\mathrm{ocp}}(\theta^{\mathrm{p}}_{\mathrm{ss}}[3,k]) - U^{\mathrm{n}}_{\mathrm{ocp}}(\theta^{\mathrm{n}}_{\mathrm{ss}}[0,k])\right) \\ &\quad + \tilde{\phi}^{\mathrm{p}}_{\mathrm{e}}[3,k] + \left(\bar{R}^{\mathrm{p}}_{\mathrm{f}} i^{\mathrm{p}}_{\mathrm{f+dl}}[3,k] - \bar{R}^{\mathrm{n}}_{\mathrm{f}} i^{\mathrm{n}}_{\mathrm{f+dl}}[0,k]\right), \end{aligned} \tag{4.25}$$

where $\theta^{\mathrm{p}}_{\mathrm{ss}}[3,k]$ and $\theta^{\mathrm{n}}_{\mathrm{ss}}[0,k]$ are evaluated using Eq. (4.18).

4.6.9 *A short summary*

Table 4.2 lists the equation numbers that address nonlinear corrections for each variable. The last column addresses all variables required to compute cell voltage.

Variable	Definition	Correction	Needed for $v_{\text{cell}}[k]$
$i^{\text{r}}_{\text{f+dl}}[\tilde{x},k]$	—	—	$i^{\text{n}}_{\text{f+dl}}[0,k]$, $i^{\text{p}}_{\text{f+dl}}[3,k]$
$i^{\text{r}}_{\text{f}}[\tilde{x},k]$	—	—	$i^{\text{n}}_{\text{f}}[0,k]$, $i^{\text{p}}_{\text{f}}[3,k]$
$i^{\text{r}}_{\text{dl}}[\tilde{x},k]$	—	—	—
$\theta^{\text{r}}_{\text{ss}}[\tilde{x},t]$	$\tilde{\theta}^{\text{r}}_{\text{ss}}[\tilde{x},k]+\theta^{\text{r}}_{\text{s},0}$	Eq. (4.18)	$\tilde{\theta}^{\text{n}}_{\text{ss}}[0,k], \tilde{\theta}^{\text{p}}_{\text{ss}}[3,k]$
$\phi^{\text{r}}_{\text{s-e}}[\tilde{x},k]$	$\tilde{\phi}^{\text{r}}_{\text{s-e}}[\tilde{x},k]+U^{\text{r}}_{\text{ocp}}(\theta^{\text{r}}_{\text{s},0})$	Eq. (4.19)	$[\tilde{\phi}^{\text{n}}_{\text{s-e}}[0,k]]^*$
$\phi^{\text{n}}_{\text{s}}[\tilde{x},k]$	$\tilde{\phi}^{\text{n}}_{\text{s}}[\tilde{x},k]+0$	Eq. (4.20)	—
$\phi^{\text{p}}_{\text{s}}[\tilde{x},k]$	$\tilde{\phi}^{\text{p}}_{\text{s}}[\tilde{x},k]+v_{\text{cell}}[k]$	Eq. (4.21)	—
$\phi^{\text{r}}_{\text{e}}[\tilde{x},k]$	$\tilde{\phi}^{\text{r}}_{\text{e}}[\tilde{x},k]+\phi^{\text{n}}_{\text{e}}[0,k]$	Eq. (4.22)	$\tilde{\phi}^{\text{p}}_{\text{e}}[3,k]$
$\theta^{\text{r}}_{\text{e}}[\tilde{x},k]$	$\tilde{\theta}^{\text{r}}_{\text{e}}[\tilde{x},k]+1$	Eq. (4.23)	$\tilde{\theta}^{\text{n}}_{\text{e}}[0,k], \tilde{\theta}^{\text{p}}_{\text{e}}[3,k]$
$v_{\text{cell}}[k]$	$\phi^{\text{p}}_{\text{s}}[3,k]-\phi^{\text{n}}_{\text{s}}[0,k]$	Eq. (4.25)	—

Table 4.2: Summary of nonlinear corrections.

$\tilde{x}$	$i_{\text{f+dl}}$ [A]	i_{f} [A]	i_{dl} [A]	θ_{ss} [unitless]	θ_{e} [unitless]	ϕ_{s} [mV]	ϕ_{e} [mV]	$\phi_{\text{s-e}}$ [mV]
0	0.344	3.130	2.841	3.143×10^{-4}	3.282×10^{-3}	—	0.697	0.697
1	0.639	3.266	3.746	4.460×10^{-4}	7.630×10^{-4}	6.474×10^{-6}	0.744	0.744
2	3.809	1.573	5.309	4.557×10^{-4}	1.277×10^{-3}	1.141	0.809	0.346
3	0.574	0.960	0.744	2.303×10^{-4}	4.541×10^{-3}	1.142	0.859	0.294

Table 4.3: Charge-neutral UDDS RMS errors for each variable.

4.7 *Simulation results near a ROM setpoint*

This section presents results from a simulation of the NMC30 cell to demonstrate the fidelity of the ROM with respect to the FOM. Note that the frequency-domain match between the ROM and the TFs is very good; most of the error that we see in these simulations is due to the linearization step required to convert the nonlinear FOM into a linear discrete-time model. The nonlinear corrections are helpful to reduce these errors, but the match is not exact.

In all cases, the FOM was simulated in COMSOL and the ROM had $n = 5$ dynamic states. The simulation implemented a single charge-neutral UDDS cycle,[24] initialized at rest at 60 % SOC. Table 4.3 summarizes the RMS errors between the FOM and ROM predictions for each variable over the course of the simulation.[25]

Simulation input current and output voltage are plotted in Fig. 4.9. The maximum input-current magnitude was a 2C rate. We observe good voltage-prediction agreement between ROM and FOM.

For the remaining variables, we present graphical results for predictions at the negative-electrode/separator boundary and the positive-electrode/separator boundary. As Table 4.3 summarizes, the quality of predictions at the current collector boundaries is usually (but not always) better. In any case, we did not wish to clutter this chapter with too many figures.[26]

Results for the total interfacial lithium flux and its faradaic com-

[24] As a reminder, the urban dynamometer driving schedule (UDDS) profile has historically been used by the EPA to estimate vehicle fuel efficiency for city driving. We convert it here to a function of current versus time (assuming the characteristics of a small passenger vehicle) and use it as a battery load that is representative of what might be expected in a vehicle application.

[25] RMS voltage-prediction error is listed in Table 4.3 under the $v_{\text{cell}}[k]=\phi^{\text{p}}_{\text{s}}[3,k]$ entry.

[26] However, the code for reproducing these figures at all four locations in the cell that were considered in this simulation is available in the toolbox: `http://mocha-java.uccs.edu/BMS3/`. You are invited to reproduce and explore these results in greater detail.

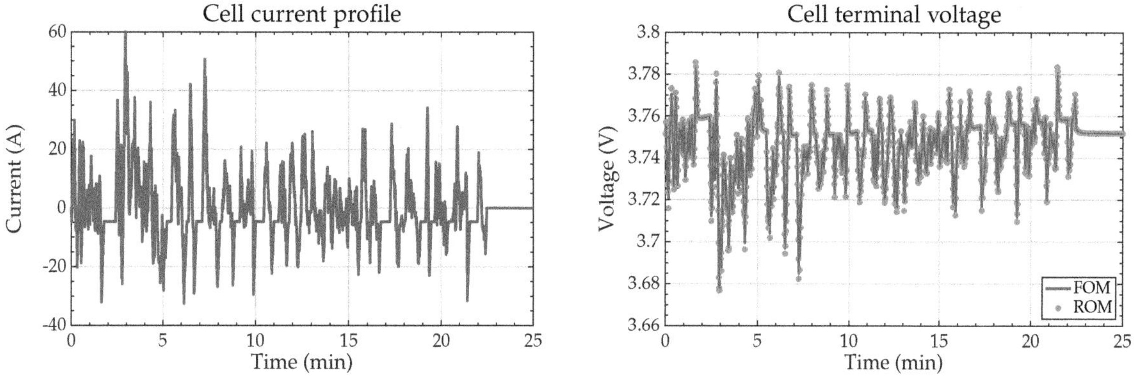

Figure 4.9: Simulation inputs and outputs for a charge-neutral UDDS profile.

ponent are presented in the top two rows of Fig. 4.10. The ROM and FOM show good visual agreement. Results for normalized lithium concentrations are presented in the bottom two rows of the figure. Finally, results for potential in the solid and in the electrolyte separately, and for the solid/electrolyte potential difference are presented in Fig. 4.11. Again, we see good visual agreement.

4.8 Simulating a cell over a wide operating range

Until now, we have created ROMs that are specialized to predict cell internal variables and voltage in the neighborhood of a specific SOC and temperature setpoint. However, since cell dynamics vary with temperature and SOC, a single model is not sufficient for applications that must give accurate predictions over a large operating window. In Vol. I of this series, we proposed a *model-blending* approach to be able to make predictions over a wide range of SOC and temperature. Since that time, we have discovered problems with model blending, and here we present an improved method that we call *output blending*.

As with model blending, we precompute ROMs ($\mathfrak{A}$, $\mathfrak{B}$, $\mathfrak{C}$, $\mathfrak{D}$ matrices) for multiple SOC and temperature setpoints. This is illustrated in Fig. 4.12 in the left frame as a grid of circle markers, where every marker represents a single ROM (out of N total) generated for a specific SOC and temperature value. The time-varying SOC and temperature of the cell define a trajectory through the set of all N ROMs, as illustrated by the blue line. At any point in time, the present SOC and temperature is surrounded by four closest models, as illustrated in the zoom view in the right side of the figure.

So far, this explanation is the same as for model blending. However, with model blending, we created time-varying $\mathfrak{A}[k]$, $\mathfrak{B}[k]$, $\mathfrak{C}[k]$, and $\mathfrak{D}[k]$ matrices based on bilinear interpolation among the static $\mathfrak{A}$, $\mathfrak{B}$, $\mathfrak{C}$, and $\mathfrak{D}$ matrices of the four nearest-neighbor models at every timestep k. There is a subtle flaw with this logic. It assumes that

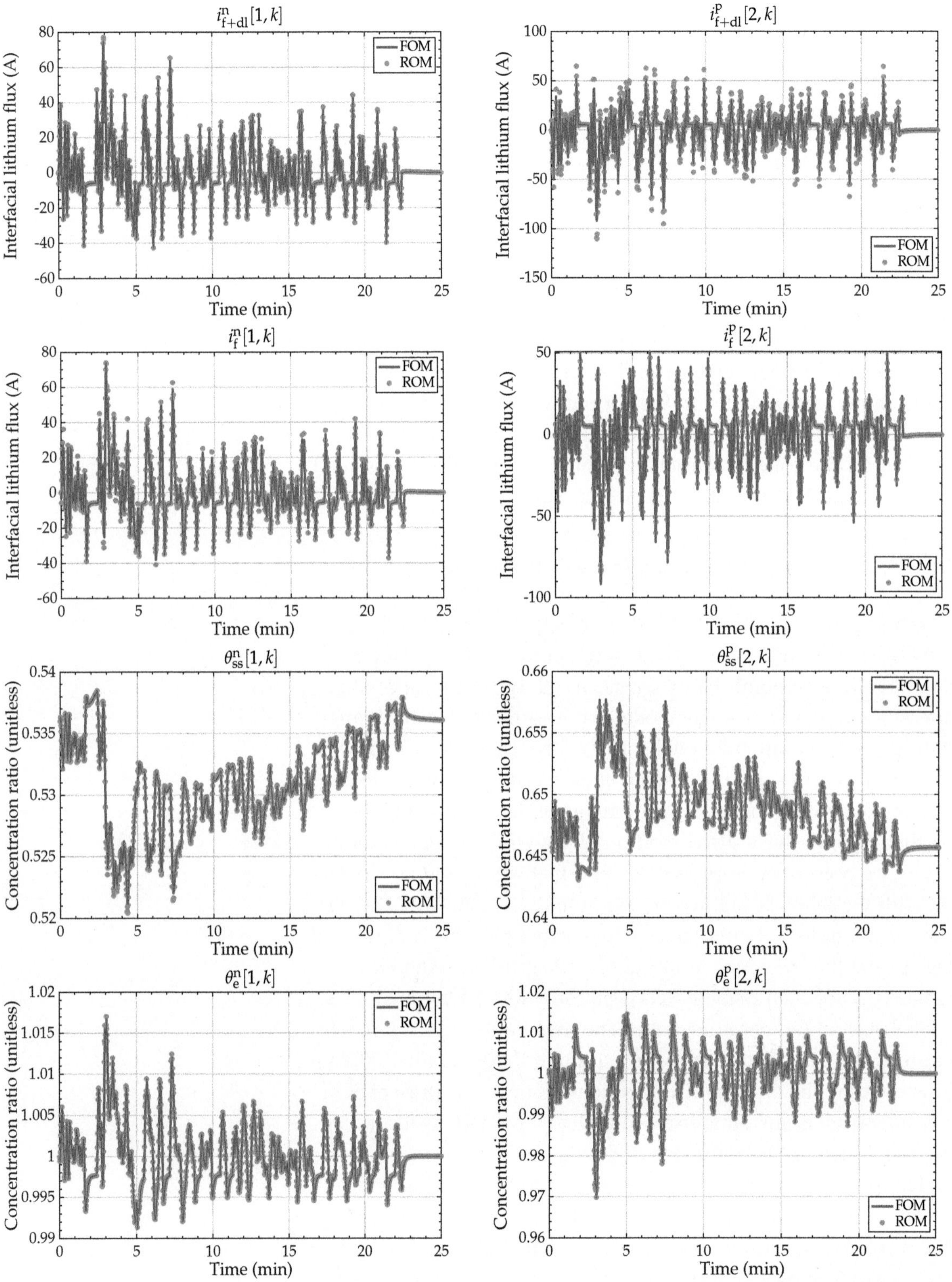

Figure 4.10: Simulation results for total and interfacial lithium flux and for normalized lithium concentrations for charge-neutral UDDS profile.

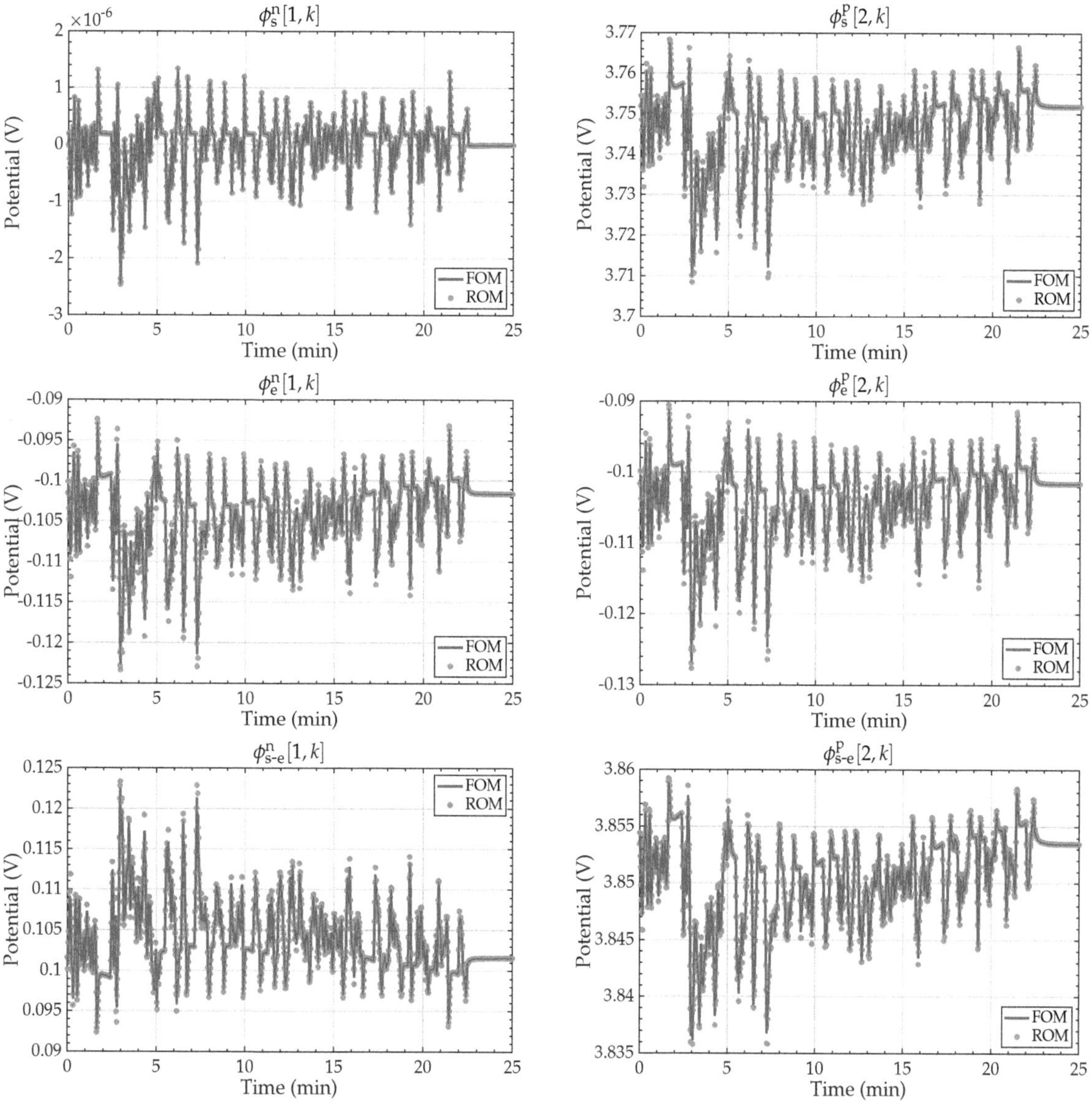

Figure 4.11: Simulation results for cell potentials for charge-neutral UDDS profile.

the $\mathfrak{X}[k]$ vector for every precomputed model has the same interpretation. That is, it assumes that the first element of $\mathfrak{X}[k]$ has the same physical meaning for all models, and so forth for all other elements. If we use an xRA to generate the models, we are not able to guarantee this. The xRA is a numeric procedure that automatically optimizes a semi-physical state for each model. We have no control over the meaning of the individual components of the state vector.[27]

To summarize, the problem with model blending is that it interpolates quantities for which there is no guarantee of physical compatibility. Output blending overcomes this problem by blending model outputs $\boldsymbol{y}[k]$, for which we *do* have a guarantee of physical compat-

[27] Model blending attempts to address this by sorting the states within the state vector $\mathfrak{X}[k]$ of every model according to the eigenvalues of the $\mathfrak{A}$ matrix. This keeps all state elements organized by their time constants. However, even this is not sufficient to ensure a compatible physical interpretation of the state elements.

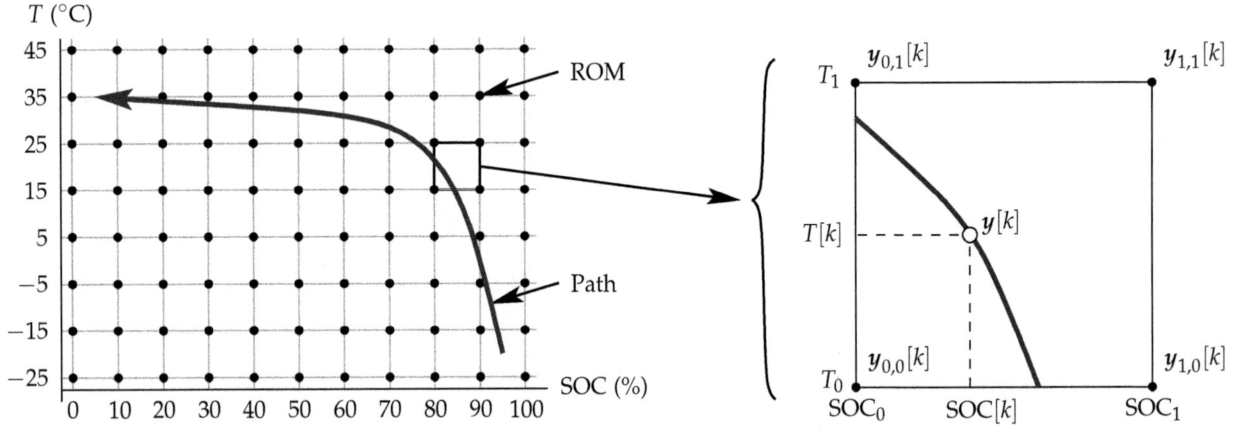

Figure 4.12: Setup for output blending.

ibility. Even if the individual ROMs have incompatible state vectors $\mathfrak{X}[k]$, their outputs are all organized in the same format to predict the same sets of linearized variables. With output blending, every model has its own unique state vector, which is updated every time step.[28] Linear outputs $y_{0,0}[k]$, $y_{0,1}[k]$, $y_{1,0}[k]$, and $y_{1,1}[k]$ are computed only for the four nearest-neighbor models to the present operating condition (see the zoom in Fig. 4.12). These four outputs are blended together to make a prediction of the cell's present linearized output. Then, we apply nonlinear corrections to this linear output.

[28] Models might even have different numbers of states. This could be handy if some portions of the operating range require more states in order to model their behaviors well.

Specifically, the real-time output-blending steps are:

Step 1: Determine the present cell SOC $z[k]$ and temperature $T[k]$.

Step 2: Update state vectors of *all* precomputed ROMs as:

$$\mathfrak{X}_j[k+1] = \mathfrak{A}_j x_j[k] + \mathfrak{B}_j i_{\text{app}}[k].$$

where the new subscript j denotes a specific ROM among the N precomputed ROMs in the model space.

Notice that we could implement this update in a "`for`" loop over all models, but if all models have the same number of states we can also use a compact matrix structure, which is convenient for implementation in programming environments such as MATLAB. Since each model's $\mathfrak{A}$ matrix is diagonal and $\mathfrak{B}_j = \mathbf{1}_{(n+1)\times 1}$ if a model has n transient states plus one integrator state, we note that:

$$\mathfrak{X}_j[k+1] = \text{diag}(\mathfrak{A}_j) \odot \mathfrak{X}_j[k] + \mathbf{1}_{(n+1)\times 1} i_{\text{app}}[k],$$

where "$\odot$" is a Hadamard (point-wise) product, or "`.*`" in MATLAB. Then all $\mathfrak{A}$-matrix diagonals can be combined in a single $(n+1) \times N$ matrix and all states can be combined in an $(n+1) \times N$ matrix as well. In MATLAB, we first build "big $\mathfrak{A}$" and "big $\mathfrak{X}$" matrices:

$$[\text{big } \mathfrak{A}] = \left[\text{diag}(\mathfrak{A}_1) \;\vdots\; \text{diag}(\mathfrak{A}_2) \;\vdots\; \cdots \;\vdots\; \text{diag}(\mathfrak{A}_N)\right]$$

$$[\text{big } \mathfrak{X}][k] = \left[\mathfrak{X}_1[k] \;\vdots\; \mathfrak{X}_2[k] \;\vdots\; \cdots \;\vdots\; \mathfrak{X}_N[k]\right],$$

where each column represents a model linearized around one setpoint. The state equations of all models are updated using a Hadamard product in MATLAB, mathematically written as:

$$[\text{big } \mathfrak{X}][k+1] = [\text{big } \mathfrak{A}] \odot [\text{big } \mathfrak{X}][k] + \mathbf{1}_{(n+1)\times N} i_{\text{app}}[k].$$

We can visualize this operation as illustrated in Fig. 4.13.

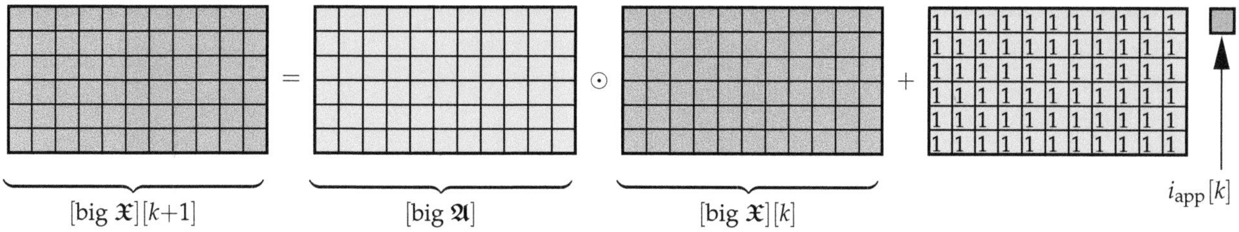

Figure 4.13: State update for output-blending method.

Step 3: Compute the linear outputs for only the four closest precomputed ROMs as:

$$y_j[k] = \mathfrak{C}_j \mathfrak{X}_j[k] + \mathfrak{D}_j i_{\text{app}}[k].$$

The structure of matrices $\mathfrak{C}$ and $\mathfrak{D}$ of a $n+1$ states and q output system are denoted as:

$$\mathfrak{C} = \begin{bmatrix} c_{11} & c_{12} & \cdots & c_{1n} & \text{res}_1^* \\ c_{21} & c_{22} & \cdots & c_{2n} & \text{res}_2^* \\ c_{31} & c_{32} & \cdots & c_{3n} & \text{res}_3^* \\ \vdots & \vdots & \ddots & \vdots & \vdots \\ c_{q1} & c_{q2} & \cdots & c_{qn} & \text{res}_q^* \end{bmatrix},$$

$$\mathfrak{D} = \left[\lim_{s\to\infty} G_1(s), \;\; \lim_{s\to\infty} G_2(s), \;\; \cdots \;\; \lim_{s\to\infty} G_q(s)\right]^T.$$

Since the C matrices do not have a diagonal form, we are unable to perform Hadamard products as previously.

Step 4: Blend the linear outputs for only the four models linearized at setpoints closest to the present (z, T) operating point. This is illustrated in Fig. 4.14, where the cell's present SOC is marked "SOC," and SOC_0 and SOC_1 are the SOC setpoints of the closest precomputed models that bracket the present SOC: $\text{SOC}_0 \leq \text{SOC} \leq \text{SOC}_1$. The cell's present temperature is marked on the figure as "T" and T_0 and T_1 are the temperature setpoints of the closest precomputed models that bracket T: $T_0 \leq T \leq T_1$. We define:

$$\lambda = \frac{\text{SOC} - \text{SOC}_0}{\text{SOC}_1 - \text{SOC}_0} \quad \text{and} \quad \tau = \frac{T - T_0}{T_1 - T_0};$$

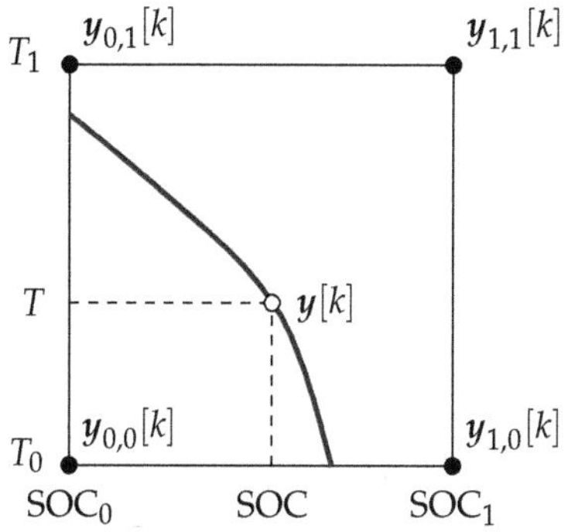

Figure 4.14: Bilinear interpolation to blend nearest-neighbor outputs.

then if $\bar{\lambda} = 1 - \lambda$ and $\bar{\tau} = 1 - \tau$,

$$\begin{aligned} \boldsymbol{y}[k] &= \bar{\tau}(\bar{\lambda}\boldsymbol{y}_{0,0}[k] + \lambda\boldsymbol{y}_{1,0}[k]) + \tau(\bar{\lambda}\boldsymbol{y}_{0,1}[k] + \lambda\boldsymbol{y}_{1,1}[k]) \\ &= \gamma_{0,0}\boldsymbol{y}_{0,0}[k] + \gamma_{0,1}\boldsymbol{y}_{0,1}[k] + \gamma_{1,0}\boldsymbol{y}_{1,0}[k] + \gamma_{1,1}\boldsymbol{y}_{1,1}[k] \qquad (4.26) \\ &= \sum_{j=1}^{N} \gamma_j \boldsymbol{y}_j[k]. \qquad (4.27) \end{aligned}$$

Notice that although we must update the state vector of all models at every time step, we are not required to compute the output vector of any model that is not being used for the present calculation of $\boldsymbol{y}[k]$. That is, in the set of weights $\{\gamma_j\}$ for $1 \le j \le N$ in Eq. (4.27), only the four closest models have nonzero weights, denoted as $\gamma_{0,0}$, $\gamma_{0,1}$, $\gamma_{1,0}$, and $\gamma_{1,1}$ in Eq. (4.26), reducing required computation.

Step 5: Apply the nonlinear corrections summarized in Table 4.2 to $\boldsymbol{y}[k]$ to produce nonlinear outputs at this point in time.

Step 6: Repeat Steps 1 through 5 for every iteration k.

4.9 *Simulation results over a wide operating range*

4.9.1 *Constant-current discharge profile*

This section presents results from a simulation of the NMC30 cell to compare model blending and output blending applied to a long-duration constant-current discharge, and to demonstrate the ROM fidelity with respect to the FOM. Note once again that the frequency-domain matches between the ROMs and the TFs are very good; the majority of the error that we see in these simulations is due to the linearization step required to convert the nonlinear FOM into TFs. The nonlinear corrections are helpful to reduce these errors, but the match is not exact.

In all cases, the FOM was simulated in COMSOL and the ROM had $n = 12$ dynamic states.[29] ROMs were generated between 0 % SOC and 100 % SOC in 2 % increments. The first simulation we consider implemented a 300 min constant-current C/5 discharge, initialized at rest at 100 % SOC. Table 4.4 summarizes the RMS errors between the FOM and ROM predictions for each variable over the simulation.

The simulation overall inputs and outputs are plotted in Fig. 4.15. The "ROM-OutB" voltage was computed using output blending, and the "ROM-MdlB" voltage was computed using model blending. We observe good agreement between both the model-blended and output-blended ROMs and the FOM for this simulation.[30]

Results for the total interfacial lithium flux and its faradaic component are presented in the top two rows of Fig. 4.16. These results

[29] This large number of states is not necessary to achieve good results, but was chosen because it amplifies the problems we have seen with the model-blending approach with respect to the output-blending approach. Output blending is much more robust to different model orders, while model blending tends to work best for low model orders.

[30] RMS voltage error for the entire simulation is listed in Table 4.4 under the $v_{\text{cell}}[k] = \phi_s^{\text{p}}[3,k]$ entry. Note that RMS error for the first 275 min is only 1.627 mV for output blending and 1.797 mV for model blending. The biggest contribution to the RMS voltage-prediction error in the table is due to the final several minutes of the discharge, at very low cell SOC.

$\tilde{x}$	i_{f+dl} [A]	i_f [A]	i_{dl} [A]	θ_{ss} [unitless]	θ_e [unitless]	ϕ_s [mV]	ϕ_e [mV]	$\phi_{s\text{-}e}$ [mV]
0	0.693	0.750	0.219	4.230×10^{-3}	1.028×10^{-3}	—	4.373	4.373
1	1.258	1.313	0.588	1.642×10^{-2}	2.435×10^{-4}	1.282×10^{-5}	4.376	4.377
2	1.195	0.244	1.424	1.055×10^{-3}	3.179×10^{-4}	13.003	4.377	1.408
3	0.252	0.311	5.987×10^{-2}	1.044×10^{-3}	1.226×10^{-3}	11.362	4.382	1.388
0	1.761	1.779	0.220	3.121×10^{-2}	1.027×10^{-2}	—	4.414	4.414
1	3.688	4.092	1.330	8.937×10^{-2}	1.404×10^{-3}	3.437×10^{-5}	4.475	4.476
2	7.241	2.768	9.814	1.325×10^{-3}	2.662×10^{-3}	13.112	4.506	1.490
3	0.952	0.740	0.387	1.356×10^{-3}	1.417×10^{-2}	11.472	4.628	1.510

Table 4.4: C/5 discharge RMS errors for each variable (top half of table is for output blending; bottom half is for model blending).

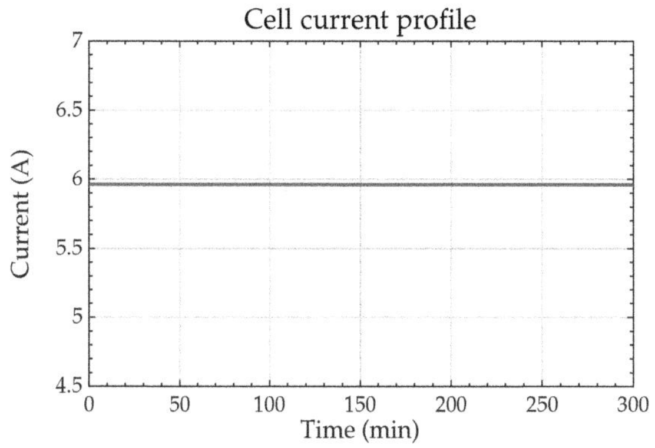

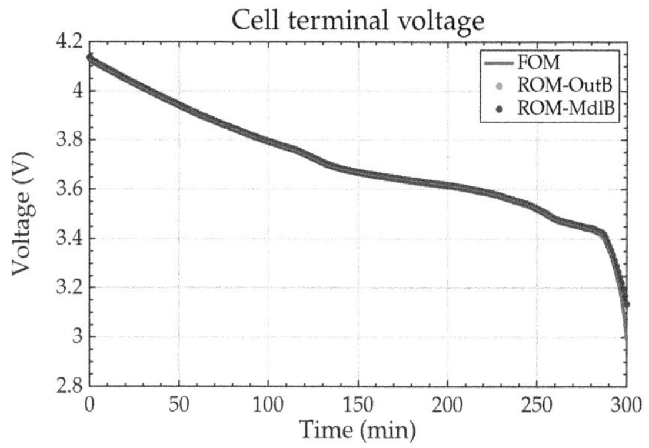

Figure 4.15: Simulation inputs and outputs for a C/5 discharge profile.

are very interesting. We notice that the model-blended ROM is not at all reliable but that output-blended ROM is giving quite good predictions, especially in the positive electrode. Examination of the COMSOL simulations shows that the FOM nonlinearities are causing large gradients across $\tilde{x}$ in $i^{n}_{f+dl}[\tilde{x}, k]$ that are not predicted well by the linearized ROM.[31] This remains an area where future investigations are warranted to seek better high-fidelity but low computational complexity modeling approaches. The unexpected inconsistencies of the model-blended ROM occur when the simulation is transitioning between models having incompatible state vectors. Effectively, the state from the model we are transitioning away from is not supplying a compatible initialization for the state of the model to which we are transitioning. Since the models are stable, we expect the transient bump to be temporary, but the time constants of the model are such that the transients still cause noticeable errors in the predictions.

Results for normalized lithium concentrations are presented in the bottom two rows of Fig. 4.16. We see that the general features are described well, but that both the model-blended and output-blended

[31] The FOM predictions also have oscillatory components that are fundamentally caused by moving concentration-profile boundaries through the electrodes as the cell is discharged. In these simulations, $\bar{D}_s$ is SOC-dependent, and so the electrode impedances also have moving boundaries. As lithium is forced into/out of an electrode, the impedance boundaries cause flux reflections, setting up an oscillating waveform. Simulations without SOC-dependent $\bar{D}_s$ do not exhibit this oscillatory behavior.

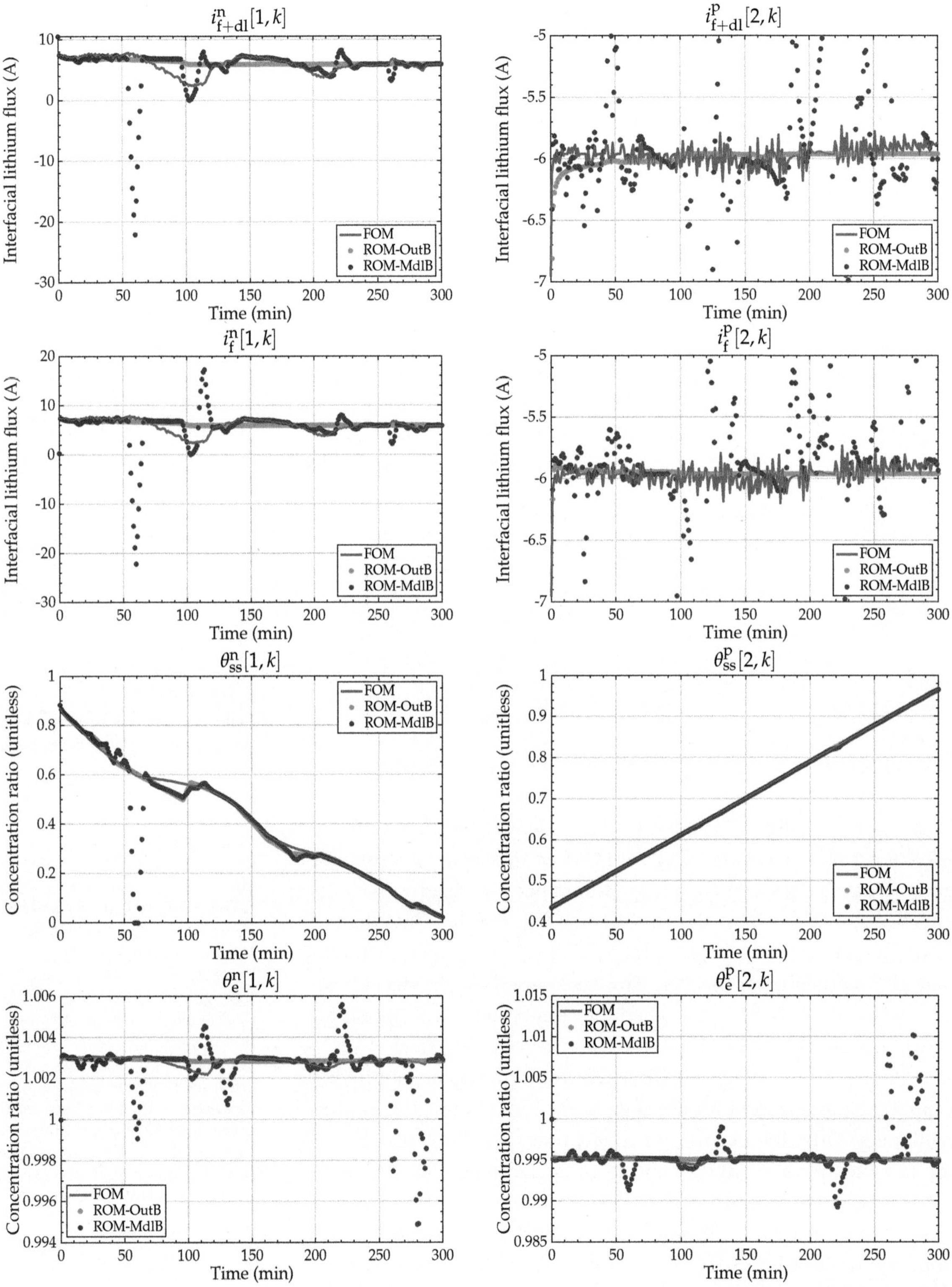

Figure 4.16: Simulation results for total and interfacial lithium flux and for normalized lithium concentrations for C/5 discharge profile.

$\tilde{x}$	$i_{\text{f+dl}}$ [A]	i_{f} [A]	i_{dl} [A]	θ_{ss} [unitless]	θ_{e} [unitless]	ϕ_{s} [mV]	ϕ_{e} [mV]	$\phi_{\text{s-e}}$ [mV]
0	2.594	3.512	1.224	1.220×10^{-2}	2.935×10^{-3}	—	1.422	1.422
1	6.028	6.837	11.562	4.121×10^{-2}	5.124×10^{-4}	5.250×10^{-5}	1.512	1.512
2	37.898	13.232	50.983	5.491×10^{-4}	8.400×10^{-4}	16.584	1.515	1.531
3	6.971	7.155	1.162	3.781×10^{-4}	3.678×10^{-3}	16.044	1.907	1.025

Table 4.5: Long-duration UDDS RMS errors (output blending).

ROMs struggle to capture the details of $\theta_{\text{ss}}^{\text{n}}[1,k]$. The largest prediction errors are when the solid surface concentration is traversing a flat region in the electrode's OCP curve, where large differences in solid concentration produce very small differences in potential. This causes large gradients in solid concentration across the thickness of the electrode and through the cross section of each particle. This effect is amplified by the SOC-dependent $\bar{D}_{\text{s}}^{\text{r}}$ relationship used in these simulations. The linearized ROM is not able to capture these effects with high accuracy.

Results for potential in the solid and in the electrolyte, and for the solid/electrolyte potential difference are presented in Fig. 4.17. Here, the output-blended ROM outperforms the model-blended ROM, especially for $\phi_{\text{s}}^{\text{n}}$.

4.9.2 *Dynamic current profile*

Finally, we present results from a simulation of the NMC30 cell to demonstrate the fidelity of the output-blended ROM with respect to the FOM applied to a long-duration sequence of charge-depleting UDDS profiles. In all cases, the FOM was simulated in COMSOL and the ROM had $n = 5$ dynamic states. ROMs were generated between 0 % SOC and 100 % SOC in 2 % increments. Here, we present results only for the output-blending method since the figures become too difficult to interpret if model-blending results are included.[32]

This simulation considers a 175 min sequence comprising seven charge-depleting UDDS cycles with maximum magnitude of a 5C rate, initialized at rest at 95 % SOC. Table 4.5 summarizes the RMS errors between the FOM and ROM predictions for each variable over the course of the simulation.

The simulation overall inputs and outputs are plotted in Fig. 4.18. We observe good agreement between voltage predictions from the output-blended ROM and FOM for this simulation.[33]

Results for the total interfacial lithium flux and its faradaic component are presented in the top two rows of Fig. 4.19. These results are more as expected than were those in Fig. 4.16. The dynamic nature of the UDDS profile tends to reduce steady-state concentration gradients across the thickness of electrodes, so the linearized

[32] But once again, you are invited to run the code from `http://mocha-java.uccs.edu/BMS3/` to plot other results that you might find interesting.

[33] RMS voltage error is under the $\phi_{\text{s}}^{\text{p}}[3,k]$ entry in Table 4.4; RMS error for the first 160 min is only 5.464 mV.

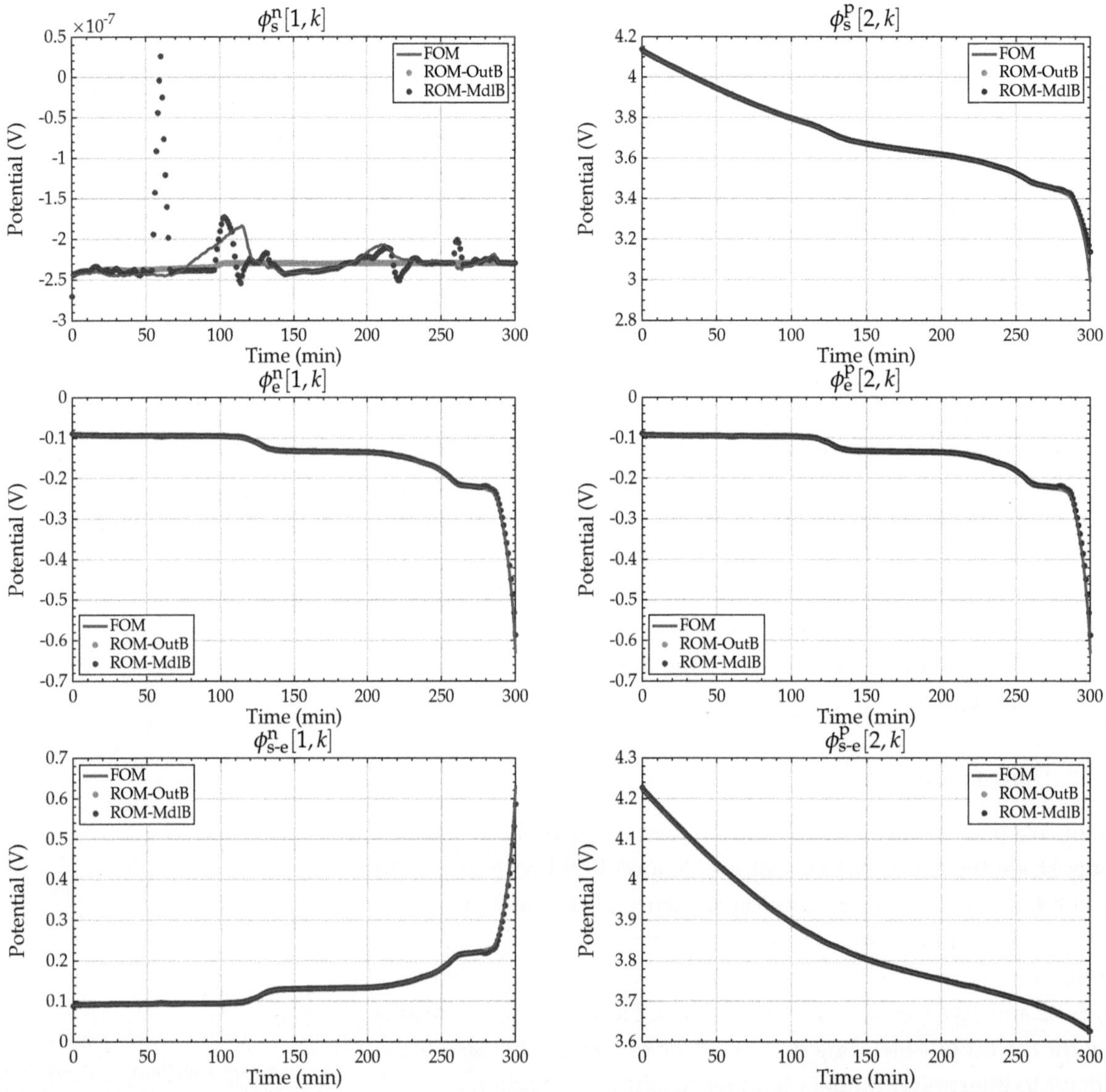

Figure 4.17: Simulation results for cell potentials for C/5 discharge profile.

ROM predictions are much closer to the FOM calculations than for a constant-current event. Results for normalized lithium concentrations are presented in the bottom two rows. We see again that the ROM struggles somewhat to capture the details of $\theta_{ss}^n[1,k]$, but that the general features are described well. The other variables are predicted very well by the ROM.

Finally, results for potential in the solid and in the electrolyte, and for the solid/electrolyte potential difference are presented in Fig. 4.20. All of these results are very good.

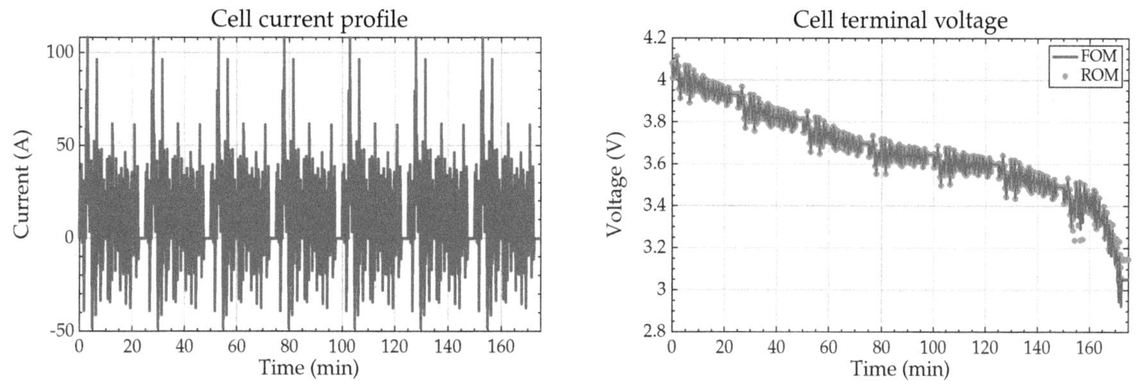

Figure 4.18: Simulation inputs and outputs for a long-duration UDDS profile.

4.10 *Simulating constant voltage and constant power*

Until this point, we have considered that the input to the ROMs is the applied current $i_{\text{app}}[k]$ and that the ROMs produce $v_{\text{cell}}[k]$ as output (as well as predictions of the cell's internal electrochemical variables). We now modify this approach somewhat to be able to use voltage or power as input to the ROMs instead. This allows us, for example, to simulate constant-voltage or constant-power conditions.

4.10.1 *Simulating constant voltage*

In order to implement a simulation where voltage is the input (i.e., cell voltage is held constant for one or more time intervals), we take an approach similar to the one presented in Vol. II of this series. In that case, we used an equivalent-circuit model, which happened to simplify the analysis since all terms in its voltage prediction that were not dependent on the present value of state were linear in the applied current. The physics-based ROMs complicate the process since their voltage calculation is nonlinear in $i_{\text{app}}[k]$.

The idea is as follows: we compute, iteration by iteration, the exact required applied current for that iteration to achieve the desired terminal voltage. The net effect is that $v_{\text{cell}}[k]$ *appears* to be the input since we are adjusting the actual model input to enforce the desired $v_{\text{cell}}[k]$, making $i_{\text{app}}[k]$ (and the electrochemical variables) the output.

To see how to do this, note that we can write present voltage as a nonlinear function of the present state and present input:

$$v_{\text{cell}}[k] = g_v(\mathfrak{X}[k], i_{\text{app}}[k]).$$

The present state $\mathfrak{X}[k]$, however, is not a function of the present input. It is a function only of prior inputs. So, the present state simply creates a bias point for cell voltage, and the present applied current modulates the present actual voltage around this bias.

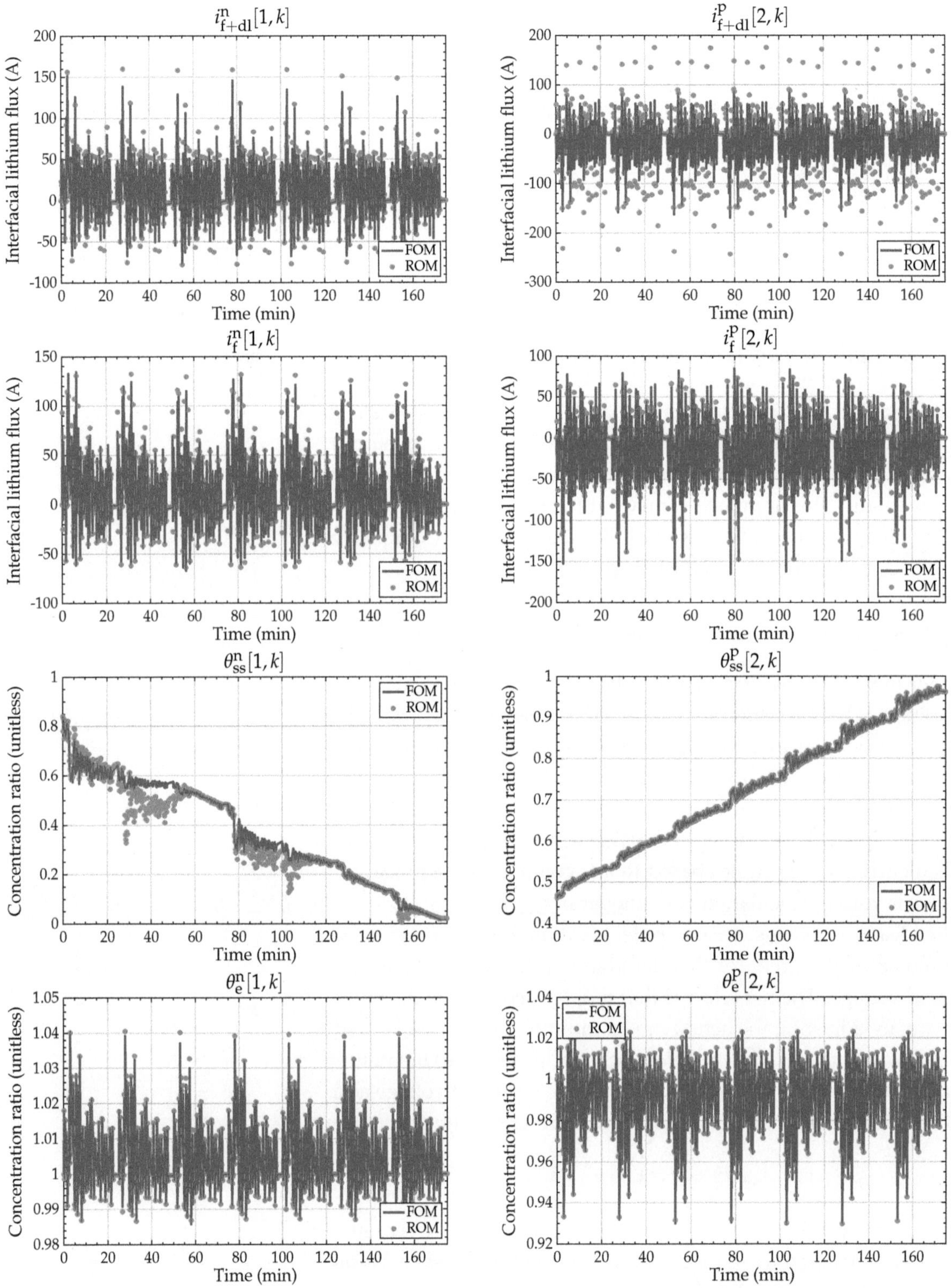

Figure 4.19: Simulation results for total and interfacial lithium flux and for normalized lithium concentrations for the long-duration UDDS profile.

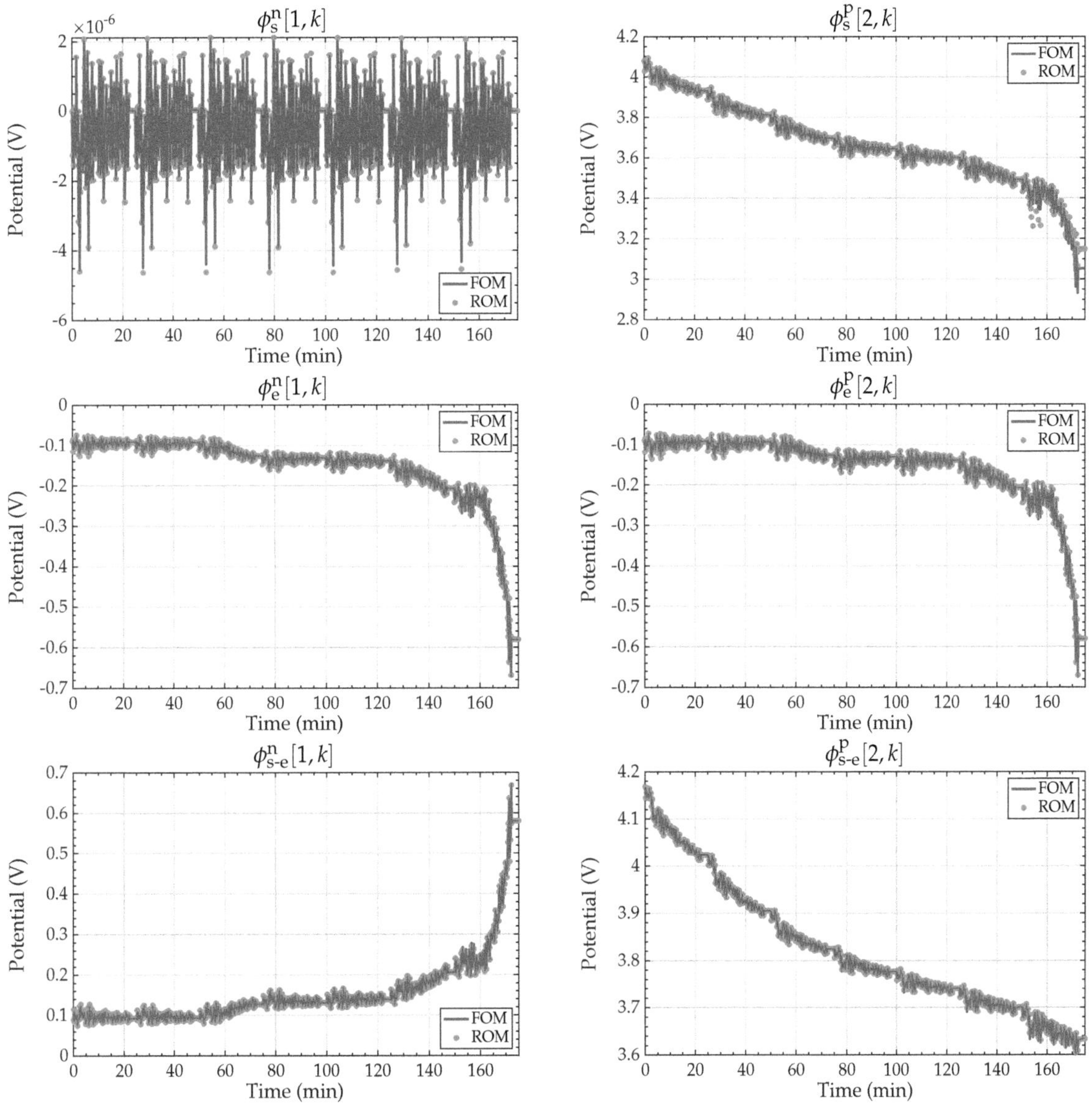

Figure 4.20: Simulation results for cell potentials for long-duration UDDS profile.

We can think of the operation of computing $i_{\text{app}}[k]$ as a kind of inverse of $g_v(\cdot)$ for the particular present value of $\mathfrak{X}[k]$:

$$i_{\text{app}}[k] = g_v^{-1}(v_{\text{cell}}[k]; \mathfrak{X}[k]).$$

How might we compute this inverse? Digging into the equations that compute $v_{\text{cell}}[k]$, we have not been able to find a closed-form expression. So, instead, we use a line-search nonlinear optimization to seek the value $i_{\text{app}}[k]$ that causes $v_{\text{cell}}[k] = v_{\text{desired}}[k]$. We will look at an example shortly.

4.10.2 *Simulating constant power*

We can use a similar strategy to implement a simulation where power is the input (i.e., cell power is held constant for one or more time intervals). Recall that power is equal to applied current multiplied by terminal voltage. In terms of the analysis to date, $p_{\text{cell}}[k] = v_{\text{cell}}[k]i_{\text{app}}[k]$. To simulate constant power, we use nonlinear optimization to search for $i_{\text{app}}[k]$ that causes $p_{\text{cell}}[k] = p_{\text{desired}}[k]$.

Note that there will be two solutions! Both solutions satisfy $p_{\text{cell}}[k] = p_{\text{desired}}[k]$; however, one solution computes a negative cell voltage. We need to be careful to find the solution that produces positive $v_{\text{cell}}[k]$. Once we recognize this detail, the solution is straightforward: if we desire negative power, we constrain $i_{\text{app}}[k]$ to be negative; if we desire positive power, we constrain $i_{\text{app}}[k]$ to be positive.

The following MATLAB code uses this logic to implement constant-current constant-voltage (CC/CV) or constant-power constant-voltage (CP/CV) charging of a cell. In the simulation, `Pmode` is true if the default applied input is power, `pk`, and it is false if the default input is current, `ik`. The maximum cell voltage, when we transition between CC or CP and CV, is defined by `Vmax` (the code also works for CP/CV or CC/CV discharges, where the transition is defined by `Vmin`).

```
for k = 0:length(ik)-1
  if Pmode % input to simulation is applied power
    if pk(k+1) > 0 % search for ik to meet power spec...
      % initialize search using prior value of cell voltage
      ik(k+1) = fmincon(@(x) ...
                  (pk(k+1) - x*simStep(x,Tk(k+1),cellState))^2,...
                  pk(k+1)/Vcell,[],[],[],[],0,Inf,[],options);
    else
      ik(k+1) = fmincon(@(x) ...
                  (pk(k+1) - x*simStep(x,Tk(k+1),cellState))^2,...
                  pk(k+1)/Vcell,[],[],[],[],-Inf,0,[],options);
    end
  end

  % now, attempt to update with user-supplied value of Iapp
  [Vcell,newCellState] = simStep(ik(k+1),Tk(k+1),cellState);

  % switch from current mode to constant-voltage mode
  if Vcell > Vmax
    ik(k+1) = fminbnd(@(x) (Vmax - simStep(x,Tk(k+1),cellState))^2,...
                min(0,ik(k+1)),0);
    [Vcell,newCellState] = simStep(ik(k+1),Tk(k+1),cellState);
  end
  if Vcell < Vmin
    ik(k+1) = fminbnd(@(x) (Vmin - simStep(x,Tk(k+1),cellState))^2,...
                0,max(0,ik(k+1)));
    [Vcell,newCellState] = simStep(ik(k+1),Tk(k+1),cellState);
  end
  cellState = newCellState;
end
```

The `simStep` function implements one timestep of the output-blended ROM. However, very importantly, `simStep` does not update the state of the ROM. The outputs from `simStep` are the voltage that would be achieved if this input were applied to the cell as well as the updated cell state that would be achieved by this input.

The first part of the loop checks to see whether the user is requesting a CP input. If so, the `fmincon` function is invoked to minimize the difference between the achieved cell power and requested cell power, where the input current is constrained to be between $-\infty$ and 0 or between 0 and ∞, depending on the sign of the requested cell power. The output of the `fmincon` optimization is a value of cell input current `ik`, but this input current is not yet applied to the cell.

The next step performs a trial application of `ik` to the cell, producing a predicted cell voltage `Vcell` and state `newCellState`. Again, this input is not yet applied permanently to the cell.

The next step checks to see whether `Vcell` is higher than `Vmax` or lower than `Vmin`. If so, then we do not want to apply `ik` to the cell—its magnitude is too large. Instead, we use the CV method to hold the cell voltage at its desired limit. We use `fminbnd` to perform a line search for the value of `ik` that obtains either `Vmax` or `Vmin`.

At last, we have the value of `ik` that we desire to apply to the cell. This is stored in the `ik` vector for later analysis and plotting. The value of cell state `newCellState` corresponding to this `ik` is now stored as the present cell state `cellState`, making the application of `ik` for this timestep permanent.

Example: Charge from 50% to 100% SOC using CC/CV and CP/CV

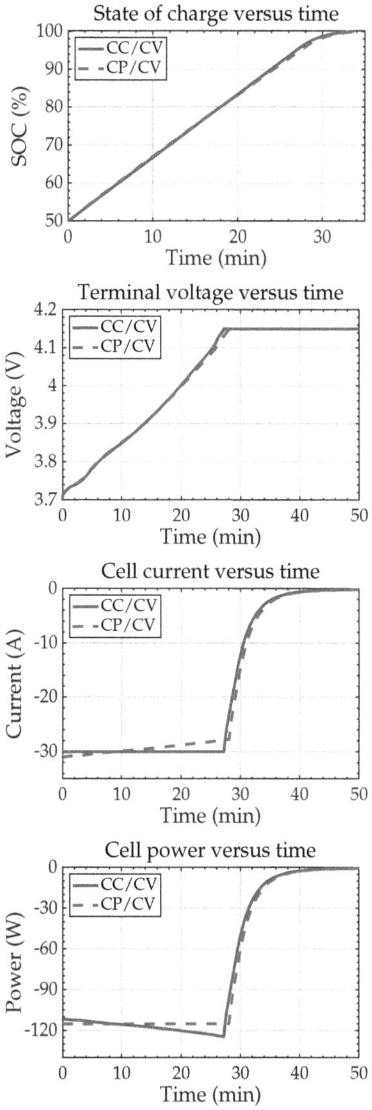

Figure 4.21: An example of CC/CV and CP/CV charging.

Fig. 4.21 shows example output from this code. The cell was initialized with SOC equal to 50 %. Either constant current or constant power was requested from the cell until the maximum voltage limit of 4.2 V was reached. After that point, the simulation logic automatically switched over to a CV mode.

The value of power for the CP/CV simulation was chosen to make the two simulations roughly comparable. Therefore, we see very little difference in the evolution of SOC and voltage over time. Small differences are seen in the figures showing current and power versus time. As expected, the first segment of the simulation demonstrates constant current for the CC/CV simulation and constant power for the CP/CV simulation.

4.11 *Simulating battery packs*

Until now, our focus has been on simulating battery *cells*. Now, we briefly consider simulating battery *packs* comprising many cells.[34]

[34] The content of this section has been adapted from: Gregory L. Plett and M. Scott Trimboli, "Simulating multicell battery packs using physics-based reduced-order models," in *Proceedings of the 35th International Electric Vehicle Symposium and Exhibition (EVS35)*, Oslo, Norway, June 2022.

The framework that we have developed to this point—especially the new concepts for CV simulations—can be extended readily to add this new capability.

4.11.1 *Series-connected cells*

Cells connected in series each experience the same current. If all cells have the same initial state and identical parameters, then all cells have exactly the same state and voltage at all times, so we need to simulate only one cell (the others will be identical). This is not generally true, however, so we can simulate all cells' dynamics by keeping separate state and parameter information for every cell, updating each cell's state once per sample interval. We can also include a per-cell interconnect resistance term R_{int} that describes terminal resistances and weld resistances and so forth when we compute pack voltage. If there are N_s cells wired in series, we compute pack voltage $v_{\text{pack}}[k]$ based on cell voltages for all cells and the interconnect resistance as:

$$v_{\text{pack}}[k] = \left(\sum_{j=1}^{N_s} v_{\text{cell},j}[k] \right) - N_s R_{\text{int}} i_{\text{app}}[k]. \qquad (4.28)$$

4.11.2 *Logically modular battery packs*

We might need to build battery packs from multiple cells for a variety of reasons. To achieve high power, we require either high voltage or high current (or both). Since parasitic resistive power losses scale with the square of the magnitude of the applied current, it is often preferable to construct a high-voltage pack by wiring many cells in series. Such series-connected packs are common for low-energy, high-power applications, such as HEV. On the other hand, high-energy applications such as EV usually have cells and even entire sub-packs that are connected in parallel to increase the pack's total capacity.

When constructing battery packs having cells wired in series and/or in parallel, it is common to employ a modular approach. One extreme is where all cells in a module are connected in parallel, forming a parallel cell module (PCM). These PCMs are then themselves wired in series to construct the battery pack. This is illustrated in the top frame of Fig. 4.22. Another extreme is where all cells in a module are connected in series, forming a series cell module (SCM). These SCMs are then themselves wired in parallel to construct a pack. This is illustrated in the bottom frame of Fig. 4.22. The features of these design choices are described in Vol. II of this series.

We would like to be able to simulate battery packs comprising either PCMs or SCMs using physics-based reduced-order models. If

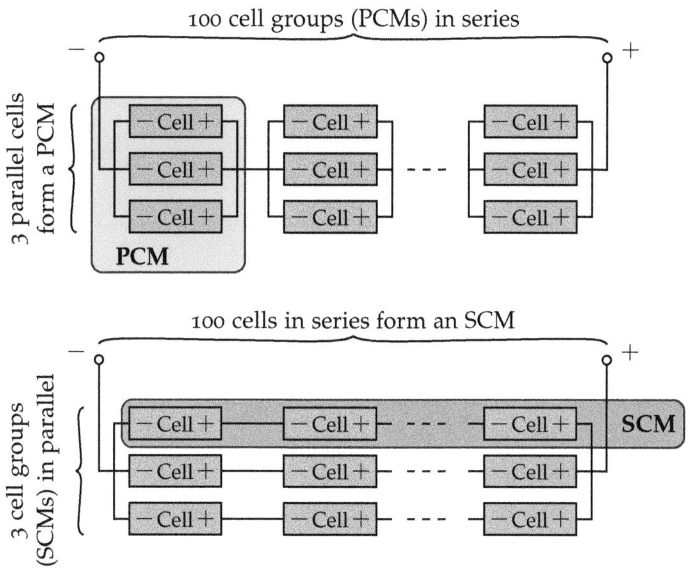

Figure 4.22: Two different approaches to modularizing 300 cells in a high-power battery pack: PCM versus SCM.

cells differ in their internal states and/or parameter values, we must simulate all cells individually and then combine the results to be able to predict the evolution of all electrochemical variables and voltages over time. But how do we do so?

4.11.3 *Simulating PCM-based packs*

We begin by considering how to simulate a single PCM. At every point in time, a cell's voltage depends in part on its state vector $\mathfrak{X}[k]$, which itself depends only on prior inputs $i_{\text{app}}[m]$ for $m < k$. The present sample of current $i_{\text{app}}[k]$ does not affect the value of $\mathfrak{X}[k]$ (but of course it will affect $\mathfrak{X}[m]$ for $m > k$). The remaining part of the cell's voltage depends directly on $i_{\text{app}}[k]$. For example, the ROM's linear outputs are directly connected to $i_{\text{app}}[k]$ via the $\mathfrak{D}$-matrix term.

Therefore, we consider that a cell's voltage comprises a fixed part that does not depend on the present cell current, and a variable part that does depend on present cell current. We can model cells in parallel as drawn in Fig. 4.23, where the voltage source in each branch is the fixed part, and the (nonlinear) resistor current is the variable part. By Kirchhoff's voltage law (KVL), all terminal voltages for cells connected in parallel must be equal. By Kirchhoff's current law (KCL), the sum of branch currents within a PCM must equal the total battery-pack current. When we applied these concepts to equivalent-circuit models in Vol. II, we found that we could solve for all branch-current and voltage values in closed form. However, due to the nonlinear voltage equation in our physics-based ROMs, we cannot, and thus must use nonlinear optimization again.

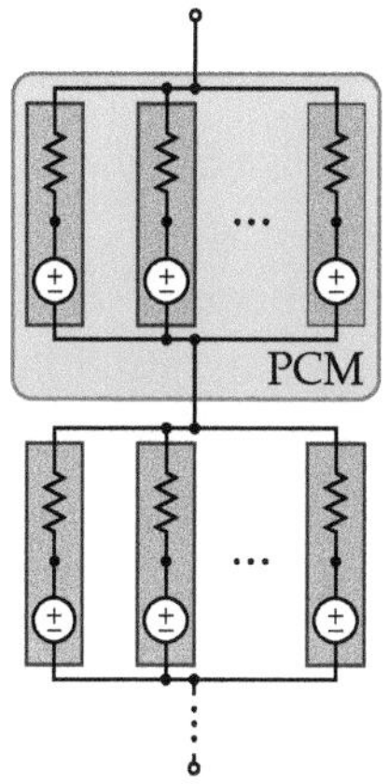

Figure 4.23: Considering how to simulate a PCM.

The procedure is as follows. We assume that total pack applied current $i_{\text{app}}[k]$ is given. Then, for every PCM:

1. We initialize PCM cell currents $i_{\text{app},j}[k] = i_{\text{app}}[k]/N_\text{p}$ for all cells j in the PCM.
2. We compute cell voltages $v_{\text{cell},j}[k]$ for each PCM cell using the cell's individual ROM and $i_{\text{app},j}[k]$.
3. We revise cell currents and loop from Step 2 until the maximum absolute difference between all cell voltages in the PCM is minimized, with the constraint that $\sum_{j=1}^{N_\text{p}} i_{\text{app},j}[k] = i_{\text{app}}[k]$.

The "revise" part of this process in Step 3 is accomplished via a nonlinear optimization, such as `fmincon` in MATLAB.

This procedure gives us all cell currents and all PCM voltages. We then update the state of each cell using its individual ROM and $i_{\text{app},j}[k]$ and proceed to the next iteration. Battery-pack voltage is computed by summing PCM voltages, in a similar fashion to Eq. (4.28), but where $v_{\text{cell},j}[k]$ is replaced by the common cell voltage of all cells in the jth PCM.

4.11.4 *Simulating SCM-based packs*

The approach to simulating an SCM is very similar to simulating a PCM. Each cell in an SCM has its own fixed and variable parts. All "fixed" parts sum to give an equivalent voltage source; all "variable" parts sum to give an equivalent (nonlinear) resistance.

This is illustrated in Fig. 4.24, where the top part of the figure shows the actual structure of an SCM-based battery pack and the lower part of the figure shows the equivalent form that we analyze, where each SCM collapses to something that looks like a single high-voltage cell (where resistances are nonlinear).

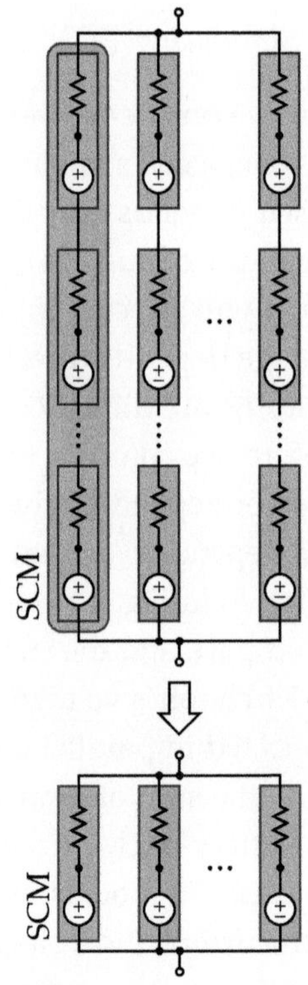

Figure 4.24: Considering how to simulate SCM.

The procedure for simulating a battery pack based on SCMs is similar to the one for simulating a PCM. We assume that total pack applied current $i_{\text{app}}[k]$ is given. Then

1. We initialize SCM currents $i_{\text{app},j}[k] = i_{\text{app}}[k]/N_\text{p}$ for all SCMs j in the pack.
2. We compute cell voltages $v_{\text{cell},j}[k]$ for each cell in the pack using the cell's individual ROM and its input current $i_{\text{app},j}[k]$.
3. We compute SCM voltage by summing voltages of all cells in that SCM in a similar fashion to Eq. (4.28).
4. We revise SCM currents and loop from Step 2 until the maximum absolute difference between all SCM voltages is minimized, with the constraint that $\sum_{j=1}^{N_\text{p}} i_{\text{app},j}[k] = i_{\text{app}}[k]$.

Again, the "revise" part of this process in Step 4 is done via nonlinear optimization. This procedure gives us all branch currents and the pack voltage. We update the state of each cell using its individual ROM and $i_{\text{app},j}[k]$ and proceed to the next iteration. In this case, the pack voltage is equal to the common voltage shared by all SCMs, except perhaps for the inclusion of a pack-level interconnect-resistance.

4.11.5 *Example simulations of PCM-based and SCM-based packs*

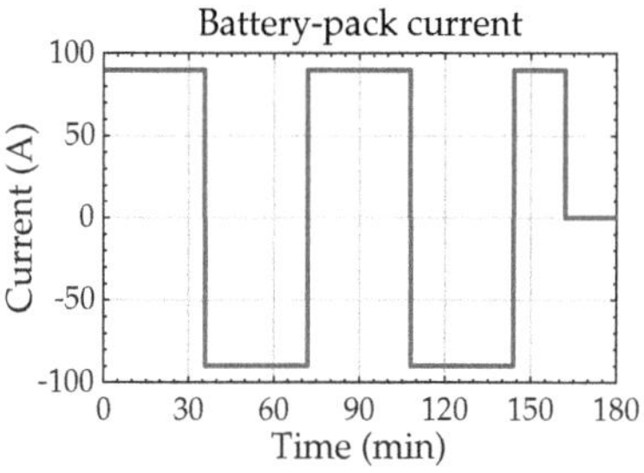

Figure 4.25: Battery-pack current for PCM and SCM simulations.

We briefly consider a simulation comparison of a PCM-based battery pack versus an SCM-based pack. Both packs had $N_s = 3$ and $N_p = 3$. The nine cells comprising the packs had individual ROMs that were generated by slightly perturbing the parameter values of the standard NMC30 cell, giving each ROM somewhat different characteristics. All cells were initialized to 80 % SOC. The pack input current, shown in Fig. 4.25, cycled the cells in the pack at a nominal 1C rate twice between nominal SOCs of 80 % and 20 % before a final discharge to bring the nominal SOC of each cell to 50 %. The pack then rested.

PCM-based battery-pack results are shown in Fig. 4.26. The voltages of all cells within the same PCM are identical so there are only three distinct voltage profiles for this simulation. However, every cell's input current can be different, depending on each cell's present state and parameters, which is what we observe in the simulation.

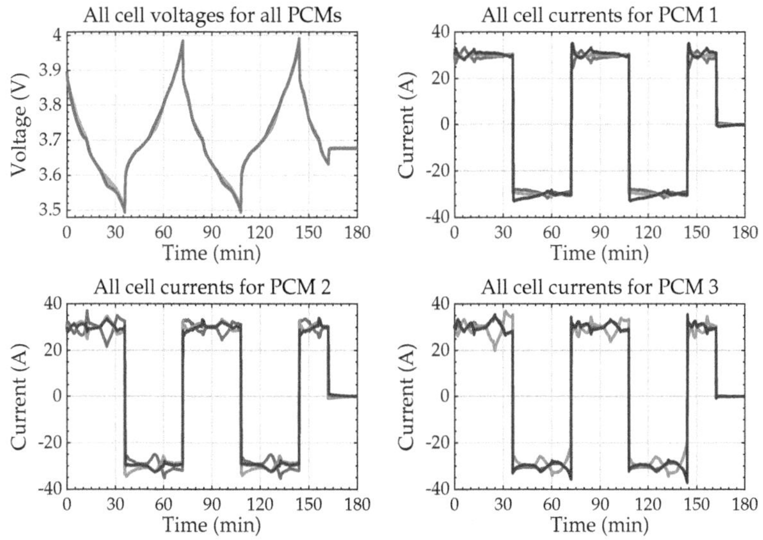

Figure 4.26: Simulation results for a PCM-based battery pack with $N_s = N_p = 3$.

SCM-based battery-pack results are shown in Fig. 4.27. In this case, the currents for all cells within the same SCM are identical so there are only three distinct current profiles for this simulation. However, every cell's voltage can be different, depending on each cell's present state and parameters, which is what we observe.

Comparing Figs. 4.26 and 4.27, we make an interesting observation. At time equal to 144 min, the overall battery-pack input current

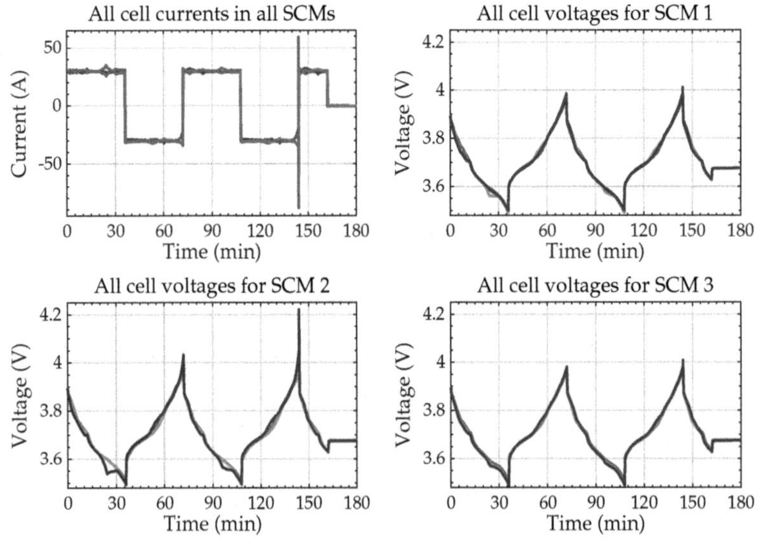

Figure 4.27: Simulation results for an SCM-based battery pack with $N_s = N_p = 3$.

switches from charging to discharging. By this point, the cells have approached a steady-state condition where the results of this cycle are different from those of the original cycle since the initial transients caused by forcing all cells to be initialized to the same SOC has died out. There is no marked difference between the PCM-simulation results at this moment and the transition at time 72 min; however, there is a large current spike in the SCM-based simulation results.

In a PCM, the parallel connection of all cells tends to average out differences in individual cell characteristics naturally since their voltages are forced to be equal. In an SCM, there is no such cell-to-cell averaging. The overall SCM voltages must be equal, but the individual cell voltages may be quite different. We conclude that without real-time balancing, the SCM architecture appears to be more prone to larger cell-voltage swings than does the PCM architecture.

The primary benefit of using physics-based models versus equivalent-circuit models is that we can also investigate cell internal variables. One variable of particular interest is the side-reaction overpotential η_s, which predicts the rate of SEI-layer growth and the onset of lithium plating.[35] To avoid lithium plating, this value must never become negative. Fig. 4.28 compares values of η_s for the PCM-based pack to those of the SCM-based pack for the same simulations. Since the SCM-based architecture does not naturally average cell differences the same way as does the PCM-based pack, we see greater fluctuation in η_s for the SCM pack. For these particular simulations, there is no risk of lithium plating for either PCM or SCM. However, we conclude that the SCM-based pack is more susceptible to lithium plating than the PCM-based pack since its side-reaction overpotential

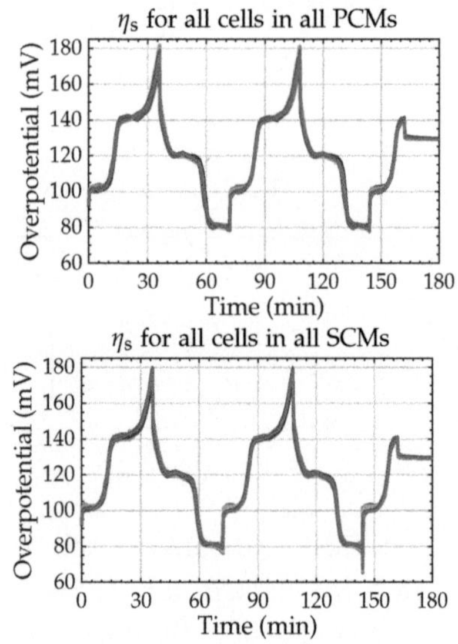

Figure 4.28: Illustrating differences between side-reaction overpotentials based on pack architecture.

[35] We will discuss side-reaction overpotential in more detail in Chap. 6.

has greater variation if there are large parameter differences between different cells in the pack.

4.12 *MATLAB toolbox*

Before closing the chapter, we return to the overview of the MATLAB toolbox that we began to discuss in Chap. 2.[36] Fig. 4.29 illustrates the components of the toolbox, highlighting the functionality added in this chapter. The most critical elements to notice at this point are the portions that generate ROMs using the HRA and that generate COMSOL models, and simulations of the FOM using COMSOL and of the ROMs using different blending methods `[XX]Blend`.[37]

[36] See: `http://mocha-java.uccs.edu/BMS3/`.

[37] As described below, `[XX]Blend` can either be `nonBlend`, `mdlBlend`, or `outBlend`.

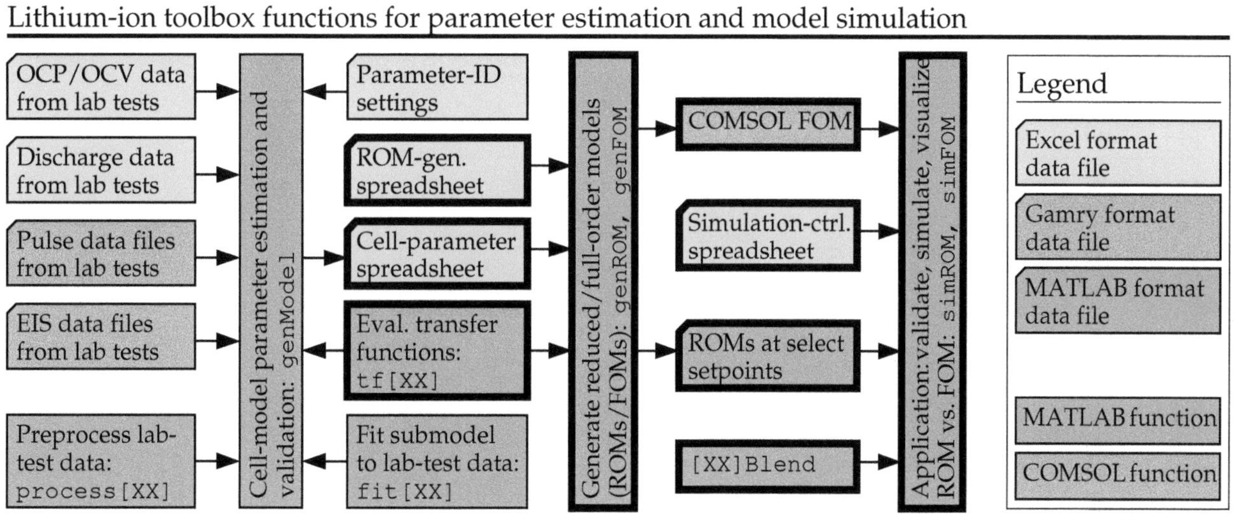

Figure 4.29: Functionality of MATLAB toolbox, highlighting the focus of this chapter.

HRA tuning values are stored in custom-format Excel spreadsheets. Fig. 4.30 shows an example. In this case, we wish to create ROMs for only one temperature, 25 °C, but for 51 SOC setpoints between 0 % and 100 % in increments of 2 % (MATLAB vector-creation notation is adopted). The sample period Δt is set to 1 s, the number of poles n_1 used by the HRA is 45, and the number of non-integrator poles in the final model is $n = 5$. The ω_k vector used by the HRA to sample the TFs is specified as a multiple of the Nyquist frequency of the final ROM. In this case, the multiplier vector comprises one point at 0 and 499 points evenly spaced on a logarithmic scale between 10^{-5} and 10^{0}. When converting continuous-time to discrete-time frequency responses, $k_{\max} = 150$, and the oversizing parameter i used when creating the HRA's internal matrices was selected to be 50. None of these values have been optimized—they seem to work well but perhaps a different tuning would work better. The final part of the spreadsheet lists the TFs to be included when building the ROM.

Each TF has "s" as its first input (the complex frequency vector), then the set of $\tilde{x}$ locations for which we want to compute outputs, then a mandatory "%s" term that the code uses internally to pass other parameter values.

See *Instructions* tab for more information

Environmental

Parameter	Code Name	Value	Unit
Operating Temperature	T	25	[°C]
Operating State of Charge Range	SOC	0:2:100	[%]

xRA Method

Parameter	Code Name	Value	Unit
Final model sample period	Tsamp	1	[s]
Number of initial non-integrator poles	n1	45	[unitless]
Number of final non-integrator poles	n	5	[unitless]
HRA - Frequency samples in [0..1]	HRA_W	[0, logspace(-5,0,499)]	[unitless]
HRA - Kmax for adding frequency aliases	HRA_Kmax	150	[unitless]
HRA - Oversizing parameter	HRA_ii	50	[unitless]

Transfer Function

Parameter	Code Name	Value	Unit
	tflist	tfPhie(s,1:3,%s)	[Ohm]
		tfPhis(s,1:2,%s)	[Ohm]
		tfPhise(s,0:3,%s)	[Ohm]
		tfIfdl(s,0:3,%s)	[unitless]
		tfIf(s,0:3,%s)	[unitless]
		tfIdl(s,0:3,%s)	[unitless]
		tfThetae(s,0:3,%s)	[1/A]
		tfThetass(s,0:3,%s)	[1/A]

Figure 4.30: HRA tuning variables in an Excel spreadsheet.

MATLAB code `loadXRA` loads the spreadsheet into MATLAB's workspace:

```
>> xraData = loadXRA('defaultHRA.xlsx')

xraData =

  struct with fields:

          T: 25
        SOC: [1 x 51 double]
      Tsamp: 1
         n1: 45
          n: 5
      HRA_W: [1 x 500 double]
   HRA_Kmax: 150
     HRA_ii: 50
         tf: {7 x 1 cell}
```

To create a ROM (possibly containing many sub-ROMs at many different temperature/SOC setpoints), use `genROM`.[38]

```
>> cellData = loadCellParams('cellDoyle.xlsx'); % load cell data
>> xraData = loadXRA('defaultHRA.xlsx'); % load XRA tuning data
>> ROM = genROM(cellData,xraData,'HRA')  % for example, using HRA

ROM =

  struct with fields:
```

[38] See comment in code in `example2.m` in the `CH4` folder of the toolbox to see how to set a debug flag that causes the HRA to plot the discrete-time frequency-response match between the TFs and the ROMs.

```
    ROMmdls: [1 x 51 struct]
   cellData: [1 x 1 struct]
    xraData: [1 x 1 struct]
     tfData: [1 x 1 struct]
```

ROM generation can take some time, so `genROM` displays status updates periodically.

To execute a ROM, we must specify the initial cell SOC as well as a profile of current and temperature versus time. This is done using another custom-format Excel spreadsheet, illustrated in Fig. 4.31. One column is time, another is current, and another is the cell's measured temperature. A final entry specifies the initial SOC.[39]

[39] This particular spreadsheet also includes a cell capacity entry that is used to scale the current.

See *Instructions* tab for more information

Time [s]	Current [A]	Temperature [°C]
0	0	25
1	-4.694370732	25
2	0.717162439	25
3	6.145697561	25
4	10.08035122	25
5	16.46195122	25
6	20.78370732	25
7	23.42853659	25
8	1.260790244	25
9	5.898526829	25

Initial cell SOC [%]
60

This is a single UDDS charge-neutral cycle, max rate = 2C

Cell capacity [Ah]
30

Figure 4.31: Runtime input-current profile in an Excel spreadsheet.

We load this spreadsheet using MATLAB function `loadInput`:

```
>> simData = loadInput('inputUDDS.xlsx')

simData =

  struct with fields:

    time: [1501 x 1 double]
    Iapp: [1501 x 1 double]
       T: [1501 x 1 double]
      Ts: 1
    SOC0: 60
```

This can now be used with `simROM` to simulate the cell for this input profile. This function can either perform output blending (`outBlend`), model blending (`mdlBlend`), or no blending (`nonBlend`):

```
ROMout = simROM(ROM,simData,'outBlend')

ROMout =

  struct with fields:

       blending: 'output-blending'
       negIfdl: [1501 x 2 double]
         negIf: [1501 x 2 double]
        negIdl: [1501 x 2 double]
       negPhis: [1501 x 1 double]
      negPhise: [1501 x 2 double]
```

```
   negThetass: [1501 x 2 double]
     negIfdl0: [1501 x 1 double]
       negIf0: [1501 x 1 double]
      negEta0: [1501 x 1 double]
    negPhise0: [1501 x 1 double]
  negThetass0: [1501 x 1 double]
      posIfdl: [1501 x 2 double]
        posIf: [1501 x 2 double]
       posIdl: [1501 x 2 double]
      posPhis: [1501 x 1 double]
     posPhise: [1501 x 2 double]
   posThetass: [1501 x 2 double]
     posIfdl3: [1501 x 1 double]
       posIf3: [1501 x 1 double]
      posEta3: [1501 x 1 double]
  posThetass3: [1501 x 1 double]
         Phie: [1501 x 4 double]
       Thetae: [1501 x 4 double]
        Vcell: [1501 x 1 double]
      cellSOC: [1501 x 1 double]
       negSOC: [1501 x 1 double]
       posSOC: [1501 x 1 double]
         time: [1501 x 1 double]
         Iapp: [1501 x 1 double]
            T: [1501 x 1 double]
         Ifdl: [1501 x 4 double]
           If: [1501 x 4 double]
         Phis: [1501 x 2 double]
        Phise: [1501 x 4 double]
      Thetass: [1501 x 4 double]
        xLocs: [1 x 1 struct]
     cellData: [1 x 1 struct]
```

The output comprises all variables simulated at all points in time in the simulation. This input profile was 1500 seconds long (1501 outputs recorded from $t = 0 \ldots 1500$); data were recorded at $\tilde{x} = \{0, 1, 2, 3\}$, as appropriate, for every variable of interest.

If COMSOL is available with LiveLink for MATLAB, then the toolbox can also build COMSOL FOMs via `genFOM`:

```
>> FOM = genFOM(cellData)

FOM =

COMSOL Model Object
Name: Lithium-ion cell
Tag: LiIonCell
Identifier: root
```

The FOM can then be simulated using `simFOM` as:

```
[FOM,FOMout] = simFOM(FOM,simData)

FOM =

COMSOL Model Object
Name: Lithium-ion cell
Tag: LiIonCell
Identifier: root
```

```
FOMout =

  struct with fields:

         negIfdl: [1501 x 21 double]
          negIf: [1501 x 21 double]
         negIdl: [1501 x 21 double]
        negPhis: [1501 x 21 double]
       negPhise: [1501 x 21 double]
     negThetass: [1501 x 21 double]
       negIfdl0: [1501 x 1 double]
         negIf0: [1501 x 1 double]
        negIdl0: [1501 x 1 double]
      negPhise0: [1501 x 1 double]
    negThetass0: [1501 x 1 double]
           Phie: [1501 x 61 double]
         Thetae: [1501 x 61 double]
          xLocs: [1 x 1 struct]
        posIfdl: [1501 x 21 double]
          posIf: [1501 x 21 double]
         posIdl: [1501 x 21 double]
        posPhis: [1501 x 21 double]
       posPhise: [1501 x 21 double]
     posThetass: [1501 x 21 double]
       posIfdl3: [1501 x 1 double]
         posIf3: [1501 x 1 double]
        posIdl3: [1501 x 1 double]
      posPhise3: [1501 x 1 double]
    posThetass3: [1501 x 1 double]
              T: [1501 x 1 double]
           time: [1501 x 1 double]
           Iapp: [1501 x 1 double]
          Vcell: [1501 x 1 double]
         negSOC: [1501 x 101 double]
         posSOC: [1501 x 101 double]
           Ifdl: [1501 x 42 double]
             If: [1501 x 42 double]
            Idl: [1501 x 42 double]
           Phis: [1501 x 42 double]
          Phise: [1501 x 42 double]
        Thetass: [1501 x 42 double]
```

The output comprises all variables simulated at all points in time in the simulation. Note that variables are recorded at 61 $\tilde{x}$ locations across the cell: there are 21 (each) locations in the negative and positive electrode-regions; the remainder of the points are in the separator region. The `xLocs` variable records the value of these locations.

4.13 *Where to from here?*

In this chapter, you have learned how to:

- Convert TFs into ROMs using an improved method;
- Add nonlinear corrections to simulate a cell using ROMs created for a single SOC/temperature linearization setpoint;
- Simulate over a wide operational range using output blending;
- Simulate constant-voltage and constant-power profiles;

- Simulate multicell battery packs.

You now have the background knowledge that enables you to take a physical cell and go through all the steps leading up to creating a ROM that describes that particular cell and then simulating the ROM to predict internal electrochemical variables and cell voltage.

But what should we do with the ROM? The next topics move on to explore concepts in SOC, SOH, and SOF/SOP estimation, as well as fast-charge controls.

4.A *Summary of variables*

The following list itemizes all parameters and variables introduced in this chapter.

- A and $\mathfrak{A}$, state-space-model state-transition matrices (for transient states only, and for transient plus integrator states, respectively; see Eqs. (4.1) and (4.13)).
- B and $\mathfrak{B}$, state-space-model input matrices (for transient states only, and for transient plus integrator states, respectively; see Eqs. (4.1) and (4.13)).
- C and $\mathfrak{C}$, state-space-model output matrices (for transient states only, and for transient plus integrator states, respectively; see Eqs. (4.2) and (4.13)).
- D and $\mathfrak{D}$, state-space-model direct-feedthrough matrices (identical for models having only transient states and also for models having transient and integrator states; see Eqs. (4.2) and (4.13)).
- Δt [s], sampling period of discrete-time model.
- $G(s)$, vector TF implemented by state-space model.
- $\bar{G}(s)$, vector TF $G(s)$ combined with ZOH (see Eq. (4.3)).
- i, the oversizing parameter used by the HRA (see Eq. (4.7)).
- $k_{\max}$, number of aliased frequency spectra when converting continuous-time to discrete-time frequency response (see Eq. (4.6)).
- $\mathcal{O}_i$, extended observability matrix, used by HRA (see Eq. (4.7)).
- $\mathcal{T}_i$, a Toeplitz matrix, part of the development of the HRA (see Eq. (4.7)).
- $x[k]$ and $\mathfrak{X}[k]$, time-varying state of state-space model (transient states only, and transient plus integrator states, respectively; see Eqs. (4.1) and (4.13)).
- $y[k]$, linear outputs from state-space model (see Eqs. (4.1) and 4.15). No notational distinction is made between outputs from an integrator-included or transient-only model; $y[k]$ is interpreted by context.

5
Electrochemical Internal Variables Estimation

Until this point, we have presented topics related to implementing physics-based models of a commercial lithium-ion battery cell. These have built on the concepts taught in Vol. I of this series: refining the model to include double-layer and constant-phase-element effects, refining the TFs to eliminate an unnecessary assumption previously made in their derivations, using these models in conjunction with data collected in the laboratory to estimate the parameter values of a commercial cell, and then implementing improved reduced-order versions of these models in efficient time-domain simulations.

From this point forward, we change direction to present topics related to implementing battery-management algorithms using these physics-based ROMs. We will build on the concepts taught in Vol. II of this series: showing how to estimate the internal electrochemical state of a cell, how to diagnose and predict degradation over the life of the cell, how to compute aggressive but safe fast-charge profiles, and how to compute dynamic power limits as the cell operates. These topics are presented iconically in Fig. 5.1.

Vol. II showed that a BMS must produce real-time SOC estimates for all battery-pack cells and that accurate, conservative, voltage-based power-limit estimates could be made if estimates of all ele-

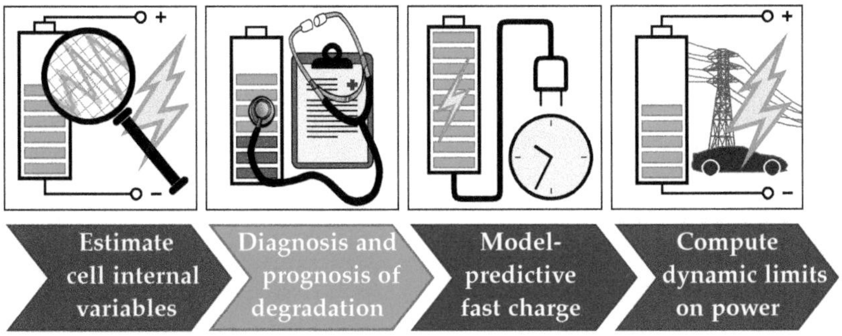

Figure 5.1: Topics in lithium-ion battery controls that we cover in this volume.

ments of the ECM state vector (plus confidence bounds) were available. The same principle is true when using electrochemical models.

This chapter focuses on making real-time estimates of the ROM's state[1] (with associated confidence bounds) and leveraging these estimates to compute estimates of SOC and any desired set of cell internal electrochemical variables,[2] also with confidence bounds. These estimates will be used in later chapters in computations needed for fast charge and for computing power-limits estimates.

[1] Note that the model's state is distinct from its parameters. The state $x[k]$ changes quickly but the parameters either do not change at all or change very slowly due to aging.

[2] For example, normalized lithium concentrations θ_s, θ_e, potentials ϕ_s, ϕ_e, fluxes i_{f+dl}, and so forth.

As in Vol. II, we present the model-based approach illustrated in Fig. 5.2. The ROM is used in a real-time algorithm to adapt a state estimate: measurements of the electrical current flowing through the physical cell are propagated through the cell ROM, updating the ROM's internal state; the ROM is used to predict cell voltage, which is compared to the voltage measured from the physical cell; the difference between measured and predicted voltages is used to adapt the ROM's state, which serves as an estimate of the cell's state.

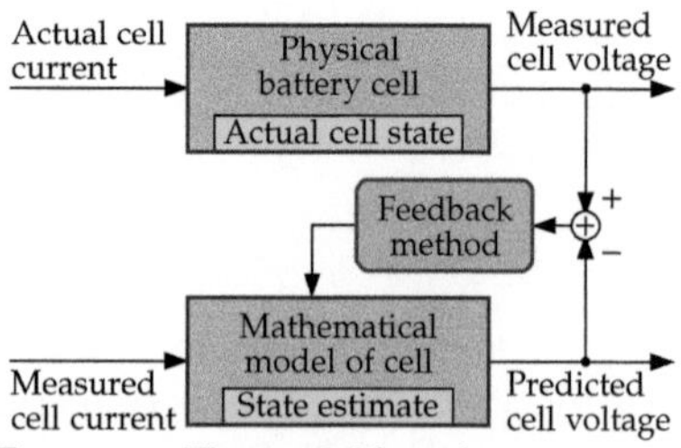

Figure 5.2: The model-based state-estimation approach.

The ROM's state vector does not directly give us estimates of the cell's internal electrochemical variables since the ROM's state vector is itself only a semi-physical quantity. We must add steps to compute the estimates we desire to find of cell internal electrochemical variables (and confidence bounds on these estimates) from the ROM's state vector.

In this chapter, we will first review the topic of sequential probabilistic inference (SPI), and will add a necessary additional step to the process for application to the ROM. We will then review approaches for estimating the mean and covariance of a random variable that is produced as the output of a nonlinear function of an input random variable. We will apply these two concepts to the output-blended ROM to develop an internal-variables estimator.[3]

[3] The earliest work in this area was by Stetzel, who investigated both EKF and SPKF with model-blended physics-based ROMs in:

- Kirk D. Stetzel, "Model-based estimation of battery cell internal physical state," Master's thesis, University of Colorado Colorado Springs, 2014.

Together with some colleagues, he published his EKF results in:

- Kirk D. Stetzel, Lukas L. Aldrich, M. Scott Trimboli, and Gregory L. Plett, "Electrochemical state and internal variables estimation using a reduced-order physics-based model of a lithium-ion cell and an extended Kalman filter," *Journal of Power Sources*, 278:490–505, 2015.

Miguel et al. were the first to investigate xKF with output-blended ROMs. Some of their results are published in:

- E. Miguel, Gregory L. Plett, M. Scott Trimboli, I. Lopetegi, Laura Oca, Unai Iraola, and E. Bekaert, "Electrochemical model and sigma point kalman filter based online oriented battery model," *IEEE Access*, 9:98072–98090, 2021.

The topics that we cover initially review some important concepts from Chap. 3 in Vol. II in summary form. You may wish to refer back to that volume for more detail. The primary intent of this review is to allow us to springboard off of these known topics to learn how to apply the methods using the ROM.

5.1 *Review of sequential probabilistic inference*

When applying model-based estimation, we start by assuming a general, possibly nonlinear, cell model:

$$x[k] = f(x[k-1], u[k-1], w[k-1]) \tag{5.1}$$

$$y[k] = h(x[k], u[k], v[k]), \tag{5.2}$$

where $u[k]$ is a known (deterministic/measured) input signal, $x[k]$ is the model's state, $w[k]$ is a process-noise random input, and $v[k]$ is sensor-noise random input.[4,5] Notice that the process-noise input has influence on the value of $x[k]$ whereas the sensor-noise input does not. Since the model's state is influenced by a random input, we must treat it as a random variable in our analysis.

Equations (5.1) and (5.2) define a nonlinear state-space model form, adopted from control-systems theory. With appropriate design and parameterization of $f(\cdot)$ and $h(\cdot)$, it can be applied to describe many dynamic systems, not only lithium-ion battery cells.[6] Consequently, by representing our cell models in this standard form, we can import estimation and control methods from control-systems theory that are designed to operate with models having this structure.

For our application, $u[k] = i_{\text{app}}[k]$, $w[k]$ is measurement error on the current sensor, and $v[k]$ is measurement error on the voltage sensor. We run into a notational conflict when defining $y[k]$, however. The literature in control systems very commonly uses $y[k]$ to denote the measured output of a system (i.e., cell voltage). This is in fact what we intend to represent by $y[k]$ at this point. However, we have already denoted the intermediate linear outputs of the ROM as $y[k]$ in Chap. 4. A further complication is that the $y[k]$ in Eq. (5.2) is a scalar quantity if it represents cell voltage, whereas the $y[k]$ in Chap. 4 is a vector quantity representing all the ROM linear outputs.

We choose to retain compatibility with the control-systems literature for the time being, while we review the SPI solution, since this is a theory that has broad application beyond cell state estimation. That is, we think about $y[k]$ in an abstract, general way at this point. When we apply the theory more specifically to our ROMs, we will then be more careful to distinguish between the linear intermediate outputs of the ROMs and the nonlinear voltage prediction of the ROMs. At that point, we will use $y[k]$ specifically to denote the linear intermediate outputs, as in Chap. 4. We will use $v_{\text{cell}}[k]$ to denote the nonlinear voltage prediction.

SPI is a process for estimating a dynamic system's present state $x[k]$ using a dynamic model and measurements of the system's noisy input and output over time. We define the set of all output measurements until timestep k to be $\mathbb{Y}[k]$. That is, $\mathbb{Y}[k] = \{y[0], \cdots, y[k]\}$. These measurements give us glimpses inside the true system. Based on the measurements and a model, we estimate the system's state. However, the existence of process- and sensor-noise randomness means we can never compute the state exactly.

This is illustrated conceptually in Fig. 5.3. The system has a true state vector that evolves over time, but we are unable to observe or measure its values directly. Even if we know the exact values of the deterministic input signal $u[k]$, we cannot make a perfect prediction of the state due to the unknown values of process noise. Therefore,

[4] At this point, the nonlinear model and its state are used in a generic way (not specific to the ROM). Later, we will apply SPI directly to the problem of estimating a ROM's state. At that time, we will make a distinction between the ROM's transient states $x[k]$ and its integrator-included states $\mathfrak{X}[k]$, but not before.

[5] This overall topic depends heavily on certain topics in probability theory, covered in detail Vol. II. We won't devote space to reviewing these topics here, but you may wish to do so on your own.

[6] In Eqs. (5.1) and (5.2), the (possibly nonlinear) functions $f(\cdot)$ and $h(\cdot)$ may even be time-varying, but we generally omit the time dependency from the notation for ease of understanding.

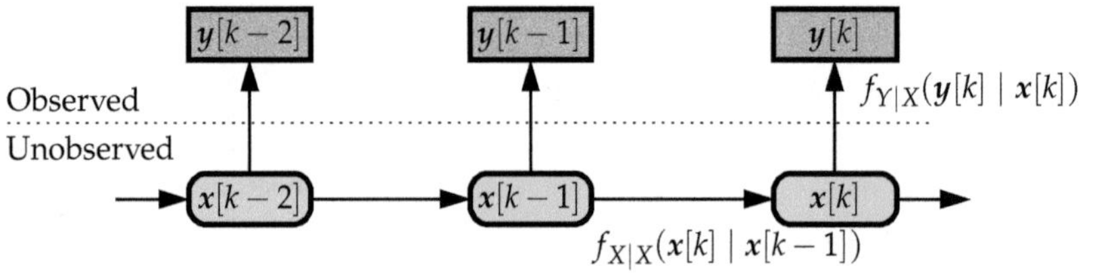

Figure 5.3: The sequential-probabilistic-inference concept.

we model the evolution of the state from one timestep to the next using the conditional probability density function $f_{X|X}(x[k] \mid x[k-1])$ where the uncertainty being modeled is that of the random process noise. We *are* able to observe the output of the system, but we recognize that it is not a deterministic function of the state due to the unknown values of measurement noise. Therefore, we model the connection between the state and the measured output using the conditional probability density function $f_{Y|X}(y[k] \mid x[k])$ where the uncertainty being modeled is that of the random measurement noise.

The SPI solution was derived in detail in Vol. II. We will not duplicate that content here;[7] instead, we summarize its steps. In the notation that we use,

[7] Note that in order to simplify the computations required to estimate the state, we assumed that both $f_{X|X}(x[k] \mid x[k-1])$ and $f_{Y|X}(y[k] \mid x[k])$ have Gaussian (normal) distributions. For this reason, our specialized solution is sometimes referred to as the Gaussian sequential-probabilistic-inference solution. It is possible to develop solutions without this assumption, leading to "particle-filter" methods. Particle filters can give better state estimates than the extended and sigma-point Kalman filters we review in this chapter if the system under observation has very nonlinear dynamics, but at the cost of much higher computational complexity. Lithium-ion cell dynamics are only mildly nonlinear under most operating conditions and probably do not warrant the use of particle filters for most BMS applications.

- A superscript minus sign $(-)$ indicates a predicted quantity based only on past measurements.
- A superscript plus sign $(+)$ indicates an estimated quantity based on both past and present measurements.
- A hat $(\hat{\ })$ symbol indicates a predicted or estimated quantity.
- A tilde $(\tilde{\ })$ symbol indicates an error: the difference between a true and either a predicted or estimated quantity.
- A Σ symbol is used to denote the correlation between the two arguments in its subscript (autocorrelation if only one is given):

$$\boldsymbol{\Sigma}_{xy} = \mathbb{E}[\boldsymbol{x}\boldsymbol{y}^T] \qquad \text{and} \qquad \boldsymbol{\Sigma}_{x} = \mathbb{E}[\boldsymbol{x}\boldsymbol{x}^T].$$

- Further, if the arguments $\tilde{x}$ and $\tilde{y}$ are zero mean, the same symbol represents covariance:

$$\boldsymbol{\Sigma}_{\tilde{x}\tilde{y}} = \mathbb{E}[\tilde{\boldsymbol{x}}\tilde{\boldsymbol{y}}^T] = \mathbb{E}[(\tilde{\boldsymbol{x}} - \mathbb{E}[\tilde{\boldsymbol{x}}])(\tilde{\boldsymbol{y}} - \mathbb{E}[\tilde{\boldsymbol{y}}])^T].$$

We seek a state estimate that minimizes the mean-squared error,

$$\hat{x}^{\text{MMSE}}[k](\mathbb{Y}[k]) = \arg\min_{\hat{x}[k]} \left(\mathbb{E}\left[\, \|x[k] - \hat{x}^+[k]\|_2^2 \mid \mathbb{Y}[k] \right] \right).$$

For this metric, we found that the optimal state estimate is $\hat{x}^+[k] = \mathbb{E}\big[x[k] \mid \mathbb{Y}[k]\big]$. If, as we assume, all random variables have a Gaussian distribution, this can be computed via a predict/correct mechanism:

$$\hat{x}^+[k] = \hat{x}^-[k] + \boldsymbol{L}[k]\tilde{y}[k].$$

The generic Gaussian sequential-probabilistic-inference recursion computes:

$$\hat{x}^+[k] = \hat{x}^-[k] + L[k]\big(y[k] - \hat{y}[k]\big) = \hat{x}^-[k] + L[k]\tilde{y}[k]$$
$$\Sigma^+_{\tilde{x}}[k] = \Sigma^-_{\tilde{x}}[k] - L[k]\Sigma_{\tilde{y}}[k]L^T[k],$$

where,

$$\begin{aligned}
\hat{x}^-[k] &= \mathbb{E}\big[x[k] \mid \mathbb{Y}[k-1]\big] & \Sigma^-_{\tilde{x}}[k] &= \mathbb{E}\big[(x[k] - \hat{x}^-[k])(x[k] - \hat{x}^-[k])^T\big] = \mathbb{E}\big[(\tilde{x}^-[k])(\tilde{x}^-[k])^T\big]\\
\hat{x}^+[k] &= \mathbb{E}\big[x[k] \mid \mathbb{Y}[k]\big] & \Sigma^+_{\tilde{x}}[k] &= \mathbb{E}\big[(x[k] - \hat{x}^+[k])(x[k] - \hat{x}^+[k])^T\big] = \mathbb{E}\big[(\tilde{x}^+[k])(\tilde{x}^+[k])^T\big]\\
\hat{y}[k] &= \mathbb{E}\big[y[k] \mid \mathbb{Y}[k-1]\big] & \Sigma_{\tilde{y}}[k] &= \mathbb{E}\big[(y[k] - \hat{y}[k])(y[k] - \hat{y}[k])^T\big] = \mathbb{E}\big[(\tilde{y}[k])(\tilde{y}[k])^T\big]\\
L[k] &= \mathbb{E}\big[(x[k] - \hat{x}^-[k])(y[k] - \hat{y}[k])^T\big]\Sigma^{-1}_{\tilde{y}}[k] = \Sigma^-_{\tilde{x}\tilde{y}}[k]\Sigma^{-1}_{\tilde{y}}[k].
\end{aligned}$$

Note that this is a linear recursion, even if the system is nonlinear.

Table 5.1: The generic Gaussian sequential-probabilistic-inference recursive solution (adapted from Table 3.1 in Vol. II).

In this equation, the state prediction is $\hat{x}^-[k] = \mathbb{E}\big[x[k] \mid \mathbb{Y}[k-1]\big]$ and the state prediction error is $\tilde{x}^-[k] = x[k] - \hat{x}^-[k]$.[8] Measurement innovation (what is new or unexpected in measurement) is $\tilde{y}[k] = y[k] - \hat{y}[k]$, where $\hat{y}[k] = \mathbb{E}\big[y[k] \mid \mathbb{Y}[k-1]\big]$. The estimator gain $L[k]$ is a function of $\Sigma^+_{\tilde{x}}[k]$, which may be computed as:

$$\Sigma^+_{\tilde{x}}[k] = \Sigma^-_{\tilde{x}}[k] - L[k]\Sigma_{\tilde{y}}[k]L^T[k].$$

Overall, the output of this process has two components:

1. *The state estimate.* Every iteration, we compute our best guess of the present state value, which is $\hat{x}^+[k]$.
2. *The covariance estimate.* The covariance matrix $\Sigma^+_{\tilde{x}}[k]$ gives the uncertainty of $\hat{x}^+[k]$, and can be used to compute error bounds.

The SPI solution is summarized in Table 5.1, reproduced from Vol. II.

[8] Error is always "truth minus prediction" or "truth minus estimate." We can never compute error in practice, since the truth value is not known. But, we can prove statistical results using this definition that lead to an algorithm for estimating the truth using measurable values.

5.2 *The eight-step process*

Implementing the SPI solution requires organizing the calculations of Table 5.1 into logical steps and finding ways to implement those steps. Here, we review how the solution can be divided into two main steps—a prediction step and an update step—each having three substeps. Then, we present an additional step—having two substeps—that is required due to the ROM's structure to estimate a cell's internal electrochemical variables.

General step 1a: State prediction time update.

Each time step, we first compute an updated prediction of $x[k]$ based on prior measurements and the system model:

$$\hat{x}^-[k] = \mathbb{E}\big[x[k] \mid \mathbb{Y}[k-1]\big] = \mathbb{E}\big[f(x[k-1], u[k-1], w[k-1]) \mid \mathbb{Y}[k-1]\big].$$

General step 1b: Prediction-error covariance time update.

Next, we calculate the state's prediction-error covariance matrix $\Sigma_{\tilde{x}}^-[k]$ based on prior information and the system model as:

$$\Sigma_{\tilde{x}}^-[k] = \mathbb{E}\big[(\tilde{x}^-[k])(\tilde{x}^-[k])^T\big],$$

where $\tilde{x}^-[k] = x[k] - \hat{x}^-[k]$.

General step 1c: Predict system output (i.e., cell voltage).

The final prediction substep is to predict the system's output using prior information:

$$\hat{y}[k] = \mathbb{E}\big[y[k] \mid \mathbb{Y}[k-1]\big] = \mathbb{E}\big[h(x[k], u[k], v[k]) \mid \mathbb{Y}[k-1]\big].$$

General step 2a: Estimator gain matrix $L[k]$.

General step 1 uses only prior measurements, so all computations are predictions. General step 2 updates the predictions based on a measurement of the system output at the present time to create a state estimate. This is the feedback mechanism of the solution. As part of this feedback, we will need to determine the estimator's gain matrix $L[k]$, which is defined to be:

$$L[k] = \Sigma_{\tilde{x}\tilde{y}}^-[k]\Sigma_{\tilde{y}}^{-1}[k].$$

General step 2b: State estimate measurement update.

We now apply a feedback mechanism to compute the state estimate from its predicted value and the *innovation* $y[k] - \hat{y}[k]$ which compares predicted and measured output:

$$\hat{x}^+[k] = \hat{x}^-[k] + L[k](y[k] - \hat{y}[k]).$$

General step 2c: Estimation-error covariance measurement update.

The final step in the standard SPI solution is to compute the covariance of the estimation error as:[9]

$$\Sigma_{\tilde{x}}^+[k] = \mathbb{E}\big[(\tilde{x}^+[k])(\tilde{x}^+[k])^T\big] = \Sigma_{\tilde{x}}^-[k] - L[k]\Sigma_{\tilde{y}}[k]L^T[k].$$

Additional step 3a: Estimate internal variables.

The standard SPI solution defines only steps 1a through 2c. It computes estimates of the model state vector and confidence intervals on those estimates. However, the state of our ROM is not directly physical. We desire to leverage the state estimate to compute estimates of physical internal electrochemical variables, which we model via the vector generically denoted as $z[k]$ as nonlinear functions of the state and the input:[10]

$$z[k] = g_z(x[k], u[k]).$$

Then, step 3a estimates the values of these variables as:

$$\hat{z}^+[k] = \mathbb{E}\big[z[k] \mid \mathbb{Y}[k]\big] = \mathbb{E}\big[g_z(x[k], u[k]) \mid \mathbb{Y}[k]\big].$$

[9] This covariance matrix is required by step 1b of the solution, but it has the added significant benefit of being useful to compute confidence intervals on the state estimate. That is, we have high confidence that the true $x[k]$ lies within $\hat{x}^+[k] \pm 3\sqrt{\text{diag}\,(\Sigma_{\tilde{x}}^+[k])}$.

[10] Note that in our notation the scalar quantity $z[k]$ is cell SOC but the vector quantity $z[k]$ is any set of cell internal electrochemical variables that we might wish to compute.

Prediction

Step 1a: State prediction time update
$\hat{x}^-[k] = \mathbb{E}\big[x[k] \mid \mathbb{Y}[k-1]\big] = \mathbb{E}\big[f(x[k-1], u[k-1], w[k-1]) \mid \mathbb{Y}[k-1]\big]$.

Step 1b: Prediction-error covariance time update
$\Sigma^-_{\tilde{x}}[k] = \mathbb{E}\big[(\tilde{x}^-[k])(\tilde{x}^-[k])^T\big] = \mathbb{E}\big[(x[k] - \hat{x}^-[k])(x[k] - \hat{x}^-[k])^T\big]$.

Step 1c: Predict system output
$\hat{y}[k] = \mathbb{E}\big[y[k] \mid \mathbb{Y}[k-1]\big] = \mathbb{E}\big[h(x[k], u[k], v[k]) \mid \mathbb{Y}[k-1]\big]$.

Correction

Step 2a: Estimator gain matrix
$L[k] = \Sigma^-_{\tilde{x}\tilde{y}}[k]\Sigma_{\tilde{y}}[k]^{-1}$.

Step 2b: State estimate measurement update
$\hat{x}^+[k] = \hat{x}^-[k] + L[k](y[k] - \hat{y}[k])$.

Step 2c: Estimation-error covariance measurement update
$\Sigma^+_{\tilde{x}}[k] = \Sigma^-_{\tilde{x}}[k] - L[k]\Sigma_{\tilde{y}}[k]L[k]^T$.

Extension

Step 3a: Estimate internal variables
$\hat{z}^+[k] = \mathbb{E}\big[z[k] \mid \mathbb{Y}[k]\big] = \mathbb{E}\big[g_z(x[k], u[k]) \mid \mathbb{Y}[k]\big]$.

Step 3b: Internal variables estimation-error covariance
$\Sigma^+_{\tilde{z}}[k] = \mathbb{E}\big[(\tilde{z}^+[k])(\tilde{z}^+[k])^T\big]$.

Figure 5.4: The eight steps of the modified sequential-probabilistic-inference solution.

Additional step 3b: Internal variables estimation-error covariance.
Finally, we compute the covariance:

$$\Sigma^+_{\tilde{z}}[k] = \mathbb{E}\big[(\tilde{z}^+[k])(\tilde{z}^+[k])^T\big].$$

KEY POINT: The estimator output comprises the estimate $\hat{z}^+[k]$ and its error covariance $\Sigma^+_{\tilde{z}}[k]$. As a result, we have high confidence that the true $z[k]$ lies within $\hat{z}^+[k] \pm 3\sqrt{\text{diag}(\Sigma^+_{\tilde{z}}[k])}$.

The estimator then waits until the next sample interval, updates k, and proceeds to step 1a.

The solution is summarized in Fig. 5.4. To implement these steps, we must find procedures to compute the expected values in the equations. Extended Kalman filters (EKFs) and sigma-point Kalman filters (SPKFs) use analytic and numeric linearization, respectively, to do so. We will present both approaches applied to output-blended, but first we review the structure of those ROMs and introduce notation that will help show how to reduce computational requirements.

5.3 *Setup for xKF with output-blended models*

Before applying xKF[11] for state and internal-variables estimation, we review the output-blending method and refine some notation. Recall that each ROM has its own state vector, modeled as:

[11] Whenever the distinction between EKF and SPKF is not important, we refer to the family of nonlinear Kalman filters as xKF.

$$\mathfrak{X}_j[k] = \mathfrak{A}_j\mathfrak{X}_j[k-1] + \mathfrak{B}_j u[k-1],$$

for all $1 \le j \le N$, where $\mathfrak{X}_j[k]$ is the state of integrator-augmented model j, $\mathfrak{A}_j$ and $\mathfrak{B}_j$ are the constant state-equation matrices for precomputed integrator-augmented model j, and N is the number of precomputed models. Since all models will have identical integral states, it would be redundant to attempt to estimate this value independently for each one. Instead, the integration state should be factored out of all models and estimated separately.

For example, if $\mathfrak{X}_j[k]$ is 5×1, when we remove the integration state and retain only transient states, we have a 4×1 $\boldsymbol{x}_j[k]$. Similarly, the 5×5 $\mathfrak{A}_j$ matrix becomes a 4×4 $\boldsymbol{A}_j$ and the 5×1 $\mathfrak{B}_j$ vector becomes a 4×1 $\boldsymbol{B}_j$. Then the transient states of model j are updated via:

$$\boldsymbol{x}_j[k] = \boldsymbol{A}_j \boldsymbol{x}_j[k-1] + \boldsymbol{B}_j u[k-1].$$

The scalar integral state $x_0[k]$ common to all models is updated as:

$$x_0[k] = x_0[k-1] + u[k-1].$$

Considering all N models together, this update is the same as:[12]

$$\underbrace{\begin{bmatrix} \boldsymbol{x}_1[k] \\ \vdots \\ \boldsymbol{x}_N[k] \\ x_0[k] \end{bmatrix}}_{\mathbf{X}[k]} = \underbrace{\begin{bmatrix} \boldsymbol{A}_1 & & & \mathbf{0} \\ & \ddots & & \\ & & \boldsymbol{A}_N & \\ \mathbf{0} & & & 1 \end{bmatrix}}_{\mathbf{A}} \underbrace{\begin{bmatrix} \boldsymbol{x}_1[k-1] \\ \vdots \\ \boldsymbol{x}_N[k-1] \\ x_0[k-1] \end{bmatrix}}_{\mathbf{X}[k-1]} + \underbrace{\begin{bmatrix} \boldsymbol{B}_1 \\ \vdots \\ \boldsymbol{B}_N \\ 1 \end{bmatrix}}_{\mathbf{B}} u[k-1]. \tag{5.3}$$

[12] Notice that this equation defines somewhat nonstandard notation. $\mathbf{X}[k]$ is a vector quantity, but we use a capital letter—normally used only for matrices—to emphasize that this is a block-vector.

Next, output blending uses bilinear interpolation to combine the linear outputs of the models corresponding to the four setpoints closest to the present operating SOC and temperature. The linear ROM output vector is computed as (per Eqs. (4.26) and (4.27)):

$$\boldsymbol{y}[k] = \gamma_{0,0}\boldsymbol{y}_{0,0}[k] + \gamma_{0,1}\boldsymbol{y}_{0,1}[k] + \gamma_{1,0}\boldsymbol{y}_{1,0}[k] + \gamma_{1,1}\boldsymbol{y}_{1,1}[k] \tag{5.4}$$

$$= \sum_{j=1}^{N} \gamma_j \boldsymbol{y}_j[k]. \tag{5.5}$$

Notice that although we must update the *state vector* of all models every time step, we are not required to compute the *output vector* of any model that is not being used for the present calculation of $\boldsymbol{y}_k$. That is, in the set of weighting constants $\{\gamma_j\}$ for $1 \le j \le N$ in Eq. (5.5), only the four closest models have nonzero weights, denoted as $\gamma_{0,0}$, $\gamma_{0,1}$, $\gamma_{1,0}$, and $\gamma_{1,1}$ in Eq. (5.4). Recognizing this fact enables us to reduce the required computations of xKF considerably.

The linear outputs described by Eq. (5.5) could be computed using the full-sized state vectors as:

$$\boldsymbol{y}_j[k] = \mathfrak{C}_j \boldsymbol{x}_j[k] + \mathfrak{D}_j u[k].$$

We can also use integrator-removed state vectors if we recall that C_j is the first n columns of $\mathfrak{C}_j$ and define C_0 to be the final column. Then, also recalling that $D_j = \mathfrak{D}_j$,

$$y_j[k] = \left[\begin{array}{c:c} C_j & C_0 \end{array}\right] \begin{bmatrix} x_j[k] \\ \hdashline x_0[k] \end{bmatrix} + D_j u[k].$$

The C_0 vector comprises the residues of the integration state for each linearized model output. Most entries in C_0 are zero since most electrochemical variables do not integrate the input current. The residues are weak functions of SOC and temperature; we choose to compute them using analytic expressions during execution rather than relying on C_0 from precomputed ROMs. Even so, we will retain C_0 as a placeholder in our notation for now.

Expanding Eq. (4.27), the output-blended final result is then:

$$y[k] = \sum_{i=1}^{N} \gamma_j \left(C_j x_j[k] + C_{0,j} x_0[k] + D_j u[k] \right). \tag{5.6}$$

The summation is written for $1 \le j \le N$ but since only four values of γ_j are nonzero, computation is reduced by considering only the nonzero terms in the summation. Internal electrochemical variables and cell voltage are computed by applying nonlinear corrections to $y[k]$. For example, we compute the nonlinear variables as:

$$z[k] = g_z(y[k], u[k]). \tag{5.7}$$

We further combine Eqs. (5.6) and (5.7) to compute cell voltage:

$$v_{\text{cell}}[k] = g_v(\mathbf{X}[k], u[k]).$$

5.4 *EKF and SPKF principles*

We wish to apply SPI to the problem of estimating a cell's internal states and variables. The challenge is that some of our model equations are nonlinear. When the inputs to these equations are random variables having known mean and covariance, we desire to compute the mean and covariance of the outputs of these equations. However, there is no general solution that does so; instead, we must make approximations. The EKF and SPKF approximate the mean and covariance of the output of a nonlinear function in different ways. This section reviews the EKF and SPKF approaches, which were developed in detail in Vol. II.

5.4.1 *The extended Kalman filter (EKF)*

The EKF makes two simplifying assumptions when adapting the SPI equations to a nonlinear system:

1. When computing means of the output of a nonlinear function, EKF assumes $\mathbb{E}[\text{fn}(x)] \approx \text{fn}(\mathbb{E}[x])$, which is not true in general;
2. When computing covariance estimates, EKF uses Taylor-series expansion to linearize the system equations around the present operating point.

We will apply the EKF assumptions to the equations for an output-blended ROM in Sect. 5.5. Here, we see generic examples of how the two assumptions are applied.

Consider a nonlinear function computing $y = f(x)$. The EKF approach approximates the expected value of y as follows:[13]

$$\begin{aligned}\bar{y} &= \mathbb{E}[f(x)] \\ &\approx f(\mathbb{E}[x]) = f(\bar{x}).\end{aligned}$$

[13] Notice that we use the bar symbol ($\bar{\cdot}$) to indicate a random-variable's mean (expected) value.

To see how the EKF approach approximates the covariance of y, we first make an approximation for $\tilde{y}$:

$$\tilde{y} = y - \bar{y} = f(x) - f(\bar{x}).$$

The first term is expanded as a Taylor series around $\bar{x}$:

$$y \approx f(\bar{x}) + \underbrace{\left.\frac{\mathrm{d}f(x])}{\mathrm{d}x}\right|_{x=\bar{x}}}_{\text{Defined as } \widehat{A}} \tilde{x},$$

where $\tilde{x} = x - \bar{x}$. This gives $\tilde{y} \approx \widehat{A}\tilde{x}$. Substituting this to find the covariance, the EKF approach calculates:

$$\Sigma_{\tilde{y}} = \mathbb{E}[\tilde{y}\tilde{y}^T] \approx \widehat{A}\Sigma_{\tilde{x}}\widehat{A}^T.$$

These two basic approaches are used throughout the EKF derivation.

5.4.2 *The sigma-point Kalman filter (SPKF)*

The SPKF uses weighted averages of function output values corresponding to carefully chosen function input values to estimate means and covariances. It chooses a set of sigma points $\mathcal{X}$ so that the weighted mean and covariance of $\mathcal{X}$ exactly matches the mean $\bar{x}$ and covariance $\Sigma_{\tilde{x}}$ of the random variable that is the input to the nonlinear function. The nonlinear function is then evaluated for each of these points, producing a transformed set of sigma points $\mathcal{Y}$. The mean $\bar{y}$ and covariance $\Sigma_{\tilde{y}}$ of the random variable y are then approximated by the mean and covariance of these transformed points $\mathcal{Y}$.

Specifically, if input random variable x has dimension L, mean $\bar{x}$, and covariance $\Sigma_{\tilde{x}}$, then $p+1 = 2L+1$ sigma points are generated as:

$$\mathcal{X} = \left\{\bar{x}, \bar{x} + \gamma\sqrt{\Sigma_{\tilde{x}}}, \bar{x} - \gamma\sqrt{\Sigma_{\tilde{x}}}\right\}. \tag{5.8}$$

The notation in Eq. (5.8) is mathematical shorthand and requires some explanation. First, braces $\{\cdot\}$ are used to underscore the fact that $\mathcal{X}$ is a set of vectors. We will find it convenient to store this set in a compact form as a matrix, where every column of the matrix is one of the members of the set; nonetheless, $\mathcal{X}$ is technically a set.

The members of $\mathcal{X}$ are indexed from 0 to p. The zeroth element of $\mathcal{X}$ is the mean $\bar{x}$ of the pdf being modeled. The next L elements of the set are written compactly as $\bar{x} + \gamma\sqrt{\boldsymbol{\Sigma}_{\tilde{x}}}$. In this notation, the matrix square root $\boldsymbol{R} = \sqrt{\boldsymbol{\Sigma}}$ computes a result such that $\boldsymbol{\Sigma} = \boldsymbol{R}\boldsymbol{R}^T$. Usually, the efficient Cholesky decomposition is used, resulting in a lower-triangular square matrix $\boldsymbol{R}$ of same dimension as $\boldsymbol{\Sigma}_{\tilde{x}}$.[14] In the equation, γ is a weighting constant that can be adjusted to tune the performance of the sigma-point method.[15]

[14] Take care: MATLAB, by default, returns an upper-triangular matrix. The `'lower'` optional argument must be used to arrive at the correct result.

[15] Note that the γ tuning factor of the sigma-point methods is different from the γ_j output-blending weighting variables.

So, $\bar{x}$ is a vector and $\gamma\sqrt{\boldsymbol{\Sigma}_{\tilde{x}}}$ is a matrix. They are of incompatible dimensions to be added, so the notation $\bar{x} + \gamma\sqrt{\boldsymbol{\Sigma}_{\tilde{x}}}$ makes no sense per standard linear algebra. Instead, what the notation *means* is that the vector $\bar{x}$ is added separately to every column of $\gamma\sqrt{\boldsymbol{\Sigma}_{\tilde{x}}}$ to produce a resulting matrix of the same size as $\boldsymbol{\Sigma}_{\tilde{x}}$. The L columns of this output matrix comprise sigma points 1 through L in $\mathcal{X}$.

Similarly, the final L sigma points of $\mathcal{X}$ are denoted as "$\bar{x} - \gamma\sqrt{\boldsymbol{\Sigma}_{\tilde{x}}}$" which means "subtract the columns of $\gamma\sqrt{\boldsymbol{\Sigma}_{\tilde{x}}}$ from $\bar{x}$ to make L sigma points." These are the elements in $\mathcal{X}$ indexed from $L+1$ to $2L$.

The weighted mean and covariance of the elements of $\mathcal{X}$ are equal to the original mean and covariance of x for some $\{\gamma, \alpha_i^{(m)}, \alpha_i^{(c)}\}$ via:

$$\bar{x} = \sum_{i=0}^{p} \alpha_i^{(m)} \mathcal{X}_i \qquad \text{and} \qquad \boldsymbol{\Sigma}_{\tilde{x}} = \sum_{i=0}^{p} \alpha_i^{(c)} (\mathcal{X}_i - \bar{x})(\mathcal{X}_i - \bar{x})^T,$$

where $\mathcal{X}_i$ is the ith vector member of the set $\mathcal{X}$, and both $\alpha_i^{(m)}$ and $\alpha_i^{(c)}$ are real scalars where $\alpha_i^{(m)}$ and $\alpha_i^{(c)}$ must both sum to one. The $\alpha_i^{(m)}$ are weighting constants used when computing the mean and the $\alpha_i^{(c)}$ are weighting constants used when computing the covariance. They are *tuning parameters* of the sigma-point methods.

The various sigma-point methods differ only in the choices taken for these weighting constants. Table 5.2 lists values used by the two most popular methods, the unscented Kalman filter (UKF) and the central-difference Kalman filter (CDKF). The original derivations of these two methods were quite different but the final steps are essentially identical. CDKF has only one tuning parameter h, so implementation is simpler. It also has marginally higher theoretic accuracy when the distributions are indeed Gaussian. However, UKF has more tuning parameters, so can be made to work better in practice when the distributions are not Gaussian.

Each one of the input random-variable sigma points $\mathcal{X}_i$ in the set $\mathcal{X}$ is passed through the nonlinear function $f(\cdot)$ to produce a

Method	γ	$\alpha_0^{(m)}$	$\alpha_k^{(m)}$	$\alpha_0^{(c)}$	$\alpha_k^{(c)}$
UKF	$\sqrt{L+\lambda}$	$\frac{\lambda}{L+\lambda}$	$\frac{1}{2(L+\lambda)}$	$\frac{\lambda}{L+\lambda}+(1-\alpha^2+\beta)$	$\frac{1}{2(L+\lambda)}$
CDKF	h	$\frac{h^2-L}{h^2}$	$\frac{1}{2h^2}$	$\frac{h^2-L}{h^2}$	$\frac{1}{2h^2}$

Table 5.2: Constants for the sigma-point methods. $\lambda = \alpha^2(L+\kappa) - L$ is a scaling parameter. Note that this $(10^{-2} \le \alpha \le 1)$ is different from $\alpha_k^{(m)}$ and $\alpha_k^{(c)}$. κ is either 0 or $3-L$. β incorporates prior information. h may take any positive value. For Gaussian RVs, $\beta = 2$ or $h = \sqrt{3}$.

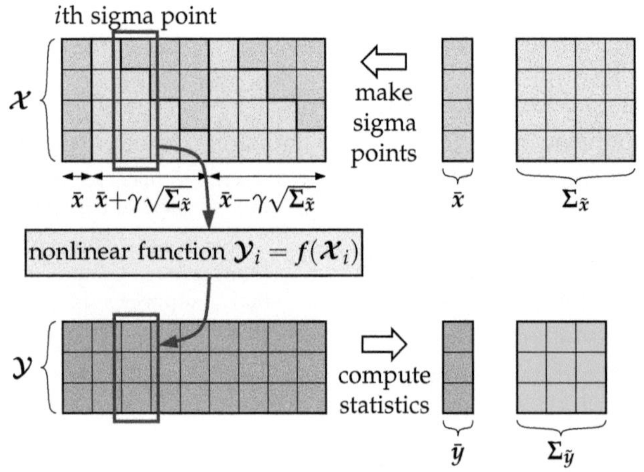

Figure 5.5: Visualizing the sigma-point approach.

corresponding output sigma point, $\mathcal{Y}_i = f(\mathcal{X}_i)$. Then the output mean and covariance are computed as well:

$$\bar{y} = \sum_{i=0}^{p} \alpha_i^{(m)} \mathcal{Y}_i \qquad \text{and} \qquad \Sigma_{\tilde{y}} = \sum_{i=0}^{p} \alpha_i^{(c)} (\mathcal{Y}_i - \bar{y})(\mathcal{Y}_i - \bar{y})^T. \tag{5.9}$$

Fig. 5.5 illustrates the overall process. On the top right, we start with the mean vector and covariance matrix of the input random variable. In the example, $\bar{x}$ is a 4-vector and $\Sigma_{\tilde{x}}$ is a 4×4 matrix. From these inputs, we create $2L+1 = 9$ sigma points, which are stored compactly as the columns of a 4×9 matrix. The zeroth sigma point is equal to $\bar{x}$ so is drawn with the same shading as the original $\bar{x}$. The next L sigma points are equal to the columns of $\gamma\sqrt{\Sigma_{\tilde{x}}}$ added to $\bar{x}$. Since $\sqrt{\Sigma_{\tilde{x}}}$ is lower triangular, all values above the diagonal are zero, and so the result when adding it to $\bar{x}$ differs from $\bar{x}$ only in the lower-triangular region. This is why the elements above the diagonal in the figure are drawn using the same shading as $\bar{x}$, but the lower-triangular elements are drawn with the same shading as $\Sigma_{\tilde{x}}$. The final L columns are computed as $\bar{x} - \gamma\sqrt{\Sigma_{\tilde{x}}}$ and stored in a similar way.

Next, each of the sigma points $\mathcal{X}_i$ in the set $\mathcal{X}$ is individually passed through the nonlinear function to produce a corresponding output sigma point $\mathcal{Y}_i$. These output sigma points form the set $\mathcal{Y}$ and are collected together in a matrix for convenient storage. In the figure, we have emphasized that the function output need not have the same dimension as the function input. In this case, input 4-

vectors produce output 3-vectors. The weighting constants $\alpha_i^{(m)}$, $\alpha_i^{(c)}$, and γ, as well as number of sigma points $p + 1 = 2L + 1$ are inherited from the dimension of the input x.

Finally, the output mean and covariance are computed from the sigma points in $\mathcal{Y}$ using Eq. (5.9). In this case, $\bar{y}$ is a 3-vector and $\Sigma_{\tilde{y}}$ is a 3×3 matrix. Note also that the elements corresponding to $\mathcal{Y}$ are drawn using a similar color scheme to those of $\mathcal{X}$, but with a different shade to emphasize that they are different quantities.

5.5 *EKF with the output-blended model*

Having reviewed the basic approaches that EKF and SPKF take to approximate the statistics of a random variable, we seek to apply both methods to the output-blended ROM. We begin with the EKF method, showing how to implement each equation in Fig. 5.4.

Step 1a: State prediction time update.

The first EKF step, every iteration, is to predict the present value of the state using only prior information. We begin by defining:

$$\widehat{\mathbf{X}}^-[k] = \mathbb{E}[\mathbf{X}[k] \mid v_{\text{cell}}[0] \dots v_{\text{cell}}[k-1]].$$

We modify our state equation to consider process noise $w[k]$. We assume that this is white Gaussian zero-mean current-sensor noise added directly to $u[k]$:

$$\mathbf{X}[k] = \mathbf{A}\mathbf{X}[k-1] + \mathbf{B}\Big(u[k-1] + w[k-1]\Big).$$

When evaluating this expectation, we find that some computational simplifications (with respect to EKF on a generic nonlinear system) result from the state equation being linear for the ROM we are using. The state prediction we wish to compute is:

$$\begin{aligned}\widehat{\mathbf{X}}^-[k] &= \mathbb{E}\big[\mathbf{X}[k] \mid \mathbb{V}_{\text{cell}}[k-1]\big] \\ &= \mathbb{E}\big[\mathbf{A}\mathbf{X}[k-1] + \mathbf{B}\left(u[k-1] + w[k-1]\right) \mid \mathbb{V}_{\text{cell}}[k-1]\big]. \\ &= \mathbf{A}\widehat{\mathbf{X}}^+[k-1] + \mathbf{B}u[k-1],\end{aligned}$$

where $\widehat{\mathbf{X}}^+[k-1] = \mathbb{E}[\mathbf{X}[k-1] \mid v_{\text{cell}}[0] \dots v_{\text{cell}}[k-1]]$ is the prior state estimate produced by the EKF. Notice that we are updating the predicted state for all models, not only the models that are presently being blended to produce a voltage estimate.

This can be implemented on a per-model basis, $1 \le j \le N$:[16]

$$\begin{aligned}\hat{x}_j^-[k] &= A_j\hat{x}_j^+[k-1] + B_j u[k-1] \\ \hat{x}_0^-[k] &= \hat{x}_0^+[k-1] + u[k-1].\end{aligned}$$

[16] Noticing that $\mathbf{A}$ is diagonal and that $\mathbf{B}$ is a vector of unit values can greatly simplify the computational complexity of an implementation in many of the xKF steps. For example, Hadamard products can be used to compute quantities like $\mathbf{A}\widehat{\mathbf{X}}^+[k-1]$.

Step 1b: Prediction-error covariance time update.

The next step of the EKF, every iteration, is to compute the covariance of the state prediction error. This describes the EKF's uncertainty regarding the quality of its state prediction. Conversely, it describes in an inverse way the confidence that the EKF has in the quality of its state prediction. We define:

$$\boldsymbol{\Sigma}_{\tilde{\mathbf{X}}}^{-}[k] = \mathbb{E}[(\mathbf{X}[k] - \widehat{\mathbf{X}}^{-}[k])(\mathbf{X}[k] - \widehat{\mathbf{X}}^{-}[k])^{T}],$$

which leads to:

$$\boldsymbol{\Sigma}_{\tilde{\mathbf{X}}}^{-}[k] = \mathbf{A}\boldsymbol{\Sigma}_{\tilde{\mathbf{X}}}^{+}[k-1]\mathbf{A}^{T} + \mathbf{B}\Sigma_{\widetilde{w}}\mathbf{B}^{T},$$

where $\boldsymbol{\Sigma}_{\tilde{\mathbf{X}}}^{+}[k-1] \in \mathbb{R}^{(Nn+1)\times(Nn+1)}$ is the prior covariance matrix of the state estimation error produced by the EKF and $\Sigma_{\widetilde{w}} \in \mathbb{R}$ is the process-noise covariance.

This can also be implemented on a per-model basis, $1 \leq j \leq N$:

$$\begin{aligned}
\boldsymbol{\Sigma}_{\tilde{x}_j}^{-}[k] &= \boldsymbol{A}_j\boldsymbol{\Sigma}_{\tilde{x}_j}^{+}[k-1]\boldsymbol{A}_j^{T} + \boldsymbol{B}_j\Sigma_{\widetilde{w}}\boldsymbol{B}_j^{T} \\
\Sigma_{\tilde{x}_0}^{-}[k] &= \Sigma_{\tilde{x}_0}^{+}[k-1] + \Sigma_{\widetilde{w}}.
\end{aligned}$$

Step 1c: Predict system output.

We next predict the voltage we will measure. That is, we compute:

$$\hat{v}_{\text{cell}}[k] = \mathbb{E}\big[v_{\text{cell}}[k] \mid \mathbb{V}_{\text{cell}}[k-1]\big].$$

Before proceeding, we modify the model's voltage equation to add white Gaussian zero-mean measurement noise $v[k]$:

$$v_{\text{cell}}[k] = g_v(\mathbf{X}[k], u[k]) + v[k]. \tag{5.10}$$

We can simplify the remaining EKF steps by recalling that $v_{\text{cell}}[k]$ depends only on the four models combined by output blending, not on the entire model set. Therefore, we define a new vector,

$$\mathbf{X}_\gamma[k] = \begin{bmatrix} x_{0,0}[k] \\ x_{0,1}[k] \\ x_{1,0}[k] \\ x_{1,1}[k] \\ x_0[k] \end{bmatrix},$$

where the subscripts "0,0" (etc.) denote the model being blended in the same fashion as used to describe y in Eq. (5.4). Then

$$v_{\text{cell}}[k] = g_v(\mathbf{X}_\gamma[k], u[k]) + v[k],$$

and since the noise is zero-mean and additive,

$$\hat{v}_{\text{cell}}[k] = \mathbb{E}\big[g_v(\mathbf{X}_\gamma[k], u[k]) \mid \mathbb{V}_{\text{cell}}[k-1]\big].$$

Since $g_v(\cdot)$ is nonlinear, we now implement EKF Assumption 1:

$$\hat{v}_{\text{cell}}[k] \approx g_v(\widehat{\mathbf{X}}_\gamma^-[k], u[k]),$$

where $\widehat{\mathbf{X}}_\gamma^-$ is a subset of $\widehat{\mathbf{X}}^-$ from step 1a. This means that:

- We use the four individual state estimates from $\widehat{\mathbf{X}}^-$ to produce four different linear-variable estimates $\hat{\boldsymbol{y}}_{0,0}$, $\hat{\boldsymbol{y}}_{0,1}$, $\hat{\boldsymbol{y}}_{1,0}$, and $\hat{\boldsymbol{y}}_{1,1}$.
- We use output blending to combine these four estimates into a single linearized output vector $\hat{\boldsymbol{y}}$.
- We apply nonlinear corrections to $\hat{\boldsymbol{y}}$ to compute $\hat{v}_{\text{cell}}$.

Step 2a: Estimator gain matrix.

We have now predicted the present state and measurement and have computed the state-prediction-error covariance using only prior information. The next steps update the prediction using present information to find the state estimate and its uncertainty. A key feature is the computation of a time-varying estimator gain matrix $\boldsymbol{L}[k] = \boldsymbol{\Sigma}^-_{\widetilde{\mathbf{X}}_\gamma \tilde{v}_{\text{cell}}}[k]\Sigma^{-1}_{\tilde{v}_{\text{cell}}}[k]$. We begin by computing the covariance of the voltage-prediction error, $\Sigma_{\tilde{v}_{\text{cell}}}[k]$, by employing EKF Assumption 2.

First, we compute $\widehat{\boldsymbol{C}}_{0,0}$, $\widehat{\boldsymbol{C}}_{0,1}$, $\widehat{\boldsymbol{C}}_{1,0}$, and $\widehat{\boldsymbol{C}}_{1,1}$, and $\widehat{C}_0$ (see Sect. 5.B for derivative calculations):[17]

$$\widehat{\boldsymbol{C}}_j = \frac{\mathrm{d}g_v(\mathbf{X}_\gamma[k], u[k])}{\mathrm{d}\boldsymbol{x}_j[k]} \quad \text{and} \quad \widehat{C}_0 = \frac{\mathrm{d}g_v(\mathbf{X}_\gamma[k], u[k])}{\mathrm{d}x_0[k]},$$

where all derivatives are evaluated at $\mathbf{X}_\gamma[k] = \widehat{\mathbf{X}}_\gamma^-[k]$. Then

$$\Sigma_{\tilde{v}_j}[k] = \widehat{\boldsymbol{C}}_j \boldsymbol{\Sigma}_{\tilde{\boldsymbol{x}}_j} \widehat{\boldsymbol{C}}_j^T + \Sigma_{\tilde{v}}, \quad \text{and} \quad \Sigma_{\tilde{v}_0}[k] = \widehat{C}_0^2 \Sigma_{\tilde{x}_0} + \Sigma_{\tilde{v}}.$$

Next, we compute the cross covariance $\boldsymbol{\Sigma}^-_{\widetilde{\mathbf{X}}_\gamma \tilde{v}_{\text{cell}}}[k]$, again using EKF Assumption 2. We find,

$$\boldsymbol{\Sigma}^-_{\widetilde{\mathbf{X}}_\gamma \tilde{v}_j}[k] = \widehat{\boldsymbol{C}}_j \boldsymbol{\Sigma}_{\tilde{\boldsymbol{x}}_j}, \quad \text{and} \quad \Sigma^-_{\widetilde{\mathbf{X}}_\gamma \tilde{v}_0}[k] = \widehat{C}_0 \Sigma_{\tilde{x}_0}.$$

This allows us to compute

$$\boldsymbol{L}_j[k] = \widehat{\boldsymbol{C}}_j \boldsymbol{\Sigma}_{\tilde{\boldsymbol{x}}_j} \left(\widehat{\boldsymbol{C}}_j \boldsymbol{\Sigma}_{\tilde{\boldsymbol{x}}_j} \widehat{\boldsymbol{C}}_j^T + \Sigma_{\tilde{v}} \right)^{-1}$$

for $\boldsymbol{L}_{0,0}[k]$, $\boldsymbol{L}_{0,1}[k]$, $\boldsymbol{L}_{1,0}[k]$, $\boldsymbol{L}_{1,1}[k]$, and $L_0[k]$.

Step 2b: State estimate measurement update.

The state prediction is now updated using the measured value of voltage to become a state estimate. This is done only for the four models that contribute to the output-blended voltage equation. For these models,

$$\hat{\boldsymbol{x}}_j^+[k] = \hat{\boldsymbol{x}}_j^-[k] + \boldsymbol{L}_j[k]\left(v_{\text{cell}}[k] - \hat{v}_{\text{cell}}[k]\right)$$
$$\hat{x}_0^+[k] = \hat{x}_0^-[k] + L_0[k]\left(v_{\text{cell}}[k] - \hat{v}_{\text{cell}}[k]\right).$$

For all other models, $\hat{\boldsymbol{x}}_j^+[k] = \hat{\boldsymbol{x}}_j^-[k]$.

[17] Since cell voltage is a scalar quantity, the derivative $\widehat{C}_0 = \mathrm{d}g_v(\mathbf{X}_\gamma[k], u[k])/\mathrm{d}x_0[k]$ used in step 2 is also a scalar quantity. This is different from how $\widehat{C}_0$ is defined and used in step 3.

Step 2c: Estimation-error covariance measurement update.

Next, the estimation-error covariance matrix is updated so all necessary values for the next iteration are computed. For the four models contributing to the output-blended voltage,

$$\Sigma^{+}_{\tilde{x}_j}[k] = \Sigma^{-}_{\tilde{x}_j}[k] - L_j[k]\Sigma_{\tilde{v}_j}[k]L_j^T[k]$$
$$\Sigma^{+}_{\tilde{x}_0}[k] = \Sigma^{-}_{\tilde{x}_0}[k] - L_0[k]\Sigma_{\tilde{v}_{\text{cell}}}[k]L_0^T[k].$$

For the remaining models, $\Sigma^{+}_{\tilde{x}_j}[k] = \Sigma^{-}_{\tilde{x}_j}[k]$.

AT THIS POINT, we have updated the state estimate and its covariance (uncertainty) matrix. In many cases, we would also like to compute estimates of the cell's electrochemical internal variables and their uncertainties. In this case, we add an additional major step having two substeps.

Step 3a: Estimate internal variables.

Cell internal variables $z[k]$ are nonlinear functions of the state. We again use EKF Assumption 1 to estimate their values, computing:

$$\begin{aligned}\hat{z}^{+}[k] &= \mathbb{E}\left[g_z(\mathbf{X}_\gamma[k], u[k]) \mid \mathbb{V}_{\text{cell}}[k-1]\right]\\ &\approx g_z(\widehat{\mathbf{X}}^{+}_\gamma[k], u[k]),\end{aligned}$$

where $\widehat{\mathbf{X}}^{+}_\gamma$ is a subset of $\widehat{\mathbf{X}}^{+}$ from step 2b. This means that:

- We use the four individual state estimates from $\widehat{\mathbf{X}}^{+}$ to produce four different linear-variable estimates $\hat{y}_{0,0}$, $\hat{y}_{0,1}$, $\hat{y}_{1,0}$, and $\hat{y}_{1,1}$.
- We use output blending to combine these four estimates into a single linearized output vector $\hat{y}$.
- We apply nonlinear corrections to $\hat{y}$ to compute $\hat{z}$.

Step 3b: Internal-variables estimation-error covariance.

Finally, we compute the covariance of the internal variable (useful for computing confidence intervals of the estimate). We begin by computing the covariance of the voltage-prediction error, $\Sigma_{\tilde{v}_{\text{cell}}}[k]$, by employing EKF Assumption 2.

First, we compute $\widehat{C}_{0,0}$, $\widehat{C}_{0,1}$, $\widehat{C}_{1,0}$, and $\widehat{C}_{1,1}$, and $\widehat{C}_0$ (see Sect. 5.B for derivative calculations):[18]

$$\widehat{C}_j = \frac{\mathrm{d}g_z(\mathbf{X}_\gamma[k], u[k])}{\mathrm{d}x_j[k]} \quad \text{and} \quad \widehat{C}_0 = \frac{\mathrm{d}g_z(\mathbf{X}_\gamma[k], u[k])}{\mathrm{d}x_0[k]},$$

where all derivatives are evaluated at $\mathbf{X}_\gamma[k] = \widehat{\mathbf{X}}^{+}_\gamma[k]$. Then

$$\Sigma_{\tilde{z}_j}[k] = \widehat{C}_j\Sigma_{\tilde{x}_j}\widehat{C}_j^T \quad \text{and} \quad \Sigma_{\tilde{z}_0}[k] = \widehat{C}_0\Sigma_{\tilde{x}_0}\widehat{C}_0^T.$$

Finally,

$$\Sigma^{+}_{\tilde{z}}[k] = \Sigma_{\tilde{z}_{0,0}}[k] + \Sigma_{\tilde{z}_{0,1}}[k] + \Sigma_{\tilde{z}_{1,0}}[k] + \Sigma_{\tilde{z}_{1,1}}[k] + \Sigma_{\tilde{z}_0}[k].$$

At the output of this process, we have high confidence that the true variable is in the range $\hat{z}^{+}[k] \pm 3\sqrt{\text{diag}\left(\Sigma^{+}_{\tilde{z}}[k]\right)}$.

[18] Since the set of cell variables of interest $z[k]$ is a vector quantity, the derivative $\widehat{C}_0 = \mathrm{d}g_z(\mathbf{X}_\gamma[k], u[k])/\mathrm{d}x_0[k]$ used in step 3 is also a vector. This is different from how $\widehat{C}_0$ is defined and used in step 2.

5.6 SPKF with the output-blended model

We now describe application of SPKF to the output-blended ROM. Some simplifications again result from the state equation being linear for the ROM we are using. In fact, SPKF steps 1a and 1b are identical to EKF steps 1a and 1b; therefore, we begin with step 1c.

Step 1c: Predict system output.

We adopt the voltage equation from Eq. (5.10) in the EKF section. So, we can write:

$$\hat{v}_{\text{cell}}[k] = \mathbb{E}\big[g_v(\mathbf{X}_\gamma[k], u[k]) \mid \mathbb{V}_{\text{cell}}[k-1]\big].$$

Since $g_v(\cdot)$ is nonlinear, we now turn to the sigma-point approach to approximate the mean and uncertainty of a random variable computed using a nonlinear equation.

The mean and error-covariance of the state prediction from steps 1a and 1b are used to form a set of sigma points based on the $\widehat{\mathbf{X}}_\gamma^-$ subset of $\widehat{\mathbf{X}}^-$ (and corresponding subset $\boldsymbol{\Sigma}_{\widetilde{\mathbf{X}}_\gamma}^-$ of $\boldsymbol{\Sigma}_{\widetilde{\mathbf{X}}}^-$):[19]

$$\boldsymbol{\mathcal{X}}^-[k] = \left\{\widehat{\mathbf{X}}_\gamma^-[k], \widehat{\mathbf{X}}_\gamma^-[k] + h\sqrt{\boldsymbol{\Sigma}_{\widetilde{\mathbf{X}}_\gamma}^-[k]}, \widehat{\mathbf{X}}_\gamma^-[k] - h\sqrt{\boldsymbol{\Sigma}_{\widetilde{\mathbf{X}}_\gamma}^-[k]}\right\}. \quad (5.11)$$

[19] The total number of elements in $\boldsymbol{\mathcal{X}}^-$ is $1 + 2(4n+1) = 8n+3$ elements, indexed from 0 to $8n+2$.

These sigma points can be organized in a convenient matrix form, as illustrated in Fig. 5.6.

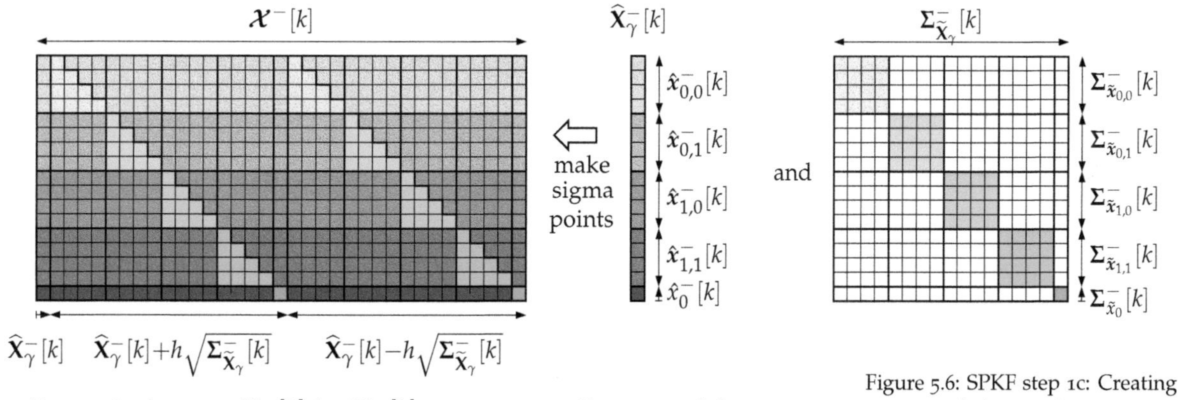

Figure 5.6: SPKF step 1c: Creating augmented sigma points.

For each element $\boldsymbol{\mathcal{X}}_i^-[k]$ in $\boldsymbol{\mathcal{X}}^-[k]$, we compute linear-model output via Eq. (5.6). Then we apply nonlinear corrections to produce voltage prediction based on that element. That is, overall we find nonlinear-output sigma points (as illustrated in Fig. 5.7):

$$\mathcal{V}_i[k] = g_v(\boldsymbol{\mathcal{X}}_i^-[k], u[k]).$$

The weighted mean of these points is the voltage prediction we are seeking to compute:

$$\hat{v}_{\text{cell}}[k] = \sum_{i=0}^{8n+2} \alpha_i^{(\text{m})} \mathcal{V}_i[k].$$

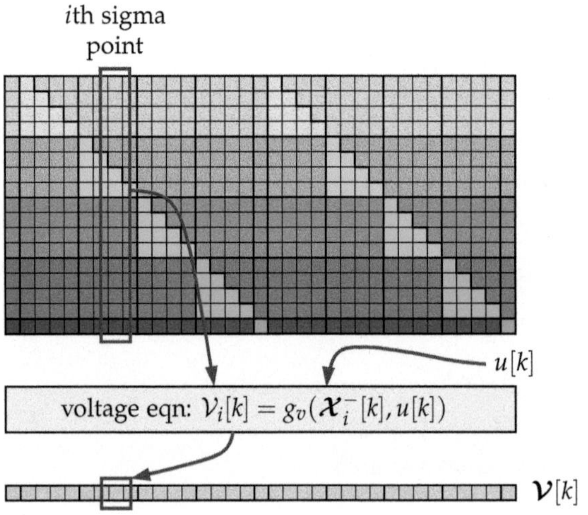

Figure 5.7: SPKF step 1c (continued).

This can be computed efficiently as an inner product between: (1) a vector $\boldsymbol{\mathcal{V}}[k]$ that stores all component values $\mathcal{V}_i$, and (2) a vector storing the tuning variables $\alpha_i^{(\mathrm{m})}$.

Step 2a: Estimator gain matrix.

We have now predicted the present state and present measurement, and have computed the covariance of the state-prediction error using only prior information. As with the EKF, the next steps update the prediction using present information to compute the state estimate and its uncertainty. A key feature is the computation of a time-varying estimator gain matrix:

$$\boldsymbol{L}[k] = \boldsymbol{\Sigma}_{\widetilde{\mathbf{X}}_\gamma \widetilde{v}_{\mathrm{cell}}}^-[k] \Sigma_{\widetilde{v}_{\mathrm{cell}}}^{-1}[k].$$

We begin by computing the voltage-prediction-error covariance:[20]

$$\Sigma_{\widetilde{v}_{\mathrm{cell}}}[k] = \sum_{i=0}^{8n+2} \alpha_i^{(\mathrm{c})} \left(\hat{v}_{\mathrm{cell}}[k] - \mathcal{V}_i[k]\right) \left(\hat{v}_{\mathrm{cell}}[k] - \mathcal{V}_i[k]\right) + \Sigma_{\widetilde{v}}.$$

[20] Note that $\Sigma_{\widetilde{v}_{\mathrm{cell}}}[k]$ on the left-hand-side is the time-varying innovation covariance and $\Sigma_{\widetilde{v}}$ on the right-hand-side is the (often time-invariant) sensor-noise covariance.

This can be computed efficiently if we define the vector $\widetilde{\boldsymbol{\mathcal{V}}}[k]$ having elements $\widetilde{\mathcal{V}}_i[k] = \hat{v}_{\mathrm{cell}}[k] - \mathcal{V}_i[k]$. Then:

$$\Sigma_{\widetilde{v}_{\mathrm{cell}}}[k] = \widetilde{\boldsymbol{\mathcal{V}}}^T[k] \operatorname{diag}(\alpha_i^{(\mathrm{c})}) \widetilde{\boldsymbol{\mathcal{V}}}[k] + \Sigma_{\widetilde{v}}.$$

We also compute the cross covariance between the state-prediction error and the voltage-prediction error:

$$\boldsymbol{\Sigma}_{\widetilde{\mathbf{X}}_\gamma \widetilde{v}_{\mathrm{cell}}}^-[k] = \sum_{i=0}^{8n+2} \alpha_i^{(\mathrm{c})} \left(\widehat{\mathbf{X}}_\gamma^-[k] - \boldsymbol{\mathcal{X}}_i^-[k]\right) \left(\hat{v}_{\mathrm{cell}}[k] - \mathcal{V}_i[k]\right).$$

If we further define the matrix $\widetilde{\boldsymbol{\mathcal{X}}}[k]$ having columns $\widetilde{\boldsymbol{\mathcal{X}}}_i[k] = \widehat{\mathbf{X}}_\gamma^-[k] - \boldsymbol{\mathcal{X}}_i^-[k]$, then this can be computed as:

$$\boldsymbol{\Sigma}_{\widetilde{\mathbf{X}}_\gamma \widetilde{v}_{\mathrm{cell}}}^-[k] = \widetilde{\boldsymbol{\mathcal{X}}}[k] \operatorname{diag}(\alpha_i^{(\mathrm{c})}) \widetilde{\boldsymbol{\mathcal{V}}}[k].$$

After these are computed, the state-estimator gain matrix can be found as:

$$\boldsymbol{L}[k] = \boldsymbol{\Sigma}^{-}_{\widetilde{\mathbf{X}}_\gamma \widetilde{v}_{\text{cell}}}[k] \left(\Sigma_{\widetilde{v}_{\text{cell}}}[k]\right)^{-1}.$$

Step 2b: State estimate measurement update.

The state prediction is now updated using the measured value of voltage to produce a state estimate. This is done for only the four models that contribute in the output-blended voltage equation. For these models,

$$\hat{\boldsymbol{x}}_j^+[k] = \hat{\boldsymbol{x}}_j^-[k] + \boldsymbol{L}_j[k]\left(v_{\text{cell}}[k] - \hat{v}_{\text{cell}}[k]\right)$$
$$\hat{x}_0^+[k] = \hat{x}_0^-[k] + L_0[k]\left(v_{\text{cell}}[k] - \hat{v}_{\text{cell}}[k]\right).$$

For all other models, $\hat{\boldsymbol{x}}_j^+[k] = \hat{\boldsymbol{x}}_j^-[k]$.

Step 2c: Estimation-error covariance measurement update.

To complete the process, the estimation-error covariance matrix is updated so that all necessary values for the next iteration are computed. For the four models contributing to the output-blended voltage,

$$\boldsymbol{\Sigma}^+_{\tilde{\boldsymbol{x}}_j}[k] = \boldsymbol{\Sigma}^-_{\tilde{\boldsymbol{x}}_j}[k] - \boldsymbol{L}_j[k]\Sigma_{\tilde{v}_j}[k]\boldsymbol{L}_j^T[k]$$
$$\Sigma^+_{\tilde{x}_0}[k] = \Sigma^-_{\tilde{x}_0}[k] - L_0[k]\Sigma_{\tilde{v}_{\text{cell}}}[k]L_0^T[k].$$

For the remaining models, $\boldsymbol{\Sigma}^+_{\tilde{\boldsymbol{x}}_j}[k] = \boldsymbol{\Sigma}^-_{\tilde{\boldsymbol{x}}_j}[k]$.

At this point, we have updated the state estimate and its covariance. In many cases, we would also like to compute estimates of the cell's electrochemical internal variables and their uncertainties. In this case, we add an additional major step having two substeps.

Step 3a: Estimate internal variables.

The cell's internal variables $\boldsymbol{z}[k]$ are nonlinear functions of the state, so once again, we must use the sigma-point method to find their estimates and uncertainties. To estimate the internal variables, we will compute new sigma points based on the present state estimate and uncertainty:

$$\boldsymbol{\mathcal{X}}^+[k] = \left\{\widehat{\mathbf{X}}^+_\gamma[k], \widehat{\mathbf{X}}^+_\gamma[k] + h\sqrt{\boldsymbol{\Sigma}^+_{\widetilde{\mathbf{X}}_\gamma}[k]}, \widehat{\mathbf{X}}^+_\gamma[k] - h\sqrt{\boldsymbol{\Sigma}^+_{\widetilde{\mathbf{X}}_\gamma}[k]}\right\}. \quad (5.12)$$

Each of these sigma points comprises only those states that participate in computing the cell voltage: $\boldsymbol{\mathcal{X}}_i^+[k]$ has dimension $4n+1$. Then, for each of the sigma points in the set $\boldsymbol{\mathcal{X}}_i^+$, we produce an output sigma point:

$$\boldsymbol{\mathcal{Z}}_i[k] = g_z(\boldsymbol{\mathcal{X}}_i^+[k], u[k]).$$

Note that $\mathcal{Z}_i[k]$ has the same size and organization as the ROM linear-output vector $y[k]$; output blending and all nonlinear corrections are applied to every element. Finally, we compute the estimate of the variable of interest as:

$$\hat{z}^+[k] = \sum_{i=0}^{8n+2} \alpha_i^{(m)} \mathcal{Z}_i[k].$$

Step 3b: Internal-variables estimation-error covariance.

- Finally, we compute the covariance of the internal variables (useful for computing confidence intervals of the estimate):

$$\Sigma_{\tilde{z}}^+[k] = \sum_{i=0}^{8n+2} \alpha_i^{(c)} (\hat{z}^+[k] - \mathcal{Z}_i^+[k])(\hat{z}^+[k] - \mathcal{Z}_i^+[k])^T.$$

At the output of this process, we have high confidence that the true variables are in the range $\hat{z}^+[k] \pm 3\sqrt{\text{diag}\left(\Sigma_{\tilde{z}}^+[k]\right)}$.

COMMENT: During operation of either xKF, only four precomputed models are used to predict cell voltage at any point in time.

- So only those same four models are updated using the measured-voltage feedback in steps 2b and 2c.
- However, all models are updated in steps 1a and 1b.

Step 1b increases the uncertainty of the predictions of model states, so this uncertainty tends to grow over time for all models. Step 2c decreases the uncertainty of the state estimates for only the four models that are updated in any timestep. As a consequence, since step 2c is not executed for most models in any timestep, the state-estimation uncertainty tends to grow over time for most models. The state-estimation uncertainty tends to decrease over time only for the four models being updated by step 2.

As the actual cell being monitored has a time-varying temperature and SOC, we expect that different sets of four precomputed models will contribute to the voltage prediction at different points in time. Any time that this set of models changes, we will begin to blend in the effect of a precomputed model that possibly has a large covariance (large uncertainty of its states) since that model has not been updated using steps 2b and 2c for some time. That is, we might expect a singularity when either SOC and/or temperature change such that we blend together a different set of four precomputed models from the set we have been using most recently.

We do see this singularity as a kind of scalloping in the error bounds of estimates, but the degree of this effect is low since both SOC and temperature change relatively slowly. When we start to blend in a new model, the γ_j weighting or blending factor associated with that model is near zero, and so the large uncertainty of that model does not have significant impact on the voltage prediction.

5.7 *Example of xKF code in operation*

The remainder of this chapter focuses on demonstrating the xKF code in operation for the three load profiles from Chap. 4 applied to the NMC30 cell. They implement variants of both EKF and SPKF that use both model blending and output blending.[21,22] In each case, the xKF initial SOC was set to 95 % of its true value to demonstrate the xKF's ability to recover from a poor initial estimate via its built-in feedback mechanism.

SOC-estimation results are presented in Fig. 5.8. The left column shows true SOC versus time for the three load profiles as well as the estimates using EKF model blending ("EKF MdlB"), EKF output blending ("EKF OutB"), SPKF model blending ("SPKF MdlB"), and SPKF output blending ("SPKF OutB"). Since the estimates are very close to the truth after an initial startup transient during which the xKFs converge to the neighborhood of the correct SOC, the right column plots SOC estimation error, which is equal to true SOC minus estimated SOC. The figures in the right column also show the $\pm 3\sigma$ confidence bounds as thin lines. We desire for all errors to be small and always to be encompassed by the confidence bounds.

A general conclusion is that all xKFs worked very well for estimating SOC (RMSE $< 1\,\%$). Results from model-blended EKF and SPKF were nearly identical to each other; results from output-blended EKF and SPKF were also nearly identical to each other. Output-blended xKF produced SOC estimates that were very slightly better than those for model-blended xKF. SOC-estimation error is always within xKF confidence bounds except for a few points at the very beginning of the simulations (due to the intentionally incorrect initialization of SOC). These results are very encouraging.

5.7.1 *Charge-neutral UDDS internal-variable estimation results*

We now consider xKF internal-variables estimation results for the charge-neutral UDDS profile. The figures in this section show estimates for each variable at the cell's two internal electrode/separator boundaries since variables at those locations are most difficult to estimate well. The left column of each figure shows the true value of the signal we wish to estimate along with the four estimates; the right column shows estimation errors and confidence bounds.

The top two rows of Fig. 5.9 present results for solid surface concentration, $\theta^{\mathrm{r}}_{\mathrm{ss}}[\tilde{x},k]$. The estimates are reasonable and the confidence bounds are fair. We will see, in general, that closed-loop xKF estimates of internal variables are similar to the open-loop predictions. The exception is when the xKF is initialized with a bad SOC esti-

[21] Application of xKF to model-blended ROMs is not discussed here, but you may find details in: Kirk D. Stetzel, "Model-based estimation of battery cell internal physical state," Master's thesis, University of Colorado Colorado Springs, 2014, and in Kirk D. Stetzel, Lukas L. Aldrich, M. Scott Trimboli, and Gregory L. Plett, "Electrochemical state and internal variables estimation using a reduced-order physics-based model of a lithium-ion cell and an extended Kalman filter," *Journal of Power Sources*, 278:490–505, 2015.

[22] You can reproduce the results by running the example code in the toolbox; it implements EKF and SPKF for both model-blending and output-blending assumptions.

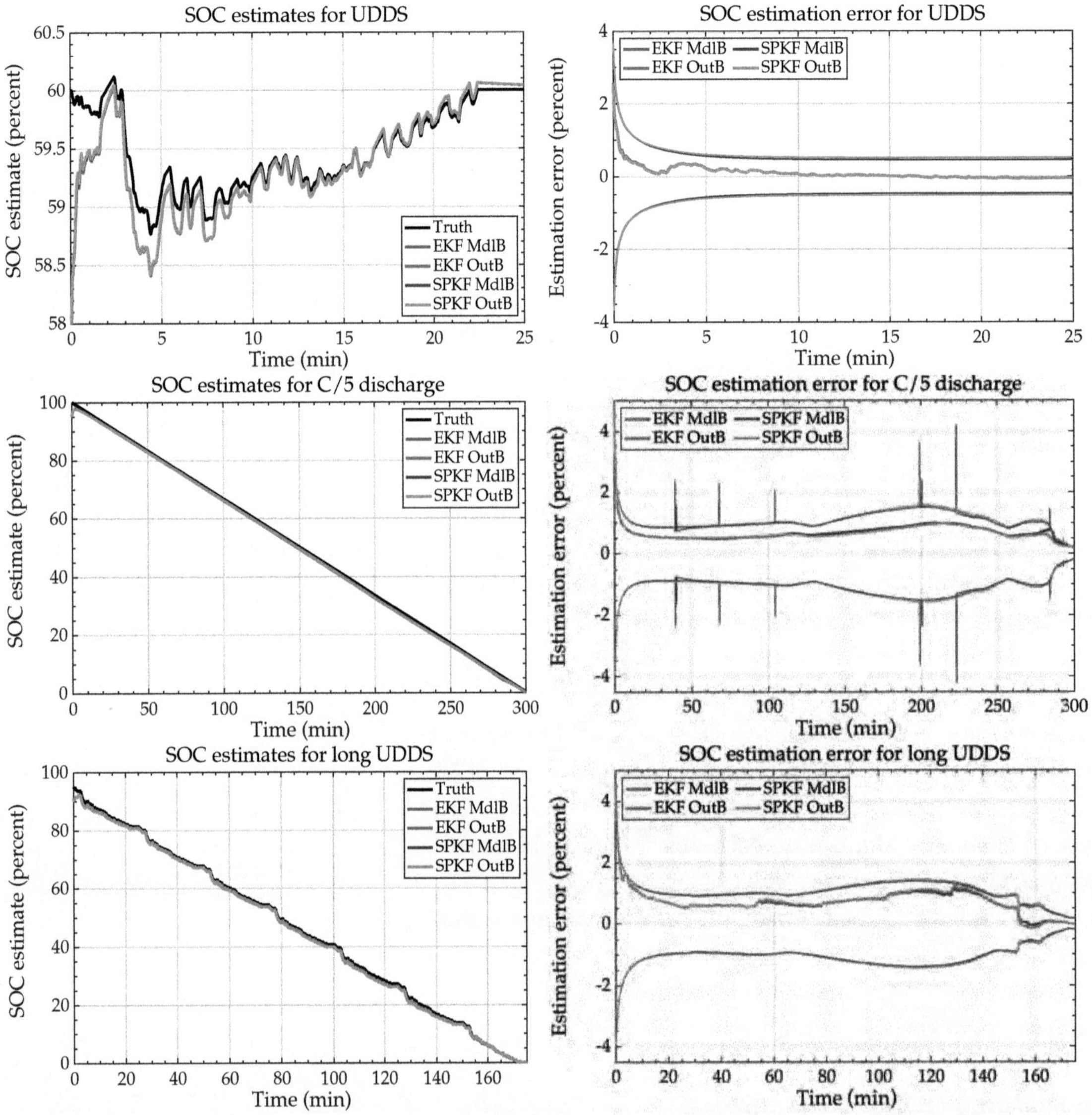

Figure 5.8: SOC-estimation results for all three scenarios.

mate; then the xKF recovers quickly from the bad initial estimate to the trajectory that aligns with the open-loop predictions. The lower two rows of the figure show results for the electrolyte concentration ratios. Once again, estimates are reasonable. SPKF produces tighter confidence bounds than EKF.

The top two rows of Fig. 5.10 show results for solid potential; the lower two rows show results for electrolyte potential. The top two rows of Fig. 5.11 show results for interphase potential difference. The lower two rows present results for total interfacial flux, $i^{\mathrm{r}}_{\mathrm{f+dl}}[\tilde{x},k]$. These results are generally good; however, we see that the confidence bounds are not always as reliable as they were when estimating SOC

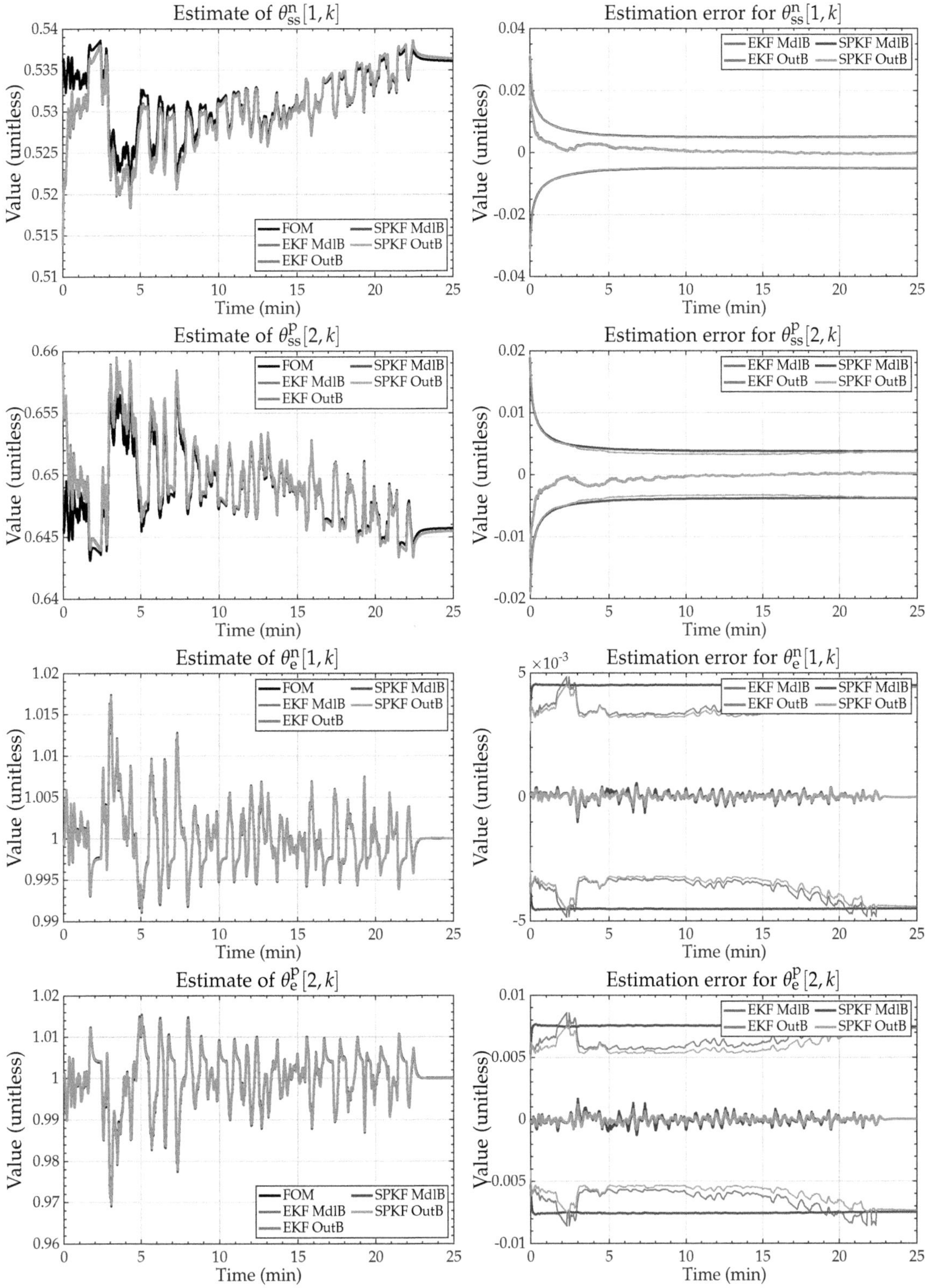

Figure 5.9: xKF results for normalized lithium concentrations for a charge-neutral UDDS profile.

due to occasional breaching of the confidence bounds at points of peak input current.

5.7.2 *C/5 discharge profile internal-variable estimation results*

Next, we look at results corresponding to the C/5 discharge profile. Results are presented in the same format as in the preceding section.

Estimates of solid surface concentrations are presented in the top two rows of Fig. 5.12 and estimates of electrolyte concentration ratio are presented in the bottom two rows. In this case, we see a significant failure of the model-blended method at just after 50 min into the simulation. This corresponds with the same failure observed in open-loop simulations using model blending in Chap. 4. While still imperfect, the output-blended xKF results are far more reliable. Notice also the prominent scalloping of the confidence bounds produced by the output-blending method. This is caused by blending from one model set to another as SOC changes during the simulation.

Estimates of cell internal potentials are presented in Fig. 5.13, and estimates of interphase potential difference and total lithium flux are shown in Fig. 5.14. Output blending is clearly better than model blending for many of these variables. We notice that the feedback of the xKF does not have much influence on reducing errors on total lithium flux, evidenced by observing that its closed-loop estimates are similar to the open-loop predictions from Chap. 4 and also by the wide confidence bounds produced by the xKFs: this is because cell voltage is a weak function of $i^{\mathrm{r}}_{\mathrm{f+dl}}[\tilde{x},k]$, and so the feedback mechanism in the xKF: (1) has no real reason to change $i^{\mathrm{r}}_{\mathrm{f+dl}}[\tilde{x},k]$ if there is a voltage-prediction error since it will not change the voltage prediction significantly, and (2) might risk overcorrecting $i^{\mathrm{r}}_{\mathrm{f+dl}}[\tilde{x},k]$ since errors in this variable do not contribute substantially to cell voltage error. The xKF will preferentially correct states and electrochemical variables that have significant impact on the voltage prediction. So, if the open-loop model of a variable to which voltage is insensitive is poor, the xKF will not do much to improve its predictions.

5.7.3 *Long-duration UDDS profile internal-variable estimation results*

For completeness, internal-variable estimation results using the xKFs with the long-duration charge-depleting UDDS profiles are presented in Figs. 5.15 through 5.17. Observations based on these figures are similar to those we have already seen.

Overall, we see significant benefits to using output blending instead of model blending; we do not see much difference between the performance of EKF and SPKF.[23]

[23] Our own preference is to use SPKF since we believe it is simpler to implement without introducing coding errors (we recognize that this is a matter of opinion).

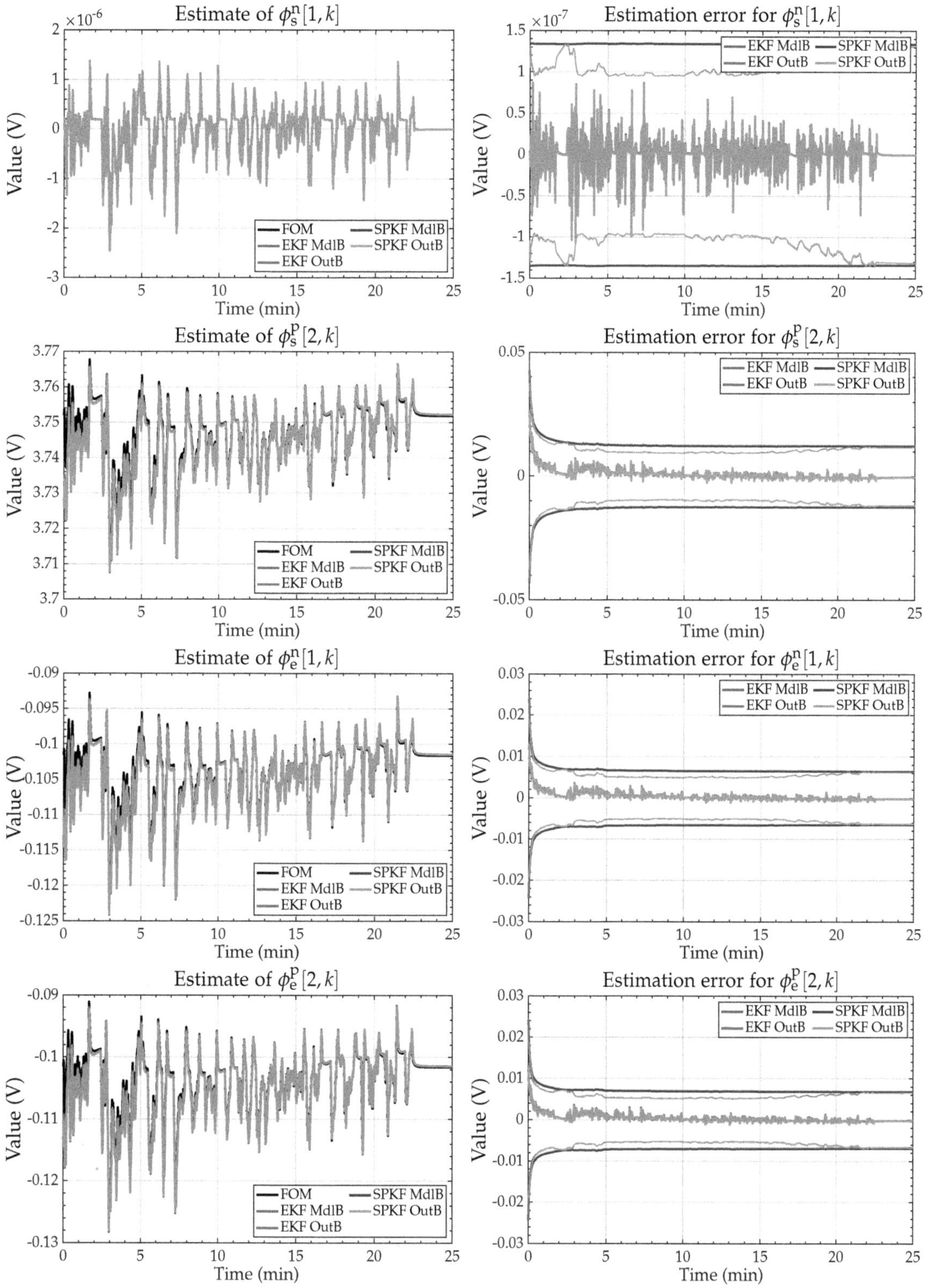

Figure 5.10: xKF results for cell potentials for a charge-neutral UDDS profile.

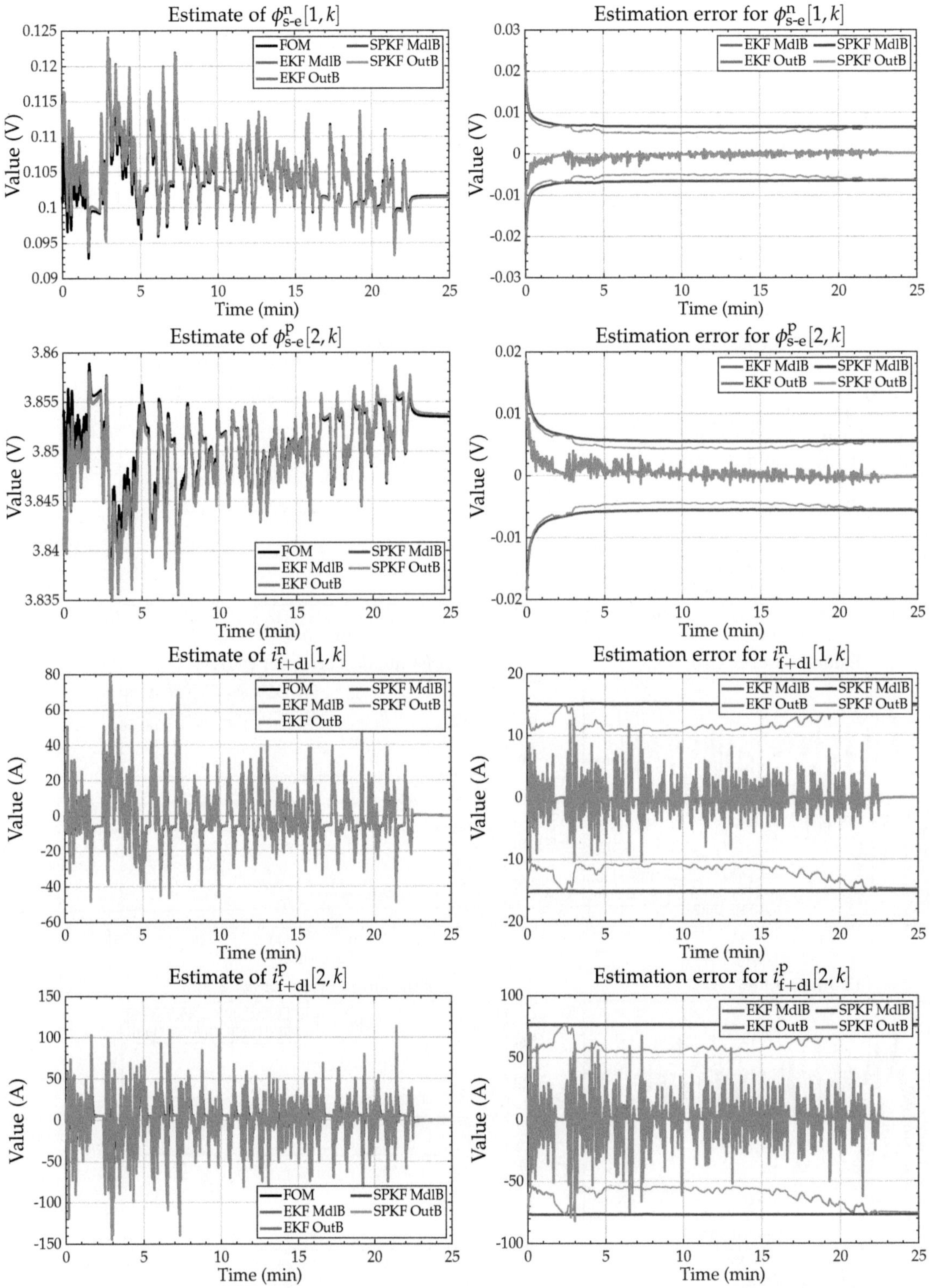

Figure 5.11: xKF results for interphase potential difference and total lithium flux for a charge-neutral UDDS profile.

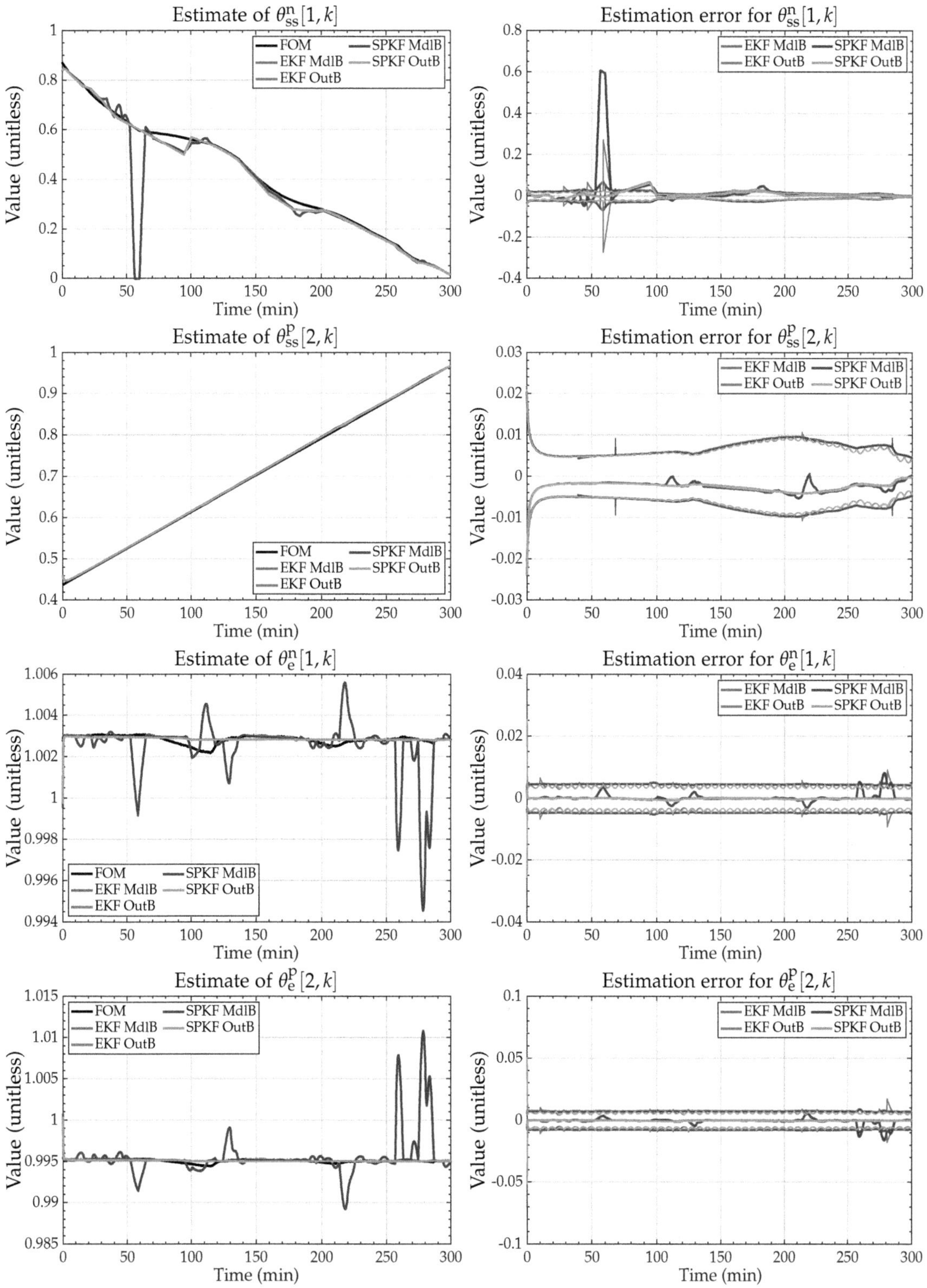

Figure 5.12: xKF results for normalized lithium concentrations for a C/5 discharge profile.

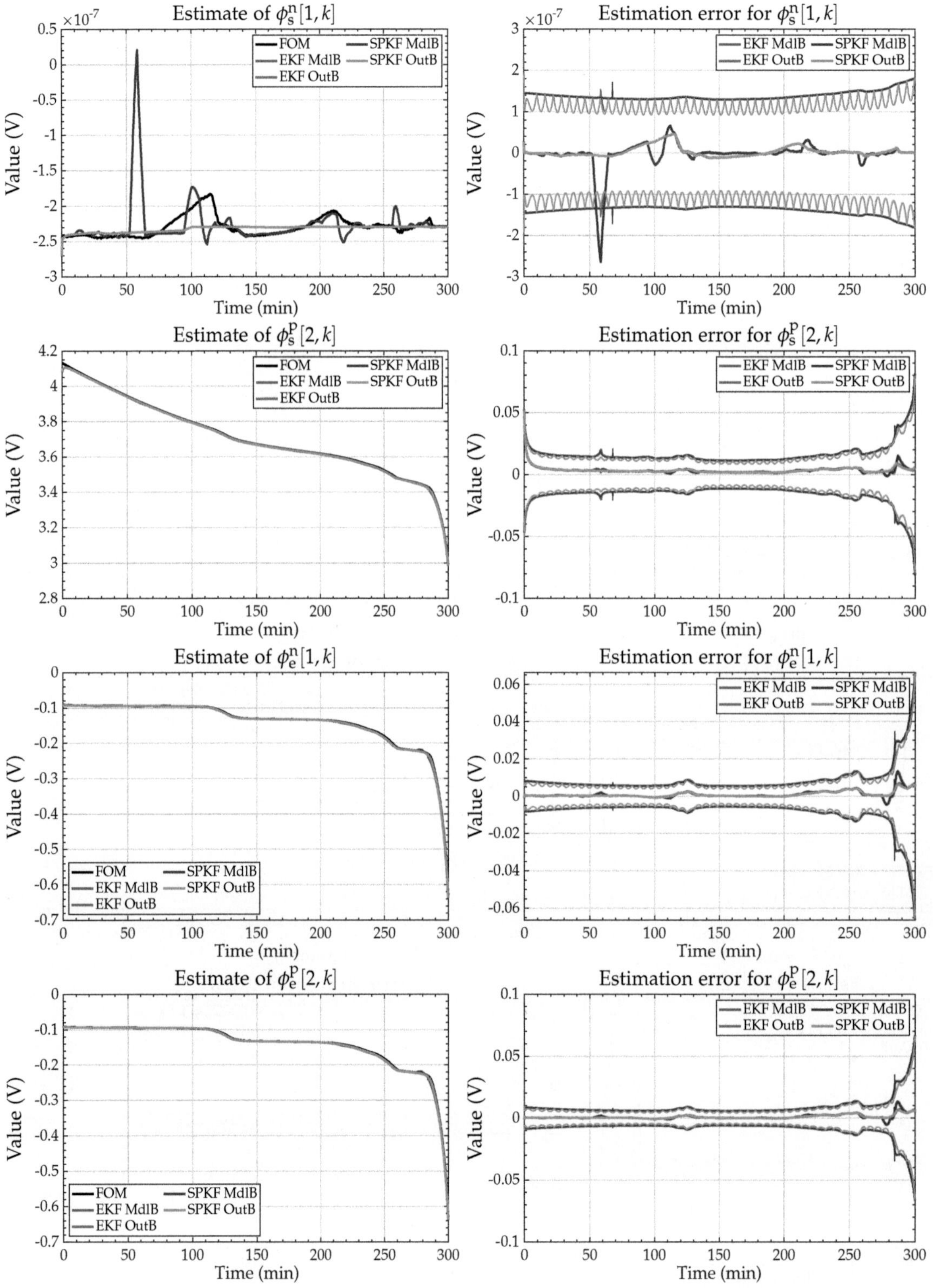

Figure 5.13: xKF results for cell potentials for a C/5 discharge profile.

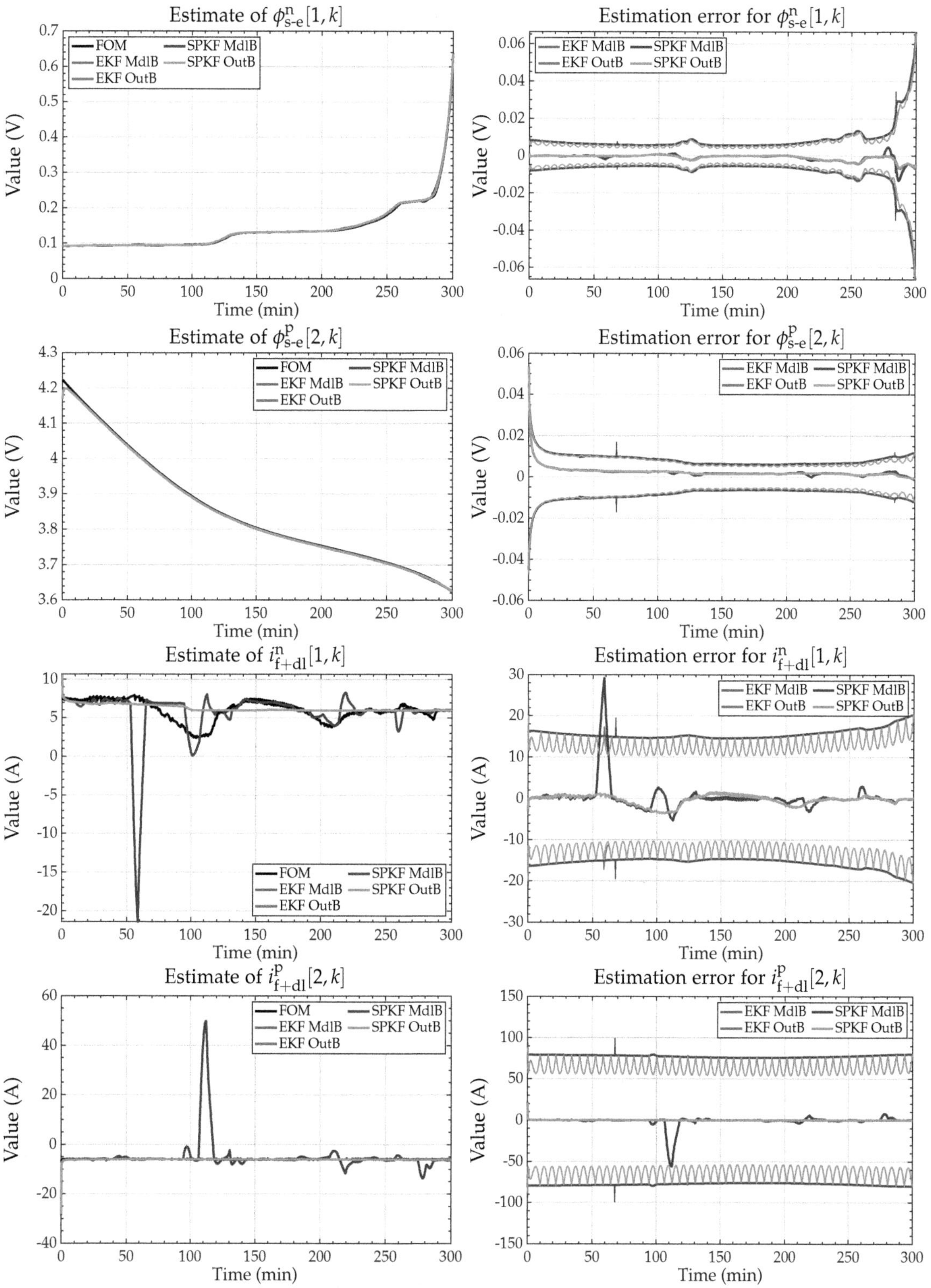

Figure 5.14: xKF results for interphase potential difference and total lithium flux for a C/5 discharge profile.

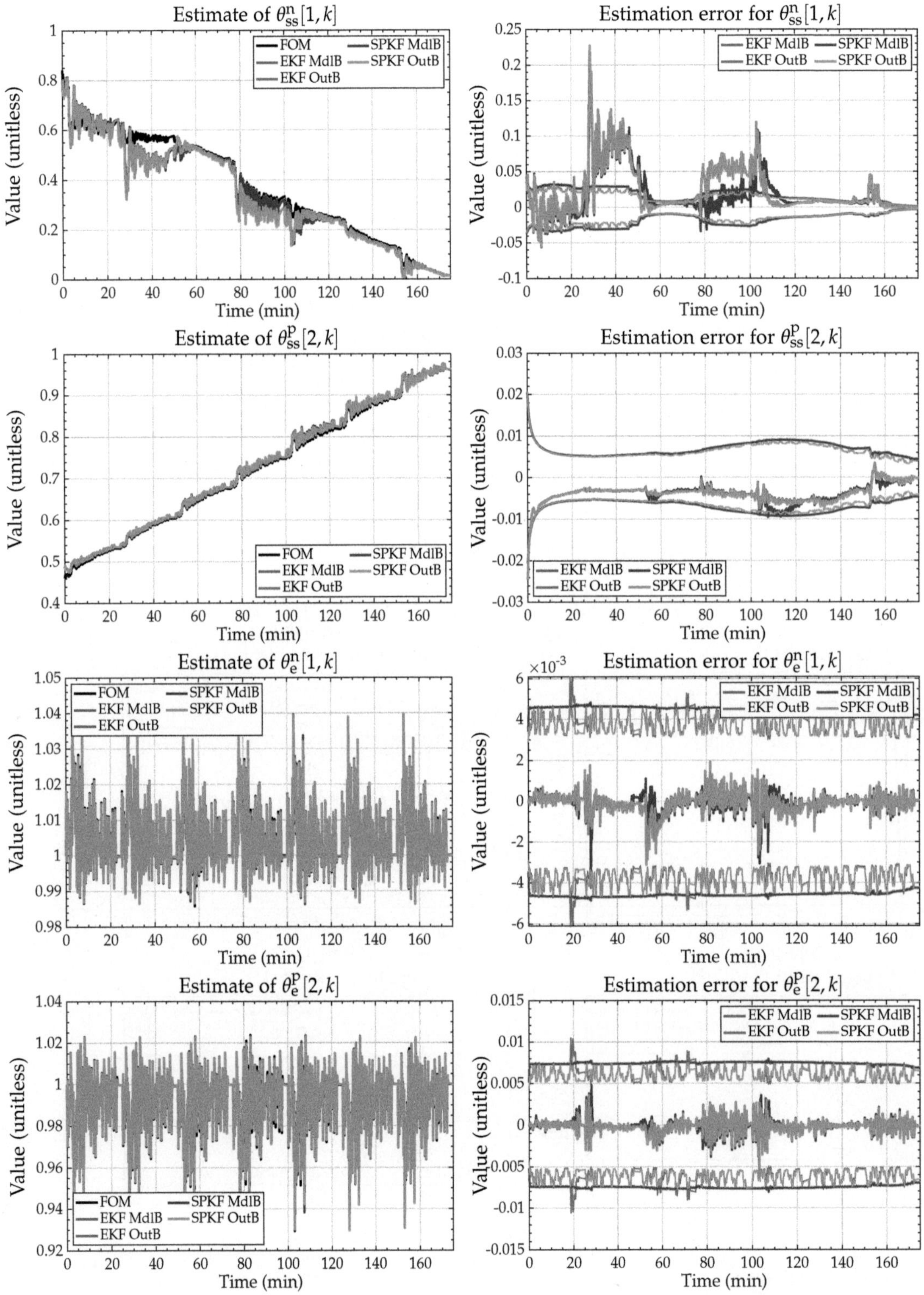

Figure 5.15: xKF results for normalized lithium concentrations for a long-duration UDDS profile.

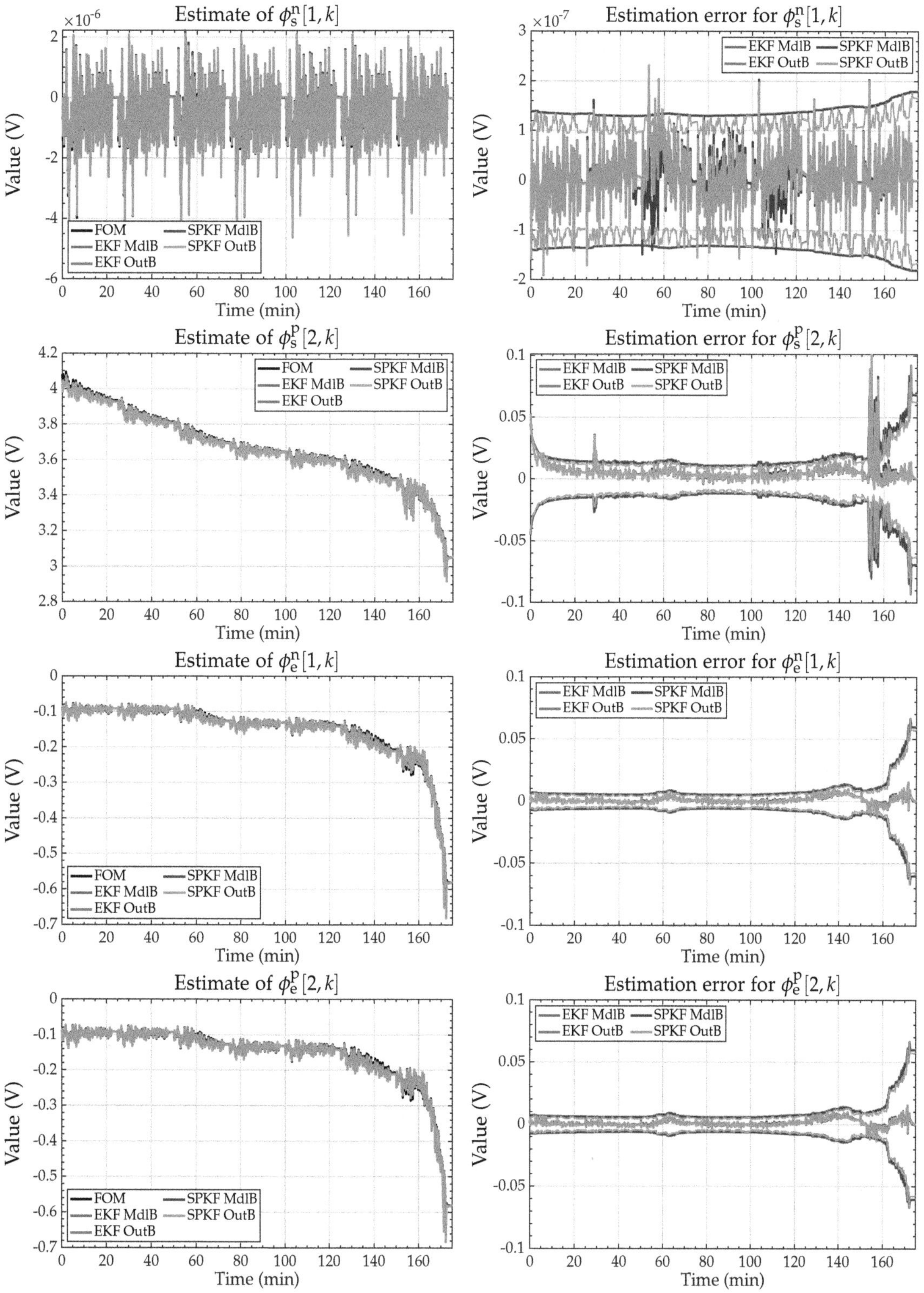

Figure 5.16: xKF results for cell potentials for a long-duration UDDS profile.

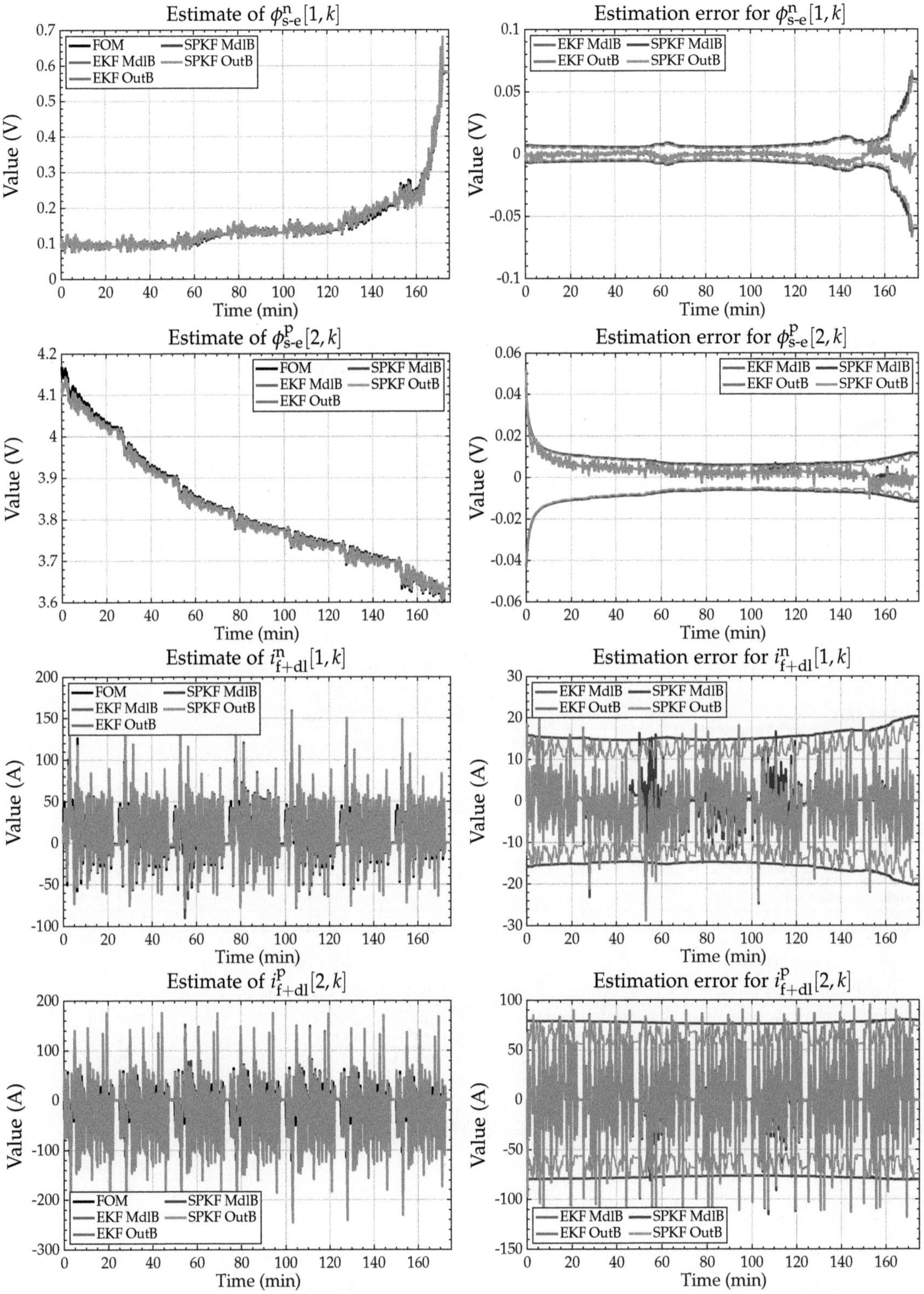

Figure 5.17: xKF results for interphase potential difference and total lithium flux for a long-duration UDDS profile.

5.8 MATLAB toolbox

Before closing this chapter, we mention the components of the MATLAB toolbox that pertain to the developments of this chapter.[24] Note that Chaps. 5–8 of this book are applications of the models developed in Chaps. 1–4, as illustrated in Fig. 5.18. The state-estimation components of the toolbox are encapsulated by the MATLAB functions `initKF` and `iterEKF` (for the EKF method) or `iterSPKF` (for SPKF).

[24] See: `http://mocha-java.uccs.edu/BMS3/`.

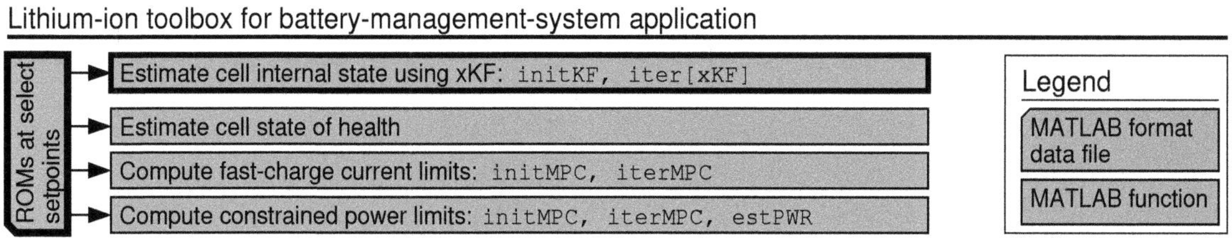

Figure 5.18: Functionality of MATLAB physics-based BMS toolbox, highlighting the focus of this chapter.

To run an xKF, we must first initialize the algorithm. This is done by calling `initKF`.

```
% First, set up covariances of initial state, current-sensor noise,
% and voltage-sensor noise (n=8 for this ROM)
SigmaX0 = diag([ones(1,5) 2e6]); % uncertainty of initial state
SigmaW = 1e2;   % uncertainty of current sensor, state equation
SigmaV = 1e-3;  % uncertainty of voltage sensor, output equation
% Set the filename where the ROMs generated by the HRA are stored
ROMfile    = 'ROM_NMC30_HRA.mat'; % generated ROM
% Set the filename of the FOM simulation output ("truth" data)
FOMoutFile = 'OUT_NMC30_UDDS.mat'; % FOM simulated output
% Set the initial true SOC of the simulation
SOC0 = 60;

% Set the blending method: either 'MdlB' or 'OutB'
blend = 'OutB';

% Load the data files
load(ROMfile,'ROM'); % Load the HRA-produced ROMs
load(FOMoutFile,'FOMout'); % Load the "truth" data

% Initialize the xKF (you can set "SOC0 = []" if you do not know
% the true initial SOC; the xKF will estimate it for you)
xkfData = initKF(SOC0,SigmaX0,SigmaV,SigmaW,blend,ROM);
```

Then we update the filter once for every measurement. We may use either `iterEKF` (for the EKF method) or `iterSPKF` (for SPKF).

```
% reserve storage for SPKF results... including voltage and SOC (the "+2")
zkEst = NaN(spkfData.nz+2,length(FOMout.Iapp));
zkBound = zkEst;

for k = 1:length(FOMout.Iapp)
  vk = FOMout.Vcell(k);
  ik = FOMout.Iapp(k);
  Tk = FOMout.T(k); % degC
  % The next line assumes we are using SPKF.
  % If we want to use EKF, replace "iterSPKF" with "iterEKF"
```

```
  [zk,zbk,xkfData] = iterSPKF(vk,ik,Tk,xkfData);
  zkEst(:,k)=zk; zkBound(:,k)=zbk;
end
% Output data structure has one column for every simulation timestep
% Each column is organized in same order as "y" outputs of ROMs, plus two
% additional rows: voltage and SOC
```

The toolbox has two examples that show how to run the xKFs and plot sample results.

5.9 *Where to from here?*

You have now learned how to implement EKF and SPKF on the output-blended ROM to estimate states, internal electrochemical variables, and confidence bounds thereon. The most important applications of these estimates are:

- To be able to predict what conditions would lead a cell in its present condition to age and fail prematurely. This is one of the principal topics of the next chapter.
- To compute power limits which, if obeyed, guarantee that these conditions will never occur. This is the topic of Chaps. 7 (in the context of fast charge) and 8 (in the context of computing dynamic power limits).

5.A *Summary of variables*

- $\alpha_i^{(c)}$, sigma-point weighting values when determining covariance.
- $\alpha_i^{(m)}$, sigma-point weighting values when determining mean.
- $\mathbf{A}$, state-transition matrix for combined states of all models (note that $\mathbf{A}$ is typeset in an upright font, not italic, and is different from $\mathfrak{A}_j$, which is the full state-transition matrix of the jth linear model being simulated, and A_j, which is the state-transition matrix of the jth linear model being simulated after integrator state removed).
- $\mathbf{B}$, input matrix for combined states of all models (note that $\mathbf{B}$ is typeset in an upright font, not italic, and is different from $\mathfrak{B}_j$, which is the full input matrix of the jth linear model being simulated, and B_j, which is the input matrix of the jth linear model being simulated after integrator state removed).
- $\mathfrak{C}_j$, full output matrix of the jth linear model being simulated.
- C_j, output matrix of the jth linear model being simulated after integrator state removed.
- C_0, output-matrix column for the integrator state common to all linear models.
- D_j, direct-feedthrough matrix of the jth linear model being simulated.

- $\mathbb{E}[\cdot]$, expected value of the argument.
- $\boldsymbol{f}(\boldsymbol{x}[k], u[k], w[k])$, nonlinear state-transition function producing $\boldsymbol{x}[k+1]$.
- γ_j, blending factor for jth ROM in output-blending approach.
- $g_v(\boldsymbol{x}[k], u[k])$, nonlinear function converting state to voltage $v_{\text{cell}}[k]$.
- $\boldsymbol{g}_z(\boldsymbol{x}[k], u[k])$, nonlinear function converting state to electrochemical variables $z[k]$.
- $\boldsymbol{L}[k]$, estimator gain.
- Σ, correlation between the two arguments in its subscript (covariance if the arguments are zero mean).
- $u[k]$, generic model input ($i_{\text{app}}[k]$ in the ROMs of this chapter).
- $v[k]$, measurement (voltage-sensor) noise.
- $\boldsymbol{\mathcal{V}}$, set of sigma points representing possible cell voltages.
- $v_{\text{cell}}[k]$, voltage.
- $\mathbb{V}_{\text{cell}}[k] = \{v_{\text{cell}}[0], \cdots, v_{\text{cell}}[k]\}$, sequence of voltage measurements up to time k.
- $w[k]$, process (current-sensor) noise.
- $x_0[k]$, the state of the integrator, shared in common by all models, at time k.
- $\mathbf{X}[k]$, the combined states of all models at time k (note that $\mathbf{X}$ is typeset in an upright font, not italic, and is different from $\boldsymbol{\mathfrak{X}}_j[k]$, which is the state of the jth single model (including integrator), and $\boldsymbol{x}_j[k]$, which is the state of the jth single model, without the integrator).
- $\mathbf{X}_\gamma[k]$, the combined states of all models being blended presently at time k (note that $\mathbf{X}_\gamma$ is typeset in an upright font, not italic).
- $\boldsymbol{\mathcal{X}}$, set of sigma points representing possible system states.
- $\mathbb{Y}[k] = \{y[0], \cdots, y[k]\}$, sequence of measurements up to time k.
- $\tilde{\boldsymbol{y}}[k]$, the innovation (difference between measurement and its prediction).
- $\boldsymbol{y}[k]$, linear output of ROM after blending but before nonlinear corrections.
- $\boldsymbol{y}_j[k]$, linear output of the jth linear ROM (before blending and nonlinear corrections).
- $\boldsymbol{\mathcal{Z}}$, set of sigma points representing possible values of internal electrochemical variables of interest.
- $\boldsymbol{z}[k]$, the vector of electrochemical variables being estimated.

5.B EKF derivative matrices

5.B.1 Computing derivatives of cell voltage with respect to states

Implementing step 2a of the EKF requires computing derivative matrices $\widehat{C}_{0,0}$, $\widehat{C}_{0,1}$, $\widehat{C}_{1,0}$, and $\widehat{C}_{1,1}$, and $\widehat{C}_0$:[25]

[25] Since cell voltage is a scalar quantity, the derivative $\widehat{C}_0 = \mathrm{d}g_v(\mathbf{X}_\gamma[k], u[k])/\mathrm{d}x_0[k]$ used in Step 2a is also a scalar quantity. This is different from how $\widehat{C}_0$ is defined and used in Step 3b.

$$\widehat{C}_j = \frac{\mathrm{d}g_v(\mathbf{X}_\gamma[k], u[k])}{\mathrm{d}x_j[k]} \quad \text{and} \quad \widehat{C}_0 = \frac{\mathrm{d}g_v(\mathbf{X}_\gamma[k], u[k])}{\mathrm{d}x_0[k]}.$$

We expand these derivatives in terms of the electrochemical variables that are needed to compute voltage:

$$\frac{\mathrm{d}g_v(\mathbf{X}_\gamma[k], u[k])}{\mathrm{d}x_j[k]} = \frac{\partial g_v(\mathbf{X}_\gamma[k], u[k])}{\partial z[k]} \frac{\mathrm{d}z[k]}{\mathrm{d}x_j[k]}.$$

From Chap. 4, we know that

$$v_{\mathrm{cell}}[k] = (\eta^{\mathrm{p}}[3,k] - \eta^{\mathrm{n}}[0,k]) + \tilde{\phi}_{\mathrm{e}}^{\mathrm{p}}[3,k] + \left(U_{\mathrm{ocp}}^{\mathrm{p}}(\theta_{\mathrm{ss}}^{\mathrm{p}}[3,k]) - U_{\mathrm{ocp}}^{\mathrm{n}}(\theta_{\mathrm{ss}}^{\mathrm{n}}[0,k])\right)$$
$$+ \left(\bar{R}_{\mathrm{f}}^{\mathrm{p}} i_{\mathrm{f+dl}}^{\mathrm{p}}[3,k] - \bar{R}_{\mathrm{f}}^{\mathrm{n}} i_{\mathrm{f+dl}}^{\mathrm{n}}[0,k]\right).$$

Therefore, to find the derivative matrices, we must be able to find the derivative of each of these terms with respect to a model state. We also need to remember that the output-blended linear model outputs are a weighted summation of multiple models (see Eq. (5.6)):

$$y[k] = \sum_{j=1}^{N} \gamma_j \left(C_j x_j[k] + C_{0,j} x_0[k] + D_j u[k]\right).$$

Therefore, there will generally be γ_j terms in each derivative.

Denote the row of C_j corresponding to a particular electrochemical variable "var" at location $\tilde{x}$ as $[C_j^{\mathrm{var}[\tilde{x}]}]$. Then, for example,

$$\frac{\mathrm{d}i_{\mathrm{f+dl}}^{\mathrm{n}}[0,k]}{\mathrm{d}x_j[k]} = \gamma_j \left[C_j^{i_{\mathrm{f+dl}}^{\mathrm{n}}[0]}\right] \quad \text{and} \quad \frac{\mathrm{d}i_{\mathrm{f+dl}}^{\mathrm{p}}[3,k]}{\mathrm{d}x_j[k]} = \gamma_j \left[C_j^{i_{\mathrm{f+dl}}^{\mathrm{p}}[3]}\right].$$

We choose to use a simplified model of overpotential for the EKF update (this simplification does not seem to cause significant differences in results compared with a full evaluation of the derivative),

$$\eta^{\mathrm{r}}[\tilde{x},k] \approx \bar{R}_{\mathrm{ct}}^{\mathrm{r}} i_{\mathrm{f}}^{\mathrm{r}}[\tilde{x},k] = \frac{RT}{\bar{i}_0 F} i_{\mathrm{f}}^{\mathrm{r}}[\tilde{x},k].$$

Therefore,

$$\frac{\mathrm{d}\eta^{\mathrm{n}}[0,k]}{\mathrm{d}x_j[k]} = \gamma_j \bar{R}_{\mathrm{ct}}^{\mathrm{n}} \left[C_j^{i_{\mathrm{f}}^{\mathrm{n}}[0]}\right] \quad \text{and} \quad \frac{\mathrm{d}\eta^{\mathrm{p}}[3,k]}{\mathrm{d}x_j[k]} = \gamma_j \bar{R}_{\mathrm{ct}}^{\mathrm{p}} \left[C_j^{i_{\mathrm{f}}^{\mathrm{p}}[3]}\right].$$

The derivatives:

$$\frac{\mathrm{d}U_{\mathrm{ocp}}^{\mathrm{r}}(\theta_{\mathrm{ss}}^{\mathrm{r}}[\tilde{x},k])}{\mathrm{d}x_j[k]} = \frac{\partial U_{\mathrm{ocp}}^{\mathrm{r}}}{\partial \theta_{\mathrm{ss}}^{\mathrm{r}}} \frac{\mathrm{d}\theta_{\mathrm{ss}}^{\mathrm{r}}[\tilde{x},k]}{\mathrm{d}x_j} = \gamma_j [U_{\mathrm{ocp}}^{\mathrm{r}}]' \left[C_j^{\tilde{\theta}_{\mathrm{ss}}^{\mathrm{r}}[\tilde{x}]}\right],$$

where $[U_{\mathrm{ocp}}^{\mathrm{r}}]'$ is evaluated at the present operating point.

In summary, $\widehat{C}_{0,0}$, $\widehat{C}_{0,1}$, $\widehat{C}_{1,0}$, and $\widehat{C}_{1,1}$, all of which have dimension $1 \times n$, are computed via:

$$\widehat{C}_j = \gamma_j \Bigg[\bar{R}_{\mathrm{ct}}^{\mathrm{p}} \left[C_j^{i_{\mathrm{f}}^{\mathrm{p}}[3]}\right] - \bar{R}_{\mathrm{ct}}^{\mathrm{n}} \left[C_j^{i_{\mathrm{f}}^{\mathrm{n}}[0]}\right] + \left[C_j^{\tilde{\phi}_{\mathrm{e}}^{\mathrm{p}}[3]}\right] + [U_{\mathrm{ocp}}^{\mathrm{p}}]' \left[C_j^{\tilde{\theta}_{\mathrm{ss}}^{\mathrm{p}}[3]}\right]$$
$$- [U_{\mathrm{ocp}}^{\mathrm{n}}]' \left[C_j^{\tilde{\theta}_{\mathrm{ss}}^{\mathrm{n}}[0]}\right] + \bar{R}_{\mathrm{f}}^{\mathrm{p}} \left[C_j^{i_{\mathrm{f+dl}}^{\mathrm{p}}[3]}\right] - \bar{R}_{\mathrm{f}}^{\mathrm{n}} \left[C_j^{i_{\mathrm{f+dl}}^{\mathrm{n}}[0]}\right] \Bigg].$$

The term relating to the integrator depends on C_0, the column of $\mathfrak{C}$ containing integration residues. The only portion of the output equation that has integration dynamics is $\theta_{\text{ss}}^{\text{r}}$. Therefore, using similar "[·]" notation for the rows of C_0, the scalar quantity $\widehat{C}_0$ is:

$$\widehat{C}_0 = [U_{\text{ocp}}^{\text{p}}]'\left[C_0^{\tilde{\theta}_{\text{ss}}^{\text{p}}[3]}\right] - [U_{\text{ocp}}^{\text{n}}]'\left[C_0^{\tilde{\theta}_{\text{ss}}^{\text{n}}[0]}\right].$$

5.B.2 *Computing derivatives of electrochemical variables with respect to states*

Implementing step 3b of the EKF requires computing the matrices:[26]

$$\widehat{C}_j = \frac{\mathrm{d}g_z(\mathbf{X}_\gamma[k], u[k])}{\mathrm{d}x_j[k]} \quad \text{and} \quad \widehat{C}_0 = \frac{\mathrm{d}g_z(\mathbf{X}_\gamma[k], u[k])}{\mathrm{d}x_0[k]}.$$

If $z[k]$ has dimensions $q \times 1$, then the dimensions of $\widehat{C}_{0,0}$, $\widehat{C}_{0,1}$, $\widehat{C}_{1,0}$, and $\widehat{C}_{1,1}$, are $q \times n$ and the dimension of $\widehat{C}_0$ is $q \times 1$. We compute these matrices one row at a time, where each row is the component of the derivative relating to the corresponding element in $z[k]$.

We consider the derivatives with respect to non-integrator states first. Similar to what we saw in the previous section, derivatives of elements of $z[k]$ corresponding to $i_{\text{f+dl}}^{\text{r}}[\tilde{x}, k]$ are computed as:

$$\frac{\mathrm{d}i_{\text{f+dl}}^{\text{r}}[\tilde{x}, k]}{\mathrm{d}x_j[k]} = \gamma_j\left[C_j^{i_{\text{f+dl}}^{\text{r}}[\tilde{x}]}\right].$$

We apply the same principle to find derivatives corresponding to $i_{\text{f}}^{\text{r}}[\tilde{x}, k]$ and $i_{\text{dl}}^{\text{r}}[\tilde{x}, k]$:

$$\frac{\mathrm{d}i_{\text{f}}^{\text{r}}[\tilde{x}, k]}{\mathrm{d}x_j[k]} = \gamma_j\left[C_j^{i_{\text{f}}^{\text{r}}[\tilde{x}]}\right] \quad \text{and} \quad \frac{\mathrm{d}i_{\text{dl}}^{\text{r}}[\tilde{x}, k]}{\mathrm{d}x_j[k]} = \gamma_j\left[C_j^{i_{\text{dl}}^{\text{r}}[\tilde{x}]}\right].$$

When computing derivatives corresponding to $\theta_{\text{ss}}^{\text{r}}[\tilde{x}, k]$, recall:

$$\begin{aligned}\theta_{\text{ss}}^{\text{r}}[\tilde{x}, k] &= \tilde{\theta}_{\text{ss}}^{\text{r}}[\tilde{x}, k] + \theta_{\text{s},0}^{\text{r}} \\ &= \left[\tilde{\theta}_{\text{ss}}^{\text{r}}[\tilde{x}, k]\right]^* - \frac{[\theta_{100}^{\text{r}} - \theta_0^{\text{r}}]\Delta t}{3600Q - \bar{C}_{\text{dl,eff}}^{\text{r}}\left|\theta_{100}^{\text{r}} - \theta_0^{\text{r}}\right|[U_{\text{ocp}}^{\text{r}}]'} x_0[k] + \theta_{\text{s},0}^{\text{r}},\end{aligned}$$

where $\left[\tilde{\theta}_{\text{ss}}^{\text{r}}[\tilde{x}, k]\right]^*$ corresponds to the integrator-removed debiased $\theta_{\text{ss}}^{\text{r}}$. We simplify the integration residue somewhat, trusting the EKF to adapt to track any error introduced by doing so:

$$\theta_{\text{ss}}^{\text{r}}[\tilde{x}, k] \approx \left[\tilde{\theta}_{\text{ss}}^{\text{r}}[\tilde{x}, k]\right]^* - \frac{[\theta_{100}^{\text{r}} - \theta_0^{\text{r}}]\Delta t}{3600Q} x_0[k] + \theta_{\text{s},0}^{\text{r}}. \tag{5.13}$$

So (because the integrator state is not considered in this derivative),

$$\frac{\mathrm{d}\theta_{\text{ss}}^{\text{r}}[\tilde{x}, k]}{\mathrm{d}x_j[k]} = \gamma_j\left[C_j^{[\tilde{\theta}_{\text{ss}}^{\text{r}}]^*[\tilde{x}]}\right].$$

[26] Since the set of cell variables of interest $z[k]$ is a vector quantity, the derivative $\widehat{C}_0 = \mathrm{d}g_z(\mathbf{X}_\gamma[k], u[k])/\mathrm{d}x_0[k]$ used in Step 3b is also a vector. This is different from how $\widehat{C}_0$ is defined and used in Step 2a. That is, these matrices have the same names as those used in Step 2a, but different definitions. Usage should be clear by context.

Derivatives corresponding to $\phi^{\mathrm{r}}_{\mathrm{s\text{-}e}}[\tilde{x},k]$ are made in a similar way. Recall,

$$\phi^{\mathrm{r}}_{\mathrm{s\text{-}e}}[\tilde{x},k] = [\tilde{\phi}^{\mathrm{r}}_{\mathrm{s\text{-}e}}[\tilde{x},k]]^* + U^{\mathrm{r}}_{\mathrm{ocp}}(\theta^{\mathrm{r}}_{\mathrm{s,avg}}[k]). \tag{5.14}$$

So, since $\theta^{\mathrm{r}}_{\mathrm{s,avg}}[k]$ depends only on the integrator state x_0,

$$\frac{\mathrm{d}\phi^{\mathrm{r}}_{\mathrm{s\text{-}e}}[\tilde{x},k]}{\mathrm{d}\boldsymbol{x}_j[k]} = \gamma_j\left[C_j^{[\tilde{\phi}^{\mathrm{r}}_{\mathrm{s\text{-}e}}]^*[\tilde{x}]}\right].$$

Derivatives corresponding to $\phi^{\mathrm{n}}_{\mathrm{s}}[\tilde{x},k]$ are simply:

$$\frac{\mathrm{d}\phi^{\mathrm{n}}_{\mathrm{s}}[\tilde{x},k]}{\mathrm{d}\boldsymbol{x}_j[k]} = \gamma_j\left[C_j^{\tilde{\phi}^{\mathrm{n}}_{\mathrm{s}}[\tilde{x}]}\right].$$

In the positive electrode, we must add the derivatives of $v_{\mathrm{cell}}[k]$ with respect to the state. These derivatives were computed in the first part of this appendix and will not be repeated here. So,

$$\frac{\mathrm{d}\phi^{\mathrm{p}}_{\mathrm{s}}[\tilde{x},k]}{\mathrm{d}\boldsymbol{x}_j[k]} = \gamma_j\left[C_j^{\tilde{\phi}^{\mathrm{p}}_{\mathrm{s}}[\tilde{x}]}\right] + \frac{\mathrm{d}g_v(\mathbf{X}_\gamma[k],u[k])}{\mathrm{d}\boldsymbol{x}_j[k]}.$$

Derivatives corresponding to $\phi^{\mathrm{n}}_{\mathrm{e}}[\tilde{x},k]$ are computed by recognizing:

$$\phi^{\mathrm{r}}_{\mathrm{e}}[\tilde{x},k] = \tilde{\phi}^{\mathrm{r}}_{\mathrm{e}}[\tilde{x},k] - [\tilde{\phi}^{\mathrm{n}}_{\mathrm{s\text{-}e}}[0,k]]^* - U^{\mathrm{n}}_{\mathrm{ocp}}(\theta^{\mathrm{n}}_{\mathrm{s,avg}}[k]). \tag{5.15}$$

Therefore, using prior results, and recognizing that the OCP term depends only on the integration state, we have:

$$\frac{\mathrm{d}\phi^{\mathrm{r}}_{\mathrm{e}}[\tilde{x},k]}{\mathrm{d}\boldsymbol{x}_j[k]} = \gamma_j\left[C_j^{\tilde{\phi}^{\mathrm{r}}_{\mathrm{e}}[\tilde{x}]}\right] - \gamma_j\left[C_j^{[\tilde{\phi}^{\mathrm{n}}_{\mathrm{s\text{-}e}}]^*[0]}\right].$$

Finally, the derivates corresponding to $\theta^{\mathrm{r}}_{\mathrm{e}}[\tilde{x},k]$ are simply:

$$\frac{\mathrm{d}\theta^{\mathrm{r}}_{\mathrm{e}}[\tilde{x},k]}{\mathrm{d}\boldsymbol{x}_j[k]} = \gamma_j\left[C_j^{\tilde{\theta}^{\mathrm{r}}_{\mathrm{e}}[\tilde{x}]}\right].$$

All that remains is to find the derivative of each variable with respect to the integrator state. The default value of every entry in $\widehat{C}_0$ is zero, since most variables do not have integration dynamics. The exceptions are entries corresponding to $\theta^{\mathrm{r}}_{\mathrm{ss}}$, $\phi^{\mathrm{r}}_{\mathrm{s\text{-}e}}$, and $\theta^{\mathrm{r}}_{\mathrm{e}}$. Based on the equations Eqs. (5.13), (5.14), and (5.15) already outlined,

$$\begin{aligned}
\frac{\mathrm{d}\theta^{\mathrm{r}}_{\mathrm{ss}}[\tilde{x},k]}{\mathrm{d}x_0[k]} &= -\frac{[\theta^{\mathrm{r}}_{100}-\theta^{\mathrm{r}}_{0}]\Delta t}{3600Q}\\
\frac{\mathrm{d}\phi^{\mathrm{r}}_{\mathrm{s\text{-}e}}[\tilde{x},k]}{\mathrm{d}x_0[k]} &= -[U^{\mathrm{r}}_{\mathrm{ocp}}]'\frac{[\theta^{\mathrm{r}}_{100}-\theta^{\mathrm{r}}_{0}]\Delta t}{3600Q}\\
\frac{\mathrm{d}\phi^{\mathrm{r}}_{\mathrm{e}}[\tilde{x},k]}{\mathrm{d}x_0[k]} &= [U^{\mathrm{r}}_{\mathrm{ocp}}]'\frac{[\theta^{\mathrm{r}}_{100}-\theta^{\mathrm{r}}_{0}]\Delta t}{3600Q}.
\end{aligned}$$

6
Diagnosis and Prognosis of Degradation

Until this point, we have focused our attention on developing a single parameterized mathematical model that describes a fresh (beginning-of-life) cell and on using that model to assist with estimating the cell's internal electrochemical variables. In reality, we must recognize that a single fixed cell model may not be sufficient: there will be cell-to-cell variations even among fresh cells; additionally, a cell's characteristics will evolve over time as the cell ages. Since a BMS must operate reliably under these conditions, it will need to be able to compensate for variations such as these.

This chapter focuses on aging-related factors, assuming that any BMS that can accommodate the changes in cell dynamics due to aging can also accommodate the relatively smaller cell-to-cell variations that arise due to imperfect manufacturing tolerances.

As illustrated in the roadmap of Fig. 6.1, we have chosen to address this topic after the main modeling and state-estimation chapters but before those that consider how a BMS might compute advisory limits on battery-pack operation. These future chapters require background knowledge on how lithium-ion cells degrade in order to optimize usage to prevent premature aging and failure while still maximizing the performance delivered by the battery pack. The top-

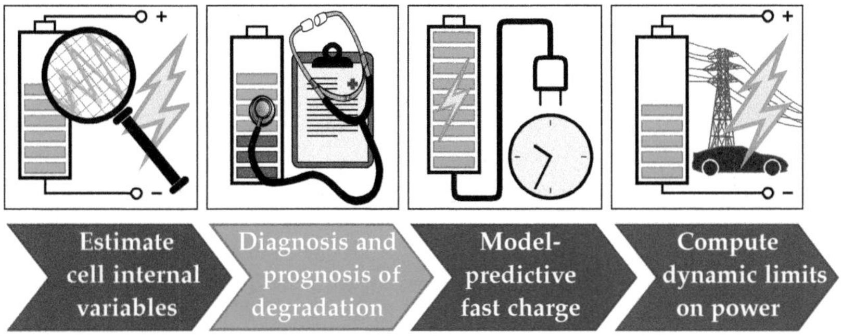

Figure 6.1: Topics in lithium-ion battery controls that we cover in this volume.

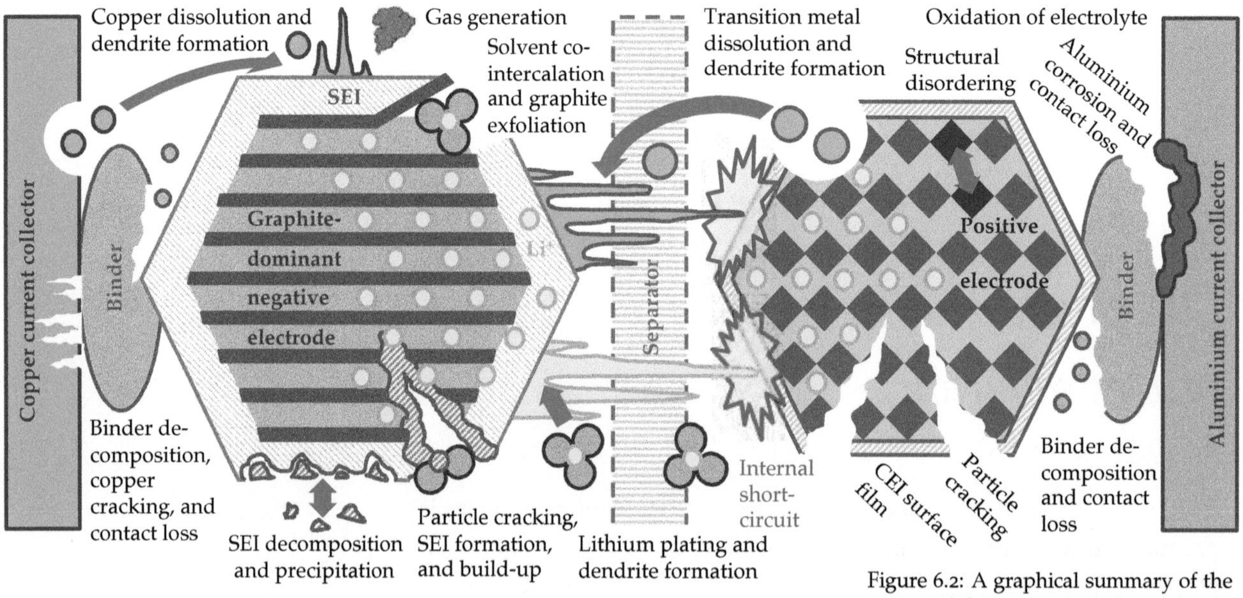

Figure 6.2: A graphical summary of the primary lithium-ion degradation mechanisms. Modified from: Christoph R. Birkl, Matthew R. Roberts, Euan McTurk, Peter G. Bruce, and David A. Howey, "Degradation diagnostics for lithium ion cells," *Journal of Power Sources*, 341:373–386, 2017.

ics leading to this point are foundational; furthermore, the remainder of this volume will build on those presented in this chapter.

After a review of cell degradation indicators, we organize the discussion of a BMS's response to cell aging under the categories of *diagnostics* and *prognostics*. These topics of study are still immature and are the focus of many research teams at this point in history. We will put forward some preliminary ideas and results, recognizing that there is much work still to do to convert them into mature, robust, and practical real-time methods.

6.1 *Degradation indicators*

There are multiple physical processes by which lithium-ion cells are known to degrade. Many of these were discussed in Vol. II of this series and Fig. 6.2 provides a graphical summary of the primary known lithium-ion degradation mechanisms. There is abundant literature that describes these mechanisms qualitatively, and an actively growing volume of research that seeks to quantify the individual mechanisms and the interactions between them.

Ideally, lithium-ion cells would last forever. Lithium would cycle between electrodes without any loss in capacity as the cell is charged and discharged, and the dynamics governing a cell's voltage response to an input-current profile would not change over time. However, this is not what we observe in practice; instead, we notice decreasing total capacity and changes to cell dynamics—especially manifesting as increases in cell resistance—over time. These two degradation indicators are known as capacity fade and power fade.[1]

[1] As resistance increases, an aged cell encounters a minimum-voltage limit during discharge earlier than would a fresh cell; therefore, available discharge power decreases or fades over time.

6.1.1 *Capacity fade*

It is straightforward to illustrate the two fundamental degradation modes leading to capacity fade. We do so with reference to Fig. 6.3, which shows a simplified depiction of ideal-cell behavior as a baseline scenario. In the figure, both the negative and positive electrodes have sixteen sites that could hold lithium. Presently, four negative-electrode sites and five positive-electrode sites are shown to be occupied. When the cell is fully charged, there will be nine occupied sites in the negative electrode and no occupied sites in the positive electrode. When the cell is fully discharged, there will be no occupied sites in the negative electrode, and nine occupied sites in the positive electrode.[2] The total capacity of the cell is equal to the minimum of the number of storage sites in the negative electrode, the number of storage sites in the positive electrode, and the amount of lithium that can be cycled. In this example, the total capacity is the minimum of 16, 16, and 9, which yields a total capacity of nine lithium atoms.

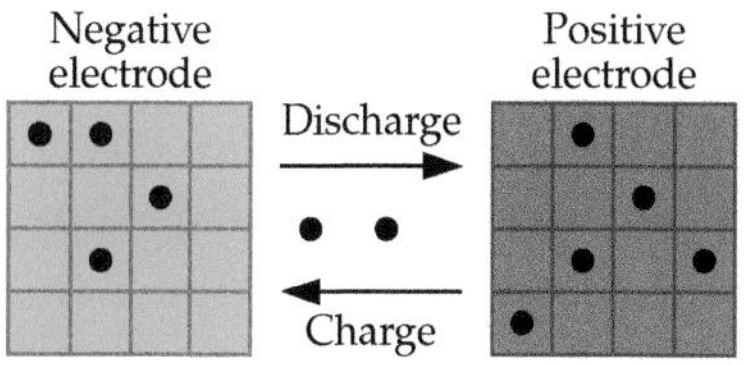

Figure 6.3: Ideal lithium-ion cell operation.

[2] This is a simplification. Neither electrode is ever completely full or completely empty during operation. Instead, a cell is considered fully charged or fully discharged when its open-circuit voltage reaches predetermined maximum and minimum levels.

Continuing the example, a side reaction is any undesired electrochemical or chemical process that consumes lithium permanently so that it is no longer able to cycle between the electrodes. This degradation mode is known as loss of lithium inventory (LLI), and is illustrated in Fig. 6.4 where one lithium atom is shown being consumed by a side reaction as the cell is being charged. The total capacity has now been reduced to the minimum of 16, 16, and 8, which yields a total capacity of eight lithium atoms.

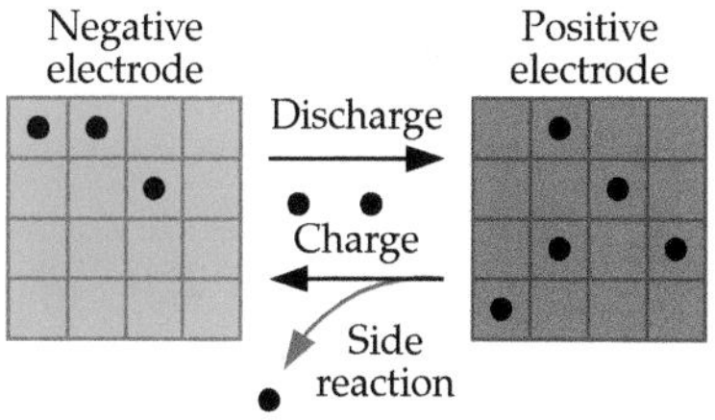

Figure 6.4: Loss of lithium inventory due to side reaction.

Structural deterioration can also reduce capacity by eliminating lithium storage sites from one (or both) of the electrodes, perhaps due to a collapse of part of the crystal structure of the electrode itself. This degradation mode is known as loss of active material (LAM), and is illustrated in Fig. 6.5 where damage to the positive electrode is shown as a white scar. In this example, six sites in the positive electrode that previously could have held lithium are now unavailable to do so in the future. The total capacity is now the minimum of 16, 10, and 9, which yields a total capacity of nine lithium atoms. Notice that the loss of six storage sites in this example is invisible to our method for accounting for total capacity even though actual damage has occurred. However, if the structure continues to deteriorate and more sites are lost, we will begin to observe capacity loss since the number of sites will become fewer than the number of lithium atoms that are available to cycle between the electrodes.

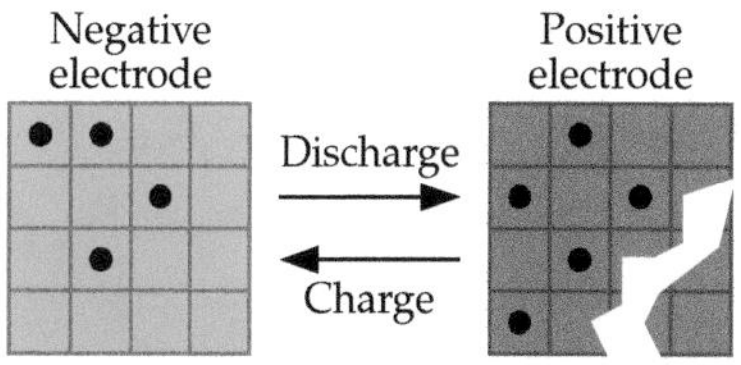

Figure 6.5: Loss of active material due to particle fracture.

When an electrode's structure deteriorates, some lithium may be trapped in the part of the structure that is lost such that it is no longer free to cycle back and forth as the cell is charged and discharged; in this case, capacity is lost due to both LLI and LAM occurring at the same time. In Fig. 6.6, the lithium atom in the lower-right corner of the positive electrode is trapped by the structural collapse and is unavailable for cycling; the positive electrode has eleven undamaged storage sites, but one is isolated from cycling, so there are only ten useable storage sites. So, the total capacity is the minimum of 16, 10, and 8, yielding a final total capacity of 8 lithium atoms.

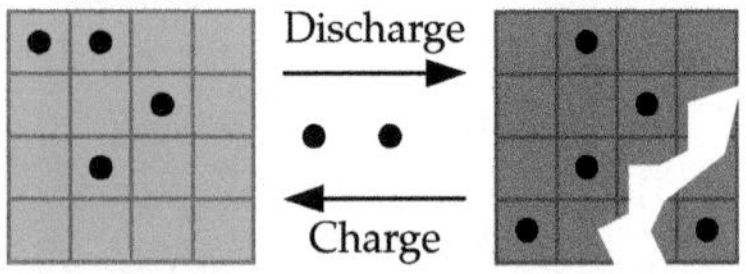

Figure 6.6: Loss of active material *and* loss of lithium inventory due to particle fracture.

In physics-based models, cell total capacity can be expressed (in ampere hours, expressed using standard parameter values) as:[3]

$$Q = \underbrace{\frac{\varepsilon_s^n A L^n c_{s,\max}^n F}{3600}}_{Q^n} \underbrace{\left|\theta_{100}^n - \theta_0^n\right|}_{\Delta\theta_s^n} = \underbrace{\frac{\varepsilon_s^p A L^p c_{s,\max}^p F}{3600}}_{Q^p} \underbrace{\left|\theta_{100}^p - \theta_0^p\right|}_{\Delta\theta_s^p}. \qquad (6.1)$$

[3] Recall that cell total capacity is defined as the quantity of ampere hours that must be moved from the negative electrode to the positive electrode to change cell OCV from $v_{\max}$ to $v_{\min}$. Total capacity expressed using negative-electrode parameters is equal to total capacity expressed using positive-electrode parameters since the quantity of lithium that must leave the negative electrode is assumed to be the same as the quantity that must be added to the positive electrode in this exercise.

We now understand that total capacity is a moving target due to LLI and LAM and so a physical cell's total capacity would actually change (somewhat) if we were to implement this discharging procedure in a laboratory setting. However, we can still express total capacity in these terms as a conceptual quantity describing the cell's characteristic at a specific point in time. It essentially states, "If no additional degradation were to occur during the process of moving cell OCV from $v_{\max}$ to $v_{\min}$, this is the quantity of discharged ampere hours we would observe while doing so."

Until this point, we have treated all parameters in Eq. (6.1) as being constant. However, we now recognize that as a cell ages and capacity is lost, some of these parameter values must change. In this chapter, we treat current-collector area A, electrode thickness L^r, and maximum lithiation $c_{s,\max}^r$ as being constant to first order; F and 3600 are physical constants and do not change. Therefore, capacity loss must be represented by changes to the solid-material volume fractions ε_s^r and to the electrode operating boundaries θ_{100}^r and θ_0^r.

LAM in one electrode directly decreases the volume fraction ε_s^r for that electrode. Therefore, in terms of lumped parameters, we will see a decrease in Q^r for that electrode. However, the loss of active material in one electrode does not cause loss of active material in the other electrode; nevertheless, Eq. (6.1) must still hold. Therefore, we conclude that if Q^r decreases in one electrode, then $\Delta\theta_s^r = \left|\theta_{100}^r - \theta_0^r\right|$ must decrease in the other electrode. This is in fact what we observe, as we will explore in more detail later in the chapter.

It is less obvious from Eq. (6.1) how capacity is lost due to LLI. Since the electrode materials are not themselves damaged by LLI, Q^r must remain constant. Therefore, we conclude that $\left|\theta_{100}^r - \theta_0^r\right|$ must decrease in both electrodes. Again, we will see later in the chapter that this is exactly what happens.

Summarizing to this point, capacity loss is caused either by LLI or LAM (or both). It is possible to explain this capacity loss by tracking changes to Q^r, θ_{100}^r, and θ_0^r.

6.1.2 *Power fade*

While capacity fade can be neatly categorized in terms of LLI and/or LAM, it is less simple to categorize power fade. Anything that increases cell resistance is a cause of power fade. And there are a lot of factors that can do so.

In Chap. 3, we saw that the very-short-duration pulse resistance of a cell is a function of 13 parameter values (more if an MSMR model is used). If any of these parameters change, cell resistance will change. However, when considering power fade, we are not usually concerned only with the very-short-duration pulse resistance, but rather with resistance to meaningful discharge or charge events, which are typically seconds or minutes in duration. Therefore, changes to the value of nearly any parameter in the cell model could lead to observed power fade.

So it is worth reviewing the list of model parameters and considering which may change due to known degradation mechanisms such as those illustrated in Fig. 6.2. And to do so carefully, we should conduct this exercise using standard parameters in case the parameter-lumping process introduced in Chap. 1 should somehow obscure how some parameters change due to aging.

We begin by investigating what happens when ε_s^r changes, as we already know that the volume fraction of active material will decrease due to any mechanism that causes LAM. A decrease to ε_s^r:[4]

- Decreases cell total capacity Q directly, as shown in Eq. (6.1).
- Decreases electrode lumped conductivity $\bar{\sigma}^r = \sigma_{\text{eff}}^r A/L^r$ because $\sigma_{\text{eff}}^r = \sigma^r(\varepsilon_s^r)^{\text{brug}}$ decreases.
- Increases film resistance $\bar{R}_f^r = R_f^r/(a_s^r A L^r)$ and double-layer resistance $\bar{R}_{\text{dl}}^r = R_{\text{dl}}^r/(a_s^r A L^r)$ since $a_s^r = 3\varepsilon_s^r/R_s^r$ decreases.[5]
- Decreases reaction-rate constants $\bar{k}_{0,j}^r = a_s^r A L^r F k_{\text{norm},0,j}^r$ since a_s^r decreases.

Therefore, a decrease to ε_s^r causes changes to Q, $\bar{\sigma}^r$, $\bar{R}_f^r$, $\bar{R}_{\text{dl}}^r$, $\bar{C}_{\text{dl}}^r$, $\bar{k}_{0,j}^r$, and possibly $\bar{D}_s^r$ in the model. As we have already seen, it will also cause changes to θ_{100} and θ_0 in the opposite electrode. Most of these changes increase cell resistance.

Certain kinds of degradation are known to cause changes to porosity ε_e^r. For example, when a solid–electrolyte interphase (SEI) film grows on the negative-electrode particles, solvent in the electrolyte is consumed and part of the volume previously occupied by electrolyte becomes filled with the SEI film. A similar process occurs in the positive electrode, although probably to a lesser extent, forming and growing a cathode–electrolyte interphase (CEI) surface film and reducing porosity. A decrease to ε_e^r:

- Decreases electrolyte lumped conductivity $\bar{\kappa}^r = \kappa_{\text{eff}}^r A/L^r$ because effective conductivity $\kappa_{\text{eff}}^r = \kappa^r(\varepsilon_e^r)^{\text{brug}}$ decreases.
- Decreases the quantity of lithium in the electrolyte:

$$\bar{q}_e^r = \frac{\varepsilon_e^r c_{e,0} A L^r F}{3600(1 - t_+^0)}.$$

- Decreases electrolyte diffusivity $\bar{D}_e^r = D_{\text{e,eff}}^r A c_{e,0}/(L^r(1 - t_+^0))$ because $D_{\text{e,eff}}^r = D_e(\varepsilon_e^r)^{\text{brug}}$ decreases.[6]

[4] It is possible that a decrease to ε_s^r also causes an increase in electrode solid diffusivity $\bar{D}_s = D_s/R_s^2$ because the effective radius of the particles decreases (which would tend to decrease cell resistance). However, it is likely that the actual diffusion length does not change when active material is lost; the material is simply unavailable for lithiation.

[5] For similar reasons, a decrease to ε_s^r will also decrease double-layer capacitance $\bar{C}_{\text{dl}}^r = a_s^r A L^r C_{\text{dl}}^r$. However, the time constant of the double layer may not change since $\tau_{\text{dl}} = \bar{R}_{\text{dl}}^r \bar{C}_{\text{dl}}^r = R_{\text{dl}}^r C_{\text{dl}}^r$; that is, a_s cancels from the computation. If the time constant were ever observed to change, that would indicate that the intrinsic resistivity or capacitance of the double layer had changed due to aging.

[6] However, note that $\bar{D}_e^r$ is not part of the LPM, so any changes to its value may not be observable. Instead, we have $\bar{\psi} = F\bar{D}_e^r/(T\bar{\kappa}^r)$. Since the changes to $\bar{D}_e^r$ and $\bar{\kappa}^r$ are assumed to be proportional, we assume that $\bar{\psi}$ does not change with aging.

Therefore, a decrease to ε_e^r causes changes to $\bar{\kappa}^r$, $\bar{q}_e^r$, and $\bar{D}_e^r$. All of these lead to increased resistance.

Other mechanical factors cause particle-to-particle contact loss due to binder decomposition and particle-to-current-collector contact loss due to delamination of the electrode from the current collector and current-collector corrosion. These decrease the conductivity of the electrode taken as a whole (effectively reducing A) and increase contact resistance R_c.

There are even some degradation mechanisms that can lead to a short-term *decrease* in cell resistance. For example, electrode swelling and shrinking caused by repeated insertion and removal of lithium can crack the SEI film, making it locally easier for lithium to intercalate and deintercalate, thus reducing $\bar{R}_f^n$. However, cracks in the SEI expose fresh graphite to the electrolyte and further SEI film quickly grows, restoring the prior level of resistance.

In summary, any aging mechanism that leads to a decrease in either ε_s^r or ε_e^r will tend to increase cell resistance indirectly. Some mechanisms operate directly on terms in the model such as $\bar{R}_f^r$, increasing resistance in a more direct way.

6.2 *Diagnosis versus prognosis*

From this discussion, we conclude that nearly all of the LPM parameters can change as the cell ages.[7] So, we need to consider carefully how a BMS should respond to aging in battery cells to provide the best benefit and protections over the entire life of the battery pack.

First, we recognize that BMS state estimates and control decisions are likely to degrade over time if the BMS does not take into account cell aging but instead uses fresh-cell models in its algorithms at all states of age. We would expect improved estimates if the BMS always had available to it parameterized models of every battery-pack cell that accurately described their dynamics at their present states of age.

So, the first important BMS response to aging is one that we will call *diagnostics*. We use this term to refer to methods that provide up-to-date aged-cell parameterized models of every cell in a battery pack.[8] This definition is broader than we find in some prognostics and health management (PHM) literature, where diagnostics is often used more narrowly to relate to "processes of detection and isolation of faults and failures."[9] Instead of focusing on cell failure, we use the term almost in a medical sense, to refer to determining what is going wrong with this cell even (and especially) if it has not yet failed utterly. That is, we are not interested in postmortem diagnostics of failed cells, but rather in diagnostics that apply to aged but still-useable cells.

[7] Except perhaps $\bar{\psi}$ and α^r. It is also possible that the MSMR-model parameter values of electrode OCP remain unchanged, but we have encountered no literature that discusses this in one way or another.

[8] This is an advanced proxy to simpler state-of-health estimation.

[9] Canh Ly, Kwok Tom, Carl S. Byington, Romano Patrick, and George J. Vachtsevanos, "Fault diagnosis and failure prognosis for engineering systems: A global perspective," in *2009 IEEE International Conference on Automation Science and Engineering*. IEEE, 2009, pp. 108–115.

Second, we recognize that the control actions a BMS should take with an aged cell are likely to be different from those that apply to a fresh cell. To be able to compute optimized controls, we must take the present state of aging into consideration in our calculations of near-term degradation that we anticipate would be caused by any proposed control action. Then we can optimize those actions to minimize the anticipated level of incremental degradation.

So, the second BMS response to aging involves aspects that we will call *prognostics*. We use this term to refer to the predicted near-future condition of the cell if some proposed current or power profile were to be applied to it. Again, our use of the term is broader than is commonly encountered in the PHM literature, where we find definitions such as "prognostics are the processes of predicting a future state (of reliability) based on current and historic conditions, or the remaining useful life (RUL) of components or systems."[10] Here, as we will discuss shortly, we are less concerned with predicting RUL—which also has to do with predicting the point in time of complete failure of the battery pack—than with predicting quantities such as the amount of short-term future degradation.

[10] Ibid.

In summary, we refer to diagnostics as methods to determine the present SOH as specifically as possible. We would like to find model parameter values that accurately describe every cell's dynamics at its present state of aging. These diagnostics include updating a total-capacity estimate as is common with SOH estimators, but also ideally updating many or all of the remaining LPM parameter values as well. We refer to prognostics as predictive of the future condition of the cell, including how some or all of the LPM parameters are expected to change if some user-provided current or power profile were to be applied. We will see in Chaps. 7–8 that these predictions can be used in optimizations when implementing fast-charge and dynamic-power-limit algorithms, for example.

In the next sections, we provide a survey of some diagnostics and prognostics methods. But, before we do so, we pose two questions:

- Regarding diagnostics: How important is it for a BMS to have accurate aged-cell models of all of its cells versus a single fresh-cell model? It turns out that this may not be as necessary as we might expect, especially if the BMS uses an adaptive model-based state-estimation method such as SPKF. For example, Miguel et al. conducted a simulation study where they applied an SPKF to both fresh cells and aged cells using only a fresh-cell model in the state estimator.[11] They found that the state estimates did degrade somewhat for the aged cells, but perhaps not enough to be a concern. The adaptivity of the SPKF overcame much of the anticipated state-prediction errors due to voltage-prediction errors caused

[11] E. Miguel, Gregory L. Plett, M. Scott Trimboli, I. Lopetegi, Laura Oca, Unai Iraola, and E. Bekaert, "Electrochemical model and sigma point Kalman filter based online oriented battery model," *IEEE Access*, 9:98072–98090, 2021.

by embedded model versus actual-cell model mismatch when computing state estimates. This result is encouraging because it implies that our diagnostics need neither be comprehensive nor perfect for the identified aged-cell models to be useful.

- Regarding prognostic: How important is it for a BMS to be able to compute RUL? There is perhaps more current debate on this topic. We agree that it is important for some applications—such as sizing a battery when designing a battery pack—to be able to predict RUL in simulation studies under anticipated usage. However, our opinion at this point is that it is not necessary for an embedded BMS in most applications to be able to predict far-future RUL. It may be valuable to notify the battery operators that a failure is likely in the next minutes or days so that they can take any necessary actions to protect themselves and ancillary equipment. However, we feel that it is of little use to be able to make real-time statements such as "the battery is expected to have an RUL of 8.75 years," especially given the uncertainties in making such far-future predictions.[12]

[12] We imagine that long-term forecasting may be more desirable for some applications than others, particularly if there are opportunities to take remedial action to extend life. Large battery installations such as those employed for grid storage could be examples of this.

6.3 *BMS diagnostics*

When discussing diagnostics, there are different levels of information that might be helpful. We have already discussed the degradation indicators of capacity fade and power fade. These can be diagnosed using the SOH algorithms presented in Vol. II to find up-to-date estimates of cell total capacity Q and equivalent-series resistance R_0. Several of the methods presented in that volume, such as the approximate weighted total least squares (AWTLS) method to estimate total capacity, and the simple voltage-differencing method to estimate resistance are model-agnostic and so can be used regardless of whether the main BMS algorithms use ECMs or PBMs.

But it would be helpful if we could glean more insight into what caused the capacity and power fade. We might further be interested in quantifying the levels of different degradation modes, such as LLI or LAM, or even in quantifying the changes to other model parameter values.

One of the simplest diagnostics reported in the literature is an application that monitors changes to the cell's OCV versus SOC relationship and the corresponding shifts in the electrode stoichiometric operating windows. Sect. 6.4 discusses how the OCV relationship can be used as a diagnostic. Then, Sect. 6.5 shows how some researchers have developed real-time methods to track the present stoichiometry of both electrodes separately in real time, which can aid with this kind of diagnostic. Sect. 6.6 considers how we might adjust other model parameter values to keep up with an aging cell. Finally, our discussion of diagnostics concludes with Sect. 6.7, which presents an alternative to adapting model parameters, which may prove to be more robust over the aging evolution of the cell.

6.4 *Changes to OCV, θ_0^r, and θ_{100}^r as a diagnostic*

In this section, we focus on showing how a cell's OCV versus SOC relationship changes subtly over time as the cell ages. This observation has led to diagnostics proposed by some researchers.[13] For the moment, we assume that we have lab-quality OCV relationships that we can examine, but will consider real-time application in Sect. 6.5.

[13] See, for example:

- Hannah M. Dahn, A. J. Smith, J. C. Burns, D. A. Stevens, and J. R. Dahn, "User-friendly differential voltage analysis freeware for the analysis of degradation mechanisms in Li-ion batteries," *Journal of The Electrochemical Society*, 159(9):A1405, 2012.
- Christoph R. Birkl, Matthew R. Roberts, Euan McTurk, Peter G. Bruce, and David A. Howey, "Degradation diagnostics for lithium ion cells," *Journal of Power Sources*, 341:373–386, 2017.

6.4.1 *Changes to OCV due to LLI*

When studying the effects of LLI and LAM, it is good to keep Eq. (6.1) in mind. With the LLI degradation mode, neither of the electrode absolute capacities Q^r change. Therefore, any observed change in total capacity must come from a change in the electrode operating boundaries, and so LLI must cause a decrease in $\Delta\theta_s^r$ in both electrodes.

To see how this comes about, we consider a thought experiment. Let us assume that a cell is initially resting at 100 % SOC prior to losing lithium in the negative electrode. At this point, $\theta_s^r = \theta_{100}^r$. This is illustrated in Fig. 6.7 by the blue markers.

The loss of lithium inventory in the negative electrode causes θ_s^n to decrease as lithium is extracted from the electrode to take part in side reactions. This is illustrated by the red marker in Fig. 6.7 for the negative electrode. The value of θ_s^p does not change. This combination causes $U_{\rm ocp}^n(\theta_s^n)$ to increase while $U_{\rm ocp}^p(\theta_s^p)$ remains constant. Overall, cell voltage decreases.

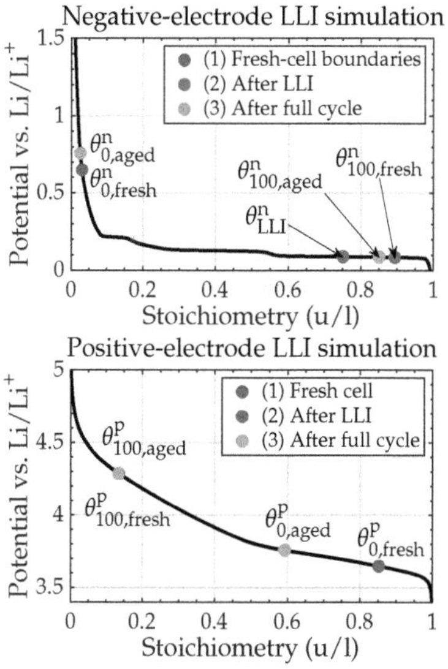

Figure 6.7: Illustrating the movement of electrode operating boundaries due to LLI. Note that θ_{100}^p moves somewhat to the left (it decreases in value).

We will not notice the impact of LLI on electrode operating boundaries until we perform a full charge and discharge of the cell. Since LLI has decreased cell voltage, we must add charge to the cell to bring its OCV back up to $v_{\rm max}$, which is our definition for 100 % SOC. Adding charge decreases θ_s^p slightly, which was already at its prior lower limit of θ_{100}^p. Therefore, θ_{100}^p moves somewhat left in the figure due to LLI in the negative electrode and $U_{\rm ocp}^p(\theta_{100}^p)$ increases versus its prior value. Since $U_{\rm ocp}^p(\theta_{100}^p)$ is greater than before, $U_{\rm ocp}^n(\theta_{100}^n)$ must also be greater than before aging for the OCV to equal $v_{\rm max}$, meaning that θ_{100}^n is less than before aging due to LLI.

When we discharge the cell to $v_{\rm min}$, moving an equal amount of charge out of the negative electrode and into the positive electrode, we discover that θ_0^r in both electrodes become lower than their prior

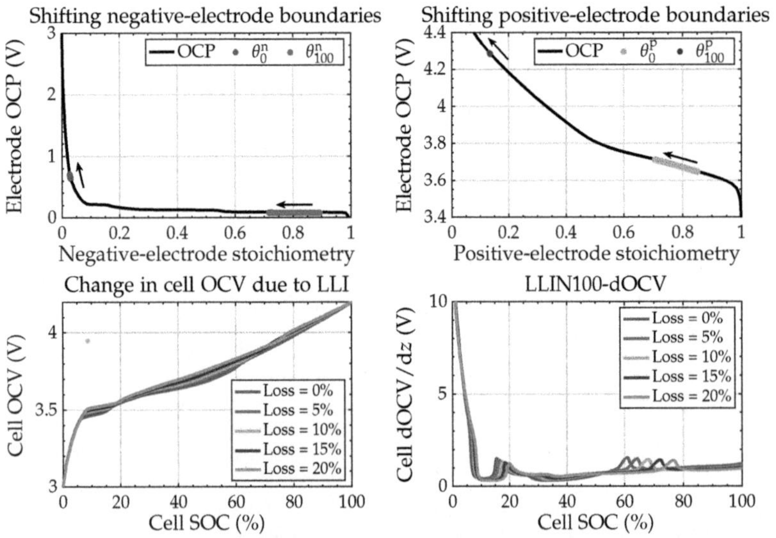

Figure 6.8: Movement in electrode operating boundaries and consequent changes in cell OCV and differential OCV due to LLI.

values. That is, all boundaries decrease, but the dominant effect is that the window width $\Delta\theta_s^r$ shrinks in both electrodes.

LLI at other conditions has an identical effect. For example, LLI in the positive electrode when the cell is initially resting at 100 % SOC causes θ_s^p to decrease. This brings about a voltage increase, and we must move θ_{100}^n downward to compensate to achieve $v_{\max}$.

Fig. 6.8 shows results from a simulation study that explores the movement of the electrode operating boundaries due to LLI for a graphite//NMC cell. Several different LLI scenarios were investigated (LLI in the negative electrode or positive electrodes, and LLI at several different initial SOC points). The results from the different scenarios are visually (and perhaps actually) indistinguishable, so we present only a single set of figures here.

The top two plots show how the operating boundaries change in both electrodes over many cycles. The arrows indicate the directions that the boundaries move. The bottom-left figure shows the evolution of the cell's OCV relationship versus total-capacity loss in percent, and the bottom-right figure shows changes in the cell's differential OCV relationship. For the case where all aging is due to LLI, we see noticeable changes to the OCV and differential OCV curves that might be used to diagnose the level of degradation in the cell.

6.4.2 *Changes to OCV due to LAM*

While we find that changes to OCV due to LLI do not appear to depend upon the electrode in which the lithium is lost or the cell SOC at which the loss occurs, we encounter a different outcome with LAM. To illustrate, we consider two different scenarios.

In the first scenario, some small amount of nonlithiated active material is lost when the cell is initially resting at 100 % SOC. Since no lithium inventory is lost, cell voltage does not change. Therefore, the cell remains at 100 % SOC and neither $\theta_{100}^{\mathrm{n}}$ nor $\theta_{100}^{\mathrm{p}}$ change. However, since $\varepsilon_{\mathrm{s}}^{\mathrm{r}}$ decreases, the absolute capacity Q^{r} for that electrode decreases. To balance the operational capacities of both electrodes, by Eq. (6.1), the value of $\Delta\theta_{\mathrm{s}}^{\bar{\mathrm{r}}}$ for the opposite electrode $\bar{\mathrm{r}}$ must decrease, forcing the OCP range of the opposite electrode to shrink as well. In order to maintain cell voltage limits between v_{min} and v_{max}, the value of $\Delta\theta_{\mathrm{s}}^{\mathrm{r}}$ must increase for the region that loses the active material. This is illustrated in Fig. 6.9 for the case when LAM occurs in the negative electrode when the cell is initially resting at 100 % SOC.

In the second scenario, some active material is lost when the cell is initially resting at 0 % SOC. We again consider losing nonlithiated material, so the cell does not lose lithium inventory at the same time. Since no lithium is lost, cell voltage does not change. Therefore, the cell remains at 0 % SOC and neither θ_0^{n} nor θ_0^{p} change. However, since $\varepsilon_{\mathrm{s}}^{\mathrm{r}}$ decreases, the theoretical capacity Q^{r} for that electrode decreases. To balance the operational capacities of both electrodes, by Eq. (6.1), the value of $\Delta\theta_{\mathrm{s}}^{\bar{\mathrm{r}}}$ for the opposite electrode $\bar{\mathrm{r}}$ must decrease, forcing the OCP range of the opposite electrode to shrink. In order to maintain cell voltage limits between v_{min} and v_{max}, the value of $\Delta\theta_{\mathrm{s}}^{\mathrm{r}}$ must increase for the region that loses the active material. This is illustrated in Fig. 6.10 for the case when LAM occurs in the positive electrode when the cell is initially resting at 0 % SOC.

In both scenarios, we find that the value of $\theta_{\mathrm{s}}^{\mathrm{r}}$ does not change for the value of cell SOC at which the active material is lost, $\Delta\theta_{\mathrm{s}}^{\mathrm{r}}$ for the electrode that loses active material increases and $\Delta\theta_{\mathrm{s}}^{\bar{\mathrm{r}}}$ for the electrode that does not lose active material decreases. When active material is lost at 0 % SOC, the boundaries θ_0^{r} do not change but both boundaries $\theta_{100}^{\mathrm{r}}$ do change; when active material is lost at 100 % SOC, the boundaries $\theta_{100}^{\mathrm{r}}$ do not change but both boundaries θ_0^{r} change; when active material is lost at an arbitrary SOC, all boundaries change.

Figs. 6.11 and 6.12 show results of two simulation scenarios. In the first figure, active material is lost in the negative electrode when the cell is initially resting at 100 % SOC; in the second figure, active material is lost in the positive electrode when the cell is initially resting at 0 % SOC. The arrows show the direction of movement of the stoichiometric operating boundaries in each electrode. The simulation results confirm our discussion regarding how the operating boundaries are expected to move, and provide further insight into the changes to cell OCV and differential OCV.

Unlike the LLI case, each LAM scenario results in different changes to OCV. However, assuming that electrode OCP relationships them-

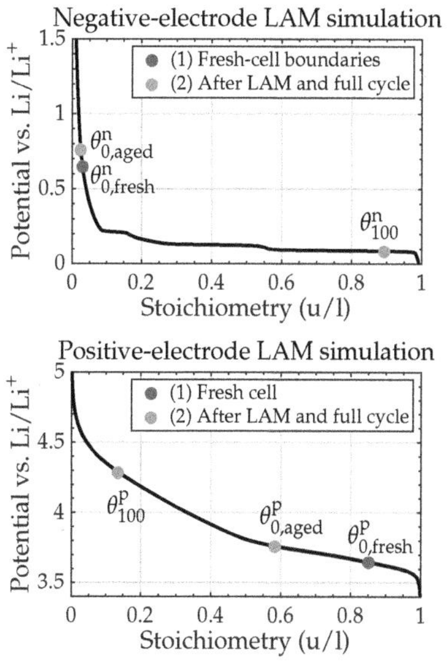

Figure 6.9: Illustrating the movement of electrode operating boundaries due to LAM in the negative electrode when the cell is initially resting at 100 % SOC. Note that neither $\theta_{100}^{\mathrm{n}}$ nor $\theta_{100}^{\mathrm{p}}$ change in this scenario.

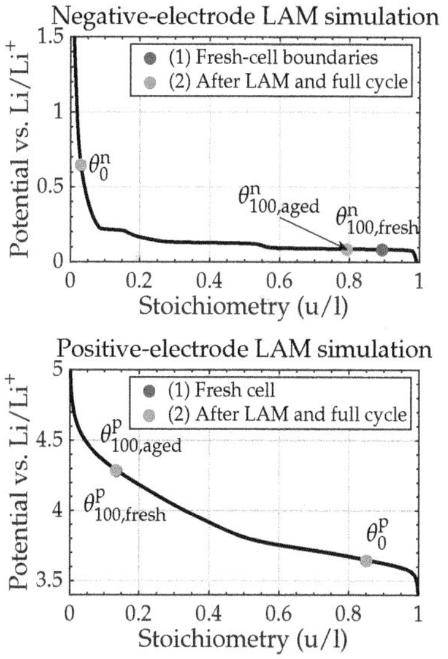

Figure 6.10: Illustrating the movement of electrode operating boundaries due to LAM in the positive electrode when the cell is initially resting at 0 % SOC. Note that neither θ_0^{n} nor θ_0^{p} change in this scenario, and that $\theta_{100}^{\mathrm{p}}$ moves somewhat to the left (it decreases in value).

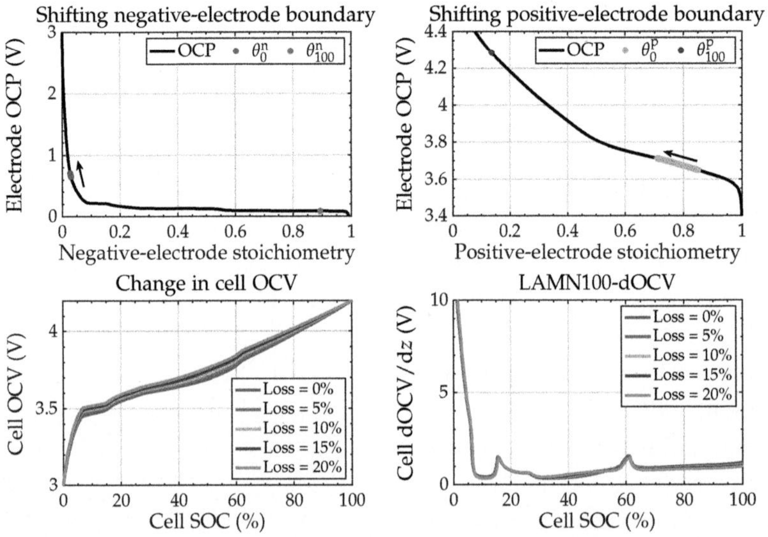

Figure 6.11: Movement in electrode operating boundaries and consequent changes in cell OCV and differential OCV due to LAM in the negative electrode when the cell is initially resting at 100 % SOC.

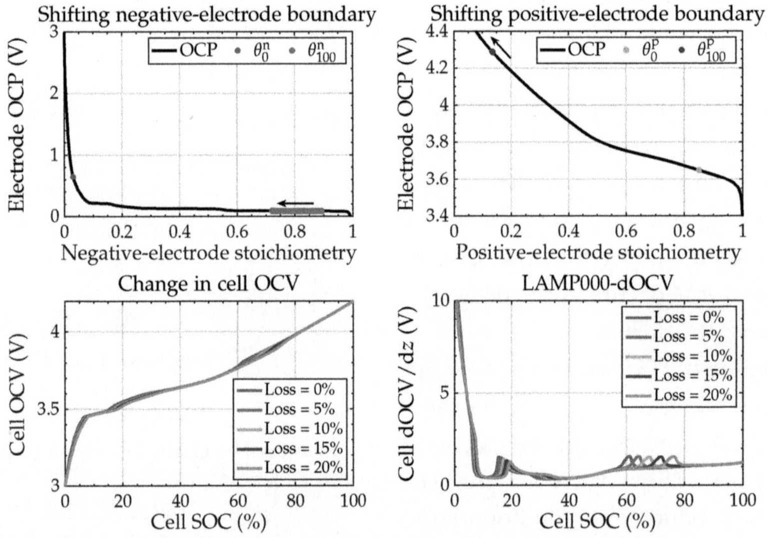

Figure 6.12: Movement in electrode operating boundaries and consequent changes in cell OCV and differential OCV due to LAM in the positive electrode when the cell is initially resting at 0 % SOC.

selves do not change as a cell ages, we can fit electrode operating boundaries θ_0^r and θ_{100}^r to match electrode OCP relationships to the aged OCV relationship—much like we did for a fresh cell in Chap. 3—to determine how the stoichiometric windows have changed due to age. Per Birkl et al., we may also be able to determine how much LLI and LAM have occurred, and in which electrodes. However, to do so requires an accurate estimate of the present cell OCV relationship, which requires laboratory equipment and specialized testing. In the next section, we review a method that may be able to accomplish similar objectives in a practical BMS in real time.

6.5 *Loosely coupled tracking of θ_s^r in both electrodes*

We now recognize that the stoichiometric operating windows of both electrodes change as a cell ages. Using lab-quality OCV data, we can update estimates of θ_0^r and θ_{100}^r. However, this approach is not feasible for a real-time BMS application. Another complication is that the change in boundaries for one electrode may be quite different from the corresponding change in the other electrode, depending on whether LLI or LAM has occurred, and at what level of SOC.

Allam and Onori have proposed a real-time architecture that is able to estimate θ_s^r reliably in both electrodes.[14,15] Their method uses separate models of negative- and positive-electrode dynamics that are loosely coupled. Each model employs a closed-loop estimator of the state dynamics for the electrode it describes directly, and an open-loop predictor of the dynamics of the opposite electrode. The models are interconnected by providing the closed-loop estimate from one as the input to the open-loop predictor of the other. By adapting the state estimate of both electrodes separately, they are allowed to adjust somewhat independently of one another and track θ_s^r correctly for both electrodes, even when the operating windows change.

[14] Anirudh Allam and Simona Onori, "An interconnected observer for concurrent estimation of bulk and surface concentration in the cathode and anode of a lithium-ion battery," *IEEE Transactions on Industrial Electronics*, 65(9):7311–7321, 2018; and Anirudh Allam and Simona Onori, "Online capacity estimation for lithium-ion battery cells via an electrochemical model-based adaptive interconnected observer," *IEEE Transactions on Control Systems Technology*, 29(4):1636–1651, 2020.

[15] Consequently, we imagine that it could be extended to adjust θ_0^r and θ_{100}^r as well.

To be more specific, the positive-electrode model is defined in a continuous-time state-space form as:

$$
\begin{aligned}
\dot{\hat{x}}_1(t) &= \boldsymbol{A}_{11}\hat{x}_1(t) + \boldsymbol{B}_1 u(t) + \mathbf{fn}_1(v_{\text{cell}}(t) - \hat{v}_1(t)) \\
\dot{\hat{x}}_{2,\text{ol}}(t) &= \boldsymbol{A}_{22}\hat{x}_2(t) + \boldsymbol{B}_2 u(t) \\
\hat{v}_1(t) &= h_1(\hat{x}_1(t), u(t)) - h_2(\hat{x}_{2,\text{ol}}(t), u(t)) - R_0 u(t),
\end{aligned}
$$

where $x_1(t)$ is the state of the positive-electrode model, $x_2(t)$ is the state of the negative-electrode model, $u(t)$ is the input current, $v_{\text{cell}}(t)$ is the actual cell voltage, and $\hat{v}_1(t)$ is the prediction of cell voltage computed by this model. The matrices $\boldsymbol{A}_{11}$, $\boldsymbol{A}_{22}$, $\boldsymbol{B}_1$, and $\boldsymbol{B}_2$ are provided by the models of the two electrodes. The function $\mathbf{fn}_1(\cdot)$ is a possibly nonlinear feedback calculation. The function $h_1(\cdot)$ computes the positive-electrode OCP plus the positive-electrode Butler–Volmer overpotential and the function $h_2(\cdot)$ computes the negative-electrode OCP plus the negative-electrode Butler–Volmer overpotential.

This model adjusts the positive-electrode state estimate using a feedback mechanism and computes an open-loop prediction of the negative-electrode state using the most recently available closed-loop estimate from the opposite observer. It predicts cell voltage (lumping all electrolyte dynamics into the Ohmic resistance term R_0) and uses the voltage-prediction error to adapt its own state estimate.

The negative-electrode model computes a symmetric relationship:

$$
\dot{\hat{x}}_{1,\text{ol}}(t) = \boldsymbol{A}_{11}\hat{x}_1(t) + \boldsymbol{B}_1 u(t)
$$

$$\dot{\hat{x}}_2(t) = A_{22}\hat{x}_2(t) + B_2u(t) + \mathbf{fn}_2(v_{\text{cell}}(t) - \hat{v}_2(t))$$
$$\hat{v}_2(t) = h_1(\hat{x}_{1,\text{ol}}(t), u(t)) - h_2(\hat{x}_2(t), u(t)) - R_0u(t).$$

Here, the positive-electrode state evolves according to an open-loop law and the negative-electrode state evolves according to a feedback law. Fig. 6.13 illustrates the coupling between the two observers.

Allam and Onori described the electrodes using single-particle models and used a sliding-mode control law to update the state estimates, which allowed them to prove convergence properties for their estimators. We believe that more comprehensive ROMs similar to those developed in Chap. 4 and nonlinear Kalman filters similar to those presented in Chap. 5 could also be substituted, if desired. This is a topic of future research.

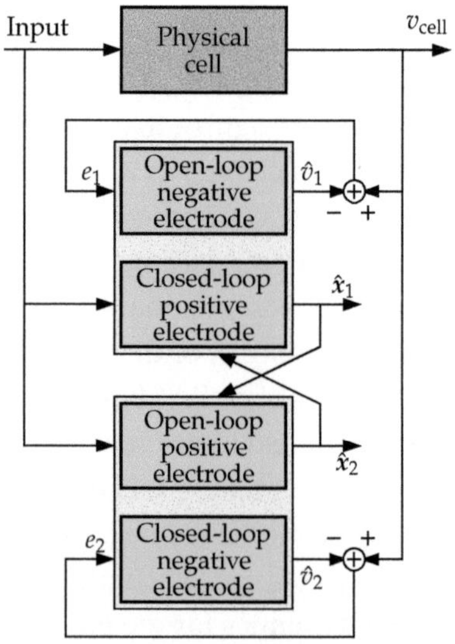

Figure 6.13: Interconnected observer design for concurrent estimation of negative-electrode and positive-electrode bulk and surface stoichiometries. Figure adapted from Allam and Onori, 2018.

6.6 *Adapting a model to track cell variations and aging*

An advantage of the Allam and Onori method is its simplicity. Using a fixed fresh-cell model, the BMS can estimate electrode bulk and surface stoichiometries accurately. But it stands to reason that we would expect better performance (e.g., faster convergence, tighter confidence bounds, more accurate estimates of internal electrochemical variables) if—instead of using a fixed fresh-cell model—we were to use models that evolved to describe more accurately the present dynamics of every battery-pack cell at its current state of age.

There are two basic approaches to doing so; we discuss one now and the other in Sect. 6.7. The first approach is illustrated in Fig. 6.14. It adapts the parameters of a cell model so that the model's predictions match the physical cell's input/output (current/voltage) behaviors. This basic idea was also discussed in Vol. II, where we introduced dual and joint Kalman filters that simultaneously adjusted the states and parameters of the cell model.

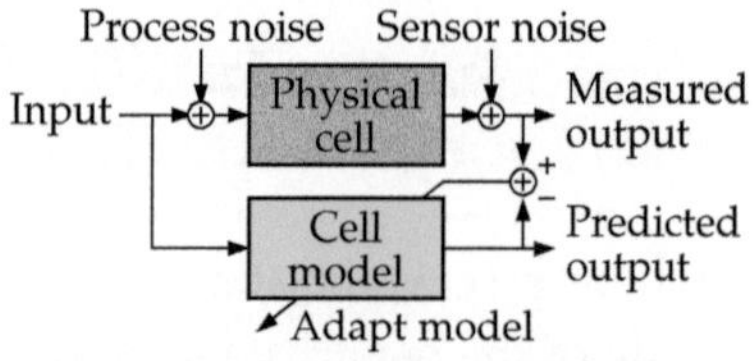

Figure 6.14: The first of two approaches to track variations in cell dynamics.

These methods can produce satisfactory results with ECMs if the Kalman filters are tuned appropriately. However, we find that this can be a very risky approach due to the vastly different timescales of state and parameter evolution. An inherent difficulty arises in separating the influence of the slowly aging parameters from the rapidly moving state, which is exacerbated when there are more states and parameters to estimate than there are outputs, or when the parameters and outputs are loosely related. There is typically not enough information available to ensure that updated parameter values are realistic and that internal states are correctly estimated, even if the model minimizes some metric of voltage-prediction error.

An example might help illustrate the problem. With an ECM, we know that voltage is modeled as OCV minus dynamic polarization (perhaps including hysteresis) minus the ohmic voltage drop, where the latter is modeled as $R_0 i_{app}(t)$. When the model voltage prediction does not match the measured voltage, the joint or dual Kalman filter will attempt to modify the model's state or parameter values so that voltage-prediction error decreases. The problem is that the easiest way to make voltage agree at this moment is to modify R_0 since it is directly connected to the voltage prediction. If we look at the evolution of R_0 versus time for a poorly tuned dual or joint xKF, we will see that it has a very noisy time series that makes no physical sense; further, we will see that the xKF makes no meaningful updates to the the states and other model parameters.

To fix this, we would need to assign extremely small process-noise covariance to the R_0 model, which causes the xKF to believe that R_0 is essentially constant, forcing the xKF to adjust other states or parameters instead. Similarly, we would need to use very small noise-covariance values for all the parameters so that they do not adapt as rapidly as the states. It is very difficult to tune such an xKF so that it produces stable results.

A modification to this idea is to monitor the input/output signals of the cell over a longer period of time and adjust the parameter values less frequently. This is essentially what we ended up doing with the different total-capacity estimation algorithms in Vol. II. Or we could choose to modify only some subset of model parameters that lead to more stable behaviors. The Allam and Onori method does this by effectively modifying only the OCP operating windows.

Another clever approach combines a fixed fresh-cell physical model with a machine-learning component such as a neural network (NN).[16] All dynamic effects are captured by the physical model; the NN simply computes a static nonlinear function of its inputs to improve model predictions. Two examples are illustrated in Fig. 6.15. In the top figure, the physical cell model produces a voltage estimate $v_{LPM}(t)$ and its own internal state estimate $\hat{x}(t)$. The NN computes a correction factor based on $i_{app}(t)$ and $\hat{x}(t)$. The correction factor is added to $v_{LPM}(t)$ to produce the overall hybrid voltage prediction, $v_{hybrid}(t)$. The parameters of the NN are adapted using standard machine-learning methods such that $v_{hybrid}(t)$ agrees with $v_{cell}(t)$ as closely as possible. In the bottom figure, the physical cell model simply produces its own internal state estimate and the NN directly computes $v_{hybrid}(t)$ based on $\hat{x}(t)$ and $i_{app}(t)$.

Both of these methods promise to be more stable than if we were to adjust the state and parameters of the physical model directly. The big advantage is that all dynamics are in the fixed physical model. The physical model does not adapt in this scenario, and so state updates will not lose physical meaning. The neural-network component

[16] Hao Tu, Scott Moura, and Huazhen Fang, "Integrating electrochemical modeling with machine learning for lithium-ion batteries," in *2021 American Control Conference (ACC)*. IEEE, 2021, pp. 4401–4407; and Hao Tu, Scott Moura, Yebin Wang, and Huazhen Fang, "Integrating physics-based modeling with machine learning for lithium-ion batteries," *Applied Energy*, 329:120289, 2023.

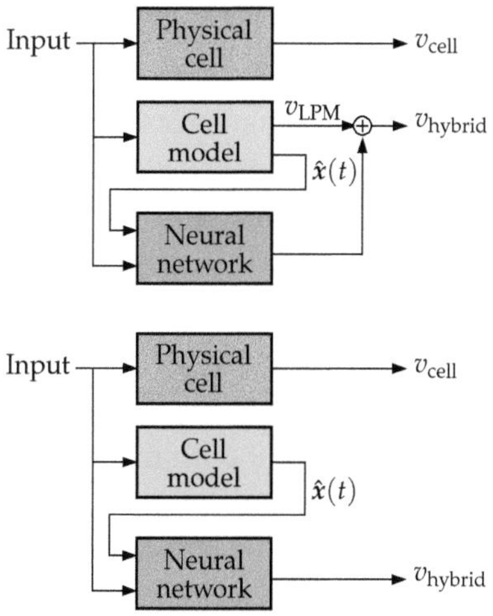

Figure 6.15: Adapting a neural network to improve model predictions. Figure adapted from the Tu et al. arXiv preprint.

simply acts to improve the estimates by providing a static (i.e., memoryless) nonlinear correction factor that can adapt to model the cell at different states of age.

6.7 *Selecting a model that describes present dynamics*

We have now discussed one approach to track cell dynamics—by adapting parameter values in a cell model. We now consider a second approach, illustrated in Fig. 6.16. This approach to parameter updating requires the construction of alternate parameter sets that seek to represent the cell in a variety of aged conditions. The task is then to identify which one of the parameter sets best represents the cell at a given moment in time. Unlike the parameter adaptation method, the preaged models used by this approach can be constrained to guarantee realistic parameter values. However, additional computation and storage requirements are imposed on the BMS as each model must be stored and evaluated whenever the parameter-update process is conducted. A trade-off is then required between the number of possible aged models that can be represented and the minimization of computational requirements.

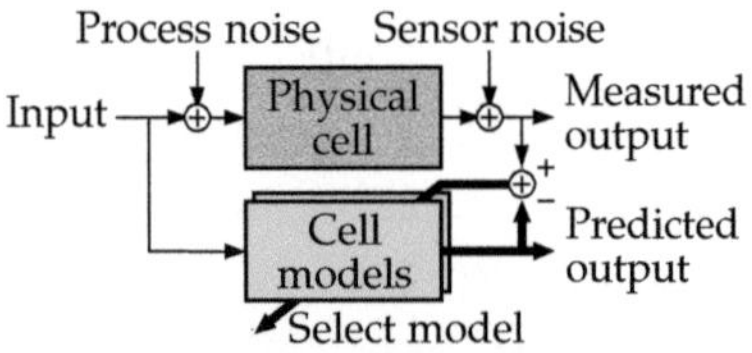

Figure 6.16: The second of two approaches to track variations in cell dynamics.

An implementation of this model-selection approach was proposed by Smiley and Plett[17] who used an interacting multiple model Kalman filter (IMM-KF)[18] to track the aging of a cell over life. A single fresh-cell model is combined with a collection of models that each has its parameter values modified to reflect different kinds and degrees of aging. The IMM executes multiple KFs in parallel, and blends their state estimates based on how closely each model matches the observed input/output dynamics of the cell. A novel output of the IMM is an estimate indicating which of the preaged models best describes the cell at its present state of age, giving some additional diagnostic information regarding how the cell might have aged to reach its current condition.

[17] Adam Smiley and Gregory L. Plett, "An adaptive physics-based reduced-order model of an aged lithium-ion cell, selected using an interacting multiple-model Kalman filter," *Journal of Energy Storage*, 19:120–134, 2018.

[18] Efim Mazor, Amir Averbuch, Yakov Bar-Shalom, and Joshua Dayan, "Interacting multiple model methods in target tracking: a survey," *IEEE Transactions on aerospace and electronic systems*, 34(1):103–123, 1998.

Fig. 6.17 illustrates the components of the IMM-KF. Each row of the figure represents a Kalman filter. There are M rows—one per model in the model set. To the left, we see the prior state and covariance estimates for each model, as well as a new vector $\mu_j[k-1]$, which is an estimate of the probability that the battery cell was best described by the jth model at the prior timestep.

The IMM first performs an interaction step: At any point in time, the aging status of the cell will be somewhat different from the aging status at a prior point in time. So, the IMM-KF method depends on knowledge of a transition-probability matrix p_{ij} that describes the likelihood of moving from model i to model j between timesteps. The interaction step of the IMM-KF blends prior state estimates and error

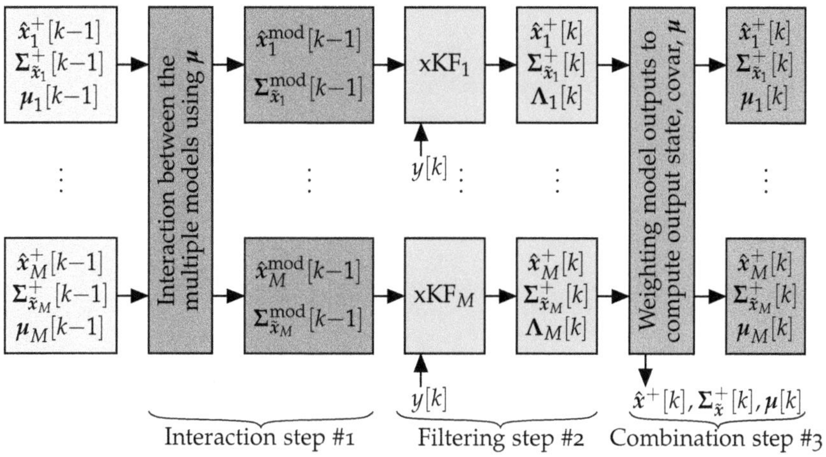

Figure 6.17: The IMM Kalman filter approach. Figure adapted from Smiley and Plett.

covariances of all models according to their transition probabilities to make modified state estimates, and covariances.

After interaction, each individual KF executes a single timestep update computing its own state estimate, error covariance, and an assessment of the likelihood of the measured voltage given the assumed model. The combination step computes an overall state estimate, covariance, and probability mass function (pmf) for the relative likelihood that cell dynamics correspond to each model.

To apply the IMM to the BMS model-selection problem, we must first develop models that are representative of a cell at different states of age. Fig. 6.18 shows an example matrix of models comprising a single fresh-cell model and 20 additional aged-cell models. The aged models represent a hypothetical cell at levels of retained capacity equal to 92.5 %, 85 %, 77.5 %, and 70 % of the original capacity. They also represent cells having degraded entirely due to LLI or entirely due to LAM or due to some blend of the two. In addition to the aged models, we must provide a model-transition matrix $\boldsymbol{p}$ containing the probability of transitioning from model i to model j:

$$p_{ij} = \Pr\left(m[k] = j \mid m[k-1] = i\right), \tag{6.2}$$

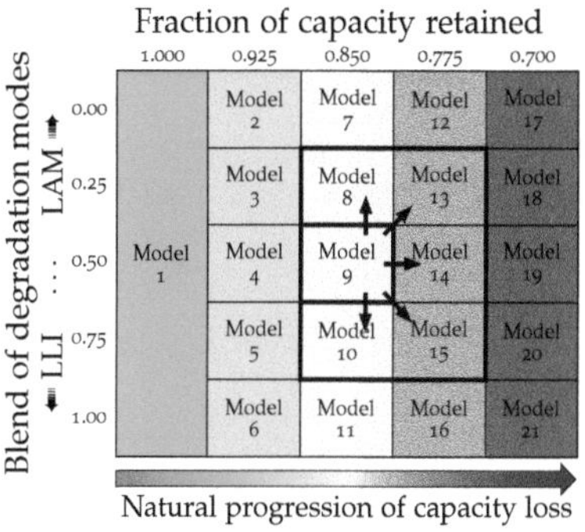

Figure 6.18: Natural progression of aging causes evolution in the model that best represents the cell at its current state of age.

where $m[k]$ is the model that best describes the cell at time k. The figure shows the assumption made by Smiley and Plett, that a cell cannot regain lost capacity, and so the transitions are never toward the left in the model space. Generally, the values on the diagonal of p_{ij} are very close to 1, indicating that it is most likely that the model that describes a cell does not change from one timestep to another. Models that are adjacent to the present model have small positive values in p_{ij}; models that are not adjacent have zeros values in p_{ij}.

The next subsections elaborate on some details of the IMM implementation.

6.7.1 *Interaction*

The estimates and covariances from the M independent xKFs (from the prior timestep) are blended together to produce the inputs to the M independent xKFs for this timestep. The interaction step begins by computing $\mu_{i|j}[k-1]$, the conditional probability of having been in mode i at the prior timestep $k-1$ given that the system is currently in mode j and the set of all prior measurements $\mathbb{Y}[k-1]$,[19]

$$\mu_{i|j}[k-1] = \Pr(m[k-1] = i \mid m[k] = j, \mathbb{Y}[k-1]).$$

[19] We use the term "mode m" to mean that the mth model is the one that describes the presently observed dynamics the best.

By assuming that the transition from one operating mode to another can be described as a Markov-chain process, movement from one mode to another is described by a static transition probability matrix, defined a priori. If $m[k]$ is the mode that the cell is operating in at a given sampling time k, this transition-probability matrix is expressed as p_{ij}, as in Eq. (6.2). If the number of models in the model space is equal to M, the p_{ij} matrix has dimensions of $M \times M$, and the values of the elements can be determined based on a statistical analysis of how mode transitions have occurred in past observations, or based on how mode transitions are expected to occur when past observations are unavailable.

Note that a version of Bayes' rule that applies is:

$$\Pr(A \mid B, C) = \frac{\Pr(B \mid A, C)\Pr(A \mid C)}{\Pr(B \mid C)}.$$

So, we have:

$$\begin{aligned}\mu_{i|j}[k-1] &= \Pr(m[k-1] = i \mid m[k] = j, \mathbb{Y}[k-1]) \\ &= \frac{\Pr(m[k] = j \mid m[k-1] = i, \mathbb{Y}[k-1])\Pr(m[k-1] = i \mid \mathbb{Y}[k-1])}{\Pr(m[k] = j \mid \mathbb{Y}[k-1])} \\ &= \frac{1}{\bar{c}_j} p_{ij}\mu_i[k-1],\end{aligned}$$

where $\mu_i[k-1]$ is the output pmf from the IMM from the prior time iteration and where the normalizing constants $\bar{c}_j$ are found as:

$$\bar{c}_j = \Pr\left(m[k] = j \mid \mathbb{Y}[k-1]\right) = \sum_{i=1}^{M} p_{ij}\mu_i[k-1]. \tag{6.3}$$

Then, we blend together state outputs from the prior step according to the probability that they could have contributed to the state at this timestep. The modified input state to xKF j at timestep k is:

$$\begin{aligned}\hat{x}_j^{\text{mod}}[k-1] &= \mathbb{E}\left[x[k-1] \mid m[k] = j, \mathbb{Y}[k-1]\right] \\ &= \sum_{i=1}^{M} \hat{x}_i^+[k-1]\Pr(m[k-1] = i \mid m[k] = j, \mathbb{Y}[k-1)\end{aligned}$$

$$= \sum_{i=1}^{M} \hat{x}_i^+[k-1]\mu_{i|j}[k-1].$$

After a lengthy but straightforward derivation, the covariance of $\hat{x}_j^{\text{mod}}[k-1]$ can be found to be:

$$\begin{aligned}\Sigma_{\tilde{x}_j}^{\text{mod}}[k-1] &= \mathbb{E}\left[(x[k-1] - \hat{x}_j^{\text{mod}}[k-1])(x[k-1] - \hat{x}_j^{\text{mod}}[k-1])^T\right] \\ &= \sum_{i=1}^{M} \left\{\Sigma_{\tilde{x}_i}^+[k-1] + \left[\hat{x}_i^+[k-1] - \hat{x}_j^{\text{mod}}[k-1]\right]\right. \\ &\qquad \left.\left[\hat{x}_i^+[k-1] - \hat{x}_j^{\text{mod}}[k-1]\right]^T\right\} \mu_{i|j}[k-1].\end{aligned}$$

The first term is the mixture of the prior covariances, and the second term is due to the spread of the means of the individual filters.

6.7.2 *Filtering*

Once the state and covariance values produced by the interaction step are available for each model j, one xKF iteration is evaluated for each model to produce a new state estimate $\hat{x}_j^+[k]$ and covariance $\Sigma_{\tilde{x}_j}^+[k]$. The xKF algorithm must be augmented to calculate the likelihood of the current system measurement being observed given that the system is currently in mode j. This likelihood, $\Lambda_j[k]$, is found as:

$$\begin{aligned}\Lambda_j[k] &= \mathcal{N}(y[k] - \hat{y}_j[k], \Sigma_{\tilde{y}_j}[k]) \\ &= \frac{1}{(2\pi)^{n/2}\left|\Sigma_{\tilde{y}_j}[k]\right|^{1/2}} \exp\left(-\frac{1}{2}(y[k] - \hat{y}_j[k])^T \Sigma_{\tilde{y}_j}^{-1}[k](y[k] - \hat{y}_j[k])\right).\end{aligned}$$

In other words, $\Lambda_j[k]$ is found by first defining the Gaussian distribution of the output estimate under the assumption that the output estimate and true output measurement are identical. The value of the distribution is then found for the innovation (the difference between the estimate and the measured value). Thus, the smaller the magnitude of the innovation and the greater the confidence in the output estimate, the larger the value of $\Lambda_j[k]$.

6.7.3 *Combination*

The final step in each IMM iteration is to determine the probability to associate with each model to form a pmf, as well as to combine the results of each individual Kalman filter into a single weighted output. We first compute the a posteriori probability of being in mode j at time k:

$$\mu_j[k] = \Pr(m[k] = j \mid \mathbb{Y}[k]) = \Pr(m[k] = j \mid \mathbb{Y}[k-1], y[k])$$

$$
\begin{aligned}
&= \frac{f(y[k] \mid m[k] = j, \mathbb{Y}[k-1]) \Pr(m[k] = j \mid \mathbb{Y}[k-1])}{\Pr(y[k])} \\
&= \frac{1}{c} \Lambda_j[k] \sum_{i=1}^{M} \Pr(m[k] = j \mid m[k-1] = i, \mathbb{Y}[k-1]) \\
&\qquad \times \Pr(m[k-1] = i \mid \mathbb{Y}[k-1]) \\
&= \frac{1}{c} \Lambda_j[k] \sum_{i=1}^{M} p_{ij} \mu_i[k-1] \\
&= \frac{1}{c} \Lambda_j[k] \bar{c}_j,
\end{aligned}
$$

where $\bar{c}_j$ has already been computed by Eq. (6.3) and c is a normalizing constant that ensures that all $\mu_j[k]$ sum to one. Using the newly found $\mu_j[k]$ values as weighting factors, the total output of the IMM is provided as the pmf $\mu_j[k]$ and

$$
\hat{x}^+[k] = \sum_{j=1}^{M} \hat{x}_j^+[k] \mu_j[k]
$$

$$
\Sigma_{\tilde{x}}^+[k] = \sum_{j=1}^{M} \left\{ \Sigma_{\tilde{x}_j}^+[k] + [\hat{x}_j^+[k] - \hat{x}^+[k]][\hat{x}_j^+[k] - \hat{x}^+[k]^T \right\} \mu_j[k].
$$

That is, the state is computed as a weighted combination of the individual filter state estimates. Likewise, the covariance is computed as a weighted combination of the individual filter covariance estimates. The output of the IMM gives state estimate and covariance like in Chap. 5 but also the pmf $\mu_j[k]$ estimating which model best describes the cell at this moment. The latter can be helpful as a diagnosis of the total impact of the history of operation of the cell.

Fig. 6.19 shows an example of the evolution of μ_j versus time in the IMM. We might consider using this pmf to select the model from the model space that best describes the input/output observations. In the top two plots, the index j that maximizes μ_j correctly identifies the true model. In the lower plot, it instead identifies a nearest-neighbor model having nearly identical input/output dynamics. Only toward the very end of the simulation does the IMM appear to recognize its error and begin to correct the values of μ_j. The nearest-neighbor model incorrectly selected at first is likely to be sufficient for BMS application; however, another paper by Smiley et al. showed how to postprocess the IMM outputs to give improved model selections, and those methods could be employed, if desired.[20]

While model selection using an IMM can work well, it is a computationally intensive approach. We find that it tends to work best when the cell voltage dynamics are very different from one model to another, which isn't always the case. State estimates are always good but the method does not always select the correct model from the

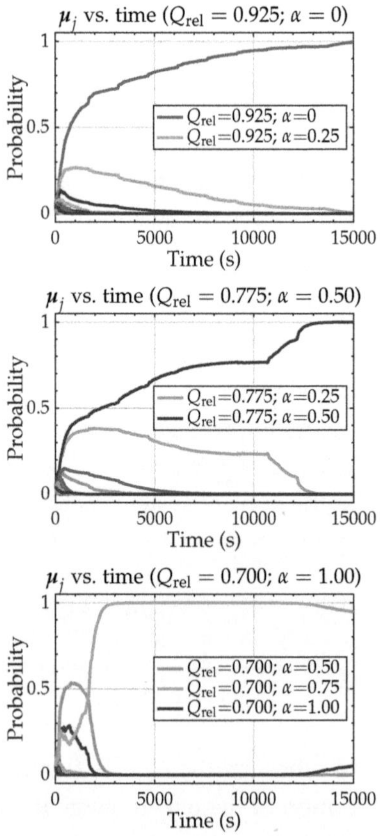

Figure 6.19: Example evolution of μ_j versus time for an IMM-SPKF application. The true aged-cell relative capacity Q_{rel} and blend of degradation modes α are shown in the figure titles. The pmf values for all models are plotted and those having highest probability are identified in the legends. In these results, $\alpha = 0$ means that all degradation is due to LAM and $\alpha = 1$ means that all degradation is due to LLI. Intermediate α indicates a blend of LAM and LLI.

[20] Adam J. Smiley, Willie K. Harrison, and Gregory L. Plett, "Postprocessing the outputs of an interacting multiple-model Kalman filter using a Markovian trellis to estimate parameter values of aged Li-ion cells," *Journal of Energy Storage*, 27:101043, 2020.

model set; as we have seen, it sometimes selects a nearest-neighbor model having similar dynamics. More work could be done to optimize the design of the model set to maximize the distance between nearest neighbors in a dynamic sense.

6.8 *BMS prognostics*

Now that we have quickly surveyed some BMS diagnostics topics, we turn our attention to prognostics. Broadly speaking, prognostics involve predicting a future condition of the cell. In this chapter, we introduce a 1D lumped-parameter thermal model and a degradation model as prognostics. These will be used in Chaps. 7–8 when implementing BMS controls.

6.9 *Cell 1D thermal model*

A BMS can measure present cell surface temperatures, but if it must predict the future thermal effect of a proposed control action, we require a thermal model.[21] In that sense, then, a thermal model provides prognostic information.

This section considers adding a description of heat generation and temperature evolution to the cell model. Cell dynamics and degradation mechanisms are highly temperature-dependent and being able to predict temperatures is important to BMS operation. Also, although a BMS can often measure the surface temperature of one or more cells in the battery pack directly, it is impractical to place temperature sensors on all cells. A thermal model can predict cell core temperatures and can provide software virtual temperature sensors for cells that are not monitored directly. Finally, it can be valuable to predict temperatures that might be reached due to proposed near-future power demands when computing power limits to keep temperature within desired bounds.

[21] This section is adapted from: Aloisio Kawakita de Souza, Gregory L. Plett, and M. Scott Trimboli, "A physics-based electrode-level thermal model for advanced battery management," in *Proceedings of the 35th International Electric Vehicle Symposium and Exhibition (EVS35)*, Oslo, Norway, June 2022.

6.9.1 *Converting thermal model into lumped-parameter form*

Vol. I of this series derived a physics-based 1D thermal model using standard DFN parameters. Here, we convert that model to use inputs from the LPM. As a reminder, the 1D model was:[22]

$$\frac{\partial(\rho^{\mathrm{r}} c_{\mathrm{p}}^{\mathrm{r}} T^{\mathrm{r}})}{\partial t} = \nabla \cdot (\lambda^{\mathrm{r}} \nabla T^{\mathrm{r}}) + q^{\mathrm{r}}, \tag{6.4}$$

where ρ^{r} is the material density of cell region r, $c_{\mathrm{p}}^{\mathrm{r}}$ is the specific heat, $T^{\mathrm{r}}(x,t)$ is the temperature, λ^{r} is the thermal conductivity, and q^{r} is the heat-generation rate. The units of T^{r} are kelvin, and the units of q^{r} are $\mathrm{W\,m^{-3}}$.

[22] We omit discussion of the boundary conditions of this PDE here; we assume convective heat transfer at the cell surface in the simplified models, but this could be extended as described in Vol. I.

The heat-generation term itself summed different heat sources:

$$q^{\mathrm{r}} = q_{\mathrm{i}}^{\mathrm{r}} + q_{\mathrm{r}}^{\mathrm{r}} + q_{\mathrm{e}}^{\mathrm{r}} + q_{\mathrm{s}}^{\mathrm{r}} + q_{\mathrm{c}}^{\mathrm{r}},$$

where each term is summarized as:

- Irreversible heat generation due to chemical reactions: $q_{\mathrm{i}}^{\mathrm{r}} = a_{\mathrm{s}}^{\mathrm{r}} F j_j^{\mathrm{r}} \eta_j^{\mathrm{r}}$ for each chemical reaction j that takes place at the solid/electrolyte interface;
- Reversible heat generation due to change in entropy: $q_{\mathrm{r}}^{\mathrm{r}} = a_{\mathrm{s}}^{\mathrm{r}} F j_j^{\mathrm{r}} T \frac{\partial U_{\mathrm{ocp},j}^{\mathrm{r}}}{\partial T}$ for each chemical reaction j that takes place at the interface;
- Joule heating due to electrical potential gradient in the solid: $q_{\mathrm{s}}^{\mathrm{r}} = \sigma_{\mathrm{eff}}^{\mathrm{r}}(\nabla \phi_{\mathrm{s}}^{\mathrm{r}} \cdot \nabla \phi_{\mathrm{s}}^{\mathrm{r}})$;
- Joule heating due to electrochemical potential gradient in the electrolyte: $q_{\mathrm{e}}^{\mathrm{r}} = \kappa_{\mathrm{eff}}^{\mathrm{r}}(\nabla \phi_{\mathrm{e}}^{\mathrm{r}} \cdot \nabla \phi_{\mathrm{e}}^{\mathrm{r}}) + \kappa_{\mathrm{D,eff}}^{\mathrm{r}}(\nabla \ln c_{\mathrm{e}}^{\mathrm{r}} \cdot \nabla \phi_{\mathrm{e}}^{\mathrm{r}})$;
- Heat generation due to contact resistance, $q_{\mathrm{c}}^{\mathrm{r}} = i_{\mathrm{app}}^2 R_{\mathrm{c}}^{\mathrm{r}}$.[23]

[23] Note the different units: q_{c} has units $\mathrm{W\,m^{-2}}$. This heat-generation term applies only to the current collector to electrode contact region, and is specified per unit area rather than per unit volume.

Following the method from Chap. 1, we can convert Eq. (6.4) into a lumped-parameter version. The net operation is to multiply both sides of the equation by AL^{r}, giving:

$$\frac{\partial(\bar{c}^{\mathrm{r}} T^{\mathrm{r}})}{\partial t} = \nabla \cdot (\bar{\lambda}^{\mathrm{r}} \nabla T^{\mathrm{r}}) + \bar{q}^{\mathrm{r}},$$

where the units of temperature remain kelvin, and the units of heat generation $\bar{q}^{\mathrm{r}}$ are watts. In the following, we assume a single reaction at the interface, but this can be regeneralized to multiple reactions if desired.

The irreversible heat generation due to de-/intercalation of lithium is written in lumped-parameter form as:

$$\bar{q}_{\mathrm{i}}^{\mathrm{r}} = i_{\mathrm{f}}^{\mathrm{r}} \eta^{\mathrm{r}}, \tag{6.5}$$

where $\mathrm{r} \in \{\mathrm{n}, \mathrm{p}\}$. The reversible heat generation due to a change in entropy during de-/intercalation is:

$$\bar{q}_{\mathrm{r}}^{\mathrm{r}} = i_{\mathrm{f}}^{\mathrm{r}} T \frac{\partial U_{\mathrm{ocp}}^{\mathrm{r}}}{\partial T}, \tag{6.6}$$

where $\mathrm{r} \in \{\mathrm{n}, \mathrm{p}\}$ and $\partial U_{\mathrm{ocp}}^{\mathrm{r}} / \partial T$ is the partial molar entropy term that relates changes in $U_{\mathrm{ocp}}^{\mathrm{r}}$ to changes in temperature. The heat generation due to polarization caused by the ionic resistance of the electrolyte is written as:

$$\bar{q}_{\mathrm{e}}^{\mathrm{r}} = \bar{\kappa}^{\mathrm{r}} \left(\nabla_{\tilde{x}} \phi_{\mathrm{e}}^{\mathrm{r}} \cdot \nabla_{\tilde{x}} \phi_{\mathrm{e}}^{\mathrm{r}} \right) + \bar{\kappa}^{\mathrm{r}} \bar{\kappa}_D T \left(\nabla_{\tilde{x}} \ln \theta_{\mathrm{e}}^{\mathrm{r}} \cdot \nabla_{\tilde{x}} \phi_{\mathrm{e}}^{\mathrm{r}} \right), \tag{6.7}$$

where $\mathrm{r} \in \{\mathrm{n}, \mathrm{s}, \mathrm{p}\}$, and the heat generation due to polarization of the electronic resistance of the solid material is written as:

$$\bar{q}_{\mathrm{s}} = \bar{\sigma}^{\mathrm{r}} \left(\nabla_{\tilde{x}} \phi_{\mathrm{s}}^{\mathrm{r}} \cdot \nabla_{\tilde{x}} \phi_{\mathrm{s}}^{\mathrm{r}} \right). \tag{6.8}$$

Finally, heat generation due to the polarization caused by contact resistance between the current collectors and active material is:

$$\bar{q}_c = R_c i_{\text{app}}^2.$$

We adopt a simplified form of the heat-flux model from Lin et al.,[24] which models the temperature evolution caused by the internal heat generated during discharge and charge processes at only two points: at the cell core and surface. The features of this model are illustrated in Fig. 6.20. This model makes the further simplifying assumptions that the internal and surface temperatures are uniform and the heat generation is uniformly distributed within the cell jelly roll. The simplified lumped thermal model is described as follows:

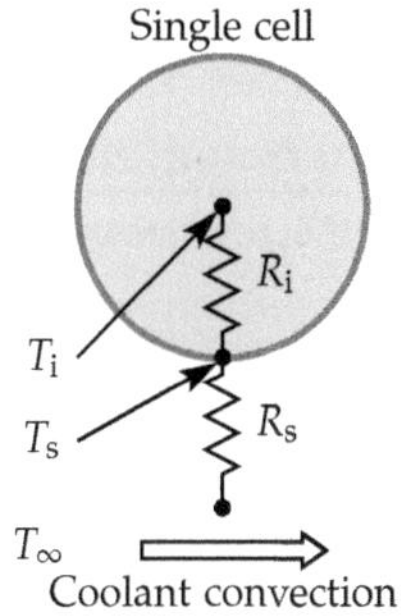

Figure 6.20: A schematic illustrating the simplified heat-flux model, identifying the internal core, surface, and ambient temperatures.

$$C_i \frac{dT_i}{dt} = \bar{q} + \frac{(T_s - T_i)}{R_i} \tag{6.9}$$

$$C_s \frac{dT_s}{dt} = \frac{(T_i - T_s)}{R_i} + \frac{(T_\infty - T_s)}{R_s}, \tag{6.10}$$

where T_i, T_s, and T_∞ are the internal core temperature, the surface temperature, and the ambient cooling temperatures. Constants C_i and C_s are the lumped battery internal and surface heat capacities, respectively; R_i is the thermal resistance between the cell core and the surface; and R_s is the thermal resistance between the cell surface and the ambient environment.[25] Since our interest is in an embedded-processing BMS application, we convert Eqs. (6.9) and (6.10) into a discrete-time state-space form:[26]

$$\begin{bmatrix} T_i[k+1] \\ T_s[k+1] \end{bmatrix} = \begin{bmatrix} \left(1 - \frac{\Delta t}{R_i C_i}\right) & \frac{\Delta t}{R_i C_i} \\ \frac{\Delta t}{R_i C_s} & \left(1 - \frac{(R_i + R_s)\Delta t}{R_i R_s C_s}\right) \end{bmatrix} \begin{bmatrix} T_i[k] \\ T_s[k] \end{bmatrix} + \begin{bmatrix} \frac{\Delta t}{C_i} & 0 \\ 0 & \frac{\Delta t}{R_s C_s} \end{bmatrix} \begin{bmatrix} \bar{q}[k] \\ T_\infty[k] \end{bmatrix}. \tag{6.11}$$

Parameter temperature-dependence is implemented using the Arrhenius relationship, as discussed in Chap. 3.

6.9.2 *Physics-based reduced-order model*

Eq. (6.11) is already in a form that can be implemented directly in a BMS. It remains only to find a way to compute the heat-generation rate $\bar{q}[k]$ using computed values from the LPM.[27] The individual components of Eqs. (6.5) and (6.6) are already implemented by the ROM; they simply need to be computed as described in Chap. 4 and multiplied together to produce predictions of $\bar{q}_i$ and $\bar{q}_r$ at different points in the electrodes. However, to compute Eqs. (6.7) and (6.8), we require knowledge of the gradients of ϕ_e, ϕ_s, and $\ln(\theta_e)$. We will

[24] Xinfan Lin, Hector E. Perez, Shankar Mohan, Jason B. Siegel, Anna G. Stefanopoulou, Yi Ding, and Matthew P. Castanier, "A lumped-parameter electro-thermal model for cylindrical batteries," *Journal of Power Sources*, 257:1–11, 2014.

[25] The paper by Lin et al. describes laboratory experimental techniques and data-processing methods to estimate these parameter values.

[26] Since we assume temperature to evolve relatively slowly compared with the sample period Δt, we use Euler's rule to discretize the ODEs instead of something more complicated. Simulation results presented later in this section appear to justify this choice.

[27] Note that the heat-generation rate term due to polarization caused by the contact resistance $\bar{q}_c$ is not dependent on spatial location and cell region. As a result, we need only obtain ROMs for the other four heat-generation rate terms.

follow the same approach as Chap. 4 to find these; namely, we will develop TFs for these gradients and then use an xRA to convert the TFs to a state-space form for real-time evaluation. In this subsection, we will quickly show how to compute Eqs. (6.5) and (6.6) if the ROM already produces the gradient terms that we require; in the following subsection we will show how to derive the necessary gradient TFs.

We start with the reversible and irreversible heat-generation terms. The ROM for irreversible heat generation at the electrode level in the negative and positive electrodes is given by:

$$\bar{q}_{\mathrm{i}}^{\mathrm{r}}[\tilde{x},k] = i_{\mathrm{f}}^{\mathrm{r}}[\tilde{x},k]\eta[x,k],$$

where, recalling Eq. (4.24),[28]

$$\eta^{\mathrm{r}}[\tilde{x},k] = \frac{2}{f}\mathrm{asinh}\left(\frac{i_{\mathrm{f}}^{\mathrm{r}}[\tilde{x},k]}{2\sum_{j=1}^{J^{\mathrm{r}}}\bar{k}_{0,j}^{\mathrm{r}}\sqrt{(X_j^{\mathrm{r}}-x_j^{\mathrm{r}}[\tilde{x},k])^{\omega_j^{\mathrm{r}}}(x_j^{\mathrm{r}}[\tilde{x},k])^{\omega_j^{\mathrm{r}}}\theta_{\mathrm{e}}^{\mathrm{r}}[\tilde{x},k]}}\right).$$

Similarly, for the reversible heat generation rate term, we have:

$$\bar{q}_{\mathrm{r}}^{\mathrm{r}}[\tilde{x},k] = i_{\mathrm{f}}^{\mathrm{r}}[\tilde{x},k]T_{\mathrm{i}}[k]\frac{\partial U_{\mathrm{ocp}}(\theta_{\mathrm{ss}}^{\mathrm{r}}[\tilde{x},k])}{\partial T},$$

where the temperature $T_{\mathrm{i}}[k]$ is updated every timestep by the thermal model.

Since we have already derived the TFs for several necessary variables—namely, those for $\tilde{\phi}_{\mathrm{s}}$, $\tilde{\phi}_{\mathrm{e}}$, and $\tilde{\theta}_{\mathrm{e}}$—we can find the gradient TFs by taking their derivatives with respect to spatial variable $\tilde{x}$. We will show how to do so in the next subsection. From the gradient TFs, we use an xRA to implement $\nabla_{\tilde{x}}\tilde{\phi}_{\mathrm{s}}[\tilde{x},k]$, $\nabla_{\tilde{x}}\tilde{\phi}_{\mathrm{e}}[\tilde{x},k]$, and $\nabla_{\tilde{x}}\tilde{\theta}_{\mathrm{e}}[\tilde{x},k]$.[29] Assuming that these signals are available at the output of the ROM, $\bar{q}_{\mathrm{s}}$ in the negative and positive electrodes is written as:

$$\bar{q}_{\mathrm{s}}^{\mathrm{r}}[\tilde{x},k] = \bar{\sigma}^{\mathrm{r}}\nabla_{\tilde{x}}\tilde{\phi}_{\mathrm{s}}^{\mathrm{r}}[\tilde{x},k]\nabla_{\tilde{x}}\tilde{\phi}_{\mathrm{s}}^{\mathrm{r}}[\tilde{x},k].$$

Similarly, the ROM for $\bar{q}_{\mathrm{e}}$ at the separator and negative/positive electrodes is written as:

$$\bar{q}_{\mathrm{e}}[\tilde{x},k] = \bar{\kappa}\left(\nabla_{\tilde{x}}\phi_{\mathrm{e}}[\tilde{x},k]\nabla_{\tilde{x}}\phi_{\mathrm{e}}[\tilde{x},k]\right) + \bar{\kappa}_D\left(\nabla_{\tilde{x}}\theta_{\mathrm{e}}[\tilde{x},k]\nabla_{\tilde{x}}\phi_{\mathrm{e}}[\tilde{x},k]\right).$$

We note that so far we have developed ROMs for the heat generation rate terms only for a given spatial location $\tilde{x}$. Therefore, in order to calculate the average heat generation rate $\bar{q}[k]$ in the separator and in the negative and positive electrodes, we must generate ROMs at different locations within the cell and then perform integration over the respective region, which yields:

$$\bar{q}_j^{\mathrm{r}}[k] = \int_0^1 \bar{q}_j[z,k]\,\mathrm{d}z$$

[28] Also recall that the x_j^{r} are computed by first finding the electrode surface potential based on $\theta_{\mathrm{ss}}^{\mathrm{r}}$ via:

$$U[\tilde{x},k] = U_{\mathrm{ocp}}^{\mathrm{r}}(\theta_{\mathrm{ss}}^{\mathrm{r}}[\tilde{x},k]).$$

Then we compute:

$$x_j^{\mathrm{r}}[\tilde{x},k] = \frac{X_j^{\mathrm{r}}}{1+\exp[f(U[\tilde{x},k]-U_j^{0,\mathrm{r}})/\omega_j^{\mathrm{r}}]}.$$

[29] Notice that we do not mention the $\nabla\ln(\theta_{\mathrm{e}})$ term. The debiased lumped electrolyte concentration was previously defined as $\tilde{\theta}_{\mathrm{e}}(\tilde{x},t) = \theta_{\mathrm{e}}(\tilde{x},t) - \theta_{\mathrm{e},0}(\tilde{x},t)$, where $\theta_{\mathrm{e},0}$ is the equilibrium electrolyte concentration, which is equal to 1 in the LPM. Linearizing $\ln(\theta_{\mathrm{e}})$ via Taylor-series expansion gives:

$$\ln(\theta_{\mathrm{e}}(\tilde{x},t)) = \ln(1+\tilde{\theta}_{\mathrm{e}}(\tilde{x},t)) \approx \tilde{\theta}_{\mathrm{e}}(\tilde{x},t).$$

As a result of linearization, we must instead find $\nabla\ln(\theta_{\mathrm{e}}) \approx \nabla\tilde{\theta}_{\mathrm{e}}(\tilde{x},s)$.

where subscript $j \in \{\mathrm{i}, \mathrm{r}, \mathrm{e}, \mathrm{s}\}$ and the spatial variable is normalized using the same notation as in Chap. 2:

$$z = \begin{cases} \tilde{x}, & \text{in the negative electrode} \\ \tilde{x} - 1, & \text{in the separator region} \\ 3 - \tilde{x}, & \text{in the positive electrode.} \end{cases}$$

Finally, the total heat-generation rate is computed as the sum of the average heat-generation rate terms at each cell region in addition to the Ohmic contact resistance heat generation rate term:

$$\bar{q}[k] = \bar{q}^{\mathrm{n}}[k] + \bar{q}^{\mathrm{s}}[k] + \bar{q}^{\mathrm{p}}[k] + R_{\mathrm{c}} i_{\mathrm{app}}[k], \tag{6.12}$$

where:

$$\begin{aligned} \bar{q}^{\mathrm{n}}[k] &= \bar{q}^{\mathrm{n}}_{\mathrm{i}}[k] + \bar{q}^{\mathrm{n}}_{\mathrm{r}}[k] + \bar{q}^{\mathrm{n}}_{\mathrm{s}}[k] + \bar{q}^{\mathrm{n}}_{\mathrm{e}}[k] \\ \bar{q}^{\mathrm{s}}[k] &= \bar{q}^{\mathrm{s}}_{\mathrm{e}}[k] \\ \bar{q}^{\mathrm{p}}[k] &= \bar{q}^{\mathrm{p}}_{\mathrm{i}}[k] + \bar{q}^{\mathrm{p}}_{\mathrm{r}}[k] + \bar{q}^{\mathrm{p}}_{\mathrm{s}}[k] + \bar{q}^{\mathrm{p}}_{\mathrm{e}}[k]. \end{aligned}$$

6.9.3 *Deriving the gradient TFs*

All that remains to describe the heat-generation model is a derivation of the gradient TFs, which we present here.

Gradient TF of electrolyte stoichiometry, $\nabla_{\tilde{x}} \tilde{\theta}_{\mathrm{e}}$

The TF of the debiased electrolyte concentration for the cell's electrode regions was given by Eq. (2.64), which we repeat here, written in terms of spatial variable z:

$$\frac{\widetilde{\Theta}^{\mathrm{r}}_{\mathrm{e}}(z,s)}{I_{\mathrm{app}}(s)} = c^{\mathrm{r}}_1(s) e^{\Lambda^{\mathrm{r}}_1 (z-1)} + c^{\mathrm{r}}_2(s) e^{-\Lambda^{\mathrm{r}}_1 z} + c^{\mathrm{r}}_3(s) e^{\Lambda^{\mathrm{r}}_2 (z-1)} + c^{\mathrm{r}}_4(s) e^{-\Lambda^{\mathrm{r}}_2 z},$$

and for the separator region as Eq. (2.67), repeated here:

$$\frac{\widetilde{\Theta}^{\mathrm{s}}_{\mathrm{e}}(z,s)}{I_{\mathrm{app}}(s)} = c^{\mathrm{s}}_1(s) e^{\Lambda^{\mathrm{s}}_1 (z-1)} + c^{\mathrm{s}}_2(s) e^{-\Lambda^{\mathrm{s}}_1 z}.$$

Beginning with these, we find the gradient transfer function in the negative electrode, written in terms of spatial variable $\tilde{x}$, to be:

$$\begin{aligned} \frac{\nabla_{\tilde{x}} \widetilde{\Theta}^{\mathrm{n}}_{\mathrm{e}}(\tilde{x},s)}{I_{\mathrm{app}}(s)} &= c^{\mathrm{n}}_1(s) \Lambda^{\mathrm{n}}_1(s) e^{\Lambda^{\mathrm{n}}_1(s)(\tilde{x}-1)} - c^{\mathrm{n}}_2(s) \Lambda^{\mathrm{n}}_1(s) e^{-\Lambda^{\mathrm{n}}_1(s)\tilde{x}} \\ &\quad + c^{\mathrm{n}}_3(s) \Lambda^{\mathrm{n}}_2(s) e^{\Lambda^{\mathrm{n}}_2(s)(\tilde{x}-1)} - c^{\mathrm{n}}_4(s) \Lambda^{\mathrm{n}}_2(s) e^{-\Lambda^{\mathrm{n}}_2(s)\tilde{x}}, \end{aligned}$$

and in the positive electrode to be:

$$\begin{aligned} \frac{\nabla_{\tilde{x}} \widetilde{\Theta}^{\mathrm{p}}_{\mathrm{e}}(\tilde{x},s)}{I_{\mathrm{app}}(s)} &= -c^{\mathrm{p}}_1(s) \Lambda^{\mathrm{p}}_1(s) e^{\Lambda^{\mathrm{p}}_1(s)(2-\tilde{x})} + c^{\mathrm{p}}_2(s) \Lambda^{\mathrm{p}}_1(s) e^{-\Lambda^{\mathrm{p}}_1(s)(3-\tilde{x})} \\ &\quad - c^{\mathrm{p}}_3(s) \Lambda^{\mathrm{p}}_2(s) e^{\Lambda^{\mathrm{p}}_2(s)(2-\tilde{x})} + c^{\mathrm{p}}_4(s) \Lambda^{\mathrm{p}}_2(s) e^{-\Lambda^{\mathrm{p}}_2(s)(3-\tilde{x})}, \end{aligned}$$

and finally in the separator to be:

$$\frac{\nabla_{\tilde{x}} \widetilde{\Theta}_{\mathrm{e}}^{\mathrm{s}}(\tilde{x},s)}{I_{\mathrm{app}}(s)} = c_1^{\mathrm{s}}(s)\Lambda_1^{\mathrm{s}}(s)e^{\Lambda_1^{\mathrm{s}}(s)(\tilde{x}-2)} - c_2^{\mathrm{s}}(s)\Lambda_1^{\mathrm{s}}(s)e^{-\Lambda_1^{\mathrm{s}}(s)(\tilde{x}-1)}.$$

Gradient TF of electrolyte potential, $\nabla_{\tilde{x}} \phi_{\mathrm{e}}$

The debiased lumped electrolyte potential was previously defined as $\tilde{\phi}_{\mathrm{e}}(\tilde{x},t) = \phi_{\mathrm{e}}(\tilde{x},t) - \phi_{\mathrm{e},0}(0,t)$. Then the gradient we seek to find is:

$$\nabla_{\tilde{x}} \phi_{\mathrm{e}}(\tilde{x},t) = \nabla_{\tilde{x}} \tilde{\phi}_{\mathrm{e}}(\tilde{x},t) + \nabla_{\tilde{x}} \phi_{\mathrm{e},0}(0,t).$$

Since $\phi_{\mathrm{e},0}(0,t)$ is not dependent on $\tilde{x}$, $\nabla_{\tilde{x}} \phi_{\mathrm{e}}(\tilde{x},t) = \nabla_{\tilde{x}} \tilde{\phi}_{\mathrm{e}}(\tilde{x},t)$.

To find this gradient, recall that the debiased electrolyte-potential TF in the negative electrode was defined as:

$$\frac{\widetilde{\Phi}_{\mathrm{e}}^{\mathrm{n}}(\tilde{x},s)}{I_{\mathrm{app}}(s)} = \frac{\left[\widetilde{\Phi}_{\mathrm{e}}^{\mathrm{n}}(\tilde{x},s)\right]_1}{I_{\mathrm{app}}(s)} + \frac{\left[\widetilde{\Phi}_{\mathrm{e}}^{\mathrm{n}}(\tilde{x},s)\right]_2}{I_{\mathrm{app}}(s)}, \tag{6.13}$$

where the two individual parts were stated in Eqs. (2.93) and (2.94) in terms of spatial variable z. Taking the gradient of Eq. (6.13), using Eqs. (2.93) and (2.94), and expressing the final result in terms of $\tilde{x}$ yields:

$$\frac{\nabla_{\tilde{x}} \widetilde{\Phi}_{\mathrm{e}}^{\mathrm{n}}(\tilde{x},s)}{I_{\mathrm{app}}(s)} = \frac{\left[\nabla_{\tilde{x}} \widetilde{\Phi}_{\mathrm{e}}^{\mathrm{n}}(\tilde{x},s)\right]_1}{I_{\mathrm{app}}(s)} + \frac{\left[\nabla_{\tilde{x}} \widetilde{\Phi}_{\mathrm{e}}^{\mathrm{n}}(\tilde{x},s)\right]_2}{I_{\mathrm{app}}(s)},$$

where:

$$\begin{aligned}\frac{\left[\nabla_{\tilde{x}} \widetilde{\Phi}_{\mathrm{e}}^{\mathrm{n}}(\tilde{x},s)\right]_1}{I_{\mathrm{app}}(s)} &= -\left(j_1^{\mathrm{n}} \frac{e^{\Lambda_1^{\mathrm{n}}(\tilde{x}-1)} - e^{-\Lambda_1^{\mathrm{n}}}}{\bar{\kappa}^{\mathrm{n}}\Lambda_1^{\mathrm{n}}} + j_2^{\mathrm{n}} \frac{-e^{-\Lambda_1^{\mathrm{n}}\tilde{x}} + 1}{\bar{\kappa}^{\mathrm{n}}\Lambda_1^{\mathrm{n}}} \right. \\ &\quad \left. + j_3^{\mathrm{n}} \frac{e^{\Lambda_2^{\mathrm{n}}(\tilde{x}-1)} - e^{-\Lambda_2^{\mathrm{n}}}}{\bar{\kappa}^{\mathrm{n}}\Lambda_2^{\mathrm{n}}} + j_4^{\mathrm{n}} \frac{-e^{-\Lambda_2^{\mathrm{n}}\tilde{x}} + 1}{\bar{\kappa}^{\mathrm{n}}\Lambda_2^{\mathrm{n}}} \right) \\ \frac{\left[\nabla_{\tilde{x}} \widetilde{\Phi}_{\mathrm{e}}^{\mathrm{n}}(\tilde{x},s)\right]_2}{I_{\mathrm{app}}(s)} &= -\bar{\kappa}_D T \left(\Lambda_1^{\mathrm{n}} \left(c_1^{\mathrm{n}} e^{\Lambda_1^{\mathrm{n}}(\tilde{x}-1)} - c_2^{\mathrm{n}} e^{-\Lambda_1^{\mathrm{n}}\tilde{x}} \right) \right. \\ &\quad \left. + \Lambda_2^{\mathrm{n}} \left(c_3^{\mathrm{n}} e^{\Lambda_2^{\mathrm{n}}(\tilde{x}-1)} - c_4^{\mathrm{n}} e^{-\Lambda_2^{\mathrm{n}}\tilde{x}} \right) \right).\end{aligned}$$

The electrolyte-potential TF for the separator region also has two spatially varying components, which were expressed in terms of spatial variable z in Eqs. (2.96) and (2.97). Taking the gradient of these terms and expressing the overall electrolyte-potential gradient in terms of $\tilde{x}$ yields:

$$\frac{\nabla_{\tilde{x}} \widetilde{\Phi}_{\mathrm{e}}^{\mathrm{s}}(\tilde{x},s)}{I_{\mathrm{app}}(s)} = -\frac{1}{\bar{\kappa}^{\mathrm{s}}} - \bar{\kappa}_D T \Lambda_1^{\mathrm{s}} \left(c_1^{\mathrm{s}} e^{\Lambda_1^{\mathrm{s}}(\tilde{x}-2)} - c_2^{\mathrm{s}} e^{-\Lambda_1^{\mathrm{s}}(\tilde{x}-1)} \right).$$

Finally, the electrolyte-potential TF for the positive electrode also has two spatially varying components, which were expressed in

Eqs. (2.99) and (2.100). Taking the gradient of these terms and expressing the overall electrolyte-potential gradient in terms of $\tilde{x}$ yields:

$$\frac{\nabla_{\tilde{x}}\widetilde{\Phi}_{\text{e}}^{\text{p}}(\tilde{x},s)}{I_{\text{app}}(s)} = \frac{\left[\nabla_{\tilde{x}}\widetilde{\Phi}_{\text{e}}^{\text{p}}(\tilde{x},s)\right]_1}{I_{\text{app}}(s)} + \frac{\left[\nabla_{\tilde{x}}\widetilde{\Phi}_{\text{e}}^{\text{p}}(\tilde{x},s)\right]_2}{I_{\text{app}}(s)},$$

where:

$$\begin{aligned}\frac{\left[\nabla_{\tilde{x}}\widetilde{\Phi}_{\text{e}}^{\text{p}}(\tilde{x},s)\right]_1}{I_{\text{app}}(s)} &= -\frac{1}{\bar{\kappa}^{\text{p}}} + \left(\text{j}_1^{\text{p}}\frac{e^{\Lambda_1^{\text{p}}(2-\tilde{x})}-1}{\bar{\kappa}^{\text{p}}\Lambda_1^{\text{p}}} + \text{j}_2^{\text{p}}\frac{e^{-\Lambda_1^{\text{p}}}-e^{-\Lambda_1^{\text{p}}(3-\tilde{x})}}{\bar{\kappa}^{\text{p}}\Lambda_1^{\text{p}}}\right.\\ &\quad\left.+\text{j}_3^{\text{p}}\frac{e^{\Lambda_2^{\text{p}}(2-\tilde{x})}-1}{\bar{\kappa}^{\text{p}}\Lambda_2^{\text{p}}} + \text{j}_4^{\text{p}}\frac{e^{-\Lambda_2^{\text{p}}}-e^{-\Lambda_2^{\text{p}}(3-\tilde{x})}}{\bar{\kappa}^{\text{p}}\Lambda_2^{\text{p}}}\right),\\ \frac{\left[\nabla_{\tilde{x}}\widetilde{\Phi}_{\text{e}}^{\text{p}}(\tilde{x},s)\right]_2}{I_{\text{app}}(s)} &= \bar{\kappa}_D T\left(\Lambda_1^{\text{p}}\left(c_1^{\text{p}}e^{\Lambda_1^{\text{p}}(2-\tilde{x})} - c_2^{\text{p}}e^{-\Lambda_1^{\text{p}}(3-\tilde{x})}\right)\right.\\ &\quad\left.+\Lambda_2^{\text{p}}\left(c_3^{\text{p}}e^{\Lambda_2^{\text{p}}(2-\tilde{x})} - c_4^{\text{p}}e^{-\Lambda_2^{\text{p}}(3-\tilde{x})}\right)\right).\end{aligned}$$

Gradient TF of solid potential, $\nabla_{\tilde{x}}\phi_{\text{s}}$

The debiased lumped solid potential was previously defined as $\tilde{\phi}_{\text{s}}(\tilde{x},t) = \phi_{\text{s}}(\tilde{x},t) - \phi_{\text{s}}(0,t)$. Then, the gradient we seek to find is:

$$\nabla_{\tilde{x}}\phi_{\text{s}}(\tilde{x},t) = \nabla_{\tilde{x}}\tilde{\phi}_{\text{s}}(\tilde{x},t) + \nabla_{\tilde{x}}\phi_{\text{s}}(0,t).$$

Since $\phi_{\text{s}}(0,t)$ is not dependent on $\tilde{x}$, $\nabla_{\tilde{x}}\phi_{\text{s}}(\tilde{x},t) = \nabla_{\tilde{x}}\tilde{\phi}_{\text{s}}(\tilde{x},t)$.

The TFs for $\tilde{\phi}_{\text{s}}(\tilde{x},t)$ in the negative and positive electrodes were expressed in Eqs. (2.90) and (2.91), respectively. Taking the gradient of these TFs, and expressing the final result in terms of $\tilde{x}$ yields:

$$\begin{aligned}\frac{\nabla_{\tilde{x}}\widetilde{\Phi}_{\text{s}}^{\text{n}}(\tilde{x},s)}{I_{\text{app}}(s)} &= -\frac{1}{\bar{\sigma}^{\text{n}}} + \text{j}_1^{\text{n}}\frac{e^{\Lambda_1^{\text{n}}(\tilde{x}-1)}-e^{-\Lambda_1^{\text{n}}}}{\bar{\sigma}^{\text{n}}\Lambda_1^{\text{n}}} + \text{j}_2^{\text{n}}\frac{1-e^{-\Lambda_1^{\text{n}}\tilde{x}}}{\bar{\sigma}^{\text{n}}\Lambda_1^{\text{n}}}\\ &\quad+\text{j}_3^{\text{n}}\frac{e^{\Lambda_2^{\text{n}}(\tilde{x}-1)}-e^{-\Lambda_2^{\text{n}}}}{\bar{\sigma}^{\text{n}}\Lambda_2^{\text{n}}} + \text{j}_4^{\text{n}}\frac{1-e^{-\Lambda_2^{\text{n}}\tilde{x}}}{\bar{\sigma}^{\text{n}}\Lambda_2^{\text{n}}}\\ \frac{\nabla_{\tilde{x}}\widetilde{\Phi}_{\text{s}}^{\text{p}}(\tilde{x},s)}{I_{\text{app}}(s)} &= -\frac{1}{\bar{\sigma}^{\text{p}}} - \text{j}_1^{\text{p}}\frac{e^{\Lambda_1^{\text{p}}(2-\tilde{x})}-e^{-\Lambda_1^{\text{p}}}}{\bar{\sigma}^{\text{p}}\Lambda_1^{\text{p}}} - \text{j}_2^{\text{p}}\frac{1-e^{-\Lambda_1^{\text{p}}(3-\tilde{x})}}{\bar{\sigma}^{\text{p}}\Lambda_1^{\text{p}}}\\ &\quad-\text{j}_3^{\text{p}}\frac{e^{\Lambda_2^{\text{p}}(2-\tilde{x})}-e^{-\Lambda_2^{\text{p}}}}{\bar{\sigma}^{\text{p}}\Lambda_2^{\text{p}}} - \text{j}_4^{\text{p}}\frac{1-e^{-\Lambda_2^{\text{p}}(3-\tilde{x})}}{\bar{\sigma}^{\text{p}}\Lambda_2^{\text{p}}}.\end{aligned}$$

6.9.4 *Simulation results and discussion*

We are now ready to present results from a time-domain simulation of the 1D reduced-order thermal model to demonstrate its performance as compared to a FOM simulated in COMSOL.[30] Both the ROM and FOM were fully coupled in the sense that temperature at any point in time is fed back to modify the temperature-dependent

[30] We use simulation in this section to validate the accuracy of the ROM's predictions of FOM variables since simulation gives access to the internal heat-generation terms of the model, which are not possible to measure in a laboratory experiment. The assumption is that the FOM adequately matches the behavior of a physical cell; if the ROM matches the FOM and the FOM matches the cell, then the ROM also matches the cell.

The paper by Lin et al. cited earlier in this section has already validated that the FOM captures very well the evolution of core and surface temperature of a cylindrical cell that they characterized in their laboratory, and we do not repeat their data here.

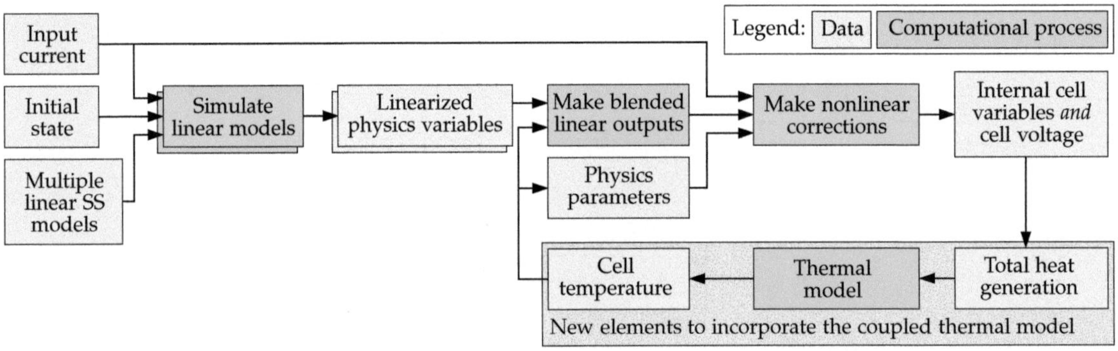

Figure 6.21: Time-domain simulation scheme.

parameters and dynamics for future times. Fig. 6.21 shows how this was implemented in the ROM simulation scheme.

Cell parameters for the simulations were for the NMC30 cell. A set of 8th-order electrochemical ROMs was created in 2 % SOC and 10 °C temperature increments using the HRA. Heat-generation ROMs were created for the same SOC and temperature setpoints at intervals of $\Delta\tilde{x} = 0.2$ within the three cell regions. Simulations employed output blending to capture time-varying behavior and used the New York city cycle (NYCC) input current profile to discharge the battery from 95 % to about 75 % SOC.

Comparisons between ROM and FOM voltage, temperature and total heat-generation rate appear in Fig. 6.22. Note that the total heat generation rate in the figure is the sum of the average heat-generation rate terms calculated using Eq. (6.12) and is the input to the thermal model in Eq. (6.11).

A comparison between ROM and FOM for the individual average

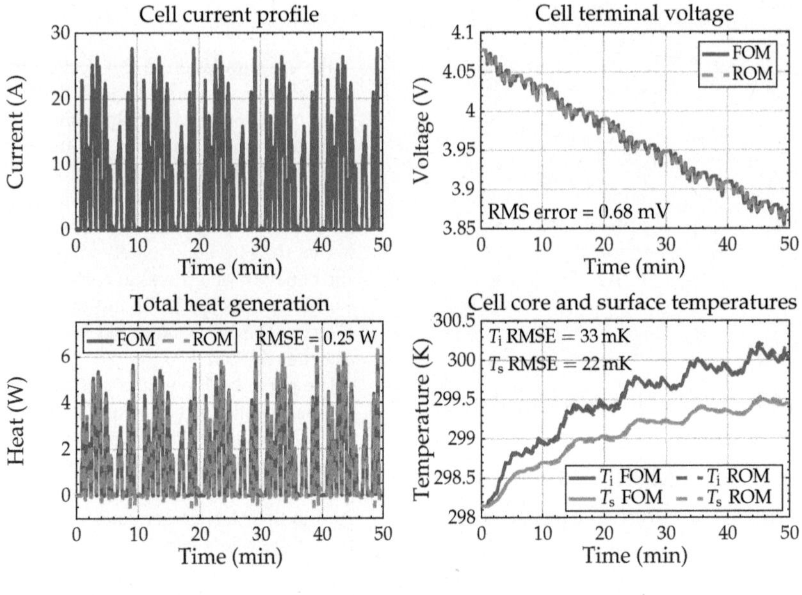

Figure 6.22: Comparison of time-domain simulations of the ROM and FOM for a NYCC profile with an initial SOC of 95 %.

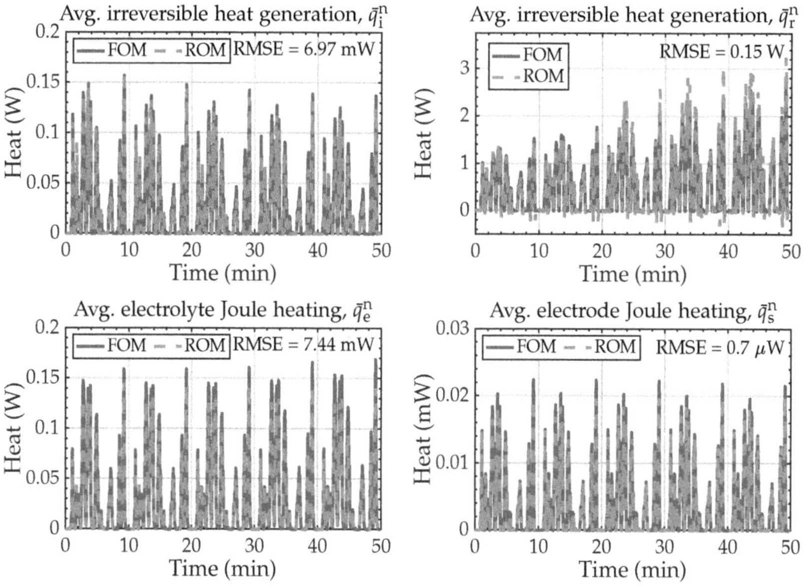

Figure 6.23: Average heat-generation-rate terms in the negative electrode.

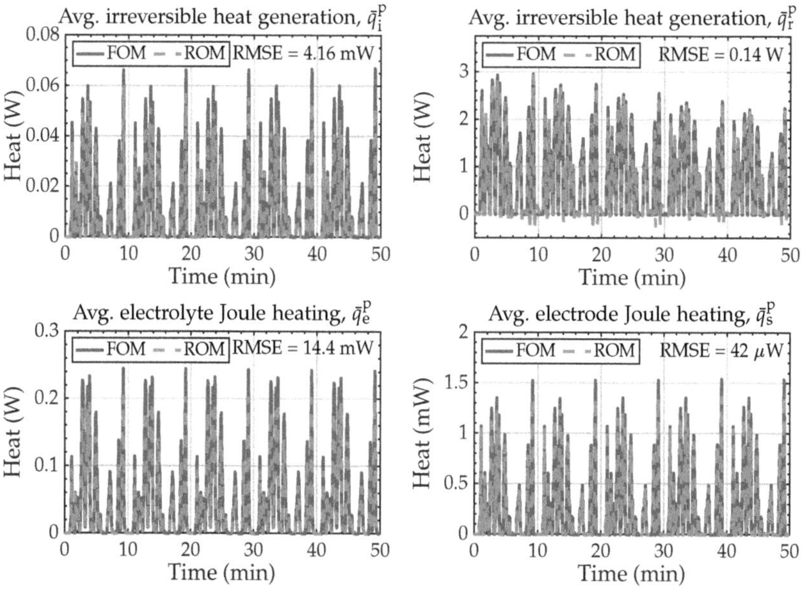

Figure 6.24: Average heat-generation-rate terms in the positive electrode.

heat-generation rate terms in the negative and positive electrodes appear in Figs. 6.23 and 6.24, respectively. Since the heat generation due to polarization caused by the contact resistance does not depend on any electrochemical variable or spatial location, its result is omitted here (but accounted for in the total heat generation of Fig. 6.22). We see in Figs. 6.23 and 6.24 that $\bar{q}_i$, $\bar{q}_e$, and $\bar{q}_r$ are the main contributors to temperature rise while $\bar{q}_s$ is comparatively small, indicating contributions from this term can be neglected. Overall, we conclude that the coupled electrochemical/thermal ROM does a good job of predicting both voltage and temperature.

Electrode

SEI kinetics
$$i_{sei} = i_0 \exp\left(\alpha n f \eta_{sei}\right)$$

Pore clogging
$$\frac{\partial \varepsilon_e^n}{\partial t} = \beta_1 i_{sei}$$

Diffusion into SEI
$$J_{diff} = D \frac{\Delta c_{solvent}}{\delta}$$

SEI

Electron tunneling
$$I_{tl} = P \frac{dN}{dt} e$$

Active material dissolution
$$\frac{\partial \varepsilon_s}{\partial t} = \frac{i_{0,diss} \exp(n f \eta_{diss})}{nFQ}$$

Loss of active material
$$\frac{\partial \varepsilon_s}{\partial t} = \beta_2 \left(\frac{\sigma_{h,max} - \sigma_{h,min}}{\sigma_{yield}}\right)^{\frac{1}{m}}$$

Surface cracking
$$\frac{\partial A_{cr}}{\partial N} = \beta_3 \left(x_{max} - x_{min}\right)^2$$

Lithium plating
$$i_{pl} = i_{0,pl} \exp\left(\alpha n f \eta_{pl}\right)$$

- Active material
- Current collector
- SEI layer
- Electrolyte solvent
- Lithium ion
- Electron

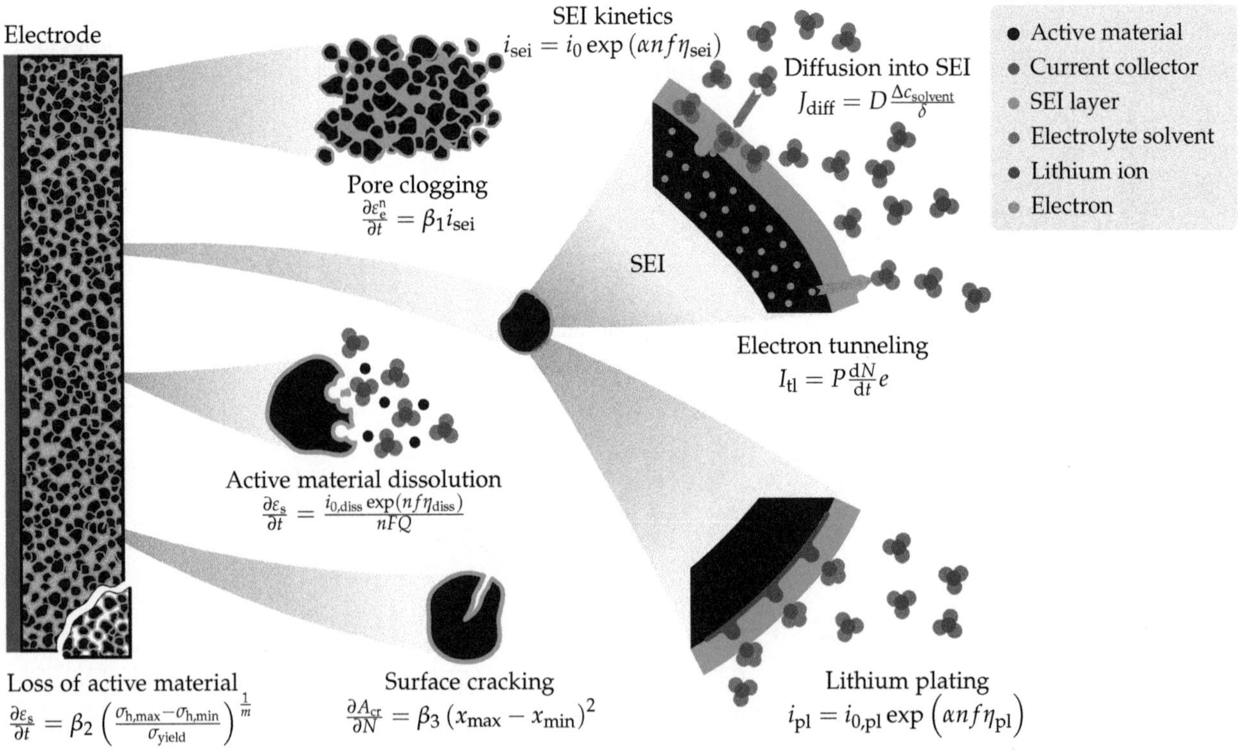

Figure 6.25: Graphical illustration of several degradation mechanisms with typical equations modeling each mechanism. Modified from Jorn M. Reniers, Grietus Mulder, and David A. Howey, "Review and performance comparison of mechanical-chemical degradation models for lithium-ion batteries," *Journal of The Electrochemical Society*, 166(14):A3189–A3200, 2019. Original figure by Dan Lovell.

6.10   Cell degradation models

Another type of prognostic comprises models that enable predicting the near-future degradation that would occur if some proposed profile of input current or power were to be applied to the cell at its present operating condition. Fig. 6.2 showed in graphical form many degradation mechanisms that are known to take place in lithium-ion cells, but not all of these have been modeled, and the interactions between them are not yet well understood.

Fortunately, we are beginning to see an acceleration in the publication of quantitative physics-based models of lithium-ion cell degradation mechanisms.[31] Fig. 6.25 illustrates some mechanisms for which models have appeared in the literature. These include LLI via SEI-layer growth and lithium plating and LAM via various mechanisms.

Whenever we encounter a new article modeling a degradation mechanism that we would like to implement in a BMS prognostic model, we must first convert that model to a lumped-parameter form, then simplify the model to be implementable in a BMS, and finally identify the parameter values of the model. In the next subsections, we will illustrate this process with an example.

[31] But many of these are primarily interested in long-term predictions to determine cell RUL under certain assumed operating regimes. As previously mentioned, we believe that the BMS should be more concerned with short-term predictions to aid with power-limit calculations that extend over the next few seconds or minutes. We propose using diagnostics methods to provide models of the cells' present conditions to use as an initialization for prognostics methods that predict near-future conditions of the cells.

6.10.1 *Full-order model of SEI layer growth*

When SEI grows on negative-electrode particles, it comprises two distinctive parts, the inner layer and outer layer. The inner layer is composed of inorganic salts which form a dense structure on the electrode surface, while the outer layer is composed of organic lithium salts which have a highly porous structure. Solvent molecules are able to diffuse through the outer layer but cannot penetrate the inner layer. The inner layer is a good insulator, but electrons can tunnel from the electrode surface to the boundary between the SEI layers if the inner layer is thin. The solvent is reduced when electrons and solvent molecules arrive at the boundary together, forming additional dense and porous SEI. Both layers continue to grow over time.

Different models of SEI-layer growth are illustrated in Fig. 6.25, each making assumptions regarding the rate-limiting process involved. Ramadass et al.[32] assumed that the reaction kinetics between electrons and solvent limited the rate of growth; Safari et al.[33] assumed that diffusion of the solvent through the outer layer was the limiting factor; and Li et al.[34] assumed that electron tunneling was the rate-limiting term.

In Vol. II, we showed how to make a 0D approximation to the Ramadass model,[35] and we used that work as our example of the process required to convert a literature aging model for BMS use. As a review, when side reactions take place at the particle surface, the local intercalation reaction rate must be modified to include an extra rate variable modeling the side reaction. The total intercalation current $i_{\text{f+s}}^{\text{r}}(\tilde{x},t)$ is given by a sum of the faradaic rate and the side-reaction rate: $i_{\text{f+s}}^{\text{r}}(\tilde{x},t) = i_{\text{f}}^{\text{r}}(\tilde{x},t) + i_{\text{s}}^{\text{r}}(\tilde{x},t)$. In this equation, $i_{\text{s}}^{\text{n}}(\tilde{x},t) \le 0$ and we assume that $i_{\text{s}}^{\text{p}}(\tilde{x},t) = 0$.

Recall, $i_{\text{f}}^{\text{r}}(\tilde{x},t)$ is the faradaic de/intercalation flux modeled via the MSMR equation:[36]

$$
\begin{aligned}
i_{\text{f}}^{\text{r}}(\tilde{x},t) &= \sum_{j=1}^{J^{\text{r}}} i_{\text{f},j}^{\text{r}}(\tilde{x},t) \\
i_{\text{f},j}^{\text{r}}(\tilde{x},t) &= i_{0,j}^{\text{r}}(\tilde{x},t)\left(\exp\left((1-\alpha_j^{\text{r}})f\eta^{\text{r}}(\tilde{x},t)\right) - \exp\left(-\alpha_j^{\text{r}}f\eta^{\text{r}}(\tilde{x},t)\right)\right) \\
i_{0,j}^{\text{r}}(\tilde{x},t) &= \bar{k}_{0,j}^{\text{r}}\left(X_j^{\text{r}} - x_j^{\text{r}}(\tilde{x},t)\right)^{\omega_j^{\text{r}}(1-\alpha_j^{\text{r}})}\left(x_j^{\text{r}}(\tilde{x},t)\right)^{\omega_j^{\text{r}}\alpha_j^{\text{r}}}\left(\theta_{\text{e}}^{\text{r}}(\tilde{x},t)\right)^{1-\alpha_j^{\text{r}}} \\
\eta^{\text{r}}(\tilde{x},t) &= \phi_{\text{s}}^{\text{r}}(\tilde{x},t) - \phi_{\text{e}}^{\text{r}}(\tilde{x},t) - U_{\text{ocp}}^{\text{r}}\left(\theta_{\text{ss}}^{\text{r}}(\tilde{x},t)\right) - \bar{R}_{\text{f}}^{\text{r}} i_{\text{f+dl}}^{\text{r}}(\tilde{x},t).
\end{aligned}
$$

The kinetics of the side reaction are assumed to be irreversible and are described using a Tafel equation:[37]

$$
\begin{aligned}
i_{\text{s}}^{\text{n}}(\tilde{x},t) &= -\bar{i}_{\text{s},0}\exp\left(-\alpha_{\text{s}}f\eta_{\text{s}}^{\text{n}}(\tilde{x},t)\right) \qquad (6.14)\\
\eta_{\text{s}}^{\text{n}}(\tilde{x},t) &= \phi_{\text{s}}^{\text{n}}(\tilde{x},t) - \phi_{\text{e}}^{\text{n}}(\tilde{x},t) - U_{\text{s}} - \bar{R}_{\text{f}}^{\text{n}} i_{\text{f+s}}^{\text{n}}(\tilde{x},t),
\end{aligned}
$$

[32] P. Ramadass, Bala Haran, Parthasarathy M. Gomadam, Ralph White, and Branko N. Popov, "Development of first principles capacity fade model for li-ion cells," *Journal of the Electrochemical Society*, 151(2):A196, 2004.

[33] M. Safari, M. Morcrette, A. Teyssot, and C. Delacourt, "Multimodal physics-based aging model for life prediction of li-ion batteries," *Journal of The Electrochemical Society*, 156(3):A145, 2008.

[34] Dongjiang Li, Dmitry Danilov, Zhongru Zhang, Huixin Chen, Yong Yang, and Peter H. L. Notten, "Electron tunneling based sei formation model," *ECS Transactions*, 62(1):1, 2014.

[35] A similar 0D reduced-order version of the Safari et al. model can be found in: Gregory L. Plett, "Reduced-order multi-modal model of SEI layer growth for management and control of lithium-ion batteries," in *2017 IEEE Conference on Control Technology and Applications (CCTA)*. IEEE, 2017, pp. 389–395.

[36] In the summations over the MSMR galleries in many of the following equations, we omit writing the lower and upper summation bounds. In all cases, the variable j iterates from $j = 1$ to $j = J^{\text{r}}$.

[37] While the SEI-growth side reaction is considered irreversible, other side reactions may be at least partially reversible and would need to be modeled differently (e.g., using a Butler–Volmer equation). One example is lithium plating, which appears to be reversible for a short period of time. Lithium can plate on the negative-electrode's surface during charge; if the current is immediately reversed, lithium can be stripped during discharge and the total capacity lost due to lithium plating is recovered. If metallic lithium remains on the electrode surface for some time without reversing current, it will react with solvent in the electrolyte to form SEI, after which it can no longer be stripped. In this case, capacity is irreversibly lost. Lithium plating is probably best modeled as a two-step process for this reason, but doing so is beyond the scope of this chapter.

where $\bar{i}_{s,0}$ is the side-reaction exchange rate, U_s is an the equilibrium potential of the reaction, and α_s is a unitless (cathodic) transfer coefficient for solvent reduction. Note that we assume there is only one reaction that forms the SEI layer instead of considering multiple galleries j. The superscripts "n" on the three SEI-model parameters are omitted because they are negative-electrode quantities by default.

6.10.2 *Eliminating redundant parameters*

In this model, two key parameters, $\bar{i}_{s,0}$ and U_s, turn out not to be independently identifiable. We show this by substituting the overpotential term $\eta_s^n(\tilde{x},t)$ back into Eq. (6.14), which yields:

$$\begin{aligned} i_s^n(\tilde{x},t) &= -\bar{i}_{s,0}\exp\left(-\alpha_s f\left(\phi_s^n(\tilde{x},t) - \phi_e^n(\tilde{x},t) - U_s - \bar{R}_f^n i_{f+s}^n(\tilde{x},t)\right)\right) \\ &= -\underbrace{\left[\bar{i}_{s,0}\exp\left(\alpha_s f U_s\right)\right]}_{\bar{k}_{s,0}(T)}\exp\left(-\alpha_s f\left(\phi_s^n(\tilde{x},t) - \phi_e^n(\tilde{x},t)\right.\right. \\ &\qquad \left.\left. -\bar{R}_f^n i_{f+s}^n(\tilde{x},t)\right)\right). \end{aligned}$$

We explore the term in the square brackets further, assuming that $\bar{i}_{s,0}$ is temperature-dependent according to an Arrhenius form, and expanding $f = F/(RT)$ for clarity:

$$\begin{aligned} \bar{k}_{s,0}(T) &= \bar{i}_{s,0,\text{ref}}\exp\left(\frac{\alpha_s F}{RT}U_s\right)\exp\left(\frac{E_{\bar{i}_{s,0}}}{R}\left(\frac{1}{T_{\text{ref}}} - \frac{1}{T}\right)\right) \\ &= \bar{i}_{s,0,\text{ref}}\exp\left(\frac{E_{\bar{i}_{s,0}}}{R}\left(\frac{1}{T_{\text{ref}}} - \frac{1}{T}\right) + \frac{\alpha_s F}{RT}U_s + \frac{\alpha_s F}{RT_{\text{ref}}}U_s - \frac{\alpha_s F}{RT_{\text{ref}}}U_s\right) \\ &= \bar{i}_{s,0,\text{ref}}\exp\left(\frac{\alpha_s F}{RT_{\text{ref}}}U_s\right)\exp\left(\frac{E_{\bar{i}_{s,0}}}{R}\left(\frac{1}{T_{\text{ref}}} - \frac{1}{T}\right) - \frac{\alpha_s F U_s}{R}\left(\frac{1}{T_{\text{ref}}} - \frac{1}{T}\right)\right) \\ &= \underbrace{\bar{i}_{s,0,\text{ref}}\exp\left(\frac{\alpha_s F}{RT_{\text{ref}}}U_s\right)}_{\bar{k}_{s,0,\text{ref}}}\exp\left(\frac{\overbrace{(E_{\bar{i}_{s,0}} - \alpha_s F U_s)}^{E_{\bar{k}_{s,0}}}}{R}\left(\frac{1}{T_{\text{ref}}} - \frac{1}{T}\right)\right). \end{aligned}$$

Overall, independent definitions of $\bar{i}_{s,0}$ and U_s can now be completely ignored, and the Tafel equation is reformulated as:

$$\begin{aligned} i_s^n(\tilde{x},t) &= -\bar{k}_{s,0}\exp\left(-\alpha_s f\left(\phi_s^n(\tilde{x},t) - \phi_e^n(\tilde{x},t) - \bar{R}_f^n i_{f+s}^n(\tilde{x},t)\right)\right) \\ \bar{k}_{s,0} &= \bar{k}_{s,0,\text{ref}}\exp\left(\frac{E_{\bar{k}_{s,0}}}{R}\left(\frac{1}{T_{\text{ref}}} - \frac{1}{T}\right)\right). \end{aligned}$$

Aging experiments must then be designed to expose three parameters in the negative-electrode: $\bar{k}_{s,0}$, $E_{\bar{k}_{s,0}}$, and α_s. Next, we will reformulate the time-domain models and create a reduced-order structure by making several assumptions.

6.10.3 *An algebraic solution for* $i_s(t) \approx i_s(\tilde{x}, t)$

We simplify the model for BMS implementation in the same way as presented in Vol. II. Namely, we first make three assumptions in addition to those from Ramadass:

1. The anodic and cathodic charge-transfer coefficients of the intercalation and side reactions are all equal ($\alpha_j^n = \alpha_j^p = \alpha_s = 0.5$).
2. The cell is always in a quasi-equilibrium state, allowing the intercalation reaction rate constant $i_{0,j}^r(\tilde{x}, t)$ to be approximated by:

$$i_{0,j}^r(\tilde{x}, t) \approx i_{0,j}^r(t) = \bar{k}_{0,j}^r \sqrt{(X_j^r - x_j^r(t))^{\omega_j^r} (x_j^r(t))^{\omega_j^r} (1)}$$

$$x_j^r(t) = \frac{X_j^r}{1 + \exp[f(U_{ocp}(\theta_{ss}^r(t)) - U_j^{0,r})/\omega_j^r]},$$

where $\theta_e^r(\tilde{x}, t) \approx \theta_{e,0}^r = 1$.
3. Intercalation (faradaic) rate $i_f^r(\tilde{x}, t)$ and side-reaction rate $i_s^r(\tilde{x}, t)$ are uniform over the electrode for a small time interval Δt, where the values are given by $i_f^r[k]$ and $i_s^r[k]$ at the kth discrete-time interval. This allows us to write the total interfacial rate $i_{f+s}^r[k]$ as:

$$i_{f+s}^n[k] = i_f^n[k] + i_s^n[k] = i_{app}[k], \quad \text{and} \quad i_{f+s}^p[k] = i_f^p[k] = -i_{app}[k].$$

With the above assumptions, a reduced-order degradation model is readily formulated. In the following content, we omit the negative-electrode superscript "n" for brevity.

First, recall that $i_f(\tilde{x}, t)$ is represented by the MSMR model; based on Assumption 1, we find the overpotential as:

$$\begin{aligned}
i_f(\tilde{x}, t) &= \sum_{j=1}^{J} i_{0,j}(\tilde{x}, t) \left\{ \exp\left(\frac{f}{2}\eta(\tilde{x}, t)\right) - \exp\left(\frac{-f}{2}\eta(\tilde{x}, t)\right) \right\} \\
&= 2\sum_{j=1}^{J} i_{0,j}(\tilde{x}, t) \sinh\left(\frac{f}{2}\eta(\tilde{x}, t)\right) \\
\eta(\tilde{x}, t) &= \frac{2}{f}\operatorname{asinh}\left(\frac{i_f(\tilde{x}, t)}{2\sum i_{0,j}(\tilde{x}, t)}\right).
\end{aligned}$$

The side-reaction and intercalation overpotentials can be related:

$$\begin{aligned}
\eta(\tilde{x}, t) &= [\phi_s(\tilde{x}, t) - \phi_e(\tilde{x}, t) - \bar{R}_f i_{f+s}(\tilde{x}, t)] - U_{ocp}(\theta_{ss}(\tilde{x}, t)) \\
\eta_s(\tilde{x}, t) &= \phi_s^n(\tilde{x}, t) - \phi_e^n(\tilde{x}, t) - \bar{R}_f^n i_{f+s}^n(\tilde{x}, t) \\
&= \eta(\tilde{x}, t) + U_{ocp}(\theta_{ss}(\tilde{x}, t)).
\end{aligned}$$

Therefore, the side-reaction rate can be written as:

$$\begin{aligned}
i_s(\tilde{x}, t) &= -\bar{k}_{s,0} \exp\left(-\frac{f}{2}\left(\eta(\tilde{x}, t) + U_{ocp}(\theta_{ss}(\tilde{x}, t))\right)\right) \\
&= -\bar{k}_{s,0} \exp\left(\operatorname{asinh}\left(\frac{-i_f(\tilde{x}, t)}{2\sum i_{0,j}(\tilde{x}, t)}\right) - \frac{f}{2}U_{ocp}(\theta_{ss}(\tilde{x}, t))\right). \quad (6.15)
\end{aligned}$$

The film resistance and solid/electrolyte potentials cancel from the calculation, which infers that we don't need the full ideal-cell LPM in order to simulate the side reactions.

Implementing Assumption 3 and rearranging Eq. (6.15):

$$\begin{aligned} i_s(\tilde{x},t) &= -\bar{k}_{s,0} \exp\left(\operatorname{asinh}\left(\frac{i_s(\tilde{x},t) - i_{app}}{2\sum i_{0,j}(\tilde{x},t)} \right) - \frac{f}{2} U_{ocp}\left(\theta_{ss}(\tilde{x},t)\right) \right) \\ &= -\bar{k}_{s,0} \exp\left(\operatorname{asinh}\left(\frac{-i_{app}}{2\sum i_{0,j}(\tilde{x},t)} + \frac{1}{2\sum i_{0,j}(\tilde{x},t)} i_s(\tilde{x},t) \right) \right. \\ &\qquad \left. - \frac{f}{2} U_{ocp}(\theta_{ss}(\tilde{x},t)) \right). \end{aligned}$$

If we assume that local variations for all variables can be ignored,

$$i_{0,j}(\tilde{x},t) \approx i_{0,j}(t), \quad i_s(\tilde{x},t) \approx i_s(t), \quad \text{and} \quad \theta_{ss}(\tilde{x},t) \approx \theta_{ss}(t).$$

We consider each variable to be constant over some small time interval Δt, where the value is denoted as, for example, $i_s[k]$ for the kth interval. Subsequently, the continuous-time formulation is then converted to discrete time as (we let $i_0[k] = \sum i_{0,j}[k]$):

$$i_s[k] = -\bar{k}_{s,0} \exp\left(\operatorname{asinh}\left(\frac{-i_{app}[k]}{2i_0[k]} + \frac{i_s[k]}{2i_0[k]} \right) - \frac{f}{2} U_{ocp}\left(\theta_{ss}[k]\right) \right). \quad (6.16)$$

This equation is implicit since $i_s[k]$ appears on both sides of the equation. We solve for $i_s[k]$ using the same method used in Vol. II. The key identity that helps to simplify the formula is:

$$\exp\left(\operatorname{asinh}(x)\right) = x + \sqrt{x^2+1},$$

and a quadratic form of Eq. (6.16) is then revealed. The root is easily solved by adopting the quadratic formula.

Deviation details are omitted here since Vol. II lists the stepwise procedure. Only the final algebraic result is shown:

$$\begin{aligned} i_s[k] &= A \exp\left(\operatorname{asinh}\left(B + C i_s[k]\right)\right) \\ &= \frac{AB + A\sqrt{B^2 + (1-2CA)}}{1-2CA}, \end{aligned}$$

where,

$$\begin{aligned} i_0[k] &= \sum i_{0,j}[k] = \sum \bar{k}_{0,j} \sqrt{\left(X_j - x_j[k]\right)^{\omega_j} \left(x_j[k]\right)^{\omega_j}} \\ x_j[k] &= \frac{X_j}{1 + \exp[f(U_{ocp}\left(\theta_{ss}[k]\right) - U_j^0)/\omega_j]} \\ A &= -\bar{k}_{s,0} \exp\left(-\frac{f}{2} U_{ocp}\left(\theta_{ss}[k]\right) \right) \end{aligned}$$

$$B = -\frac{i_{\text{app}}[k]}{2i_0[k]}$$
$$C = \frac{1}{2i_0[k]}.$$

A key observation is that in order to simulate $i_s[k]$ the following parameters are required: (1) universal constants F and R, and (2) intercalation and side-reaction rate constants $\bar{k}^{\text{n}}_{0,j}$ and $\bar{k}_{\text{s},0}$, respectively. The variables needed to compute the time-varying output for every time sample k are: (1) temperature T, (2) applied current $i_{\text{app},k}$, (3) the negative-electrode OCP function $U^{\text{n}}_{\text{ocp}}$, and (4) the negative-electrode surface stoichiometry $\theta^{\text{n}}_{\text{ss}}[k]$.

6.10.4 *Identifying SEI model parameter values*

We have now reformulated the Ramadass SEI model to eliminate redundant parameters and have simplified its computation by deriving a 0D reduced-order model. But before we implement this model, we must still estimate $\bar{k}_{\text{s},0}$ and $E_{\bar{k}_{\text{s},0}}$ (we have assumed that $\alpha_{\text{s}} = 0.5$). These constants may be found by conducting calendar-life tests at different temperatures and periodically measuring lost capacity via a full calibrated discharge and charge cycle. A subtle point is that summing the lithium ion loss predicted by $i_s[k]$ over time is not the same as the measured total-capacity loss—some capacity loss is recovered by the shifting stoichiometric operating windows in both electrodes. It is also possible to use voltage measurements during the calendar-life test to verify the applicability of the Ramadass, Safari, or Li methods. However, cell voltage does not decrease due to SEI side-reactions alone, and this must be taken into account. Lu et al. discuss these issues in detail, including how to estimate the SEI model parameter values from calendar-life data.[38]

[38] The content of this section has been adapted from: Dongliang Lu, M. Scott Trimboli, Yujun Wang, and Gregory L. Plett, "Modeling voltage decay during calendar-life aging," *Journal of The Electrochemical Society*, 169(12):120515, 2022.

6.11 *Where to from here?*

This chapter has provided a high-level survey of some important topics in cell diagnosis and prognosis. This field is rapidly evolving, and high-quality articles are being published at an accelerating rate by researchers around the world. There remains a need to improve the speed, accuracy, robustness, and specificity of diagnostics, and to validate the results in meaningful ways. There is also a need to expand the number of degradation mechanisms that are modeled to improve prognosis. Physics-based models are required to do both of these well. O'Kane et al. write, "Empirical models may capture [degradation] behavior, but they can offer insights into how to prevent degradation only when combined with physics-based models."[39]

[39] Simon EJ O'Kane, Weilong Ai, Ganesh Madabattula, Diego Alonso-Alvarez, Robert Timms, Valentin Sulzer, Jacqueline Sophie Edge, Billy Wu, Gregory J. Offer, and Monica Marinescu, "Lithium-ion battery degradation: how to model it," *Physical Chemistry Chemical Physics*, 24(13):7909–7922, 2022.

Another significant omission in the literature is advice regarding how to combine the models of different degradation mechanisms when their effects are known to be coupled. The only article of which we are aware that makes an attempt to do so is the work by O'Kane et al.[40] We look forward to seeing more works like this in the future.

[40] Ibid.

However, even the simplified diagnostics and prognostics that we have discussed are a good starting point. Chaps. 7–8 will particularly use the prognostics models to inform BMS computations that provide advisory limits of operation during charging and dynamic usage.

6.A Summary of variables

- $\Delta\theta_{\mathrm{s}}^{\mathrm{r}} = \left|\theta_{100}^{\mathrm{r}} - \theta_{0}^{\mathrm{r}}\right|$ [unitless], the range of stoichiometry used in electrode r between SOC of 0 % and 100 % (see Eq. (6.1)).
- η_{s} [A], side-reaction overpotential.
- i_{s} [A], side-reaction current.
- $\boldsymbol{\Lambda}_j$, the likelihood in the IMM framework of the present voltage measurement given that the cell is described by model j.
- μ_i, the pmf output by the IMM framework summarizing the probability that the cell is best described by model i at the current timestep.
- $\mu_{i|j}$, the conditional probability in the IMM framework of having been in mode i at the prior timestep given that the system is currently in mode j.
- p_{ij}, the probability in the IMM framework of transitioning from model i to model j (see Eq. (6.2)).
- $Q^{\mathrm{r}} = \frac{\varepsilon_{\mathrm{s}}^{\mathrm{n}} A L^{\mathrm{n}} c_{\mathrm{s,max}}^{\mathrm{n}} F}{3600}$ [Ah], the theoretical capacity of electrode r, if the entire stoichiometric range from 0 to 1 could be used (see Eq. (6.1)).
- $\bar{q}^{\mathrm{r}}$ [W], total heat generation in cell region r.
- $\bar{q}_{\mathrm{c}}^{\mathrm{r}}$ [W], heat generation due to contact resistance.
- $\bar{q}_{\mathrm{e}}^{\mathrm{r}}$ [W], Joule heat generation in cell region r in the electrolyte.
- $\bar{q}_{\mathrm{i}}^{\mathrm{r}}$ [W], irreversible heat generation in cell region r.
- $\bar{q}_{\mathrm{r}}^{\mathrm{r}}$ [W], reversible (entropic) heat generation in cell region r.
- $\bar{q}_{\mathrm{s}}^{\mathrm{r}}$ [W], Joule heat generation in the solid (electrode) in cell region r.
- T_{i} [K], temperature at the interior (core) of the cell.
- T_{s} [K], temperature at the surface of the cell.
- T_{∞} [K], ambient temperature.

7
Optimal Fast Charge

So far, we've seen how physics-based models of lithium-ion dynamics can accurately predict externally measured indicators of ideal battery behavior. We have further seen how physics-based models of degradation mechanisms can forecast a loss of performance or shortening of battery lifetime. In this and the next chapter, we introduce two practical applications of these physics-based methods by way of advanced controls: (1) optimal fast charge, and (2) calculation of operational power limits. Present state-of-the-art battery controls based on ECMs address primarily short-term performance objectives using voltage-based design limits. Although simple ECMs and straightforward voltage limits provide adequate performance for many applications, the ability to bring a lithium-ion battery to a desired state-of-charge in the shortest time possible is ultimately limited by internal electrochemical effects. Exceeding certain current rates and voltages can cause irreversible damage and capacity loss—degrading long-term performance. The widely used constant-current constant-voltage (CCCV) charge profile is based on conservative voltage limits—and may not take full advantage of cell's true operating range! Surprisingly, little has been done to address battery performance or lifetime by exploiting internal electrochemical quantities.

Our progress so far can again be visualized using the roadmap of Fig. 7.1. In the first half of this book, we described a process for

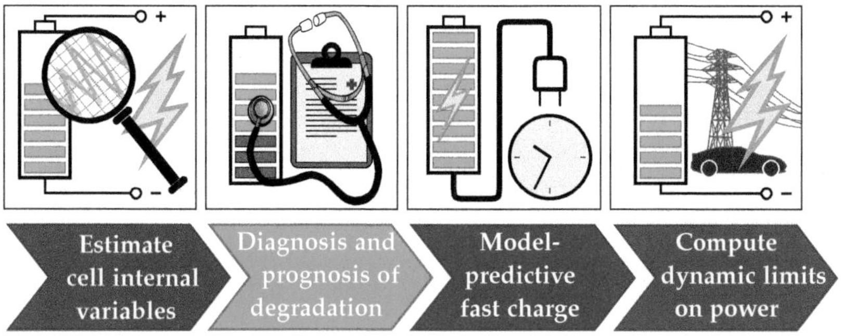

Figure 7.1: Topics in lithium-ion battery controls that we cover in this volume.

finding an LPM of a specific lithium-ion cell. In the second half, we present applications of such an LPM. To this point, we have shown how to use nonlinear Kalman filters to estimate a cell's internal electrochemical variables and have discussed the topics of diagnosis and prognosis of degradation.

In this chapter, we show potential benefits of an electrochemical-based approach by applying model predictive control (MPC) to the fast-charge of a lithium ion battery. The proposed method:

- Uses the computationally compact reduced-order, electrochemical model introduced in Chap. 4.
- Imposes hard constraints on internal cell variables.
- Computes step-wise optimal applied maximum and/or minimum value of current to bring about desired result.

We hypothesize that battery cell performance—and lifetime—can be extended by designing controls allowing cell performance up to but not exceeding true electrochemical limits.

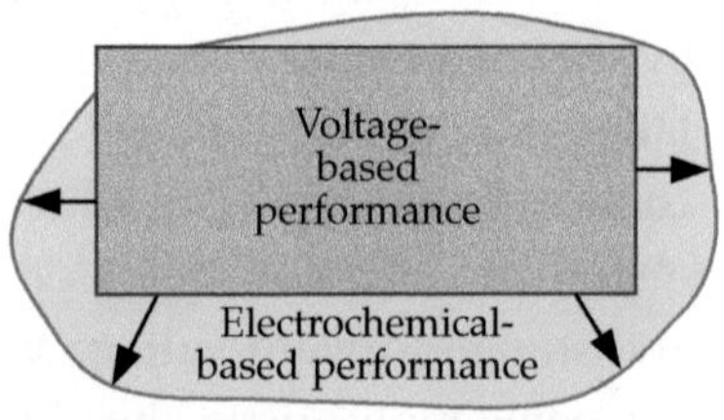

Figure 7.2: Performance envelope for a lithium-ion cell.

Fig. 7.2 portrays a notional performance envelope for a lithium-ion battery cell. State-of-the-art BMS algorithms that rely on simplified ECMs and empirically obtained voltage limits give performance defined by the inner box. Physics-based approaches can theoretically move the performance boundary outward by removing inherent conservatism by incorporating insights gleaned from knowledge of the values of internal electrochemical variables that is not available from ECMs. This enables operation right to the true electrochemical performance limits of the cell. Additionally, it is possible that empirically derived limits can result in potentially harmful operation outside of the true safe region because, again, the empirical models are unaware of the true internal electrochemical state of the cell and may, at times, allow damaging and unsafe operation.

Only recently have BMS engineers considered electrochemical models as the basis for BMS controls. Reasons for this are varied, but are mainly attributed to: (1) the high complexity of physics-based models and (2) a lack of understanding of them. This series on battery-management systems seeks to overcome these obstacles. Researchers over the past decade or so have made great strides in gaining a deeper understanding of the abstruse world of lithium-ion electrochemistry. Volume I and the first chapters of this volume have presented summaries of this research in terms of models that describe lithium-ion cell operation and how to develop parameterized models for specific cells. In addition, fresh methods of model-order reduction—including those presented in this volume in Chap. 4—have enabled efficient computation of complicated physics-based models that operate much faster than real time. These advances have now made it possible to apply powerful advanced control techniques to lithium-ion BMS design. The remainder of this chapter will examine the application of one such method—MPC—to enable cell-level fast charge in a health-conscious manner. As one important example, we shall employ MPC to compute an optimal fast-charge current profile that avoids conditions known to induce lithium-plating.

7.1 *Fast-charge limitations*

Battery fast-charging protocols are typically governed by a cell's upper voltage limit. But as previously pointed out, an exact limit on cell terminal voltage is not a criterion a BMS algorithm designer ought to be directly concerned about. Manufacturer-supplied voltage limits are merely indirect indicators of something else—actual causes of aging stem from electrochemical and mechanical processes occurring within the cell. It turns out that cell terminal voltage alone is a poor—but measurable—indicator of the true state of these processes. The real issue when designing methods for fast-charge is not a cell's terminal voltage, but rather the amount of damage expected to occur at different charging levels at any particular point in time caused by following various charging protocols. So, charging rates should ideally be calculated in real-time in order to effect a tradeoff between time-to-charge and the resulting degradation experienced by the cell. Fig. 7.3 shows a notional charge scheme limited strictly by voltage. In this hypothetical case, the actual internal cell electrochemical degradation parameter never reaches its limit P_{max}, indicating that our voltage-limited scheme is conservative. However, as shown in Fig. 7.4, if we charge instead to the *true* electrochemical-based degradation limit, then the maximum voltage specified on a cell manufacturer's data sheet may be exceeded, but without harm!

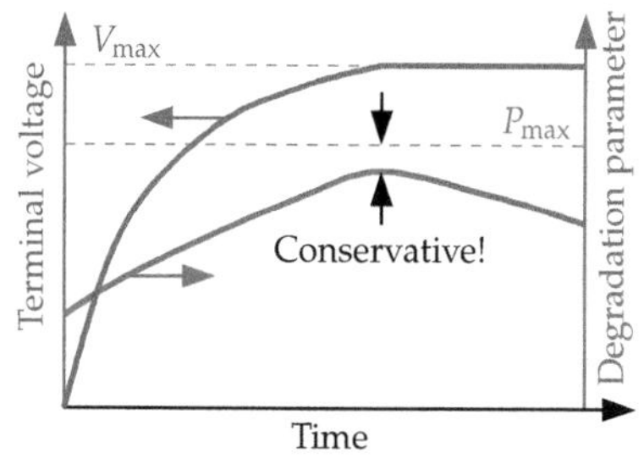

Figure 7.3: External voltage-limited charge.

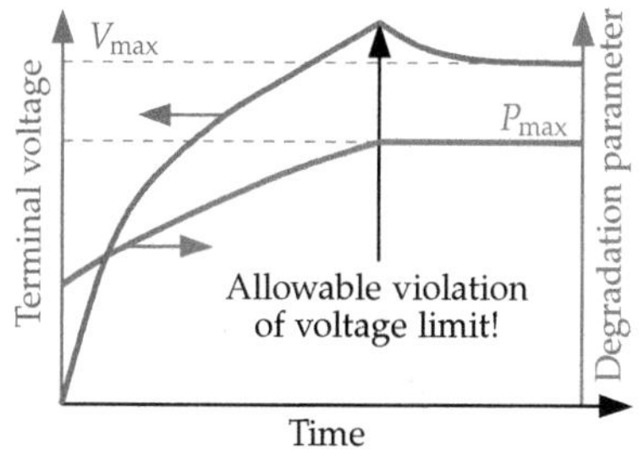

Figure 7.4: Internal electrochemical-limited charge.

Chap. 6 introduced the idea that lithium-ion battery cells are prone to a wide and complex array of degradation mechanisms. Indeed, there is no single mechanism responsible for total-capacity fade and power loss; rather, cell degradation arises from the intricate coupling of a variety of factors.[1] This makes it difficult to isolate specific causes, and research on cell degradation remains active. The good news is that we do not need complete knowledge of all of these mechanisms and how they apply to our specific cell to be able to implement advanced BMS controls. If we are able to model only the most prevalent and damaging of them, we can make big improvements over the state of the art.

[1] A good review can be found in J. Vetter, P. Novák, M.R. Wagner, C. Veit, K-C Möller, J.O. Besenhard, M. Winter, M. Wohlfahrt-Mehrens, C. Vogler, and A. Hammouche, "Ageing mechanisms in lithium-ion batteries," *Journal of Power Sources*, 147(1):269–281, 2005.

One of the most significant single contributors to capacity fade if a cell is improperly charged is lithium plating, which can severely impact both battery lifetime as well as performance. Metallic lithium assumes a variety of structural forms on the electrode surface, one of which can produce dendrites that can introduce a short-circuit safety concern. Since the conditions that bring about this degradation mode are relatively well understood (and amenable to mathematical modeling), lithium plating provides an excellent starting point for the development of advanced charge-control strategies aiming to improve battery safety and performance. Before we can incorporate lithium plating concerns into our control approach, however, we require a mathematical model of its behavior, which we introduced in Chap. 6 and will utilize in the following sections.

7.1.1 Lithium plating

A model for lithium plating was introduced in Volume II and a model of SEI layer growth was introduced in Chap. 6 of this volume. In principle, both models could be used as the basis for a control strategy aimed at mitigating the effects of these degradation mechanisms. For purposes of this chapter, we focus on lithium plating, where the conditions bringing about its formation are particularly easy to specify and thus include in our MPC procedure. First, we shall recapture the salient points of the lithium plating model as they relate to the battery control problem.

Lithium plating occurs due to an electrochemical side reaction at the negative electrode that may occur during battery charging; the effect is most acute at cold temperatures where diffusion of lithium within the electrode solid particles is generally slower. During a fast-charge event, local surface overpotential can cause lithium ions from the electrolyte to combine with electrons from the external circuit and plate solid metallic lithium metal onto the graphite negative-electrode surface.[2]

This metallic lithium adversely coats the surface of the electrode, effectively reducing lithium inventory and inhibiting the intercalation of lithium ions into the graphite particles. The associated reaction occurs when the solid–electrolyte potential difference at the electrode surface drops below 0 V. This condition is more prone to occur at high states-of-charge where the OCP of the negative electrode is relatively low (less than 0.1 V). The undesired result of this reaction is that capacity is irreversibly lost.[3] Additionally, lithium metal will further catalyze SEI growth and can form a metallic annealing site that promotes growth of metal dendrites, which can penetrate the separator and eventually lead to a potentially dangerous cell short circuit.

The lithium-plating model summarized here originates from the work of Arora et al.[4] and assumes for the full order cell model that

[2] An excellent treatment of the electrochemical process of lithium plating is given by Thomas Waldmann, Björn-Ingo Hogg, and Margret Wohlfahrt-Mehrens, "Li plating as unwanted side reaction in commercial li-ion cells–a review," *Journal of Power Sources*, 384:107–124, 2018.

[3] Lithium plating on charge can be accompanied by the reverse process of *lithium stripping* on discharge which returns some lithium inventory as was discussed in Chap. 6. For this development, we assume the amount of lithium stripped is negligible with respect to the amount plated. A more comprehensive degradation model accounting for lithium stripping could improve the utility of the approach.

[4] Pankaj Arora, Marc Doyle, and Ralph E White, "Mathematical modeling of the lithium deposition overcharge reaction in lithium-ion batteries using carbon-based negative electrodes," *Journal of The Electrochemical Society*, 146(10):3543–3553, 1999.

the intercalation kinetics of lithium at normalized spatial location $\tilde{x}$ in the negative electrode is expressed using a standard Butler–Volmer equation,[5]

$$i_f(\tilde{x},t) = i_0\left[\exp\left(\frac{(1-\alpha)F}{RT}\eta(\tilde{x},t)\right) - \exp\left(-\frac{\alpha F}{RT}\eta(\tilde{x},t)\right)\right],$$

driven by the overpotential,

$$\eta(\tilde{x},t) = \phi_s(\tilde{x},t) - \phi_e(\tilde{x},t) - U_{ocp}(\theta_{ss}) - \bar{R}_f i_f(\tilde{x},t),$$

and where i_0 is the exchange current density given by,

$$i_0 = \bar{k}_0(\theta_e)^{1-\alpha}(1-\theta_{ss})^{1-\alpha}(\theta_{ss})^{\alpha}.$$

In this and the following development, we omit the negative-electrode superscript "n" for clarity. Here, $U_{ocp}(\theta_{ss})$ gives the equilibrium potential as a function of the normalized solid phase concentration at the particle surface. The side-reaction rate of lithium deposition leading to irreversible lithium loss is then given by:

$$i_{sr}(\tilde{x},t) = \min\left(0, i_{0,sr}\left[\exp\left(\frac{(1-\alpha_s)F}{RT}\eta_s(\tilde{x},t)\right) - \exp\left(-\frac{\alpha_s F}{RT}\eta_s(\tilde{x},t)\right)\right]\right),$$

driven by the side reaction overpotential,

$$\eta_s(\tilde{x},t) = \phi_s(\tilde{x},t) - \phi_e(\tilde{x},t) - \bar{U}_s - F\bar{R}_f i_{sr}(\tilde{x},t), \qquad (7.1)$$

and where the side-reaction exchange current density is,

$$i_{0,sr} = \bar{k}_{0,sr}(\theta_e)^{1-\alpha}.$$

A key result of this expression is the following condition:[6]

> *The side reaction occurs only at spatial locations in the negative electrode where $\eta_s(\tilde{x},t) < 0$.*

As discussed in Vol. II, the spatial distribution of overpotential across the negative electrode governs the region in which lithium plating will occur. It has been shown[7] that the profile of overpotential with respect to electrode dimension is monotonic and approximately parabolic in shape, reaching its lowest value at the electrode/separator boundary. In addition, during charge, the local value of $\eta_s(\tilde{x},t)$ decreases over time—again decreasing more quickly near the separator. This geometry is portrayed in Fig. 7.5.

For the purposes of fast-charge control, we shall be concerned only with the overpotential value computed at the extreme dimension of the negative electrode (i.e., $x = L^n$ or equivalently $\tilde{x} = 1$). Computation of the time-dependent overpotential value from the ROM output equation is addressed in Sect. 7.6.

[5] For simplicity in this treatment, we shall assume a single Butler–Volmer equation for the entire negative electrode; the method can be extended for the MSMR model as well.

[6] While the 0 V value is a generally accepted threshold for the onset of lithium plating, it has been shown to occur at values slightly offset from this value in practice, depending in part on the dynamic condition of the cell. Some modern work that seeks to model this behavior can be found in:

- Daniel Baker and Mark Verbrugge, "Modeling overcharge at lithiated-graphite porous electrodes: plating and dissolution of lithium," *Journal of The Electrochemical Society*, 167(10):100508, 2020.
- Mark W. Verbrugge, Xingcheng Xiao, and Daniel R. Baker, "Experimental and theoretical examination of low-current overcharge at lithiated-graphite porous electrodes," *Journal of The Electrochemical Society*, 167(8):080523, 2020.

[7] Roger D. Perkins, Alfred V. Randall, Xiangchun Zhang, and Gregory L. Plett, "Controls oriented reduced order modeling of lithium deposition on overcharge," *Journal of Power Sources*, 209:318–325, July 2012.

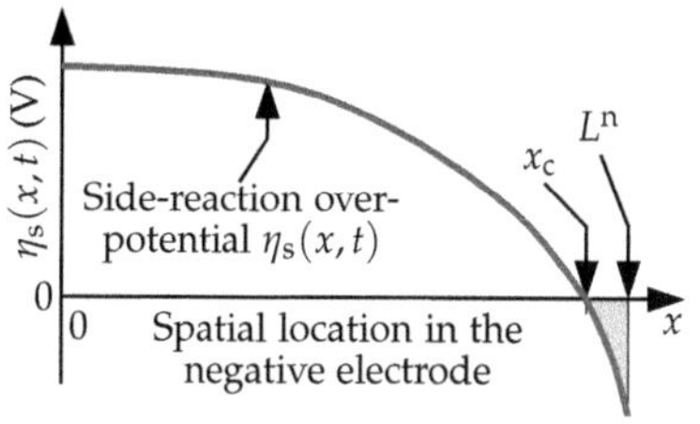

Figure 7.5: Plot of overpotential versus negative electrode dimension.

We are nearly in a position to combine our reduced order ideal-cell model with a degradation model for lithium plating to formulate an advanced fast-charge control strategy. But first, we introduce a real-time predictive control method, MPC, which we shall use to enforce the conditions required to avoid lithium plating.

7.2 *Model predictive control*

MPC belongs to a special class of computer-based control algorithms that make use of an explicit process model to predict the future response of a system and base control actions on those predictions. A process model is here defined as a mathematical representation of the process we wish to control (i.e., a model of lithium-ion cell dynamics which includes a description of lithium plating). The method employs a look-ahead strategy that attempts to compute a *best* sequence of future inputs that will result in a best output response for the objective at hand. Technically speaking, the term *MPC* refers to a family of control algorithms that adopt this approach; in what follows we shall introduce one such implementation.[8]

Generally speaking, MPC can be configured for either continuous- or discrete-time implementation. Since our aim is real-time embedded control implemented by an advanced BMS, we consider here only the discrete-time setting and adopt the state-space representation introduced in Chap. 4.[9]

Fig. 7.6 illustrates the basic idea behind MPC. At this point, the illustration is for a generic control problem where we desire to move a system output $y(t)$ to a reference value. We will adapt this idea to the lithium plating fast-charge problem later. In the figure, future system behavior is influenced by a sequence of future manipulated input variables indicated as a discrete-time sequence of $u[k]$ values, of which only the first is implemented operationally. The input sequence is computed by optimizing the future output sequence in some meaningful sense. For example, the figure shows an output signal $y(t)$ attempting to reach a desired set-point in the shortest possible time. A trivial solution to this problem can be obtained by applying near-infinite input energy thus bringing about near-perfect alignment of our output trajectory with the set-point trace. However, reality suggests we must trim our input energy in order to respect the constraints imposed by real-world system capabilities. This effects a necessary trade-off between the zero-output-error solution and the realistic input-energy solution depicted in the plot.

MPC is carried out according to the receding horizon principle, whereby an optimal control solution is computed for the duration of a finite horizon, yet only the first value of the input solution is

[8] The standard MPC implementation described here is adapted from: Liuping Wang, *Model Predictive Control System Design and Implementation Using MATLAB®*. Springer Verlag, 2009.

[9] MPC can be implemented using a variety of system model structures. Early versions used finite impulse response (FIR) and transfer function (input-output) models; the latter leading to highly complex computational forms. State-space models are computationally compact and have the advantage that all information required for n-step-ahead prediction is contained in the present value of the state vector $x[k]$, thus eliminating the need to store past input and output values.

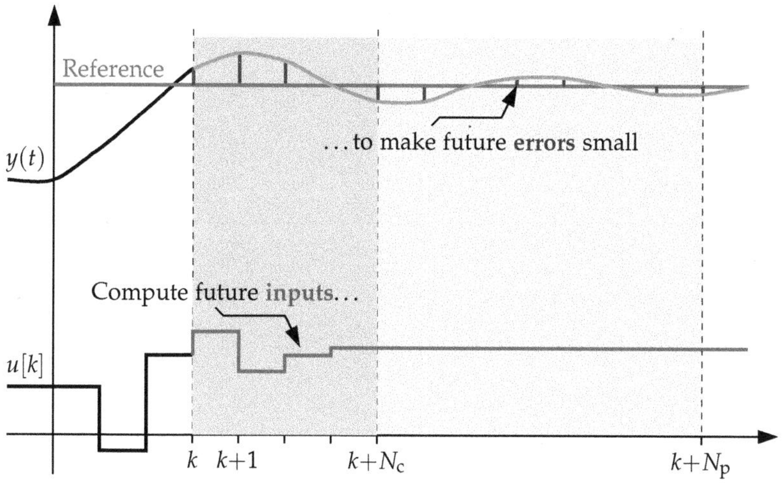

Figure 7.6: MPC: The basic idea.

used in the implementation of control. The entire solution process is carried out at each discrete sample point, each time taking the first value in the newly computed sequence and discarding the rest.

This can be viewed notionally as a finite-length window moving forward in time, much like a flashlight beam illuminating a fixed distance ahead as you move steadily forward. That is, imagine that you are attempting to navigate a dark room, having only the illumination of a flashlight to help guide you. At any point in time, you mentally process what you are able to see and you plan a path forward. You take one step of that path and now the flashlight illuminates an area that exposes more of the surroundings than were visible before you took that step. So, you discard the rest of your previously planned path and replan a path forward using this new information. You take one step of this new planned path and iterate the process.

Predictive control methods—properly applied—can be extremely effective and deliver excellent overall performance due to their reliance on optimality criteria. Nonetheless, the true value of MPC in advanced controls lies in its ability to handle hard constraints in real time during system operation. Indeed, MPC forces a dynamic system to conform exactly to multiple constraints imposed on designated problem variables. It is this aspect that makes MPC particularly appealing for the battery control problem, where respecting certain parameter limits can be shown to influence both instantaneous and long-term cell performance.

The next sections will develop the classic MPC algorithm.[10] First we give a general solution to the unconstrained problem and then treat the case incorporating hard constraints.

[10] Alternative formulations of MPC have emerged in the literature. The dual-mode method, for example, assumes an infinite horizon linear quadratic regulator (LQR) control action for the unconstrained solution. In this text we shall focus on what we term the classic form of MPC tailored for state-space system models. An excellent treatment of dual-mode MPC can be found in: Basil Kouvaritakis and Mark Cannon, *Model Predictive Control: Classical, Robust and Stochastic*. Springer Science & Business Media, 2016.

7.3 MPC: The classic algorithm

We begin by first defining two important time horizons: (1) the *prediction horizon, N_p,* and (2) the *control horizon, N_c;* these parameters establish the degrees of freedom available to effect a solution to our control problem.

- PREDICTION HORIZON: How far into the future to predict system behavior.
- CONTROL HORIZON: How long into the future over which to optimize control action.

Since we are dealing with discrete-time systems, these horizons are typically denoted by integers signifying a specified number of sampling intervals. For example, if the sampling rate is 10 Hz then the sampling interval is $T_s = 0.1$ s and a prediction horizon of, say, 5 s would correspond to $N_p = 50$. The selection of an appropriate prediction horizon length can have an important effect on MPC controller performance, although the design guidance for its selection is largely heuristic, noting that longer prediction horizons generally tend to stabilize controller performance.[11] Theoretically, as the prediction horizon approaches infinity, the solution to the unconstrained MPC optimization can be shown to approach a linear quadratic optimal solution, which is guaranteed to be stable. It would thus seem prudent to select long prediction horizons to leverage this stability property. Rough guidelines suggest choosing a value of N_p that captures 80 % of the dominant system rise time, although in practice many systems perform satisfactorily using prediction horizons well below this value. Of course the trade-off is a resource one—larger horizons demand larger computational effort. So in practice it is important to strike a balance between low computational complexity and good controller performance, and this normally requires a bit of trial and error.

[11] An interesting paper addressing this topic is: Mohammed Alhajeri and Masoud Soroush, "Tuning guidelines for model-predictive control," *Industrial & Engineering Chemistry Research,* 59(10):4177–4191, 2020.

Regarding the control horizon, we normally require $N_c \le N_p$ for simplicity of computation. It turns out, for systems similar to the lithium-ion battery that exhibit relatively slow dynamics, controller performance is largely insensitive to choice of N_c. For most implementations, we shall use a control horizon in the range $2 \le N_c \le 5$.

Since we've established that $N_c \le N_p$, it is interesting to consider what happens with future input variables that lie outside of the control horizon range. In typical implementations, we assume that the control input is held constant at its final value $u[k + N_c]$ beyond the control horizon and up to the length of the prediction horizon. The decision variables in our predictive-control optimization problem are thus the N_c future control choices, $u[k]$, $u[k+1]$,..., $u[k + N_c]$.[12]

[12] Other strategies exist for assuming control actions beyond the control horizon. For example, the aforementioned work of Kouvaritakis et al. proposes a linear quadratic regulator solution for the control action in dual-mode MPC, while an exponential control profile is proposed in the work of: Marcelo A. Xavier, Aloisio Kawakita de Souza, and M. Scott Trimboli, "A split-future MPC algorithm for lithium-ion battery cell-level fast-charge control," *IFAC-PapersOnLine*, 53(2):12459–12464, 2020.

To see in more detail how MPC works, let's consider a linearized, discrete-time, state-space model of a general multi-input multi-output (MIMO) system:

$$\begin{aligned} \boldsymbol{x}[k+1] &= \boldsymbol{A}[k]\boldsymbol{x}[k] + \boldsymbol{B}[k]\boldsymbol{u}[k] \\ \boldsymbol{y}[k] &= \boldsymbol{C}[k]\boldsymbol{x}[k] + \boldsymbol{D}[k]\boldsymbol{u}[k], \end{aligned} \tag{7.2}$$

where $\boldsymbol{x}[k] \in \mathbb{R}^{n\times 1}$, $\boldsymbol{u}[k] \in \mathbb{R}^{m\times 1}$, $\boldsymbol{y}[k] \in \mathbb{R}^{q\times 1}$ are as defined in Chap. 4, namely, the state vector, the input vector, and the output vector, respectively. $\boldsymbol{A}[k]$, $\boldsymbol{B}[k]$, $\boldsymbol{C}[k]$, and $\boldsymbol{D}[k]$ are (possibly time-varying) matrices defining the state-space model, and index k is the sampling instant. We note that allowance for time-varying state matrices is important here in that the physics-based ROM that forms the basis of our linearized model can be expected to vary over the operational space of the battery charging profile. We saw in Chap. 4 that cell dynamics vary with temperature and SOC and thus a blending technique is necessary to describe behavior over a wide operating range. In our MPC implementation, we shall incorporate the output blending method introduced in Sect. 4.8.[13]

[13] Of course, at the expense of sacrificing accuracy and performance, one could choose to implement a very simple model comprising a single constant (time-invariant) set of state matrices.

Classical forms of MPC construct the dynamic state equation in terms of the control-input increment $\Delta\boldsymbol{u}[k+1] = \boldsymbol{u}[k+1] - \boldsymbol{u}[k]$ instead of the control input $\boldsymbol{u}[k]$. This formulation implies the inclusion of integral action within the feedback loop and caters directly for eliminating steady-state error. In order to achieve this form, we first define an augmented state vector:

$$\boldsymbol{\chi}[k] = \left[\boldsymbol{x}^T[k] \;\vdots\; \boldsymbol{u}^T[k]\right]^T, \tag{7.3}$$

and rewrite the state equations as:

$$\begin{aligned} \boldsymbol{\chi}[k+1] &= \tilde{\boldsymbol{A}}[k]\boldsymbol{\chi}[k] + \tilde{\boldsymbol{B}}[k]\Delta\boldsymbol{u}[k] \\ \boldsymbol{y}[k] &= \tilde{\boldsymbol{C}}[k]\boldsymbol{\chi}[k], \end{aligned} \tag{7.4}$$

where we formulate the augmented state matrices as:

$$\tilde{\boldsymbol{A}}[k] = \begin{bmatrix} \boldsymbol{A}[k] & \boldsymbol{B}[k] \\ \boldsymbol{0}_{m\times n} & \boldsymbol{I}_{m\times n} \end{bmatrix}, \quad \tilde{\boldsymbol{B}} = \begin{bmatrix} \boldsymbol{0}_{n\times m} \\ \boldsymbol{I}_{m\times m} \end{bmatrix}, \quad \tilde{\boldsymbol{C}}[k] = \left[\boldsymbol{C}[k] \;\vdots\; \boldsymbol{D}[k]\right].$$

Note that real-time implementation of the MPC algorithm requires the timely update of the constituent state matrices at the start of each calculation step. This in turn requires that we compute the appropriate output-blended solution for the set $\boldsymbol{A}[k]$, $\boldsymbol{B}[k]$, $\boldsymbol{C}[k]$, and $\boldsymbol{D}[k]$ based on the current state value $\boldsymbol{x}[k]$.

Future values of the augmented state vector can now be obtained by propagating the state equations recursively:

$$\boldsymbol{\chi}[k+1] = \tilde{\boldsymbol{A}}\boldsymbol{\chi}[k] + \tilde{\boldsymbol{B}}\Delta\boldsymbol{u}[k]$$

$$\begin{aligned}
\chi[k+2] &= \tilde{A}\chi[k+1] + \tilde{B}\Delta u[k+1] \\
&= \tilde{A}^2\chi[k] + \tilde{A}\tilde{B}\Delta u[k] + \tilde{B}\Delta u[k+1] \\
&\;\;\vdots \\
\chi[k+N_\text{p}] &= \tilde{A}^{N_\text{p}}\chi[k] + \tilde{A}^{N_\text{p}-1}\tilde{B}\Delta u[k] + \tilde{A}^{N_\text{p}-2}\tilde{B}\Delta u[k+1] \\
&\quad \cdots + \tilde{A}^{N_\text{p}-N_\text{c}}\tilde{B}\Delta u[k+N_\text{c}-1].
\end{aligned}$$

Similarly, applying the output equation from Eq. (7.4) we can compute future output values as:

$$\begin{aligned}
y[k+1] &= \tilde{C}\tilde{A}\chi[k] + \tilde{C}\tilde{B}\Delta u[k] \\
y[k+2] &= \tilde{C}\tilde{A}\chi[k+1] + \tilde{C}\tilde{B}\Delta u[k+1] \\
&= \tilde{C}\tilde{A}^2\chi[k] + \tilde{C}\tilde{A}\tilde{B}\Delta u[k] + \tilde{C}\tilde{B}\Delta u[k+1] \\
&\;\;\vdots \\
y[k+N_\text{p}] &= \tilde{C}\tilde{A}^{N_\text{p}}\chi[k] + \tilde{C}\tilde{A}^{N_\text{p}-1}\tilde{B}\Delta u[k] + \tilde{C}\tilde{A}^{N_\text{p}-2}\tilde{B}\Delta u[k+1] \\
&\quad \cdots + \tilde{C}\tilde{A}^{N_\text{p}-N_\text{c}}\tilde{B}\Delta u[k+N_\text{c}-1].
\end{aligned}$$

At this point, we define special vectors containing the set of future outputs and future control input increments, respectively, as:

$$\begin{aligned}
\underrightarrow{Y}[k+1] &= \begin{bmatrix} y^T[k+1] & y^T[k+2] & \cdots & y^T[k+N_\text{p}+1 \end{bmatrix}^T \\
\Delta\underrightarrow{U}[k+1] &= \begin{bmatrix} \Delta u^T[k+1] & \Delta u^T[k+2] & \cdots & \Delta u^T[k+N_\text{c}] \end{bmatrix}^T,
\end{aligned}$$

where we have introduced the underscore arrow notation to denote sequences of future variables.

The output prediction equation can now be written in the more compact form,

$$\underrightarrow{Y}[k+1] = \Phi[k]\tilde{A}[k]\chi[k] + G[k]\Delta\underrightarrow{U}[k+1], \tag{7.5}$$

where matrices $\Phi[k] \in \mathbb{R}^{(N_\text{p}\cdot q)\times(n+m)}$ and $G[k] \in \mathbb{R}^{(N_\text{p}\cdot q)\times(N_\text{c}\cdot m)}$ are defined as:

$$\Phi[k] = \begin{bmatrix} \tilde{C}[k] \\ \tilde{C}[k]\tilde{A}[k] \\ \vdots \\ \tilde{C}[k]\tilde{A}^{n_p-1}[k] \end{bmatrix}$$

$$G[k] = \begin{bmatrix} \tilde{C}[k]\tilde{B}[k] & 0 & \cdots & 0 \\ \tilde{C}[k]\tilde{A}[k]\tilde{B}[k] & \tilde{C}[k]\tilde{B}[k] & \cdots & 0 \\ \vdots & \vdots & \ddots & \\ \tilde{C}[k]\tilde{A}^{N_\text{p}-1}[k]\tilde{B}[k] & \tilde{C}[k]\tilde{A}^{N_\text{p}-2}[k]\tilde{B}[k] & \cdots & \tilde{C}[k]\tilde{A}^{N_\text{p}-N_\text{c}}[k]\tilde{B}[k] \end{bmatrix},$$

and compute the free and forced response components, respectively, of the predicted output.

Since the original dynamic system is nonstrictly causal (i.e., it has a nonzero D-term), this formulation assumes that at time sample k, the control input $\boldsymbol{u}[k]$ was computed at previous time sample $k-1$. This information, together with the current state vector value $\boldsymbol{x}[k]$, is used to compute the set of future input increments $\Delta\underrightarrow{\boldsymbol{U}}[k+1]$ from which the next control input will be computed, namely $\boldsymbol{u}[k+1]$.[14] This process is carried out according to the receding-horizon principle.

Now that we have an expression for the predicted output in terms of the future input decision variables, we next fashion a cost function that penalizes the error between future values of the output (i.e., vector $\underrightarrow{\boldsymbol{Y}}[k+1]$) and future values of a reference. For convenience, define the $(q \cdot N_p) \times 1$ reference vector:

$$\boldsymbol{R}[k+1] = \left[\boldsymbol{r}^T[k+1, \boldsymbol{r}^T[k+2], \ldots, \boldsymbol{r}^T[k+N_p+1]\right]^T,$$

where each of the elements $\boldsymbol{r}[j]$ are $q \times 1$ vectors stipulating a references value for each of the q outputs. Recall that for realizability (i.e., to avoid the trivial solution), the cost function must place a nonzero penalty on the energy in the input, in this case defined by the magnitude of the elements of the vector of future input increments, $\Delta\underrightarrow{\boldsymbol{U}}[k+1]$.

The scalar cost function to be minimized can be written as a quadratic form:

$$\begin{aligned} J[k] = &\left(\underrightarrow{\boldsymbol{R}}[k+1] - \underrightarrow{\boldsymbol{Y}}[k+1]\right)^T \left(\underrightarrow{\boldsymbol{R}}[k+1] - \underrightarrow{\boldsymbol{Y}}[k+1]\right) \\ &+ \Delta\underrightarrow{\boldsymbol{U}}^T[k+1]\bar{\boldsymbol{R}}\Delta\underrightarrow{\boldsymbol{U}}[k+1], \end{aligned}$$

where the vector multiplications bring about a nonnegative sum-of-squares cost measure. Here, $\bar{\boldsymbol{R}}$ denotes a weighting matrix that trades off the penalty on the control input increment energy with that on the predicted output error. In many cases, there is little justification for differentially weighting the individual input sequence elements; so we normally take $\bar{\boldsymbol{R}} = \rho \boldsymbol{I}$ where ρ is a scalar weight and $\boldsymbol{I}$ is an identity matrix of square dimension $N_c \cdot m$. In this context, weighting element ρ serves as another tuning parameter in the optimization problem.

We now use the prediction equation, Eq. (7.5), and substitute for $\underrightarrow{\boldsymbol{Y}}[k+1]$ to rewrite the cost function as:

$$\begin{aligned} J[k] = &\left(\underrightarrow{\boldsymbol{R}}[k+1] - \boldsymbol{\Phi}[k]\tilde{\boldsymbol{A}}[k]\boldsymbol{\chi}[k]\right)^T \left(\underrightarrow{\boldsymbol{R}}[k+1] - \boldsymbol{\Phi}[k]\tilde{\boldsymbol{A}}[k]\boldsymbol{\chi}[k]\right) \\ &- 2\left(\underrightarrow{\boldsymbol{R}}[k+1] - \boldsymbol{\Phi}[k]\tilde{\boldsymbol{A}}[k]\boldsymbol{\chi}[k]\right)^T G[k]\Delta\underrightarrow{\boldsymbol{U}}[k+1] \end{aligned}$$

[14] This is a nontrivial point for the battery control problem. Since the battery exhibits an internal ohmic resistance, there is a near-instantaneous voltage change with the application of current. This implies the presence of a direct feedthrough term (D-matrix) in the state-space model. Most classic forms of MPC neglect the D-matrix; however, in our development we adopt the modification of Ordys and Pike to cater for this term. For details, see: A.W. Ordys and A.W. Pike, "State space generalized predictive control incorporating direct through terms," in *Proceedings of the 37th IEEE Conference on Decision and Control*, vol. 4. IEEE, 1998, pp. 4740–4741.

$$+\underrightarrow{\Delta \mathcal{U}}^T[k+1]\left(G^T[k]G[k]+\rho I\right)\underrightarrow{\Delta \mathcal{U}}[k+1], \tag{7.6}$$

where $J[k]$ is an explicit function of the unknown future input sequence $\underrightarrow{\Delta \mathcal{U}}[k+1]$; all other values in the expression are updated at each sample time. We shall first develop the solution for the case where all problem variables are unconstrained. We have previously noted that most meaningful predictive-control problems include constraints; in fact, this is the principal reason for using a real-time model-predictive control strategy in the first place. Nevertheless, during operation, the algorithm will often traverse regions of the functional space where no constraints are active. In these instances, the optimal unconstrained solution is the correct solution.

7.4 *Optimal MPC solution: unconstrained case*

For notational simplicity in what follows, we shall drop dependence on the sampling index k except where needed for clarity. Our task is now to find the future input sequence $\underrightarrow{\Delta \mathcal{U}}$ that minimizes our cost function J as defined by Eq. (7.6). Since J is a quadratic form, a global optimum is found as the solution to the stationarity condition:

$$\frac{\partial J}{\partial \underrightarrow{\Delta \mathcal{U}}} = -2G^T\left(\underrightarrow{\mathcal{R}} - \Phi\tilde{A}\chi[k]\right) + 2\left(G^TG+\rho I\right)\underrightarrow{\Delta \mathcal{U}} = 0. \tag{7.7}$$

By construction of the quadratic form, matrix $G^TG+\rho I$ is positive definite; thus, the solution to the stationarity condition is a unique minimizing one. Solving Eq. (7.7) for $\underrightarrow{\Delta \mathcal{U}}$ yields the optimal (unconstrained) control sequence as the $N_c \cdot m \times 1$ vector:

$$\underrightarrow{\Delta \mathcal{U}}^* = \left(G^TG+\rho I\right)^{-1}G^T\left(\underrightarrow{\mathcal{R}} - \Phi\tilde{A}\chi[k]\right). \tag{7.8}$$

Note that by application of the receding-horizon principle, only the first element of the optimal sequence $\underrightarrow{\Delta \mathcal{U}}^*$ is implemented, namely, $\Delta u^*[k+1]$; the process starts over and repeats at the next time sample. Practitioners of control theory will recognize Eq. (7.8) as a time-varying full state feedback controller, functionally dependent on the augmented state vector $\chi[k]$ as:

$$\begin{aligned}
\chi[k+1] &= \tilde{A}[k]\chi[k] + \tilde{B}[k]\Delta u[k+1] \\
&= \tilde{A}[k]\chi[k] - \tilde{B}[k]\left(G^T[k]G[k]+\rho I\right)^{-1}G^T[k]\Phi[k]\tilde{A}[k]\chi[k] \\
&= \tilde{A}[k]\chi[k] - \tilde{B}[k]K[k]\chi[k],
\end{aligned}$$

where the time-varying state-feedback gain is given by,

$$K[k] = \left(G^T[k]G[k]+\rho I\right)^{-1}G^T[k]\Phi[k]\tilde{A}[k].$$

This expression implies that fundamentally, unconstrained MPC implements a closed-loop feedback system, and furthermore, for systems where the state-space model is assumed to be constant over all time, stability may be assured by careful selection of tuning parameters that give a stable set of closed-loop eigenvalues for the constant matrix $\tilde{A} - \tilde{B}K$. Stability assurances for the time-varying case are less straightforward.

So far, we have not considered applying constraints to the system state or output. When problem constraints are active, however, the above solution is no longer valid and we therefore must turn to alternative strategies to find the correct constrained solution.

7.5 *Optimal MPC solution: constrained case*

Practical control problems impose constraints to reflect real-world system operation. Basically speaking, constraints may be imposed on three classes of parameters: (1) control inputs; (2) system outputs; and (3) state variables. With respect to the battery control problem, our primary cell-level control input is the applied current (i.e., $\boldsymbol{u}[k] = i_{\text{app}}[k]$);[15] this implies that $m = 1$ and that $u[k]$ is a scalar. Hard constraints on input current are normally stipulated by the nature of the charging infrastructure and are imposed accordingly. That is, charging electronics will have a maximum rated current that is independent of cell performance limits and MPC must be informed of these hard limits on $u[k]$.

[15] Ambient or flow temperature may also be used as a control input to regulate cell core temperature. In this case, we would adopt a 2×1 input vector $\boldsymbol{u}[k] = \begin{bmatrix} i_{\text{app}}[k] & T_{\text{flow}}[k] \end{bmatrix}^T$ and implement a multivariable form of MPC. This is beyond the scope of this chapter and will not be addressed here.

Output variables can be further subdivided into both measurable and nonmeasurable outputs of interest. For the battery, we generally consider cell voltage and surface temperature to be measurable outputs since they are readily obtained with sensors. The class of unmeasurable outputs are perhaps the most interesting from an advanced controls point of view as they may harbor information directly related to phenomena that can harm a cell. These might include lithium concentration ratios or potentials at various locations in the cell, for example. Forcing constraints on such indicators can address the true causes of cell degradation and will form the basis of the control approach to follow. Clearly, unmeasurable outputs cannot be acquired from sensors, but must be deduced using state-estimation techniques as discussed in Chap. 5.

The final class of parameter constraints deals with the dynamic state variables. Depending on the underlying model structure, state variables may or may not have a meaningful physical interpretation. In the case that they do, it is always possible to create a corresponding output variable by proper selection of an output C-matrix, in which case they are handled in exactly the same manner as the unmeasurable outputs above.

7.5.1 *Quadratic programming*

The constrained optimization result can be found by solving a quadratic programming (QP) problem, defined as one in which the cost function is quadratic and the constraint functions are all linear. QP problems can be solved systematically in a finite number of steps and have been shown to be robust and computationally efficient. Although quadratic programming routines are generally available (e.g., MATLAB's `quadprog`), we shall present an implementable algorithm herein; this provides the ability to access the code to control error handling and make any necessary algorithm modifications.

For generality, and to be consistent with the prevailing literature, we shall denote the objective function J in the QP problem in the general form:

$$J[k] = \frac{1}{2}\underrightarrow{\Delta u}^T[k+1]H[k]\underrightarrow{\Delta u}[k+1] + g^T[k]\underrightarrow{\Delta u}[k+1] + f[k], \qquad (7.9)$$

where the decision vector $\underrightarrow{\Delta u}[k+1]$ contains the degrees of freedom for the solution. Symmetric Hessian matrix $H[k]$, vector $g[k]$, and scalar $f[k]$ are compatible with the objective function defined in Eq. (7.6) where:

$$\begin{aligned} H[k] &= 2\left(G^T[k]G[k] + \rho I\right) \\ g[k] &= -2\left(\underrightarrow{R}[k+1] - \Phi[k]\tilde{A}[k]\chi[k]\right)^T G[k] \\ f[k] &= \left(\underrightarrow{R}[k+1] - \Phi[k]\tilde{A}[k]\chi[k]\right)^T \left(\underrightarrow{R}[k+1] - \Phi[k]\tilde{A}[k]\chi[k]\right). \end{aligned}$$

Note that when $H[k[$ is positive definite, then $x^*[k]$ is a unique global minimizer.

Additionally, we shall stack the linear constraint set that we wish to impose on the solution into a convenient linear algebraic inequality as:

$$M\underrightarrow{\Delta u}[k+1] \le \gamma[k], \qquad (7.10)$$

where each constraint equation can be expressed row-wise as:

$$m_i^T\underrightarrow{\Delta u}[k+1] \le \gamma_i[k].$$

Here, m_i^T are the rows of M and $\gamma_i[k]$ is the corresponding ith element of $\gamma[k]$.

For general application to the battery control problem, we consider constraints imposed on: (1) the input, (2) the measured output, and (3) designated unmeasured outputs. In what follows, we shall consider a scalar-valued input variable and a vector-valued output variable.

We first consider the simplest case: constraints imposed on the input increment, $\Delta u[k]$. Enforcing a constraint on the input increment effectively limits the rate at which the input can be changed. Defining hard bounds on the input increment gives:

$$\Delta u_{\min}[k] \leq \Delta u[k] \leq \Delta u_{\max}[k].$$

In classic MPC, input constraints are generally applied to all future input increments, however this choice is left up to the controls designer as the mathematical approach is unaffected by this choice. We next build up a matrix representation that stacks the individual constraint equations imposed at each future sample point as:

$$\left[\begin{array}{ccccc} 1 & 0 & 0 & \cdots & 0 \\ 0 & 1 & 0 & \cdots & 0 \\ 0 & 0 & 1 & \cdots & 0 \\ \vdots & \vdots & \vdots & \ddots & \vdots \\ 0 & 0 & 0 & 0 & 1 \\ \hdashline -1 & 0 & 0 & \cdots & 0 \\ 0 & -1 & 0 & \cdots & 0 \\ 0 & 0 & -1 & \cdots & 0 \\ \vdots & \vdots & \vdots & \ddots & \vdots \\ 0 & 0 & 0 & \cdots & 1 \end{array}\right] \begin{bmatrix} \Delta u[k+1] \\ \Delta[k+2] \\ \Delta[k+3] \\ \vdots \\ \Delta[k+N_c] \end{bmatrix} \leq \left[\begin{array}{c} \Delta u_{\max}[k] \\ \Delta u_{\max}[k] \\ \Delta u_{\max}[k] \\ \vdots \\ \Delta u_{\max}[k] \\ \hdashline \Delta u_{\min}[k] \\ \Delta u_{\min}[k] \\ \Delta u_{\min}[k] \\ \vdots \\ \Delta u_{\min}[k] \end{array}\right],$$

where we have flipped the sign on the lower half in order to pose the minimum bound as a left-hand inequality. This can be written in more compact form as:

$$\begin{bmatrix} \boldsymbol{I}_{N_c} \\ -\boldsymbol{I}_{N_c} \end{bmatrix} \underrightarrow{\boldsymbol{\Delta u}}[k+1] \leq \begin{bmatrix} \boldsymbol{e}_{N_c} \cdot \Delta u_{\max} \\ \boldsymbol{e}_{N_c} \cdot \Delta u_{\max} \end{bmatrix},$$

where $\boldsymbol{I}_{N_c}$ denotes an N_c-dimensioned identity matrix and $\boldsymbol{e}_{N_c} \in \mathbb{R}^{N_c \times 1}$ is a vector of ones. This results in a form identical to Eq. (7.10) above.

Next, we consider constraints imposed directly on the input magnitude $u[k]$ by defining the bounds:

$$u_{\min}[k] \leq u[k] \leq u_{\max}[k].$$

As before, note that the bounds $u_{\min}[k]$ and $u_{\max}[k]$ are, in general, time-varying. For simplicity of the present development, we shall consider the hard bounds on $u[k]$ to be constant over the control horizon. We proceed as before to build a vector-valued inequality by stacking the individual constraint equations for each future sample point as:

$$\begin{bmatrix} 1 & 0 & 0 & \cdots & 0 \\ 1 & 1 & 0 & \cdots & 0 \\ 1 & 1 & 1 & \cdots & 0 \\ \vdots & \vdots & \vdots & \ddots & \vdots \\ 1 & 1 & 1 & \cdots & 1 \\ \hline -1 & 0 & 0 & \cdots & 0 \\ -1 & -1 & 0 & \cdots & 0 \\ -1 & -1 & -1 & \cdots & 0 \\ \vdots & \vdots & & \ddots & \vdots \\ -1 & -1 & -1 & \cdots & -1 \end{bmatrix} \begin{bmatrix} \Delta u[k+1] \\ \Delta[k+2] \\ \Delta[k+3] \\ \vdots \\ \Delta[k+N_c] \end{bmatrix} \le \begin{bmatrix} u_{\max} - u[k] \\ u_{\max} - u[k] \\ u_{\max} - u[k] \\ \vdots \\ u_{\max} - u[k] \\ \hline u_{\min} + u[k] \\ u_{\min} + u[k] \\ u_{\min} + u[k] \\ \vdots \\ u_{\min} + u[k] \end{bmatrix},$$

or more compactly,

$$\begin{bmatrix} E_1 \\ -E_1 \end{bmatrix} \underrightarrow{\Delta \mathcal{U}} \le \begin{bmatrix} e_{N_c} \cdot u_{\max} \\ e_{N_c} \cdot u_{\min} \end{bmatrix} + \begin{bmatrix} -e_{N_c} \\ e_{N_c} \end{bmatrix} u[k],$$

where we've introduced the matrix $E_1 \in \mathbb{R}^{N_c \times N_c}$ as a lower triangular matrix of ones. Note that in this formulation, respective values of $\gamma[k]$ are updated each time step using the previously computed input magnitude.

Output constraints can be addressed by starting with the matrix form of the output equation:

$$\underrightarrow{Y}[k+1] = \Phi \tilde{A} \chi[k] + G \underrightarrow{\Delta \mathcal{U}}[k+1],$$

where we fashion bounds as:

$$y_{\min} \le y[k] \le y_{\max}.$$

We can express this in equivalent vector form as:

$$e_{N_p} y_{\min} \le \underrightarrow{Y}[k+1] \le e_{N_p} y_{\max},$$

and proceeding as above we obtain the constraint inequality:

$$\begin{bmatrix} G \\ -G \end{bmatrix} \underrightarrow{\Delta \mathcal{U}}[k+1] \le \begin{bmatrix} e_{N_p} y_{\max} \\ -e_{N_p} y_{\min} \end{bmatrix} - \begin{bmatrix} \Phi \tilde{A} \chi[k] \\ -\Phi \tilde{A} \chi[k] \end{bmatrix},$$

where here e_{N_p} now represents an $(N_p + 1)$-length vector of ones. Note that both constraint inequalities introduced above have time-varying elements on their right-hand sides, requiring that we update their values at each sampling point.

Additionally, as mentioned above, we may construct constraint inequalities for the state variables by defining them as special outputs. For example,

$$y_z[k] = C_z x[k],$$

by constructing the appropriate output matrix, C_z. From here, we simply proceed exactly as we did for output constraints above. Both outputs and state variables can be used to enforce hard constraints on other problem variables of interest (e.g., cell temperature) that are available in the selected model.

In most practical control problems, we will have a combination of constraints imposed on different problem variables. This case is handled by stacking the relevant constraint matrices and vectors into a combined form. For example, if our problem imposed both input-magnitude and output constraints, we would assemble these as:

$$\begin{bmatrix} E_1 \\ -E_1 \\ G \\ -G \end{bmatrix} \underrightarrow{\Delta u} \le \begin{bmatrix} e_{N_c} \cdot u_{\max} \\ e_{N_c} \cdot u_{\min} \\ e_{N_p} y_{\max} \\ -e_{N_p} y_{\min} \end{bmatrix} + \begin{bmatrix} -e_{N_c} \\ e_{N_c} \\ \mathbf{0} \\ \mathbf{0} \end{bmatrix} u[k] + \begin{bmatrix} \mathbf{0} \\ \mathbf{0} \\ \boldsymbol{\Phi}\tilde{\boldsymbol{A}}\chi[k] \\ -\boldsymbol{\Phi}\tilde{\boldsymbol{A}}\chi[k] \end{bmatrix},$$

where by appropriate definition, we arrive at the form of Eq. (7.10),

$$\boldsymbol{M}\Delta\boldsymbol{U} \le \gamma.$$

Summarizing to this point, we have seen that the fundamental optimization problem addressed by MPC can be written as:

$$\min_{\underrightarrow{\Delta u}} J[k]$$

subject to linear constraints, $\boldsymbol{M}\underrightarrow{\Delta u} \le \gamma$. It turns out that solving this constrained-optimization problem in real time can be made significantly easier by solving its corresponding dual form instead.

A dual method can be used in a quadratic-optimization problem to identify constraints that are not active in optimization so that they can be systematically eliminated from the solution process. This method explicitly incorporates the original constraints into the dual formulation. Assuming feasibility (i.e., that there exists a candidate solution $\boldsymbol{x}$ such that $\boldsymbol{M}\boldsymbol{x} < \gamma$), then the primal problem is equivalent to:

$$\max_{\boldsymbol{\lambda} \ge \mathbf{0}} \min_{\boldsymbol{x}} \left[\frac{1}{2}\boldsymbol{x}^T \boldsymbol{H}\boldsymbol{x} + \boldsymbol{g}^T\boldsymbol{x} + \boldsymbol{\lambda}^T(\boldsymbol{M}\boldsymbol{x} - \gamma) \right],$$

where here we have adjoined the constraints to the cost function with a vector of Lagrange multipliers, $\boldsymbol{\lambda}$.[16] The minimum of this augmented cost function over $\boldsymbol{x}$ is now unconstrained and is attained by:

$$\boldsymbol{x}^o = -\boldsymbol{H}^{-1}(\boldsymbol{g} + \boldsymbol{M}^T\boldsymbol{\lambda}).$$

Substituting $\boldsymbol{x}^o$ into the above expression gives the dual problem (see App. 7.B for the complete derivation):

$$\max_{\boldsymbol{\lambda} \ge \mathbf{0}} \left(-\frac{1}{2}\boldsymbol{\lambda}^T\boldsymbol{P}\boldsymbol{\lambda} - \boldsymbol{\lambda}^T\boldsymbol{k} - \frac{1}{2}\boldsymbol{g}^T\boldsymbol{H}^{-1}\boldsymbol{g} \right),$$

[16] Note that the vector inequality $\boldsymbol{\lambda} \ge \mathbf{0}$ means that all individual elements λ_i of $\boldsymbol{\lambda}$ must satisfy $\lambda_i \ge 0$, $\forall i$.

where:

$$\begin{aligned} P &= MH^{-1}M^T \\ k &= \gamma + MH^{-1}g. \end{aligned}$$

Hence the dual problem turns out to be another quadratic-programming problem, only now with λ as the decision variable instead of x and with the original problem constraints baked in to the dual formulation. Switching sign, the dual problem is made equivalent to the minimization problem:

$$\min_{\lambda \geq 0} \left(\frac{1}{2}\lambda^T P \lambda + \lambda^T k + \frac{1}{2} g^T H^{-1} g \right).$$

The solution to this new quadratic programming problem can be accomplished using an iterative variant of the well-known Gauss–Seidel routine developed by Clifford Hildreth.[17] Hildreth's algorithm employs a primal-dual active set method to solve a linear system of equations using element-by-element reduction, which avoids matrix inversion. The algorithm is guaranteed to converge under some mild conditions and has been shown to be robust in practice. Details of the Hildreth method are presented in App. 7.C.

[17] Clifford Hildreth, "A quadratic programming procedure," *Naval research logistics quarterly*, 4(1):79–85, 1957.

7.6 *MPC fast charge of lithium-ion cells*

Application of MPC to the fast-charge problem will utilize the state-space ROM generated using the subspace reduction methods introduced in Chap. 4. In its most general form, we can write the state-space model as:[18]

[18] This formulation was originally presented in: Marcelo A. Xavier, Aloisio K. De Souza, Kiana Karami, Gregory L. Plett, and M. Scott Trimboli, "A computational framework for lithium ion cell-level model predictive control using a physics-based reduced-order model," *IEEE Control Systems Letters*, 5(4):1387–1392, 2020.

$$\begin{aligned} x[k+1] &= A[k]x[k] + B[k]i_{\text{app}}[k] \\ \tilde{y}[k] &= C[k]x[k] + D[k]i_{\text{app}}[k] \\ y[k] &= g(\tilde{y}[k], i_{\text{app}}[k]) \\ v[k] &= g_v(\tilde{y}[k], i_{\text{app}}[k]). \end{aligned}$$

In this formulation, vector $\tilde{y}[k]$ is the linearized version of the electrochemical variables of interest emerging directly from the ROM. State-matrices are time-varying and are updated with new values using the output blending method discussed in Chap. 4. Note that individual electrochemical variables are computed in the physics-based model relative to normalized spatial coordinates across the electrode, $\tilde{x}$. We shall use this linearized model to generate the prediction dynamics required by MPC and as the basis for the nonlinear Kalman filter that will ultimately provide estimates of the internal electrochemical state of the cell. Importantly, we shall use the estimated variables supplied by the model to enforce hard limits during operation. Note that although the basic ROM is configured in linear state-space form, key outputs of interest are *not* linear functions of the state variables and must be reconfigured to obtain linear approximate forms.

In order to construct a more accurate simulation of battery behavior, however, nonlinear corrections are applied to appropriate linearized variables; these are computed via the function, $g(\cdot)$ to give corrected output values in $\boldsymbol{y}[k]$. Cell terminal voltage is computed as a nonlinear function of electrochemical parameters, $g_v(\cdot)$.

Leveraging the results of Chap. 4, we now recap SOC and cell-voltage calculations.

7.6.1 *State-of-Charge (SOC)*

Cell SOC defined in terms of the negative electrode is given by:

$$z[k] = \frac{\theta^{\mathrm{n}}_{\mathrm{s,avg}}[k] - \theta^{\mathrm{n}}_{0}}{\theta^{\mathrm{n}}_{100} - \theta^{\mathrm{n}}_{0}},$$

where $\theta^{\mathrm{n}}_{100}$ and θ^{n}_{0} give the electrode stoichiometry at 100 % and 0 % SOC, respectively for the negative electrode. Additionally, if the cell model has no double layer effect,[19] then the electrode state of charge $\theta^{\mathrm{n}}_{\mathrm{s,avg}}$ can be computed as:

$$\theta^{\mathrm{n}}_{\mathrm{s,avg}}[k] = \theta^{\mathrm{n}}_{\mathrm{s,0}} + \theta^{\mathrm{res0,n}}_{\mathrm{ss}} x_0[k], \tag{7.11}$$

where $\theta^{\mathrm{n}}_{\mathrm{s,0}}$ is the initial stoichiometry, $x_0[k]$ is the integrator state, and $\theta^{\mathrm{res0,n}}_{\mathrm{ss}}$ is the integrator residue for the appropriate TF. Initial electrode stoichiometry is determined from:

$$\theta^{\mathrm{n}}_{\mathrm{s,0}} = \theta^{\mathrm{n}}_{0} + z_0 \left(\theta^{\mathrm{n}}_{100} - \theta^{\mathrm{n}}_{0}\right),$$

where z_0 is the initial cell SOC. Performing appropriate substitutions, we can express the SOC as:

$$z[k] = C_{z,}[k]\boldsymbol{x}[k] + z_0,$$

where we have constructed output matrix $C_z[k] = \begin{bmatrix} 0 & 0 & \cdots & c_{n_x}[k] \end{bmatrix}$ where $c_{n_x}[k]$ is a function of $\theta^{\mathrm{n}}_{\mathrm{s,avg}}[k]$ and multiplies the integrator state $x_i[k]$.[20]

[19] Lithium flux $i_{\mathrm{f+dl}}$ is divided between faradaic flux, which changes the electrode SOC, and double-layer flux, which does not. However, during fast-charge, the double-layer charges to a steady-state level *very* quickly, and since its capacity is negligible with respect to the capacity of the electrode, Eq. (7.11) is still approximately true and can be considered to be accurate for the fast-charge problem.

[20] In our formulation, we assume that the integrator state occupies the last position within the state vector $\boldsymbol{x}[k]$.

7.6.2 *Cell voltage*

Cell terminal voltage is a somewhat complicated function of a subset of the cell variables contained in the output vector $\boldsymbol{y}_k$:

$$\begin{aligned} v[k] &= (\eta^{\mathrm{p}}[3,k] - \eta^{\mathrm{n}}[0,k]) + \left(\phi^{\mathrm{p}}_{\mathrm{e}}[3,k] - \phi^{\mathrm{n}}_{\mathrm{e}}[0,k]\right) \\ &\quad + \left(U^{\mathrm{p}}_{\mathrm{ocp}}\left(\theta^{\mathrm{p}}_{\mathrm{ss}}[3,k]\right) - U^{\mathrm{n}}_{\mathrm{ocp}}\left(\theta^{\mathrm{n}}_{\mathrm{ss}}[0,k]\right)\right) - \left(\bar{R}^{\mathrm{p}}_{\mathrm{f}} i^{\mathrm{p}}_{\mathrm{f+dl}}[3,k] - \bar{R}^{\mathrm{n}}_{\mathrm{f}} i^{\mathrm{n}}_{\mathrm{f+dl}}[0,k]\right). \end{aligned}$$

By linearizing this equation around the present operating point, rearranging terms, and making appropriate substitutions, we can reformulate as an affine relationship:

$$v[k] = C_v[k]\boldsymbol{x}[k] + D_v[k]i_{\text{app}} + b_v[k].$$

7.6.3 *Lithium-plating side-reaction overpotential*

From Sect. 7.1 we recall that the negative-electrode side-reaction overpotential, η_{s}, can serve as an indicator of the onset of lithium plating. Here we formulate the expressions we'll need for its calculation in the fast-charge algorithm. In Chap. 1 we reformulated the full-order DFN P2D physics-based PDE model of a lithium-ion cell, and then created from it a computationally compact ROM using the xRA method in Chap. 4. The output of that ROM provides the electrochemical quantities needed to determine the discrete-time values of $\eta_{\text{s}}[\tilde{x}, k]$ for the simple case where $\bar{R}_{\text{f}} = 0$. Starting from Eq. (7.1), we make the following approximation:

$$\eta_{\text{s}}[\tilde{x}, k] \approx \phi_{\text{s}}[\tilde{x}, k] - \phi_{\text{e}}[\tilde{x}, k] = \phi_{\text{s-e}}[\tilde{x}, k], \tag{7.12}$$

where we have further assumed $\bar{U}_{\text{s}} = 0$, as is the case for lithium plating. Thus the ROMs can be used to produce values for the electrochemical variables required to monitor for the lithium plating condition.[21]

In the present case, we have the phase-potential difference variable for the negative electrode,

$$\phi_{\text{s-e}}[\tilde{x}, k] = \tilde{\phi}_{\text{s-e}}[\tilde{x}, k] + U_{\text{ocp}}[\theta_{\text{s},0}^{\text{r}}],$$

where $\tilde{\phi}_{\text{s-e}}[\tilde{x}, k]$ is obtained directly from the system linear output vector $\tilde{\boldsymbol{y}}[k]$, and where the negative electrode open-circuit potential function is evaluated at the initial local state of charge and applied as a nonlinear correction. In Chap. 4 it was shown that by treating the integrator pole separately in transfer function $\widetilde{\Phi}_{\text{s-e}}(z, s) / I_{\text{app}}(s)$ we can obtain the more accurate expression,

$$\phi_{\text{s-e}}[\tilde{x}, k] = [\tilde{\phi}_{\text{s-e}}[\tilde{x}, k]]^{*} + U_{\text{ocp}}[\theta_{\text{s,avg}}[k]].$$

Therefore, for purposes of advanced physics-based MPC, we can approximate the condition for avoiding lithium plating as,

$$\phi_{\text{s-e}}[\tilde{x}, k] \geq 0.$$

Next, we need to calculate the solid-electrolyte phase potential difference from the xRA-produced state-space model so that it is made available in proper form to the algorithm. A state-space representation can be written as,

$$\phi_{\text{s-e}}[k, \tilde{x}] = C_{\phi_{\text{s-e}}}[k]\boldsymbol{x}[k] + U_{\text{ocp}}[\theta_{\text{s,avg}}^{\text{n}}[k]], \tag{7.13}$$

[21] Note here that the ROMs are actually supplied to the EKF to compute estimates of the required variables that will then be introduced to MPC for constraint handling.

where we have introduced the output matrix $C_{\phi_{\text{s-e}}}$ which linearly combines the elements of the state vector to generate the linear portion of the output variable $\phi_{\text{s-e}}[k, \tilde{x}]$.[22] Substituting Eq. (7.13) into Eq. (7.12), and renaming the output matrix gives the expression for the side-reaction overpotential we'll use in the fast-charge algorithm:[23]

$$\eta_s[k, \tilde{x}] = C_{\eta_s}[k]\boldsymbol{x}[k] + U_{\text{ocp}}[\theta^{\text{n}}_{\text{s,avg}}[k]].$$

[22] In App. 5.B, these were denoted as $[C^{\phi_{\text{s-e}}[\tilde{x}]}]$. We change the notation slightly here for a more compact presentation.

[23] Spatial variable $\tilde{x}$ is not to be confused with state vector $\boldsymbol{x}$; the latter is vector-valued and indicated in bold type.

7.6.4 *State-space prediction model*

Combining the results from above, we can now define an augmented state-space prediction model for the physics-based ROM as:

$$\begin{aligned}
\chi[k+1] &= \tilde{\boldsymbol{A}}[k]\chi[k] + \tilde{\boldsymbol{B}}[k]\triangle i_{\text{app}}[k+1] \\
v[k] &= \tilde{\boldsymbol{C}}_v[k]\chi[k] + b_v[k] \\
z[k] &= \tilde{\boldsymbol{C}}_z[k]\chi[k] + z_0 \\
\eta_s[k] &= \tilde{\boldsymbol{C}}_{\eta_s}[k]\chi[k] + U_{\text{ocp}}[\theta^{\text{n}}_{\text{s,avg}}[k]],
\end{aligned}$$

where the augmented state vector is given by,

$$\chi[k] = \begin{bmatrix} \boldsymbol{x}^T[k] \\ i_{\text{app}}[k] \end{bmatrix}.$$

Appropriate MPC prediction matrices $\boldsymbol{\Phi}$ and $\boldsymbol{G}$ are next built according to the procedure in Sect. 7.5, enabling predicted sequences of SOC, cell voltage, and side-reaction overpotential to be computed as:

$$\begin{aligned}
\underset{\rightarrow}{z}[k+1] &= \boldsymbol{\Phi}_z\tilde{\boldsymbol{A}}[k]\chi[k] + \boldsymbol{G}_z\boldsymbol{\Delta i}_{\text{app}}[k+1] + \boldsymbol{e}_{N_p}z_0 \\
\underset{\rightarrow}{v}[k+1] &= \boldsymbol{\Phi}_v\tilde{\boldsymbol{A}}[k]\chi[k] + \boldsymbol{G}_v\boldsymbol{\Delta i}_{\text{app}}[k+1] + \boldsymbol{e}_{N_p}b_v[k] \\
\underset{\rightarrow}{\boldsymbol{\eta}_s}[k+1] &= \boldsymbol{\Phi}_{\eta_s}\tilde{\boldsymbol{A}}[k]\chi[k] + \boldsymbol{G}_{\eta_s}\boldsymbol{\Delta i}_{\text{app}}[k+1] + \boldsymbol{e}_{N_p}U_{\text{ocp}}[\theta^{\text{n}}_{\text{s,avg}}[k]].
\end{aligned}$$

The construction of the $\boldsymbol{\Phi}$ and $\boldsymbol{G}$ matrices utilizes common $\tilde{\boldsymbol{A}}[k]$ and $\tilde{\boldsymbol{B}}[k]$ state and input matrices, but separate output $\tilde{\boldsymbol{C}}$ matrices corresponding to the output equations for each required variable. Now that we have the prediction model in place, we next turn toward constructing the needed constraint matrices.

7.6.5 *Defining constraints on PBM variables*

The linear-constraint inequality $\boldsymbol{M}\underset{\rightarrow}{\boldsymbol{\Delta i}}_{\text{app}}[k+1] \le \gamma[k]$ introduced in Sect. 7.5 is built from the bounds defined above as:

$$\underbrace{\begin{bmatrix} \boldsymbol{M}_i \\ \boldsymbol{M}_z \\ \boldsymbol{M}_v \\ \boldsymbol{M}_{\eta_s} \end{bmatrix}}_{M} \underset{\rightarrow}{\boldsymbol{\Delta i}}_{\text{app}}[k+1] \le \underbrace{\begin{bmatrix} \gamma_i[k] \\ \gamma_z[k] \\ \gamma_v[k] \\ \gamma_{\eta_s}[k] \end{bmatrix}}_{\gamma},$$

where:

$$
\begin{aligned}
\boldsymbol{M}_i &= \begin{bmatrix} \boldsymbol{E}_1 \\ -\boldsymbol{E}_1 \end{bmatrix} \\
\boldsymbol{\gamma}_i[k] &= \begin{bmatrix} \boldsymbol{e}_{N_c}\left(i_{\max} - i_{\text{app}}[k]\right) \\ -\boldsymbol{e}_{N_c}\left(i_{\min} - i_{\text{app}}[k]\right) \end{bmatrix} \\
\boldsymbol{M}_v &= \boldsymbol{G}_v \\
\boldsymbol{M}_v &= \boldsymbol{e}_{N_p} v_{\max} - \boldsymbol{\Phi}_v \tilde{\boldsymbol{A}}[k]\boldsymbol{\chi}[k] - \boldsymbol{e}_{N_p} b_v[k] \\
\boldsymbol{M}_z &= \boldsymbol{G}_z \\
\boldsymbol{\gamma}_z[k] &= \boldsymbol{e}_{N_p} z_{\max} - \boldsymbol{\Phi}_z \tilde{\boldsymbol{A}}[k]\boldsymbol{\chi}[k] - \boldsymbol{e}_{N_p} z_0 \\
\boldsymbol{M}_{\eta_s} &= -\boldsymbol{G}_{\eta_s} \\
\boldsymbol{\gamma}_{\eta_s}[k] &= -\boldsymbol{e}_{N_p} \eta_{s,\min} + \boldsymbol{\Phi}_{\eta_s} \tilde{\boldsymbol{A}}[k]\boldsymbol{\chi}[k] + \boldsymbol{e}_{N_p} U_{\text{ocp}}[\theta^{\text{n}}_{\text{s,avg}}[k]].
\end{aligned}
$$

7.6.6 Setting up the fast-charge problem

At this point, we have most of the algorithmic machinery in place to execute the MPC-EKF fast-charge problem. What remains is to formulate the optimization objective that will bring about a fastest-time-to-charge profile. Strictly speaking, an optimal fast-charge is obtained as the solution to a minimum-time optimal control problem posed as:

$$\min_{\underrightarrow{\boldsymbol{\Delta i}}_{\text{app}}[k+1]} \left\{ T_{\text{charge}} \right\}$$

subject to constraints. In other words, the cost function is $J[k] = T_{\text{charge}}$, defined as the total time interval required to bring the cell from an initial state-of-charge z_0 at time t_0 to a final state-of-charge z_f at time t_f. Optimization is carried out over the decision vector $\underrightarrow{\boldsymbol{\Delta i}}_{\text{app}}[k+1]$; that is, the sequence of future input increments of applied current, and must adhere to all stipulated constraints. The optimal MPC solution $\underrightarrow{\boldsymbol{\Delta i}}^{*}_{\text{app}}[k+1]$ describes the N_p-length input current profile (i.e., the $\Delta i_{\text{app}}[k]$ sequence) that achieves this objective.[24] Interestingly, when constraints are imposed simply on the input current and cell terminal voltage, the solution approaches the well-known constant-current-constant-voltage (CCCV) profile.

In general, optimal minimum-time problems are notoriously difficult to solve and are generally not amenable to real-time computation.[25] So for our implementation, we instead fashion what we shall term a pseudo-minimum-time problem:[26]

$$\min_{\underrightarrow{\boldsymbol{\Delta i}}_{\text{app}}[k+1]} J[k],$$

subject to appropriate constraints, where here we define the cost

[24] Recall that although the optimization produces the entire optimal input sequence, only the first element $\Delta i_{\text{app}}[k]$ is supplied in accordance with the receding-horizon principle.

[25] For an appreciation of the difficulty in solving minimum-time optimal control problems, see Arthur Earl Bryson, *Applied optimal control: optimization, estimation and control*. CRC Press, 1975.

[26] See Marcelo A. Xavier and M. Scott Trimboli, "Lithium-ion battery cell-level control using constrained model predictive control and equivalent circuit models," *Journal of Power Sources*, 285:374–384, 2015 for a more complete discussion of the pseudo-minimum-time problem.

function as:

$$J[k] = \left(\underset{\rightarrow}{\boldsymbol{R}}[k+1] - \underset{\rightarrow}{\boldsymbol{Y}}[k+1]\right)^T \left(\underset{\rightarrow}{\boldsymbol{R}}[k+1] - \underset{\rightarrow}{\boldsymbol{Y}}[k+1]\right) + \rho \Delta \underset{\rightarrow}{\boldsymbol{U}}^T[k+1] \Delta \underset{\rightarrow}{\boldsymbol{U}}[k+1].$$

The output is defined to be the instantaneous SOC $y[k] = z[k]$, and the reference is the desired final SOC charge value, $r[k] = z_f$. In essence, the cost function is designed to minimize the normed distance between the measured SOC and the output reference while penalizing the magnitude of the input current rate. As such, the optimal solution seeks to bring the SOC to its target value as quickly as possible. By detuning the input weighting penalty ρ to an arbitrarily small (but nonzero) value, the result can be shown to approach the min-time solution.

7.7 *MPC implementation*

In our configuration, the state-space physics-based ROM produced in Chap. 4 will serve as the MPC prediction model. This model embodies the dynamics needed to describe the internal electrochemical variables of interest. But we cannot rely on the model alone to inform our controller since any error in initial conditions—or model accuracy—will propagate as errors in the corresponding variable estimates. Instead we employ an extended Kalman filter (EKF) to incorporate measurable outputs and statistical representations of process uncertainty into the computation of estimates of internal variables. The principles of Kalman filtering were presented in Chap. 5 and the reader is encouraged to review that material for details of the process. Since both MPC and Kalman filtering rely on an underlying model of the process, it is natural (and computationally efficient) to use the same physics-based ROM to drive both methods. Fig. 7.7 presents a notional implementation combining a Kalman filter estimator with MPC control.

The diagram is centered around the linearized state-space model, comprising the basic set of $\boldsymbol{A}[k]$, $\boldsymbol{B}[k]$, $\boldsymbol{C}[k]$, and $\boldsymbol{D}[k]$ matrices generated from the output-blending procedure applied to stored model sets precomputed by the xRA process of Chap. 4. Note that the model is updated at every sample increment to reflect changes in the operational state. At the top of the figure is our lithium-ion cell—the actual process to be controlled. Sensors capture externally available quantities (normally voltage and temperature, although only voltage is depicted in the figure)[27] and supply these to the Kalman-filtering algorithm at each time step in order to compute the estimate of the battery's state, $\hat{x}^+[k]$. Note that when cell surface temperature is

[27] We point out here that cell temperature is an important and limiting factor for fast-charge, and should be included in the constraint set of any practical fast-charge problem. For ease of exposition, we consider only voltage as the measured output in this development. A full treatment of the electro-thermal case can be found in: Aloisio Kawakita de Souza, Gregory L. Plett, and M. Scott Trimboli, "Lithium-ion battery charging control using a coupled electro-thermal model and model predictive control," in *2020 IEEE Applied Power Electronics Conference and Exposition (APEC)*. IEEE, 2020, pp. 3534–3539.

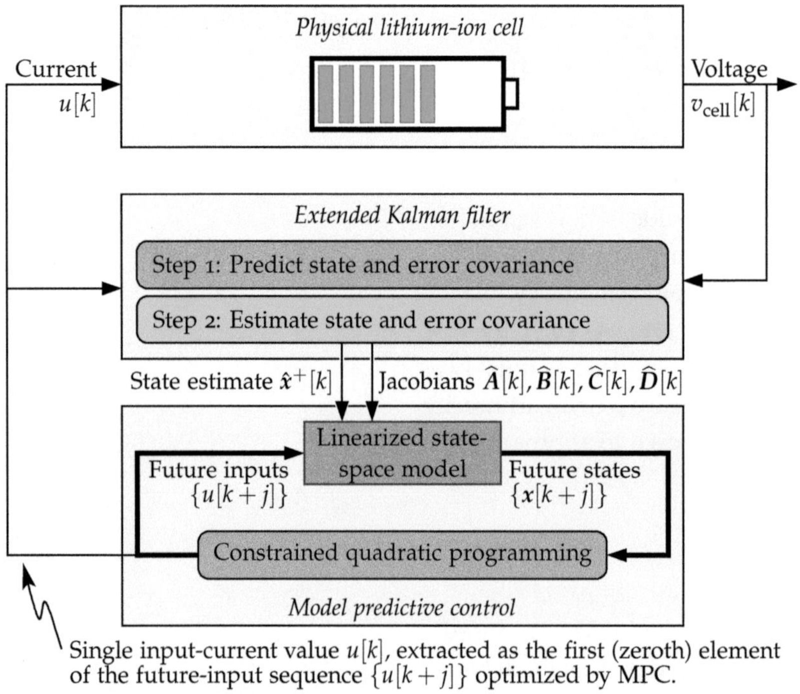

Figure 7.7: Combined MPC-EKF architecture with measured terminal voltage output to the Kalman filter.

supplied, the Kalman filter can be used to estimate the corresponding core (or worst-case) internal temperature on which constraints should be enforced. The algorithm further updates the Jacobian matrices $\hat{A}[k]$, $\hat{B}[k]$, $\hat{C}[k]$, and $\hat{D}[k]$ used by the EKF to compute process covariance (see Chap. 5 for details).

This state estimate is then supplied to the MPC algorithm to initialize its N_p-ahead prediction of the state dynamics using the same state-matrices employed by the EKF. The trajectory of future state values is indicated by the bold black line on the right-hand side of the MPC block in the figure. Proceeding into the MPC algorithm, a constrained quadratic program is executed (e.g., the Hildreth algorithm) to produce an optimized sequence of future input values, $\{u[k+j]\}$. In accordance with the receding-horizon principle, the first of these values is implemented (and delivered to the EKF), while the others are discarded. This completes one loop through the combined MPC-EKF algorithm.

It is important to point out here that the algorithm just described operates on the central estimate provided by the EKF, although the principles of Kalman filtering tell us that this is only the statistically most likely value with respect to the modeled statistical uncertainty. When using such estimates to impose bounds on system variables, it is suggested to fashion those limits instead on statistical bounding envelopes surrounding the central parameter estimates. The EKF computes these bounds from its own moving estimate of covariance values; a 95% confidence bound is a typical choice. While this imposes a somewhat conservative bound, it is nonetheless a sensible bound that effects a good compromise between performance and safety. Fig. 7.8 shows how we might bound a notional measure of cell degradation on estimated Kalman-filter statistical limits.

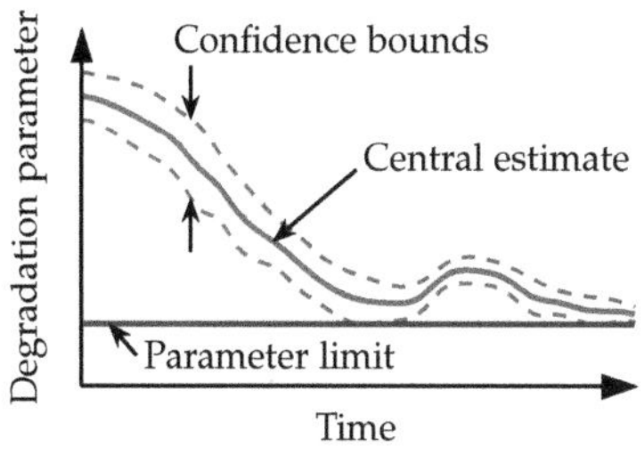

Figure 7.8: Kalman filter confidence bounds.

7.8 *Example of MPC-EKF fast charge*

The physics-based fast-charge method is next demonstrated by applying the MPC-EKF code to the same NMC30 cell model introduced in earlier chapters. The first case establishes a baseline of performance by imposing limits on maximum applied current and cell terminal voltage. Maximum charge current is constrained to a magnitude of 59.72 A which corresponds to a 2C charge rate and maximum cell voltage is set to 4.1 V. Additionally, a threshold of 0.08 V is defined for the side-reaction overpotential as a conservative limit for the onset of lithium plating. This example requests a fast charge from an initial SOC of 10 % to a target value of 95 %. Here, MPC is tuned with relatively short horizons, $N_p = 5$ and $N_c = 2$, assuring compact prediction matrices and efficient computation.

Fig. 7.9 shows simulation results where the MPC charge current is applied to the nonlinear model. As expected, when constraining only current and voltage, the optimal solution gives the CCCV charge profile. Despite adhering to both limits, however, the side-reaction overpotential limit is breached approximately 800 s into the charge and extends to nearly 1500 s, suggesting that the conditions for lithium plating may be attained well within terminal voltage limits.

Fig. 7.10 presents the same simulation conditions with the addition of a constraint imposed directly on the side-reaction overpotential as estimated by the EKF. It is immediately apparent that the current profile is markedly different from the CCCV version. In particular, MPC foresees the impending violation of the side-reaction overpotential constraint and forces the current to back off such that the limit is observed. Subsequent current calculations enable the fast charge to continue while respecting all imposed constraints. Note that in this case, the voltage limit is again reached just prior to achieving the SOC target. It is also important to note that lifetime-extending operation within electrochemical limits for this example comes at the cost of some charging time.

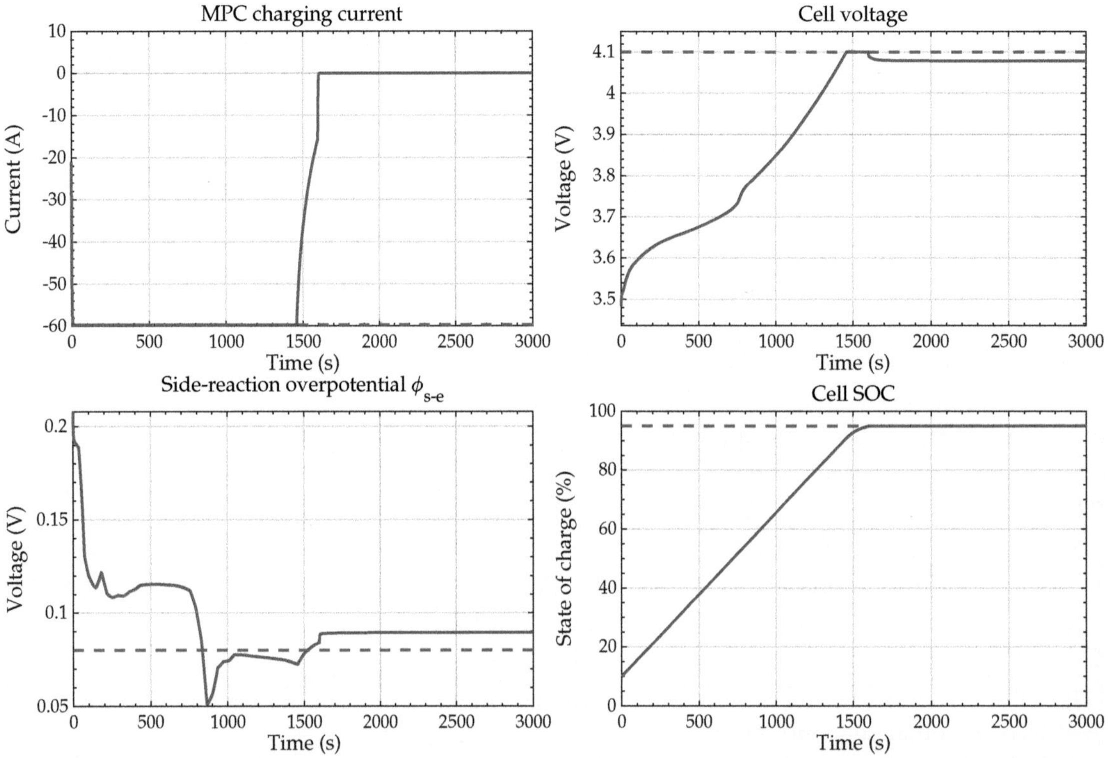

Figure 7.9: Case 1: Constraints on applied current and terminal voltage.

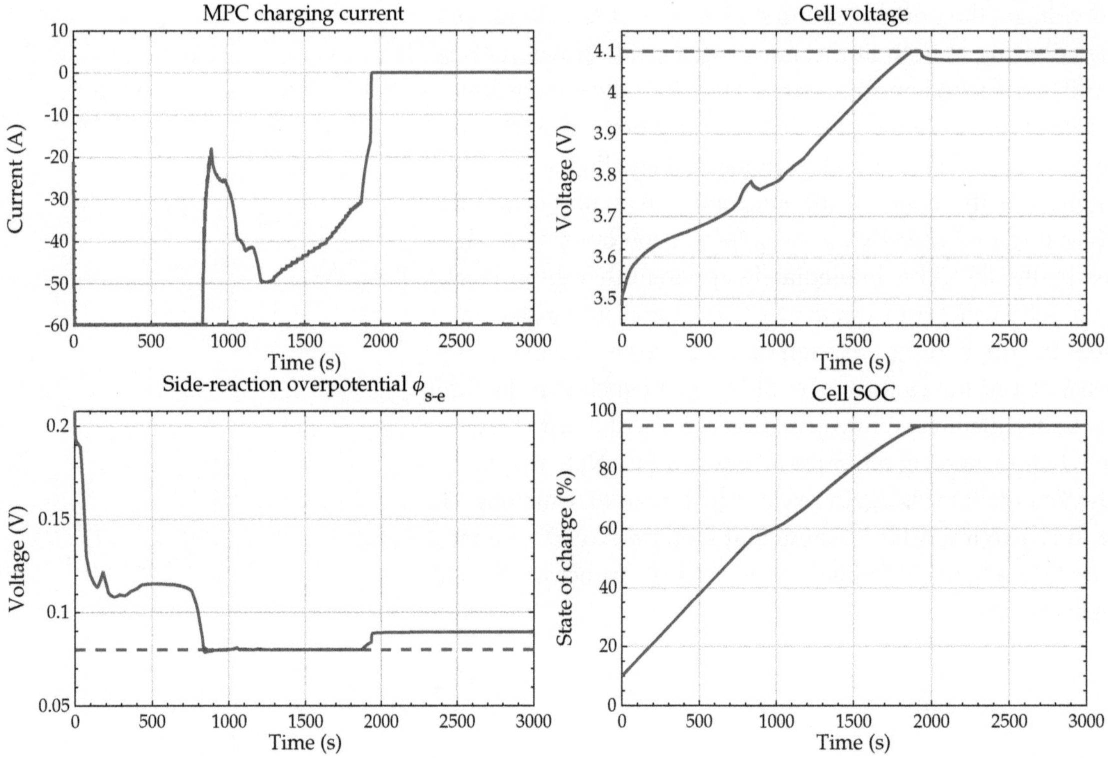

Figure 7.10: Case 2: Constraints on applied current, terminal voltage, and side-reaction overpotential.

7.9 *Experimental validation*

Direct experimental validation of the proposed techniques is not easily obtained, due mainly to the inaccessibility of internal electrochemical properties in situ. Nonetheless, a number of indirect methods have been devised to show experimentally the viability of the electrochemically constrained fast-charge idea. Of note, recent work by Sieg et al.[28] experimentally investigates the fast charge of a commercial 51 Ah lithium-ion pouch cell using a control method similar in principle to that presented here. In particular, the authors assemble three-electrode test cells of a high-energy graphite anode pouch cell using lithium metal as a reference electrode in order to measure the overpotential value at the negative electrode-electrolyte interface. They employ a current-control strategy different from what is presented here, but having the same objective of maintaining the side-reaction overpotential above 0 V. A variety of test setups are utilized; the setup that most closely approximates the fast-charge problem presented in this chapter is shown to extend battery lifetime by more than 30% as measured by normalized capacity obtained by a C/10 discharge. The controlled cell reached a remaining capacity of 80 % after 1800 cycles, compared to 1350 cycles for a CC–CV reference.

[28] Johannes Sieg, Jochen Bandlow, Tim Mitsch, Daniel Dragicevic, Torben Materna, Bernd Spier, Heiko Witzenhausen, Madeleine Ecker, and Dirk Uwe Sauer, "Fast charging of an electric vehicle lithium-ion battery at the limit of the lithium deposition process," *Journal of Power Sources*, 427:260–270, 2019.

Yin and Choe[29] further demonstrate the viability of the proposed method using an experimental battery-in-the-loop configuration and a 40 Ah high-energy pouch cell. Their work features a multiobjective optimization scheme based on a reduced-order physics-based model where they seek an optimal charging profile for fast-charge that alleviates capacity fade. As in our approach, they consider the side-reaction overpotential (as estimated by a nonlinear Kalman filter) for mitigation of lithium-plating. They also consider temperature effects, which as noted previously, are easily accommodated in the approach outlined above, as well as discharge pulses to promote lithium stripping. Upon cycling, the controlled case reaches 95 % of original capacity after 140 cycles, whereas the reference case (CC–CV) reached the same level after just 25 cycles.

[29] Yilin Yin and Song-Yul Choe, "Actively temperature controlled health-aware fast charging method for lithium-ion battery using nonlinear model predictive control," *Applied Energy*, 271:115232, 2020.

In summary, this chapter has outlined an advanced battery control approach that exploits the rich information content of a physics-based model. Unmeasurable quantities are estimated using a nonlinear Kalman filter applied to a computationally compact reduced-order form of the full electrochemical description. Finally, MPC is employed to obtain optimal charge current profiles that adhere to hard constraints chosen to mitigate certain degradation effects and thereby extend life. The potential viability of this approach is inferred from experimental results obtained for strategies similar in principle to that presented here.

7.10 MATLAB toolbox

The MPC-EKF algorithms featured in this chapter have been incorporated into the MATLAB toolbox that accompanies this volume. As with Chaps. 5–6, this chapter and the next are applications of the reduced order physics-based models developed in Chap. 4, as indicated in Fig. 7.11. The model-predictive control components of the toolbox are encapsulated by the MATLAB functions `initMPC` and `iterMPC`.

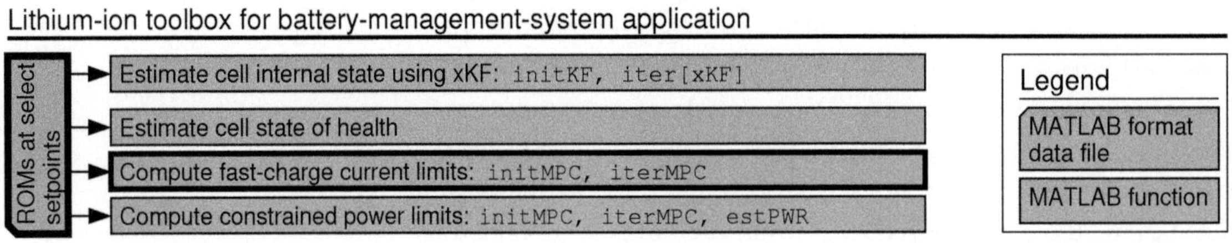

Figure 7.11: Functionality of MATLAB physics-based BMS toolbox, highlighting the focus of this chapter.

To run an MPC simulation, we must first initialize the algorithm. This is done by calling `initMPC`.

```
% Establish MPC tuning parameters:
mpcData.Np = Np;           % Prediction horizon
mpcData.Nc = Nc;           % Control horizon
mpcData.ref = targetSOC; % Target value for final SOC
mpcData.Sigma = tril(ones(mpcData.Nc,mpcData.Nc));
mpcData.Nsim = Nsim;       % Number of simulation points
mpcData.uk_1 = 0;          % Initialization value
mpcData.SOCk_1 = 0;
mpcData.DUk_1 = 0;
mpcData.maxHild = 100;     % Maximum iterations for Hildreth
mpcData.lambda = [];
mpcData.SOC0 = SOC0;       % Initialization value
mpcData.Ts = 1;            % Sampling interval [sec]
Iapp_max  = -ROM.cellData.function.const.Q(0.5)*Crate; % Max current
Iapp_min  = 0;

% Establish MPC constraints
mpcData.const.constraints = constraints; % Enter desired conditions
if constraints(1) == 1
    mpcData.const.u_max =   Iapp_min;   % Minimum applied current
    mpcData.const.u_min =   Iapp_max;   % Maximum applied current
    mpcData.const.du_max =  50;         % Maximum increment
    mpcData.const.du_min = -50;         % Minimum increment
end
if constraints(2) == 1
    mpcData.const.v_min = V_min;
    mpcData.const.v_max = V_max;
end
if constraints(3) == 1
    mpcData.const.phise_min = Phise_min;
end
```

After initialization, we execute the MPC algorithm to compute the constrained optimal solution at each sampling interval using `iterMPC`.

```
function [uk, mpcData] = iterMPC(xk,xk_1,cellState, mpcData)
%ITERMPC This function computes one iteration of the MPC charging protocol
% Inputs:
%   xk:        present state vector
%   xk_1:      previous timestep state vector
%   cellState: structure, contains cell information from simStep
%   mpcData:   structure, contains MPC information
% Outputs:
%   uk:        optimal input current
%   mpcData:   structure, contains updated MPC information

% Load MPC data
Np = mpcData.Np;
Nc = mpcData.Nc;
uk_1 = mpcData.uk_1;
Sigma = mpcData.Sigma;
Ref = mpcData.ref*ones(Np,1)/100;

% Obtain matrices from Kalman Filter
Am = cellState.MPC.A;
Bm = cellState.MPC.B;
Csoc = cellState.MPC.Csoc;
Dsoc = cellState.MPC.Dsoc;

% Compute SOC prediction matrices
[Phi_soc,G_soc] = predMat(Am,Bm,Csoc,Dsoc,Np,Nc);

% Build augmented state vector
dx = [(xk - xk_1); mpcData.SOCk_1];

% Compute values needed by hildreth
F = -2*G_soc'*(Ref - Phi_soc*dx );
[~,S,~] = svd(G_soc'*G_soc);
[m,n] = size(S);
Ru = (norm(F,2)/(2*mpcData.const.du_max*sqrt(Nc)))-(S(m,n));
Ru = Ru*eye(mpcData.Nc,mpcData.Nc);
mpcData.Ru = Ru;

E = 2*(G_soc'*G_soc + Ru);
DU = -E\F;

[M,gamma] = constraintsMPC(dx,cellState,mpcData);

% Check if constraints are violated; if so, run Hildreth
if sum(M*DU - gamma > 0) > 0
    [DU,lambda,nexec] = hildreth(E,F,M,gamma,mpcData.lambda,mpcData.
      maxHild);
    mpcData.lambda = lambda;
    mpcData.nexec = nexec;
else
    mpcData.nexec = 0;
end

du = DU(1);
uk = du + mpcData.uk_1;

mpcData.uk_1 = uk;
```

```
mpcData.xk_1 = xk;
mpcData.SOCk_1 = cellState.MPC.prior_SOC;
mpcData.DUk_1 = DU;
end
```

7.11 *Where to from here?*

This chapter presented a practical implementation of MPC combined with EKF state estimation to bring about the safe fast charge of a lithium-ion cell within limits imposed by internal battery states. Use of statistical confidence bounds enhances safety of operation. The most important implications of this approach are:

- Computationally compact physics-based models can form the basis of an effective battery management and control system.
- Kalman filter estimation can be used in synchronization with predictive control to produce a workable algorithm for real-time application.

It turns out that predictive fast-charge methods may also be used to manage energy flow into and out of a battery cell during standard operation; this has important implications for power limits estimation and will be the topic of Chap. 8.

7.A *Summary of variables*

- $b_v[k]$, affine term in voltage output equation.
- $i_{\text{app}}[k]$, applied input current value [A].
- k, sampling index.
- $J[k]$, quadratic programming objective function.
- N_{c}, MPC control horizon.
- N_{p}, MPC prediction horizon.
- T_{charge}, the charge-time interval $t_f - t_0$ [s].
- T_{s} , sampling period of discrete-time model [s].
- $u_{\max}$, input variable maximum bound.
- $u_{\min}$, input variable minimum bound.
- $\Delta u_{\max}$, input-increment variable maximum bound.
- $\Delta u_{\min}$, input-increment variable minimum bound.
- η_s, side-reaction overpotential [V].
- ρ, input increment weighting factor for MPC cost function.
- $\boldsymbol{A}[k]$, time-varying state-transition matrix of state-space model.
- $\tilde{\boldsymbol{A}}[k]$, time-varying state-transition matrix of augmented state-space model.
- $\boldsymbol{B}[k]$, time-varying input matrix of state-space model.
- $\tilde{\boldsymbol{B}}[k]$, time-varying input matrix of augmented state-space model.

- $C[k]$, time-varying output matrix of state-space model.
- $\tilde{C}[k]$, time-varying output matrix of augmented state-space model.
- $C_z[k]$, time-varying SOC output matrix.
- $C_{\eta_s}[k]$, time-varying side-reaction overpotential output matrix.
- $C_v[k]$, time-varying voltage output matrix.
- $C_{\phi_{\text{s-e}}}[k]$, time-varying phase potential difference output matrix.
- $D[k]$, time-varying direct-feedthrough matrix of state-space model.
- $D_v[k]$, time-varying voltage direct-feedthrough matrix.
- E_1, a lower triangular matrix of ones.
- $H[k]$, symmetric Hessian matrix in quadratic program objective function.
- $G[k]$, time-and-input-dependent prediction matrix for MPC.
- M, matrix whose rows contain linear constraint equation coefficients.
- $P[k]$, time-varying Hessian matrix of dual quadratic program objective function.
- $\Phi[k]$, time-and-state-dependent prediction matrix for MPC.
- e_N, an $N \times 1$ dimension vector of ones.
- $f[k]$, time-varying scalar value in quadratic program objective function.
- $g[k]$, time-varying gradient vector in quadratic program objective function.
- $k[k]$, time-varying gradient vector in dual quadratic program objective function.
- $\Delta u[k]$, input-increment vector for MPC.
- $\chi[k]$, MPC augmented state-vector.
- $x[k]$, time-varying state vector of state-space model.
- $\tilde{y}[k]$, vector of linearized system outputs from ROM.
- $y[k]$, vector of nonlinear-corrected system outputs.
- $\gamma[k]$, vector of time-varying bounds in linear constraint equations.

7.B Derivation of the dual optimization problem

Starting with the primal problem:

$$J^* = \max_{\lambda \geq 0} \min_{x} \left[\frac{1}{2} x^T H x + g^T x + \lambda^T (Mx - \gamma) \right],$$

we solve for the unconstrained minimizer via the stationarity condition:

$$\frac{\partial J}{\partial x} = Hx + g + M^T \lambda = 0,$$

giving the solution,

$$x^o = -H^{-1}(g + M^T \lambda).$$

Substituting x^o into the primal problem, we obtain:

$$\begin{aligned} J^* = \max_{\lambda \ge 0} \Big\{ & \frac{1}{2}\left(g + M^T\lambda\right)^T H^{-1} H H^{-1}\left(g + M^T\lambda\right) \\ & - \left(g + M^T\lambda\right)^T H^{-1} g + \lambda^T \left(M\left(-H^{-1}(g + M^T\lambda)\right) - \gamma\right)\Big\}. \end{aligned}$$

Expanding and cancelling terms:

$$\begin{aligned} J^* &= \max_{\lambda \ge 0} \Big\{ \frac{1}{2}\left(\lambda^T M + g^T\right) H^{-1}\left(g + M^T\lambda\right) \\ &\quad - \left(\lambda^T M + g^T\right) H^{-1} g + \lambda^T\left[-MH^{-1}\left(g + M^T\lambda\right) - \gamma\right]\Big\} \\ &= \max_{\lambda \ge 0} \Big\{ \frac{1}{2}\left[\lambda^T M H^{-1}\left(g + M^T\lambda\right) + g^T H^{-1}\left(g + M^T\lambda\right)\right] \\ &\quad - \lambda^T M H^{-1} g - g^T H^{-1} g + \lambda^T\left[-MH^{-1}g - MH^{-1}M^T\lambda - \gamma\right]\Big\} \\ &= \max_{\lambda \ge 0} \Big\{ \frac{1}{2}\lambda^T M H^{-1} g + \frac{1}{2}\lambda^T M H^{-1} M^T \lambda \\ &\quad + \frac{1}{2} g^T H^{-1} g + \frac{1}{2} g^T H^{-1} M^T \lambda - \lambda^T M H^{-1} g - g^T H^{-1} g \\ &\quad - \lambda^T M H^{-1} g - \lambda^T M H^{-1} M^T \lambda - \lambda^T \gamma \Big\}. \end{aligned}$$

And finally, we achieve:

$$J^* = \max_{\lambda \ge 0} \left\{ -\lambda^T M H^{-1} g - \frac{1}{2}\lambda^T M H^{-1} M^T \lambda - \frac{1}{2} g^T H^{-1} g - \lambda^T \gamma \right\}.$$

7.C *Hildreth's algorithm*

Hildreth's quadratic programming procedure is a simple variant of the Gauss–Seidel method for solving a linear set of equations. Gauss–Seidel is an iterative technique for solving equations of the general form:

$$Ax = b,$$

for a square matrix A. It proceeds by decomposing A as:

$$A = L_A + U_A,$$

where L_A is a lower triangular matrix and U_A is strictly upper triangular. By doing so, we can write:

$$\begin{aligned} Ax &= b \\ (L_A + U_A)x &= b \\ L_A x &= -U_A x + b. \end{aligned}$$

Gauss–Seidel then solves for x on the left-hand side using previously computed values of x on the right-hand side:

$$x^{(k+1)} = L_A^{-1}\left(-U_A x^{(k)} + b\right).$$

By exploiting the triangular nature of L_A and U_A we avoid inversion and compute the elements of x by substitution,

$$x_i^{(k+1)} = \frac{1}{a_{ii}}\left[-\sum_{j=1}^{i-1} a_{ij}x_j^{(k+1)} - \sum_{j=i+1}^{n} a_{ij}x_j^{(k)} + b_i\right].$$

The Hildreth procedure utilizes this method to solve for the individual elements of λ that satisfy the dual optimization problem,

$$\min_{\lambda \geq 0} J = \left(\frac{1}{2}\lambda^T P\lambda + \lambda^T k + \frac{1}{2}g^T H^{-1} g\right).$$

The stationarity condition is obtained from:

$$\frac{\partial J}{\partial \lambda} = P\lambda + k = 0,$$

which gives the linear matrix equation:

$$P\lambda = -k.$$

In particular, the method computes the individual elements λ_i by setting $\frac{\partial J}{\partial \lambda_i} = 0$ while holding all other components fixed at their previously obtained values. Positive λ_i values are retained while negative values are replaced by $\lambda_i = 0$. The Hildreth procedure is computed according to the following algorithm:

$$\lambda_i^{(m+1)} = \max\left(0, w_i^{(m+1)}\right),$$

where:

$$w_i^{(m+1)} = -\frac{1}{P_{ii}}\left[\sum_{j=1}^{i-1} P_{ij}\lambda_j^{(m+1)} + \sum_{j=i+1}^{n_\lambda} P_{ij}\lambda_j^{(m)} + k_i\right],$$

for $i = 1, 2, \ldots, n_\lambda$. Here, P_{ij} is the ijth element of the matrix $P = MH^{-1}M^T$, k_i is the ith element of the vector k, and n_λ is the length of λ.

Some features of the algorithm:

- λ is systematically varied one component at a time and the λ_i elements chosen to minimize the objective function.
- The objective function can be regarded as a quadratic function in each λ_i component.
- Previously calculated values of λ_i are incorporated into subsequent calculations.
- The Hildreth procedure, when convergent, approaches the solution asymptotically.
- If constraint matrix M has full row rank, then P is positive definite and the problem has a unique solution.

8

Computing Dynamic Power Limits

In the final chapter of this book we present another important application of advanced battery management that leverages dynamic models and predictive methods: power limit estimation. Among the most important functions of a BMS is the management of energy within the battery pack during vehicle operation. Two interrelated quantities exist to facilitate this task: (1) state of energy (SOE), and (2) state of power (SOP). Definitions for these two quantities are not hard and fast, and some ambiguity persists in their application. In this chapter we will attempt to bring some clarity to this discussion and illustrate one important example of using a predictive framework with physics-based models to generate online estimates of SOP.

One final look at the roadmap of Fig. 8.1 shows that our journey is nearly complete. We have developed an identifiable version of the dynamic equations in the form of a lumped-parameter DFN model, derived linearized transfer functions for cell internal electrochemical variables, and generated a full-cell impedance model. This formulation served as the basis for a model order reduction technique giving rise to a computationally compact ROM harboring key electrochemical dynamic information. In Chapter 7 we applied MPC methods to obtain an optimizing fast-charge profile guided by underlying electrochemical limits. In this chapter, we extend the fast-charge concept to develop a physics-based approach to battery SOP estimation.

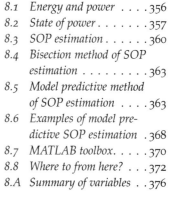

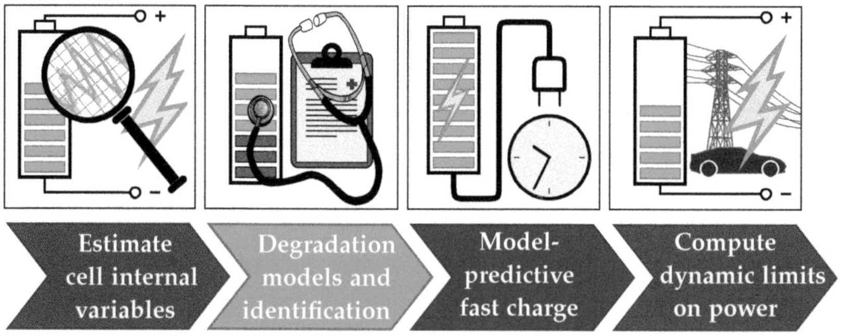

Figure 8.1: Topics in lithium-ion battery controls that we cover in this volume.

Similar to SOC, both SOE and SOP (in their various definitions) are not directly measurable with sensors, and thus can only be estimated from available externally defined quantities. Volume II of this series outlined a method for SOP estimation using ECMs and provided general background on SOP methods. It is not the intent of this chapter to duplicate that material; instead, the interested reader is encouraged to review Volume II to gain a broader view of the topic. Here, we shall build on the earlier development to construct a comprehensive framework for understanding SOP and then introduce a physics-based application in line with the topic of this book.

8.1 *Energy and power*

Volume II introduced the idea that careful management of energy flow into—and out of—a battery is of paramount concern to a BMS. Recall that energy is defined as the ability of the battery to do work, and is a total quantity typically measured in Wh or kWh for vehicle applications. A measure of remaining total energy defines a fuel gauge equivalent for a BEV and relates to the vehicle's available driving range. A state of energy measure is often used to capture this quantity. In general practice, however, SOC is used often instead as it trends in the same sense as SOE and is easier to estimate. Although they are fundamentally related, the relationship that links SOC with SOE is not straightforward. In a mathematical context, SOC is obtained via the integral of applied current, whereby SOE is computed using the integral of the product of applied current and battery terminal voltage, which complicates its evaluation.[1] Nonetheless, a variety of methods exist for estimation of SOE.[2]

It is important to note that total energy is to be distinguished from available energy, in that the latter takes into account the manner in which the energy is removed from the battery. We mention this here because it provides a direct connection to the idea behind state of power. Whereas energy content defines useable range, it is the rate at which this energy is transported which ultimately influences a vehicle's performance. The time-rate-of change of energy is defined as power, and it is on this topic we will focus the remainder of this chapter.

Since energy within a battery is stored as charge, power thus relates to the movement of charge into and out of a battery cell, and this dynamic is governed by its time-dependent current-voltage behavior. When we speak of a high-power cell, for example, we generally mean a battery where current can be drawn off (or inserted) at a high rate, which when multiplied by its instantaneous voltage, delivers a correspondingly high value of power. The converse is true for

[1] More precisely, total energy is related to SOC as $E = \left[\int_{z_{\min}}^{z} OCV(\xi)d\xi\right] \times Q$ for a single cell. It is more complicated to compute for a battery pack (see Vol. II) and is not equivalent to available energy.

[2] See, for example, Guangzhong Dong, Xu Zhang, Chenbin Zhang, and Zonghai Chen, "A method for state of energy estimation of lithium-ion batteries based on neural network model," *Energy*, 90:879–888, 2015.

the case where transport of charge is inhibited (by increased internal resistance, for example).

Estimating available power on charge and discharge enables a vehicle controller to manage battery utilization wisely for efficient operation of BEVs and HEVs.[3] Accurate estimation of SOP over a future time horizon assists with planning for acceleration and regenerative braking requirements, among other operational demands. The following sections outline a practical definition of SOP and present a physics-based model-predictive method for its computation.

[3] Vital van Reeven and Theo Hofman, "Multi-level energy management for hybrid electric vehicles—part i," *Vehicles*, 1(1):3–40, 2019.

8.2 *State of power*

SOP is generally defined to mean how much power—on charge or discharge—can be sustained over a specified time period into the future.[4] Sometimes termed state of available power (SOAP)[5] the term lends itself to a variety of interpretations in the literature. In some cases, it is used to refer to a *peak* power level available at a given moment in time. More generally it is used in a predictive sense, as in the definition introduced above. The premise here is that we wish to supply an upcoming load demand (e.g., accelerate, climb a hill, or supply charge from regenerative braking) and must inform the vehicle controller of what it can safely expect from the battery at its present state. This is a critical element in HEV energy utilization decisions and is equally important in BEVs to ensure the battery is operated in a safe manner. Thus in our development we shall define SOP as the maximum *constant* power level that can be supplied by or introduced to a battery cell over a defined power horizon, which we term ΔT. We shall next introduce a visual representation of SOP to facilitate our understanding of these terms.

[4] Volume II employed the term *power limits* for this quantity; here we synonymously adopt the widely used term state of power to describe the same metric.

[5] Wladislaw Waag, Christian Fleischer, and Dirk Uwe Sauer, "Adaptive on-line prediction of the available power of lithium-ion batteries," *Journal of Power Sources*, 242:548–559, 2013.

8.2.1 *Voltage-current plot*

It is insightful to generate simplified SOP estimates based on standard current profiles such as constant-current (CC), constant-voltage (CV), or constant-current-constant voltage (CCCV). It is important to note that these cases violate the constant power (CP) ideal and SOP estimates made using these will generally depart from the true value. This is easy to see as follows. A nonzero constant applied current brings about an inevitable change in voltage due to the fact that charge is either entering or leaving the cell electrodes and the SOC-OCV relationship is monotonic. Conversely, using the same reasoning, maintaining constant voltage requires a changing value of applied current (or zero current). Therefore, it is not possible to achieve constant power while holding either the voltage or the applied current constant; in fact, both must change simultaneously. Thus, any SOP calculations based on the simplified assumption of either constant current or constant voltage are necessarily conservative.

To illustrate the impact of these assumptions and provide a useful framework for understanding general SOP behavior, we employ a simple voltage-current (VI) plot, where current is displayed on the abscissa and voltage on the ordinate axis. For visual simplicity, only positive values of current are used; charge or discharge cases will be evident by context.

Each point on the plot represents an instantaneous power value computed as the product $P(t) = v(t) \times i(t)$. Contours of constant power are drawn for reference. A notional V-I plot is shown in Fig. 8.2. The purple trace indicates the trajectory of instantaneous power for a cell discharge event starting at an equilibrium voltage, V_0. Sudden application of discharge current I_1 brings about an instantaneous drop in cell voltage to V_1 due to ohmic resistance; the absolute value of the slope is equal to the ohmic resistance value, R_0. The corresponding instantaneous power level is $P_1 = V_1 \times I_1$. In this example, the discharge current is controlled so as to maintain a constant power level—denoted by alignment with the constant power contour traversed between points $P_1 = V_1 \times I_1$ and $P_2 = V_2 \times I_2$.

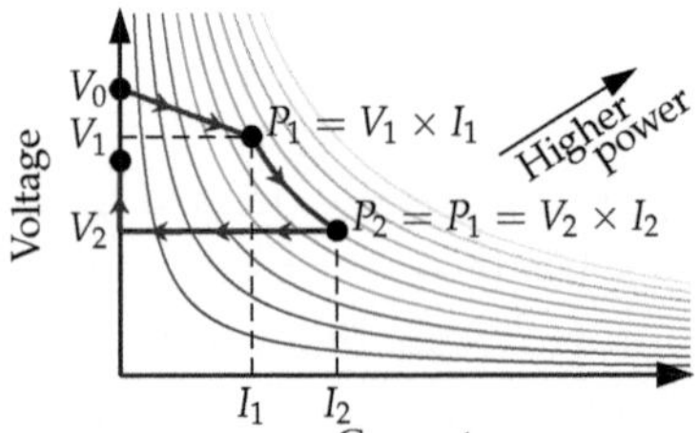

Figure 8.2: Voltage-current (V-I) plot.

8.2.2 *Constant-current-constant-voltage discharge*

We first demonstrate a long duration discharge power trajectory for a battery cell where maximum and minimum limits are imposed on both current and voltage and a CCCV discharge strategy is employed. Starting from an initial SOC $z(0)$, maximum instantaneous power is achieved immediately at the maximum applied current level I_{max} (we assume the cell is initially at equilibrium with $i(0) = 0$). Assuming we reach I_{max} without encountering any other limit, then in order to sustain this power level into the future we need to maintain a constant voltage at the applied current I_{max}. But since OCV is monotonically decreasing with SOC, voltage must drop for a sustained current level. This means that our trajectory moves downward on the VI plot and instantaneous power continuously decreases until reaching the lower voltage limit V_{min}.

Another possibility is that we reach the SOC bound z_{min} before reaching V_{max}. In this case, $i(t)$ will immediately drop to zero and the voltage will relax to equilibrium. Fig 8.3 shows a representative VI plot depicting a simulation result for the instantaneous discharge power behavior of a 3 Ah 18650 NMC cell. Starting at an initial SOC of $z(0) = 0.6$ and corresponding equilibrium voltage of 3.83 V, a maximum rated discharge current of 10 A is continuously applied

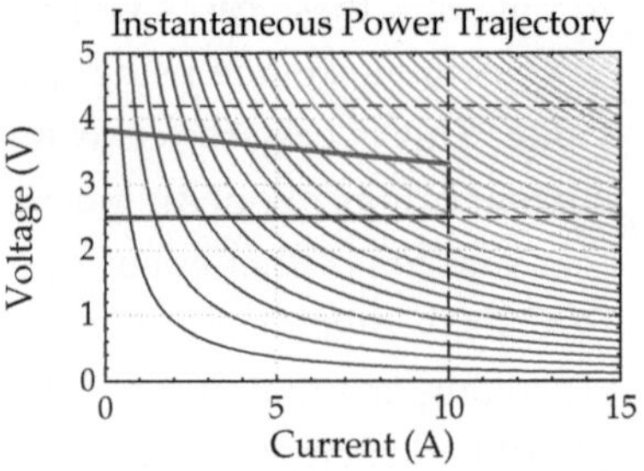

Figure 8.3: Discharge power V-I plot for CCCV.

until the minimum voltage limit of 2.5 V is reached. The current then reduces to maintain the lower voltage bound as just described. Note that for this example instantaneous maximum power $P_{\text{max}} = 33.1\,\text{W}$ is achieved immediately upon application of maximum current, after which instantaneous power continuously decreases to $P_{\text{inst}} = 25.0\,\text{W}$ when the lower voltage bound is reached. This value is simply $P = V_{\text{min}} \times I_{\text{max}}$. Upon reaching V_{min} the current continuously decreases to zero proportionate with the instantaneous power.

Figures 8.4, 8.5, and 8.6 show current, voltage, and instantaneous power plots corresponding to the CCCV discharge case presented above. We note here that constant discharge current results in decreasing voltage and a correspondingly decreasing power level, implying that the CC assumption cannot bring about a sustained constant power level and that power limit estimates derived with this assumption will exhibit conservatism, especially at longer power horizons.

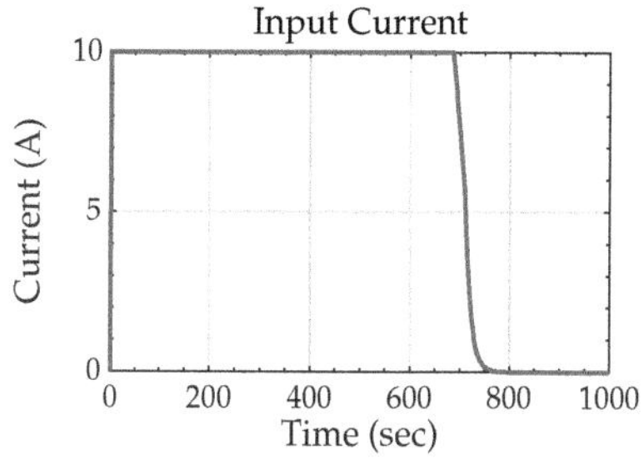

Figure 8.4: Discharge power input current for CCCV.

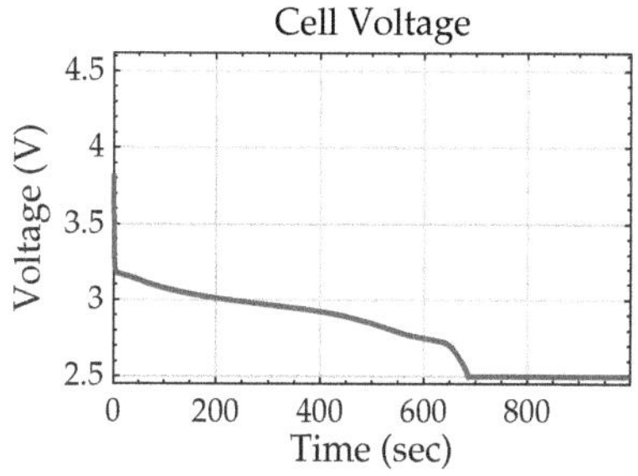

Figure 8.5: Discharge power cell voltage for CCCV.

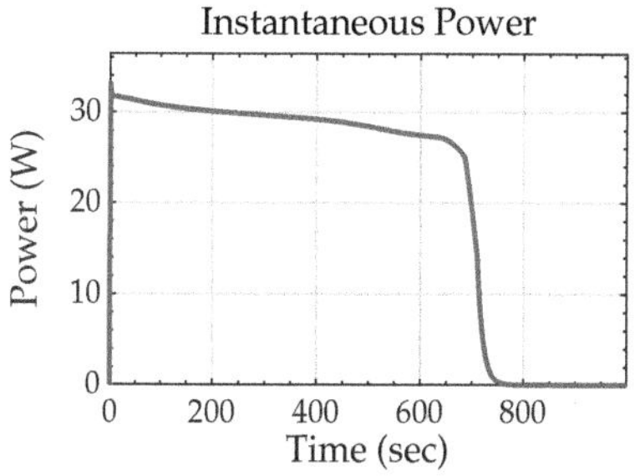

Figure 8.6: Discharge instantaneous power for CCCV.

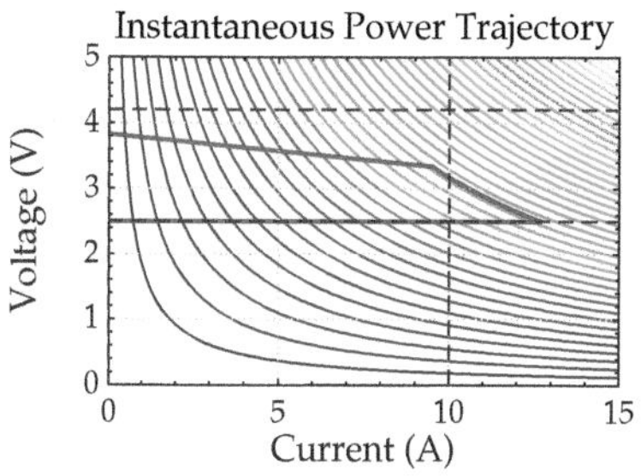

Figure 8.7: Discharge power V-I plot for CPCV.

8.2.3 *Constant-power-constant-voltage discharge*

We next demonstrate a maximum discharge power trajectory where once again maximum and minimum limits are imposed on both current and voltage. This time a constant-power-constant voltage (CPCV) strategy is employed. Figure 8.7 depicts the corresponding V-I portrait where it is easy to see that the CPCV strategy follows a constant power contour until reaching a specified limit. In this case, both I_{max} and V_{min} are encountered at the same time, after which the CV current is applied. We note here that the maximum constant power derived SOP value is 25.0 W, which is the same as the minimum instantaneous power value obtained at I_{max} in the CCCV case.

Figures 8.8, 8.9, and 8.10 show current, voltage, and instantaneous power plots corresponding to the CPCV discharge case presented above. It is clear from this simulation that constant power generally requires both current and voltage to change in opposition to one another, thus reaffirming that assumptions of CC (or CV) made to simplify SOP calculation will normally result in conservative SOP estimates.

The preceding examples suggest a surprisingly simple calculation for the maximum constant discharge power over a long duration power horizon (i.e., one measured in minutes), namely, $P = V_{\text{min}} \times I_{\text{max}}$. Let's explore this situation a bit further. We've assumed for our examples that the only relevant limits are imposed on cell voltage and discharge current (and, practically speaking, SOC). Furthermore, we assumed we could achieve the maximum current level without encountering any other limits. Note, however, that if the current limit is high enough, the initial cell voltage low enough, and/or the cell's internal resistance is high enough, then $V_{\min}$ may be encountered well before reaching $I_{\max}$. We can picture this as a steepening of the initial slope in the V-I trajectory (which is proportional to cell resistance) causing the plot to intersect the $V_{\min}$ bound prior to reaching the $I_{\max}$ boundary. This of course results in an immediate curtailment of instantaneous power which cannot be held constant for any length of time without violating the voltage limit. In this case, once $V_{\min}$ is reached, the current (and the power) must continue to decrease in order to maintain the constant minimum voltage value.

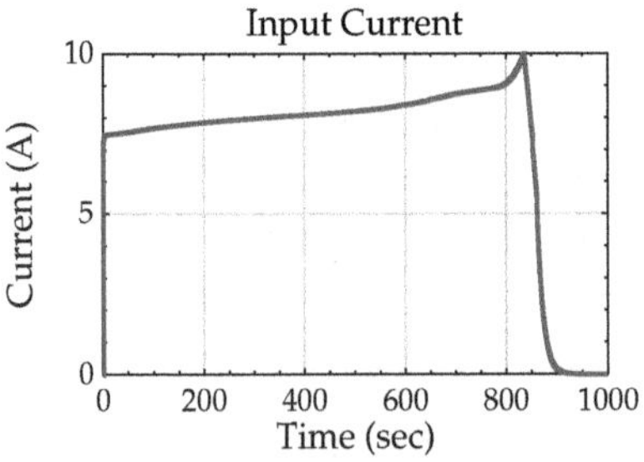

Figure 8.8: Discharge power input current for CPCV.

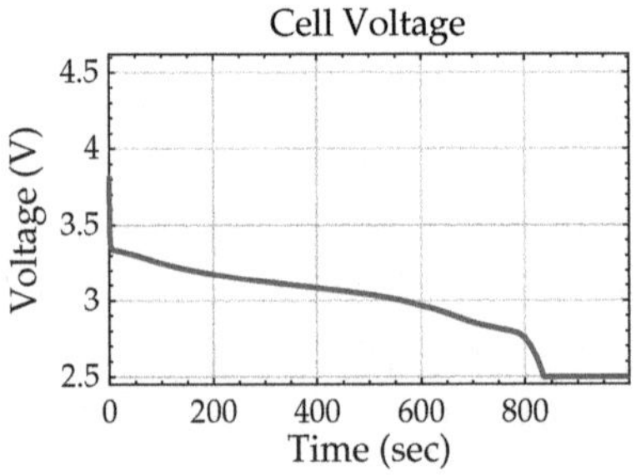

Figure 8.9: Discharge power cell voltage for CPCV.

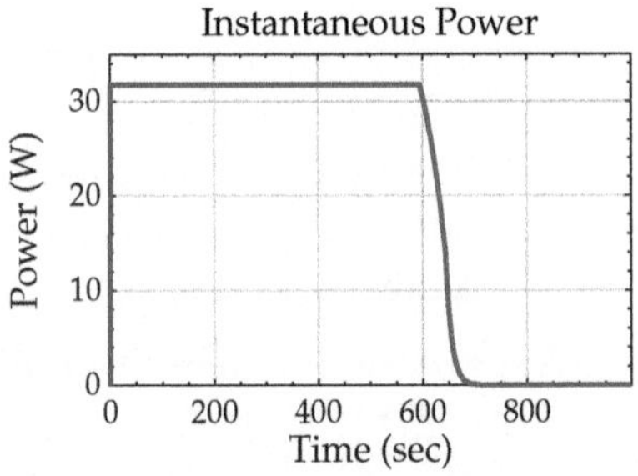

Figure 8.10: Discharge instantaneous power for CPCV.

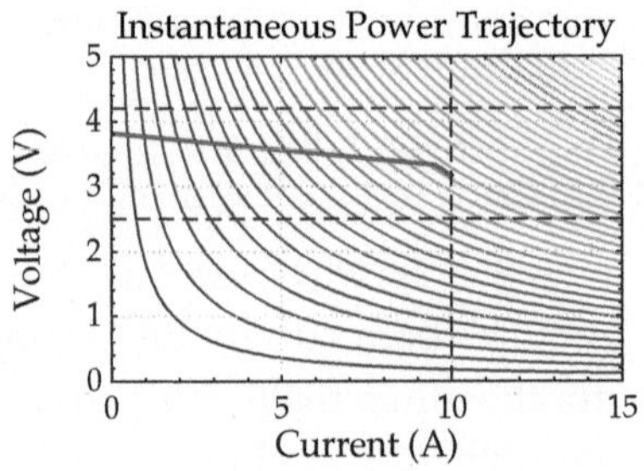

Figure 8.11: Discharge power V-I plot: short duration.

In most vehicle applications we are not interested in long duration horizons, but typically consider power prediction on the order of seconds, with 10 to 20 sec a commonly used range for xEVs. Sustaining power over a short horizon means we are generally less likely to encounter a lower voltage limit on discharge, especially at high values of initial SOC. In this case, we must consider the upper current limit as the bounding event for maximum sustained power. Turning again to the V-I plot, we seek a trajectory that intersects the largest constant power contour it can follow prior to encountering the maximum current limit within that specified time interval. An example is shown in Fig 8.11. Clearly this becomes a rather complicated calculation to make and one can quickly appreciate the desire to find suitable approximations.

Our discussion so far suggests that cell internal resistance will ultimately govern short-term power availability for a battery cell. However, we have purposely oversimplified the problem by only placing bounds on external measurable quantities cell voltage and applied current. In reality, the issue is far more complicated.

8.3 *SOP estimation*

Practically speaking, it is not straightforward to compute SOP as defined here since it involves a constrained nonlinear optimization. Most authors instead resort to alternative formulations of SOP, which are easier to obtain yet characterize the true SOP in some meaningful way. In some applications, it may be important to predict peak power, which we define here as the maximum instantaneous value of power attainable over a future time period. This is distinctly different from our definition of SOP; nevertheless we note here that the maximum constant power over ΔT in fact approaches the maximum instantaneous power as $\Delta T \to 0$. This idea is illustrated in Fig 8.12.

So far, we have discussed a basic approach for maximizing instantaneous power, whereas our ultimate interest is in maximizing predicted power. In the sections that follow, we shall leverage this paradigm as a basis for formulating predictive power limit estimation strategies and comparing their behavior. State of power cannot be measured by direct sensing and therefore must be determined from measurable information, normally voltage, current, and temperature. Since the measurements themselves are considered fixed, we are left to select: (1) a mathematical model, and (2) a computational method that we shall use to determining our SOP estimate. Fig. 8.13 illustrates this idea.

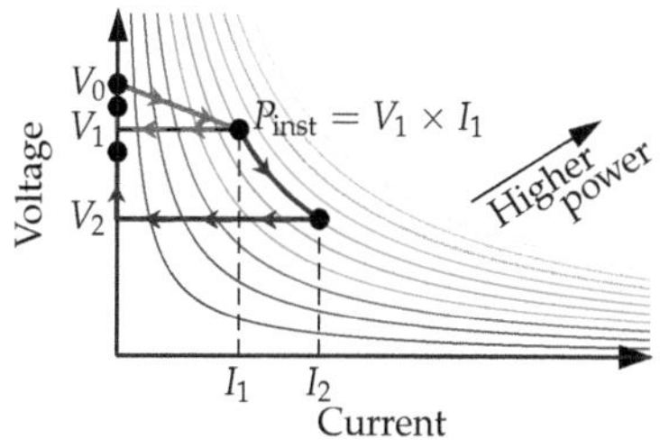

Figure 8.12: V-I plot for instantaneous power.

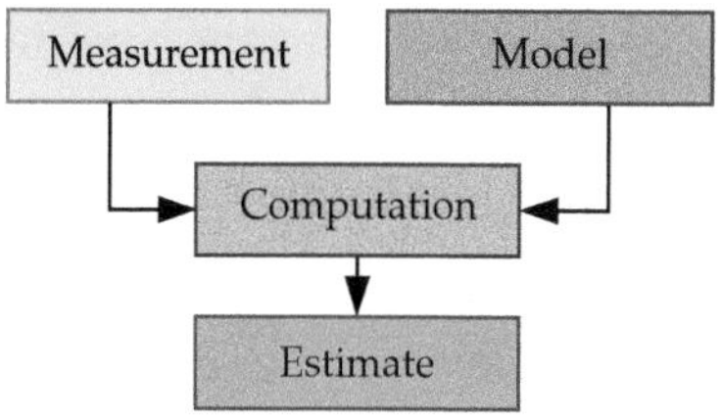

Figure 8.13: SOP estimation method: basic approach.

The following development shall make use of the following notational conventions:

$v[k]$	Cell terminal voltage at time k
$z[k]$	Cell state-of-charge at time k
$P[k]$	Cell instantaneous power at time k
$i[k]$	Cell applied current at time k

For ease of plotting, we shall occasionally refer to the maximum value of both charge or discharge current as the positive value I_{max}. In other words, we shall treat the charge/discharge current as its absolute value. The basic idea behind the computation of dynamic power limits can then be stated as follows: Starting at time index k, compute a value $P_{limit}[k]$ according to the predictions of $v[k]$ and $i[k]$ over the time interval defined by $\{k, k+1, \ldots, k+k_{\Delta T}\}$, where $k_{\Delta T}$ is the number of discrete-time time steps in real-time interval ΔT seconds. Since instantaneous power is computed as $P[k] = v[k] \times i[k]$, a computation of maximum power is ultimately limited by bounds on certain cell-level quantities that may be adversely affected by the rapid movement of current. In this sense, the maximum power problem resembles that for fast charge. For simplicity, we shall only consider here limits imposed on: current, voltage, and state-of-charge; these are of course relevant for the case of ECMs. Nonetheless, the concepts easily generalize to the PBM case with inclusion of hard bounds on other electrochemical quantities of interest. We shall illustrate the idea of maximum power by simple example: first examining the case of instantaneous discharge power, after which we'll address the charge case.

In general, a predictive estimate of available power is preferred over an instantaneous estimate to allow for load-schedule optimization over a near-term horizon. In short, we wish to know how much power can be sustained over a specified time into the future, termed the *power horizon*, which we shall denote mathematically by ΔT. Since the constant power level sustained must be less than or equal to the maximum instantaneous power, the predictive estimate can be considered conservative in some sense. A common value for the power horizon used by the automotive industry is $\Delta T = 10$ sec; we shall use this value in subsequent calculations. We note further that power limits are a dynamic measure—a function of the evolving operational environment of the battery. Thus, power limits must be updated periodically in order to supply timely and useful load-planning information—this update rate is typically more frequently than once every ΔT. Hence, our computation methodology must include a prediction component that can supply needed model variables out to the length of the power horizon. Volume II introduced one very effective method that utilized a state-space equivalent circuit model (ECM) and a polytopic[6] set of hard constraints to compute power limits. We shall term this the bisection method for power limits estimation (PL-BSM). PL-BSM assumes a constant level of input current ΔT seconds into the future for its estimates; we shall review the essential elements of this method in the next section.

[6] The term *polytopic* refers to constraints that appear as a set of parallel hyperplanes in the parameter space.

While effective, we might wonder if PL-BSM is perhaps a bit too conservative in that it fails to consider the dynamic nature of the predicted response within the power prediction window. To that end, we shall investigate an alternative method (PL-MPC) that makes use of the full dynamic profile and compare this to PL-BSM. Furthermore, PL-BSM considered only empirical ECMs and relied mostly on voltage-based performance limits. So again here we may wonder if a physics-based model (PBM), harboring electrochemical parameter information, might bring about a more accurate power limits estimate. We'll introduce this improved PL-BSM approach in this chapter. Finally, we'll combine the PBM with PL-MPC to explore the potential of computing dynamically predictive power limits governed by true electrochemical quantities.

Before proceeding, we recall a few definitions:

- Discharge power: maximum discharge power that may be maintained constant for ΔT seconds without violating design limits.
- Charge power: maximum battery charge power that may be maintained constant for ΔT seconds without violating design limits.

We note that both quantities are based on the present battery state. In the material to follow, we shall develop power limits for the single-cell configuration; the methods are readily adapted to obtain pack-level estimates as discussed in Volume II.

8.4 Bisection method of SOP estimation

By way of brief review, this method of SOP estimation relies on a general state-space model of battery dynamics,

$$\begin{aligned} x[k+1] &= f(\boldsymbol{x}[k], i[k]) \\ v[k] &= h(\boldsymbol{x}[k], i[k]), \end{aligned}$$

to compute a predicted battery state at a single fixed time index $k + k_{\Delta T}$. Assuming a constant current is applied throughout the duration of the power horizon (i.e., from time index k to $k + k_{\Delta T}$), the method computes the appropriate constant current value ($i_{\text{max}}^{\text{dis,volt}}[k]$ and $i_{\text{min}}^{\text{chg,volt}}[k]$) that brings the predicted terminal voltage to a prescribed limit (v_{min} or v_{max}) ΔT seconds in the future. This is accomplished by seeking the $i[k]$ value that brings about equality in:

$$\begin{aligned} v_{\text{min}} &= h(\boldsymbol{x}[k+k_{\Delta T}], i[k]), \text{ or} \\ 0 &= h(\boldsymbol{x}[k+k_{\Delta T}], i[k]) - v_{\text{min}} \end{aligned}$$

for $i_{\text{max}}^{\text{dis,volt}}[k]$, and the $i[k]$ that brings about equality in:

$$\begin{aligned} v_{\text{max}} &= h(\boldsymbol{x}[k+k_{\Delta T}], i[k]), \text{ or} \\ 0 &= (\boldsymbol{x}[k+k_{\Delta T}], i[k]) - v_{\text{max}} \end{aligned}$$

to find $i_{\text{min}}^{\text{chg,volt}}[k]$. Solving these equations for the limiting $i[k]$ involves finding the root of a nonlinear equation, which is accomplished using the bisection search algorithm. SOC-based limits can be computed in a similar fashion using a simplified dynamic model. The interested reader is directed to Volume II, Chap. 6 for additional details.

The resultant products, $p_{\text{min}}^{\text{chg}}[k] = i_{\text{min}}^{\text{chg}}[k] \times v[k+k_{\Delta T}]$ and $p_{\text{max}}^{\text{dis}}[k] = i_{\text{max}}^{\text{dis}}[k] \times v[k+k_{\Delta T}]$ give the SOP estimates for charge and discharge, respectively.

8.5 Model predictive method of SOP estimation

Previously, in Volume II, SOP estimates were computed using ECMs together with designated limits on current, voltage, and SOC. Although a carefully identified ECM can give a very accurate picture of battery cell input-output voltage (and SOC) behavior, it cannot provide useful information about internal electrochemical properties that determine the genuine boundaries of performance. Consequently, ECM limits may be considered only indirect indicators of true physics-based limits, as was the case for fast charge. Thus, power limit estimates based solely on ECM prediction models (and current, voltage, and SOC bounds) will not, in general, be exact bounds.

Chapter 7 applied model predictive methods to a physics-inspired dynamic model in order to obtain an optimizing fast-charge profile guided by underlying electrochemical limits. In the following sections, we shall leverage the same fast-charge concept to develop a physics-based approach to battery SOP estimation.

8.5.1 *Shortcomings of the bisection method*

The bisection method utilizes a constant-current assumption over the power prediction interval in its computation of power limit estimates; this necessarily gives rise to conservative approximations. This is easy to see if we consider how the action of the bisection method appears on a V-I diagram, as shown in Fig. 8.14 for the charge case of Fig. 8.15 for the discharge case. First, under the assumption of constant current for the duration of ΔT, it is not possible to maintain constant voltage during the same interval simultaneously, meaning that the respective bounds v_{min} and v_{max} are enforced only at the end of the power prediction window. The resulting SOP estimate is equal to the power contour level intersected by the trajectory corner point marked $P = V_{min} \times I_{\text{max}}^{\text{dis}}$ in Fig. 8.15.[7]

An alternative method inspired by model predictive control (MPC) instead considers the full dynamic nature of the predicted trajectories. Here it is conjectured that by taking account of time-varying predictions throughout the power horizon time interval, a more accurate power limit approximation may be obtained. Next, we present the MPC method of power limit estimation.

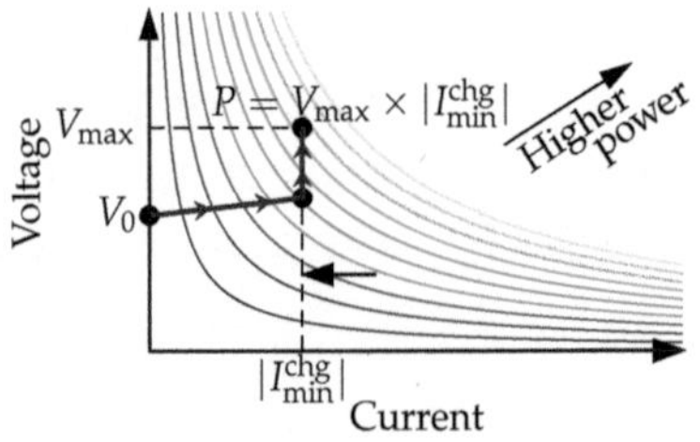

Figure 8.14: V-I plot for instantaneous charge power.

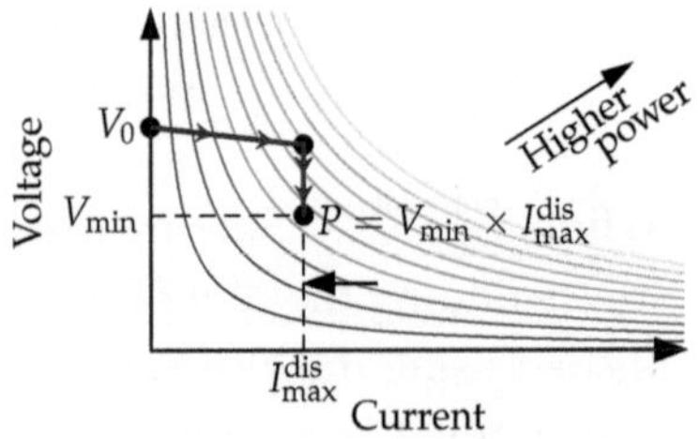

Figure 8.15: V-I plot for instantaneous discharge power.

[7] Note that the horizontal left-pointing arrows in the figures indicate the value of constant current that the bisection algorithm seeks to optimize.

8.5.2 *Basic idea*

Chapter 7 introduced MPC as a real-time computational control method that can predict dynamic behavior, optimize to a specified performance criterion, and enforce hard constraints on problem variables. There, we applied MPC as a control method to compute an optimal fast-charge profile, while enforcing hard constraints on internal electrochemical variables, thereby eliminating the conservatism inherent using voltage limits alone. Inspired by this result, it is natural to ask here whether we can adopt a similar methodology to the problem of power limits estimation; and therrefore the remainder of this section will endeavor to address this question. As with PL-BSM previously, in this approach, we shall compute an estimate of available charge and discharge power over a finite-time power horizon, ΔT. The method executes a constrained MPC algorithm at each sample point to predict an optimal future power profile and then numerically integrates the time-varying instantaneous power to compute total energy. It then averages the total energy over a finite time horizon to provide an estimate of maximum sustainable power. It is important to note here that the MPC algorithm is used here not to control the battery, but rather to inform the BMS.

The PL-BSM approach to power limit estimation, as seen previously, relies on a straight-line prediction of cell dynamics over a finite time horizon. Fig. 8.16 depicts this situation. Here, the dynamic predictor uses the system model to compute the instantaneous power $P[k] = v[k] \times i[k]$ at time value $k + k_{\Delta T}$ such that it adheres to specified voltage (and perhaps other) hard limits. This value is then used directly as an estimate of maximum sustainable power.

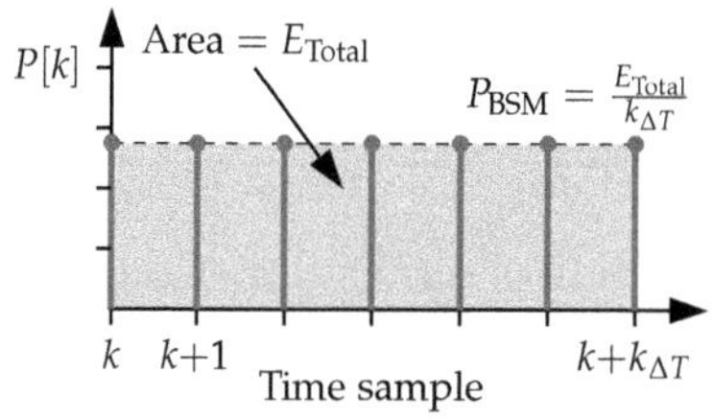

Figure 8.16: Bisection SOP approach.

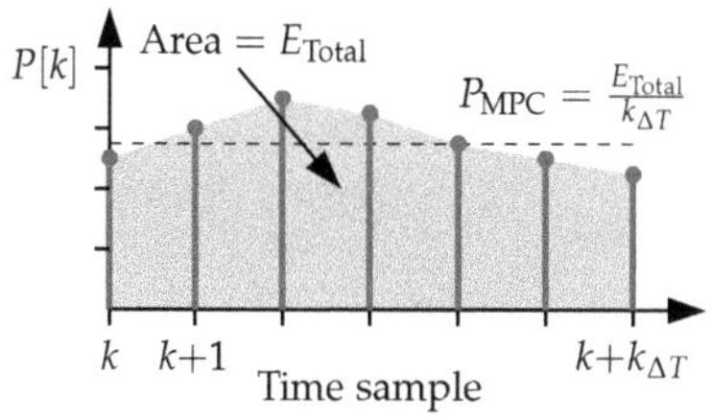

Figure 8.17: Model predictive SOP approach.

In other words, the PL-BSM method considers the area of the shaded rectangle in the figure as the total charge-discharge energy estimate over the power horizon, and distributes this energy at a constant power level, $P_{\text{BSM}} = E_{\text{Total}}/k_{\Delta T}$. By contrast, the MPC-inspired method uses a dynamic predictor—together with step-wise constrained optimization—to compute an optimal fast-charge/discharge profile that adheres to hard constraints. Fig. 8.17 depicts this new situation.

The optimal profile of instantaneous power values is numerically integrated to obtain a computation of total energy. This value reflects the largest amount of energy that can either be inserted or removed from the battery cell over the specified power horizon while respecting constraints. Taking the average of this total energy value gives an estimate of the maximum sustainable power than can be inserted/extracted from the battery upon charge/discharge respectively. It is interesting to note that in this sense, PL-MPC gives a near-optimal *instantaneous power profile* for the duration of ΔT. This means the maximum charge/discharge energy computed can be achieved by following the prescribed current profile. It is not, however, a tight bound on the sustainable energy level; this can be seen be seen by examining some simulation results.

8.5.3 *Model predictive SOP estimation*

Model predictive control was introduced in Chap. 7 in the context of the battery fast-charge problem. SOP estimation will employ precisely the same MPC computation used to solve the pseudo-minimum-time problem for fast charge, only in this case the resulting optimal input values are not implemented for control, rather they are used strictly as input to the SOP estimation procedure. The algorithm outlined here executes one optimal constrained MPC calculation at each time instant for which an SOP estimate is desired. For consis-

Algorithm 1 MPC SOP estimation.

- Initialization
  - Define initial augmented state-space model matrices $\widetilde{\boldsymbol{A}}_0$, $\widetilde{\boldsymbol{B}}_0$, $\widetilde{\boldsymbol{C}}$
  - Select MPC tuning parameters: N_p, N_c, and ρ
  - Compute corresponding initial MPC prediction matrices, $\boldsymbol{\Phi}_0$, $\boldsymbol{G}_0$
- Power prediction loop
  - For simulation sampling time $k = 1, \ldots, N_\mathrm{sim}$:
    * Update time-varying matrices ($\widetilde{\boldsymbol{A}}_k$, $\widetilde{\boldsymbol{B}}_k$, $\boldsymbol{\Theta}_k$, $\boldsymbol{G}_k$)
    * Update constraint bounds via $\boldsymbol{M}_k$, γ_k
    * Compute optimal input sequence, $\left\{\boldsymbol{\Delta}\overrightarrow{u}[k]\right\}$ which satisfies constrained fast-charge or fast-discharge objective over power horizon
    * Compute instantaneous power as $P[k] = v[k] \times i[k]$, for $k = 1 : k_{\Delta T}$
    * Compute numerical integral of $P[k]$ over the interval $k = 1 : k_{\Delta T}$ to obtain total energy, $E[k]$
    * Compute available power $P_\mathrm{chg,min}$ or $P_\mathrm{dis,max}$ at sample time k as $\frac{E[k]}{(k_{\Delta T} \times T_\mathrm{s})}$
    * Propagate state equations

tency with the predictive SOP framework, the prediction horizon index N_p is selected so that the MPC time horizon $N_\mathrm{p} \times T_\mathrm{s}$ is less than the power horizon ΔT. Of course, this is not a requirement, but rather a convenient choice that promotes computational efficiency.[8]

The MPC-based power estimation algorithm is outlined in Algorithm 1.

[8] It is of course possible to design alternative MPC strategies for this computation which use different horizon lengths (e.g., a shrinking horizon). Numerical experiments show some advantages may be gained, but overall, results are similar.

8.5.4 *Adaptive input weighting*

One challenge associated with implementing the MPC method of SOP estimation regards the selection of the input weighting factor, ρ. For a stand-alone optimization problem, there exist ad hoc rules (along with trial and error) that give a suitable value for the input weighting factor. If chosen too small, overly aggressive input actions may result; too large and the response is often sluggish. Furthermore, it has been observed that some values of ρ in constrained quadratic programming problems may lead to slow convergence or worse yet, infeasibility conditions.

Real-time SOP estimation will generally encounter a wide variety of operational conditions that may give rise to very different optimization problems at each SOP sample time. It is impractical to expect that a single value of ρ will be satisfactory in all situations.

In order to address this difficulty, our method incorporates a simple adaptive input weighting strategy designed to ensure the unconstrained input solution (which is supplied to the quadratic program) is approximately of the same size—in a vector norm sense—as the constraint bound.[9] One can think of this idea as analogous to a warm start for the Hildreth algorithm. The conjecture here is that by starting the constrained optimization algorithm close to the constraint boundary, the algorithm will be better behaved and give more consistent constrained solutions. Although this will not be proven here, experience shows good agreement with the conjecture. The algorithm is constructed as follows.

[9] This idea was first introduced in Aloisio Kawakita de Souza, Gregory L. Plett, and M. Scott Trimboli, "A model predictive control-based state of power estimation algorithm using adaptive weighting," in *International Electric Vehicle Symposium and Exhibition (EVS35)*, 2022.

Utilizing the vector-matrix 2-norm we can apply the triangle inequality to express an upper bound of the unconstrained input solution as,

$$\begin{aligned}\left\|\underset{\rightarrow}{\Delta u}^*_{k+1}\right\| &\le \frac{1}{2}\left\|H^{-1}g\right\| \\ &\le \frac{1}{2}\left\|H^{-1}\right\|\left\|g\right\| \\ &\le \frac{1}{2}\left\|\left(G^TG+\rho I_{N_c}\right)^{-1}\right\|\left\|g\right\|. \end{aligned} \tag{8.1}$$

Ideally, we seek a bound on the magnitude of the largest element of the input sequence contained in the vector $\underset{\rightarrow}{\Delta u}^*_{k+1}$, but this is cumbersome. Instead, the vector 2-norm gives us a sensible aggregate measure of input size, is relatively easy to compute, and serves as an adequate basis on which to guide the selection of the input weighting value ρ.[10]

[10] Other norms could be used for computational convenience; the Frobenius norm, defined for a matrix/vector as the square root of the sum of the squares of its elements, is an obvious choice.

In order to obtain an absolute magnitude-based algorithm, we rewrite Eq. (8.1) by equating the right-hand side upper bound directly to the constraint value, yielding,

$$\left\|\left(G^TG+\rho I_{N_c}\right)^{-1}\right\| = \frac{2\Delta u_{\max}\sqrt{N_c}}{\|g\|}$$

where $\Delta u_{\max}$ is the maximum allowable magnitude of the input increment. Rearranging, we have

$$\underline{\sigma}\left(G^TG+\rho I_{N_c}\right) = \frac{\|g\|}{2\Delta u_{\max}\sqrt{N_c}}$$

where $\underline{\sigma}(\cdot)$ denotes the minimum singular value. Solving for ρ we obtain the result:

$$\rho = \frac{\|g\|}{2\Delta u_{\max}\sqrt{N_c}} - \underline{\sigma}\left(G^TG+\rho I_{N_c}\right).$$

In practice, the input weighting factor ρ is recalculated at each sample point where a constrained solution to the quadratic optimization problem is required within the power prediction loop.

8.6 Examples of model predictive SOP estimation

The model predictive-based SOP estimation method is next demonstrated by applying both the BSM algorithm and the previous MPC-EKF code to the same NMC30 cell model used in the fast-charge example of Chapter 7. For these examples, a 9-state ROM (8 electrochemical states and one integrator state) was generated from the NMC30 FOM using the HRA model reduction method introduced in Chap. 4.

Case 1 establishes a baseline of performance by imposing hard limits on maximum applied current and cell terminal voltage. Maximum current is constrained to a magnitude of 150 A for both charge and discharge, which corresponds to a 5 C rate. Maximum and minimum cell voltage limits are is set to 4.2 V and 3.3 V, respectively. The simulation utilizes a sequence of charge-depleting UDDS profiles applied over a total time of 42,000 sec (11.67 hours) as depicted in Fig. 8.18. Accompanying Fig. 8.19 shows the corresponding cell voltage resulting from application of this profile.

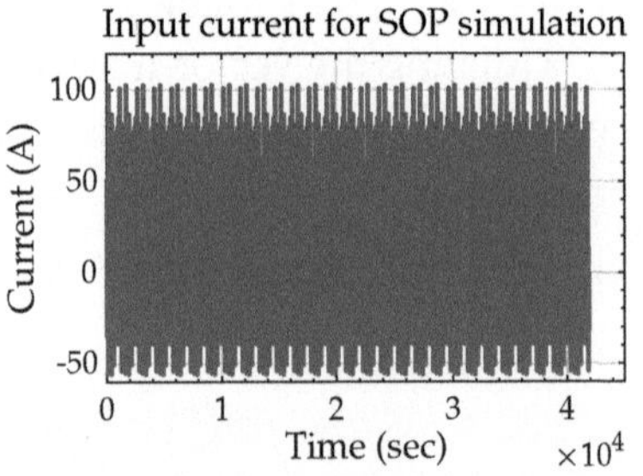

Figure 8.18: UDDS input profile for SOP simulation.

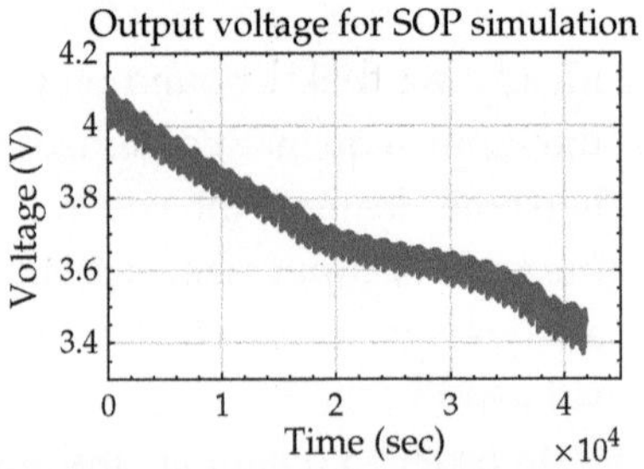

Figure 8.19: Output cell voltage for SOP simulation.

Figure 8.20 shows the resulting SOP estimates computed for a power horizon ΔT of 10 sec on the left and 20 sec on the right, where, consistent with our applied current convention, charge power is negative-valued. In order to ensure compact prediction matrices and efficient computation, MPC horizons of $N_p = 3$ and $N_c = 2$ were selected.[11] As one would expect, available discharge power diminishes with decreasing SOC and terminal voltage. Note that power estimates become increasingly erratic at very low SOC values where linearized model behavior is generally less accurate. Available charge power also exhibits a steady decline aligned with a corresponding decrease in battery voltage levels. In this case, the SOP estimate is essentially the product of the maximum applied charge current and the voltage that can be attained over the power horizon from the given operating point with application of constant maximum current. It's not too surprising that both BSM and MPC algorithms give very similar results for this particular case, generally agreeing within 1 % over most of the simulation. Computed estimates diverge more significantly at the SOC extremes where BSM tends to overestimate the available power. This discrepancy is more apparent at the longer power horizon, where we also can observe a larger decrease in available discharge power as SOC approaches very low values.

[11] Numerical experiments indicated only marginal improvements in the SOP estimate were obtained with larger values of these parameters.

Case 2 demonstrates the viability of using estimates of internal

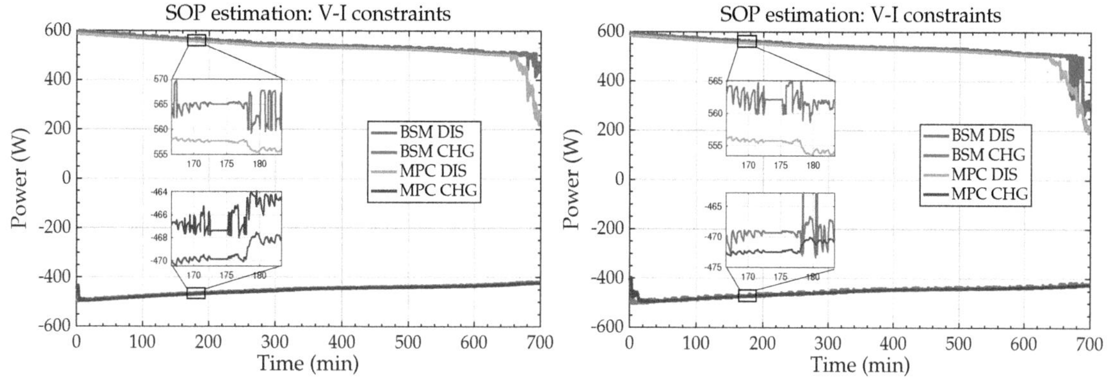

Figure 8.20: Case 1: Constraints on applied current and terminal voltage.

electrochemical states to refine SOP values further. For this set of simulations, the constraint on cell voltage is removed and replaced by constraints on solid surface concentration (θ_{ss}) for the discharge case, and overpotential ($\phi_{s\text{-}e}$) for the charge case. Both quantities are denoted for the negative electrode. Fig. 8.21 shows simulation results for BSM and MPC SOP algorithms computed for the same power horizons used in Case 1 for comparison. The available charge power shows some interesting differences from what we observed in the previous case. Namely, starting around $16,500$ sec, we note a relative decrease in charge power. This is brought about by the algorithm anticipating that the overpotential $\phi_{s\text{-}e}$ is about to reach its lower bound, and risk lithium plating at the anode. As a consequence of this, allowable current levels are decreased in this operating region thus lowering the available power estimate. We further note that the BSM approach struggles to capture this effect correctly since the overpotential value hovers very close to its lower bound. MPC, on the other hand, foresees this impending breach before it occurs and takes protective action.

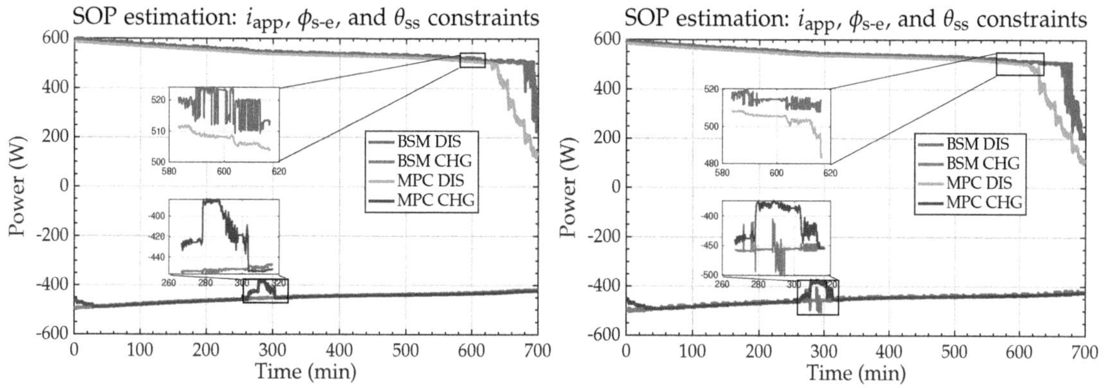

Figure 8.21: Case 2: Constraints on applied current, $\phi_{s\text{-}e}$, and θ_{ss}.

8.7 *MATLAB toolbox*

The SOP algorithms featured in this chapter have also been incorporated in the MATLAB toolbox that accompanies this volume. Flow diagrams depicting the overall structure of both BSM and MPC SOP estimation methods are illustrated in Figs. 8.22 and 8.23, respectively; a simulation implementation is shown for reference.

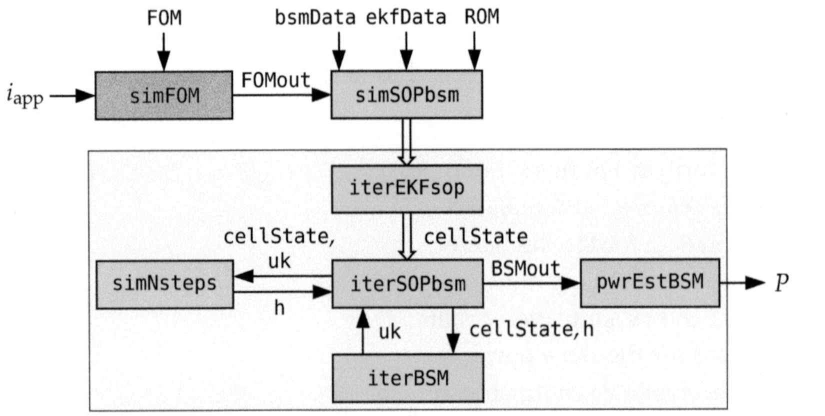

Figure 8.22: SOP BSM.

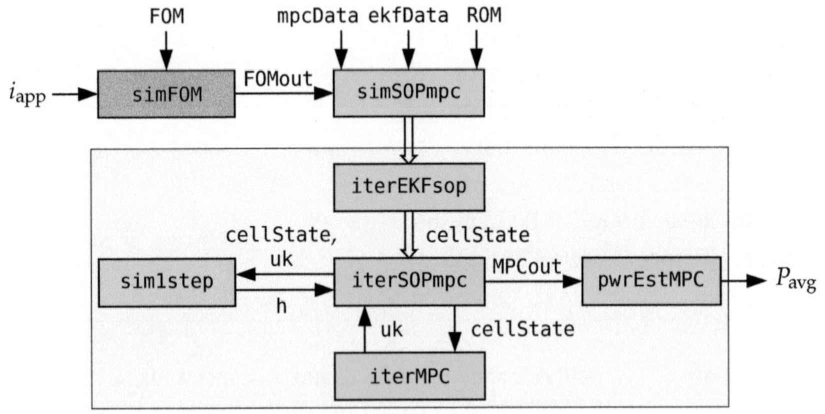

Figure 8.23: SOP MPC.

The first step involves generation of the full-order model output for a designated input profile, i_{app}, accomplished here via the FOM simulation block `simFOM`. This step can be carried out by any of a number of available numerical simulation packages specialized to full-order battery modeling. Output variables contained in the data structure `FOMout` then serve as input to the `simSOPx` routines initiating either BSM or MPC algorithms. Note also that `simSOPx` utilizes data supplied by structures `bsmData`, `ekfData`, as well as the appropriate `ROM`. Function `simSOPx` orchestrates the remainder of the SOP estimation process by first calling a Kalman filter routine to generate estimates of the cell's internal state and provide updates to the state matrices. From this point on, the two methods operate differently, executing the algorithms detailed in the previous sections. The main code for `simSOPbsm` is shown.

```
function [SOP_BSMresults_DIS,SOP_BSMresults_CHG] = simSOPbsm(bsmData,
  FOMout,ekfData,ROM)
%SIMSOBPBSM  This function carries out the steps required to implement the
   bisection method for SOP estimation
% Inputs:
%   bsmData:    data structure containing bisection parameters
%   FOMout:     data structure containing FOM simulation output
%   ekfData:    data structure containing extended Kalman filter output
%   ROM:        data structure defining ROM
% Outputs:
%   SOP_BSMresults_x:   data structure containing BSM SOP results

% Set up preliminaries
Nsim = length(FOMout.Iapp);
bsmData_DIS = bsmData;
bsmData_CHG = bsmData;
initVariables;              % function to initialize data variables

% Run main program
for k = 1:Nsim
   vk = FOMout.Vcell(k);    % cell voltage [V]
   uk = FOMout.Iapp(k);     % applied current [A]
   Tk = FOMout.T(k);        % temperature [C]

% Run Kalman filter
   [zk,zbk,ekfData,Xind] = iterEKF_SOP(vk,uk,Tk,ekfData);
   zkEst(:,k) = zk; zkBound(:,k) = zbk;
   ekfData.k = k;
   gamma(:,k)=Xind.gamma; theT(:,k)=Xind.theT; theZ(:,k)=Xind.theZ;

% DISCHARGE CASE
   if zk(end) > bsmData.DISref
      hdk = @(x) iterSOPbisect(x,ROM,ekfData,bsmData_DIS,'DIS');
      [idbk(k) BSMout] = bisect(hdk,0,bsmData.const.u_max,10e-5,k);
      SOP_BSMresults_DIS = storing(BSMout,SOP_BSMresults_DIS,k);
   else
    idbk(k) = 0;
   end

   P_dis(k) = bsmData.const.v_min*idbk(k);

% CHARGE CASE
   if zk(end) < bsmData.CHGref
      hck = @(x) iterSOPbisect(x,ROM,ekfData,bsmData_CHG,'CHG');
      [icbk(k) BSMout] = bisect(hck,bsmData.const.u_min,0,10e-5,k);
      SOP_BSMresults_CHG = storing(BSMout,SOP_BSMresults_CHG,k);
   else
      icbk(k) = 0;
   end

   P_chg(k) = bsmData.const.v_max*icbk(k);

end
```

8.8 *Where to from here?*

We have now reached the end of this three-volume series on battery-management systems. What an incredible journey! In the first volume, we developed both empirical and physics-based lithium-ion cell models; in the second volume, we explored BMS algorithms based on empirical equivalent-circuit models; and finally, in this volume, we investigated ways to make BMS algorithms based on PBMs practical. To do so, we first had to overcome several roadblocks.

- We discovered that one challenge to using physics-based models in a BMS is that not all of the parameters of the DFN model are identifiable from current/voltage measurements. So, we applied a parameter-lumping process to the equations comprising the model to generate the reformulated LPM. All of the parameters of the LPM are identifiable, which means that they can be estimated from current/voltage measurements, at least in principle.
- We then learned that a second impediment to using physics-based models is that the original DFN-style model does not describe all lithium-ion cell dynamics observed in experimental data. So, we enhanced the DFN model by replacing the standard Butler–Volmer kinetics with the more general MSMR model. We added a description of an electrical double layer to the electrode/electrolyte interface. We generalized the description of this double layer and also of solid diffusion by incorporating CPEs in the model, and we added a physics-based model of SOC dependence to electrode solid diffusivity. Due to the addition of CPEs, this model can be evaluated in closed form only in the frequency domain (but we learned later how to make a high-fidelity approximation in the time domain). To assist with computing the model in the frequency domain, we developed a set of exact closed-form equations (TFs) that allow computing the frequency response of any electrochemical variable at any spatial location in the cell very quickly. By combining several TFs, we also derived a closed-form expression of the impedance of the cell, which can be visualized using Nyquist diagrams, which are commonly used to display cell impedance measured via EIS.
- A third obstacle to using physics-based models is the need to estimate all model parameter values. We focused on nonteardown approaches to doing so. Even though it is mathematically possible to estimate all parameter values of the enhanced LPM from current/voltage data, the sheer number of parameters that must be estimated and the fact that nonlinear optimization is required makes the problem very challenging. We proposed a divide-and-conquer approach that determines subsets of parameter values us-

ing specialized targeted laboratory tests. First, OCP testing allows us to estimate OCP versus stoichiometry for both electrodes. Second, OCV testing determines the electrode operating boundaries and cell capacity. Third, discharge testing improves the calibration of the positive-electrode OCP function versus absolute stoichiometry and makes finding OCP of both electrodes possible without cell teardown. Fourth, pulse testing exposes the parameters of the potential and kinetics equations in the LPM. Fifth, EIS testing determines the majority of the remaining parameter values. And finally, sixth, discharge testing combined with the pseudo-steady-state model allows us to resolve the few remaining electrolyte parameter values.

- The fourth challenge to using physics-based models is the fact that they are computationally very intensive. Volume I of this series showed how to use a subspace-projection method named the discrete-time realization algorithm (DRA) to convert TFs to discrete-time state-space ROMs that can be simulated very quickly. This volume presented an improved subspace-projection method named the hybrid realization algorithm (HRA) that performs the same operation much more quickly and requires much less memory. Substeps of the HRA enable validation of the discrete-time frequency response of the ROM versus the original TFs, which gives insight into how well the final model is approximating each of the TFs that it is seeking to reproduce. We also compared the model-blending approach from Volume I to a new output-blending approach to using multiple ROMs to provide robust predictions over a wide operational window, and discovered that the output-blending approach is often much better. Finally, we saw ways to use a set of ROMs describing multiple cells having distinct parameter values to simulate battery packs comprising those cells wired in parallel and series.

Having addressed these fundamental impediments to being able to use physics-based models in a BMS, we began to look at applications of these models to satisfy BMS-algorithm requirements.

- Volume II of this series introduced the nonlinear EKF and SPKF as methods to estimate cell SOC using circuit models. In this volume, we discovered that we can also apply EKF and SPKF to estimate the state of a physics-based ROM. But since the ROM describes much more than simply cell SOC or voltage, we can now also estimate any of the electrochemical variables of the cell at any spatial location(s) of interest. To do so, we needed to add a step 3 to the xKF framework, greatly enhancing the predictive capabilities of the methods. We also saw how to apply xKF to an output-blended model, to make accurate state and internal-variables predictions over a wide operational range.

- We explored some concepts in BMS diagnostics and prognostics. We paid special attention to how a cell's OCV relationship changes as the cell ages and how this can give us information regarding the kinds and levels of degradation that are occurring. We discussed some real-time approaches to adapting or selecting a model that best describes the cell under observation at its present point of age. And we presented several thermal and degradation models in lumped-parameter format, for use in power-limits calculations.
- Having developed the tools for lithium-ion battery modeling and state estimation, we next examined some real-time applications of advanced battery management methods. To that end, we outlined a fast-charge strategy employing model predictive control to compute an optimal charge current profile that respects hard constraints on problem variables of interest. Numerical simulations demonstrated the viability of the approach and its potential to extend battery lifetime by mitigating true mechanisms of cell degradation.
- Finally, we examined the important topic of energy management within a battery from the standpoint of a state of power measure. In addition to the remaining energy a battery contains, we considered the implications of how quickly that energy is added to or removed from a battery in terms of its power utilization. Such a measure is a governing factor in vehicle performance. Again utilizing and MPC framework, we developed an advanced real-time SOP estimation algorithm and demonstrated its operation via numerical simulation.

This book has presented our best understanding at this point in time of best practices relating to physics-based modeling of lithium-ion cells and of physics-based BMS algorithms. But this field is developing at a rapid pace. Our own research team and teams at other universities and research institutes are making steady progress. The community has not yet converged on any best modeling methods or practices, or best battery-management approaches. We fully expect that this field will continue to evolve, driven by the need to support the significant investments being made in large-scale battery-based installations (and hence the urgency of making them work safely, optimally, and robustly for as long a service life as possible).

- We envision ongoing refinements to the fundamental lithium-ion PDEs, perhaps improving descriptions of cell dynamics at low SOCs, low temperatures, and high C-rates. (We also expect to see the approaches presented in this series start to be applied to other battery types, featuring new anodes such as lithium-metal and silicon, solid electrolytes, and new cathode materials like sulphur.)

- We anticipate advances in nonteardown methods to estimate model parameter values. We are pleased with the OCV and teardown-based OCP estimation methods, but there is considerable room for improvement in nonteardown-based OCP estimation. We know that OCV should equal the positive electrode's OCP minus the negative electrode's OCP, but experimental errors accumulate and there is room for refining these estimates to achieve better agreement at the cell level. We have found that the pulse-resistance test provides accurate parameter estimates for some cells but not for others.[12] Since the pulse-resistance test is the only one of those that we have presented that is designed to expose the parameters describing the nonlinearities in the cell, different tests must be developed that somehow use a variety of C-rates to actuate the cell in ways that enable estimating these parameter values efficiently and accurately.
- We have a strong affinity to the HRA-based ROM, especially for simulations of fresh cells, but we admit that it does have some shortcomings. The most limiting of these is that the HRA is a numeric method that obscures the connection between physical parameter values of the PDEs in the final ROMs. It is an open problem to determine how to adjust the coefficients of these ROMs in a physically meaningful way as the cells being managed by the BMS age. We have presented some of our thoughts on how to approach this problem in this volume and even have some preliminary results along these lines, but there remains significant work to do. One possibility is to replace the ROM with a single-particle model (SPM) or enhanced single-particle model (SPMe) for some applications; the methods of Chaps. 1 to 3 in this volume still apply to that case, but the approaches of Chaps. 4 to 8 would need to be redeveloped for the different model equations of the SPM.
- Research on physics-based fast-charge and state-of-power estimation shows great promise, but is still in its infancy. It remains to perform comprehensive experimental validation of these techniques, which may point the way to further refinements. At the time of this writing, other topics worthy of investigation in this rapidly emerging field include new models of lithium plating/stripping, pulse-charging techniques, as well as advances to improve performance/robustness of MPC for nonlinear systems.

In closing, we hope that this series has been helpful to your understanding of lithium-ion battery cells and how they can be managed by a BMS that uses either empirical or physics-based models. We entrust the next steps to you to apply these methods, refine them,

[12] We think that perhaps $\bar{R}_{\text{dl}}$ is too small on some cells for the data quality to be sufficient to find good estimates of the values of the kinetics and potential equations.

and even replace them with transformational new approaches. Best wishes to you as you do so!

8.A *Summary of variables*

- E_{Total}, total energy [J].
- $E[k]$, total energy computed at time instant k [W].
- $i[k]$, applied current at time instant k [A].
- $i_{\text{min}}^{\text{chg,volt}}[k]$, constant charge current value that brings voltage to a limit over ΔT [A].
- $i_{\text{max}}^{\text{dis,volt}}[k]$, constant discharge current value that brings voltage to a limit over ΔT [A].
- I_{max}, maximum applied current [A].
- I_{min}, minimum applied current [A].
- $k_{\Delta T}$, integer time index for power horizon (i.e., $\Delta T = k_{\Delta T} \times T_s$).
- $P[k]$, cell power at time instant k [W].
- P_{avg},average power over a defined time interval [W].
- $P_{\text{chg,min}}$, SOP estimate of available charge power [W].
- $P_{\text{dis,max}}$, SOP estimate of available discharge power [W].
- P_{inst}, instantaneous power [W].
- P_{max}, maximum instantaneous power [W].
- T_s, sampling interval [sec].
- $v[k]$, cell terminal voltage at time instant k [V].
- V_{max}, maximum allowable cell voltage [V].
- V_{min}, minimum allowable cell voltage [V].
- ΔT, power horizon [sec].
- $\underline{\sigma}(\boldsymbol{M})$, minimum singular value of matrix $\boldsymbol{M}$.

About the Authors

Gregory L. Plett is a professor of electrical and computer engineering at the University of Colorado Colorado Springs (UCCS). His research since 2001 has focused on applications of control-systems theory to the management of high-capacity battery systems such as found in hybrid and electric vehicles. Current investigations include physics-based reduced-order modeling of ideal lithium-ion and lithium-metal dynamics; system identification of physics-based model parameters using only current–voltage input–output data; physics-based reduced-order modeling of degradation mechanisms in electrochemical cells; estimation of cell internal state and degradation state; and state-of-charge, state-of-health, and state-of-life estimation. He offers courses in control systems and in battery modeling and management to support this research. Dr. Plett holds a bachelor of engineering degree in computer systems engineering from Carleton University (Ontario, Canada) and master of science and PhD degrees in electrical engineering from Stanford University (Stanford, California). He is a senior member of the Institute of Electrical and Electronic Engineers and life member of the Electrochemical Society.

M. Scott Trimboli is an associate professor of electrical and computer engineering at UCCS where he presently conducts research in modeling and advanced control of battery systems. Previous research involved application of smart controls to multifunctional materials to achieve vibration suppression, work conducted as an exchange scientist with the German Aerospace Research Establishment (DLR) in Göttingen, Germany. Current investigations focus on applying model-predictive control using physics-based reduced-order models to achieve improved performance and life extension for lithium-ion batteries. He teaches courses in modern and robust control theory including model predictive control. Dr. Trimboli received his bachelor's degree in engineering science from the United States Air Force Academy (Colorado Springs, Colorado) and master of science in engineering mechanics from Columbia University (New York, New

York). He holds a DPhil degree in control engineering from the University of Oxford (Oxford, United Kingdom).

The research that Profs. Plett and Trimboli perform is both theoretical and empirical: the UCCS Charging, Heating, and Aging for Resolving Governing Equations (CHARGE) Laboratory houses equipment to test cells, modules, and battery packs and is home to custom battery-management-system projects, which enable cutting-edge research in advanced but practical algorithm prototyping.

Index

Artech House Power Engineering Library

Naoki Matsumura, Series Editor

Advanced Technology for Smart Buildings, James Sinopoli

The Advanced Smart Grid: Edge Power Driving Sustainability, Second Edition, Andres Carvallo and John Cooper

Applications of Energy Harvesting Technologies in Buildings, Joseph W. Matiko and Stephen Beeby

Battery Management Systems, Volume I: Battery Modelings, Gregory L. Plett

Battery Management Systems, Volume II: Equivalent-Circuit Methods, Gregory L. Plett

Battery Management Systems, Volume III: Physics-Based Methods, Gregory L. Plett and M. Scott Trimboli

Battery Management Systems for Large Lithium Ion Battery Packs, Davide Andrea

Battery Power Management for Portable Devices, Yevgen Barsukov and Jinrong Qian

Big Data Analytics for Connected Vehicles and Smart Cities, Bob McQueen

Design and Analysis of Large Lithium-Ion Battery Systems, Shriram Santhanagopalan, Kandler Smith, Jeremy Neubauer, Gi-Heon Kim, Matthew Keyser, and Ahmad Pesaran

Designing Control Loops for Linear and Switching Power Supplies: A Tutorial Guide, Christophe Basso

Electric Power System Fundamentals, Salvador Acha Daza

Electric Systems Operations: Evolving to the Modern Grid, Second Edition, Mani Vadari

Energy Harvesting for Autonomous Systems, Stephen Beeby and Neil White

Energy IoT Architecture: From Theory to Practice, Stuart McCafferty

Energy Storage Technologies and Applications, C. Michael Hoff

GIS for Enhanced Electric Utility Performance, Bill Meehan

IEC 61850 Demystified, Herbert Falk

IEC 61850: Digitizing the Electric Power Grid, Alexander Apostolov

Introduction to Power Electronics, Paul H. Chappell

Introduction to Power Utility Communications, Harvey Lehpamer

IoT Technical Challenges and Solutions, Arpan Pal and Balamuralidhar Purushothaman

Lithium-Ion Batteries and Applications: A Practical and Comprehensive Guide to Lithium-Ion Batteries and Arrays, from Toys to Towns, Volume 1, Batteries, Davide Andrea

Lithium-Ion Batteries and Applications: A Practical and Comprehensive Guide to Lithium-Ion Batteries and Arrays, from Toys to Towns, Volume 2, Applications, Davide Andrea

Lithium-Ion Battery Failures in Consumer Electronics, Ashish Arora, Sneha Arun Lele, Noshirwan Medora, and Shukri Souri

Microgrid Design and Operation: Toward Smart Energy in Cities, Federico Delfino, Renato Procopio, Mansueto Rossi, Stefano Bracco, Massimo Brignone, and Michela Robba

Plug-in Electric Vehicle Grid Integration, Islam Safak Bayram and Ali Tajer

Power Grid Resiliency for Adverse Conditions, Nicholas Abi-Samra

Power Line Communications in Practice, Xavier Carcelle

Power System State Estimation, Mukhtar Ahmad

Practical Battery Design and Control, Naoki Matsumura

Renewable Energy Technologies and Resources, Nader Anani

Robust Battery Management System Design with MATLAB®, Balakumar Balasingam

Signal Processing for RF Circuit Impairment Mitigation in Wireless Communications, Xinping Huang, Zhiwen Zhu, and Henry Leung

The Smart Grid as An Application Development Platform, George Koutitas and Stan McClellan

Smart Grid Redefined: Transformation of the Electric Utility, Mani Vadari

Sustainable Power, Autonomous Ships, and Cleaner Energy for Shipping, John Erik Hagen

Synergies for Sustainable Energy, Elvin Yüzügüllü

A Systems Approach to Lithium-Ion Battery Management, Phil Weicker

Telecommunication Networks for the Smart Grid, Alberto Sendin, Miguel A. Sanchez-Fornie, Iñigo Berganza, Javier Simon, and Iker Urrutia

A Whole-System Approach to High-Performance Green Buildings, David Strong and Victoria Burrows